NAYFEH—Perturbation Methods
NAYFEH and MOOK—Nonlinear Oscillations
ODEN and REDDY—An Introduction to the Mathematical
PASSMAN—The Algebraic Structure of Group Rings
PETRICH—Inverse Semigroups
PIER—Amenable Locally Compact Groups
PRENTER—Splines and Variational Methods
RAO—Measure Theory and Integration
RIBENBOIM—Algebraic Numbers
RICHTMYER and MORTON—Difference Methods for Initial-Value Problems, 2nd Edition
RIVLIN—The Chebyshev Polynomials
ROCKAFELLAR—Network Flows and Monotropic Optimization
RUDIN—Fourier Analysis on Groups
SAMELSON—An Introduction to Linear Algebra
SCHUMAKER—Spline Functions: Basic Theory
SHAPIRO—Introduction to the Theory of Numbers
SIEGEL—Topics in Complex Function Theory
Volume 1—Elliptic Functions and Uniformization Theory
Volume 2—Automorphic Functions and Abelian Integrals
Volume 3—Abelian Functions and Modular Functions of Several Variables
STAKGOLD—Green's Functions and Boundary Value Problems
STOKER—Differential Geometry
STOKER—Nonlinear Vibrations in Mechanical and Electrical Systems
STOKER—Water Waves
TURÁN—On A New Method of Analysis and Its Applications
WHITHAM—Linear and Nonlinear Waves
WOUK—A Course of Applied Functional Analysis
ZAUDERER—Partial Differential Equations of Applied Mathematics

MEASURE THEORY AND INTEGRATION

MEASURE THEORY AND INTEGRATION

M. M. RAO
Professor of Mathematics
University of California

A Wiley-Interscience Publication

JOHN WILEY & SONS

New York · Chichester · Brisbane · Toronto · Singapore

Library of Congress Cataloging in Publication Data:
Rao, M. M. (Malempati Madhusudana), Date.
Measure theory and integration.

(Pure and applied mathematics)
"A Wiley-Interscience publication."
Bibliography: p.
Includes indexes.
1. Measure theory. 2. Integrals, Generalized.
I. Title. II. Series: Pure and applied mathematics
(John Wiley & Sons)
QA312.R34 1987 515.4′2 87-2005
ISBN 0-471-82822-X

Printed in the United States of America

10 9 8 7 6 5 4 3 2

To My Brother

MUKUNDA RAO

Whose Life Was Cut Short
So Suddenly

PREFACE

This book presents a detailed exposition of the general theory of measure and integration. It is meant to be a text for a first year graduate course, often given under such titles as "Measure Theory", "Integration", "Real Analysis", or "Measure and Integral". The material is unified from various sets of notes, and of experience gained, from my frequent teaching of such a class since 1960.

Generally the subject is approached from two points of view as evidenced from the standard works. Traditionally one starts with measure, then defines the integral and develops the subject following Lebesgue's work. Alternatively one can introduce the integral as a positive linear functional on a vector space of functions and get a measure from it, following the method of Daniell's. Both approaches have their advantages, and eventually one needs to learn both methods. As the preponderance of existing texts indicates, the latter approach does not easily lead to a full appreciation of the distinctions between the (sigma) finite, localizable, and general measures, or their impact on the subject. On the other hand, too often the former approach appears to have little motivation, rendering the subject somewhat dry. Here I have tried to remedy this by emphasizing the positive and minimizing the negative aspects of these methods, essentially following the natural growth of the subject in its presentation. This book covers all the standard theory and includes several contemporary results of interest for different applications.

Each topic is introduced with ample motivation. I start with an abstraction of lengths, areas, volumes and other measurements of known geometric figures and develop the basic ideas of Lebesgue in $\mathbb{R}^n$. This is then used as a model and a reference for the general study leading to the Carathéodory process. The measure approach as a basic step is especially natural in such areas as functional analysis, probability and statistics, and ergodic theory, whereas reference to Lebesgue's method keeps in view the applications to differential equations and mathematical physics among others. I now indicate some features of the present treatment and contrast it with earlier works.

The Carathéodory process, which here takes center stage and helps in an efficient presentation, was effectively used earlier by Dunford and Schwartz

(1958), by Zaanen (1967), and more recently by Sion (1968, 1973). In addition, inner measures have a special role in several types of extension procedures. This is particularly true in obtaining regular extensions of topological measures. It was indicated by Royden (1968), but the full potential is utilized and emphasized here. In the context of topological measures, I have presented the Henry extension theorem and used it later in shortening and illuminating the structure of some other results. (See, e.g., Theorems 6.4.7–8 for novel applications.)

Inclusion of image measures and vague convergence is discussed for sequences. For instance, Skorokhod's representation (cf. Theorem 3.3.5) in this context is of interest in probability and Fourier analysis. A few results given in Section 4.3, on integration of not necessarily measurable functions, exhibit the power of Carathéodory's process and also help in simplifying some arguments for product integrals later in Chapter 6, while enlarging the scope of applications of Lebesgue's limit theorems. An account of L^p-spaces is included in Sections 4.5 and 5.5, illustrating the methods of integration. Then signed measures and the Vitali-Hahn-Saks theorem find a natural place there. Further a detailed treatment of differentiation of set functions is given. The Radon-Nikodým theorem is presented with multiple proofs and shown to imply the Jordan-Hahn decomposition. This exhibits a deeper equivalence between these two theorems since each is also shown to be provable independently of the other and deducible from one another. The localizability concept introduced earlier is utilized to establish Segal's theorem on the equivalence of the Radon-Nikodým property for μ, with the dual of $L^1(\mu)$ as $L^\infty(\mu)$. Also absolutely continuous and completely monotone real functions on the line are treated. Only Zaanen (1967) had considered an extended discussion of the Radon-Nikodým theorem. However, localizability is also found to be interesting in product measure theory. (See, e.g., Exercises 6.2.7 and 6.2.8.)

Infinite product measures are given an extended treatment. I include the Kolmogorov-Bochner, Prokhorov, Tulcea, and Fubini-Jessen theorems. Their relation with two martingale convergence results is established. In the earlier works, only Hewitt and Stromberg (1965) have considered an aspect of this theory. These results find an important place in the current work on stochastic analysis. As useful applications, Bochner's representation theorem for continuous positive definite functions on the line, and a realization of an abstract Hilbert space as a subspace of an $L^2(\mu)$-space are presented (Section 6.5).

A novel treatment is an inclusion of Choquet's capacity theorem for analytic sets from which one obtains the Daniell integration as a consequence. This approach was indicated by Meyer (1966), and a comprehensive account is given here. Recently Jacobs (1978) also considered Choquet's theorem, but my purpose is to obtain Daniell's results easily and quickly from the former. Next an elementary proof of the lifting theorem, due to T. Traynor (1974), is included. This result vividly shows the facility and problems created by sets of measure zero in the Lebesgue theory, in addition to its intrinsic importance. Finally the interplay of topology and measure is expounded in Chapters 9 and

10. Here regular measures, on locally compact and some general topological spaces, as well as Pettis's theorem on extensions of a measure from a lattice (usually of compact sets) to the σ-algebras generated by them, the Riesz-Markov theorem, and an integral representation of local functionals of Gel'fand-Vilenkin on compactly based continuous function spaces are presented. Topologies induced by a measure, the Stone isomorphism theorem of a measure algebra, and some applications as well as a treatment of the Haar measure find a place here.

I have presented both the classical and some contemporary topics often used in the current mathematical activity. Indeed, almost all the measure and integration theory needed by probabilists as well as functional analysists, and in particular most of what is needed for my earlier books (1979, 1981, 1984), is found here. I hope it will be useful to others in similar applications in which measure and integration play an important role.

The book is intended primarily as a text for a year's or a semester's course on contemporary real analysis. The following suggestions are offered for this purpose. Omitting a few special topics, such a standard analysis course is covered by the *first six* chapters.

A respectable course for a semester (or a two quarter) length class is obtained by the selection: Chapter I, Sections 2.1–2.3, 2.6, 3.1, 3.2, 4.1, 4.3, 4.5, 5.1; the first two results of Section 5.2; Section 5.3; the first half of Section 5.5 and Sections 6.1 and 6.2. If any time is left one can cover Chapter 7 for either of the above two classes. However, Chapter 9 can be studied immediately after the first four chapters, with only a reference to the Radon-Nikodým thorem, or by omitting Theorem 9.3.5. For a year's course, it is possible to cover all the first seven chapters. Chapters 8, 9 are essentially independent and can be taken up in any order (after Chapter 4) and then Chapter 10 may be appended.

There is more than enough material for a year's course, even with selected omission of certain sections, according to one's tastes. However, the treatment throughout is considerably detailed with alternative arguments (including some repetitions of notation and definitions to ease a search by the reader), keeping the student's needs in mind. Therefore, the book is also suitable for a self-study.

A prerequisite for this text is a knowledge of advanced calculus such as that found in Bartle (1976) or Rudin (1976). Essentially everything else is detailed here. A short appendix presents some results from topology and set theory with references. I have included many exercises (over 400) of varying difficulty at the end of each section, and those which are less simple are provided with hints. As the study progresses, the reader is expected to gain sophistication, and in any case, some of the more advanced topics can be skipped in a first reading.

The numbering system is standard: m.n.p denotes the chapter (m), the section (n), and the proposition, definition, or exercise, etc. (p). In a given chapter, m is omitted and in a section, m.n is also omitted.

The material is influenced by the many texts used before, but I should especially like to acknowledge that my point of view has shifted from the traditional one with the appearance of Dunford and Schwartz (1958) at the beginning of my career. This and that of Sion's books (1968, 1973) have strengthened my belief in the efficacy of the Carathéodory process even for pedagogical purposes. Also, the reactions of my audiences have encouraged me in this approach.

The preparation of the manuscript over the past two years has been facilitated by a year's UCR sabbatical leave, spent at the Institute for Advanced Study during 1984–1985, partially supported by an ONR contract. Typing of my handwritten and difficult manuscript, and its revision, was patiently carried out by Mrs. Eva Stewart. This preparation was helped by a UCR Academic Senate grant. Joseph Sroka and Derek Chang assisted me in proof-reading and preparation of indexes. To all these people and institutions I wish to express my deep appreciation.

M. M. Rao

Riverside, California
May 1987

CONTENTS

MEASURE THEORY AND INTEGRATION

1

INTRODUCTION AND PRELIMINARIES

After a discussion of the need for a general study of measures and integration, an overview of the subject is sketched. Then some preliminaries on set operations and an abstraction of the latter for an extended analysis are indicated. To motivate the general study, the Lebesgue measure on the line and a few of its properties are discussed.

1.1. MOTIVATION AND OUTLOOK

The concepts of length, area, volume, mass, and weight are familiar in measuring the sizes of various geometrical objects, and are usually treated in the study of elementary calculus. Typically these are nonnegative values attached to certain elementary figures such as intervals, rectangles, spheres, or balls. When the objects are more complicated, then we assign the corresponding numerical values by approximating them, whenever possible, with the above types of figures. Since from experience it is found that these approximations cannot always consist of a finite number of these elementary objects, we introduce the Riemann-Darboux sums and the consequent integrals. However, the latter are obtained only for certain fairly simple, "smooth" figures. More complicated objects appear in real problems, and they must be measured and assigned suitable numerical values. This leads to analyses of objects which are composed of (or obtained from) elementary figures in the sense of using sums (or unions), differences (or intersections and complements), and other decompositions. These ideas motivate an abstraction and use of set theoretical operations on them, to study new and nonelementary figures. This will be our basic step—the establishment of a domain of operations for all measurements—and naturally one introduces an algebraic structure into such a class, suggestively calling them algebras of measurable sets.

The modern approach here is not to begin with the geometric structures of figures, but to make a general study of the algebraic properties of these objects. Specialization and geometry will come later. We shall motivate this in the next section with intervals on the number space $\mathbb{R}^n$, $n \geq 1$.

Having thus isolated a class of sets (a neutral word for all these objects), we next proceed to study volumes, areas, and their extensions for figures defined by various functions. If these are bounded and continuous, the Riemann-Darboux or Riemann-Stieltjes integrals of our early work in calculus provide solutions. However, problems in applications are not always definable with such "smooth" or elementary functions. Consequently we want to study the properties of functions which do not depend on the continuity hypotheses to start with, but which can be specialized later to these nicer classes. To use another neutral word, we start to analyze the structure of all functions which are candidates for measuring the properties of the figures, and call them collectively measurable functions. Naturally these are closely related to (and largely determined by) the abovementioned measurable classes of sets. Then one proceeds to define an integral for these functions on those sets. But generally both these "measurable" classes of sets and functions depend on the measuring instrument (for length, area, volume, mass, etc.) prescribed by the problem being investigated. The common properties of such a method are again abstracted into what are suggestively called measure functions. These will provide measures for the simplest (or elementary) figures, in sufficient number, with a natural consistency property, and then one needs to extend them to all measurable sets, in the appropriate sense, and perhaps enlarge that class if possible. This is again a nontrivial task; the necessary study starts in Chapter 2 and continues in various forms for the rest of the book. Our central concern is thus a study of measure functions and their properties on classes of measurable sets. (Measure function is often abbreviated to measure.)

Once measures, measurable functions, and sets are at our disposal, we proceed to the basic task of integration of such (suitable) measurable functions on measurable sets relative to certain measure functions. The problem here is to include everything we already know about Riemann-Stieltjes integrals. However, this turns out to be unwieldly, and an important (but still very large) subclass of the former integrals (namely the so-called absolutely continuous ones) will be generalized with a method of Lebesgue's. The thus obtained integral is profound, and it is found sufficient for most of the current work in analysis. That study will start in Chapter 4, and will occupy our attention for the rest of this book. The non-absolutely-continuous aspect of the Riemann integral is also useful for some work, and has been extended by Perron, Denjoy, and others. We shall not consider such theories in this book, since most of the modern analysis lives, as it were, with Lebesgue's theory and especially with its abstract version due to Carathéodory, which we study in detail.

Our general treatment, without specializations or additional restrictions, will continue through Chapter 6, since this account should be useful for a very broad spectrum of applications in which the theory of integration plays a key role. These include the areas of functional analysis, probability and statistics, harmonic analysis, differential equations, mathematical economics, mechanics, and others. For some of the latter applications one can start with the concept of linear functionals, without mention of measure, and proceed to the theory of integration following the method of Daniell. From this a measure theory can be developed later. Since this procedure will not be convenient for the broadest areas of application mentioned above, we proceed first with measures and then go to integration. However, in the current framework we can present the abstract capacity theory due to Choquet, as a natural flow of ideas, and with this, in Chapter 7, we obtain the Daniell integral as an easy consequence. Since the capacity theory has important applications in the modern (i.e., axiomatic) potential theory, our procedure has therefore additional advantages.

After the standard material (including product integrals) is covered, we turn to some specialization when the spaces have a topology, and then certain interesting regularity properties of measure functions are treated in Chapters 9 and 10. Using the work of Chapter 8 on the lifting theorem, it is shown that one can often introduce a topology in an abstract measure space, and this will clarify the subject in relation to the original topology, if any. In Chapter 9 we give different approaches to the Riesz-Markov theorem and finally a brief indication of the possibility of introducing a Hausdorff topology for a class of measurable groups is given in Chapter 10.

1.2. THE SPACE $\mathbb{R}^n$ AS A MODEL

We start with the basic set $\mathbb{R}^n$, called the *number space* if it is regarded as a vector space without using its inner product structure, or the *Euclidean n-space* if its metric property is also considered. This will serve as a model for all the work that follows, besides being important in its own right. In the number space $\mathbb{R}^n$, a subset A is called an *interval*, or a *rectangle* (or a *box*) if it is of the form

$$A = \{x = (x_1, \ldots, x_n) \in \mathbb{R}^n : a_i < x_i \leqslant b_i,\ i = 1, \ldots, n\}, \qquad (1)$$

where $-\infty < a_i \leqslant b_i \leqslant \infty$, with "$< b_i$" if $b_i = +\infty$. When both the end points a_i and b_i are included in the interval, it will be termed a *closed interval*; and when both are excluded, an *open interval*. The set in (1) may be termed a right-closed (and left-open) interval. Thus if A_1, A_2 are intervals of any of the above kinds, then it is evident that their intersection $A_1 \cap A_2$ is of the same kind. However, neither their difference $A_1 - A_2$ nor their union $A_1 \cup A_2$ need be an interval of the same kind. But the intervals of the type (1) have nicer properties, and they turn out to be convenient in our work. Note that, by definition, $\mathbb{R}^n$ is simultaneously open, closed, and of the type (1). The same is true of the empty set, denoted $\varnothing$. Hereafter we use the word *rectangle* for sets of the form (1).

Let us begin with this special structure:

Proposition 1. *Let A, B be two rectangles in $\mathbb{R}^n$ [of the type (1)].*

(i) *If $A \subset B$, then there exist rectangles A_k, $1 \leqslant k \leqslant m$, such that*

$$A = A_0 \subset A_1 \subset \cdots \subset A_m = B,$$

with $A_k - A_{k-1}$, $k = 1, \ldots, m$, being a rectangle [of the type (1)].

(ii) *If A, B are arbitrary (without the inclusion relation), then $B - A$ can be expressed as a finite disjoint union of rectangles [of the type (1) again].*

Proof. (i): First consider the special case that $B = \mathbb{R}^n$. Then letting $A_0 = A$, define A_k as follows:

$$A_1 = \{(x_1, \ldots, x_n) \colon a_1 < x_1 < \infty,\ a_i < x_i \leqslant b_i,\ i = 2, \ldots, n\},$$

$$A_2 = \{(x_1, \ldots, x_n) \colon a_i < x_i \leqslant b_i,\ i = 2, \ldots, n\}.$$

In general, for $j \geqslant 2$ define

$$A_{2j-1} = \{(x_1, \ldots, x_n) \colon a_j < x_j < \infty,\ a_i < x_i \leqslant b_i,\ i = j+1, \ldots, n\},$$

$$A_{2j} = \{(x_1, \ldots, x_n) \colon a_i < x_i \leqslant b_i,\ i = j+1, \ldots, n\},$$

and set $A_{2n} = \mathbb{R}^n$. Then $A = A_0 \subset A_1 \subset \cdots \subset A_{2n} = \mathbb{R}^n$, and each A_k is a rectangle. On the other hand for $j \geqslant 1$,

$$A_{2j} - A_{2j-1} = \{(x_1, \ldots, x_n) \colon x_j \leqslant a_j,\ a_i < x_i \leqslant b_i,\ i = j+1, \ldots, n\},$$

and

$$A_{2j+1} - A_{2j}$$
$$= \{(x_1, \ldots, x_n) \colon b_{j+1} < x_{j+1} < \infty,\ a_i < x_i \leqslant b_i,\ i = j+2, \ldots, n\}. \quad (2)$$

Since each of these differences is a rectangle (draw a picture in the plane), it follows that the conclusion holds if $B = \mathbb{R}^n$. (Thus $m = 2n$ works.)

In the general case that $B \subset \mathbb{R}^n$ (properly), since $A \subset B \subset \mathbb{R}^n$, by the special case there exist rectangles A_k^* such that for some m

$$A = A_0^* \subset A_1^* \subset \cdots \subset A_m^* = \mathbb{R}^n, \qquad A_k^* - A_{k-1}^* \text{ a rectangle.}$$

Set $A_i = A_i^* \cap B$. Then, by our earlier observation on intersections, A_i is a rectangle, and since $B \subset \mathbb{R}^n$, there exists a $k_0 \leqslant m$ such that $A_{k_0}^* \supset B$. Hence $A_{k_0} = B$ $(= A_{k_0+1} = \cdots = A_m)$, and $A = A_0 \subset A_1 \subset \cdots \subset A_{k_0} = B$. But

$$A_k - A_{k-1} = A_k^* \cap B - A_{k-1}^* \cap B = (A_k^* - A_{k-1}^*) \cap B \quad (3)$$

is a rectangle. Hence (i) holds in the general case also.

(ii): Since $B - A$ is generally not a rectangle, consider $\tilde{A} = A \cap B$. Then $\tilde{A} \subset B$ and $B - A = B - A \cap B = B - \tilde{A}$. By part (i), there exist rectangles A_k such that $A_0 = \tilde{A} \subset A_1 \subset \cdots \subset A_m = B$, and $B_k = A_k - A_{k-1}$, $k = 1, \ldots, m$, are disjoint rectangles. It follows that

$$B - A = B - \tilde{A} = \bigcup_{k=1}^{m} B_k, \tag{4}$$

is a finite disjoint union of rectangles, completing the proof.

It should be observed that the special type of definition of a rectangle in (1) played an important role in the properties (2)–(4).

The above proposition implies the following interesting algebraic fact: *Let* $\mathcal{S}$ *be the class of all rectangles in* $\mathbb{R}^n$ *defined by* (1). *Then we have* (i) $A, B \in \mathcal{S} \Rightarrow A \cap B \in \mathcal{S}$, (ii) $\varnothing \in \mathcal{S}$, $\mathbb{R}^n \in \mathcal{S}$, (iii) $A, B \in \mathcal{S}$, $A \subset B$ $\Rightarrow$ *there exist* $A_k \in \mathcal{S}$ *such that* $A = A_0 \subset A_1 \subset \cdots \subset A_m = B$ *and* $A_k - A_{k-1} \in \mathcal{S}$.

A collection $\mathcal{S}$ of subsets, with the properties noted, plays a basic role in the study of measure theory, since we can easily attach a measure (area, volume etc.) to such elements.

Let us now introduce a larger class of sets than those of $\mathcal{S}$.

Definition 2. A subset A of $\mathbb{R}^n$ is called *elementary* if it is a finite disjoint union of rectangles, i.e., $A = \bigcup_{i=1}^{k} A_i$, $A_i \cap A_j = \varnothing$ if $i \neq j$, and $A_i \in \mathcal{S}$, $i = 1, \ldots, k$ (k depending on A). Let $\mathcal{E}$ denote the class of all elementary sets of $\mathbb{R}^n$.

The algebraic structure of $\mathcal{E}$ is even nicer than that of $\mathcal{S}$.

Proposition 3. *The class* $\mathcal{E}$ *is closed under finite unions, intersections, differences, and (hence) symmetric differences. In symbols* $\{A, B\} \subset \mathcal{E} \Rightarrow \{A^c, B^c, A \cup B, A \cap B, A - B, A \triangle B\} \subset \mathcal{E}$. *Moreover,* $\mathcal{S} \subset \mathcal{E}$ *properly. (As usual,* A^c *is the complement of* A.)

Proof. Let A, B be a pair of elementary sets. Then there exist disjoint collections $\{A_k : 1 \leqslant k \leqslant m_1\}$ and $\{B_j : 1 \leqslant j \leqslant m_2\}$ of rectangles (hence elements of $\mathcal{S}$) such that $A = \bigcup_{k=1}^{m_1} A_k$, $B = \bigcup_{j=1}^{m_2} B_j$. Then

$$\begin{aligned} A \cap B &= \left(\bigcup_{k=1}^{m_1} A_k \right) \cap \left(\bigcup_{j=1}^{m_2} B_j \right) \\ &= \bigcup_{k=1}^{m_1} \bigcup_{j=1}^{m_2} (A_k \cap B_j), \end{aligned}$$

by the elementary set calculus. But $A_k \cap B_j = A_{kj} \in \mathcal{S}$, and $\{A_{kj} : 1 \leqslant k \leqslant$

$m_1, 1 \leqslant j \leqslant m_2\}$ is a disjoint collection of $m_1 m_2$ sets in $\mathscr{S}$. Hence $A \cap B \in \mathscr{E}$, by definition.

Since $B - A = B \cap A^c$, where $A^c = \mathbb{R}^n - A$ (the complement of A), it suffices to show that $A^c \in \mathscr{E}$, because of the preceding paragraph. However, $A^c = \mathbb{R}^n - A = \bigcup_{k=1}^m C_k$ for disjoint $C_k \in \mathscr{S}$ by Proposition 1(ii). Hence A^c is an elementary set, and so $B - A \in \mathscr{E}$.

Next consider $A \cup B = A \cup (B - A)$. The right side sets are disjoint, and each is elementary. Hence they may be represented as

$$A = \bigcup_{k=1}^{m_1} A_k, \quad B - A = \bigcup_{k=1}^{m_2} D_k, \qquad A_k \in \mathscr{S}, \quad D_k \in \mathscr{S},$$

and the whole collection $\{A_k, D_j : 1 \leqslant k \leqslant m_1, 1 \leqslant j \leqslant m_2\} \subset \mathscr{S}$ is pairwise disjoint. But letting $E_k = A_k$ for $1 \leqslant k \leqslant m_1$, and $= D_j$ for $m_1 + j, 1 \leqslant j \leqslant m_2$, we get

$$A \cup B = \bigcup_{k=1}^{m_1+m_2} E_k, \qquad E_k \in \mathscr{S} \text{ (disjoint)},$$

so that $A \cup B \in \mathscr{E}$.

Finally, by the definition of symmetric difference operation $\vartriangle$,

$$A \vartriangle B = (A - B) \cup (B - A),$$

and by the preceding two paragraphs, $A \vartriangle B \in \mathscr{E}$. The last statement is obvious.

This result (and its proof) can be phrased in a more appealing form as

Corollary 4. *If $\mathscr{E}$ is the collection of all elementary sets formed from $\mathscr{S}$, then it is closed under finite unions and complements (hence also under differences and intersections).*

Another useful consequence of the proposition is given by

Corollary 5. *If $A_i \in \mathscr{S}$, $i = 1, \ldots, m$, are arbitrary, then $A = \bigcup_{i=1}^m A_i \in \mathscr{E}$, so that $A = \bigcup_{i=1}^m \tilde{A}_i$, $\tilde{A}_j \in \mathscr{S}$ and disjoint.*

The above classes of sets form the first (and basic) building blocks to introduce a measure function for many subsets of $\mathbb{R}^n$ which are geometrically complicated. We illustrate this point with the fundamental Lebesgue length (or volume) function. Thus consider $A \in \mathscr{S}$ so that it can be represented as (1):

$$A = \{(x_1, \ldots, x_n) : a_i < x_i \leqslant b_i, i = 1, \ldots, n\}.$$

Note that this may be regarded as the Cartesian product of the n intervals $(a_i, b_i] = \{x_i \in \mathbb{R}: a_i < x_i \leqslant b_i\}$, so that the measure (or volume) of A, say $m(A)$, is given from geometric properties by

$$m(A) = \prod_{i=1}^{n} (b_i - a_i). \tag{5}$$

If some $a_i = b_i$, then the volume is zero, and if $a_i < b_i$ for all i, $a_{j_0} = -\infty$, or $b_{j_0} = +\infty$ for some $i = j_0$, then $m(A) = +\infty$. In case $B \in \mathscr{E}$ is any elementary set having a representation $B = \bigcup_{j=1}^{k} B_j$, $B_j \in \mathscr{S}$ disjoint, then it is natural to define the measure (or volume) of B, $\overline{m}(B)$, as

$$\overline{m}(B) = \sum_{j=1}^{k} m(B_j). \tag{6}$$

This will be meaningful only if $\overline{m}$ does not depend on the particular representation of B. That fact is a consequence of

Proposition 6. *The function $\overline{m}$ on $\mathscr{E}$ given by* (6) *is uniquely defined, nonnegative, additive, monotone, and coincides with m on $\mathscr{S}$. Thus $A \subset B$ in $\mathscr{E}$ implies $0 \leqslant \overline{m}(A) \leqslant \overline{m}(B)$, and $\{C, D\} \subset \mathscr{E}$, $C \cap D = \varnothing$ implies*

$$\overline{m}(C \cup D) = \overline{m}(C) + \overline{m}(D), \tag{7}$$

and $\overline{m}(E) = m(E)$ for all $E \in \mathscr{S}$.

Proof. Let $A = \bigcup_{i=1}^{k} P_i = \bigcup_{j=1}^{r} Q_i$, $P_i \in \mathscr{S}$, disjoint, and $Q_j \in \mathscr{S}$ also disjoint. Since $Q_j \subset A$, $P_i \subset A$, $1 \leqslant i \leqslant k$, $1 \leqslant j \leqslant r$, we have the computation

$$\begin{aligned}
\overline{m}(A) &= \sum_{i=1}^{k} m(P_i) = \sum_{i=1}^{k} m(P_i \cap A) \\
&= \sum_{i=1}^{k} m\left(P_i \cap \bigcup_{j=1}^{r} Q_j\right) = \sum_{i=1}^{k} m\left(\bigcup_{j=1}^{r} (P_i \cap Q_j)\right) \\
&= \sum_{i=1}^{k}\left[\sum_{i=1}^{r} m(P_i \cap Q_j)\right]
\end{aligned}$$

[since $m(\cdot)$ is additive on disjoint rectangles, by the definition (5)]

$$\begin{aligned}
&= \sum_{j=1}^{r} \sum_{i=1}^{k} m(P_i \cap Q_j) \\
&= \sum_{j=1}^{r} m\left(\bigcup_{i=1}^{k} (P_i \cap Q_j)\right) = \sum_{j=1}^{r} m\left(\left(\bigcup_{i=1}^{k} P_i\right) \cap Q_j\right) \\
&= \sum_{j=1}^{r} m(A \cap Q_j) = \sum_{j=1}^{r} m(Q_j).
\end{aligned}$$

Thus $\overline{m}(\cdot)$ does not depend on the particular representation of A, and is uniquely defined.

Next let A, B be disjoint elementary sets such that $A = \bigcup_{j=1}^{k} P_j$ and $B = \bigcup_{j=1}^{r} Q_j$, where all the P_j and the Q_i are also mutually disjoint rectangles. Then

$$A \cup B = \bigcup_{i=1}^{k+r} S_i, \qquad S_i = \begin{cases} P_i, & 1 \leqslant i \leqslant k, \\ Q_j, & i = k+j, \quad 1 \leqslant j \leqslant r. \end{cases}$$

Hence

$$\begin{aligned} \overline{m}(A \cup B) &= \sum_{i=1}^{k+r} m(S_i) = \sum_{i=1}^{k} m(P_i) + \sum_{j=1}^{r} m(Q_j) \\ &= \overline{m}(A) + \overline{m}(B), \end{aligned}$$

which is (7).

Finally, if $A \subset B$ are elementary sets, then $B - A$, A are disjoint elementary sets by Proposition 3. Hence by (7), since $\overline{m}$ (and m) is nonnegative,

$$\overline{m}(B) = \overline{m}(A \cup (B - A)) = \overline{m}(A) + \overline{m}(B - A) \geqslant \overline{m}(A).$$

Thus $\overline{m}$ is monotone, and it is clear that $\overline{m}(A) = m(A)$ if $A \in \mathscr{S}$. This proves the result.

Remark. In view of the last part, we may use the same symbol m for the measure of the elements of $\mathscr{E}$ as well as those of $\mathscr{S}$. Also note that we have not yet used the topological (or metric) properties of $\mathbb{R}^n$.

To use the special properties of $m(\cdot)$ given by (5) *and* the metric properties of $\mathbb{R}^n$, we recall the following result from advanced calculus. (*Our topologies will be Hausdorff unless stated otherwise.*)

Theorem 7. *For a subset A of the Euclidean space $\mathbb{R}^n$, the following statements are equivalent:*

(i) *A has the Heine-Borel property, i.e., every open covering of A admits a finite covering.*

(ii) *A has the Bolzano-Weierstrass property, i.e., every infinite subset of A has a limit point in A.*

(iii) *A is sequentially compact, i.e., every infinite sequence in A has a convergent subsequence with limit in A.*

(iv) *A is closed and bounded, i.e., A has all its limit points in itself and A is contained in some ball of radius $r > 0$.*

(v) *A enjoys the finite intersection property, i.e., if a family of closed subsets of A is such that any finite subcollection has a point in common, then the whole family has at least one point in common.*

A proof of this result is found in any standard advanced calculus text. [See, for instance, Bartle (1976, Section 11).] A set having any one of the above five equivalent properties will be called *compact*. To give some feeling for this concept, we sketch the argument for one of the equivalences (it is valid for any topological space Ω):

Proof of (i) $\Leftrightarrow$ (v). Suppose (i) holds, and let $\{F_\alpha: \alpha \in I\}$ be a closed system of subsets of A having the finite intersection property. Let $U_\alpha = A \cap F_\alpha^c$ so that each U_α is relatively open by definition. (Here F_α^c stands as usual for the complement of F.) By the finite intersection property of the F_α, $A \supset \bigcap_{i=1}^n F_{\alpha_i} \neq \varnothing$ for each $\{\alpha_1, \dots, \alpha_n\} \subset I$. Hence,

$$A \cap \left(\bigcap_{i=1}^n F_{\alpha_i}\right)^c = A \cap \left(\bigcup_{i=1}^n F_{\alpha_i}^c\right) = \bigcup_{i=1}^n U_{\alpha_i} \subset A - \{x\},$$

where $x \in \bigcap_{i=1}^n F_{\alpha_i}$. So no finite subcollection of the (relatively) open sets can cover A. By (i) therefore, $\bigcup_{\alpha \in I} U_\alpha \subset A - \{x_0\}$ for some $x_0 \in A$. Consequently (taking complements) we get $x_0 \in A \cap \bigcap_{\alpha \in I} F_\alpha$ (for the whole collection), and so (v) holds.

For the opposite direction suppose (v) holds. Let $\{U_\alpha: \alpha \in I\}$ be any open covering of A. Then $F_\alpha = A \cap U_\alpha^c$ is (relatively) closed. By hypothesis, since $\{F_\alpha: \alpha \in I\}$ is a (relatively) closed subcollection such that $\bigcap_{\alpha \in I} F_\alpha = A \cap A^c = \varnothing$, there must be a finite subcollection $\{F_{\alpha_1}, \dots, F_{\alpha_m}\}$ with an empty intersection. Thus

$$\varnothing = \bigcap_{i=1}^m F_{\alpha_i} = A \cap \bigcap_{i=1}^m U_{\alpha_i}^c = A \cap \left(\bigcup_{i=1}^m U_{\alpha_i}\right)^c.$$

Hence $\{U_{\alpha_1}, \dots, U_{\alpha_m}\}$ covers A, and (i) follows.

We now use the topological (i.e., metric) nature of $\mathbb{R}^n$, in showing that $\overline{m}$ (and m) of Proposition 6 inherits a stronger "continuity" (or countable additivity) property, of great interest in the theory. The precise statement is given by

Theorem 8. *The measure function $\overline{m}$ on the class of elementary sets $\mathscr{E}$ of $\mathbb{R}^n$ is countably additive in the sense that if $A_n \in \mathscr{E}$, disjoint, and $A = \bigcup_{n=1}^\infty A_n$ is also in $\mathscr{E}$, then*

$$\overline{m}(A) = \sum_{n=1}^{\infty} \overline{m}(A_n). \tag{8}$$

Proof. The argument here, and in all such results of this kind, is to show that the left side is no larger than the right side and vice versa.

First note that $\bigcup_{k=1}^{n} A_k \in \mathscr{E}$ and hence $A - \bigcup_{k=1}^{n} A_k = \bigcup_{k=n+1}^{\infty} A_k \in \mathscr{E}$, by Proposition 3. Then by the (finite) additivity of $\overline{m}$ established in Proposition 6,

$$\begin{aligned}\overline{m}(A) &= \overline{m}\left(\bigcup_{k=1}^{n} A_k\right) + \overline{m}\left(\bigcup_{k=n+1}^{\infty} A_k\right) && \text{since these are disjoint}\\ &= \sum_{k=1}^{n} \overline{m}(A_k) + \overline{m}\left(\bigcup_{k=n+1}^{\infty} A_k\right) && \text{by (7)}\\ &\geqslant \sum_{k=1}^{n} \overline{m}(A_k) && \text{since } \overline{m}(\cdot) \text{ is nonnegative.}\end{aligned}$$

Letting $n \to \infty$, and noting that the left side is independent of n, it follows that

$$\overline{m}(A) \geqslant \sum_{k=1}^{\infty} \overline{m}(A_k). \tag{9}$$

Note that (9) is always true for any nonnegative additive measure function, and no topological properties are used. However, the opposite inequality does not obtain without such an additional hypothesis.

If $\overline{m}(A_k) = +\infty$ for at least one k, or if the right side of (8) is infinite, then the result follows from (9). So suppose that $\sum_{k=1}^{\infty} \overline{m}(A_k) < \infty$. Since if $\overline{m}(A) = 0$, (8) is true and trivial by (9), let $\overline{m}(A) > 0$; then choose an elementary set E such that its closure $\overline{E} \subset A$ and such that with $\alpha = \overline{m}(A)$,

$$\alpha \geqslant \overline{m}(E) > \alpha - \frac{\epsilon}{2}. \tag{10}$$

That this is possible is seen as follows. By definition $A = \bigcup_{k=1}^{m} I_k$, where $I_i \in \mathscr{S}$. Thus $I_k = \{(x_1, \ldots, x_n): a_{kj} < x_j \leqslant b_{kj},\ j = 1, \ldots, n\}$, $k = 1, \ldots, m$, without equality if $b_{kj} = \infty$ for any j. Then we can choose a rectangle whose closure (i.e., both boundaries included) is contained in I_k but whose measure (= volume) satisfies the inequality (10). In other words, if A itself is a rectangle and $0 < \alpha < m(A)$, then a rectangle E can be found (nonuniquely in general) to satisfy $\overline{E} \subset A$ and (10). If A is elementary, then we repeat this procedure finitely many times to get $E = \bigcup_{i=1}^{m} E_i \in \mathscr{E}$, $\overline{E}_i \subset I_i$, satisfying (10).

Next, since $\overline{m}(A_k) < \infty$, choose a set $J_k \in \mathscr{E}$ such that its interior $\mathring{J}_k$ contains A_k and

$$\overline{m}(J_k) < \overline{m}(A_k) + \frac{\epsilon}{2^{k+1}}. \tag{11}$$

This is done as above, except that the right end points (boundaries) are enlarged so that the open sets $\mathring{J}_k$ contain the corresponding subsets A_k. Thus one has

$$\overline{E} \subset A \subset \bigcup_{k=1}^{\infty} A_k \subset \bigcup_{k=1}^{\infty} \mathring{J}_k. \tag{12}$$

Since $\overline{E}$ is a finite union of closed and bounded rectangles, it is compact and is covered by the open sets $\{\mathring{J}_k: k \geqslant 1\}$. Hence by Theorem 7, $\overline{E}$ can be covered by a finite subcollection. Writing this as

$$E \subset \overline{E} \subset \bigcup_{k=1}^{r} \mathring{J}_k \subset \bigcup_{k=1}^{r} J_k \tag{13}$$

for some $1 \leqslant r < \infty$, we note that the last union in (13) is in $\mathscr{E}$. Then by Corollary 5, this can be disjunctified to get

$$\bigcup_{k=1}^{r} J_k = J_1 \cup (J_2 - J_1) \cup \cdots \cup \left(J_r - \bigcup_{i=1}^{r-1} J_i \right),$$

and hence

$$\begin{aligned} \overline{m}(E) \leqslant \overline{m}\left(\bigcup_{k=1}^{r} J_k \right) &= \sum_{k=1}^{r} \overline{m}\left(J_k - \bigcup_{i=1}^{r-1} J_i \right) \\ &\leqslant \sum_{k=1}^{r} \overline{m}(J_k) \qquad \text{since } \overline{m} \text{ is monotone} \\ &\leqslant \sum_{k=1}^{\infty} \overline{m}(A_k) + \frac{\epsilon}{2} \qquad \text{by (11).} \end{aligned} \tag{14}$$

It follows from (10) and (14) that

$$\alpha \leqslant \sum_{k=1}^{\infty} \overline{m}(A_k) + \epsilon. \tag{15}$$

Since $\epsilon > 0$ is arbitrary and $0 < \alpha < \overline{m}(A)$ is any number, we get on letting $\epsilon \downarrow 0$ and then $\alpha \uparrow \overline{m}(A)$, that

$$\overline{m}(A) \leqslant \sum_{k=1}^{\infty} \overline{m}(A_k). \tag{16}$$

Hence (8) follows from (9) and (16), and the proof is complete.

The details are spelled out here to show how the topological structure of $\mathbb{R}^n$ intervenes critically in the inequality (16) so that $\overline{m}$ on $\mathscr{E}$ is really countably additive. Here the continuity of the volume measure was also used. These two aspects will now be abstracted, and the same method of proof applied in those cases. This (continuity) property is called "regularity" and will be generalized later.

EXERCISES

0. Let $\mathbb{R}^n$ be the number space and $\mathcal{O}(\mathcal{C})$ be the class of all of its open (closed) intervals. Show that

(a) $\mathcal{O}$ is closed under finite intersections and arbitrary unions, but Proposition 1 is not true for it;

(b) $\mathcal{C}$ is closed under finite unions and arbitrary intersections, but Proposition 1 is not true for it.

1. Let $\mathcal{S}$ be the class of all rectangles of $\mathbb{R}^n$ of the form (1). If $\tilde{\mathcal{E}} = \{A \subset \mathbb{R}^n: A = \bigcup_{k=1}^{\infty} P_k,\ P_k \in \mathcal{S}$ (P_k need not be disjoint)$\}$, show that $\tilde{\mathcal{E}}$ is closed under countable unions, finite intersections, and finite differences, and that $\tilde{\mathcal{E}} \supset \mathcal{E}$. The same holds if $\mathcal{S}$ contains only proper rectangles (so that $\mathbb{R}^n \notin \mathcal{S}$). [Here and below $\mathcal{E}$ is the same as in Definition 2.]

2. Let $\tilde{\mathcal{E}}$ be as above. Show that $A \in \tilde{\mathcal{E}} \Rightarrow A = \bigcup_{k=1}^{\infty} Q_k,\ Q_k \in \mathcal{S},\ Q_k \cap Q_{k'} = \varnothing$ if $k \neq k'$. Hence deduce that an additive set function $\tilde{m}$ can be unambiguously defined on $\tilde{\mathcal{E}}$ such that $\tilde{m}(A) = \overline{m}(A)$, if $A \in \mathcal{E}$. Verify that $\tilde{m}$ is also countably additive on $\tilde{\mathcal{E}}$, so that it is an extension of m and $\overline{m}$ to this larger collection.

3. Let $\Omega = \{x = (x_1, x_2, \ldots): x_i = 0$ or 1 for all $i\}$. Let $\mathcal{S} = \{I_n \subset \Omega: I_n$ consist of those $x \in \Omega$ whose first n components $x_1, \ldots, x_n$ have a prescribed form, $n \geqslant 1\}$. We regard $\varnothing, \Omega$ as members of $\mathcal{S}$. Thus I_2 may be the set of all x in Ω such that either $x_1 = 1$ and $x_2 = 0$ or $x_1 = 0$ and $x_2 = 1$. Let $I_0 = \Omega$. Verify that the family $\mathcal{S}$ has the following properties: If $I, J \in \mathcal{S}$, then (i) $I \cap J \in \mathcal{S}$; (ii) if $I \subset J$, then there exist disjoint $I_1, I_2, \ldots, I_k$ in $\mathcal{S}$ such that $J - I = \bigcup_{i=1}^{k} I_i$. (In some descriptions, Ω is called the coin tossing (or binary) space, since 1 can stand for head and 0 for tail in an infinite sequence of tosses of a coin.)

4. A prototype of our length (or volume) function can be given as follows. Let Ω be a compact Hausdorff space, and $\mathcal{E}$ be a class of subsets of Ω such that (i) $\varnothing, \Omega$ are in $\mathcal{E}$, (ii) $\mathcal{E}$ is closed under finite unions and differences. Suppose $m: \mathcal{E} \to \mathbb{R}$ is an additive nonnegative function [i.e. $m(A \cup B) = m(A) + m(B)$, A, B disjoint in $\mathcal{E}$; $0 \leqslant m(A) < \infty$, $A \in \mathcal{E}$] that is "regular", i.e., for each $A \in \mathcal{E}$ and $\epsilon > 0$, there exist G, F in $\mathcal{E}$ such that $\overline{F}$, the closure of F, $\subset A \subset \mathring{G}$, the interior of G, and for each $D \in \mathcal{E}$, $D \subset G - F$, we have $m(D) < \epsilon$. Show that $m(\cdot)$ is countably additive on $\mathcal{E}$. [This is a particular case of a result due to A. D. Alexandrov.]

5. Let $\Omega = [0,1]$ and $F_1 = \Omega - (\frac{1}{3}, \frac{2}{3})$, so that we have deleted the open middle third of the unit interval Ω. Similarly delete open intervals $(\frac{1}{9}, \frac{2}{9})$

and $(\frac{7}{9}, \frac{8}{9})$ from F_1, and call it F_2. Proceeding in this way we delete at the nth stage 2^{n-1} open intervals from F_{n-1}, each of length 3^{-n}, so that $\Omega \supset F_1 \supset \cdots \supset F_n$, and let $F = \cap_{n=1}^{\infty} F_n$. Then F is called the *Cantor set*. If the class $\tilde{\mathscr{E}}$ of Problem 1 (and also $\mathscr{E}$) is considered for this Ω, then show that each open interval of Ω is in $\tilde{\mathscr{E}}$ (but not in $\mathscr{E}$), and then $F \in \tilde{\mathscr{E}}$. (Note that $(a, b) = \cup_{n \geqslant n_0}(a, b - 1/n]$.) Deduce that $\tilde{m}(F) = 0$, where the measure $\tilde{m}$ [derived from the length function $m(\cdot)$] is as in Exercise 2. However, F is closed and contains the numbers $0, 1, \frac{1}{3}, \frac{2}{3}, \frac{1}{9}, \frac{2}{9}, \frac{7}{9}, \frac{8}{9}, \ldots$. More generally, if we let $\xi = \sum_{k=1}^{\infty} 3^{-k} a_k$, $\eta = \sum_{k=1}^{\infty} 2^{-k} b_k(a_k)$, where $a_k = 0, 2$, and where $b_k(0) = 0$, $b_k(2) = 1$, $k \geqslant 1$, then show that $\xi \in F$, $\eta \in \Omega$ and that the correspondence $\xi \leftrightarrow \eta$ is a bijection between F and Ω. [Thus F has the same cardinality as Ω, whereas $\tilde{m}(\Omega) = 1$ and $\tilde{m}(F) = 0$. This indicates the intricacy of measure theory. The Cantor set will be very useful for both illustrations and counterexamples in our development.]

6. There is yet another way of defining a class $\mathscr{E}'$ of sets and then an extension of the same measure function m. Thus let $\mathscr{I}$ be the class of all intervals of $\mathbb{R}^n$, which may be closed, open, or half-closed (so $\mathscr{I}$ does *not* have the properties of $\mathscr{S}$ used in Exercises 1, 2 or that of Proposition 1 or 2). Let $\mathscr{E}'$ be the class of all finite disjoint unions of elements of $\mathscr{I}$. Show that $\mathscr{E}'$ is closed under finite unions, intersections, and differences. If $I \in \mathscr{I}$, then let $m(I)$ be the usual volume expression, which is the same as in (5). For each $A \in \mathscr{E}'$ (so $A = \cup_{j=1}^{r} I_j$, disjoint union) define $m'(A) = \sum_{j=1}^{r} m(I_j)$. Show that m' is well defined on $\mathscr{E}'$, nonnegative, and countably additive, as in Exercise 2. (Note that $\mathscr{E}' \neq \mathscr{E}$ and $\mathscr{E}' \not\subset \tilde{\mathscr{E}}$. But we show later that all these three collections generate the same class of "Borel" sets, and thus there are different routes to arrive at this class.)

1.3. ABSTRACTION OF THE SALIENT FEATURES

In the preceding work most of what we have done did not depend on the topology or even the linear structure of $\mathbb{R}^n$. An exception for this is Theorem 2.8. To understand the problem clearly, we abstract the main properties and state them for a general point set. Not only will this identify the key features of the measure function, but the theory applies to much more general spaces than $\mathbb{R}^n$, leading to some natural extensions. We first introduce some terminology.

Let Ω be a nonempty point set, and let $\mathscr{S}$ and $\mathscr{E}$ be classes of subsets of Ω. Then we have

Definition 1. The class $\mathscr{E}$ is a *ring* of subsets of Ω if

(i) $\varnothing \in \mathscr{E}$,

(ii) $\{A, B\} \subset \mathscr{E} \Rightarrow A \cup B \in \mathscr{E}$ and $A - B \in \mathscr{E}$.

A ring $\mathscr{E}$ is an *algebra* if also $\Omega \in \mathscr{E}$. [The words "clan" and "field" are also used in the literature for "ring" and "algebra" respectively.]

Definition 1′. The class $\mathscr{S}$ is a *semiring* of subsets of Ω if

(i) $\varnothing \in \mathscr{S}$,
(ii) $\{A, B\} \subset \mathscr{S} \Rightarrow A \cap B \in \mathscr{S}$, and
(iii) $\{A, B\} \subset \mathscr{S}$ and $A \subset B \Rightarrow$ there are A_i, $0 \leqslant i \leqslant m$, in $\mathscr{S}$ such that $A = A_0 \subset A_1 \subset \cdots \subset A_m = B$ with the property that $A_i - A_{i-1} \in \mathscr{S}$.

A semiring $\mathscr{S}$ is a *semialgebra* if also $\Omega \in \mathscr{S}$.

Since $A \cap B = A - (A - B)$, a ring is thus closed under finite intersections. Hence Propositions 2.1 and 2.3 can be restated as

Proposition 2. *The class $\mathscr{S}$ of all rectangles of $\mathbb{R}^n$ is a semialgebra, and the class $\mathscr{E}$ of all elementary sets formed from the elements of $\mathscr{S}$ is an algebra. Moreover, $\mathscr{E}$ is the smallest algebra that contains $\mathscr{S}$.*

Proof. Only the last statement needs a proof. Indeed, if $\mathscr{E}_0$ is the smallest algebra containing $\mathscr{S}$, then clearly $\mathscr{E}_0 \subset \mathscr{E}$. On the other hand, if $A \in \mathscr{E}$, then $A = \bigcup_{i=1}^r A_i$, $A_i \in \mathscr{S}$, disjoint. Hence $A_i \in \mathscr{E}_0$, and so $\bigcup_{i=1}^r A_i = A \in \mathscr{E}_0$, since it is closed under finite unions. Thus $\mathscr{E} \subset \mathscr{E}_0$, and so $\mathscr{E} = \mathscr{E}_0$, as asserted.

If $\mathscr{F}$ is a class of subsets of Ω, then the (semi-) ring (or an algebra) *generated by* $\mathscr{F}$ is the smallest such family $\mathscr{S}(\mathscr{F})$ [or $\mathscr{E}(\mathscr{F})$] which contains $\mathscr{F}$. Such a generated family is simply the intersection of all (semi-) rings or algebras containing $\mathscr{F}$. These rings and algebras always exist, since, for instance, the power set $\mathscr{P} = 2^\Omega$ of Ω is one such. Just as a length (or volume) function $m(\cdot)$ on rectangles or intervals is given from prior considerations, one may assume a measure function given on a "primitive" class of sets as follows:

Definition 3. An *additive function* μ on a semiring $\mathscr{S}$ into $\mathbb{R}$ is a mapping such that

(i) $\mu(E) \geq 0$ for each $E \in \mathscr{S}$,
(ii) if $A = \bigcup_{i=1}^r E_i$, $E_i \in \mathscr{S}$, disjoint, then $\tilde{\mu}$ on $\mathscr{E}$ is defined by $\tilde{\mu}(A) = \sum_{i=1}^r \mu(E_i)$, and
(iii) $\mu(\varnothing) = 0$.

$\tilde{\mu}$ is called a *measure function* (or simply a *measure*) on $\mathscr{E}(\mathscr{S}) \to \overline{\mathbb{R}}^+$ if the following hold:

(a) $\tilde{\mu}$ is additive,
(b) $A_n \in \mathscr{E}$, A_n disjoint, and $A = \bigcup_{n=1}^{\infty} A_n \in \mathscr{E} \Rightarrow \tilde{\mu}(A) = \sum_{n=1}^{\infty} \tilde{\mu}(A_n)$, where the series is convergent or $= +\infty$.

Here (b) is called *countable additivity* or *σ-additivity* of $\tilde{\mu}$. (Since $\tilde{\mu} = \mu$ on $\mathscr{S}$, we write μ for both classes indifferently.)

As an example, $\mu = m$ (the length or volume) is a measure on the ring of elementary sets of $\mathbb{R}^n$ (cf. Theorem 2.8). This important special case is called *Lebesgue measure*. Note that by the result of Proposition 6, $\tilde{\mu}$ given by (i) and (ii) of the above definition is unambiguous. We shall see, in Section 2.6, that with each continuous monotone increasing function $f: \mathbb{R} \to \mathbb{R}$, it is possible to let $\mu_f: \mathscr{S} \to \mathbb{R}$, where $\mu_f([a, b]) = f(b) - f(a)$. This again defines a measure, called the *Lebesgue-Stieltjes* measure, $f(x) = x$ corresponding to Lebesgue's length function, occasionally termed *linear Lebesgue measure*.

For many problems it is necessary to measure sets that are more complicated than those of the elementary class $\mathscr{E}$. The general collection for which our methods admit a natural extension can be stated as

Definition 4. An algebra (ring) of sets $\mathscr{A}$ of a point set Ω is called a *σ-algebra* (*σ-ring*) if $A_n \in \mathscr{A}$, $n = 1, 2, \ldots \Rightarrow \bigcup_{n=1}^{\infty} A_n \in \mathscr{A}$ also. A ring $\mathscr{A}$ is a *δ-ring* if $A_n \in \mathscr{A} \Rightarrow \bigcap_{n=1}^{\infty} A_n \in \mathscr{A}$. [The word "tribe" is also used for "σ-ring", and sometimes "σ-field" for "σ-algebra", in the literature.]

Example A. The elementary class $\mathscr{E}$ of $\mathbb{R}^n$ considered above is an algebra but not a σ-algebra. If $\tilde{\mathscr{E}}_0 = \{A \in \tilde{\mathscr{E}}: \tilde{m}(A) < \infty\}$, $\tilde{\mathscr{E}}$, $\tilde{m}$ as in Exercise 2.1, then $\tilde{\mathscr{E}}_0$ is a δ-ring but not a σ-ring.

Example B. Another simple one is $\Omega = [0, 1]$, and $\mathscr{A}_0$ the class of all sets that are finite or have finite complements. This is an algebra but not a σ-algebra and not a σ-ring.

Example C. If in the above $\mathscr{A}_1$ is the collection of countable sets or those that have countable complements, then $\mathscr{A}_1$ is a σ-algebra.

The importance and intricacy of these larger classes will be seen later. For instance, it will become clear that the σ-algebras generated by (i) the semiring or algebra of rectangles in $\mathbb{R}^n$, (ii) the elementary sets $\mathscr{E}$, (iii) $\tilde{\mathscr{E}}$ of Exercise 2.1, and (iv) $\mathscr{E}'$ of Exercise 2.6 are all identical. Thus when we show in the next chapter how these Lebesgue measure functions have unique extensions to their generated σ-algebras, it will follow that the different starting classes of measuring sets lead to the same σ-rings; but this result is not evident initially.

Because of the importance of set operations in our subject, we now introduce methods that aid in simplifying some computations. Thus if $A \subset \Omega$ is a set, then the *indicator function*, denoted $\chi_A: \Omega \to \mathbb{R}$, is defined as $\chi_A(\omega) = 1$ if $\omega \in A$, and $= 0$ if $\omega \in \Omega - A$. Also, if $A_1, A_2, \ldots$ is any sequence of subsets of Ω, then we define their lim sup and lim inf as

$$\limsup_n A_n = \overline{\lim_n} A_n = \text{the set of points that belong to infinitely many } A_n,$$

$$\liminf_n A_n = \underline{\lim_n} A_n = \text{the set of points that belong to all but finitely many } A_n.$$

Clearly

$$\underline{\lim_n} A_n \subset \overline{\lim_n} A_n,$$

and if there is equality here, we say that the sequence has a limit, denoted $\lim_n A_n$. It is not difficult to see that

$$E = \overline{\lim_n} A_n = \bigcap_{k \geq 1} \bigcup_{n \geq k} A_n, \tag{1}$$

$$F = \underline{\lim_n} A_n = \bigcup_{k \geq 1} \bigcap_{n \geq k} A_n. \tag{2}$$

These can be stated alternatively as follows:

$$\chi_E = \overline{\lim_n} \chi_{A_n}, \qquad \chi_F = \underline{\lim_n} \chi_{A_n}, \tag{3}$$

where the limits are in terms of the calculus of real functions. Thus the sequence of sets $\{A_n: n \geq 1\}$ has a limit A iff (if and only if) the corresponding function sequence $\{\chi_{A_n}: n \geq 1\}$ has the limit χ_A. Similarly if A, B are two subsets of Ω and $A \vartriangle B$ is their symmetric difference, then

$$\chi_{A \vartriangle B} = |\chi_A - \chi_B|. \tag{4}$$

In this abstract setting, let us introduce some further terminology regarding measure functions on rings. If $\mathscr{E}$ is a ring and μ on $\mathscr{E}$ is a measure function, as in Definition 3, then it is called *countably subadditive*, or *σ-subadditive*, if $A_n \in \mathscr{E}, \bigcup_{n \geq 1} A_n \in \mathscr{E} \Rightarrow$

$$\mu\left(\bigcup_{n=1}^{\infty} A_n\right) \leq \sum_{n=1}^{\infty} \mu(A_n). \tag{5}$$

If the opposite inequality holds, then it is called *countably* (or σ-) *superadditive*.

Lemma 5. *If $\mu \geqslant 0$ on a ring $\mathscr{E}$ is finitely additive, $\mu(\varnothing) = 0$, and countably subadditive, then it is σ-additive, so that it is a measure.*

Proof. Since $\mathscr{E}$ is a ring, if $A_n \in \mathscr{E}$, disjoint, and $\bigcup_{n \geqslant 1} A_n = A \in \mathscr{E}$, then $\bigcup_{k=1}^{n} A_k \in \mathscr{E}$ and $\bigcup_{k=n+1}^{\infty} A_k = A - \bigcup_{k=1}^{n} A_k \in \mathscr{E}$. So by the finite additivity of μ we have

$$\mu(A) = \mu\left(\bigcup_{k=1}^{n} A_k\right) + \mu\left(\bigcup_{k=n+1}^{\infty} A_k\right)$$

(since the displayed sets are disjoint)

$$\geqslant \sum_{k=1}^{n} \mu(A_k), \qquad \text{since} \quad \mu \geqslant 0 \quad \text{and is additive.}$$

Letting $n \to \infty$, we get

$$\mu(A) \geqslant \sum_{n=1}^{\infty} \mu(A_n). \tag{6}$$

But by the σ-subadditivity hypothesis

$$\mu(A) \leqslant \sum_{n=1}^{\infty} \mu(A_n). \tag{7}$$

These two inequalities imply μ is σ-additive. Since $\mu(\varnothing) = 0$, by Definition 3, μ is a measure, as asserted.

Comparing this with Theorem 2.8, we see that the first part, (6), on σ-superadditivity has the same argument, but the verification of σ-subadditivity was the hard part of that theorem which depended on the topological properties of the space. That property is *assumed* here. In general, it cannot be deduced from the other hypotheses in the abstract case.

Since $\mu(\Omega) = +\infty$ is possible [as in the case that $m(\mathbb{R}^n) = +\infty$], we did not assume that μ is real valued. It is *extended nonnegative real valued*, i.e., $\mu(A) \in \overline{\mathbb{R}}^+ = \mathbb{R}^+ \cup \{+\infty\}$. For this reason, we consider the extended real line $\overline{\mathbb{R}} = \mathbb{R} \cup \{+\infty, -\infty\}$. The order structure of $\mathbb{R}$ is extended to $\overline{\mathbb{R}}$. Namely:

1. $-\infty < +\infty$ and $-\infty < x < +\infty$; (i.e. $x \in \mathbb{R}$)
2. $\pm\infty + x = x \pm \infty = \pm\infty$, $x \in \mathbb{R}$;
3. $x \cdot (+\infty) = +\infty = (+\infty) \cdot x = (-\infty) \cdot (-\infty)$ if $x \in \overline{\mathbb{R}}^+ - \{0\}$, and $x > 0 \Rightarrow x \cdot (-\infty) = -x \cdot (+\infty) = (+\infty) \cdot (-x) = (-\infty) \cdot (+\infty) = -\infty$;

4. if $x < 0$ then $x \cdot (+\infty) = (+\infty) \cdot x = -\infty$ and $x \cdot (-\infty) = (-\infty) \cdot x = +\infty$; and
5. $0 \cdot (\pm\infty) = (\pm\infty) \cdot 0 = 0$.

The last property is not the usual indeterminate quantity, familiar in other parts of analysis. It is explicitly assumed here, since this proves to be convenient for measures. However, it makes the distributive and cancellation laws invalid in $\overline{\mathbb{R}}$, and we use them only when restricted to $\mathbb{R}$.

With the above conventions, intervals, $I \subset \overline{\mathbb{R}}$ are defined with end points real or $\pm\infty$. The usual limit properties and topology of $\mathbb{R}$ extend without any difficulty. For instance, $(a, +\infty]$ and $[-\infty, b)$ are taken as the neighborhoods of $+\infty$ and $-\infty$ respectively with a, b in $\mathbb{R}$. The other neighborhoods of $\mathbb{R}$ are used for $\overline{\mathbb{R}}$ without change.

Having thus abstracted the salient features of the model $\mathbb{R}^n$, we shall begin, in the next chapter, a general study of measures. Also, classes of sets which can be measured are considered and we then proceed to functions which may be integrated with such measures on different types of (measurable) sets.

EXERCISES

0. (a) Let A, B be subsets of a nonempty set Ω. If $A - B = A$, $B - A = B$, find $A \cap B$.

(b) Let A, B, C be subsets of Ω. Find the relation (comparison) between $(A \cup B) \cap (A \cup C)$ and $A \cup (B \cap C)$.

(c) Let A_n, $n \geq 1$, be a sequence of subsets of Ω. Verify that $\underline{\lim}_n A_n \subset \overline{\lim}_n A_n$.

(d) Let $A_1, \ldots, A_n$ be disjoint subsets of Ω, with $\Omega = \bigcup_{i=1}^n A_i$. If $a_i = \chi_{A_i}$, $\bar{a}_i = 1 - a_i$, then establish the following identity:

$$\sum_{i=1}^{n} a_i = \sum_{j=0}^{n} j b_j,$$

where $b_j = \Sigma a_{i_1} a_{i_2} \ldots a_{i_j} \bar{a}_{i_{j+1}} \ldots \bar{a}_{i_n}$, the summation ranging over all permutations $(i_1, \ldots, i_n)$ of $(1, 2, \ldots, n)$ divided into two groups of j and $(n - j)$ items.

1. Establish the identities (1), (2), (3), and (4).

2. Using (4), verify the following relations for any subsets A, B, C of a given nonempty set Ω, with $\vartriangle$ as the symmetric difference:

(a) $A \vartriangle (B \vartriangle C) = (A \vartriangle B) \vartriangle C$, $A \cap (B \vartriangle C) = (A \cap B) \vartriangle (A \cap C)$.

(b) $A \vartriangle B = B \vartriangle A = A^c \vartriangle B^c$, $(A \vartriangle B) \cap (C \vartriangle A) = A \vartriangle (B \cup C) - (B \vartriangle C)$.

(c) Given A, B, there is a unique $D \subset \Omega$ such that $A \vartriangle D = B$. (Try $D = A \vartriangle B$.)

3. Let $\{A_n: n \geqslant 1\}$ and $\{B_n: n \geqslant 1\}$ be sequences of subsets of a nonempty set Ω such that $A_n \subset A_{n+1}$, $B_n \supset B_{n+1}$. Show that $\lim_n A_n = A$, $\lim_n B_n = B$ exist. If $C_{2n-1} = A_n$, $C_{2n} = B_n$, $n \geqslant 1$, show that $\lim_n C_n$ exists iff $A = B$, and in any case $\liminf_n C_n = A \cap B$.

4. Abstract measures arise in a number of different ways. We indicate a few examples here. Let Ω be a set of an infinite number of points, and $\mathscr{P}$ be its power set. Then establish the following assertions:

(a) If $\mu(A) = 0$ for $A = \varnothing$, and $= +\infty$ for $A \neq \varnothing$, $A \in \mathscr{P}$, then μ is a measure on $\mathscr{P}$.

(b) If $\mathscr{E}(\subset \mathscr{P})$ is a collection of sets which are finite, or which have finite complements, and $\mu(A) = 0$ for A finite and $= \alpha$ otherwise, where $A \in \mathscr{E}$, then $\mu: \mathscr{E} \to \mathbb{R}^+$ is not additive for $0 < \alpha < \infty$. What happens if $\alpha = 0$ or $\alpha = +\infty$?

(c) If $\mu(A)$ is the number of points in A, $A \in \mathscr{P}$ $(= +\infty$ if A is not finite), then μ is a measure on $\mathscr{P}$.

5. Let $\{\mathscr{A}_i: i \in I\}$ be a collection of σ-rings on Ω. Show that $\mathscr{A} = \bigcap_{i \in I} \mathscr{A}_i$ is a σ-ring, but $\mathscr{A}_{i_1} \cup \mathscr{A}_{i_2}$ need not even be a ring. If, on the other hand, the index set I is ordered by $\leqslant$ (e.g., $I = \mathbb{N}$, the integers), and $\mathscr{A}_{i_1} \subset \mathscr{A}_{i_2}$ iff $i_1 \leqslant i_2$, verify that $\bigcup_{i \in I} \mathscr{A}_i$ is a ring but need not be a σ-ring if I is infinite.

6. A class $\mathscr{L}$ of subsets of Ω containing $\varnothing$ is called a (σ-) *lattice* if it is closed under (countable) unions and (countable) intersections. Demonstrate that (i) a (σ-) ring is a (σ-) lattice but not conversely, (ii) a σ-lattice need not be a δ-ring, or conversely, and (iii) a σ-lattice which is a ring is a σ-ring. Give an example of a lattice which is not a ring.

7. Let $\mathscr{E}$ be an algebra of subsets of Ω, and $\mu: \mathscr{E} \to \mathbb{R}^+$ be additive. Show that μ is then a measure (i.e., σ-additive) iff for each sequence $A_n \in \mathscr{E}$ such that $A_n \supset A_{n+1}$, one has $\bigcap_{n=1}^{\infty} A_n = \varnothing \Rightarrow \lim_{n \to \infty} \mu(A_n) = 0$. Given an example showing that the statement need not hold when $\mathscr{E}$ is only a ring instead of being an algebra. Verify, however, that the "if" part holds for rings even when $\mathbb{R}^+$ is replaced by $\overline{\mathbb{R}}^+$, and that for any A, B in $\mathscr{E}$, $\mu(A \cup B) = \mu(A) + \mu(B)$ (even if μ is only additive but nonnegative).

2

MEASURABILITY AND MEASURES

A deeper study of a large class of sets that can be measured by a given abstract measure on a ring is given here. For this purpose the fundamental Carathéodory process is employed; and the extension procedure for measures, their properties, and the concrete Lebesgue-Stieltjes measures are studied in detail. A brief discussion of metric (outer) measures is also included. This chapter contains the main work on the first phase of a generalization of calculus pointed out in the discussion of Section 1.1.

2.1. MEASURABILITY AND CLASS PROPERTIES

Since a σ-ring is closed under countable unions, differences, and (hence) countable intersections (as noted in Section 1.3), they form natural domains for (general) measure functions, if the latter can be defined on them. Let us note some advantages of such a definition.

(In this book, a *set* is a *nonempty* collection of objects or points, unless the contrary is stated.)

Proposition 1. *Let $\mathscr{T}$ be a σ-ring of an abstract nonvoid point set Ω, and suppose μ on $\mathscr{T}$ is a measure.* [*Thus it is σ-additive and nonnegative with $\mu(\varnothing) = 0$.*] *Then μ has the following properties on $\mathscr{T}$*:

(i) $A_n \in \mathscr{T}$, $A_n \subset A_{n+1}$, $n \geq 1 \Rightarrow \lim_{n\to\infty}\mu(A_n) = \mu(\lim_{n\to\infty} A_n)$;

(ii) $A_n \in \mathscr{T}$, $A_n \supset A_{n+1}$, $n \geq 1$, *and* $\mu(A_{n_0}) < \infty$ *for some* $n_0 \Rightarrow \lim_{n\to\infty}\mu(A_n) = \mu(\lim_{n\to\infty} A_n)$;

(iii) $A_n \in \mathscr{T} \Rightarrow \mu(\liminf_{n\to\infty} A_n) \leq \liminf_{n\to\infty}\mu(A_n)$, *and if also* $\mu(\sup_{n \geq n_0} A_n) < \infty$ *for some* n_0, *then* $\mu(\limsup_{n\to\infty} A_n) \geq \limsup_{n\to\infty}\mu(A_n)$;

(iv) $A_n \in \mathscr{T}$ *and* $\sum_{n=1}^{\infty}\mu(A_n) < \infty \Rightarrow \mu(\limsup_{n\to\infty} A_n) = 0$;

(v) $A_n \in \mathscr{T}$ *and* $A = \bigcup_{n=1}^{\infty} A_n \Rightarrow \mu(A) \leq \sum_{n=1}^{\infty}\mu(A_n)$.

All these properties of μ are simple but extremely useful in our work.

Proof. (i): Since the A_n are increasing, they have a limit. In fact

$$\underline{\lim_n} A_n = \bigcup_{k\geq 1}\bigcap_{n\geq k} A_n = \bigcup_{k\geq 1} A_k = \bigcap_{k\geq 1}\bigcup_{n\geq k} A_n = \overline{\lim_n} A_n. \tag{1}$$

Thus $A = \lim_n A_n = \bigcup_{n=1}^{\infty} A_n \in \mathscr{T}$ (since $\mathscr{T}$ is a σ-ring). Being a measure, μ is increasing, and $A \supset A_k \Rightarrow \mu(A) \geq \mu(A_k)$. If $\mu(A_k) = \infty$ for some k, then the result holds. So let $\mu(A_k) < \infty$ for all k. Set $A_0 = \varnothing$, and $B_k = A_k - A_{k-1}$, $k \geq 1$. Then $B_k \in \mathscr{T}$, and disjoint. Also $A = \bigcup_{k=1}^{\infty} B_k$, $A_k = B_k \cup A_{k-1}$, both being disjoint unions. Hence $\mu(A_k) = \mu(B_k) + \mu(A_{k-1})$, so that we have

$$\mu(B_k) = \mu(A_k) - \mu(A_{k-1}), \quad \text{since} \quad \mu(A_{k-1}) < \infty, \tag{2}$$

and by σ-additivity

$$\mu(A) = \sum_{k=1}^{\infty} \mu(B_k) = \lim_{n\to\infty} \sum_{k=1}^{n} [\mu(A_k) - \mu(A_{k-1})], \qquad \text{by (2)}$$

$$= \lim_{n\to\infty} \mu(A_n), \qquad \text{since} \quad \mu(A_0) = 0.$$

(ii): Now $A_n \supset A_{n+1}$, so that again the sequence has a limit, since

$$\overline{\lim_n} A_n = \bigcap_{k\geq 1}\bigcup_{n\geq k} A_n = \bigcap_{k\geq 1} A_k = \bigcup_{k\geq 1}\bigcap_{n\geq k} A_n$$

$$= \underline{\lim_n} A_n = A \quad \text{(say)}. \tag{3}$$

Given $\mu(A_{n_0}) < \infty$, consider $C_k = A_{n_0} - A_k$, for $k \geq n_0$. Then $C_k \in \mathscr{T}$, $C_k \subset C_{k+1}$, and by (1) $\lim_{k\to\infty} C_k = A_{n_0} - A \in \mathscr{T}$. Hence $A \subset A_{n_0}$ and

$$\mu(A_{n_0}) - \mu(A) = \mu(A_{n_0} - A), \qquad \text{as in (2)},$$

$$= \mu\left(\lim_{k\to\infty} C_k\right)$$

$$= \lim_{k\to\infty} \mu(C_k), \qquad \text{by (1)}$$

$$= \lim_{k\to\infty} \left[\mu(A_{n_0}) - \mu(A_k)\right], \qquad \text{as in (2)}$$

$$= \mu(A_{n_0}) - \lim_{n\to\infty} \mu(A_n). \tag{4}$$

Canceling the finite number $\mu(A_{n_0})$ on both sides of (4), one gets (ii).

We note at once that this result will be false if $\mu(A_{n_0}) < \infty$ is dropped. For instance, let $\mu(\cdot) = m(\cdot)$, the Lebesgue measure on $\mathbb{R}$, $A_n = (n, \infty)$. Then

$\mu(A_n) = +\infty$, all n, and $\lim_n A_n = \varnothing$. So $0 = \mu(\lim_n A_n) < \lim_{n\to\infty}\mu(A_n) = +\infty$.

(iii): Let $B_k = \bigcap_{n\geqslant k} A_n$. Then we have

$$A = \underline{\lim}_n A_n = \bigcup_{k\geqslant 1}\bigcap_{n\geqslant k} A_n = \bigcup_{k=1}^{\infty} B_k,$$

and since $B_k \subset B_{k+1}$,

$$\mu(A) = \lim_{k\to\infty}\mu(B_k) = \underline{\lim}_{k\to\infty}\mu(B_k), \qquad \text{by (i)}$$

$$\leqslant \underline{\lim}_{k\to\infty}\mu(A_k), \qquad \text{since } B_k \subset A_n, \quad n \geqslant k.$$

Similarly, if $C_k = \bigcup_{n\geqslant k} A_n$, then $C_n \supset C_{n+1}$, $C_k = \sup_{n\geqslant k} A_n$, and

$$\tilde{A} = \overline{\lim}_n A_n = \bigcap_{k\geqslant 1}\bigcup_{n\geqslant k} A_n$$

$$= \bigcap_{k=1}^{\infty} C_k \qquad (\in \mathscr{T}, \text{ since } \mathscr{T} \text{ is a } \sigma\text{-ring}).$$

Hence $\mu(C_{n_0}) < \infty \Rightarrow$

$$\mu(\tilde{A}) = \lim_{k\to\infty}\mu(C_k), \qquad \text{by (ii)}$$

$$= \overline{\lim}_{k\to\infty}\mu(C_k)$$

$$\geqslant \overline{\lim}_{n\to\infty}\mu(A_n), \qquad \text{since } A_n \subset C_k, \quad n \geqslant k.$$

This shows (iii) is true.

(iv): Using the preceding part,

$$\mu\Big(\sup_{n\geqslant k} A_n\Big) = \mu\Big(\bigcup_{n\geqslant k} A_n\Big)$$

$$\leqslant \sum_{n\geqslant k}\mu(A_n), \qquad \text{by (v) to be proved below.}$$

The right side series converges by hypothesis. Hence letting $k \to \infty$ and using (ii), we get

$$0 \leqslant \mu\Big(\limsup_{n\to\infty} A_n\Big) = \lim_{k\to\infty}\mu\Big(\sup_{n\geqslant k} A_n\Big)$$

$$\leqslant \lim_{k\to\infty}\sum_{n\geqslant k}\mu(A_n) = 0.$$

(v): Finally, let $A = \bigcup_{n \geq 1} A_n (\in \mathcal{T})$, and consider sets $D_1 = A_1$, $D_2 = A_2 - A_1$, and inductively $D_n = A_n - \bigcup_{k=1}^{n-1} A_k$ for $n \geq 2$. Then $D_n \in \mathcal{T}$, disjoint, and $A = \bigcup_{n=1}^{\infty} D_n$. Also $D_n \subset A_n$. Hence

$$\mu(A) = \sum_{n=1}^{\infty} \mu(D_n)$$

$$\leq \sum_{n=1}^{\infty} \mu(A_n), \qquad \text{since } \mu \text{ is monotone.}$$

This yields (v), hence (iv), and the proof is complete.

In view of the above proposition, it will be useful to know how σ-algebras (σ-rings) can be "constructively" obtained from algebras (rings). We already know (cf. Proposition 1.3.2) that the ring generated by a semiring is the class of all finite disjoint unions of sets from the latter. The corresponding assertion for σ-rings is not so simple. We present two types of characterizations both of which are useful in applications.

Definition 2. Let $\mathcal{A}$ be a nonempty class of subsets for Ω. Then

(1) it is a *monotone class* if $A_n \in \mathcal{A}$ and $\{A_n: n \geq 1\}$ is a monotone sequence (i.e., either $A_n \subset A_{n+1}$ or $A_n \supset A_{n+1}$, $n \geq 1$), then $\lim_n A_n \in \mathcal{A}$;

(2) it is a *π- (or product) class* if $A, B \in \mathcal{A} \Rightarrow A \cap B \in \mathcal{A}$;

(3) it is a *λ- (or lattice) class* if (a) $A, B \in \mathcal{A}$, $A \cap B = \varnothing \Rightarrow A \cup B \in \mathcal{A}$; (b) $A, B \in \mathcal{A}$, $A \supset B \Rightarrow A - B \in \mathcal{A}$; $\Omega \in \mathcal{A}$; and (c) $A_n \in \mathcal{A}$, $A_n \subset A_{n+1} \Rightarrow \bigcup_{n=1}^{\infty} A_n \in \mathcal{A}$.

Of these classes, the first one is traditional and the last two were introduced by E. B. Dynkin in 1959 for applications in probability theory. Note that, if $\mathcal{A}$ is a λ-class, then by (3)(b) it has both $\varnothing, \Omega$, and if $A \in \mathcal{A}$ then $A^c = \Omega - A \in \mathcal{A}$. However, it need not be closed under finite unions. In case it is a π-class, then by (3)(a) it is also closed under finite unions, so that with (3)(c), it will become a σ-algebra.

We have the following characterizations of a σ-algebra.

Proposition 3. *The following assertions hold:*

(a) *If $\mathcal{A}$ is a ring (algebra) of sets of a point set Ω, then the smallest σ-ring (algebra) containing $\mathcal{A}$ is the same as the smallest monotone class containing $\mathcal{A}$. (Thus both generated classes are the same.)*

(b) *If $\mathcal{A}$ is a λ-class and $\mathcal{B}$ is a π-class of sets of Ω, and $\mathcal{A} \supset \mathcal{B}$, then $\mathcal{A} \supset \sigma(\mathcal{B})$, where $\sigma(\mathcal{B})$ is the σ-algebra generated by $\mathcal{B}$.*

Proof. The argument in both cases is similar, but involves a nontrivial trick which becomes one of the special (and standard) tools in the subject.

(a): Let $M(\mathscr{A})$ be the smallest monotone class containing the ring $\mathscr{A}$, and $\sigma(\mathscr{A})$ be the σ-ring generated by $\mathscr{A}$. Since each σ-ring is closed under monotone limits, it follows that $M(\mathscr{A}) \subset \sigma(\mathscr{A})$. For the opposite inclusion, since $\sigma(\mathscr{A})$ is the smallest such σ-ring, it suffices to show $M(\mathscr{A})$ is a σ-ring. But if $M(\mathscr{A})$ is a ring, then $A_n \in M(\mathscr{A}) \Rightarrow \bigcup_{n=1}^{\infty} A_n = \bigcup_{k=1}^{\infty}(\bigcup_{i=1}^{k} A_i) \in M(\mathscr{A})$, which makes it a σ-ring (cf. Definition 1.3.4). To show the ring property we need the following unmotivated but crucial idea. Consider for each $A \subset \Omega$, a class

$$\mathscr{A}_A = \{B \subset \Omega \colon B - A,\ A - B, \text{ and } A \cup B \text{ are in } M(\mathscr{A})\}. \tag{5}$$

The class of $\mathscr{A}_A$ may be empty, but if it is nonempty, then by symmetry of the definition in A and B, $B \in \mathscr{A}_A$ iff $A \in \mathscr{A}_B$. Further it is closed under monotone limits, since $B_n \in \mathscr{A}_A$, monotone, implies

$$B_n - A,\ A - B_n, \text{ and } A \cup B_n \text{ are monotone, and are in } M(\mathscr{A}), \tag{6}$$

since $M(\mathscr{A})$ is a monotone class. Thus $\mathscr{A}_A$ is a monotone class. Now if A, B are in $\mathscr{A}$, which is a ring, $A - B$, $B - A$, $A \cup B$ are also in $\mathscr{A} \subset M(\mathscr{A})$, so $B \in \mathscr{A}_A$ and hence $\mathscr{A} \subset \mathscr{A}_A$ for each $A \in \mathscr{A}$. Since $\mathscr{A}_A$ is shown above to be a monotone class, it follows that $M(\mathscr{A}) \subset \mathscr{A}_A$, $A \in \mathscr{A}$. Hence, if $B \in M(\mathscr{A})$, then $B \in \mathscr{A}_A$ for all $A \in \mathscr{A}$, and by symmetry $A \in \mathscr{A}_B$ for all $B \in M(\mathscr{A})$, which means that for all A, B in $M(\mathscr{A})$, $A \in \mathscr{A}_B$ and $B \in \mathscr{A}_A$. Thus $A - B$, $B - A$, $A \cup B$ are in $M(\mathscr{A})$ and it is a ring. This shows that $M(\mathscr{A}) = \sigma(\mathscr{A})$.

(b): We prove the second part with an analogous argument. Let $\mathscr{A}_0$ be the smallest λ-class containing the π-class $\mathscr{B}$. If we show that $\mathscr{A}_0$ is also a π-class, then it is a σ-algebra containing $\mathscr{B}$, so $\sigma(\mathscr{B}) \subset \mathscr{A}_0 \subset \mathscr{A}$, the last since $\mathscr{A}_0$ is minimal. Hence we only need to show that $\mathscr{A}_0$ is a π-class. For this we again use a construction as in (5). Thus consider

$$\mathscr{A}_1 = \{A \subset \Omega \colon A \cap B \in \mathscr{A}_0 \text{ for all } B \in \mathscr{B}\}. \tag{7}$$

It is seen that $\mathscr{B} \subset \mathscr{A}_1$, since $\mathscr{B} \subset \mathscr{A}_0$ by construction. Moreover $A_n \in \mathscr{A}_1$, $A_n \subset A_{n+1} \Rightarrow \lim_n (A_n \cap B) \in \mathscr{A}_0$ for all $B \in \mathscr{B}$, since $\mathscr{A}_0$ is a λ-class. Hence $\lim_n A_n \in \mathscr{A}_1$. But $\Omega \in \mathscr{A}_1$ and $A_1, A_2 \in \mathscr{A}_1$, disjoint, $\Rightarrow A_i \cap B \in \mathscr{A}_0$, disjoint for $i = 1, 2$, and $B \in \mathscr{B}$. So $(A_1 \cup A_2) \cap B = (A_1 \cap B) \cup (A_2 \cap B) \in \mathscr{A}_0 \Rightarrow A_1 \cup A_2 \in \mathscr{A}_1$. Similarly $A_1 \supset A_2 (A_i \in \mathscr{A}_1) \Rightarrow (A_1 - A_2) \cap B = (A_1 \cap B) - (A_2 \cap B) \in \mathscr{A}_0$, $B \in \mathscr{B}$, so $A_1 - A_2 \in \mathscr{A}_1$. Thus $\mathscr{A}_1$ is a λ-class, and by the minimality, $\mathscr{A}_1 \supset \mathscr{A}_0 \supset \mathscr{B}$. So $A \in \mathscr{A}_0 \Rightarrow A \in \mathscr{A}_1$ and $A \cap B \in \mathscr{A}_0$ for all $B \in \mathscr{B}$.

Now to see that $\mathscr{B}$ can be replaced by $\mathscr{A}_0$ consider the new class

$$\mathscr{A}_2 = \{A \subset \Omega \colon A \cap B \in \mathscr{A}_0 \text{ for all } B \in \mathscr{A}_0\}. \tag{8}$$

It follows from the preceding that $\mathscr{B} \subset \mathscr{A}_2$, and we repeat the above argument to show that $\mathscr{A}_2$ is also a λ-class containing $\mathscr{B}$. Since $\mathscr{A}_0$ is the smallest such, we must have $\mathscr{A}_0 \subset \mathscr{A}_2$. Hence $A_1, A_2 \in \mathscr{A}_0 \Rightarrow A_1 \in \mathscr{A}_2$ and $A_1 \cap A_2 \in \mathscr{A}_0$. In other words $\mathscr{A}_0$ is a λ-class. This completes the proof of (b), and hence of the proposition.

We may wish to have a clearer idea of a σ-ring than that given by the above proposition, i.e., one which can be obtained by a step by step procedure starting with a given collection. Typically this should involve adding the countable unions and then the countable intersections of the resulting construction, and repeating the process. This can indeed be done, but then we must know as to how far should this process continue. Can we start with the (countable) intersection operation and proceed to the unions, and will the end result be the same? These are nontrivial problems whose solutions involve a use of the calculus of ordinal numbers. The simplest, yet nonelementary, treatment of such a "constructive" procedure is given in detail by F. Hausdorff (1962, 3rd ed., Section 18), and its employment in other contexts, especially in generalizations of Lebesgue integrals, is treated in S. Saks (1964, 2nd ed., Chapter 8, Sections 4, 5). We omit its discussion here.

Definition 4. Let $\mathscr{G}$ be the class of all open sets of $\mathbb{R}^n$. The σ-algebra generated by $\mathscr{G}$ (i.e., the smallest σ-algebra containing $\mathscr{G}$), is called the *Borel σ-algebra* of $\mathbb{R}^n$. More generally, the σ-algebra generated by all the open sets of a topological space is termed its *Borel σ-algebra*.

This class for $\mathbb{R}^n$ can be described easily:

Proposition 5. *Let $\mathscr{S}$ be the semiring of rectangles and $\mathscr{E}$ be the class of elementary sets formed from $\mathscr{S}$ in $\mathbb{R}^n$ (cf. Proposition 1.3.2). Let $\tilde{\mathscr{E}} = \{A \subset \mathbb{R}^n: A = \bigcup_{i=1}^{\infty} P_i,\ P_i \in \mathscr{S}\}$, $\mathscr{I}$ = the set of all intervals (which may be closed, open, or half-closed) in $\mathbb{R}^n$, and $\mathscr{E}' = \{A \subset \mathbb{R}^n: A = \bigcup_{i=1}^{m} Q_i,\ Q_i \in \mathscr{I}$, disjoint$\}$. Then the σ-algebras generated by each of the classes, $\mathscr{S}$, $\mathscr{I}$, $\mathscr{E}$, $\tilde{\mathscr{E}}$ and $\mathscr{E}'$ are identical with the Borel σ-algebra of $\mathbb{R}^n$.*

Proof. Let $\sigma(\mathscr{A})$ denote the σ-algebra generated by the class $\mathscr{A}$, and let $\mathscr{B}$ stand for the Borel σ-algebra of $\mathbb{R}^n$. Since

$$\mathscr{S} \subset \mathscr{E} \subset \sigma(\mathscr{S}), \qquad \mathscr{S} \subset \tilde{\mathscr{E}} \subset \sigma(\mathscr{S}), \tag{9}$$

it follows that $\sigma(\mathscr{S}) = \sigma(\mathscr{E}) = \sigma(\tilde{\mathscr{E}})$ $[\sigma(\sigma(\mathscr{S})) = \sigma(\mathscr{S})]$. Similarly,

$$\mathscr{I} \subset \mathscr{E}' \subset \sigma(\mathscr{I}), \tag{10}$$

and hence $\sigma(\mathscr{I}) = \sigma(\mathscr{E}')$. On the other hand, we have

$$\{\mathbf{x} = (x_1, \ldots, x_n) \colon a_i < x_i \leqslant b_i, 1 \leqslant i \leqslant n\}$$

$$= \bigcap_{k \geqslant 1} \left\{\mathbf{x} \colon a_i < x_i < b_i + \frac{1}{k}, 1 \leqslant i \leqslant n\right\},$$

$$\{\mathbf{x} = (x_1, \ldots, x_n) \colon a_i < x_i < b_i, 1 \leqslant i \leqslant n\}$$

$$= \bigcup_{k \geqslant 1} \left\{\mathbf{x} \colon a_i < x_i \leqslant b_i - \frac{1}{k}, 1 \leqslant i \leqslant n\right\},$$

from which one deduces at once that

$$\mathscr{I} \subset \sigma(\mathscr{S}) \quad \text{and} \quad \mathscr{S} \subset \sigma(\mathscr{I}). \tag{11}$$

Hence $\sigma(\mathscr{I}) = \sigma(\mathscr{S}) = \sigma(\mathscr{E}') = \sigma(\mathscr{E}) = \sigma(\tilde{\mathscr{E}})$. We also have $\mathscr{S} \subset \mathscr{B}$, so that $\sigma(\mathscr{S}) \subset \mathscr{B}$, since $\mathscr{B}$ is a σ-algebra. If $G \subset \mathbb{R}^n$ is an open set, then it can be expressed as a countable union of closed intervals (even cubes). (We provide a proof of this well-known fact in a lemma below.) This implies that each open set G is in $\sigma(\mathscr{E}')$ and hence $\mathscr{B} \subset \sigma(\mathscr{E}') = \sigma(\mathscr{S})$, so that $\mathscr{B} = \sigma(\mathscr{S})$. In other words, all the σ-algebras are identical. This finishes the proof.

The lemma mentioned above follows essentially from the result that $\mathbb{R}^n$ has a countable base of intervals with rational end points and that every open set in $\mathbb{R}^n$ is a countable union of these basis elements (since $\mathbb{R}^n$ is separable in its metric, i.e., it contains a countable dense subset). In fact, if the (countable set of) points of $\mathbb{R}^n$ with integral coordinates are considered together with unit (closed) cubes at each of these points, we have $\mathbb{R}^n$ to be a countable union of *nonoverlapping* (= with disjoint interiors) cubes. Call this set $\mathscr{C}_1$. Split each side of each cube into equal parts, so that the cube is replaced by 2^n subcubes of side $\frac{1}{2}$ unit. Call the thus obtained (countable) collection $\mathscr{C}_2$. Repeating this procedure, we get at the kth stage a (countable) collection $\mathscr{C}_k$ of (closed) cubes each of side 2^{-k} and nonoverlapping. Note that a unit cube is the Cartesian product of unit intervals. One has:

Lemma 6. *If $G \subset \mathbb{R}^n$ is an open set, then it can be expressed as a countable union of closed cubes which moreover are nonoverlapping.*

Proof. Let $\mathscr{K}_1$ be the set of cubes from $\mathscr{C}_1$ that lie entirely inside the open set G. Now let $\mathscr{K}_2$ be the elements of $\mathscr{C}_2$ that are contained in G but are not subcubes of elements of $\mathscr{K}_1$. By induction let $\mathscr{K}_r$ be the elements of $\mathscr{C}_r$ which are inside G but which are not subcubes of those already selected classes, $\mathscr{K}_1, \ldots, \mathscr{K}_{r-1}$, each of which is at most countable. If $\mathscr{K}$ is the union of all these collections, then as $r \to \infty$ the side of each cube in $\mathscr{K}_r$ is reduced, and since $G \subset \mathbb{R}^n$, each point of G is in one of the elements of $\mathscr{K}_r$ for some r, so that (each element being a subset of G) $G = \bigcup\{C: C \in \mathscr{K}\}$. Since $\mathscr{K}$ is countable, we conclude that G has the stated representation.

Remark. If $n = 1$, then we can strengthen the statement of the lemma to the following: Each open set of $\mathbb{R}$ can be represented as a countable disjoint union of open intervals. But an analog of this statement does not obtain for $\mathbb{R}^n$ if $n \geqslant 2$. [cf. Bartle (1976, pp. 66 and Example 9.H)].

The preceding discussion already shows the extent of the new classes under countable operations, in contrast to the finite operations seen in the first chapter. An additional enlargement is possible when measures are associated with this process; but that becomes the next step in our search for the broadest class of sets that can be measured by a particular measure function, whereas we have not thus far used any such functions. This is done in the next section.

EXERCISES

0. Let $\{\mathscr{A}_i, i \in I\}$ be a family of classes of subsets of a set Ω, and define $\mathscr{A} = \bigcap_{i \in I} \mathscr{A}_i$, $\mathscr{B} = \bigcup_{i \in I} \mathscr{A}_i$. Show that if each $\mathscr{A}_i$ is either a ring, algebra, monotone class, lattice, π, or δ-class, then the same is true of $\mathscr{A}$. Construct examples to show that an analogous statement is false for $\mathscr{B}$ even if the index I has only two elements.

1. Let $\mathscr{A}$ be a collection of subsets of a set Ω, and $A \subset \Omega$ be a given set. Then $\mathscr{A}(A) = \{A \cap B: B \in \mathscr{A}\}$ is called the *restriction* or *trace* of $\mathscr{A}$ on A. Show that $\sigma(\mathscr{A})(A)$, the trace of $\sigma(\mathscr{A})$ on A, is the same as $\sigma(\mathscr{A}(A))$, the σ-algebra generated by $\mathscr{A}(A)$ in A.

2. Let $\mathscr{B}$ be the Borel σ-algebra of $\mathbb{R}^n$. If $A \subset \mathbb{R}^n$ and $x_0 \in \mathbb{R}^n$, then $A + x_0 = \{x_0 + y: y \in A\}$ is called the *translation* or *shift* of A by x_0 units. Show that if $A \in \mathscr{B}$, then $A + x_0 \in \mathscr{B}$. (Note that if a set is open or compact in $\mathbb{R}^n$, then its translate has the same property.)

3. Let b ($\in \mathbb{R}^n$) be a fixed vector and $T: \mathbb{R}^n \to \mathbb{R}^n$ be a nonsingular linear transformation. It may be thought of as an invertible matrix $T = (t_{ij})$ relative to a basis of the vector space $\mathbb{R}^n$. We define an *affine* mapping: For any set $A \subset \mathbb{R}^n$, define $B = TA + b = \{y = Tx + b: x \in A\} \subset \mathbb{R}^n$. Using the ideas of the preceding problem, show that, if A is a Borel set in $\mathbb{R}^n$, then B is also a Borel set. (Thus not only translations, but affine transformations leave the Borel σ-algebra in $\mathbb{R}^n$ invariant.)

4. A subset C of a topological space Ω is called a G_δ-*set* if it is a countable intersection of open sets, and an F_σ-*set* if it is a countable union of closed sets. Show that the σ-algebra $\mathscr{B}$ of $\mathbb{R}^n$ is also generated by all the G_δ-sets (or F_σ-sets) of $\mathbb{R}^n$. [First verify that each closed (open) set of $\mathbb{R}^n$ is a G_δ- (F_σ-) set. The result is true if $\mathbb{R}^n$ is replaced by any complete metric space.

In fact, let $x \in C$ (closed), and $B(x, 1/n)$ be an open ball of radius $1/n$ at x. If $U_n = \bigcup_{x \in C} B_n(x, 1/n)$, then $C \subset \bigcap_{n \geq 1} U_n$, and the opposite inclusion follows on using the completeness of the metric space under consideration.]

5. Let $\mathscr{A}$ be any collection of subsets of a point set Ω, and consider the generated σ-ring $\sigma(\mathscr{A})$. Show that a set $A \in \sigma(\mathscr{A})$ iff there is (at most) a countable subcollection $\mathscr{A}_1$ of $\mathscr{A}$ (depending on A in general) such that $A \in \sigma(\mathscr{A}_1)$. [Note that if $\mathscr{A}_i \subset \mathscr{A}$, $i \in I$, is a collection such that each $\mathscr{A}_i$ is at most *countable*, then $\{A \subset \Omega: A \in \sigma(\mathscr{A}_i) \text{ for some } i \in I\}$ is a σ-algebra containing $\mathscr{A}$.]

6. This problem is a concrete illustration of the preceding one. Let $\Omega = \mathbb{R}^{[0,1]}$, the space of all real valued functions on $[0,1]$. Let $\mathscr{A} = \{A \subset \Omega: A = \{f \in \Omega: a_i < f(t_i) \leqslant b_i, i = 1, \ldots, n\}, 0 < t_i \leqslant 1, n \geqslant 1\}$. If $A_0 = C[0,1]$, the set of all real continuous functions on $[0,1]$, show that $A_0 \notin \sigma(\mathscr{A})$. [Observe that a function whose values are prescribed at a countable set of points need not be continuous but can agree with a continuous one.]

7. Let $\mathscr{C}$ be a given ring of sets of a point set Ω, and suppose that μ_1, μ_2 are two finite measures on the σ-ring $\sigma(\mathscr{C})$ which agree on $\mathscr{C}$. Show that they must be identical. If, on the other hand, $\mathscr{C}$ is a π-class, and if μ_1, μ_2 are finite, are defined on the σ-algebra $\sigma(\mathscr{C})$, and agree on $\mathscr{C}$ as well as on the whole space Ω, then again $\mu_1 = \mu_2$ on $\sigma(\mathscr{C})$. [Use Proposition 3.]

8. Here is an extension of the preceding problem for some more functions. A measure μ on a ring $\mathscr{C}$ is called *σ-finite* if for each $A \in \mathscr{C}$, there exist $A_n \in \mathscr{C}$ such that $A \subset \bigcup_{n=1}^{\infty} A_n$ and $\mu(A_n) < \infty$, $n \geqslant 1$. (Also each such A is termed a *σ-finite* set.) Suppose μ_1, μ_2 are measures on a σ-ring $\sigma(\mathscr{C})$, where $\mathscr{C}$ is a ring, and agree on $\mathscr{C}$. If both are σ-finite on $\mathscr{C}$ show that they also agree on $\sigma(\mathscr{C})$. If, in the preceding, $\mathscr{C}$ is only a π-class, $\sigma(\mathscr{C})$ is the generated σ-algebra, and Ω is σ-finite for both μ_1, μ_2 by sets of $\mathscr{C}$ (i.e., $\Omega = \bigcup_{n \geqslant 1} A_{in}$, $A_{in} \in \mathscr{C}$, $\mu_i(A_{in}) < \infty$, $i = 1, 2$), then we again have $\mu_1 | \mathscr{C} = \mu_2 | \mathscr{C} \Rightarrow \mu_1 = \mu_2$ on $\sigma(\mathscr{C})$ where $\mu | \mathscr{C}$ is the restriction of μ to $\mathscr{C}$. [Observe that one can take a single sequence $\{A_n: n \geqslant 1\}$ with the stated properties for both measures, and consider the traces $\mathscr{C}(A_n)$, $\mu_{in} = \mu_i | \mathscr{C}(A_n)$, $i = 1, 2$, for which the result of Exercise 6 applies for each n. Then Exercise 1 implies the general case. A further extension of this will be considered in Section 3.]

2.2. THE LEBESGUE OUTER MEASURE AND THE CARATHÉODORY PROCESS

The preceding section indicated how small classes of sets, on which abstract measures are defined *a priori*, can be enlarged to have nice closure properties for set operations. But we have not discussed the crucial question that the given measures can be extended to these larger classes. The last two problems, above, show certain uniqueness properties of measures if such extensions are available. In this section we present some fundamental results with consequences to enable this extension, starting with Lebesgue's procedure and then giving a potentially powerful extension method due to C. Carathéodory.

The earliest attempts to extend the length function for geometric figures more complicated than elementary sets appear to be due to G. Peano in 1887 and C. Jordan in 1892. The basic ideas here are an extension of the formulas used for the arc length in calculus. Thus the *Jordan measure* (or *content*) of a set $A \subset \mathbb{R}$ is defined as ($(a, b] = \phi$, if $a > b$ as usual)

$$\mu_J(A) = \inf\left\{ \sum_{i=1}^{n} (b_i - a_i) : A \subset \bigcup_{i=1}^{n} (a_i, b_i], \text{disjoint} \right\}. \tag{1}$$

Then $\mu_J(\cdot) \geqslant 0$, nondecreasing and a well-defined number. While this is additive for intervals and simple figures, μ_J is *not* additive for moderately complicated sets, as the following classical illustration shows.

Example 1. Let $\Omega = [0, 1]$, and $\mathbb{Q} \subset \Omega$ be the set of rationals (any dense denumerable set suffices). Let $\mathbb{Q} \subset \bigcup_{i=1}^{n}(a_i, b_i]$ be any covering by intervals $(a_i, b_i]$. Since the closure $\overline{\mathbb{Q}}$ of $\mathbb{Q}$ is Ω, we have

$$\overline{\mathbb{Q}} = \Omega = \bigcup_{i=1}^{n} [a_i, b_i].$$

We show that $\mu_J(\mathbb{Q}) = 1$, by evaluating (1). Since $\overline{\mathbb{Q}} \subset (-\epsilon, 1 + \epsilon)$, $\epsilon > 0$, it follows that $\mu_J(\mathbb{Q}) \leqslant 1$. For the opposite inequality, with $\epsilon > 0$, let $a_i' = a_i - (\epsilon/2^{i+1})$, $b_i' = b_i + (\epsilon/2^{i+1})$ so that $(a_i', b_i') \supset (a_i, b_i]$. We may suppose (by relabeling if necessary) that $a_1' < 0 < b_1'$. Since $\epsilon > 0$, such an interval $(a_1', b_1']$ clearly exists. If $b_1' > 1$, we stop, and if $b_1' \leqslant 1$, choose $(a_2', b_2']$ such that $a_1' < a_2' < b_1' < b_2'$ and stop if $b_2' > 1$. Continuing this process, which stops at some stage $m \leqslant n$, we get

$$a_{i-1}' < a_i' < b_{i-1}' \quad \text{and} \quad 0 < b_1' < b_2' < \cdots < b_m', \quad b_m' > 1.$$

Hence

$$\sum_{i=1}^{n} (b_i' - a_i') \geqslant \sum_{i=1}^{m} (b_i' - a_i') \geqslant \sum_{i=2}^{m} (a_i' - a_{i-1}') + (b_m' - a_m')$$
$$= b_m' - a_1' \geqslant 1.$$

Consequently

$$1 \leqslant \sum_{i=1}^{n} (b_i' - a_i') = \sum_{i=1}^{n} (b_i - a_i) + \epsilon \sum_{i=1}^{n} 2^{-i} < \sum_{i=1}^{n} (b_i - a_i) + \epsilon.$$

This implies that $\mu_J(\mathbb{Q}) \geqslant 1$. Thus $\mu_J(\mathbb{Q}) = 1$. An analogous argument applies to $\mathbb{Q}^c = \Omega - \mathbb{Q}$; one gets $\mu_J(\mathbb{Q}^c) = 1$ also. Thus

$$1 = \mu_J(\Omega) = \mu_J(\mathbb{Q} \cup \mathbb{Q}^c) < \mu_J(\mathbb{Q}) + \mu_J(\mathbb{Q}^c) = 2,$$

and μ_J is not even finitely additive on simple but not elementary sets, showing thereby that the definition of μ_J by (1) is not satisfactory.

To analyze this example further, note that the topology of Ω did not help. Since in $\mathbb{R}$ each point is a G_δ-set $[\{a\} = \bigcap_{n=1}^{\infty}(a - 1/n, a + 1/n)]$, $\mathbb{Q}$ is a countable union of G_δ-sets. However, $\mathbb{Q}$ is not a G_δ-set. (See Exercise 1.3.) The complement of a G_δ in the metric space Ω being an F_σ-set, it follows that $\mathbb{Q}^c$ is a countable intersection of F_σ-sets which is not an F_σ. (In fact, it can be shown that if a set such as $\mathbb{Q}$ in a complete metric space, such as Ω, is a G_δ, then it is homeomorphic to a complete metric space. This is violated for $\mathbb{Q}$ here.) Actually E. Borel showed in 1898 that μ_J is σ-additive on a certain collection of sets including the ring of elementary sets studied in Section 1.2. Since each of the sets $\mathbb{Q}$ and $\mathbb{Q}^c$ is a Borel set (cf. Exercise 1.3), and Proposition 1.1 implies that σ-rings are the natural domains for measure functions, it follows that Jordan's definition is not adequate for extending measure functions from a ring to a σ-ring.

It remained for H. Lebesgue in 1902 to make a seemingly small but in fact a decisive modification of (1) to obtain the desired solution, by replacing the finite covers with infinite ones. He recognized and exploited the consequences already in his thesis (written under E. Borel), and spent much of his research on this. Its impact parallels the countable operations on classes of sets, and the resulting change is as profound as that implied by the work of the preceding section.

Definition 2. If $\mathscr{S}$ is the semiring of rectangles on $\mathbb{R}^n$, and $A \subset \mathbb{R}^n$ is any set, then the *Lebesgue outer measure* μ^* of A, induced by m, is given by

$$\mu^*(A) = \inf\left\{ \sum_{i=1}^{\infty} m(P_i) : A \subset \bigcup_{i=1}^{\infty} P_i,\ P_i \in \mathscr{S}, \text{disjoint} \right\}, \tag{2}$$

where $m(\cdot)$ is the volume function on rectangles of $\mathbb{R}^n$.

The terminology is justified by the fact that μ^* is nonnegative, $\mu^*(A) \leqslant \mu^*(B)$ for $A \subset B$, and especially the following:

Proposition 3. *The set function μ^* defined by* (2) *is countably subadditive, and $\mu^*(A) = m(A)$ for each elementary set $A \subset \mathbb{R}^n$. In particular $m = \mu^*|\mathscr{S}$.*

Proof. Let A, A_n, $n \geqslant 1$, be subsets of $\mathbb{R}^n$ such that $A \subset \bigcup_{n \geqslant 1} A_n$. We need to show that

$$\mu^*(A) \leqslant \sum_{n=1}^{\infty} \mu^*(A_n). \tag{3}$$

Since this is true if the right side is infinite, we may assume that the series in (3) converges so that $\mu^*(A_n) < \infty$ for each $n \geqslant 1$. Then by definition of (2), given $\epsilon > 0$, there exists a disjoint family $\{P_{ni}: i \geqslant 1\} \subset \mathscr{S}$ such that $A_n \subset \bigcup_{i=1}^{\infty} P_{ni}$ and

$$\mu^*(A_n) > \sum_{i=1}^{\infty} m(P_{ni}) - \frac{\epsilon}{2^n}. \tag{4}$$

Hence $A \subset \bigcup_{n \geqslant 1} A_n \subset \bigcup_{n \geqslant 1} \bigcup_{i \geqslant 1} P_{ni}$, so that $\{P_{ni}: n \geqslant 1, i \geqslant 1\} \subset \mathscr{S}$ is also a (disjoint) covering of A, and by (2) we get

$$\begin{aligned} \mu^*(A) &\leqslant \sum_{n=1}^{\infty} \sum_{i=1}^{\infty} m(P_{ni}) \\ &\leqslant \sum_{n=1}^{\infty} \left[\mu^*(A_n) + \frac{\epsilon}{2^n}\right], \qquad \text{by (4)} \\ &= \sum_{n=1}^{\infty} \mu^*(A_n) + \epsilon. \end{aligned} \tag{5}$$

Since $\epsilon > 0$ is arbitrary, (5) implies μ^* is σ-subadditive.

For the second assertion, let A be an elementary set, i.e., $A = \bigcup_{i=1}^{k} P_i$, $P_i \in \mathscr{S}$, disjoint. Then

$$\begin{aligned} \overline{m}(A) &= \sum_{i=1}^{k} m(P_i) \\ &\geqslant \inf\left\{ \sum_{i=1}^{\infty} m(Q_i): A \subset \bigcup_{i=1}^{\infty} Q_i,\ Q_i \in \mathscr{S}, \text{disjoint}\right\} \\ &= \mu^*(A). \end{aligned} \tag{6}$$

On the other hand, since $\overline{m}$ is σ-additive on $\mathscr{E}$ (the ring of elementary sets) by Theorem 1.2.8, and then $A \subset \bigcup_{i=1}^{\infty} Q_i$ implies $A = \bigcup_{i=1}^{\infty}(A \cap Q_i)$, $A \cap Q_i \in \mathscr{E}$, we get

$$\overline{m}(A) = \sum_{i=1}^{\infty} \overline{m}(A \cap Q_i) \leqslant \sum_{i=1}^{\infty} m(Q_i), \qquad Q_i \in \mathscr{S} \text{ (disjoint)}. \tag{7}$$

Taking infimum on all such coverings of A, one has, from (7) and (2), that $\overline{m}(A) \leqslant \mu^*(A)$. This and (6) give $\mu^*(A) = \overline{m}(A)$, $A \in \mathscr{E}$, as asserted.

Remark. It is to be noted that the σ-subadditivity of $\overline{m}$, obtained in (7), depends essentially on the topological properties of $\overline{m}$ and $\mathbb{R}^n$. In the general study of measures, this property has to be postulated so that any mention of topology can be suppressed. However, it will be seen later than the assumption of σ-additivity of μ enables us to introduce a topology in the underlying space, emphasizing the importance of this critical postulate. (See Chapters 8 and 10.)

The preceding proposition and Theorem 1.2.8 imply that μ^* is σ-additive on $\mathscr{E}$. It is now necessary to characterize a class of sets on which μ^* is σ-additive, i.e., a class that can be consistently measured by μ^*. Now, to motivate a new idea, (2) shows how to approximate the measure of a set from above, and if the set is "measurable" for a given measure, one should be able to approximate it from below (or inside) as well. This point is made precise by Lebesgue, using the already defined outer measure, as follows.

Let A, E be two sets in $\mathscr{E}$, the ring of elementary sets in $\mathbb{R}^n$. If $\overline{m}(A - E) < \infty$, then we have

$$\overline{m}(A) = \overline{m}(A - A \cap E) + \overline{m}(A \cap E),$$

and hence

$$\overline{m}(A \cap E) = \overline{m}(A) - \overline{m}(A - E). \tag{8}$$

If $\alpha = \sup_{A \in \mathscr{E}} \overline{m}(A \cap E) \leqslant \infty$, then there exist $A_n \in \mathscr{E}$ such that $\lim_{n \to \infty} \overline{m}(A_n \cap E) = \alpha$. Since $\mathscr{E}$ is a ring, we may assume that $A_n \subset A_{n+1}$ (replacing each A_k by $\bigcup_{i=1}^k A_i \in \mathscr{E}$, if necessary). Hence by using the fact that $\overline{m}(\cdot)$ is σ-additive on $\mathscr{E}$ (cf. Theorem 1.2.8) and then invoking Proposition 1.2, we get (let $A = \lim_n A_n = \bigcup_{n=1}^\infty A_n$)

$$\alpha = \lim_{n \to \infty} \overline{m}(A_n \cap E) = \overline{m}(E \cap A). \tag{9}$$

Since $\mathbb{R}^n \in \mathscr{E}$, the definition of α implies $A \supset E$; so $\overline{m}(E \cap A) = \overline{m}(E)$. Thus by (8) and (9)

$$\overline{m}(E) = \sup\{\overline{m}(A) - \overline{m}(A - E) : A \in \mathscr{E}, \overline{m}(A - E) < \infty\}. \tag{10}$$

Next we suppose that $E \subset \mathbb{R}^n$ is an arbitrary set. If μ^* is the outer measure induced by $m(\cdot)$ [or $\overline{m}(\cdot)$], as in (2), then the right side of (10) is well defined with μ^* in place of $\overline{m}$ [and $\overline{m}(A) = \mu^*(A)$, $A \in \mathscr{E}$, by Proposition 3], and that gives a new measure for E, which need not be equal to $\mu^*(E)$. It is denoted by $\mu_*(E)$, to indicate that the approximation is obtained from sets $A \cap E$, contained in E. This leads to the following concept:

Definition 4. Let $E \subset \mathbb{R}^n$ be an arbitrary set, and $\mathscr{E}$ be the ring of elementary sets. Then a function $\mu_*(\cdot)$ induced by $m(\cdot)$, called the Lebesgue *inner measure*, is given by

$$\mu_*(E) = \sup\{\overline{m}(A) - \mu^*(A - E) : A \in \mathscr{E}, \mu^*(A - E) < \infty\}, \tag{11}$$

where μ^* is the (Lebesgue) outer measure induced by $m(\cdot)$. A set $E \subset \mathbb{R}^n$ is called *Lebesgue measurable* iff $\mu_*(E) = \mu^*(E)$.

The concept of inner measure (as well as its evaluation) is decidedly cumbersome, but it is useful in extending measures from a class to a larger one. For instance, given a measure on a σ-ring $\mathscr{A}$, and a set $A \notin \mathscr{A}$, one may want to extend μ from $\mathscr{A}$ to a larger class generated by $\mathscr{A}$ and A. In such a situation we have to use this concept for an extension (and these problems arise frequently in areas such as stochastic analysis). So we present a few properties of μ_* which also motivate an alternative approach to the subject. The basic result here is:

Proposition 5. *The inner measure μ_* is nonnegative and nondecreasing, and $\mu_*(E) \leqslant \mu^*(E)$, $E \subset \mathbb{R}^n$, with equality if $E \in \mathscr{E}$. Moreover,*

$$\overline{m}(A) = \mu_*(A \cap E) + \mu^*(A \cap E^c), \qquad A \in \mathscr{E}, \quad E \subset \mathbb{R}^n. \tag{12}$$

Hence E is (Lebesgue) measurable if

$$\overline{m}(A) = \mu^*(A \cap E) + \mu^*(A \cap E^c), \qquad A \in \mathscr{E}, \tag{13}$$

and conversely every bounded Lebesgue measurable set E ($\subset \mathbb{R}^n$) satisfies the functional equation (13).

Proof. First, by the subadditivity of μ^*, we have

$$\overline{m}(A) = \mu^*(A) \leqslant \mu^*(A - A \cap E) + \mu^*(E), \qquad A \in \mathscr{E}. \tag{14}$$

It follows from (11) that $\mu_*(E) \geqslant 0$, and by (14)

$$\mu^*(E) \geqslant \overline{m}(A) - \mu^*(A - E), \qquad \text{using} \quad \mu^*(A - E) < \infty. \tag{15}$$

Now taking appropriate suprema over A, we get $\mu^*(E) \geqslant \mu_*(E)$. Note that if $E \subset F$, then subject to the conditions of (11), we have

$$\begin{aligned} \mu_*(E) &= \sup\{\overline{m}(A) - \mu^*(A - E)\} \\ &\leqslant \sup\{\overline{m}(A) - \mu^*(A - F)\} = \mu_*(F), \end{aligned}$$

so that μ_* is monotone nondecreasing, and moreover if $E \in \mathscr{E}$, since $\mu^*|\mathscr{E} = \overline{m}$ by Proposition 3, there is equality in (14) and hence in (15). Consequently $\mu^*(E) = \mu_*(E)$, $E \in \mathscr{E}$.

We now establish the key equation (12), which depends on the definition (2) of μ^*. Thus let $E \subset A \subset B$, where $A, B \in \mathscr{E}$, and E is an arbitrary set. Then $B - E = (B - A) \cup (A - E)$, a disjoint union, and $B - A \in \mathscr{E}$, so that $B - A = \bigcup_{i=1}^k Q_i$ for a finite disjoint union of rectangles Q_i ($\in \mathscr{S}$). Thus by (2)

$$
\begin{aligned}
\mu^*(B - E) &\\
&= \inf\left\{\sum_{i=1}^{\infty} m(P_i) : \bigcup_{i=1}^{\infty} P_i \supset B - E,\ P_i \in \mathscr{S},\ \text{disjoint}\right\}\\
&= \inf\left\{\sum_{i=1}^{k} m(Q_i) + \sum_{j=1}^{\infty} m(P_j) : B - A = \bigcup_{i=1}^{k} Q_i,\right.\\
&\qquad\qquad \left. A - E \subset \bigcup_{j=1}^{\infty} P_j,\ Q_i,\ P_j \in \mathscr{S},\ \text{all disjoint}\right\}\\
&= \overline{m}(B - A) + \inf\left\{\sum_{j=1}^{\infty} m(P_j) : \bigcup_{j=1}^{\infty} P_j \supset A - E,\ P_j \in \mathscr{S},\ \text{disjoint}\right\}\\
&= \overline{m}(B - A) + \mu^*(A - E). \qquad (16)
\end{aligned}
$$

If $\mu^*(A - E) \leqslant \mu^*(B - E) < \infty$, then $\overline{m}(B - A) < \infty$ by (16), and

$$\overline{m}(B) = \overline{m}(B - A) + \overline{m}(A) = \overline{m}(A) + \mu^*(B - E) - \mu^*(A - E).$$

This shows that for $B \supset A \supset E$,

$$\overline{m}(B) - \mu^*(B - E) = \overline{m}(A) - \mu^*(A - E). \qquad (17)$$

Thus the number in (17) does not depend on the sets A, B which cover E, and hence their supremum subject to $\mu^*(A - E) < \infty$, $A \in \mathscr{E}$, is independent of A, and this number is $\mu_*(E)$:

$$\mu_*(E) = \overline{m}(A) - \mu^*(A - E), \qquad A \in \mathscr{E}. \qquad (18)$$

Replacing E by $A - E$ in (18), we get (on transposition)

$$\overline{m}(A) = \mu_*(A \cap E^c) + \mu^*(A \cap E), \qquad A \in \mathscr{E}. \qquad (19)$$

This is true for all $\mu^*(A - E) < \infty$. But $\infty = \mu^*(A - E) \leqslant \mu^*(A) = \overline{m}(A)$ implies (19) is also true in the general case, so that (12) holds as stated.

Finally, if $E \subset \mathbb{R}^n$ satisfies (13), then since (12) [or (19)] must hold in any case, we have for all $A \in \mathscr{E}$

$$\mu^*(A \cap E) + \mu^*(A \cap E^c) = \overline{m}(A) = \mu_*(A \cap E) + \mu^*(A \cap E^c).$$

In particular, letting $A = \mathbb{R}^n$, we get, after cancelling $\mu^*(A \cap E^c)$,

$$\mu^*(E) = \mu_*(E),$$

and hence E is Lebesgue measurable, by Definition 4. We obtain the converse implication after establishing some key properties of μ^* in Theorem 10 below, which then will imply that a bounded set $E \subset \mathbb{R}^n$ is Lebesgue measurable in the sense of Definition 4 iff it satisfies (13). (See Corollary 11 in particular.) Thus further discussion is postponed.

The definition of outer and inner measures can be given alternatively by formulas equivalent to those of (2) and (11). With the early 1898 work of Borel, it became clear that (cf. Theorem 10 below) the length or volume function $m(\cdot)$ can be extended to all open and closed (in fact all Borel) sets in $\mathbb{R}^n$. We then show that

$$\mu^*(E) = \inf\{m'(G): G \supset E, G \text{ open}\}, \qquad E \subset \mathbb{R}^n, \tag{20}$$

and

$$\mu_*(E) = \sup\{m'(F): F \subset E, F \text{ closed}\}, \qquad F \subset \mathbb{R}^n, \tag{21}$$

where $m'(\cdot)$ is the (Borel) extension of $m(\cdot)$ to the Borel sets (cf. also Exercise 1.2.6 and Theorem 17 below). Other properties, with indications of arguments when they are not direct, are included in the exercises at the end of this section. Lebesgue showed, using his μ_*, μ^*, that the class of Lebesgue-measurable sets forms a σ-algebra containing the Borel class properly.

By now the reader has undoubtedly noted that working with inner measures is somewhat involved. Indeed after analyzing Lebesgue's successful work, C. Carathéodory developed a more elegant and powerful process in 1914. This procedure uses only outer measures. No better or more general method than this one is available even now. We therefore present this important process, and give specializations later on.

Definition 6. (1) An abstract function $\tilde{\mu}$ on $\mathcal{P}$ $(= 2^{\Omega})$, the power set of Ω, is called an *outer measure* if it is nonnegative, extended real valued; and σ-subadditive and vanishes at the empty set. Thus (i) $\tilde{\mu}(\varnothing) = 0$, (ii) $0 \leqslant \tilde{\mu}(A) \leqslant \tilde{\mu}(B)$ for $A \subset B \subset \Omega$, and (iii) $\tilde{\mu}(\bigcup_{n=1}^{\infty} A_n) \leqslant \sum_{n=1}^{\infty} \tilde{\mu}(A_n)$.

(2) For an (abstract) outer measure $\tilde{\mu}$, a set A $(\subset \Omega)$ is called (*Carathéodory* or) *$\tilde{\mu}$-measurable* iff the following system of equations holds:

$$\tilde{\mu}(E) = \tilde{\mu}(E \cap A) + \tilde{\mu}(E \cap A^c), \qquad E \in \mathcal{P}. \tag{22}$$

(3) Let $\mathcal{M}_{\tilde{\mu}}$ be the class of all $\tilde{\mu}$-measurable subsets of Ω. Then the outer measure $\tilde{\mu}$ is called *Carathéodory regular* iff for each $E \subset \Omega$ one has

$$\tilde{\mu}(E) = \inf\{\tilde{\mu}(A): A \supset E, A \in \mathcal{M}_{\tilde{\mu}}\}. \tag{23}$$

The preceding work shows that, if $\tilde{\mu} = \mu^*$, the Lebesgue outer measure on $\mathbb{R}^n$ induced by the volume function $m(\cdot)$ is an outer measure in the above sense, and each Lebesgue-measurable set is μ^*-measurable (cf. Lemma 8 below, and the remark following the proof of Proposition 5 above). It will be shown below that μ^* satisfies (23) also. Indeed, this can be established for many generated measures in the following sense:

Definition 7. Let $\varnothing \in \mathscr{A} \subset \mathscr{P}$ $(= 2^{\Omega})$ be an arbitrary subfamily of sets, and $\alpha: \mathscr{A} \to \bar{R}^+$ any mapping (i.e. a nonnegative set function) such that $\alpha(\varnothing) = 0$. Then the object $\mu^*(\cdot)$ given by

$$\mu^*(A) = \inf\left\{ \sum_{i=1}^{\infty} \alpha(A_i): A \subset \bigcup_{i=1}^{\infty} A_i,\ A_i \in \mathscr{A} \right\}, \qquad A \in \mathscr{P}, \tag{24}$$

is called a *generated outer measure* by the pair $(\alpha, \mathscr{A})$.

As usual, if there is no collection $\{A_n\}_{n \geqslant 1}$ in $\mathscr{A}$ which covers A, then we have $\inf\{\varnothing\} = +\infty$, since for $\varnothing \subset \mathbb{R}$, every real number is a lower bound (vacuously) to the empty set.

It will be shown below that a generated measure has many nice properties, and if $\mathscr{A}$ is a *(semi-)ring* and α on $\mathscr{A}$ is additive, then the $(\alpha, \mathscr{A})$-generated measure is automatically Carathéodory regular. However, a general abstract outer measure $\tilde{\mu}$ need not have such a regularity property. Note that, since $\varnothing \in \mathscr{M}_{\tilde{\mu}}$, the latter class is nonempty.

Let us show that (13) and (22) are equivalent for the Lebesgue case:

Lemma 8. *Let μ^* be the Lebesgue outer measure in $\mathbb{R}^n$. Then $\overline{m}(A) = \mu^*(A \cap E) + \mu^*(A \cap E^c)$ for all $A \in \mathscr{E}$ iff $\mu^*(A) = \mu^*(A \cap E) + \mu^*(A \cap E^c)$ for all $A \subset \mathbb{R}^n$ and any given $E \subset \mathbb{R}^n$.*

Proof. Let $E \subset \mathbb{R}^n$ be given. Suppose

$$\mu^*(A) = \mu^*(A \cap E) + \mu^*(A \cap E^c), \qquad \text{all} \quad A \subset \mathbb{R}^n, \tag{25}$$

Then it also holds for $A \in \mathscr{E}$. But $\mu^*|\mathscr{E} = \overline{m}$. Hence

$$\overline{m}(A) = \mu^*(A \cap E) + \mu^*(A \cap E^c), \qquad A \in \mathscr{E}. \tag{26}$$

Conversely, let (26) hold. If $\mu^*(A) = +\infty$, then by the subadditivity of μ^*, (25) is true trivially. So if $\mu^*(A) < \infty$, then for each $\epsilon > 0$ there exists a cover $\{P_i: i \geqslant 1\} \subset \mathscr{S} \subset \mathscr{E}$ such that

$$\mu^*(A) + \epsilon > \sum_{i=1}^{\infty} m(P_i), \qquad A \subset \bigcup_{i=1}^{\infty} P_i. \tag{27}$$

But $A \cap E \subset \bigcup_{i=1}^{\infty}(P_i \cap E)$ and $A \cap E^c \subset \bigcup_{i=1}^{\infty}(P_i \cap E^c)$, so that

$$\mu^*(A \cap E) + \mu^*(A \cap E^c) \leqslant \sum_{i=1}^{\infty} \left[\mu^*(P_i \cap E) + \mu^*(P_i \cap E^c)\right],$$

by the σ-subadditivity of μ^*,

$$= \sum_{i=1}^{\infty} m(P_i), \qquad \text{by (26)}, \quad \text{since } P_i \in \mathscr{E},$$

$$< \mu^*(A) + \epsilon, \qquad \text{by (27)}.$$

Since $\epsilon > 0$ is arbitrary,

$$\mu^*(A \cap E) + \mu^*(A \cap E^c) \leqslant \mu^*(A).$$

But the opposite inequality always holds by the subadditivity of μ^*, and hence (25) is true. Since $E \subset \mathbb{R}^n$ is arbitrary, the lemma follows.

At this point we turn to a study of measures on abstract sets without topology. It shows that a deep analysis of classes of sets that can be measured is possible if only an outer measure is available.

The first major result of this chapter is given by:

Theorem 9 (Carathéodory). *Let μ be any outer measure on $\mathscr{P}$ $(= 2^{\Omega})$ of a point set Ω in the sense of Definition* 6. *Then*

(i) *the class $\mathscr{M}_{\mu}$ of all μ-measurable subsets of Ω is a σ-algebra;*
(ii) *the restriction of μ to $\mathscr{M}_{\mu}$ is a measure, so that it is σ-additive; and μ is even σ-additive on the trace σ-algebra $\mathscr{M}_{\mu}(E)$ for each $E \subset \Omega$;*
(iii) *if $A \subset \Omega$, and $\mu(A) = 0$ then $A \in \mathscr{M}_{\mu}$.*

Further, if μ is Carathéodory regular, then

(iv) *for any $A \subset \Omega$, there exists a $B \in \mathscr{M}_{\mu}$, $A \subset B$, (B is termed a **μ-measurable cover of** A) such that $\mu(A) = \mu(B)$,*
(v) *if $A_n \subset A_{n+1}$ is any sequence of $\mathscr{P}$, then $\mu(\lim_n A_n) = \lim_n \mu(A_n)$. In case μ is a finite (Carathéodory regular) outer measure, then a subset A of Ω is in $\mathscr{M}_{\mu}$ if it satisfies the single equation*

$$\mu(\Omega) = \mu(A) + \mu(A^c) < \infty.$$

Proof. We present the proof in steps for clarity.

1. By the symmetry of the equation (22) of the measurability definition for A and A^c, it is clear that $A \in \mathscr{M}_{\mu}$ iff $A^c \in \mathscr{M}_{\mu}$. If $A \subset \Omega$ and $\mu(A) = 0$, then by the monotonicity of μ, we get

$$0 \leqslant \mu(A \cap E) \leqslant \mu(A) = 0, \qquad E \subset \Omega, \tag{28}$$

and hence

$$\mu(E) \geqslant \mu(E \cap A^c) + \mu(A \cap E), \qquad E \subset \Omega. \tag{29}$$

Since the opposite inequality always holds by subadditivity, this implies the equality in (29), and so by (22) we have $A \in \mathscr{M}_\mu$. Thus $\mathscr{M}_\mu$ contains every set $A \subset \Omega$ of μ-measure zero. Since $A = \varnothing$ is a candidate, $\varnothing^c = \Omega \in \mathscr{M}_\mu$ also.

2. In general $\mathscr{M}_\mu$ may not have anything else. But if it has A, B, then we assert that $A - B$ and $B - A$ are in $\mathscr{M}_\mu$ as well.

Indeed, let $E \subset \Omega$ be arbitrary. Since $A, B \in \mathscr{M}_\mu$, we have by (22)

$$\begin{aligned}\mu(E) &= \mu(E \cap A) + \mu(E \cap A^c)\\ &= \mu(B \cap (E \cap A)) + \mu(B^c \cap E \cap A) + \mu(E \cap A^c)\end{aligned}$$

($E \cap A$ playing a role of a test set for B)

$$\begin{aligned}&= \mu(E \cap (A - B)) + [\mu(A \cap B \cap E) + \mu(E \cap A^c)]\\ &\geqslant \mu(E \cap (A - B)) + \mu(E \cap (A - B)^c), \qquad (30)\end{aligned}$$

because

$$\begin{aligned}E \cap (A - B)^c &= E \cap (A^c \cup B) = (E \cap A^c) \cup (E \cap B)\\ &= (E \cap A^c) \cup ((E \cap B \cap A) \cup (E \cap B \cap A^c))\\ &= (E \cap A^c) \cup (E \cap B \cap A), \qquad \text{since} \quad E \cap B \cap A^c \subset E \cap A^c,\end{aligned}$$

so that

$$\mu(E \cap (A - B)^c) \leqslant \mu(E \cap A^c) + \mu(E \cap B \cap A).$$

But by the subadditivity of μ, the opposite inequality of (30) is always true, so that $A - B \in \mathscr{M}_\mu$. Interchanging A, B in the above we get $B - A \in \mathscr{M}_\mu$. So $\mathscr{M}_\mu$ is closed under differences and complementation.

3. $\mathscr{M}_\mu$ is also closed under countable disjoint unions, so it is a σ-algebra, and μ is σ-additive on $\mathscr{M}_\mu$ at the same time. For, let $A_n \in \mathscr{M}_\mu$, disjoint, $A = \bigcup_{n=1}^\infty A_n$, and $E \subset \Omega$ be any set. Then

$$\begin{aligned}\mu(E) &= \mu(E \cap A_1) + \mu(E \cap A_1^c),\\ &= \mu(E \cap A_1) + \mu(E \cap A_1^c \cap A_2) + \mu(E \cap A_1^c \cap A_2^c),\end{aligned}$$

(treating $E \cap A_1^c$ as the new test set for $A_2 \in \mathscr{M}_\mu$, and $A_1 \cap A_2 = \varnothing \Rightarrow A_2 \subset A_1^c$)

$$= \mu(E \cap A_1) + \mu(E \cap A_2) + \mu(E \cap (A_1 \cup A_2)^c).$$

Hence by induction one has

$$\mu(E) = \sum_{i=1}^{n} \mu(E \cap A_i) + \mu\left(E \cap \left(\bigcup_{i=1}^{n} A_i\right)^c\right)$$

$$\geqslant \sum_{i=1}^{n} \mu(E \cap A_i) + \mu(E \cap A^c), \qquad \text{since } A \subset \bigcup_{i=1}^{n} A_i. \tag{31}$$

Letting $n \to \infty$ in (31), and using the σ-subadditivity of μ,

$$\mu(E) \geqslant \sum_{i=1}^{\infty} \mu(E \cap A_i) + \mu(E \cap A^c) \geqslant \mu(E \cap A) + \mu(E \cap A^c). \tag{32}$$

Since the opposite inequality in (32) is always true, $A \in \mathscr{M}_\mu$. Also, replacing E with the set $E \cap A$ in (32), we get

$$\mu(A \cap E) = \sum_{i=1}^{\infty} \mu(A_i \cap E), \qquad E \in \mathscr{P}. \tag{33}$$

Thus μ is a measure on $\mathscr{M}_\mu(E)$, $E \in \mathscr{P}$, and taking $E = \Omega$, it has the same property on $\mathscr{M}_\mu$ itself.

If $B_n \in \mathscr{M}_\mu$ are not necessarily disjoint, then

$$B_1 \cup B_2 = B_1 \cup (B_2 - B_1)$$

is a disjoint union with $B_1 \in \mathscr{M}_\mu$, $B_2 \in \mathscr{M}_\mu \Rightarrow B_2 - B_1 \in \mathscr{M}_\mu$ (by step 2), so that $B_1 \cup B_2 \in \mathscr{M}_\mu$. Thus $\mathscr{M}_\mu$ is closed under finite unions. Let $A_1 = B_1$, $A_n = B_n - \bigcup_{i=1}^{n-1} B_i \in \mathscr{M}_\mu$, $n > 1$. Then A_n's are disjoint, and

$$\bigcup_{n=1}^{\infty} B_n = \bigcup_{n=1}^{\infty} A_n \in \mathscr{M}_\mu.$$

Thus $\mathscr{M}_\mu$ is a σ-algebra. This gives parts (i)–(iii) of the theorem.

We next use the hypothesis that μ is Carathéodory regular, i.e. (23) holds.

4. Each $A \subset \Omega$ has a μ-measurable cover B, and if $A_n \uparrow A$ is any sequence, then $\mu(A) = \lim_{n \to \infty} \mu(A_n)$. For, this is trivial if $\mu(A) = +\infty$, since $B = \Omega \in \mathscr{M}_\mu$ will suffice. So let $\mu(A) < \infty$. Since μ is Carathéodory regular, by definition [cf. (23)] for each $\epsilon = 1/n$ there exists a $D_n \in \mathscr{M}_\mu$ such that $D_n \supset A$ and

$$\mu(A) + \frac{1}{n} > \mu(D_n).$$

Let $B = \bigcap_{n=1}^{\infty} D_n$. Then $B \in \mathscr{M}_\mu$, since the latter is a σ-algebra, and $B \supset A$, so that

$$\mu(A) \leqslant \mu(B) \leqslant \mu(D_n) \leqslant \mu(A) + \frac{1}{n}.$$

Taking lim inf on both sides, we get $\mu(A) = \mu(B)$.

Now let $A_n \subset A_{n+1} \subset \Omega$ with $A = \lim_{n \to \infty} A_n$. By the above paragraph there exist measurable covers $B_n \supset A_n$, $n \geqslant 1$, with $\mu(A_n) = \mu(B_n)$, $B_n \in \mathcal{M}_\mu$. Since B_n's are not necessarily increasing, let $F_n = \bigcup_{i=1}^n B_i$ ($\in \mathcal{M}_\mu$). But $A_n \subset A_{n+1} \subset B_{n+1} \subset F_{n+1}$, and $A_k \subset B_k \cap B_n$, $n \geqslant k$. Also

$$\mu(A_n) \leqslant \mu(A), \qquad \text{since} \quad A_n \subset A, \quad n \geqslant 1.$$

If $\mu(A_n) = +\infty$, for some n, then $\lim_{n \to \infty} \mu(A_n) = \mu(A) = \infty$. Let $\mu(A_n) < \infty$, all $n \geqslant 1$. But

$$\mu(A_k) \leqslant \mu(B_k \cap B_n) \leqslant \mu(B_k) = \mu(A_k), \qquad k \leqslant n.$$

Hence

$$\mu(B_k) = \mu(B_k \cap B_n),\ n \geqslant k, \text{ and } B_k \cap B_n \in \mathcal{M}_\mu \Rightarrow$$

$$\begin{aligned}\mu(B_k - B_n) &= \mu(B_k - B_k \cap B_n) \\ &= \mu(B_k) - \mu(B_k \cap B_n) = 0, \qquad \text{since} \quad \mu(B_k) < \infty. \quad (34)\end{aligned}$$

This implies $\mu(A_n) = \mu(F_n)$ also, because

$$\mu(A_n) \leqslant \mu(F_n) = \mu\left(\bigcup_{k=1}^n B_k\right) \leqslant \mu(B_n) + \mu\left(\bigcup_{i=1}^n B_i - B_n\right)$$

$$\leqslant \mu(B_n) + \sum_{i=1}^n \mu(B_i - B_n)$$

(by the subadditivity of μ)

$$= \mu(B_n) = \mu(A_n), \qquad \text{using (34)}.$$

Since it was shown that $\mu | \mathcal{M}_\mu$ is a measure, and $F_n \uparrow$, one has by Proposition 1.1(i)

$$\lim_{n \to \infty} \mu(A_n) = \lim_{n \to \infty} \mu(F_n) = \mu\left(\bigcup_{n=1}^\infty F_n\right)$$

$$= \mu\left(\bigcup_{n=1}^\infty B_n\right) \geqslant \mu(A) \geqslant \mu(A_k). \qquad (35)$$

Letting $k \to \infty$ on both sides of (35), we see that there is equality throughout. This establishes (iv) and (v).

5. Finally suppose that the Carathéodory regular μ is finite, and $\mu(\Omega) = \mu(A) + \mu(A^c)$, $A \subset \Omega$. Then by step 4, there is a μ-measurable cover $B \supset A$, so that $B \in \mathscr{M}_\mu$ and

$$\mu(B) + \mu(B^c) = \mu(\Omega) = \mu(A) + \mu(A^c) < \infty.$$

Since $\mu(B) = \mu(A)$, so that $\mu(A^c) = \mu(B^c)$, and using A^c as a test set for B, we get [cf. (22)]

$$\begin{aligned}\mu(A^c) &= \mu(A^c \cap B) + \mu(A^c \cap B^c)\\ &= \mu(B - A) + \mu(B^c), \qquad \text{since} \quad B^c \subset A^c.\end{aligned}$$

Hence using $\mu(A^c) = \mu(B^c)$, we deduce that $\mu(B - A) = 0$. So $B - A \in \mathscr{M}_\mu$. But $A = B - (B - A)$, so that $A \in \mathscr{M}_\mu$, and the theorem is completely proved.

Our next task is to show that generated outer measures (in particular our Lebesgue outer measure) are Carathéodory regular. This and several other properties of such (outer) measures are given in our second major result of this chapter as follows:

Theorem 10 (C. Carathéodory, E. Hopf). *Let μ^* be a generated outer measure by a pair $(\alpha, \mathscr{A})$ where $\varnothing \in \mathscr{A} \subset 2^\Omega$, and $\alpha: \mathscr{A} \to \overline{\mathbb{R}}^+$ with $\alpha(\varnothing) = 0$, as in Definition 7. Then we assert:*

(i) *μ^* is an outer measure satisfying*

$$\mu^*(A) = \inf\{\mu^*(B): B \in \mathscr{A}_\sigma,\ B \supset A\},$$

where $\mathscr{A}_\sigma = \{B \subset \Omega: B = \bigcup_{n=1}^\infty A_n,\ A_n \in \mathscr{A}\}$.

(ii) *If there exists a sequence $\{A_n: n \geq 1\} \subset \mathscr{A}$ which covers A, i.e., $A \subset \bigcup_{n=1}^\infty A_n \in \mathscr{A}_\sigma$, then there exists a $B \in \mathscr{A}_{\sigma\delta}$ such that $A \subset B$ and $\mu^*(A) = \mu^*(B)$, where*

$$\mathscr{A}_{\sigma\delta} = \left\{B \subset \Omega: B = \bigcap_{n=1}^\infty B_n,\ B_n \in \mathscr{A}_\sigma\right\}.$$

(iii) *If, in the above, $\mathscr{A}$ is a (semi-)ring and α on $\mathscr{A}$ is finitely additive, then $\mathscr{A} \subset \mathscr{M}_{\mu^*}$ and μ^* is Carathéodory regular. In fact more is true, namely, it is regular for $\mathscr{A}_\sigma$:*

$$\mu^*(A) = \inf\{\mu(B): B \in \mathscr{A}_\sigma,\ B \supset A\}.$$

Further, $\mu^*|\mathscr{A} = \alpha$ iff α is σ-additive on $\mathscr{A}$.

If this result is granted, then we deduce immediately:

Corollary 11. *Let μ^* be the Lebesgue outer measure in $\mathbb{R}^n$ and $E \subset \mathbb{R}^n$ be a bounded set. Then E is Lebesgue measurable in the sense of Definition* 4 *iff it satisfies* (13), *i.e.*

$$\overline{m}(A) = \mu^*(A \cap E) + \mu^*(A \cap E^c) \qquad \text{for all elementary} \quad A \subset \mathbb{R}^n.$$

Proof of Corollary. Let E be any bounded subset of $\mathbb{R}^n$. Then there exists a bounded rectangle $I \supset E$, and $m(I) < \infty$. Since by the above theorem μ^* is Carathéodory regular, and by Theorem 1.2.8 $\overline{m}$ is σ-additive on the class of elementary sets $\mathscr{E}$ of $\mathbb{R}^n$, it follows by (iii) of the above theorem that $\mathscr{E} \subset \mathscr{M}_{\mu^*}$ and $\mu^*|\mathscr{E} = \overline{m}$, the extension of the volume measure of $\mathscr{E}$ (cf. Proposition 1.2.6). Hence taking $A = I$ in (12), we get

$$\begin{aligned}\overline{m}(I) = \mu^*(I) &= \mu_*(I \cap E) + \mu^*(I - E)\\ &= \mu_*(E) + \mu^*(I - E) < \infty.\end{aligned} \tag{36}$$

If E is Lebesgue measurable, then by Definition 4, $\mu_*(E) = \mu^*(E)$. Consequently (36) may be written as

$$\overline{m}(I) = \mu^*(I) = \mu^*(E) + \mu^*(I - E) < \infty. \tag{37}$$

Taking $I = \Omega$ in Theorem 9(v), (37) implies that $E \in \mathscr{M}_{\mu^*}(I) \subset \mathscr{M}_{\mu^*}$ and hence very bounded Lebesgue-measurable set is Carathéodory measurable. Since $\mathbb{R}^n$ can be expressed as a disjoint union of bounded intervals, this implies that every Lebesgue-measurable set in $\mathbb{R}^n$ is Carathéodory measurable relative to the outer measure μ^*.

On the other hand, every $E \subset \mathbb{R}^n$ satisfying (13) was already shown to be Lebesgue measurable in Proposition 5 above without any other restrictions. The corollary follows, and this also completes the converse part of (13) at the same time.

We can now proceed to the

Proof of Theorem 10. (i): To see that μ^* is an outer measure, we need to show that it is σ-subadditive, since $\mu^*(\varnothing) = 0$ and its monotonicity are clear. Thus let $A \subset \Omega$ and $A_n \subset \Omega$. If $\mu^*(A_n) = +\infty$ for some n, the result is true and trivial. So let $\mu^*(A_n) < \infty$ for each n, and $A \subset \bigcup_{n=1}^{\infty} A_n$. Then by definition of μ^*, given $\epsilon > 0$, there exist sets $B_{ni} \in \mathscr{A}$, $A_n \subset \bigcup_{i=1}^{\infty} B_{ni}$ and

$$\sum_{i=1}^{\infty} \alpha(B_{ni}) - \frac{\epsilon}{2^n} < \mu^*(A_n). \tag{38}$$

But one also has $A \subset \bigcup_{n=1}^{\infty} A_n \subset \bigcup_{n=1}^{\infty}\bigcup_{i=1}^{\infty} B_{ni}$, so that

$$\mu^*(A) \leqslant \sum_{n=1}^{\infty}\sum_{i=1}^{\infty} \alpha(B_{ni})$$

$$\leqslant \sum_{n=1}^{\infty}\left[\mu^*(A_n) + \frac{\epsilon}{2^n}\right] = \sum_{n=1}^{\infty} \mu^*(A_n) + \epsilon, \qquad \text{by (38)}.$$

Since $\epsilon > 0$ is arbitrary, the σ-subadditivity follows.

To prove the stated approximation property for any $A \subset \Omega$, note that if there is no B in $\mathscr{A}_\sigma$ containing A, then $\mu^*(A) = +\infty$, and the result is trivial again, since $\inf\{\mu^*(B): B \supset A, B \in \mathscr{A}_\sigma\} = \inf\{\varnothing\} = +\infty$. Suppose therefore that there exists a B in $\mathscr{A}_\sigma$ such that $B \supset A$. Since $B = \bigcup_{k=1}^{\infty} C_k$, $C_k \in \mathscr{A}$, and by definition

$$\mu^*(A) = \inf\left\{\sum_{k=1}^{\infty} \alpha(C_k): C_k \in \mathscr{A}, A \subset \bigcup_{k=1}^{\infty} C_k = B\right\},$$

the result follows if $\mu^*(A) = +\infty$, since any such $B \in \mathscr{A}_\sigma$ will give an approximation. Thus we may also assume that $\mu^*(A) < \infty$. Then for each $\epsilon > 0$, there is a sequence $B_n \in \mathscr{A}$ such that $A \subset B = \bigcup_{n=1}^{\infty} B_n$ and

$$\mu^*(A) + \epsilon > \sum_{n=1}^{\infty} \alpha(B_n) \geqslant \mu^*(B), \tag{39}$$

since B is obviously covered by itself (i.e., $B = B_n$, $n \geqslant 1$). But $\mu^*(A) \leqslant \mu^*(B)$, so that

$$\mu^*(A) \leqslant \inf\{\mu^*(C): C \in \mathscr{A}_\sigma, C \supset A\} \leqslant \mu^*(B). \tag{40}$$

Hence (39) and (40) imply that (i) is true in this case also.

(ii): Suppose now $A \subset \tilde{B}$, $\tilde{B} \in \mathscr{A}_\sigma$. Then by (i) above, for each $\epsilon = 1/n$, there exists a $B_n \in \mathscr{A}_\sigma$ such that

$$\mu^*(A) + \frac{1}{n} \geqslant \mu^*(B_n) \geqslant \mu^*(A). \tag{41}$$

If $B = \bigcap_{n=1}^{\infty} B_n \supset A$, then $B \in \mathscr{A}_{\sigma\delta}$ and (41) implies

$$\mu^*(A) \leqslant \mu^*(B) \leqslant \mu^*(A) + \frac{1}{n}.$$

Since n is arbitrary, this gives (ii).

(iii): In the above part, $\mathscr{A}$ was arbitrary, and we cannot infer that the approximating sets B (from $\mathscr{A}_\sigma$, $\mathscr{A}_{\sigma\delta}$, etc.) for A are μ^*-measurable. In fact, they need not be in $\mathscr{M}_{\mu^*}$ as easy counterexamples show. (One is included below as Exercise 6.) However, under the given hypothesis that $\mathscr{A}$ is a (semi-)ring and $\alpha: \mathscr{A} \to \overline{\mathbb{R}}^+$ is additive, this cannot happen. Since such an α can be uniquely extended to the ring generated by $\mathscr{A}$ (cf. Proposition 1.2.4), we may (and do) assume that $\mathscr{A}$ is a ring. Let $A \in \mathscr{A}$ be arbitrary. If $E \in \mathscr{P}$ is a test set, then we need to show only that

$$\mu^*(E) \geqslant \mu^*(E \cap A) + \mu^*(E \cap A^c), \tag{42}$$

because of the subadditivity of μ^* on the power set $\mathscr{P}$. This is obvious if $\mu^*(E) = \infty$. So let $\mu^*(E) < \infty$. Then for each $\epsilon > 0$, there exist $E_n \in \mathscr{A}$ such that $E \subset \bigcup_{n=1}^{\infty} E_n$ and

$$\mu^*(E) + \epsilon > \sum_{n=1}^{\infty} \alpha(E_n). \tag{43}$$

But then $E \cap A \subset \bigcup_{n=1}^{\infty} E_n \cap A$, $E \cap A^c \subset \bigcup_{n=1}^{\infty} E_n \cap A^c$, and $E_n \cap A \in \mathscr{A}$, $E_n - A \in \mathscr{A}$, since it is a ring. By the σ-subadditivity of μ^* on $\mathscr{P}$, shown in (i) above, we get

$$\mu^*(E \cap A) + \mu^*(E \cap A^c) \leqslant \sum_{n=1}^{\infty} \left[\alpha(E_n \cap A) + \alpha(E_n \cap A^c)\right]$$

$$= \sum_{n=1}^{\infty} \alpha(E_n)$$

(since α is additive on $\mathscr{A}$),

$$< \mu^*(E) + \epsilon, \qquad \text{by (43)}.$$

Since $\epsilon > 0$ is arbitrary, we get (42) and hence $A \in \mathscr{M}_{\mu^*}$.

But it was shown in (ii) that μ^* has the regularity relative to $\mathscr{A}_\sigma \subset \mathscr{M}_{\mu^*}$, the latter being a σ-algebra. Since the infimum is attained on a subset of $\mathscr{M}_{\mu^*}$, it follows that the same holds on $\mathscr{M}_{\mu^*}$, and μ^* is Carathéodory regular.

If $A \in \mathscr{A}$, it is covered by itself, so $\mu^*(A) \leqslant \alpha(A)$. Since $\mu^*|\mathscr{M}_{\mu^*}$ is σ-additive, $\mu^*|\mathscr{A} = \alpha$ implies α is also σ-additive on $\mathscr{A}$. On the other hand, if $\alpha(\cdot)$ is σ-additive and $A_n \in \mathscr{A}$, $A \subset \bigcup_{n=1}^{\infty} A_n$, then (since σ-additivity implies its σ-subadditivity) we have

$$\alpha(A) \leqslant \sum_{n=1}^{\infty} \alpha(A_n).$$

Taking infimum of all such sums gives $\alpha(A) \leqslant \mu^*(A)$, implying $\alpha = \mu^*|\mathscr{A}$. This completes the proof of the theorem.

A function $\bar{\mu} \geqslant 0$ on a ring $\mathscr{A}$ is said to be *complete* if $A \in \mathscr{A}$, $\bar{\mu}(A) = 0$, and $B \subset A$ implies $\bar{\mu}(B) = 0$. Such a B is called a *$\bar{\mu}$-null set*. If $\mathscr{A}$ contains all such B, it is termed a *complete ring* for $\bar{\mu}$. Thus in Theorem 9, we showed that, *for any outer measure μ, the class of μ-measurable sets $\mathscr{M}_\mu$ is a complete σ-algebra and $\mu|\mathscr{M}_\sigma$ is a complete measure.*

Remark. The last part of the above theorem implies that even when $\alpha(\cdot)$ is additive and $\mathscr{A}$ is a ring, then $\mathscr{A} \subset \mathscr{M}_{\mu^*}$ but $\mu^* \leqslant \alpha$ on $\mathscr{A}$. It is therefore natural to ask for equality conditions in the latter. Indeed, if μ^* is generated by $(\alpha, \mathscr{A})$ with only the Jordan finite sums, as in (1), instead of the Lebesgue infinite sums, as in Definition 2, and denoting this by $\tilde{\mu}$, we have a positive answer. Thus

$$\tilde{\mu}(A) = \inf\left\{ \sum_{i=1}^{n} \alpha(P_i) : A \subset \bigcup_{i=1}^{n} P_i,\ P_i \in \mathscr{A},\ n \geqslant 1 \right\}, \qquad A \subset \Omega,$$

and the $\tilde{\mu}$-(measurable) sets $\mathscr{M}_{\tilde{\mu}}$ satisfy $A \in \mathscr{M}_{\tilde{\mu}}$ iff

$$\tilde{\mu}(E) = \tilde{\mu}(E \cap A) + \tilde{\mu}(E \cap A^c), \qquad E \subset \Omega.$$

Then Theorems 9 and 10 take the following form, whose proof is left to the reader. (See Exercise 5.)

Proposition 12. *Let $\tilde{\mu}$ be the Jordan outer set function generated by a ring $\mathscr{A}$ and an additive $\alpha: \mathscr{A} \to \overline{\mathbb{R}}^+$. If $\mathscr{M}_{\tilde{\mu}}$ is the class of $\tilde{\mu}$-sets, then we have: $\mathscr{M}_{\tilde{\mu}}$ is an algebra containing $\mathscr{A}$, $\tilde{\mu}$ is monotone, $\tilde{\mu}|\mathscr{M}_{\tilde{\mu}}$ is additive, and $\tilde{\mu}|\mathscr{A} = \alpha$. Also it is complete. Moreover*

$$\tilde{\mu}(A) = \inf\{\tilde{\mu}(B) : B \supset A,\ B \in \mathscr{A}\}, \qquad A \subset \Omega. \tag{44}$$

In (44), *$\mathscr{A}$ can be replaced by $\mathscr{M}_{\tilde{\mu}}$.*

(Recall that $\inf\{\varnothing\} = +\infty$.)

We present some consequences of, and supplements to, the two basic results established above. Let us start with a simple but useful observation.

Lemma 13. *Let $\alpha: \mathscr{S} \to \overline{\mathbb{R}}^+$ be a σ-additive function with $\mathscr{S}$ as a (semi-)ring. If μ^* is a generated outer measure by $(\alpha, \mathscr{S})$, and $\mathscr{M}_{\mu^*}$ is the collection of μ^*-measurable sets, then the test class in determining the μ^*-measurable sets can be restricted to $\mathscr{S}$ from $\mathscr{P}$. Thus, if we denote*

$$\mathscr{M}_{\mu^*} = \{A \subset \Omega : \mu^*(E) = \mu(E \cap A) + \mu^*(E \cap A^c),\ E \in \mathscr{P}\}$$

and

$$\tilde{\mathscr{M}}_{\mu^*} = \{A \subset \Omega: \alpha(E) = \mu^*(E \cap A) + \mu^*(E \cap A^c),\ E \in \mathscr{S}\},$$

then $\mathscr{M}_{\mu^*} = \tilde{\mathscr{M}}_{\mu^*}$. [*Replacing* A *by* $A \cap E$ *here, one can also conclude that* $A \in \mathscr{M}_{\mu^*}$ *iff* $A \cap E \in \mathscr{M}_{\mu^*}$ *for all* $E \in \mathscr{S}$ *for which* $\alpha(E) < \infty$.]

When we compare this result with that of Theorem 10(iii), it is clear that $\mathscr{S} \subset \mathscr{M}_{\mu^*}$, the latter being a complete σ-algebra. Let $\mathscr{A}$ be the ring generated by $\mathscr{S}$. Then α can be (uniquely) extended to $\mathscr{A}$ (see Proposition 1.2.6, and the remark following its proof indicating the validity of the result generally). In view of this and the obvious fact that $\mathscr{M}_{\mu^*} \subset \tilde{\mathscr{M}}_{\mu^*}$, only the opposite inclusion has to be established. But the proof is the same as that of the converse part of Lemma 8. (The extension is also proved in Theorem 3.1.)

The point of this result is that for generated outer measures, from (semi-) rings and additive functions, measurable sets are all determined by a relatively small number of functional equations as in $\tilde{\mathscr{M}}_{\mu^*}$. This need not be true if μ^* is generated by an arbitrary pair $(\alpha, \mathscr{A})$, as in Definition 7.

The connection between arbitrary, generated, and Carathéodory regular outer measures is not difficult to establish. It is given by the following result.

Proposition 14. *Let* μ *be an arbitrary outer measure on* Ω, *and define* μ_i, *with* $\mu_0 = \mu$, *inductively on* Ω, *for* $i \geqslant 1$, *by*

$$\mu_i(A) = \inf\{\mu_{i-1}(B): B \supset A,\ B \in \mathscr{M}_{\mu_{i-1}}\}, \qquad A \subset \Omega, \tag{45}$$

where $\mathscr{M}_{\mu_i}$ *stands for the collection of* μ_i*-measurable sets in terms of Definition 6(2). Then for each* $i \geqslant 1$, μ_i *is a Carathéodory regular outer measure on* Ω, *and* $\mu_{i+1} = \mu_i$, $\mathscr{M}_{\mu_{i+1}} = \mathscr{M}_{\mu_i} \supset \mathscr{M}_{\mu_0}$. *There is equality in the latter if* μ_0 *is Carathéodory regular, and this happens iff* μ_0 *is a generated outer measure by some pair* $(\alpha, \mathscr{A})$ *where* $\mathscr{A}$ *is a (semi-)ring and* $\alpha: \mathscr{A} \to \bar{\mathbb{R}}^+$ *is additive. (However,* $\mathscr{M}_{\mu_0} = \mathscr{M}_{\mu_1}$ *does not necessarily imply that* $\mu_0 = \mu_1$ *or equivalently* μ_0 *is Carathéodory regular. See the example below.)*

Proof. To see that each μ_i is an outer measure, it suffices to verify this for μ_1. It is clear that μ_1 is monotone and nonnegative and that $\mu_1(\varnothing) = 0$. Regarding the σ-subadditivity, let A, A_n be subsets of Ω and $A \subset \bigcup_{n=1}^{\infty} A_n$. Using the by now familiar argument, we only need to consider the case that $\mu(A_n) < \infty$ for all $n \geqslant 1$, since the result is true and trivial otherwise. Then given $\epsilon > 0$, (45) implies that there are $B_n^\epsilon \in \mathscr{M}_{\mu_0}$ such that $A_n \subset B_n^\epsilon$ and

$$\mu_1(A_n) + \frac{\epsilon}{2^n} > \mu_0(B_n^\epsilon), \qquad n \geqslant 1. \tag{46}$$

But $A \subset \bigcup_{n=1}^{\infty} A_n \subset \bigcup_{n=1}^{\infty} B_n^\epsilon \in \mathscr{M}_{\mu_0}$ since $\mathscr{M}_{\mu_0}$ is a σ-algebra. Also $\mu_0 | \mathscr{M}_{\mu_0}$ is

σ-additive, *and* $\mu_1 = \mu$ on $\mathscr{M}_{\mu_0}$. Hence

$$\mu_1(A) \leqslant \mu_1\left(\bigcup_{n=1}^{\infty} B_n^{\epsilon}\right) = \mu_0\left(\bigcup_{n=1}^{\infty} B_n^{\epsilon}\right)$$

$$\leqslant \sum_{n=1}^{\infty} \mu_0(B_n^{\epsilon}), \qquad \text{since } \mu_0 \text{ is } \sigma\text{-subadditive,}$$

$$< \sum_{n=1}^{\infty} \left(\mu_1(A_n) + \frac{\epsilon}{2^n}\right), \qquad \text{by (46)}$$

$$= \sum_{n=1}^{\infty} \mu_1(A_n) + \epsilon. \tag{47}$$

Since $\epsilon > 0$ is arbitrary, μ_1 is σ-subadditive and hence is an outer measure. By induction, this shows that every μ_i, $i \geqslant 1$, is an outer measure, and $\mathscr{M}_{\mu_i}$ is a σ-algebra on which μ_i is σ-additive.

For the inclusion relation $\mathscr{M}_{\mu_i} \subset \mathscr{M}_{\mu_{i+1}}$, $i \geqslant 0$, again it suffices to consider the case that $i = 0$, the argument being the same for all the other i. Let $A \in \mathscr{M}_{\mu_0}$; it is to be shown (in view of the subadditivity of μ_1) that

$$\mu_1(E) \geqslant \mu_1(E \cap A) + \mu_1(E \cap A^c), \qquad E \subset \Omega. \tag{48}$$

Here we may assume that $\mu_1(E) < \infty$. Then, as before, given $\epsilon > 0$, there is a $B_\epsilon \in \mathscr{M}_{\mu_0}$ such that $E \subset B_\epsilon$ and

$$\mu_1(E) + \epsilon > \mu_0(B_\epsilon).$$

Hence by the monotonicity of μ_1 and the fact that $\mu_0 = \mu_1$ on $\mathscr{M}_{\mu_0}$,

$$\mu_1(E \cap A) + \mu_1(E \cap A^c) \leqslant \mu_0(B_\epsilon \cap A) + \mu_0(B_\epsilon \cap A^c)$$

$$\leqslant \mu_0(B_\epsilon), \qquad \text{since } \mu_0 \text{ is additive on } \mathscr{M}_{\mu_0}$$

$$< \mu_1(E) + \epsilon.$$

This shows that (48) is true and hence $A \in \mathscr{M}_{\mu_1}$. Since each $\mu_{i+1}|\mathscr{M}_{\mu_i} = \mu_i$, and $\mathscr{M}_{\mu_i} \subset \mathscr{M}_{\mu_{i+1}}$, we conclude that each μ_{i+1} is an extension of μ_i given by the process of (45) for $i \geqslant 0$.

We now assert that this process stops at $i = 1$, i.e., $\mu_{i+1} = \mu_i$ for $i \geqslant 1$. For this we need to show that μ_1 is a generated outer measure by a suitable pair $(\alpha, \mathscr{A})$, and deducing the desired result from Theorem 10(iii). Indeed, let $\tilde{\mu}$ be the generated outer measure by $(\mu_0, \mathscr{M}_{\mu_0})$:

$$\tilde{\mu}(A) = \inf\left\{ \sum_{n=1}^{\infty} \mu_0(B_n): B_n \in \mathscr{M}_{\mu_0}, A \subset \bigcup_{n=1}^{\infty} B_n \right\}. \tag{49}$$

If $B = \bigcup_{n=1}^{\infty} B_n$, then $B \in \mathscr{M}_{\mu_0}$, $B \supset A$ for each such covering, and hence by (45)

$$\mu_1(A) \leqslant \mu_0(B) \leqslant \sum_{n=1}^{\infty} \mu_0(B_n).$$

Consequently $\mu_1(A) \leqslant \tilde{\mu}(A)$, which results by taking the infimum of the sums in (49). On the other hand, for any $B \in \mathscr{M}_{\mu_0}$, $B \supset A$, let $B_1 = B$, $B_k = \varnothing$, $k > 1$. Then $\bigcup_{n=1}^{\infty} B_n \supset A$ and forms a covering, so that $\tilde{\mu}(A) \leqslant \mu_1(A)$ always holds by (49) and (45). This and the preceding inequalities yield that $\mu_1 = \tilde{\mu}$. Since $\tilde{\mu}$ is regular by Theorem 10(iii), we conclude that μ_1 is also Carathéodory regular and moreover it is a generated outer measure for a suitable pair $(\alpha, \mathscr{A})$, namely $\alpha = \mu_0$, $\mathscr{A} = \mathscr{M}_{\mu_0}$. Thus for each $A \subset \Omega$

$$\mu_2(A) = \inf\{ \mu_1(B): B \supset A, B \in \mathscr{M}_{\mu_1} \} = \mu_1(A). \tag{50}$$

Hence $\mu_1 = \mu_2$, and by induction $\mu_i = \mu_{i+1}$, $i \geqslant 1$. It follows that $\mathscr{M}_{\mu_i} = \mathscr{M}_{\mu_{i+1}}$, $i \geqslant 1$.

Finally, if μ_0 itself is Carathéodory regular, then in (50) μ_1, μ_2 can be replaced by μ_0, μ_1, so that $\mu_0 = \mu_1 = \mu_i$, $i > 1$, and $\mathscr{M}_{\mu_0} = \mathscr{M}_{\mu_1} = \mathscr{M}_{\mu_i}$, $i > 1$. Since each regular measure μ_1 satisfies $\mu_1 = \tilde{\mu}$ as shown above [the converse being given by Theorem 10(iii)], the proof is finished.

The parenthetical comment of the proposition is shown by the following:

Example. Let $\Omega = \{0,1\}$—the "coin tossing space". Consider a map $\mu_0(\{0\}) = p$, $\mu_0(\{1\}) = q$, $\mu_0(\Omega) = 1$ and $p + q > 1$. Then μ_0 is an outer measure on $\mathscr{P} = \{\varnothing, \{0\}, \{1\}, \Omega\}$. Moreover $\mathscr{M}_{\mu_0} = \{\varnothing, \Omega\}$. Define μ_1 by (45) on $\mathscr{P}$. Then $\mu_1(\varnothing) = 0$, $\mu_1(\{0\}) = 1$, $\mu_1(\{1\}) = 1$, and $\mu_1(\Omega) = 1$, and again $\mathscr{M}_{\mu_1} = \{\varnothing, \Omega\} = \mathscr{M}_{\mu_0}$. Thus $\mu_1 = \mu_2$ on the common class of measurable sets $\{\varnothing, \Omega\}$, but, e.g., if $p = \frac{1}{3}$, $q = \frac{3}{4}$, then $\mu_1(A) \neq \mu_2(A)$ for $A = \{0\}$ or $\{1\}$. Thus μ_0 is not regular even though μ_1 is and both have the same class of measurable sets. (Note that if we let $p = q = 1$, then $\mu_1 = \mu_0$, so that μ_0 becomes a generated outer measure.)

The preceding example admits a generalization.

Lemma 15. *Let μ be a finite outer measure on Ω [i.e., $\mu(\Omega) < \infty$]. If μ_1 is the measure function defined by (45) with $(\mu, \mathscr{M}_{\mu})$, and if $\mathscr{M}_{\mu_1}$ is the class of*

μ_1-measurable sets, then $\mathscr{M}_\mu = \mathscr{M}_{\mu_1}$ even though the functions μ_1 and μ do not necessarily agree on all subsets of Ω.

Proof. Since by the preceding proposition $\mathscr{M}_\mu \subset \mathscr{M}_{\mu_1}$ always, let $A \in \mathscr{M}_{\mu_1}$. By definition, $\mu_1(\Omega) \leqslant \mu(\Omega) < \infty$ so that $\mu_1(A) < \infty$. Now we show, a little more generally, even in the arbitrary case, that if $\mu_1(A) < \infty$, then $A \in \mathscr{M}_\mu$. Thus if $\epsilon = 1/n$ is fixed, by (45), there is a $B_n \in \mathscr{M}_\mu$ such that $A \subset B_n$ and

$$\mu_1(A) + \frac{1}{n} > \mu(B_n).$$

Let $B_0 = \bigcap_{n=1}^{\infty} B_n \supset A$. Since $\mathscr{M}_\mu$ is a σ-algebra, we have $B_0 \in \mathscr{M}_\mu$ and

$$\mu_1(A) + \frac{1}{n} > \mu(B_n) \geqslant \mu(B_0) \geqslant \mu_1(A).$$

It follows that $\mu_1(A) = \mu(B_0) < \infty$. Since $B_0 \in \mathscr{M}_\mu \subset \mathscr{M}_{\mu_1}$ and $A \in \mathscr{M}_{\mu_1}$, we have

$$\mu_1(B_0) = \mu_1(B_0 \cap A) + \mu_1(B_0 \cap A^c) = \mu_1(A) + \mu_1(B_0 - A),$$

so that

$$\mu_1(B_0 - A) = \mu_1(B_0) - \mu_1(A) = \mu(B_0) - \mu_1(A) = 0. \tag{51}$$

But then there exist $C_n \in \mathscr{M}_\mu$, $C_n \supset B_0 - A$ and

$$\frac{1}{n} = \mu_1(B_0 - A) + \frac{1}{n} > \mu(C_n) \geqslant \mu(B_0 - A).$$

So $\mu(B_0 - A) = 0$, and $B_0 - A \in \mathscr{M}_\mu$. Since $A = B_0 - (B_0 - A)$ and since $B_0 \in \mathscr{M}_\mu$, $B_0 - A \in \mathscr{M}_\mu$, and the latter is a ring (σ-algebra, in fact), it results that $A \in \mathscr{M}_\mu$. The lemma follows.

Some other easy characterizations of measurable sets for general and generated outer measures can be given, complementing the above discussion. The following result contains a pair of such statements as well as the maximality of these collections in the regular case.

Proposition 16. *Let μ be an outer measure on Ω, and $A \subset \Omega$. Then*

(i) *A is μ-measurable (i.e., $A \in \mathscr{M}_\mu$) iff for each $\epsilon > 0$ there exist $A_1^\epsilon \subset A \subset A_2^\epsilon$ such that $A_i^\epsilon \in \mathscr{M}_\mu$, $i = 1, 2$, and $\mu(A_2^\epsilon - A_1^\epsilon) < \epsilon$;*

(ii) *if μ is generated by a pair $(\alpha, \mathscr{A})$ where $\mathscr{A}$ is an algebra, $\alpha: \mathscr{A} \to \overline{\mathbb{R}}^+$ is a measure, $A \in \mathscr{M}_\mu - \mathscr{A}$, and $\mathscr{A}_1$ is the smallest ring containing $(A, \mathscr{A})$, or more generally $\mathscr{A} \subset \mathscr{A}_1 \subset \mathscr{M}_\mu$ is any algebra, then μ and μ_1^**

(the latter being generated by the pair $\mu_1 = \mu|\mathscr{A}_1$ *and* $\mathscr{A}_1$*) are the same (hence* $\mathscr{M}_\mu = \mathscr{M}_{\mu_1^*}$ *also)*;

(iii) (Kolmogorov) *for the* μ *of* (ii) *together with* $\alpha(\Omega) < \infty$, A *is* μ*-measurable iff given* $\epsilon > 0$, *there exists* $A^\epsilon \in \mathscr{A}$ *such that* $\mu(A \,\triangle\, A^\epsilon) < \epsilon$.

Proof. (i): If $A \in \mathscr{M}_\mu$, then for any $\epsilon > 0$, simply take $A_1^\epsilon = A_2^\epsilon = A$. For the converse, which is less trivial, suppose the given condition holds. Then for any $E \subset \Omega$,

$$
\begin{aligned}
\mu(E) &\leqslant \mu(E \cap A) + \mu(E \cap A^c), &&\text{by the subadditivity of } \mu \\
&\leqslant \mu(E \cap A_2^\epsilon) + \mu(E - A_1^\epsilon), &&\text{since } A_1^\epsilon \subset A \subset A_2^\epsilon \\
&= \mu(E \cap [A_1^\epsilon \cup (A_2^\epsilon - A_1^\epsilon)]) + \mu(E - A_1^\epsilon) \\
&\leqslant \mu(E \cap A_1^\epsilon) + \mu(E - A_1^\epsilon) + \mu(E \cap (A_2^\epsilon - A_1^\epsilon)) &&\text{(subadditivity of } \mu) \\
&\leqslant \mu(E) + \mu(A_2^\epsilon - A_1^\epsilon), &&\text{since } A_1^\epsilon \in \mathscr{M}_\mu \\
&< \mu(E) + \epsilon, &&\text{by the hypothesis.}
\end{aligned}
$$

Since $\epsilon > 0$ is arbitrary, there is equality throughout and so $A \in \mathscr{M}_\mu$.

(ii): Let μ be the generated outer measure by $(\alpha, \mathscr{A})$. Then it is regular, and for each $E \subset \Omega$

$$\mu(E) = \inf\{\mu(B): B \supset E,\ B \in \mathscr{M}_\mu\}. \tag{52}$$

But $\mu_1 = \mu|\mathscr{A}_1$, where $\mathscr{A} \subset \mathscr{A}_1 \subset \mathscr{M}_\mu$ is an algebra. If μ^* is the generated outer measure of $(\mu_1, \mathscr{A}_1)$ then by Theorem 10(iii), for all $E \subset \Omega$ one has

$$\mu_1^*(E) = \inf\{\mu_1^*(B): B \supset E,\ B \in \mathscr{M}_\mu\}, \qquad \text{since } (\mathrm{A}_1)_\sigma \subset \mathscr{M}_\mu.$$

But by definition $\mu_1^*(B) = \mu(B)$, $B \in \mathscr{M}_\mu$. Hence this and (52) imply $\mu_1^* = \mu$, and the result follows.

(iii): Let $A \in \mathscr{M}_\mu$, and only suppose that $\mu(A) < \infty$ [instead of the stronger condition $\mu(\Omega) < \infty$]. Then for each $\epsilon > 0$, there exists a sequence $A_n \in \mathscr{A}$, such that $A \subset \bigcup_{n=1}^\infty A_n = B$, and

$$\infty > \mu(A) + \frac{\epsilon}{2} > \sum_{n=1}^{\infty} \alpha(A_n). \tag{53}$$

So there exists n_0 $[= n_0(\epsilon)]$ such that $\sum_{k=n}^\infty \alpha(A_k) < \epsilon/2$ if $n > n_0$. Let

$A^\epsilon = \bigcup_{k=1}^{n_0} A_k \in \mathscr{A}$. Then

$$\mu(A - A^\epsilon) \leqslant \mu(B - A^\epsilon) = \mu\Big(\bigcup_{k>n_0} A_k\Big)$$

$$\leqslant \sum_{k>n_0} \mu(A_k) = \sum_{k>n_0} \alpha(A_k), \qquad \text{since} \quad \alpha = \mu|\mathscr{A}$$

$$< \epsilon/2. \tag{54}$$

Also

$$\mu(A^\epsilon - A) \leqslant \mu(B - A)$$

$$= \mu(B) - \mu(A), \qquad \text{since} \quad \mu(A) < \infty \text{ and } A \in \mathscr{M}_\mu$$

$$= \mu\Big(\bigcup_{n=1}^{\infty} A_n\Big) - \mu(A)$$

$$\leqslant \sum_{n=1}^{\infty} \mu(A_n) - \mu(A) < \frac{\epsilon}{2}, \qquad \text{by (53)}. \tag{55}$$

Hence (54) and (55) imply

$$\mu(A \vartriangle A^\epsilon) = \mu(A - A^\epsilon) + \mu(A^\epsilon - A) < \epsilon, \tag{56}$$

proving the necessity.

Conversely, suppose (56) holds with $A^\epsilon \in \mathscr{A}$, and for this part we use the full hypothesis that $\mu(\Omega) < \infty$. Then by Theorem 9(v), A is measurable if we can show that (since μ is also Carathéodory regular)

$$\mu(\Omega) = \mu(A) + \mu(A^c). \tag{57}$$

However, it is clear that since $A \vartriangle B = (A - B) \cup (B - A)$,

$$A \cup (A \vartriangle B) = B \cup (A \vartriangle B) = A \cup B,$$

so that by subadditivity of μ, for any A, B,

$$\mu(A) \leqslant \mu(B) + \mu(A \vartriangle B), \qquad \mu(A) \leqslant \mu(B) + \mu(A \vartriangle B).$$

Hence,

$$|\mu(A) - \mu(B)| \leqslant \mu(A \vartriangle B). \tag{58}$$

Thus in our case taking $B = A^{\epsilon}$ and $A \subset \Omega$, (58) implies

$$|\mu(A) - \mu(A^{\epsilon})| \leqslant \mu(A \vartriangle A^{\epsilon}) < \epsilon$$

and replacing the sets by their complements and noting that $A \vartriangle A^{\epsilon} = A^c \vartriangle (A^{\epsilon})^c$, we get

$$\left|\mu(A^c) - \mu\left((A^{\epsilon})^c\right)\right| \leqslant \mu(A \vartriangle A^{\epsilon}) < \epsilon.$$

But $\mu(A^{\epsilon}) = \alpha(A^{\epsilon})$, $\mu((A^{\epsilon})^c) = \alpha((A^{\epsilon})^c)$. Hence

$$\mu(A) \leqslant \mu(A^{\epsilon}) + \epsilon = \alpha(A^{\epsilon}) + \epsilon$$

$$\mu(A^c) \leqslant \mu\left((A^{\epsilon})^c\right) + \epsilon = \alpha\left((A^{\epsilon})^c\right) + \epsilon.$$

Adding these two inequalities and using the subadditivity of μ, one has

$$\begin{aligned}\mu(\Omega) &\leqslant \mu(A) + \mu(A^c)\\ &\leqslant \alpha(A^{\epsilon}) + \alpha\left((A^{\epsilon})^c\right) + 2\epsilon = \alpha(\Omega) + 2\epsilon\\ &= \mu(\Omega) + 2\epsilon,\end{aligned}$$

α being additive and $\alpha = \mu|\mathscr{A}$. Since ϵ is arbitrary, this implies (57), and the proof is finished.

Even though the preceding treatment of measures is quite general, the original Lebesgue (outer) measure is always in the background, guiding our development. But in the Lebesgue case, the space $\mathbb{R}^n$ has a topology, and so improvements to the above work are possible. Being a generated measure, the Lebesgue outer measure μ^* is (Carathéodory) regular, according to Theorem 10(iii). A variant of this condition and another of the approximations given in the above proposition will be included in the next result.

Theorem 17. *Let μ^* be Lebesgue outer measure on $\mathbb{R}^n$, generated by the volume function $m(\cdot)$ and the ring of elementary sets $\mathscr{E}$. Then the following statements hold*:

(i) *If $\mathscr{G}$ is the class of all open sets of $\mathbb{R}^n$, then $\mathscr{G} \subset \mathscr{M}_{\mu^*}$ and*

$$\mu^*(A) = \inf\{\mu^*(G): G \supset A, G \in \mathscr{G}\}, \qquad A \subset \mathbb{R}^n, \tag{59}$$

so that μ^* is regular relative to the (smaller) class $\mathscr{G}$.

(ii) *If $\mathscr{C}$ is the class of all compact sets of $\mathbb{R}^n$, then $\mathscr{C} \subset \mathscr{M}_{\mu^*}$ and*

$$\mu^*(A) = \sup\{\mu^*(C): C \subset A, C \in \mathscr{C}\}, \qquad A \in \mathscr{M}_{\mu^*}. \tag{59'}$$

(iii) *Every countable set* $K \subset \mathbb{R}^n$ *(is an* F_σ*-set and) satisfies* $\mu^*(K) = 0$.

(iv) $A \in \mathscr{M}_{\mu^*}$ *iff for each* $\epsilon > 0$, *there exist an open set* G_ϵ *and a closed set* F_ϵ *such that* $F_\epsilon \subset A \subset G_\epsilon$ *and* $\mu^*(G_\epsilon - F_\epsilon) < \epsilon$. *Thus* $A \in \mathscr{M}_{\mu^*}$ *iff there is a* G_δ*-set* $\tilde{G} \supset A$ *such that* $\mu^*(\tilde{G} - A) = 0$, *or an* F_σ*-set* $\tilde{F} \subset A$ *such that* $\mu^*(A - \tilde{F}) = 0$.

(v) *If* $A \in \mathscr{M}_{\mu^*}$ *and* $x \in \mathbb{R}^n$, *then* $A + x = \{y + x \colon y \in A\} \in \mathscr{M}_{\mu^*}$ *and has the same measure as* A, *i.e.,* $\mu^*(A + x) = \mu^*(A)$ *and* μ^* *is translation invariant.*

Proof. (i): Since μ^* is a generated measure, Theorem 10(iii) already shows that it is Carathéodory regular. But we need to prove the stronger assertion here. By the same result, we deduce that $\mathscr{E} \subset \mathscr{M}_{\mu^*}$, and since $\mathscr{M}_{\mu^*}$ is a σ-algebra, it follows that the Borel algebra $\mathscr{B} = \sigma(\mathscr{E}) \subset \mathscr{M}_{\mu^*}$. It is a consequence of Proposition 1.5 that $\mathscr{G} \subset \mathscr{B}$ as well as $\mathscr{C} \subset \mathscr{B}$, so that both open and closed sets of $\mathbb{R}^n$ are μ^*-measurable.

To establish (59), since $\mathscr{E}_\sigma \subset \mathscr{B}$, Theorem 10(iii) yields

$$\begin{aligned} \mu^*(A) &= \inf\{\mu^*(B) \colon B \supset A,\ B \in \mathscr{E}_\sigma\} \\ &= \inf\{\mu^*(B) \colon B \supset A,\ B \in \mathscr{B}\} \\ &\leqslant \inf\{\mu^*(G) \colon G \supset A,\ G \in \mathscr{G}\}, \end{aligned} \tag{60}$$

since $\mathscr{G} \subsetneqq \mathscr{B}$ and the infimum is taken on a smaller collection. If $\mu^*(A) = \infty$, we are done. Suppose then $\mu^*(A) < \infty$. Hence for each $\epsilon > 0$, there exist $P_n \in \mathscr{S} \subset \mathscr{E}$ such that $A \subset \bigcup_{n=1}^{\infty} \mathring{P}_n = G \subset \bigcup_{n=1}^{\infty} P_n$ ($\mathring{P}_n$ is the interior of P_n) and

$$\begin{aligned} \mu^*(A) + \epsilon &> \sum_{n=1}^{\infty} \overline{m}(P_n), && \text{as in Theorem 1.2.8} \\ &\geqslant \overline{m}\left(\bigcup_{n=1}^{\infty} P_n\right), && \text{since } \overline{m} \text{ is } \sigma\text{-subadditive,} \\ &\geqslant \mu^*(G). \end{aligned} \tag{61}$$

But $G \in \mathscr{G}$, and $\epsilon > 0$ is arbitrary. So (60) and (61) together prove (59).

(ii): We first deduce (59′) from (i) and Theorem 9 (v), for $A \in \mathscr{M}_{\mu^*}$. Unlike (59), the relation (59′) is false if $A \subset \mathbb{R}^n$ but not μ^*-measurable. This will follow later when we establish the existence of such non-μ^*-measurable sets. [The right side of (59′) gives $\mu_*(A)$, the Lebesgue inner measure of $A \subset \mathbb{R}^n$, when we allow the class $\mathscr{C}$ to be the class of all closed sets.]

Observing that $\mathbb{R}^n = \bigcup_{k=1}^{\infty} C_k$, where $C_k \subset C_{k+1}$, $C_k \in \mathscr{C}$ (a result from advanced calculus), we note that $\mathscr{C} \subset \mathscr{B} \subset \mathscr{M}_{\mu^*}$, so that $A \cap C_i \in \mathscr{M}_{\mu^*}$ and increasing. Hence by the regularity of μ^*, we can invoke Theorem 9(v), so that

$$\mu^*(A) = \lim_{k \to \infty} \mu^*(A \cap C_k). \tag{62}$$

So given $\epsilon > 0$ pick a $t \in \mathbb{R}$ such that $t + \epsilon < \mu^*(A)$. Hence there is a k_0 such that $\mu^*(A \cap C_{k_0}) > t + \epsilon$. Also $C_{k_0} - A \in \mathscr{M}_{\mu^*}$, and since C_{k_0} is compact, (hence bounded and closed, so that $C_{k_0} - A \subset C_{k_0}$), it is contained in a bounded rectangle. Thus $\mu^*(C_{k_0} - A) < \infty$. By (i) there exists an open $G \supset C_{k_0} - A$ such that

$$\mu^*(C_{k_0} - A) + \epsilon > \mu^*(G). \tag{63}$$

If $C = C_{k_0} \cap G^c$, then C is compact and

$$C = C_{k_0} - G \subset C_{k_0} - (C_{k_0} - A) = A \cap C_{k_0} \subset A.$$

Consequently,

$$\begin{aligned} \mu^*(C) &= \mu^*(C_{k_0} - G) = \mu^*(C_{k_0}) - \mu^*(C_{k_0} \cap G) \\ &\geqslant \mu^*(C_{k_0}) - \mu^*(G) \geqslant \mu^*(C_{k_0}) - \mu^*(C_{k_0} - A) - \epsilon, \qquad \text{by (63)} \\ &= \mu^*(C_{k_0} \cap A) - \epsilon > t. \end{aligned} \tag{64}$$

It follows from (64) that

$$\sup\{\mu^*(C): C \subset A, C \in \mathscr{C}\} > t.$$

Letting $t \uparrow \mu^*(A)$, this yields (59′).

(iii): Since $\overline{m}(\{x\}) = 0$ for each $x \in \mathbb{R}^n$, we have $\overline{m}(K) = 0$ and $\mu^*(K) = \overline{m}(K) = 0$.

(iv): Since $\mathscr{G} \subset \mathscr{B} \subset \mathscr{M}_{\mu^*}$ and the class of closed sets $\mathscr{F} \subset \mathscr{B} \subset \mathscr{M}_{\mu^*}$, the condition $\mu^*(G_\epsilon - F_\epsilon) < \epsilon$ with $F_\epsilon \subset A \subset G_\epsilon$ implies $A \in \mathscr{M}_{\mu^*}$ by Proposition 16(i). For the converse, suppose $A \in \mathscr{M}_{\mu^*}$. Since $\mu^*(A) = \infty$ is possible, and since $\mathbb{R}^n = \bigcup_{k=1}^{\infty} I_k$ where $I_k \subset I_{k+1}$ and I_k is a bounded rectangle, let $A_k = A \cap I_k \in \mathscr{M}_{\mu^*}$ so that $\mu^*(A_k) < \infty$. Then by Theorem 9(v), or part (i) above, for each $\epsilon > 0$ there is an open set $G_n \supset A_n$ such that $\mu(G_n - A_n) < \epsilon/2^{n+1}$. Let $H_n = \bigcup_{k=1}^{n} G_k = \mathscr{G}$, so that $A_n \subset A_{n+1} \subset G_{n+1} \subset H_{n+1}$ and $A_k \subset G_k \cap G_n$. Also

$$\mu^*(A_k) \leqslant \mu^*(G_k \cap G_n) \leqslant \mu^*(G_k) < \mu^*(A_k) + \frac{\epsilon}{2^{k+1}}. \tag{65}$$

Hence

$$\mu^*(G_k - G_n) = \mu^*(G_k) - \mu^*(G_k \cap G_n) < \frac{\epsilon}{2^{k+1}}. \tag{66}$$

Similarly

$$\mu^*(A_n) \leqslant \mu^*(H_n) \leqslant \mu^*(G_n) + \sum_{k=1}^{n} \mu^*(G_k - G_n)$$

$$\leqslant \mu^*(A_n) + \sum_{k=1}^{n} \frac{\epsilon}{2^{k+1}}, \qquad \text{by (65) and (66).}$$

It follows from this computation that

$$\mu^*(H_n - A) \leqslant \mu^*(H_n - A_n)$$

$$\leqslant \mu^*(H_n) - \mu^*(A_n) < \epsilon/2, \qquad \text{since} \quad \mu^*(A_n) < \infty.$$

Letting $n \to \infty$, and noting that $H = \lim_n H_n = \bigcup_{n=1}^{\infty} H_n$ is open, we get

$$\mu^*(H - A) < \epsilon/2. \tag{67}$$

Applying this to $A^c \in \mathscr{M}_{\mu^*}$, there exists an open set $S \supset A^c$ such that $(A \supset S^c,\ S^c \in \mathscr{F})$

$$\mu^*(A - S^c) = \mu^*(S - A^c) < \epsilon/2. \tag{68}$$

Letting $G_\epsilon = H$ and $F_\epsilon = S^c$, so that $F_\epsilon \subset A \subset G_\epsilon$, one has

$$\mu^*(G_\epsilon - F_\epsilon) \leqslant \mu^*(G_\epsilon - A) + \mu^*(A - F_\epsilon) < \epsilon, \qquad \text{by (67) and (68).}$$

If we set $\epsilon_n = 1/n$, and $\tilde{G} = \bigcap_{n=1}^{\infty} G_{\epsilon_n}$, $\tilde{F} = \bigcup_{n=1}^{\infty} F_{\epsilon_n}$, then $\tilde{G}$ is aG_δ-set and $\tilde{F}$ is an F_σ-set, $\tilde{F} \subset A \subset \tilde{G}$, and $\mu^*(\tilde{G} - \tilde{F}) = 0$. This proves (iv).

(v): By (iv), $A \in \mathscr{M}_{\mu^*}$ implies $A = \tilde{F} \cup (A - \tilde{F})$ where $N = A - \tilde{F}$ is a set in $\mathscr{M}_{\mu^*}$ [since $\mu^*(N) = 0$] and $\tilde{F}$ is an F_σ, so that it is a Borel set. But we have already remarked that a translate of a Borel set is Borel (cf. Exercise 1.2), and, by Definition 2, that $\mu^*(B + x) = \mu^*(B)$ for each Borel set $B \in \mathscr{B}$. Also it is clear that [since for any mapping f, $f(A \cup B) = f(A) \cup f(B)$, and since $f(\cdot) = f_x(\cdot) = f(\cdot + x)$ is the translation mapping here]

$$A + x = (\tilde{F} + x) \cup (N + x).$$

Hence $A + x \in \mathscr{M}_{\mu^*}$. But by (59), since $\mu^*(G + x) = \mu^*(G)$, $G \in \mathscr{G}$, we deduce at once that $\mu^*(A + x) = \mu^*(A)$. Thus the proof the theorem is complete.

In the last two parts, we have also established the following fact, which will be stated for reference.

Corollary 18. *Let μ^* be the Lebesgue outer measure on $\mathbb{R}^n$ generated by the volume $m(\cdot)$ and the ring $\mathscr{E}$. Then a set A is in $\mathscr{M}_{\mu^*}$ iff it can be represented as $A = \tilde{F} \cup N_1 = \tilde{G} \cup N_2$ where $\tilde{F}$ is an F_σ-set, $\tilde{G}$ is a G_δ-set, and $\mu^*(N_i) = 0$, so that N_i is a μ^*-null set, $i = 1, 2$.*

[This and Theorem 17 are generalized in Chapter 9.]

Considering again the example preceding Lemma 15, it is seen that the μ^*-measurable sets are $\{\varnothing, \Omega\}$, while $\mathscr{P} = \{\varnothing, \{0\}, \{1\}, \Omega\}$. Thus the sets $\{0\}$ and $\{1\}$ are not μ^*-measurable, and the only μ^*-null set is the empty set. However, for a finer measure function such as the Lebesgue measure μ^* on $\mathbb{R}^n$, all Borel sets are μ^*-measurable. The above corollary shows, on the other hand, that every μ^*- (or Lebesgue-) measurable set is a union of a Borel set and a μ^*-null set. But Exercise 1.2.5 shows that the Cantor set on $\Omega = [0, 1]$ has length zero and hence μ^*-measure zero, but it is nonempty (it is even uncountable). Thus a null set depends on the measure function and need not be empty, although an empty set is null for every measure. Now a natural question is this. How large is the class $\mathscr{M}_{\mu^*}$ of Lebesgue-measurable sets in $\mathbb{R}^n$? This is not an idle or a simple matter. In fact, we show below the existence of a non-Lebesgue-measurable set on $\mathbb{R}$ whose construction depends explicitly on the axiom of choice, and goes back to G. Vitali in 1905. It is now known that, without use of some form of the axiom of choice, it is impossible to show the existence of such a "bad" set.† The latter result has been established by R. M. Solovay (1970). In the other direction, one can construct what are called "analytic sets" that are Lebesgue measurable but not Borel sets. We discuss such a construction in Chapter 7. Here only a nonconstructive argument, based on the sizes (i.e., cardinalities) of Borel and Lebesgue classes, is given to show this distinction.

The above discussion is made precise in

Theorem 19. *Let $\Omega = [0, 1]$, and μ^* be Lebesgue outer measure determined by the length function $m(\cdot)$ and the elementary sets $\mathscr{E}$ of Ω. If $\mathscr{B} = \sigma(\mathscr{E})$ is the Borel σ-algebra of Ω, $\mathscr{M}_{\mu^*}$ the class of Lebesgue (or μ^*-) measurable sets of Ω, and $\mathscr{P}$ is its power set, then*

$$\mathscr{B} \subsetneqq \mathscr{M}_{\mu^*} \subsetneqq \mathscr{P}. \tag{69}$$

† More precisely, this statement relates to the standard classical methods of mathematics. Actually a weaker form of the choice axiom, namely the existence of ultrafilters (see Section 7.2), suffices. But this technique uses nonstandard analysis. For such a construction, see the book by M. Davis (1977, pp. 72–74). It is different and interesting if one studies nonstandard methods also.

Proof. We consider the Cantor set $F \subset \Omega$, and show that even though $F \in \mathscr{B}$, $\mu^*(F) = 0$, there are subsets of F which are not Borel sets. Recall that $F = \bigcap_{n=1}^{\infty} F_n$, where F_n is the complement of the union of 2^{n-1} deleted open middle third sets of Ω. Let $M = \Omega - F$. Consequently M is open and

$$\overline{m}(M) = \frac{1}{3} + \frac{2}{3^2} + \frac{2^2}{3^3} + \cdots = 1.$$

Also, $\overline{m}(F) = 1 - \overline{m}(M) = 0$, since F (being closed) is in $\mathscr{B}$. So $F \in \mathscr{M}_{\mu^*}$ and since μ^* was shown to be a complete measure, every subset of F is a μ^*-null set, and hence its power set $2^F \subset \mathscr{M}_{\mu^*}$. Now using some simple ideas of the cardinal number calculus, we deduce that

$$\operatorname{card}(2^F) = 2^{\operatorname{card}(F)} = 2^c,$$

where c is the cardinality of the continuum (i.e., of Ω). Since $2^F \subset \mathscr{M}_{\mu^*} \subset 2^\Omega$, we have

$$\operatorname{card}(2^F) \leqslant \operatorname{card}(\mathscr{M}_{\mu^*}) \leqslant \operatorname{card}(2^\Omega) = 2^{\operatorname{card}(\Omega)}.$$

Hence $\operatorname{card}(\mathscr{M}_{\mu^*}) = 2^c$, and it is the same as that of all the subsets of the μ^*-null set F.

On the other hand $\mathscr{B} \subset \mathscr{M}_{\mu^*}$, and we now calculate the cardinality of $\mathscr{B}$. In fact, since each Borel set is obtained by the countable operations of unions, intersections, or complements from the class $\mathscr{D} = \{(a, b)\colon a, b \text{ rationals in } \Omega\}$, $\mathscr{B}$ can be represented as

$$\mathscr{B} = \Omega^{\mathscr{D}} = \{\text{all functions } f\colon f(\mathscr{D}) \subset \Omega\}.$$

Hence,

$$\operatorname{card}(\mathscr{B}) = \operatorname{card}(\Omega^{\mathscr{D}}) = \operatorname{card}(\Omega)^{\operatorname{card}(\mathscr{D})} = 2^{\aleph_0} = c,$$

where $\aleph_0$ is the cardinality of the natural numbers. Since $c < 2^c$, it follows that $\mathscr{B} \subset \mathscr{M}_{\mu^*}$ is a proper inclusion; so there are many (in fact 2^c-c of them) Lebesgue-measurable subsets of Ω which are not Borel sets. There are the "analytic" sets, included in this class, and we shall discuss this concept in Chapter 7.

The cardinality argument is too crude to show that $\mathscr{M}_{\mu^*} \subset \mathscr{P}$ is also proper, since they both have the same number of sets, i.e., 2^c. We therefore follow Vitali's construction and exhibit a set in $\mathscr{P} - \mathscr{M}_{\mu^*}$.

Let $\mathbb{Q}$ be the set of all rationals in $\mathbb{R}$, $T_t\colon \mathbb{R} \to \mathbb{R}$ be the translation $T_t(x) = x + t$, $x \in \mathbb{R}$. If $\Gamma_t = T_t\mathbb{Q}$, then Γ_t is countable and $\Gamma_t \cap I \neq \varnothing$, where $I = (0, 1) \subset \Omega$. Note that for $t \neq t'$, either $\Gamma_t \cap \Gamma_{t'}$ is empty or $\Gamma_t = \Gamma_{t'}$. In fact, if $t - t' \in \mathbb{Q}$, then $\Gamma_t = \Gamma_{t'}$, since $\Gamma_t = t + \mathbb{Q} = t' + \mathbb{Q} = \Gamma_{t'}\mathbb{Q}$, and if

$t - t' \notin \mathbb{Q}$, then $\Gamma_t \cap \Gamma_{t'} = \varnothing$. Let $\mathscr{K} = \{\Gamma_t: t \in \mathbb{R}$, with only distinct Γ_t counted$\}$. Define a set A_0, using the axiom of choice (recalled in the Appendix) as follows:

$$A_0 = \{x \in I \cap \Gamma_t: \Gamma_t \in \mathscr{K}\},$$

i.e., we pick an element from each $I \cap \Gamma_t$ as the Γ_t vary over $\mathscr{K}$ using a choice function. (The method stops here if this axiom cannot be used.) We *claim* that $A_0 \in \mathscr{P} - \mathscr{M}_{\mu^*}$, i.e., is non-$\mu^*$-measurable.

1. If $\{r_i, i \geqslant 1\}$ is an enumeration of all (distinct) rationals in $(-1,1)$, then we assert that

$$I \subset \bigcup_{i=1}^{\infty} T_{r_i} A_0 \subset (-1,2). \tag{70}$$

For, since $T_{r_i} A_0 \subset (-1,2)$ for each r_i, the right side inclusion is clear. If $x \in I$, so that $x \in \Gamma_t$ for some t so that $x - t \in \mathbb{Q}$, and hence $t \in \Gamma_x$. Thus $\Gamma_x = \Gamma_t$, i.e., $x \in \Gamma_x$. Since $\Gamma_x \cap A_0 \neq \varnothing$, by definition, there is some $y \in \Gamma_x \cap A_0$ such that $x - y \in \mathbb{Q}$. Since $0 < x < 1$ and $0 < y < 1$, this rational must be one of the r_i in $(-1,1)$. So $x = y - r_i = T_{-r_i} y$ and $x \in T_{-r_i} A_0$ and thus the first inclusion of (70) also holds.

2. $(T_{r_i} A_0) \cap (T_{r_j} A_0) = \varnothing$ if $i \neq j$. If not, let y be in this intersection. Then $y = x + r_i = x' + r_j$ for some x, x' in A_0, so that $x - x' \in \mathbb{Q}$ and $x \in \Gamma_x = \Gamma_{x'}$. But by the definition of $\mathscr{K}$, only one point is selected from each Γ_x, so that $x = x'$ and hence $r_i = r_j$, a contradiction. Thus the translates are disjoint.

3. Each $T_{r_i} A_0$ and hence A_0 itself is nonmeasurable. For, in the opposite case, each of these will be measurable, having the same measure by Theorem 17(v). If $\mu^*(A_0) = \alpha$, then by (70)

$$\mu^*\left(\bigcup_{j=1}^{\infty} T_{r_j} A_0\right) \leqslant \mu^*((-1,2)) = 3,$$

and using the σ-additivity of μ^* on $\mathscr{M}_{\mu^*}$,

$$1 = \mu^*(I) \leqslant \mu^*\left(\bigcup_{j=1}^{\infty} T_{r_j} A_0\right) = \sum_{j=1}^{\infty} \mu^*\left(T_{r_j} A_0\right) = \sum_{j=1}^{\infty} \alpha \leqslant 3.$$

This is impossible, and hence $A_0 \notin \mathscr{M}_{\mu^*}$. Since $T_{-r_i}(T_{r_i} A_0) = A_0$, it follows also that none of the $T_{r_j} A_0$ can be in $\mathscr{M}_{\mu^*}$. This completes the proof of the theorem.

Remark. The above result allows us to draw several conclusions about Lebesgue measure, which we denote simply by μ $(= \mu^*|\mathscr{M}_{\mu^*})$. The following consequences of Theorems 17 and 19 are useful:

1. In Theorem 9(v), it was shown that for a generated (or Carathéodory regular) outer measure μ^*, if $A_n \uparrow A$ is any sequence, then one has $\mu^*(A_n) \uparrow \mu^*(A)$. This need not be true for a decreasing sequence even if $\mu^*(\Omega) < \infty$. [Compare it with Proposition 1.1(ii).] For, let $\Omega = [0,1]$, $\mu^* =$ the Lebesgue outer measure, $B_n = \bigcup_{i=n}^{\infty}(T_{r_i}A_0)$ where A_0 is the nonmeasurable set constructed above. Then $B_n \supset B_{n+1} \downarrow \varnothing$. Since $T_{r_k}(A_0) \subset B_k$, if $\alpha = \mu^*(T_{r_k}(A_0))$ ($\alpha = 0$ is not possible, since then $T_{r_k}A_0 \in \mathscr{M}_{\mu^*}$, contrary to the last part of the Theorem 19), then one has

$$0 = \mu^*(\varnothing) = \mu^*\Big(\lim_n B_n\Big) < \alpha \leqslant \lim_n \mu^*(B_n).$$

2. The lower approximation in Theorem 17(ii) need not be true for non-Lebesgue-measurable sets. Indeed, let $\underline{\mu}$ be the right side quantity of the expression (59′) for any $A \subset \Omega$. Thus in the present notation

$$\underline{\mu}(A) = \sup\{\mu(B) \colon B \subset A,\ B \in \mathscr{M}_{\mu^*}\}, \qquad A \subset \mathbb{R}^n. \tag{71}$$

Note that this is the same as the inner measure of Definition 4 with μ for $\overline{m}$ there. In view of Theorem 17(i), this is legitimate and we get the same expression (cf. Exercise 9). Thus

$$\mu_*(A) = \sup\{\mu(B) - \mu^*(B \cap A^c) \colon B \in \mathscr{M}_{\mu^*},\ \mu^*(B \cap A^c) < \infty\}. \tag{72}$$

For, by subadditivity of μ^*, we have

$$\mu(B) \leqslant \mu^*(B \cap A) + \mu^*(B \cap A^c),$$

and since the last term is finite,

$$\begin{aligned}\mu_*(A) &\leqslant \sup\{\mu^*(B \cap A) \colon \mu^*(B \cap A^c) < \infty,\ B \in \mathscr{M}_{\mu^*}\}\\ &\leqslant \sup\{\mu^*(B) \colon B \subset A,\ B \in \mathscr{M}_{\mu^*}\} = \underline{\mu}(A).\end{aligned}$$

Also, $\mu_*(\cdot)$ is monotone and $\mu_*(B) = \mu(B)$ for $B \in \mathscr{M}_{\mu^*}$ by Proposition 5 [with $\overline{m}(\cdot)$ replaced by μ now]. We thus get $\mu_*(A) \geqslant \mu(B)$, $B \subset A$, $B \in \mathscr{M}_{\mu^*}$, and, taking the supremum on all such B, it follows that $\mu_*(A) \geqslant \underline{\mu}(A)$. These two inequalities imply $\mu_* = \underline{\mu}$.

Now let $B \subset A$, $B \in \mathscr{M}_{\mu^*}$, and consider $T_{r_i}B$ which is in $\mathscr{M}_{\mu^*}$, and disjoint for different i. Taking $A = A_0$, the nonmeasurable set of Theorem 19, then

$\mu(T_{r_i}B) = \mu(B)$, one has

$$\sum_{i=1}^{\infty} \mu(T_{r_i}B) = \mu\left(\bigcup_{i=1}^{\infty} T_{r_i}B\right) \leqslant \mu(-1,2) = 3.$$

Hence $\mu(B) = 0$ for each $B \in \mathscr{M}_{\mu^*}$, $B \subset A_0$. Consequently,

$$\underline{\mu}(A_0) = \sup\{\mu(B): B \subset A,\ B \in \mathscr{M}_{\mu^*}\} = 0,$$

whereas $\mu^*(A_0) > 0$. Since A_0^c is also nonmeasurable, the same argument shows that $\underline{\mu}(A_0^c) = 0$. Now using the fact that $\mu_* = \underline{\mu}$, and $\underline{\mu}(A_0^c) = 0$, it is easily seen from (72) that $\mu^*(A_0) = 1$ on taking $B = \Omega$ there. But $\mu_*(A_0) + \mu^*(A^c) = \mu(\Omega)$ so that $\mu^*(A^c) = 1$ also. Thus (59′) need not be true for nonmeasurable sets.

We state the above remarks, for reference, in the following

Proposition 20. *Let $\mathscr{A}$ be an algebra from 2^Ω, and $\mu: \mathscr{A} \to \overline{\mathbb{R}}^+$ be a measure. If μ^* is the outer measure and μ_* the inner measure generated by $(\mu, \mathscr{A})$, then for any $A \subset \Omega$ and any $A_n \subset A_{n+1} \subset \Omega$*

$$\mu^*(A) = \inf\{\mu(B): B \supset A,\ B \in \mathscr{M}_{\mu^*}\},$$

$$\lim_n \mu^*(A_n) = \mu^*\left(\lim_n A_n\right).$$

But if $\Omega \supset \tilde{A}_n \supset \tilde{A}_{n+1}$, and $A \subset \Omega$ are any sets, then the following need **not** *hold*:

$$\lim_n \mu^*(\tilde{A}_n) = \mu^*\left(\lim_n \tilde{A}_n\right) \qquad (> \textit{may hold})$$

$$\mu^*(A) = \sup\{\mu(B): B \subset A,\ B \in \mathscr{M}_{\mu^*}\}. \quad (> \textit{may hold})$$

If $\Omega = [0,1]$, and μ^*, μ_* are the Lebesgue outer and inner measures, then there exists a set $A_0 \in 2^\Omega - \mathscr{M}_{\mu^*}$ such that $\mu^*(A_0) = \mu^*(A_0^c) = \mu(\Omega)$ $(= 1)$ and $\mu_*(A) = 0 = \mu_*(A^c)$.

For nongenerated measures (i.e., an outer measure which is not Carathéodory regular), there is no natural definition of inner measure to extend Definition 4. In such cases expression (71) is taken as a definition, and one develops the theory from there, since for the generated measures notions of the two inner measures coincide. We present some of these facts and complements as part of the exercises below. They also show why the Carathéodory process based only on outer measures is most convenient and satisfactory in many cases.

EXERCISES

0. Let $A \subset \mathbb{R}$ be a set. If $m(\cdot)$ is the Lebesgue length function on $\mathbb{R}$, denote by $m^*(\cdot)$, the outer measure generated by m. Show that
(a) $m^*(A) = \inf\{m^*(\mathscr{O}) \colon \mathscr{O} \supset A,\ \mathscr{O} \text{ open}\}$,
(b) $m^*(A) = \inf\{m^*(E) \colon E \supset A,\ E \text{ elementary}\}$,
(c) $m^*(A) = \inf\{\sum_{n=1}^{\infty} m(I_n) \colon A \subset \bigcup_{n=1}^{\infty} I_n,\ I_n$ are intervals of any of the four kinds, and need not be disjoint$\}$.

1. Let Ω be a nonempty set and $\mathscr{A}$ be the collection of $\varnothing$ and all one point sets of Ω. If $\alpha \colon \mathscr{A} \to \overline{\mathbb{R}}^+$ is defined as $\alpha(\{\omega\}) = 1$, $\omega \in \Omega$, and μ^* is generated by $(\alpha, \mathscr{A})$, determine $\mathscr{M}_{\mu^*}$, the class of μ^*-measurable sets. Show that $\mathscr{A} \subset \mathscr{M}_{\mu^*}$ and $\mu^* | \mathscr{A} = \alpha$. Check the hypothesis of Theorem 10(iii) for this example. If $\{A_n \colon n \geqslant 1\}$ is a monotone sequence of sets in Ω with A as its limit, and if A_{n_0} has at most 10^9 points for some n_0, show that $\mu^*(A_n) \to \mu^*(A)$ as $n \to \infty$.

2. Suppose that μ is an outer measure on Ω, $\mu(\Omega) < \infty$, and $\mathscr{M}_\mu$ is the corresponding class. If $\{A_n \colon n \geqslant 1\} \subset \mathscr{M}_\mu$ and $A_n \to A$, then verify that $\mu(A_n) \to \mu(A)$ as $n \to \infty$.

3. Let $\mathscr{A}$ be a ring of subsets of a point set Ω, and $\mu \colon \mathscr{A} \to \mathbb{R}^+$ be additive. Show that for A, B in $\mathscr{A}$, if $d(A, B) = \mu(A \vartriangle B)$, then $d \colon \mathscr{A} \times \mathscr{A} \to \mathbb{R}^+$ is a semimetric, so that $(\mathscr{A}, d)$ becomes a semimetric space. Here $d(A, B) = 0$ if $A = B$, but it can be zero even if $A \neq B$. Give an example of the latter possibility.

4. Let $\mathscr{A}$ be a ring and $\mu \colon \mathscr{A} \to \mathbb{R}^+$ be additive. If $A_1, \dots, A_n$ are sets from $\mathscr{A}$, establish the *Poincaré formula*:

$$\mu\left(\bigcup_{i=1}^{n} A_i\right) = \sum_{i=1}^{n} \mu(A_i) - \sum_{1 \leqslant i < j \leqslant n} \mu(A_i \cap A_j) + \sum_{1 \leqslant i < j < k \leqslant n} \mu(A_i \cap A_j \cap A_k) - \cdots + (-1)^{n-1} \mu\left(\bigcap_{i=1}^{n} A_i\right).$$

Transporting all the negative terms to the left, show that the resulting identity holds even if $\mu \colon \mathscr{A} \to \overline{\mathbb{R}}^+$. [Use induction.]

5. Complete the details of proof of Proposition 12.

6. This problem illustrates the fact that, in Theorem 10(iii), if $\mathscr{A}$ is not a semiring, then $\mathscr{A}$ need not be a subset of $\mathscr{M}_\mu$. Thus, let $\Omega = \{\alpha_1, \alpha_2, \dots\}$ and $\mathscr{A}$ be the collection of all the k-element subsets of Ω $(\alpha_i \neq \alpha_j,\ i \neq j)$ and $\varnothing \in \mathscr{A}$. Let $\mu \colon \mathscr{A} \to \mathbb{R}^+$ be defined as $\mu(A_k) =$ number of points in A_k. Let μ^* be the outer measure generated by $(\mu, \mathscr{A})$. Show that $\mathscr{A} \not\subset \mathscr{M}_{\mu^*}$ if $k \geqslant 2$. Describe $\mathscr{M}_{\mu^*}$ if $k = 2$ and 3. (If $k = 1$ this cannot happen. Why?)

7. Let μ be any outer measure on $\mathscr{P}$ $(= 2^{\Omega})$, and let $f: \Omega \to \tilde{\Omega}$ be a function where $\tilde{\Omega}$ is any other point set. Noting that the inverse mapping $f^{-1}: 2^{\tilde{\Omega}} \to 2^{\Omega}$ is well defined [$f^{-1}(\tilde{A})$ is taken as empty for $\tilde{A} \subset \tilde{\Omega} - f(\Omega)$], and preserves *all* set operations (i.e., arbitrary unions, complements, intersections—verify this), let $\tilde{\mu} = \mu \circ f^{-1}: 2^{\tilde{\Omega}} \to \mathbb{R}^+$ be the image mapping. Show that $\tilde{\mu}$ is an outer measure, and if $\mathscr{A} = \{\tilde{A} \subset \tilde{\Omega}: f^{-1}(\tilde{A}) \in \mathscr{M}_\mu\}$ then $\mathscr{A}$ is a σ-subalgebra of $\mathscr{M}_{\tilde{\mu}}$. Give examples to show where there is equality and proper inclusion between $\mathscr{A}$ and $\mathscr{M}_{\tilde{\mu}}$, even when μ is a finite measure (e.g., try $f = \chi_A$, $A \notin \mathscr{M}_\mu$).
8. Complete the details of the proof of Lemma 13.
9. Let μ be an outer measure on Ω, and μ^* be the generated measure by the pair $(\mu, \mathscr{M}_\mu)$ where $\mathscr{M}_\mu$ is, as usual, the class of μ-measurable sets. Define the inner measure μ_* and a function $\underline{\mu}$ as

$$\mu_*(A) = \sup\{\mu(B) - \mu^*(B - A)\}: B \in \mathscr{M}_\mu, \mu^*(B - A) < \infty\},$$

$$\underline{\mu}(A) = \sup\{\mu(B): B \supset A, B \in \mathscr{M}_\mu\}.$$

Show that $\mu_* = \underline{\mu}$, and that $\underline{\mu}$ (or μ_*) is σ-superadditive, but not necessarily σ-subadditive. However, it is σ-additive on the trace $\mathscr{M}_\mu(A)$ for any $A \subset \Omega$. This need not hold on $\mathscr{M}_{\mu^*}(A)$.
10. Let $\underline{\mu}$ and μ^* be the inner and outer measures generated by $(\mu, \mathscr{M}_\mu)$ as in the above exercise. Establish Proposition 5 for this pair $(\underline{\mu}, \mu^*)$, and as a consequence, deduce that

$$\mu(A_1 \cup A_2) = \underline{\mu}(A_1) + \mu^*(A_2), \qquad A_i \in \mathscr{M}_\mu, \quad i = 1, 2, \quad \text{disjoint.}$$

11. If $\mu(\cdot)$ is the Lebesgue measure on $\mathbb{R}$, and μ_*, μ^* are the inner and outer measures generated by $(\mu, \mathscr{E})$ as in Definitions 2 and 4, show that there exists a set $A \subset \mathbb{R}$ such that $\mu^*(A) = \infty$, $\mu_*(A) = \mu^*(A)$, but $A \notin \mathscr{M}_{\mu^*}$. (By Proposition 5, this cannot happen if A is also bounded.) Show however that, for the general (μ_*, μ^*) defined in Exercise 9 above, we must have

$$\{A \subset \Omega: \mu_*(A) = \mu^*(A), \mu(A) < \infty\} \subset \mathscr{M}_\mu \subset \mathscr{M}_{\mu^*}.$$

12. Let $\{\mu_i: i \in I\}$ be a collection of outer measures on Ω, and $\mathscr{F}$ be the class of all finite subsets of I, directed by inclusion. Thus for α, β in $\mathscr{F}$, $\alpha < \beta$ if $\alpha \subset \beta$. For each $\alpha \in \mathscr{F}$, define

$$\mu_\alpha(A) = \sum_{i \in \alpha} \mu_i(A), \qquad A \subset \Omega.$$

Show that μ_α is an outer measure and if $\alpha < \beta$ then $\mu_\alpha(A) \leqslant \mu_\beta(A)$,

$A \subset \Omega$. If $\mu(A) = \sup\{\mu_\alpha(A): \alpha \in \mathscr{F}\}$, verify that μ is an outer measure on Ω and that $\bigcap_{\alpha \in \mathscr{F}} \mathscr{M}_{\mu_\alpha} \subset \bigcap_{i \in I} \mathscr{M}_{\mu_i} \subset \mathscr{M}_\mu$. [Note that for any directed set of numbers $\beta_i \geqslant 0$, $\bar{\beta}_i \geqslant 0$, satisfying $i \leqslant i' \Rightarrow \beta_i \leqslant \beta_{i'}, \bar{\beta}_i \leqslant \bar{\beta}_{i'}$, we have $\sup_i(\beta_i + \bar{\beta}_i) = \sup_i \beta_i + \sup_i \bar{\beta}_i$.]

13. In producing new outer measures from a given set as in the above problem, some properties may be lost. Thus the sum of a pair of generated measures need not be a generated measure. Show this by taking $\Omega = \mathbb{R}^+$, μ_1^* as the Lebesgue outer measure, and μ_2^* as the measure generated by $(m(\cdot), \mathscr{A})$, where $\mathscr{A} = \{\varnothing, \Omega\}$ and $m(\cdot)$ is the length function. Does the situation change if $\Omega = [0, 1]$ here?

14. Let $\alpha: 2^\Omega \to \bar{\mathbb{R}}$ be a mapping such that $\alpha(\varnothing) = 0$. Define

$$\mathscr{M}_\alpha = \{B: \alpha(A) = \alpha(A \cap B) + \alpha(A - B), \quad \text{all} \quad A \subset \Omega\}.$$

Show that $\mathscr{M}_\alpha$ is an algebra and $\alpha|\mathscr{M}_\alpha(T)$ is an additive set function, where $\mathscr{M}_\alpha(T)$ is the trace of $\mathscr{M}_\alpha$ on $T \subset \Omega$. Give an example to show that $\alpha(A) = 0$ does not imply that $A \in \mathscr{M}_\alpha$.

15. (H. Steinhaus) Let $A \subset \mathbb{R}$ be a Lebesgue-measurable set and μ be the Lebesgue measure. Suppose that $0 < \mu(A) < \infty$, and $B = A - A = \{x - y: x, y \in A\}$. Using the following hints, establish that $B \supset I$ where I is an open interval symmetric about the origin. [By Theorem 17(i), for any $0 < \epsilon < 1$ there is an open set $G \supset A$ such that $\mu(A) > \epsilon\mu(G)$, and since $G \subset \mathbb{R}$ is a countable disjoint union of intervals I_n, we have $A = \bigcup_{n=0}^\infty (A \cap I_n)$, and so for some n_0, $\mu(A \cap I_{n_0}) \geqslant \epsilon\mu(I_{n_0})$. Letting $I = I_{n_0}$ and $\epsilon = \frac{3}{4}$, we get $\mu(A \cap I) \geqslant \frac{3}{4}\mu(I)$. Note that the sets $A \cap I$ and $(A \cap I) + x$, for each $|x| < \frac{1}{2}\mu(I)$, contain at least one common point, for otherwise, since $(A \cap I) \cup ((A \cap I) + x) \subset I \cup (I + x)$, which is an interval with length $< \frac{3}{2}\mu(I)$, we get the contradiction:

$$\tfrac{3}{2}\mu(I) > \mu((A \cap I) \cup (A \cap I) + x) = 2\mu(A \cap I) \geqslant \tfrac{3}{2}\mu(I).$$

Hence $|x| < \frac{1}{2}\mu(I)$, implies $y \in ((A \cap I) + x)$ for some $y \in A \cap I$, or $x - y \in A \cap I$, or $x \in B$.]

16. Let Ω be a nonempty set and μ be an outer measure on it such that $\mu(\Omega) < \infty$. If $A \subset \Omega$, $\mu(A) > 0$, define a new set function μ_A by the equation $\mu_A(B) = \mu(A \cap B)/\mu(A)$ for all $B \subset \Omega$. Verify that μ_A is an outer measure for each such A, (μ_A may be termed a "conditional outer measure", given A). Show that the μ_A-measurable class $\mathscr{M}_{\mu_A}$ contains $\mathscr{M}_\mu$ as well as $\mathscr{M}_\mu(A)$, the trace of $\mathscr{M}_\mu$ on A. Further establish $\mathscr{M}_\mu = \bigcap\{\mathscr{M}_{\mu_A}: \mu(A) > 0\}$, and characterize the class of sets $A \subset \Omega$ for which $\mathscr{M}_\mu(A) \subset \mathscr{M}_\mu$.

17. Recall that a subset S of a topological space Ω is *perfect* if it is closed and each of its points is a limit point of the set. Also, if Ω is a separable metric space, and $S \subset \Omega$ is closed, then there is a perfect set $P \subset S$ such that

$S - P$ is at most countable. (This is the Cantor-Bendixon lemma.) If $\Omega = [0,1]$, $A \subset \Omega$, and μ^* is the Lebesgue outer measure on Ω, show, with the above facts and Theorem 17, that A is μ^*-measurable iff it can be approximated by perfect sets in that for each $\epsilon > 0$, there is a perfect set $P_\epsilon \subset A$ such that $\mu^*(A - P_\epsilon) < \epsilon$. Deduce that a set $A \in \mathscr{M}_{\mu^*}$ iff

$$\mu^*(A) = \sup\{\mu^*(P) \colon P \subset A,\ P \text{ perfect in } \Omega\}.$$

18. Let $\Omega = [0,1]$ and $r \geqslant 3$ be a number. Remove an open interval of length $1/r$ symmetrically from the middle of Ω, and repeat the process with the remaining parts, just as in constructing the Cantor set, which corresponds to $r = 3$. Let the resulting "generalized Cantor set" be called K_r. Show that $\mu(K_r) = (r-3)/(r-2)$, where μ is the Lebesgue measure on Ω, and that K_r is a perfect set for every $r \geqslant 3$, K_r is *nowhere dense* in Ω, i.e., K_r^c is dense in Ω. Deduce that for each $\epsilon > 0$, there is a nowhere dense set $A \subset \Omega$ such that $\mu(A) > 1 - \epsilon$. A countable union of nowhere dense sets is a set of the *first category*. Show that there is an F_σ set $A \subset \Omega$ of the first category such that $\mu(A) = 1$.

19. The definition of inner measure could not mimic that of the outer measure, as shown by this exercise. Suppose we define

$$\tilde{\mu}_*(A) = \sup\left\{\sum_{i=1}^{\infty} \alpha(B_i) \colon B_i \in \mathscr{A},\ \bigcup_{i=1}^{\infty} B_i \subset A\right\}, \qquad A \subset \Omega,$$

where $(\alpha, \mathscr{A})$ is a pair as in Theorem 9. Verify that $\tilde{\mu}_*$ is monotone and σ-superadditive, $\tilde{\mu}_*(A) \leqslant \mu_*(A) = \sup\{\mu^*(C) \colon C \subset A,\ C \in \mathscr{M}_{\mu^*}\}$, where μ^* is the generated [by $(\alpha, \mathscr{A})$] outer measure. However, let $\Omega = [0,1]$, $\mathscr{A} =$ ring of elementary sets, and $\alpha(\cdot) = m(\cdot)$, the Lebesgue measure. If A is the set of all rationals in Ω, then show that we have $\tilde{\mu}_*(A^c) = 0 < \mu_*(A^c) = 1 - \mu^*(A) = 1$. Thus the previously known set A^c, which is Lebesgue measurable, will not be so in this new definition. (Compare this with Example 1 at the beginning of this section, and the resulting class will be much less satisfactory.)

20. In extending the Lebesgue definition of inner measure we have used a generated outer measure by $(\alpha, \mathscr{A})$, where $\mathscr{A}$ is an algebra and $\alpha \colon \mathscr{A} \to \mathbb{R}^+$ is a measure, and showed that a set A is μ^*-measurable iff $\mu_*(A) = \mu^*(A)$, where $\mu_*(A) = \alpha(E) - \mu^*(E \cap A^c)$ with $\mu^*(\Omega) < \infty$, and $E \in \mathscr{A}$, so that $\mu_*(A) = \mu^*(\Omega) - \mu^*(A^c)$. The latter definition will not produce a satisfactory theory if μ^* is not a generated outer measure, as seen here: Let $\Omega = \{1,2,3,4\}$; $\mu(\varnothing) = 0$; $\mu(\{i\}) = \mu(\{i,j\}) = 1$, $i = 1,\ldots,4$, $j = 1,\ldots,4$ $(i \neq j)$; $\mu(\{i,j,k\}) = 2$, $i \neq j \neq k$, $1 \leqslant i,j,k \leqslant 4$; and $\mu(\Omega) = 3$. Then μ is an outer measure, and $\mu_*(A) = \mu(\Omega) - \mu(A^c) = 3 - \mu(A^c)$ defines μ_* as a monotone σ-subadditive set function. Show that $\mathscr{M}_\mu$ does not have any three point sets but $\mu_*(A) = \mu(A)$ for all such sets. In other words, this equality does not signify the μ-measurability of A, in contrast to Proposition 5.

2.3. EXTENSIONS OF MEASURES TO LARGER CLASSES

It was noted in Section 1, as well as in the previous chapter, that several properties of measures can be obtained if their domains are σ-algebras (or σ-rings) and that the primary classes of sets are (semi-)rings or algebras on which the measure functions are defined. This is reasonable in view of our basic model of Section 1.2, especially Proposition 1.2.6. Here we shall establish that every such function can be extended to the σ-algebra generated by these basic (semi-)algebras. Indeed this can be done uniquely whenever the measures are somewhat restricted (e.g., σ-finite). Using the fundamental work of the preceding section, we present the extension problem for measures under different conditions.

Recall that a measure $\alpha(\cdot)$ on a ring $\mathscr{A}$ is *σ-finite* if for each $A \in \mathscr{A}$ there exist $A_n \in \mathscr{A}$, $\alpha(A_n) < \infty$, such that $A \subset \bigcup_{n=1}^{\infty} A_n$. We first establish the extension theorems for σ-finite measures. These are the most common ones in applications. However, there are important problems for which one needs to go beyond σ-finiteness. So we discuss the rationale and then consider the general case.

The classical formulation of a result which is the most often used in applications, is given in the following:

Theorem 1 (Hahn extension). *If $\mathscr{A}$ is a ring on a set Ω and $\alpha: \mathscr{A} \to \overline{\mathbb{R}}^+$ is a measure, then it has a σ-additive extension $\hat{\alpha}$ onto the σ-ring $\sigma(\mathscr{A})$. Moreover, if α is σ-finite on $\mathscr{A}$, then $\hat{\alpha}$ is also σ-finite on $\sigma(\mathscr{A})$, the extension is unique, and the completion $\hat{\sigma}(\mathscr{A})$ of $\sigma(\mathscr{A})$ under $\hat{\alpha}$ is identical with $\mathscr{M}_{\mu^*}$, where μ^* is the generated outer measure by the pair $(\alpha, \mathscr{A})$ and $\mathscr{A}$ is now also an algebra.*

Proof. The existence of an extension of α from $\mathscr{A}$ to $\sigma(\mathscr{A})$ is an immediate consequence of Theorem 2.10. In fact, let μ^* be the generated outer measure of $(\alpha, \mathscr{A})$, and $\mathscr{M}_{\mu^*}$ be the class of μ^*-sets. Then $\mathscr{M}_{\mu^*}$ is a σ-algebra containing $\mathscr{A}$, and μ^* is a measure on $\mathscr{M}_{\mu^*}$ such that $\mu^*|\mathscr{A} = \alpha$. So $\sigma(\mathscr{A}) \subset \mathscr{M}_{\mu^*}$ and let $\hat{\alpha} = \mu^*|\sigma(\mathscr{A})$. Then $\hat{\alpha}$ is an extension of α to the generated σ-algebra $\sigma(\mathscr{A})$.

We now show that, under the σ-finiteness hypothesis, this is the only extension. Such a result was already indicated in Exercise 1.8, but we add the detail here, to use it for a further generalization.

Let μ_1 be another extension of α, so that $\mu_1|\mathscr{A} = \hat{\alpha}|\mathscr{A} = \alpha$. Define the class

$$\mathscr{A}_0 = \{A \in \mathscr{A}: \hat{\alpha}(A) < \infty, \mu_1(A) < \infty\} = \{A \in \mathscr{A}: \alpha(A) < \infty\}. \quad (1)$$

Then $\mathscr{A}_0$ is a subring of $\mathscr{A}$, and both generate the same σ-ring. For clearly

$\sigma(\mathscr{A}_0) \subset \sigma(\mathscr{A})$, and if $A \in \mathscr{A}$, then by the σ-finiteness of α on $\mathscr{A}$, we have $A \subset \bigcup_{n=1}^{\infty} A_n$, $A_n \in \mathscr{A}_0 \subset \sigma(\mathscr{A}_0)$. Since $A = \bigcup_{n=1}^{\infty}(A \cap A_n)$ and $A \cap A_n \in \mathscr{A}_0$, it follows that $A \in \sigma(\mathscr{A}_0)$ so that $\mathscr{A} \subset \sigma(\mathscr{A}_0)$. Hence $\sigma(\mathscr{A}) \subset \sigma(\mathscr{A}_0)$. Consequently $\sigma(\mathscr{A}_0) = \sigma(\mathscr{A})$. This fact is used to deduce the σ-finiteness of $\hat{\alpha}$ (and μ_1) on $\sigma(\mathscr{A})$.

Consider the class

$$\mathscr{A}_1 = \left\{ A \subset \Omega \colon A \subset \bigcup_{n=1}^{\infty} A_n,\ A_n \in \mathscr{A}_0 \right\}. \tag{2}$$

Then $\mathscr{A}_1$ is a σ-ring. Indeed, if $B_n \in \mathscr{A}_1$ so that $B_n \subset \bigcup_{m=1}^{\infty} A_{nm}$, $A_{nm} \in \mathscr{A}_0$, and $\bigcup_{n=1}^{\infty} B_n \subset \bigcup_{n=1}^{\infty}\bigcup_{m=1}^{\infty} A_{nm}$, then $\bigcup_{n=1}^{\infty} B_n \in \mathscr{A}_1$. Also, $B_1 - B_2 = B_1 - (B_2 \cap B_1) \subset B_1 \subset \bigcup_{n=1}^{\infty} A_{1n}$, so that $B_1 - B_2 \in \mathscr{A}_1$. It is clear that $\mathscr{A}_0 \subset \mathscr{A}_1$, whence $\sigma(\mathscr{A}_0) \subset \mathscr{A}_1$. Thus every element of $\sigma(\mathscr{A}_0) = \sigma(\mathscr{A})$ is covered by a countable sequence of elements of $\mathscr{A}_0$, so that $\hat{\alpha}$ (as well as μ_1) is σ-finite on $\sigma(\mathscr{A})$. It remains to verify that $\hat{\alpha} = \mu_1$ on $\sigma(\mathscr{A})$. For this we use the monotone class argument.

Define a class

$$\mathscr{A}_2 = \{ A \in \sigma(\mathscr{A}) \colon \hat{\alpha}(A) = \mu_1(A) \}. \tag{3}$$

Since $\hat{\alpha}$ and μ_1 agree on $\mathscr{A}$, and are defined on $\sigma(\mathscr{A})$, we have $\mathscr{A} \subset \mathscr{A}_2 \subset \sigma(\mathscr{A})$. To show that $\mathscr{A}_2 = \sigma(\mathscr{A})$, it suffices to check, by Proposition 1.3(a), that $\mathscr{A}_2$ is closed under monotone limits. Thus let A_n in $\mathscr{A}_2$ be monotone and $A = \lim_n A_n$. In case $A_n \subset A_{n+1}$ so that $A = \bigcup_{n=1}^{\infty} A_n$, we have

$$\hat{\alpha}(A) = \lim_n \hat{\alpha}(A_n) = \lim_n \mu_1(A_n) = \mu_1(A), \tag{4}$$

since $\hat{\alpha}$ and μ_1 are measures and Proposition 1.1 applies. Hence $A \in \mathscr{A}_2$. $A_n \supset A_{n+1}$, so that $A = \bigcap_{n=1}^{\infty} A_n$, we have $A, A_1 \in \sigma(\mathscr{A})$. So by the preceding paragraph, there exist $B_n \in \mathscr{A}_0$, $n \geqslant 1$, such that

$$A \subset A_1 \subset \bigcup_{n=1}^{\infty} B_n = \bigcup_{n=1}^{\infty} C_n, \qquad C_n \subset C_{n+1},$$

where $C_n = \bigcup_{k=1}^{n} B_k \in \mathscr{A}_0$, since $\mathscr{A}_0$ is a ring. Hence

$$A = \bigcup_{n=1}^{\infty} (A \cap C_n), \qquad A \cap C_n \in \sigma(\mathscr{A}), \tag{5}$$

and $\hat{\alpha}(A \cap C_n) \leqslant \hat{\alpha}(C_n) = \alpha(C_n) < \infty$; also $A \cap C_n \subset A \cap C_{n+1}$. It suffices to verify that $\hat{\alpha}(A \cap C_n) = \mu_1(A \cap C_n)$, all n, so that (5) implies

$$\hat{\alpha}(A) = \lim_n \hat{\alpha}(A \cap C_n) = \lim_n \mu_1(A \cap C_n) = \mu_1(A),$$

or that $A \in \mathscr{A}_2$ as desired. Thus to check the above key equation, for the sets $A \cap C_n = D_0$ (say), where $D_0 \subset C_n$ ($C_n \in \mathscr{A}_0$), so that $D_0 \subset \bigcup_{n=1}^{\infty} C_n$, we have

$$\mu_1(D_0) \leqslant \sum_{n=1}^{\infty} \mu_1(C_n) = \sum_{n=1}^{\infty} \alpha(C_n), \qquad \text{since} \quad \alpha = \mu_1|\mathscr{A} = \hat{\alpha}|\mathscr{A},$$

and hence $\mu_1(D_0) \leqslant \hat{\alpha}(D_0)$ [recall $\hat{\alpha} = \mu^*|\sigma(\mathscr{A})$]. Similarly, considering $C_n - D_0 \in \sigma(\mathscr{A})$ in place of D_0 above, we get

$$\mu_1(C_n - D_0) \leqslant \hat{\alpha}(C_n - D_0).$$

On the other hand,

$$\mu_1(D_0) + \mu_1(C_n - D_0) = \mu_1(C_n) = \hat{\alpha}(C_n) = \hat{\alpha}(D_0) + \hat{\alpha}(C_n - D_0), \quad (6)$$

since $\mu_1, \hat{\alpha}$ are additive (they are measures) on $\sigma(\mathscr{A})$. This shows that the preceding inequalities must be equalities, i.e., $\mu_1(D_0) = \hat{\alpha}(D_0)$ as asserted. Thus $\hat{\alpha} = \mu_1$ on $\sigma(\mathscr{A})$, and the uniqueness, existence, and σ-finiteness of $\hat{\alpha}$ are established. From now on let $\mathscr{A}$ be an algebra and α is given on it as before.

It remains to show that $\mathscr{M}_{\mu^*}$ is the $\hat{\alpha}$-completion of $\sigma(\mathscr{A})$. Since $\mathscr{M}_{\mu^*}$ is always complete by Theorem 2.9, let α be σ-finite, and $A \in \mathscr{M}_{\mu^*}$. If $\mathscr{A}_0$ is as in (1), then $\mathscr{A}_0 \subset \mathscr{M}_{\mu^*}$. Since μ_0 is the generated outer measure by $(\alpha, \mathscr{A})$, by Theorem 2.10(ii), μ^* is $\mathscr{A}_{\sigma\delta}$-regular, so that there exists a $B \in \mathscr{A}_{\sigma\delta} \subset \sigma(\mathscr{A}) = \sigma(\mathscr{A}_0)$, $B \supset A$, and $\mu^*(A) = \mu^*(B)$. Since $\hat{\alpha} = \mu^*|\sigma(\mathscr{A})$ is σ-finite by a preceding paragraph, there exist $B_n \in \mathscr{A}_0$, $n \geqslant 1$, $B \subset \bigcup_{n=1}^{\infty} B_n$, and hence

$$A \subset B \subset \bigcup_{n=1}^{\infty} B_n, \qquad \text{or} \quad A = \bigcup_{n=1}^{\infty} (A \cap B_n). \quad (7)$$

Thus $A_n = A \cap B_n \in \mathscr{M}_{\mu^*}$, and again by Theorem 2.10(ii) there exists a $C_n \in \mathscr{A}_0$ such that $A_n \subset C_n$, $\mu^*(A_n) = \mu^*(C_n) < \infty$. Hence $\mu^*(C_n - A_n) = 0$. Consequently, if $C = \bigcup_{n=1}^{\infty} C_n \in \sigma(\mathscr{A})$, one has

$$\mu^*(C - A) = \mu^*\left(\bigcup_{n=1}^{\infty} C_n - \bigcup_{n=1}^{\infty} A_n\right) = \mu^*\left(\bigcup_{n=1}^{\infty}\left(C_n - \bigcup_{m=1}^{\infty} A_m\right)\right)$$

$$\leqslant \mu^*\left(\bigcup_{n=1}^{\infty} (C_n - A_n)\right) \leqslant \sum_{n=1}^{\infty} \mu^*(C_n - A_n) = 0.$$

If $N_1 = C - A$, then $\mu^*(N) = 0$ and $A = C - N_1$. Also $N_1 \in \mathscr{M}_{\mu^*}$, so by Theorem 2.10 there is an $N \in \mathscr{A}_{\sigma\delta}$ such that $N \supset N_1$, $0 = \mu^*(N_1) = \hat{\alpha}(N)$. Thus if we set $A_0 = C - N \in \sigma(\mathscr{A})$ and $N_2 = A - A_0$, then

$$N_2 \subset C - A_0 = C - (C - N) \subset N, \qquad A_0 \subset C - N_1 = A. \quad (8)$$

It follows that $A = A_0 \cup (A - A_0) = A_0 \cup N_2$ where $A_0 \in \sigma(\mathscr{A})$, and N_2 is a subset of a set of $\hat{\alpha}$ measure zero, i.e. $A \in \hat{\sigma}(\mathscr{A})$, the completion. Thus $\mathscr{M}_{\mu^*} = \hat{\sigma}(\mathscr{A})$, and the proof is complete.

An immediate question to ask is about the nonuniqueness of the extensions of measures on rings if the σ-finiteness assumption is dropped. Without any further restrictions, there can be no uniqueness, as the following simple example reveals.

Example (Nonuniqueness). Let $\Omega = [0,1]$, μ = Lebesgue measure on the Lebesgue σ-algebra Σ of Ω (i.e., Σ is the completed Borel σ-algebra of Ω), and let $\mathscr{A} \subset \Sigma$ be the subalgebra of all countable sets and of those with countable complements. Then $\alpha_1 = \mu|\mathscr{A}$ is a measure on $\mathscr{A}$ such that $\alpha(A) = 0$ or $+\infty$ according as A is countable or has a countable complement. Let μ_1^* be the generated measure by the pair $(\alpha_1, \mathscr{A})$. Then $\mathscr{M}_{\mu_1^*}$ is the power set of Ω, and $\mu_1^*|\Sigma$ is different from μ even though $\mu_1^*|\mathscr{A} = \alpha = \mu|\sigma(\mathscr{A})$, since μ_1^* takes only two values whereas μ takes all positive real values. In fact, if $\nu_a = a\mu_1^*$, then for each $a > 0$, ν_a is an extension of α, so that there are in fact infinitely many extensions.

A closer examination of this and similar examples reveal that the measure α on $\mathscr{A}$ is already pathological in that it has no sets of finite positive measure. Now if we insist upon having at least some sets of $\mathscr{A}$ without such pathology, can this phenomenon be eliminated? To analyze this behavior, we introduce the following:

Definition 2. Let Ω be a set of Σ, a σ-algebra on Ω. Then (Ω, Σ) is called a *measurable space*. If $\mu: \Sigma \to \overline{\mathbb{R}}^+$ is a measure then the triple (Ω, Σ, μ) is termed a (*measured space by* μ, or simply) *measure space*. The measure μ is said to have the *finite subset property* if for each $A \in \Sigma$, $\mu(A) > 0$, there exists a $B \subset A$, $B \in \Sigma$ such that $0 < \mu(B) < \infty$.

The last property of μ is sometimes called "semifiniteness", because of the following characterization.

Lemma 3. *Let (Ω, Σ, μ) be a complete measure space, i.e., Σ is complete for μ. Then μ has the finite subset property iff*

$$\mu(A) = \sup\{\mu(B): B \subset A,\ B \in \Sigma,\ \mu(B) < \infty\}, \qquad A \in \Sigma. \tag{9}$$

Proof. Let a_0 denote the right side of (9). Then clearly $0 \leqslant a_0 \leqslant \mu(A)$. To show that there is equality, we may assume that $a_0 < \infty$ since the result is true and trivial otherwise. Hence by the definition of supremum, there exist $B_n \in \Sigma$, $n \geqslant 1$, $B_n \subset A$ such that $\lim_n \mu(B_n) = a_0$. Since Σ is a ring, we may assume that $B_n \subset B_{n+1}$ so that $\lim_n \mu(B_n) = \mu(\tilde{B}) = a_0$, where $\tilde{B} =$

$\lim_n B_n (= \bigcup_{n=1}^{\infty} B_n)$ and $\tilde{B} \in \Sigma$. We now claim that $\mu(A - \tilde{B}) = 0$. In the contrary case, by the finite subset property there exists a $C \subset A - \tilde{B}$, $C \in \Sigma$, $0 < \mu(C) < \infty$. Since $\tilde{B} \cap C = \varnothing$, and $C \subset A$, we get

$$\begin{aligned} a_0 &= \sup\{\mu(B): B \subset A,\ B \in \Sigma,\ \mu(B) < \infty\} \\ &= \mu(\tilde{B}) < \mu(\tilde{B}) + \mu(C) = \mu(\tilde{B} \cup C) \\ &\leqslant \sup\{\mu(B): B \subset A,\ B \in \Sigma,\ \mu(B) < \infty\} = a_0, \end{aligned}$$

which is impossible. Hence $\mu(A - \tilde{B}) = 0$ must hold. Since $\tilde{B} \subset A$, we also have

$$\mu(A) = \mu(A - \tilde{B}) + \mu(\tilde{B}) = 0 + a_0 = a_0.$$

Hence (9) is true as stated.

For the converse, if $\mu(A) > 0$, then by (9) there certainly exists a $B \subset A$, $B \in \Sigma$, $0 < \mu(B) < \infty$. So μ must have the finite subset property. Hence the lemma holds as stated.

This result shows that the nonunique extended measures ν_a of the preceding example do not have the finite subset property. However, we may eliminate such behavior and modify the given measure space slightly, so that every measure has the finite subset property. This is done as follows. Let (Ω, Σ, μ) be a measure space and $\Sigma_0 = \{A \in \Sigma: \mu(A) < \infty\}$. Clearly Σ_0 is a ring (actually a δ-ring). Let $\tilde{\Sigma} = \{A \subset \Omega: A \cap B \in \Sigma_0 \text{ for all } B \in \Sigma_0\}$. Then $A_n \in \tilde{\Sigma} \Rightarrow (\bigcup_{n=1}^{\infty} A_n) \cap B \in \Sigma_0$, $A_1^c \cap B \in \Sigma_0$, $B \in \Sigma_0$. Hence $\tilde{\Sigma}$ is a σ-algebra. Let $\tilde{\mu}(A) = \sup\{\mu(A \cap B): B \in \Sigma_0\}$, $A \in \tilde{\Sigma}$. Then $\tilde{\mu} = \sup\{\mu_B: B \in \Sigma_0\}$ and is a measure on $\tilde{\Sigma}$ [as in Exercise 2.12, $\mu_B(\cdot) = \mu(B \cap \cdot)$], satisfying

$$\tilde{\mu}(B) = \mu(B), \qquad B \in \Sigma_0, \quad \text{and} \quad \tilde{\tilde{\Sigma}} = \tilde{\Sigma}, \quad \tilde{\tilde{\mu}} = \tilde{\mu}, \tag{10}$$

where the "tilde" operation is as defined above. Thus no new classes are obtained, and $(\Omega, \tilde{\Sigma}, \tilde{\mu})$ is a measure space with the finite subset property.

In view of the simple modification that suffices to eliminate the pathological measures noted before, it is not a real restriction to assume the finite subset property. However, this by itself is not sufficient to restore uniqueness in the extension problem for measures, as the following example illustrates.

Let $\Omega = [0, 1]$, $\mathscr{A} = \{A \subset \Omega: A \text{ is finite or } A^c \text{ is finite}\}$. Let $\alpha(A)$ = number of points in $\mathscr{A}$ if A is finite, and $+\infty$ if A is not finite. If $\tilde{\Sigma}$ is defined as above, then $\tilde{\Sigma}$ is the power set of Ω, and the extended measure $\tilde{\mu}$ is again *counting measure*, i.e., $\tilde{\mu}(A)$ = number of points in A. If μ is Lebesgue measure on Ω, and $\nu = \tilde{\mu} + \mu$, then $\tilde{\mu}|\mathscr{A} = \nu|\mathscr{A}$ but $\tilde{\mu} \neq \nu$ on $\sigma(\mathscr{A})$.

This shows that even though both ν and $\tilde{\mu}$ have the finite subset property, they are not the same after extension. Thus a further condition including

σ-finiteness is needed. *Hereafter we therefore assume that each measure has the finite subset property, unless the contrary is stated*. The desired condition, in two parts, is what is called (*strict*) *localizability*, and it plays an important role in many places, including the proof of the Radon-Nikodým theorem, differentiation theory, and the lifting theorem (see Chapters 5, 8). This concept, introduced by I. E. Segal (1954), is as follows:

Definition 4. Let (Ω, Σ, μ) be a measure space (with the finite subset property). Then:

(a) μ is called *localizable* if for any (not necessarily countable) collection $\mathscr{A} \subset \Sigma$ there exists a set $B \in \Sigma$, called the *supremum*, such that (i) for each $A \in \mathscr{A}$, $\mu(A - B) = 0$, and (ii) if $\tilde{B} \in \Sigma$ satisfies (i), then $\mu(B - \tilde{B}) = 0$.

(b) μ is *strictly localizable*, or has the *direct sum property*, if there exists a collection $\{A_j: j \in I\} \subset \Sigma$, $0 < \mu(A_j) < \infty$, (I not necessarily countable) such that (i) $\Omega - \bigcup_{j \in I} A_j \subset N$ where $\mu(N) = 0$, and (ii) if $B \in \Sigma$, $\mu(B) < \infty$, then the set $\{j \in I: \mu(A_j \cap B) > 0\}$ is at most countable.

Remark. Taking $I = N$, the natural numbers, it is clear that each σ-finite measure is strictly localizable. Further, strict localizability implies localizability (Exercise 5). Also, in Definition 4(a), $\mathscr{A}$ may be assumed to have only sets of finite measure. To see this, since the above condition clearly implies that each collection of sets of finite measure has a supremum in Σ, assume the latter property. If $\mathscr{A}$ is any collection ($\subset \Sigma$), define

$$\tilde{\mathscr{A}} = \{B \in \Sigma_0: B \subset A,\ A \in \mathscr{A}\},$$

where $\Sigma_0 = \{B \in \Sigma: \mu(B) < \infty\}$. Then $\tilde{\mathscr{A}}$ has a supremum $\tilde{B}$ in Σ, by hypothesis. We claim that $\tilde{B}$ is also a supremum of $\mathscr{A}$. Indeed, if this is false, there exists a set $A \in \mathscr{A}$ such that $\mu(A - \tilde{B}) > 0$. Then by the finite subset property, there exists a $B_1 \subset A - \tilde{B}$ where $B_1 \in \Sigma_0$ and $\mu(B_1) > 0$. But then $B_1 \subset A$, so $B_1 \in \tilde{\mathscr{A}}$ and $B_1 \cap \tilde{B} = \varnothing$. Hence $\mu(B_1 - \tilde{B}) > 0$, contradicting the fact that $\tilde{B}$ is a supremum of $\tilde{\mathscr{A}}$. Similarly, it is seen that $\tilde{B}$ is the smallest such. This proves the claim.

Thus we may restate that μ *is localizable if any collection* $\tilde{\mathscr{A}} \subset \Sigma$ *of sets of finite measure has a supremum in* Σ. We use this fact hereafter without comment.

Now Theorem 1 admits the following extension:

Proposition 5. *Let $\mathscr{A}$ be a ring on Ω, and α be a localizable measure on $\mathscr{A}$, so that $\mathscr{A}$ is complete for α. Then it has a unique extension $\hat{\alpha}$ onto the generated σ-ring $\sigma(\mathscr{A})$ on which $\hat{\alpha}$ is again localizable.*

Proof. The argument is an adaptation of that of Theorem 1 and will be sketched. As before, the existence of an extension $\hat{\alpha}$ follows from Theorem 2.10. Suppose that μ is another extension such that $\alpha = \mu|\mathscr{A}$. If $\mathscr{A}_0$ is as in (1), then $\mathscr{A}_0$ is a complete ring. Let $\hat{\sigma}(\mathscr{A})$ be the completed (for α) σ-ring $\sigma(\mathscr{A})$. Then $\hat{\sigma}(\mathscr{A}_0) \subset \hat{\sigma}(\mathscr{A})$. If $A \in \mathscr{A}$, then the collection $\{B \subset A: B \in \mathscr{A}_0\}$ has a supremum by the localizability of α on $\mathscr{A}$, and if it is denoted $\hat{B}$, then $\hat{B}$ differs from A only by an α-null set. Consequently, $\hat{B} \in \hat{\sigma}(\mathscr{A}_0)$ and hence $A = \hat{B} \cup N \in \hat{\sigma}(\mathscr{A}_0)$ where $\alpha(N) = 0$. Thus $\hat{\sigma}(\mathscr{A}_0) = \hat{\sigma}(\mathscr{A})$, where the circumflex denotes completion in both cases.

Let $\mathscr{A}_1$ be the class defined by

$$\mathscr{A}_1 = \left\{A \subset \Omega: A \subset \bigcup_{\beta} A_\beta,\ A_\beta \in \mathscr{A}_0\right\}. \tag{11}$$

It follows, as before, that $\mathscr{A}_1$ is a σ-ring, complete for $\hat{\alpha}$. [$\Omega \in \mathscr{A}_1$ if $\Omega \in \mathscr{A}$, since $\alpha(\cdot)$ is localizable on $\mathscr{A}$.] Also $\hat{\sigma}(\mathscr{A}_0) = \hat{\sigma}(\mathscr{A}) \subset \mathscr{A}_2$, and then $\hat{\alpha}$ is localizable on $\mathscr{A}_2$ ($\supset \mathscr{A}$) and hence also on $\hat{\sigma}(\mathscr{A})$. If it is shown that $\hat{\alpha}$ and μ agree on all sets of finite $\hat{\alpha}$-measure in $\hat{\sigma}(\mathscr{A})$, then (9) will imply that they agree on $\hat{\sigma}(\mathscr{A})$ itself. But on sets of finite $\hat{\alpha}$-measure, $\hat{\alpha}$ and μ agree by the same argument as in (6). Hence considering the contractions of $\hat{\alpha}$ and μ to $\sigma(\mathscr{A})$, we may conclude that they are identical there. This finishes the argument, and the proposition follows.

The localizability property is the most general concept available for the unique extension problem. However, the problem for finite measures itself plays the crucial part in the proof. But the localizability concept proves to be indispensable for some of the important parts of analysis. For instance, one shows (Chapter 9) that a Haar measure on a general locally compact group, which is essential for a study in harmonic analysis, is (strictly) localizable but not σ-finite. Hence it is necessary to include such problems into our framework.

We next establish a pair of extension theorems for which the Carathéodory process does not apply. Let us first answer a simple question. Suppose a measure space (Ω, Σ, μ) is given, and consider the class $\tilde{\Sigma} = \{A \cup N: A \in \Sigma, \mu(N) = 0\}$. What is the structure of $\tilde{\Sigma}$, and is there a unique extension of μ to $\tilde{\Sigma}$? A positive answer to this question is given by

Proposition 6. *Let (Ω, Σ, μ) be a measure space, and $\tilde{\Sigma}$ be defined as the class $\{A \cup N \subset \Omega: A \in \Sigma,\ N \subset M,\ M \in \Sigma,\ \mu(M) = 0\}$. Then $\tilde{\Sigma}$ is a σ-algebra, and if $\tilde{\mu}(A \cup N) = \mu(A)$, then $\tilde{\mu}$ is well defined and $(\Omega, \tilde{\Sigma}, \tilde{\mu})$ is a complete measure space. Moreover, $\tilde{\mu}$ is the minimal extension of μ in the sense that if $\tilde{\mu}_1$ is another complete measure on $\tilde{\Sigma}_1$ containing Σ, satisfying $\tilde{\mu}_1|\Sigma = \mu$, then $\tilde{\Sigma} \subset \tilde{\Sigma}_1$ and $\tilde{\mu} = \tilde{\mu}_1|\tilde{\Sigma}$.*

Proof. Since $\Sigma \subset \tilde{\Sigma}$, to see that $\tilde{\Sigma}$ is a σ-algebra, let $\tilde{A}_n \in \tilde{\Sigma}$, $\tilde{A}_n = A_n \cup N_n$, $N_n \subset M_n$, $\mu(M_n) = 0$ with $A_n, M_n \in \Sigma$, $n \geqslant 1$. Then

$$\bigcup_{n=1}^{\infty} \tilde{A}_n = \left(\bigcup_{n=1}^{\infty} A_n\right) \cup \left(\bigcup_{n=1}^{\infty} N_n\right) = A \cup N \qquad \text{(say)}. \tag{12}$$

Then $A \in \Sigma$, and since $\bigcup_{n=1}^{\infty} M_n \in \Sigma$, $\mu(\bigcup_{n=1}^{\infty} M_n) \leqslant \sum_{n=1}^{\infty} \mu(M_n) = 0$, it follows that $\tilde{\Sigma}$ is closed under countable unions. Also $A - N \in \tilde{\Sigma}$ if $A \in \Sigma$ and $N \subset M \in \Sigma$, $\mu(M) = 0$, because if $Q = M - N$, then

$$A - N = A \cap (M^c \cup Q) = (A - M) \cup (A \cap Q) \in \tilde{\Sigma},$$
$$\text{since} \quad A \cap Q \subset M. \tag{13}$$

Thus using the same representation,

$$\begin{aligned}\tilde{A}_1 - \tilde{A}_2 &= (A_1 \cup N_1) \cap (A_2 \cup N_2)^c = (A_1 \cap A_2^c \cap N_2^c) \cup (N_1 \cap A_2^c \cap N_2^c) \\ &= [(A_1 - A_2) - N_2] \cup N_3 \qquad \text{(say)}.\end{aligned} \tag{14}$$

Since $(A_1 - A_2) - N_2 \in \tilde{\Sigma}$ by (13), it follows by (12) that $\tilde{A}_1 - \tilde{A}_2 \in \tilde{\Sigma}$, so that it is closed under differences. But $\Omega \in \tilde{\Sigma}$. Hence $\tilde{\Sigma}$ is a σ-algebra.

That $\tilde{\mu}$ is well defined on $\tilde{\Sigma}$ and is a measure are also easy. Indeed, let $A = A_1 \cup N_1 = A_2 \cup N_2$ be two representations. Then letting $\overline{N} = N_1 \cup N_2$, we get

$$A_1 \cup (N_1 \cup N_2) = (A_2 \cup N_2) \cup N_1, \qquad \text{or} \quad A_1 \cup \overline{N} = A_2 \cup \overline{N}.$$

Hence

$$\mu(A_1) = \tilde{\mu}(A_1 \cup \overline{N}) = \tilde{\mu}(A_2 \cup \overline{N}) = \mu(A_2).$$

Consequently,

$$\tilde{\mu}(A) = \tilde{\mu}(A_1 \cup N_1) = \mu(A_1) = \mu(A_2) = \tilde{\mu}(A_2 \cup N_2).$$

Thus $\tilde{\mu}$ is well defined on $\tilde{\Sigma}$. If $\tilde{A}_n \in \tilde{\Sigma}$, $n \geqslant 1$, are disjoint, so that $A_n \subset \tilde{A}_n$, and the A_n are disjoint, where $\tilde{A}_n = A_n \cup N_n$, then

$$\begin{aligned}\tilde{\mu}\left(\bigcup_{n=1}^{\infty} \tilde{A}_n\right) &= \mu\left(\bigcup_{n=1}^{\infty} A_n\right) \\ &= \sum_{n=1}^{\infty} \mu(A_n), \qquad \text{since } \mu \text{ is a measure,} \\ &= \sum_{n=1}^{\infty} \tilde{\mu}(\tilde{A}_n).\end{aligned}$$

Hence it follows that $(\Omega, \tilde{\Sigma}, \tilde{\mu})$ is a complete measure space.

Finally, since now $\Sigma \subset \tilde{\Sigma}_1$ and for $M \in \Sigma$, $\mu(M) = 0$, and $N \subset M$ we have $\mu_1(N) = 0$, it is clear that $A_1 = A \cup N \in \tilde{\Sigma}_1 \Rightarrow \tilde{\Sigma} \subset \tilde{\Sigma}_1$. Also

$$\mu(A) = \tilde{\mu}_1(A) \leqslant \tilde{\mu}_1(A \cup N) \leqslant \tilde{\mu}_1(A) + \tilde{\mu}_1(N) = \tilde{\mu}_1(A).$$

Thus,

$$\tilde{\mu}_1(A \cup N) = \mu(A) = \tilde{\mu}(A \cup N),$$

and $\tilde{\mu}_1|\Sigma = \tilde{\mu}$, proving the minimality of $\tilde{\mu}$. This completes the proof.

Remark. It should be noted that this result does not say that a given complete measure μ on a (complete) σ-ring cannot be extended; only that every measure can be completed in a unique way by the procedure given. If μ is not Carathéodory regular relative to μ, then by Proposition 2.14 it may be extended to a complete and regular measure. The above method will also be used later for Theorem 9.2.6.

If $\Omega = \mathbb{R}^n$, $\mu =$ Lebesgue measure, then it cannot be extended further than $\mathscr{M}_{\mu^*}$, since μ^* is Carathéodory regular. However, we have seen that there exist sets $A \in \mathscr{P} - \mathscr{M}_{\mu^*}$, i.e., nonmeasurable sets. If $\tilde{\mathscr{M}}$ is the smallest σ-algebra containing A and $\mathscr{M}_{\mu^*}$, is it possible to extend μ^* to $\tilde{\mathscr{M}}$? The Carathéodory process cannot be used directly here. We now show that there *are* extensions (in general nonunique even for finite measures), using the ideas of inner measures.

Theorem 7. *Let $\mathscr{A}$ be an algebra of sets of Ω, and α be a measure. If $A \subset \Omega$, $A \notin \mathscr{A}$, and $\mathscr{B}$ is the algebra generated by $(A, \mathscr{A})$, let μ_* and μ^* be the inner and outer measures generated by $(\alpha, \mathscr{A})$. Then for each number a_0 satisfying $\mu_*(A) \leqslant a_0 \leqslant \mu^*(A)$ there exists a measure μ_0 on $\mathscr{B}$ such that* (i) $\mu_0(A) = a_0$, (ii) $\mu_0|\mathscr{A} = \alpha$, *and* (iii) *$\mu_0$ can be further extended to $\sigma(\mathscr{B})$. Moreover, every extension μ of $\alpha(\cdot)$ to $\mathscr{B}$ [or $\sigma(\mathscr{B})$] verifies $\mu(A) \in [\mu_*(A), \mu^*(A)]$ $(\subset \overline{\mathbb{R}}^+)$.*

Proof. If $\mu^*(A) = 0$, then $a_0 = 0$, and by Theorem 2.10, $\mu^*|\mathscr{M}_{\mu^*}$ is a measure. Moreover $\mathscr{A} \subset \mathscr{B} \subset \mathscr{M}_{\mu^*}$, and $\mu_0 = \mu^*|\mathscr{B}$ gives $\alpha = \mu^*|\mathscr{A} = \mu_0|\mathscr{A}$, so that $\mu_0(A) = \mu^*(A) = a_0$, as desired.

For the nontrivial case, let $\mu^*(A) > 0$ and $\mu_*(A) \leqslant a_0 \leqslant \mu^*(A)$. First we note a simple characterization of $\mathscr{B}$. Indeed, let

$$\mathscr{B}_1 = \{B = (A_1 \cap A) \cup (A_2 \cap A^c) \colon A_i \in \mathscr{A}, i = 1, 2\}. \tag{15}$$

Taking $A_1 = \Omega$, $A_2 = \varnothing$, we get $A \in \mathscr{B}_1$, and then setting $A_1 = A_2$, we see that $B = A_1 = A_2$, so that $\mathscr{A} \subset \mathscr{B}_1$. Since $\mathscr{B}$ is an algebra generated by $\mathscr{A}$ and A, it is clear that each B in $\mathscr{B}_1$ is in $\mathscr{B}$, so $\mathscr{B}_1 \subset \mathscr{B}$. On the other hand, if $B \in \mathscr{B}_1$, then

$$B^c = (A_1^c \cup A^c) \cap (A_2^c \cup A) = (A \cap A_1^c) \cup (A^c \cap A_2^c) \in \mathscr{B}_1,$$

since $A_i^c \in \mathscr{A}$. Clearly $\mathscr{B}_1$ is closed under finite unions. Thus it is an algebra, and hence contains $\mathscr{B}$. So $\mathscr{B} = \mathscr{B}_1$. We next define two set functions μ_1, μ_2 on $\mathscr{B}$ by the equations:

$$\mu_1(B) = \mu_*(B \cap A) + \mu^*(B \cap A^c), \tag{16}$$

$$\mu_2(B) = \mu^*(B \cap A) + \mu_*(B \cap A^c). \tag{17}$$

We show that μ_1, μ_2 are measures on $\mathscr{B}$ and both agree with α on $\mathscr{A}$. If $\gamma \geqslant 0$ is chosen so that $a_0 = \gamma\mu_*(A) + (1-\gamma)\mu^*(A)$, then we set $\mu_0 = \gamma\mu_1 + (1-\gamma)\mu_2$ on $\mathscr{B}$. This will prove (i), (ii); and (iii) is then an immediate consequence of the first part of Theorem 1. Thus let us establish (16) and (17).

Since both μ_* and μ^* are nonnegative and nondecreasing, so are μ_i, $i = 1, 2$, and both vanish at $\varnothing$. Hence $\mu_1(C) = \mu_2(C) = \mu_*(C) = \mu^*(C)$, $C \in \mathscr{A}$, since $\alpha = \mu^*|\mathscr{A} = \mu_*|\mathscr{A}$ by Theorem 2.10 and Proposition 2.5, the latter being valid for abstract measures also (see Exercise 2.10). It remains to verify that both μ_1, μ_2 are σ-additive on $\mathscr{B}$.

Let $B_n \in \mathscr{B}$, disjoint, and $B = \bigcup_{n=1}^{\infty} B_n \in \mathscr{B} = \mathscr{B}_1$. Then by (15) each $B_n = (A_{1n} \cap A) \cup (A_{2n} \cap A^c)$, $A_{in} \in \mathscr{A}$, $i = 1, 2$. Hence $B = ((\bigcup_{n=1}^{\infty} A_{1n}) \cap ((\bigcup_{n=1}^{\infty} A_{2n}) \cap A^c)$. However, the A_{in} need not be disjoint. Hence we disjunctify them as

$$C_{i1} = A_{i1},$$

and for $n > 1$,

$$C_{in} = A_{in} - \bigcup_{k=1}^{n-1} A_{ik}, \qquad i = 1, 2.$$

Then $C_{in} \in \mathscr{A}$, disjoint, and

$$C_{1n} \cap A = \left(A_{1n} - \bigcup_{k=1}^{n-1} A_{1k} \right) \cap A = A_{1n} \cap A, \tag{18}$$

since $B_n \cap B_m = \varnothing$ implies $A_{1n} \cap A_{1m} \cap A = \varnothing$, so $A_{1n} \cap A_{1m}^c \cap A = A_{1n} \cap A$; *and* $A_{2n} \cap A_{2m} \cap A^c = \varnothing$. Thus we also have $C_{2n} \cap A^c = A_{2n} \cap A^c$, and hence

$$B_n = (C_{1n} \cap A) \cup (C_{2n} \cap A^c), \qquad n \geqslant 1, \tag{19}$$

the C_{in}, $n \geqslant 1$, being disjoint. This implies

$$B \cap A = \bigcup_{n=1}^{\infty} (C_{1n} \cap A), \qquad B \cap A^c = \bigcup_{n=1}^{\infty} (C_{2n} \cap A^c). \tag{20}$$

But by Theorem 2.9(ii), μ^* is σ-additive on $\mathscr{M}_{\mu^*}(A)$ and hence also on

$\mathscr{A}(D)$. Thus for any $D \subset \Omega$,

$$\mu^*(B \cap D) = \sum_{n=1}^{\infty} \mu^*(C_{1n} \cap D) = \sum_{n=1}^{\infty} \mu^*(B_n \cap D), \qquad C_{1n} \in \mathscr{A}. \quad (21)$$

Using this result in Proposition 2.5, after noting that $\alpha(\cdot)$ is σ-additive on $\mathscr{A}$, we get μ_* to be σ-additive on $\mathscr{A}(D)$, and

$$\mu_*(B \cap D) = \sum_{n=1}^{\infty} \mu_*(C_{1n} \cap D) = \sum_{n=1}^{\infty} \mu_*(B_n \cap D), \qquad C_{1n} \in \mathscr{A}. \quad (22)$$

Taking $D = A$ in (21) and $D = A^c$ in (22) and adding, we get μ_2 of (17) to be σ-additive. Similarly μ_1 of (16) is a measure, as we see by taking $D = A^c$ in (21) and $D = A$ in (22) and adding. This proves the main part.

Finally, if μ is any extension of $\alpha(\cdot)$ to $\mathscr{B}$, then by definition of μ^*, we have for any disjoint sequence $\{C_n: n \geqslant 1\} \subset \mathscr{A}$ such that $A \subset \bigcup_{n=1}^{\infty} C_n$, so that $A = \bigcup_{n=1}^{\infty}(A \cap C_n)$ $(A \cap C_n \in \mathscr{B})$,

$$\mu(A) = \sum_{n=1}^{\infty} \mu(A \cap C_n) \leqslant \sum_{n=1}^{\infty} \mu(C_n).$$

Taking the infimum of all such sums, we get $\mu(A) \leqslant \mu^*(A)$. On the other hand, if $E \in \mathscr{A}$ such that $\mu^*(E - A) < \infty$, then by the above, $\mu(E - A) \leqslant \mu^*(E - A) < \infty$, so that by Proposition 2.5 (see also Exercise 2.10), we get

$$\begin{aligned}\alpha(E) - \mu^*(E - A) &\leqslant \alpha(E) - \mu(E - A) \\ &= \mu(E) - \mu(E - A) = \mu(E \cap A) \leqslant \mu(A),\end{aligned}$$

since $\alpha = \mu|\mathscr{A}$, and μ is additive on $\mathscr{B}$. Hence taking the supremum over all such E in $\mathscr{A}$, we get $\mu_*(A) \leqslant \mu(A) \leqslant \mu^*(A)$. This completes the proof of the theorem.

In case $\mu_*(A^c) = 0$ and $\mu^*(A) > 0$, so that $A \notin \mathscr{M}_{\mu^*}$ and $\alpha(\Omega) = \mu^*(\Omega) = \mu^*(A)$, by Proposition 2.5, one calls A a *thick* set. In this case (17) shows that $\mu_2(B) = \mu^*(B \cap A)$ or $\mu_2 = \alpha|\mathscr{A}(A)$, and then $(A, \sigma(\mathscr{A}(A)), \mu_2)$ is a measure space. Motivated by some probabilistic applications, this fact was established by J. L. Doob in 1930's and the above theorem contains his result. So we state it for reference as:

Corollary 8 (Doob). *Let A be a thick set of a Carathéodory regular measure space (Ω, Σ, μ). Define μ_1 on $\Sigma(A)$ by $\mu_1(A \cap B) = \mu(B)$, $B \in \Sigma$. Then μ is unambiguously defined and $(A, \Sigma(A), \mu_1)$ is a measure space. The μ_1 however is generally not Carathéodory regular, and the extension need not be unique.*

As a last extension result, we include one for the case when there is a topology for the underlying space Ω, and the measure is finite. This is due to J.-P. Henry (1969), and will be used in a variety of applications to follow. It depends on the topological properties as well as Theorem 7. Topological measure spaces will be considered exclusively in Chapter 9. We introduce the regularity of a measure in a topological context to prove Henry's theorem.

The properties of Lebesgue's outer measure given in Theorem 2.17 are now abstracted and taken as a definition for measures on general topological spaces to be well behaved. The concept is introduced as:

Definition 9. Let Ω be a topological space, and $\mathscr{B}$ be the Borel σ-algebra which by definition is generated by all open sets of Ω. An outer measure μ on Ω is called a *Radon* (outer) *measure* if:

(i) open sets are μ-measurable (i.e., are in $\mathscr{M}_\mu$);
(ii) every compact set C has finite measure (i.e., $\mu(C) < \infty$);
(iii) for every $A \subset \Omega$ (*outer regularity*)

$$\mu(A) = \inf\{\mu(G): G \supset A, G \text{ open}\};$$

(iv) for every $B \in \mathscr{B}$ (*inner regularity*)

$$\mu(B) = \sup\{\mu(C): C \subset A, C \text{ compact and closed}\}.$$

It should be noted that if $\Omega = \mathbb{R}^n$, then the above definition is exactly the result of Theorem 17 when μ is taken as our Lebesgue outer measure. Also, in (iv) of the definition the closedness is, of course, automatic if Ω is Hausdorff. In general, however, a compact set, which by definition is a set admitting a finite subcover from its open covering, need not be closed. (Sometimes this is called *quasicompactness*, if a distinction is to be emphasized.) In the following proof, if compactness is understood as *quasicompact and closed*, then the result is valid even when Ω is not a Hausdorff space. However, for most applications, the latter property is available and we prove it for that case.

With this definition we have the following

Theorem 10 (Henry's extension). *Let Ω be a topological space, $\mathscr{A}$ be a subalgebra of the Borel σ-algebra $\mathscr{B}$, and μ: $\mathscr{A} \to \mathbb{R}^+$ be a function such that* (i) *μ is finitely additive and* (ii) *it is inner regular in the sense that for each $B \in \mathscr{A}$,*

$$\mu(B) = \sup\{\mu(C): C \subset B, C \in \mathscr{A}, C \text{ compact (and closed)}\}. \quad (23)$$

Then μ can be extended (nonuniquely in general) to $\hat{\mu}$, a Radon measure on $\mathscr{B}$.

Proof. It is first noted that the inner regularity of μ implies that it is a measure, i.e., μ is σ-additive. Since $\mu(\Omega) < \infty$, it suffices for this to show that for any $B_n \in \mathscr{A}$, $B_n \downarrow \varnothing$, then $\mu(B_n) \to 0$. This is so because for any disjoint

$\{A_n: n \geqslant 1\} \subset \mathscr{A}$ with $A = \bigcup_{n=1}^{\infty} A_n \in \mathscr{A}$, we get

$$\mu(A) = \sum_{k=1}^{n} \mu(A_k) + \mu\left(\bigcup_{k>n} A_k\right) \to \sum_{k=1}^{\infty} \mu(A_k) \qquad \text{as} \quad n \to \infty,$$

since $B_n = \bigcup_{k>n} A_k \in \mathscr{A}$, $B_n \downarrow \varnothing$.

If $\epsilon > 0$ is given, choose by (23), for each B_n, a compact $C_n \in \mathscr{A}$ such that $C_n \subset B_n$ and $\mu(B_n - C_n) < \epsilon/2^{n+1}$. But then

$$\bigcap_{n=1}^{\infty} C_n \subset \bigcap_{n=1}^{\infty} B_n = \varnothing,$$

and the compactness hypothesis implies $\bigcap_{n=1}^{n_0} C_n = \varnothing$ for some $n_0 \geqslant 1$. Consequently

$$B_{n_0} = B_{n_0} - \bigcap_{k=1}^{n_0} C_k = B_{n_0} \cap \left(\bigcup_{k=1}^{n_0} C_k^c\right) = \bigcup_{k=1}^{n_0} (B_{n_0} - C_k),$$

so that for $n \geqslant n_0$

$$\mu(B_n) \leqslant \mu(B_{n_0}) \leqslant \sum_{k=1}^{n_0} \mu(B_{n_0} - C_k) < \sum_{k=1}^{\infty} \frac{\epsilon}{2^{k+1}} = \epsilon.$$

Hence μ is σ-additive on $\mathscr{A}$. Since μ is a finite measure, Theorem 1 implies that μ has a unique extension to $\sigma(\mathscr{A}) \subset \mathscr{B}$, and it is a (Radon) measure there. However, the inclusion may be proper here. So we need to proceed differently. The desired extension will be obtained by using *Zorn's lemma*, which states: If $\mathscr{L}$ is a nonempty partially ordered set such that each totally ordered subset of $\mathscr{L}$ has an upper bound in $\mathscr{L}$, then $\mathscr{L}$ contains at least one maximal element. (See the Appendix for other forms of this lemma.)

Consider the set $\mathscr{L}$ of all pairs $(\mu_i, \mathscr{A}_i)$, where $\mathscr{A} \subset \mathscr{A}_i \subset \mathscr{B}$ is an algebra and $\mu_i: \mathscr{A}_i \to \mathbb{R}^+$ is a Radon measure such that $\mu_i|\mathscr{A} = \mu$. This set is partially ordered by defining $(\mu_i, \mathscr{A}_i) \leqslant (\mu_{i'}, \mathscr{A}_{i'})$ iff $\mathscr{A}_i \subset \mathscr{A}_{i'}$ and $\mu_i = \mu_{i'}|\mathscr{A}_i$. If $\{(\mu_i, \mathscr{A}_i): i \in I\}$ is a totally ordered subset of $\mathscr{L}$, let $\tilde{\mathscr{A}} = \bigcup_{i \in I} \mathscr{A}_i$ and $\tilde{\mu}: \tilde{\mathscr{A}} \to \mathbb{R}^+$ be given by $\tilde{\mu}(A) = \mu_i(A)$ if $A \in \mathscr{A}_i$. Then $\tilde{\mathscr{A}} \subset \mathscr{B}$ is an algebra, and $\tilde{\mu}$ is additive on $\tilde{\mathscr{A}}$ and unambiguously defined. Since $A \in \tilde{\mathscr{A}} \Rightarrow A \in \mathscr{A}_i$ for some $i \in I$, and μ_i is Radon, hence inner regular, it follows that $\tilde{\mu}$ on $\mathscr{A}$ is inner regular, and by the preceding paragraph $\tilde{\mu}$ must be σ-additive. Thus $(\tilde{\mu}, \tilde{\mathscr{A}})$ is an upper bound in $\mathscr{L}$ of $\{(\mu_i, \mathscr{A}_i): i \in I\}$. Hence by Zorn's lemma, there is at least one maximal element, say $(\mu', \mathscr{A}')$, in $\mathscr{L}$ which extends $(\mu, \mathscr{A})$. If $\mathscr{A}'$ is not a σ-algebra, then the finite Radon measure μ' on $\mathscr{A}'$ can be extended to a finite Radon measure onto $\sigma(\mathscr{A}') \subset \mathscr{B}$ by Theorem 1, and this contradicts the maximality of $(\mu', \mathscr{A}')$. Consequently $\mathscr{A}'$ is a σ-algebra, and μ' is a finite Radon measure on $\mathscr{A}'$ such that $\mu'|\mathscr{A} = \mu$.

We now show that $\mathscr{A}' = \mathscr{B}$, so μ' is an extension of the desired kind. Since $\mathscr{B}$ is generated by all the open or equivalently closed sets of Ω, it suffices to show that every closed (or open) set $A \subset \Omega$ is also in $\mathscr{A}'$. This involves some additional work.

Suppose there is a closed set A of Ω which is not in $\mathscr{A}'$. Let $\mathscr{B}_1$ be the algebra generated by A and $\mathscr{A}'$. Then $\mathscr{B}_1$ is given by (15), and by Theorem 7 there exists a σ-additive $\mu_0: \mathscr{B}_1 \to \mathbb{R}^+$ extending μ' from $\mathscr{A}'$ to $\mathscr{B}_1$ and it is given by (16) and (17). Thus if $B = (B_1 \cap A) \cup (B_2 \cap A^c)$, $B_i \in \mathscr{A}'$, then

$$\mu_0(B) = \mu_*(B \cap A) + \mu^*(B \cap A^c). \tag{24}$$

We need only establish that μ_0 is inner regular. This may be shown separately for sets $B \subset A$ and $B \subset A^c$. Thus if $B \subset A^c$,

$$\mu_0(B) = \mu_*(B) = \sup\{\mu'(D): D \subset B,\ D \in \mathscr{A}'\}, \tag{25}$$

by definition of μ_* (in one of its equivalent forms). But μ' on $\mathscr{A}'$ is Radon, so

$$\mu'(D) = \sup\{\mu'(C): C \subset D, C \in \mathscr{A}', C \text{ compact}\}.$$

Substituting this in (25), it follows that μ_0 is inner regular on $\mathscr{B}_1(A^c)$. Next suppose that $B \subset A$. Then $B = A \cap B_1$, $B_1 \in \mathscr{A}'$, by (15). With the inner regularity of μ' on $\mathscr{A}'$ we can find compact $C_n \in \mathscr{A}$ such that $C_n \subset C_{n+1} \subset B_1$ and $\mu'(B_1 - C_n) \to 0$ as $n \to \infty$, since μ' is a finite measure. Hence

$$\mu_0(B - A \cap C_n) = \mu_*(A \cap B_1 - A \cap C_n) \to 0 \qquad \text{as} \quad n \to \infty, \tag{26}$$

since μ_* is σ-additive. But A is closed, so $A \cap C_n$ is compact and (26) implies that μ_0 is also inner regular on $\mathscr{B}_1(A)$. Thus $(\mu_0, \mathscr{B}_1)$ is an extension of $(\mu', \mathscr{A}')$, contradicting again the maximality of the latter. So $A \in \mathscr{A}'$ for each closed $A \subset \Omega$. It follows therefore that $\mathscr{A}' = \mathscr{B}$, and μ_0 is the desired Radon extended measure of μ on $\mathscr{A}$. This establishes the theorem.

Since the extension here is obtained through Zorn's lemma, it may not be unique. However, sufficient conditions for uniqueness can be given. (See Exercise 11 for one such.)

Remark. In the above proof, we used the fact that $\mathscr{A}$ is an algebra. The proof is not special, but the result itself is false, if $\mathscr{A}$ is a subring and not an algebra. For instance, let $\Omega = [0, 1]$, $\mathscr{A}$ be the ring of finite subsets of Ω, and $\mu: \mathscr{A} \to \mathbb{R}^+$ be counting measure. Then μ has no Radon extension to the Borel σ-algebra $\mathscr{B}$ of Ω.

EXERCISES

0. Let (Ω, Σ, μ) be the Lebesgue unit interval. If μ^* is the outer measure generated by (μ, Σ), show that $\mathscr{M}_{\mu^*} = \Sigma$. If $A \subset \Omega$ is a thick set (so

$A \notin \mathcal{M}_{\mu^*}$), can we extend μ^* to a σ-algebra $\mathcal{A} \supset \{\mathcal{M}_{\mu^*}, A\}$? [For a general case, see Exercise 3 below.]

1. Show that there exists a measurable space $(\Omega, \mathcal{A})$ and a measure μ on $\mathcal{A}$ such that μ is not σ-finite on $\mathcal{A}$ but has an extension onto a larger σ-algebra $\mathcal{B}$ containing the σ-algebra $\mathcal{A}$ such that μ is σ-finite on $\mathcal{B}$.

2. This goes in the converse direction to the above. If $(\Omega, \mathcal{A}, \mu)$ is a σ-finite measure space and $\mathcal{B} \supset \mathcal{A}$ is an algebra to which μ has a σ-additive extension, then verify that μ is σ-finite on $\mathcal{B}$ as well as on $\sigma(\mathcal{B})$.

3. Let $(\Omega, \mathcal{A}, \mu)$ be a measure space and $\mathcal{B} \supset \mathcal{A}$ be an arbitrary algebra of Ω. Show that μ has an extension $\hat{\mu}$ to $\mathcal{B}$ which is just additive. If $\mathcal{B}$ can be obtained from $\mathcal{A}$ by adjoining sets $A_1, \ldots, A_n$ such that $A_{i+1} \notin$ to the algebra generated by $(\mathcal{A}, A_1, \ldots, A_i)$, then verify that there is at least one measure $\tilde{\mu}$ on $\mathcal{B}$ which is an extension of μ. What if $\{A_n: n \geqslant 1\}$ is infinite? [Use Zorn's lemma for the first part, and Theorem 7 for the second.]

4. Complete the details of proof of Proposition 5.

5. This exercise is designed to show that a strictly localizable measure is localizable, using the following development.

(a) If (Ω, Σ, μ) is a finite complete measure space, then it is localizable. [Let $\mathcal{A} \subset \Sigma$ be any collection, and set $\mathcal{D} = \{B \in \Sigma: B$ is an upper bound of $\mathcal{A}\}$. Since $\Omega \in \mathcal{D}$, it is nonempty. If $a_0 = \inf\{\mu(B): B \in \mathcal{D}\}$, since $\mathcal{D}$ is closed under finite unions and intersections, we conclude that there exist $B_n \in \mathcal{D}$ such that $B_n \downarrow \bar{B}$ and $a_0 = \lim_n \mu(B_n) = \mu(\bar{B})$. So $\bar{B} \in \mathcal{D}$. It is the least such set, since if $\tilde{B}$ is another, then $\mu(\bar{B} \cap \tilde{B}) \geqslant a_0$, so that $\mu(\bar{B} - \bar{B} \cap \tilde{B}) = 0$. Thus $\bar{B}$ is the supremum of $\mathcal{A}$, and μ is localizable.]

(b) If (Ω, Σ, μ) is a complete σ-finite space, then it is localizable. [There exists $A_n \in \Sigma$, $A_n \uparrow \Omega$, $\mu(A_n) < \infty$. So if $\mathcal{A} \subset \Sigma$ is any collection, then $\mathcal{A}_n = \mathcal{A}(A_n) \subset \Sigma(A_n)$, the trace classes of sets on A_n, and by (a) each $\mathcal{A}_n$ has a supremum $\bar{B}_n$ in $\Sigma(A_n)$. If $\bar{B} = \bigcup_{n=1}^{\infty} \bar{B}_n \in \Sigma$, then $\bar{B}$ is the supremum of $\mathcal{A}$ in Σ, so μ is localizable.]

(c) If (Ω, Σ, μ) is strictly localizable, where $\Sigma = \mathcal{M}_\mu$, then μ is localizable. [By hypothesis there exists a disjoint class $\{A_i: i \in I\} \subset \Sigma$ such that $\Omega - \bigcup_{i \in I} A_i \subset N$, $\mu(N) = 0$, $0 < \mu(A_i) < \infty$, and for $A \in \Sigma$, $\mu(A) < \infty \Rightarrow \{i: \mu(A \cap A_i) > 0\}$ is countable. Also A is in $\mathcal{M}_\mu$ iff $A \cap B \in \mathcal{M}_\mu$ for all $B \in \Sigma_0$, the δ-ring of sets from Σ of finite μ-measure, by Lemma 2.8. Let $\mu_i = \mu|\Sigma(A_i)$. Then $(A_i, \Sigma(A_i), \mu_i)$ is a finite complete measure space, which is localizable by (a) above. Hence $\mathcal{A}(A_i) \subset \Sigma(A_i)$ has a supremum, B_i. If $\tilde{B} = \bigcup_{i \in I} B_i$, then $\tilde{B} \cap B \in \mathcal{M}_{\tilde{\mu}} = \Sigma$ for each $B \in \Sigma_0$, so that $\tilde{B}$ is μ-measurable. Next check that $\tilde{B}$ is the supremum of $\mathcal{A}$ (cf. the argument of the remark following Definition 4). It was shown in 1978 by D. H. Fremlin that there exist localizable measure spaces which are not strictly localizable, i.e., do not have the direct sum property. Thus localizability is a more general concept. Also, there exist measure spaces with the finite subset property that are not localizable, as well as localizable spaces definable without the finite subset property. The latter situation is unimportant, however.]

6. Let $\mathscr{A}$ be an algebra of sets of Ω, and $\alpha: \mathscr{A} \to \mathbb{R}^+$ be a measure. Show that $\mathscr{A}$ is a "dense" subspace of $\sigma(\mathscr{A})$ in the sense that for each $\epsilon > 0$, and $A \in \sigma(\mathscr{A})$, there exists a $B_\epsilon \in \mathscr{A}$ such that $d(A, B_\epsilon) \leqslant \epsilon$ where $d(A,B) = \mu(A \triangle B)$ (as in Exercise 2.3), is a (semi-)distance on $\sigma(\mathscr{A})$.

7. Let $\Omega = [0, a]$, $a > 0$, and μ be the Lebesgue measure on Ω. Let $A \subset \Omega$ be a nonmeasurable set such that $\mu_*(A) = 0$ and $\mu^*(A) > 0$ [so $\mu^*(A) = \mu^*(\Omega) = a$; by Theorem 2.19 such sets exist]. Let $\mathscr{B}$ be the smallest algebra containing $(A, \mathscr{M}_\mu)$, where $\mathscr{M}_\mu$ is the Lebesgue σ-algebra of measurable sets. Show that $\mu_\lambda: \mathscr{B} \to \mathbb{R}^+$ defined by $\mu_\lambda(C) = \lambda\mu(B_1) + (1 - \lambda)\mu(B_2)$ for each $C = (B_1 \cap A) \cup (B_2 \cap A^c) \in \mathscr{B}$ [cf., e.g., (15)], with $0 < \lambda < 1$, $B_i \in \mathscr{M}_\mu$, $i = 1, 2$, is a measure for each λ, and $\mu_\lambda(B) = \mu(B)$, $B \in \mathscr{M}_\mu$, $\mu_\lambda(A) = \lambda a$. Thus there are continuum-many extensions of μ in this case also.

8. Let $\mathscr{A}$ be an algebra on Ω, and μ be σ-finite on $\mathscr{A}$. If $\hat{\mu}$ is the extension of μ to $\sigma(\mathscr{A})$, and $\tilde{\mu}$ is another measure on $\sigma(\mathscr{A})$ which agrees with μ on $\mathscr{A}$, then demonstrate that μ and $\tilde{\mu}$ also agree on $\sigma(\mathscr{A})$. Verify that we may replace "σ-finite" by "localizable" everywhere above.

9. Let (Ω, Σ, μ) be a complete σ-finite measure space, and μ^* be the generated outer measure by the pair (μ, Σ). If $\mathscr{M}_{\mu^*}$ is, as usual, the class of μ^*-measurable sets, characterize the class $\mathscr{M}_{\mu^*} - \Sigma$. Do the same if "$\sigma$-finiteness" is replaced by "localizability", "direct sum property", and "finite subset property" respectively.

10. If Ω is a topological space and μ is an outer measure, then we define its *support* as the closed set

$$\operatorname{spt}(\mu) = \Omega - \bigcup\{G: \mu(G) = 0, G \text{ open}\}.$$

Suppose μ is a Radon measure in the sense of Definition 9, with support Ω_0. Show that $\mu(\Omega - \Omega_0) = 0$, i.e., the open set Ω_0^c cannot have a positive μ-measure, so that we may conclude that $\mu(\Omega) = \mu(\Omega_0)$.

11. If $\mathscr{A}$ is an algebra on a topological space Ω, containing the sets which form a base for the topology of Ω, and $\mu: \mathscr{A} \to \mathbb{R}^+$ is an inner regular, additive set function (so it is a finite Radon measure), then show that its extension to the Borel σ-algebra $\mathscr{B}$ of Ω as a Radon measure is *unique* and the extended measure has a full support, so its support has measure $\mu(\Omega)$. Prove an extension of Henry's theorem if μ on $\mathscr{A}$ is σ-finite or if Ω is countably compact, but μ is Radon on $\mathscr{A}$.

2.4. DISTINCTION BETWEEN FINITE AND INFINITE MEASURES

Thus far we have treated the subject of measures from a general point of view without making specialized assumptions, except in a few cases. Clearly the work with finite measures has fewer pathologies; and it occupies perhaps the major part of analysis. For instance, the whole probability theory rests on finite (actually, normalized so that the total space has unit mass) measures. However, for a finer analysis of the structure of certain spaces one has to consider infinite measures also. In fact, a given point set may be measured differently using different measure functions whose employment is dictated by the problem at hand. As an example, the Cantor set has Lebesgue measure zero, but it has a positive (finite) Hausdorff α-dimensional measure (to be briefly considered in the next section) and gets infinite value for the counting measure. In general Hausdorff measures are not even σ-finite. They are important in studies of exceptional sets in Fourier analysis and in an analysis of the behavior of Brownian motion paths in probability, among others. Since they are important, our general study should take these questions into account. But the theory needs modifications and special restrictions in treating such diverse types of measure functions.

For a systematic study, we have therefore classified the measures into (1) finite, (2) σ-finite, (3) strictly localizable (or having the direct sum property, also called *decomposable*), (4) localizable, (5) with the finite subset property (or semifiniteness), and (6) completely general. The last class includes the pathological two valued (0 and ∞ only) measures. Thus Lebesgue's original study dealt with the first category, namely the length measure on the unit interval. But later applications demanded the general study involving all the other classes.

In addition to the previous results and remarks on a need for additional conditions in the case of infinite measures, we now include the following simple but useful result for both cases, on turning a measure space into a metric space (cf. also Exercise 2.3).

Proposition 1. *Let μ be an outer measure on Ω with $\mathscr{M}_\mu$ as the class of μ-measurable sets. If $\varphi: \overline{\mathbb{R}}^+ \to \mathbb{R}^+$ is a bounded continuous concave function such that $\varphi(x) = 0$ iff $x = 0$, define $d: \mathscr{M}_\mu \times \mathscr{M}_\mu \to \mathbb{R}$ by*

$$d(A, B) = \varphi(\mu(A \vartriangle B)), \qquad A, B \text{ in } \mathscr{M}_\mu. \tag{1}$$

Then $(\mathscr{M}_\mu, d)$ is a complete semimetric space. In case $\mu(\Omega) < \infty$, we may take $\varphi(x) = x \in [0, \mu(\Omega)]$, the identity mapping. Moreover, the measure function μ and the operations $\cap, \cup$ are uniformly continuous in this semimetric topology.

Remark 1. If $\tilde{\mathscr{M}}_\mu = \mathscr{M}_\mu / \mathscr{N}_\mu$ is the quotient space with $\mathscr{N}_\mu$ as the ring of μ-null sets, and if we define $\tilde{d}: \tilde{\mathscr{M}}_\mu \times \tilde{\mathscr{M}}_\mu \to \mathbb{R}^+$ as

$$\tilde{d}([A],[B]) = d(A, B), \qquad [A] = A + \mathscr{N}_\mu \in \tilde{\mathscr{M}}_\mu, \tag{2}$$

then $(\tilde{\mathscr{M}}_\mu, \tilde{d})$ is shown to be a complete metric space. $\tilde{\mathscr{M}}_\mu$ is called a *measure algebra*.

Remark 2. The general form of a φ above can be taken as

$$\varphi(x) = \int_0^x f(t)\,dt, \tag{3}$$

where f is a monotone nonincreasing piecewise continuous function. Examples of φ are $\varphi(x) = \arctan x$, $x(1+x)^{-1}$, etc. Also, from (3) one has

$$\begin{aligned}\varphi(x+y) &= \int_0^x f(t)\,dt + \int_0^y f(x+t)\,dt\\ &\leqslant \varphi(x) + \int_0^y f(t)\,dt, \qquad \text{since } f \text{ is nonincreasing}\\ &= \varphi(x) + \varphi(y).\end{aligned} \tag{4}$$

Any such φ will give an "equivalent metric", and so the completeness assertion is true for any of these. If $\mu(\Omega) < \infty$, then $d(\cdot,\cdot)$ can be defined as $d(A, B) = \mu(A \,\triangle\, B)$. Thus a distinction in the nonfinite case is immediately noticeable.

Proof. The facts that $d(A, A) = 0$, $d(A, B) = d(B, A)$ are clear, and since for any A, B, C (seen at once by using indicator functions)

$$A \,\triangle\, B \subset (A \,\triangle\, C) \cup (C \,\triangle\, B), \tag{5}$$

and since μ and φ [by (4)] are subadditive, the triangle inequality obtains by (5). Thus $d(\cdot,\cdot)$ is a semimetric. To prove the completeness of $(\mathscr{M}_\mu, d)$, let $\{A_n: n \geqslant 1\} \subset \mathscr{M}_\mu$ be a Cauchy sequence, i.e., $d(A_n, A_m) \to 0$ as $n, m \to \infty$. Then we choose a subsequence $n_1 < n_2 < \cdots$ such that

$$d(A_{n_1}, A_n) < 1, \quad n > n_1, \qquad d(A_{n_2}, A_n) < 2^{-1}, \quad n > n_2 > n_1.$$

By induction, $n_{k+1} > n_k$ is chosen such that

$$d(A_{n_{k+1}}, A_n) < 2^{-k}, \qquad n > n_{k+1}. \tag{6}$$

Let $B_k = A_{n_{k+1}}$, so that $d(B_k, B_{k+1}) < 2^{-k}$. Now to produce a limit set, we consider $B = \limsup_k B_k = \bigcap_{k \geqslant 1} \bigcup_{n \geqslant k} B_n$ and claim that $d(A_n, B) \to 0$ as $n \to \infty$, proving the first part.

In fact $B \in \mathscr{M}_\mu$, since the latter is a σ-algebra. Also, for the B_n-sequence we have

$$B - B_n \subset \bigcup_{k \geqslant n+1} B_k - B_n = \bigcup_{k \geqslant n+1} \left[B_k - \bigcup_{i=n}^{k-1} B_i \right]$$

$$= \bigcup_{k=n+1}^{\infty} \left[B_k \cap \bigcap_{i=n}^{k-1} B_i^c \right] \subset \bigcup_{k=n+1}^{\infty} (B_k - B_{k-1}),$$

so that by the σ-subadditivity of μ,

$$\mu(B - B_n) \leqslant \sum_{k \geqslant n+1} \mu(B_k - B_{k-1}) = \lim_{p \to \infty} \sum_{k=n+1}^{n+p} \mu(B_k - B_{k-1}). \quad (7)$$

Similarly, B being determined by the tail part of the B_k-sequence,

$$B_n - B = B_n \cap \left(\bigcup_{m=n+1}^{\infty} \bigcap_{k \geqslant m} B_k^c \right) \subset \bigcup_{k=n+1}^{\infty} (B_n \cap B_k^c) = \bigcup_{k=n+1}^{\infty} (B_k^c \cap B_n),$$

so that the simplification of (7) applied here to B_k^c and B_n gives

$$\mu(B_n - B) \leqslant \sum_{k=n+1}^{\infty} \mu(B_k^c \cap B_{k-1}) = \lim_{p \to \infty} \sum_{k=n}^{n+p} \mu(B_k - B_{k+1}). \quad (8)$$

Adding (7) and (8),

$$\mu(B \vartriangle B_n) \leqslant \lim_{p \to \infty} \sum_{k=n+1}^{n+p} \mu(B_k \vartriangle B_{k+1}). \quad (9)$$

Since φ is continuous, monotone, and subadditive, (9) implies

$$d(B, B_n) = \varphi(\mu(B \vartriangle B_n)) \leqslant \lim_{p \to \infty} \varphi\left(\sum_{k=n+1}^{n+p} \mu(B_k \vartriangle B_{k+1}) \right)$$

$$\leqslant \lim_{p \to \infty} \sum_{k=n+1}^{n+p} \varphi(\mu(B_k \vartriangle B_{k+1})), \qquad \text{by (4)}$$

$$= \sum_{k=n+1}^{\infty} d(B_k, B_{k+1}) < \sum_{k=n+1}^{\infty} 2^{-k} = 2^{-n}.$$

Hence $\lim_n d(B, B_n) = 0$. Considering the full sequence,

$$d(B, A_n) \leqslant d(B, A_{n_{k+1}}) + d(A_{n_{k+1}}, A_n)$$

$$\leqslant 2^{-k} + 2^{-k} \to 0 \qquad \text{as } n \text{ and hence } k \to \infty.$$

Thus $(\mathscr{M}_\mu, d)$ is a complete semimetric space.

Let $[A]$ be defined as in (2). Then $A \in [A] \subset \mathscr{M}_\mu$, and A is termed a representative of the equivalence class $[A]$. If $A \sim B$ and $E \sim F$ denote $[A] = [B]$ and $[E] = [F]$, then clearly $\mu(A \vartriangle E) = \mu(B \vartriangle F)$. Indeed, we have seen [in the proof of Proposition 2.16(iii)] that for any C, D in $\mathscr{M}_\mu$, of finite measure,

$$|\mu(C) - \mu(D)| \leqslant \mu(A \vartriangle B). \tag{10}$$

Hence if $A \vartriangle E$ and $B \vartriangle F$ have finite measures, we get

$$\begin{aligned} |\mu(A \vartriangle E) - \mu(B \vartriangle F)| &\leqslant \mu((A \vartriangle E) \vartriangle (B \vartriangle F)) \\ &\leqslant \mu(A \vartriangle B) + \mu(E \vartriangle F) = 0, \end{aligned} \tag{11}$$

since the operation $\vartriangle$ is commutative and associative, and μ is subadditive on $\vartriangle$. We have also used the fact that $A \sim B$ iff $\mu(A \vartriangle B) = 0$ and similarly for $E \sim F$. Hence (11) implies that the value of μ is unchanged on equivalence classes of finite measure. But the same trivially holds if any of these sets has infinite measure. Thus $d(\cdot, \cdot)$ is unchanged, and if $\tilde{d}([A], [B]) = d(A, B)$, then $\tilde{d}$ is a distance, since $\tilde{d}([A], [B]) = 0$ iff $\varphi(\mu(A \vartriangle B)) = 0$, and this holds iff $\mu(A \vartriangle B) = 0$, so that $A \sim B$ or $[A] = [B]$. Thus $\tilde{\mathscr{M}} = \mathscr{M}_\mu / \sim$ is the space of equivalence classes, and $(\tilde{\mathscr{M}}, \tilde{d})$ is a complete metric space, as asserted in Remark 1.

If $\mu(\Omega) < \infty$, then evidently $\varphi(x) = x$ will satisfy our requirements. Finally, (10) implies for any $A, B \in \mathscr{M}_\mu$, $|\mu(A) - \mu(B)| \leqslant d(A, B)$ and hence μ is a uniformly continuous function on $\mathscr{M}_\mu$, since we now take $\varphi(x) = x$. If $d(E_1, E_2) \to 0$ and $d(F_1, F_2) \to 0$, then from the easy set operations with $\vartriangle$ we get the following two relations:

$$d(E_1 \,\eta\, F_1, E_2 \,\eta\, F_2) \leqslant d(E_1, E_2) + d(F_1, F_2), \qquad \eta = \cup, \cap, \tag{12}$$

which imply the uniform continuity of the operations of $\cup$ and $\cap$ on $(\mathscr{M}_\mu, d)$ at once, completing the proof.

The above result will be used later in some limit theorems. It should be noted how the availability of an outer measure μ on an abstract set Ω enabled us to introduce a metric topology on the space of subsets of Ω. This topology will generally be different from that of Ω if the latter has already a topology (e.g., $\Omega = \mathbb{R}^n$) and is usually coarser than the original one, since points are "collapsed" and only the elements of $\mathscr{M}_\mu$ are primary. We shall see later that there are other ways of considering a topology in an abstract measure space so as to be consistent (and dependent only on) with the measuring functions.

Let us also introduce here another concept.

Definition 2. Let (Ω, Σ, μ) be a (completed) measure space. If the associated (semi-)metric space (Σ, d) is separable, in that it has a countable dense family (in Σ) relative to the function d, then the measure μ(and hence the measure space itself) is called *separable*.

Using the separability of the metric space $\mathbb{R}^n$ and the regularity of Lebesgue measure μ on $\mathbb{R}^n$, one can show without difficulty that μ is separable. (See Exercise 3.) This concept is especially interesting when one studies the structure of function spaces over (separable) measure spaces, as will become evident in our work later (Section 4.5). In that study, one also sees the necessity for a distinction between finite and infinite measures, since without supplementary hypotheses some results on finite measure spaces do not hold in the nonfinite case. Moreover, distinctions among the σ-finite, localizable, and general cases have to be maintained.

EXERCISES

0. Let $\Omega = [0, 1]$, Σ = Borel σ-algebra of Ω, and μ = counting measure on Σ. Show that the metric space $(\tilde{\Sigma}, d)$ associated with (Σ, μ) is not compact (although Ω is compact as a subspace of $\mathbb{R}$). Does the situation improve if μ is Lebesgue measure? [Recall that a compact metric space is complete and separable.]

1. Let (Ω, Σ, μ) be a measure space and $A_n \in \Sigma$, $n \geqslant 1$, be sets such that $A_n \to A$ (i.e., $A = \limsup_n A_n = \liminf_n A_n$). Show that $\mu(A_n) \to \mu(A)$ if $\mu(\Omega) < \infty$, and the result may or may not be true if $\mu(\Omega) = +\infty$. Give examples illustrating both the latter possibilities.

2. Establish the relation (12), and complete the details on the uniform continuity of the binary operations $\cap$ and $\cup$, as stated in Proposition 1.

3. Let $\mathscr{A}$ be an algebra over a set Ω, and suppose it has a countable set of generators. Thus $\mathscr{A}$ is the smallest algebra containing the countable set $\{A_n: n \geqslant 1\}$, $A_n \subset \Omega$. If μ is a measure on $\mathscr{A}$, and μ^* is the generated outer measure by $(\mu, \mathscr{A})$, show that μ^* is separable. Taking $\mathscr{A} = \mathscr{E}$, the class of elementary sets of $\mathbb{R}^n$ (generated by the rectangles with rational coordinates), deduce that Lebesgue measure μ on $\mathbb{R}^n$ is separable.

4. The counting measure on $\Omega = [0, 1]$ and the trivial measure $\tilde{\mu}$, where $\tilde{\mu}(A) = 0$ if $A = \varnothing$ and $= +\infty$ if $A \neq \varnothing$, are not separable. Verify these statements.

5. If $\tilde{\mu}$ is a (finitely subadditive) Jordan outer measure as in Proposition 2.12, and $\mathscr{M}_{\tilde{\mu}}$ is the class of $\tilde{\mu}$-sets, then one can define a metric d_0 as in Proposition 1 above. Show that $(\mathscr{M}_{\tilde{\mu}}, d_0)$ is a (semi-)metric space but need not be complete. This can be verified if $\Omega = [0, 1]$ and $\tilde{\mu}$ is obtained from the length function as in (2.1), with $\tilde{\mu} = \mu_J$ there.

6. Let (Ω, Σ, μ) be a complete measure space. A set $A \in \Sigma$ is called an *atom* of μ if $\mu(A) > 0$ and $B \subset A$, $B \in \Sigma$ implies either $\mu(B) = 0$ or $\mu(A - B) = 0$. If there are no atoms, then μ is called a *nonatomic* or *diffuse* measure.

Verify that if $\mu(\Omega) < \infty$, then μ can have at most countably many atoms. Show, moreover, that the (semi-)metric space (Σ, d) is "convex" iff μ is diffuse. Here (Σ, d) convex means: for any A, B in Σ, there is a $C \in \Sigma$ such that if $d(A, B) > 0$, then $d(A, C) > 0$, $d(B, C) > 0$, and $d(A, B) = d(A, C) + d(C, B)$. [First show that the result holds if $\Omega = [0, a]$, $0 < a < \infty$, and μ is Lebesgue's length function. The general case is reduced to this case with the following theorem (can be proved using Section 10.1) of Carathéodory: If (Ω, Σ, μ) is as given let (Σ, d) be the associated complete (semi-)metric space. If $[0, \mu(\Omega)]$ is the finite interval with $m(\cdot)$ as the Lebesgue measure, and $\mathscr{M}$ is the corresponding Lebesgue σ-algebra and $(\mathscr{M}, \tilde{d})$ its associated (semi-)metric space, then there exists an isomorphism φ of Σ into $\mathscr{M}$ preserving countable unions, intersections, and complements, such that $\tilde{d}(\varphi(A), \varphi(B)) = d(A, B)$, and φ is onto iff μ is diffuse.]

7. If (Ω, Σ, μ) is as in the above exercise, (Σ, d) is its associated (semi-)metric space, and $A \in \Sigma$, consider the trace $\Sigma(A)$ and the restriction $d_A(\cdot)$ of d to $\Sigma(A)$. Show that $d(B \,\eta\, A, C \,\eta\, A) = d_E(B, C)$ where $\eta = \cup \Rightarrow E = A^c$, $\eta = \cap \Rightarrow E = A$, and $\eta = \vartriangle \Rightarrow E = \Omega$.

2.5. METRIC OUTER MEASURES

Motivated by another property of our basic model, the Lebesgue triple of $\mathbb{R}^n$, we consider an analysis of a class of outer measures defined on metric spaces. An early generalization of this was given by Hausdorff. Abstracting that construction, Carathéodory obtained an extension, of considerable interest in certain geometric studies, called metric outer measures. We include a brief discussion of both these results, since they are important in applications, and they show a reason for the considerations in the preceding section about non-(sigma-)finite measures which appear in analysis. Indeed, these results stand between the Lebesgue case and measures on general topological (even locally compact) spaces.

Since $\mathbb{R}^n$ has a metric, we start with its relation to the Lebesgue measure, which serves as a motivation for Carathéodory's general result on metric outer measures.

Proposition 1. *Let μ^* denote Lebesgue outer measure on $\mathbb{R}^n$, and A, B be a pair of nonempty subsets which are at a positive distance apart, i.e., if $\|\cdot\|$ stands for the metric or norm in $\mathbb{R}^n$, then the distance between A, B is positive:*

$$d(A, B) = \inf\{\|x - y\|: x \in A, y \in B\} > 0. \tag{1}$$

Then μ^* satisfies

$$\mu(A \cup B) = \mu^*(A) + \mu^*(B). \tag{2}$$

Proof. Since $\mathbb{R}^n$ is countably compact, we can cover it by a sequence $\{I_k: k \geq 1\} \subset \mathcal{S}$, the semiring of rectangles of arbitrarily small sides. Thus given $p > 0$, let $\mathcal{S}_p \subset \mathcal{S}$ be those rectangles with diameters not exceeding p [i.e., $\text{diam}(I_n) = \sup\{\|x - y\|: x, y \in I_n\} \leq p$], and let μ_p^* be the generated outer measure by the pair $(m(\cdot), \mathcal{S}_p)$, where $m(\cdot)$ is the volume function. As usual, μ^* is the generated outer measure by $(m(\cdot), \mathcal{S})$. If $A \subset \mathbb{R}^n$, then $\mu^*(A) \leq \mu_p^*(A)$, since the infimum for μ^* is taken on a larger class $\mathcal{S}$. We claim that there is equality here, or that $\mu_p^*(A) \leq \mu^*(A)$. This is to be shown only for $\mu^*(A) < \infty$. Then for each $\epsilon > 0$, there is a covering $\{I_k: k \geq 1\} \subset \mathcal{S}$ such that

$$\mu^*(A) + \frac{\epsilon}{2} \geq \sum_{k=1}^{\infty} m(I_k), \qquad A \subset \bigcup_{k=1}^{\infty} I_k. \tag{3}$$

Consider $\bar{I}_k$, which is closed and bounded, i.e., compact. But it can also be covered by a finite collection of the interiors $\mathring{I}_{kj}$ of $I_{kj} \in \mathcal{S}_p$, as noted in Theorem 1.2.8: $\bar{I}_k \subset \bigcup_{j=1}^{N_k} \mathring{I}_{kj}$, such that

$$m(I_k) + \frac{\epsilon}{2^{k+1}} > \sum_{j=1}^{N_k} m(I_{kj}). \tag{4}$$

Hence

$$\sum_{k=1}^{\infty} \sum_{j=1}^{N_k} m(I_{kj}) < \sum_{k=1}^{\infty} m(I_k) + \frac{\epsilon}{2}$$

$$< \mu^*(A) + \epsilon, \qquad \text{by (3)}. \tag{5}$$

But $\{I_{jk}: 1 \leq j \leq N_k,\ k \geq 1\} \subset \mathcal{S}_p$ is a particular cover for A, and taking infimum on all such covers, we get $\mu_p^*(A) \leq \mu^*(A) + \epsilon$, by (5). Since $\epsilon > 0$ is arbitrary, we deduce that $\mu^*(A) = \mu_p^*(A)$ for each $p > 0$.

Now let A, B be any pair of subsets of $\mathbb{R}^n$ at a distance $p > 0$ apart. Since μ^* is subadditive [$\mu^*(A \cup B) \leq \mu^*(A) + \mu^*(B)$], we show the opposite inequality, with the positive distance hypothesis. This is required only for $\mu^*(A \cup B) < \infty$, the other case being true. Then, to use the above paragraph, let $\{I_k: k \geq 1\} \subset \mathcal{S}_p$ be a covering of $A \cup B$ such that

$$\mu^*(A \cup B) + \epsilon > \sum_{k=1}^{\infty} m(I_k). \tag{6}$$

Note that no I_k can contain a point of A and also a point of B, for then the distance between these two points must be $> p$ and this is impossible for the elements I_k of $\mathcal{S}_p$. So we can divide the covering into two exclusive classes

$\{I_k': k \geqslant 1\}$ and $\{I_k'': k \geqslant 1\}$ which cover A and B respectively. Hence

$$\mu^*(A) + \mu^*(B) \leqslant \sum_{k=1}^{\infty} m(I_k') + \sum_{k=1}^{\infty} m(I_k'')$$

$$= \sum_{k=1}^{\infty} m(I_k) \leqslant \mu^*(A \cup B) + \epsilon, \qquad \text{by (6)}.$$

Since $\epsilon > 0$ is arbitrary, this shows that (2) must be true, completing the proof.

Based on this property of the Lebesgue measure, we may now introduce a general concept as:

Definition 2. Let (Ω, d) be a metric space with $d(\cdot, \cdot)$ as its metric function. Then $\mu: 2^{\Omega} \to \overline{\mathbb{R}}^+$ is called a *metric outer measure* if it is an outer measure and if, for any nonempty subsets A, B of Ω which are at a positive distance, i.e.,

$$d(A, B) = \inf\{d(x, y): x \in A,\ y \in B\} > 0,$$

we always have

$$\mu(A \cup B) = \mu(A) + \mu(B). \tag{7}$$

The preceding proposition implies that Lebesgue's measure is a metric outer measure. Moreover, we can assert the same property for all generated measures on (Ω, d), just as for the Lebesgue μ^*. A precise statement is given by the following basic result.

Theorem 3 (Carathéodory). *Let (Ω, d) be a metric space, and $\mathscr{A}$ be a class of subsets including $\varnothing$, such that each subset of Ω can be covered by a countable class of sets of $\mathscr{A}$. Let $\alpha: \mathscr{A} \to \overline{\mathbb{R}}^+$ be a mapping such that $\alpha(\varnothing) = 0$, $\mathscr{A}_n = \{A \in \mathscr{A}: \operatorname{diam}(A) = \sup\{d(x, y): x, y \in A\} \leqslant 1/n\}$. If $\mathscr{A}_n$ is also a sequential covering of Ω, let μ_n^* and μ^* be the generated measures by $(\alpha, \mathscr{A}_n)$ and $(\alpha, \mathscr{A})$ respectively. If $\mu_n^*(B) \leqslant \alpha(B)$ for each $B \in \mathscr{A}$, $n \geqslant 1$, then $\mu^* = \mu_n^*$ and μ^* is a metric outer measure on Ω. Moreover, for any metric outer measure μ, every Borel set of Ω is μ-measurable.*

Proof. The first part of the proof that μ is a metric outer measure is identical with that of Proposition 1, since (4) is now an assumption from which (5) follows, where we use the fact that $\mu_n(A) \leqslant \alpha(A)$. So we leave the details to the reader (Exercise 1). Here we establish that each Borel set is μ-measurable.

Since a Borel σ-algebra is generated by the class of all open or equivalently closed sets and the μ-measurable sets form a σ-algebra, it is sufficient to show that each generator is μ-measurable. So let $F \subset \Omega$ be a closed set. Then we need to show that

$$\mu(T) = \mu(T \cap F) + \mu(T \cap F^c), \qquad T \subset \Omega. \tag{8}$$

Since $T \cap F \subset F$, $T \cap F^c \subset F^c$, and their union is T, we show more generally that for any $A \subset F$, $B \subset F^c$,

$$\mu(A \cup B) = \mu(A) + \mu(B), \tag{9}$$

and this will prove (8), if we set $A = T \cap F$, $B = T \cap F^c$.

Let $A_k = \{x \in A: d(x, F^c) \geqslant 1/k\} \subset A_{k+1} \subset A$. Since $B \subset F^c$, it is clear that A_k and B are at a distance at least $1/k$ units apart. Consequently by (7) we get

$$\mu(A \cup B) \geqslant \mu(A_k \cup B) = \mu(A_k) + \mu(B), \qquad k \geqslant 1. \tag{10}$$

Since $\lim_k \mu(A_k) \leqslant \mu(A)$ and $\mu(A \cup B) \leqslant \mu(A) + \mu(B)$, to prove (9) it is sufficient to show in (10) that $\lim_k \mu(A_k) = \mu(A)$. Now for this it is only required to prove $\mu(A) \leqslant \lim_k \mu(A_k)$ when the right side is finite. Note that this would have followed immediately from Theorem 2.10 if μ were a generated metric measure, as in the first part, when $A = \bigcup_{n=1}^{\infty} A_n$ is shown. However, we are not assuming that μ is also generated, and so have to proceed by a different method which involves a trick.

First note that each point of $A \subset F$ is at a distance $\geqslant k^{-1}$ from F^c for some $k \geqslant 1$, so that $A \subset \bigcup_{k=1}^{\infty} A_k$ and hence $A = \bigcup_{k=1}^{\infty} A_k$. Let $C_k = A_{k+1} - A_k$. Then (the reader should draw a picture) by the triangle inequality and the fact that each point of A_k is at a distance $\geqslant 1/k$ from B, we get

$$\frac{1}{k} \leqslant d(A_k, B) \leqslant d(A_k, C_{k+1}) + d(C_{k+1}, B), \tag{11}$$

on taking suitable infima separately on A_k, C_{k+1}, and B. Hence

$$\begin{aligned} d(A_k, C_{k+1}) &\geqslant \frac{1}{k} - d(C_{k+1}, B) \\ &\geqslant \frac{1}{k} - d(A_{k+1}, B) = \frac{1}{k} - \frac{1}{k+1} > 0. \end{aligned} \tag{12}$$

Since $A = \lim_k A_k = A_k \cup (\bigcup_{j \geqslant k} C_j)$, by the σ-subadditivity of μ

$$\mu(A) \leqslant \mu(A_k) + \sum_{j=k}^{\infty} \mu(C_j), \tag{13}$$

and we show now, with a new trick, that the series on the right is convergent. It then follows on letting $k \to \infty$ in (13) that $\mu(A) \leqslant \lim_k \mu(A_k)$, as desired.

To prove the convergence of the series in (13), we consider all the even and odd suffixed terms and show that their sums converge. This is needed, because one only has $d(C_{j-1}, C_{j+1}) > 0$ for all j, which follows from (12), since $C_{j-1} \subset A_j$ by definition. Hence, using (7) for μ, we get

$$\sum_{j=1}^{n-1} \mu(C_{2j}) = \mu\left(\bigcup_{j=1}^{n-1} C_{2j}\right) \leqslant \mu(A_{2n}) \leqslant \mu(A) < \infty$$

and also

$$\sum_{j=0}^{n-1} \mu(C_{2j+1}) = \mu\left(\bigcup_{j=0}^{n-1} C_{2j+1}\right) \leqslant \mu(A_{2n+1}) \leqslant \mu(A) < \infty.$$

Hence adding these two series and letting $n \to \infty$, we get

$$\sum_{j \geqslant 1} \mu(C_j) \leqslant 2\mu(A) < \infty. \tag{14}$$

This and (13) show that (9) holds, and hence the proof is complete.

An important application of the above result is the following, which will also demonstrate its generality.

Proposition 4. *Let* $\varphi: \mathbb{R}^+ \to \overline{\mathbb{R}}^+$ *be a nondecreasing (left continuous) function,* $\varphi(x) = 0$ *iff* $x = 0$, *and let* (Ω, d) *be a metric space. For each* $\epsilon > 0$, *and for each* $A \subset \Omega$, *define*

$$H_\varphi^\epsilon(A) = \inf\left\{\sum_{i=1}^{\infty} \varphi(\delta_1(A_i)) : \delta_1(A_i) \leqslant \epsilon,\ A \subset \bigcup_{i=1}^{\infty} A_i\right\}, \tag{15}$$

where $\delta_1(A) = \sup\{d(x, y): x, y \in A\}$ *is the diameter of* A. *Then for* $0 < \epsilon' < \epsilon$, $H_\varphi^\epsilon(A) \leqslant H_\varphi^{\epsilon'}(A)$, $A \subset \Omega$, *so that* $H_\varphi(A) = \lim_{\epsilon \downarrow 0} H_\varphi^\epsilon(A)$ *exists, and* $H_\varphi(\cdot)$ *is a metric outer measure on* Ω. *Moreover,* H_φ *is outer regular relative to* $\mathscr{U}_\delta$, *where* $\mathscr{U}$ *is the class of open sets in* Ω *(so that* $\mathscr{U}_\delta$ *is the class of* G_δ*-sets of* Ω*), in the sense that for each* $A \subset \Omega$ *there is a* G_δ *set* $B \supset A$ *such that* $H_\varphi(A) = H_\varphi(B)$.

Proof. The first part leading to the assertion that $H_\varphi(\cdot)$ is a metric outer measure is similar to that of Proposition 1 and is omitted (see Exercise 2).

Regarding the outer regularity, since H_φ^ϵ is a generated outer measure, Theorem 2.10 guarantees that H_φ^ϵ is $\mathscr{U}_{\sigma\delta}(= \mathscr{U}_\delta)$-regular. To show that the same property holds for H_φ, let $n \geqslant 1$ and $\epsilon = 1/n$. Thus there is a $B_n \in \mathscr{U}_\delta$ with $H_\varphi^{1/n}(A) = H_\varphi^{1/n}(B_n)$. So $B = \bigcap_{n=1}^{\infty} B_n \in \mathscr{U}_\delta$, $B \supset A$. Now given $\eta > 0$, choose $n \geqslant 1$ such that $1/n < \eta$, and hence

$$H_\varphi^\eta(B) \leqslant H_\varphi^{1/n}(B) \leqslant H_\varphi^{1/n}(B_n)$$

$$= H_\varphi^{1/n}(A), \qquad \text{since } H_\varphi^\epsilon(\cdot) \text{ is monotone}$$

$$\leqslant H_\varphi(A), \qquad \text{since } H_\varphi^\epsilon(A)\uparrow \text{ as } \epsilon\downarrow 0.$$

Letting $\eta \downarrow 0$, and then using the fact that $B \supset A$, one has

$$H_\varphi(B) \leqslant H_\varphi(A) \leqslant H_\varphi(B),$$

and the regularity is established.

Taking $\varphi(x) = x^\alpha$, $\alpha > 0$, in the above and writing H_φ as H_α, we may state the following concept (here $\alpha > 0$ is real).

Definition 5. The metric outer measure $H_\alpha(\cdot)$ on a metric space (Ω, d), given by the above proposition, is called the *Hausdorff α-dimensional* (outer) *measure*. For any set $A \subset \Omega$, the *Hausdorff dimension* β of A is defined as

$$\beta = \beta(A) = \sup\{\alpha > 0 \colon H_\alpha(A) = +\infty\}. \tag{16}$$

Example. To understand these concepts, we shall calculate $\beta(C)$, where $C \subset [0,1]$ is the Cantor set, and show that $\beta(C) = (\log 2)/(\log 3)$.

Recalling the definition of the Cantor set from Exercise 1.2.5, if F_n is the set of 2^n closed intervals each of length 3^{-n} obtained at the nth stage by removing 2^{n-1} middle thirds from $[0,1]$, then $C = \bigcap_{n=1}^\infty F_n$. Since the left-over pieces F_{nj} at the nth stage are intervals we see that each of their lengths is the same as their diameters. Thus $m(F_{nj}) = \delta_1(F_{nj})$ and if $\epsilon > 3^{-n}$, then $\{F_{nj}, 1 \leqslant j \leqslant 2^n\}$ is an ϵ-covering of C, so that

$$H_\alpha^\epsilon(C) \leqslant \sum_{j=1}^{2^n} \left[\delta_1(F_{nj})\right]^\alpha = 2^n \times 3^{-n\alpha}.$$

Thus

$$H_\alpha(C) = \lim_{\epsilon\downarrow 0} H_\alpha^\epsilon(C) \leqslant \lim_{n\to\infty}\left(\frac{2}{3^\alpha}\right)^n = 0, 1, \text{ or } +\infty,$$

according as $2 < 3^\alpha$, $2 = 3^\alpha$, or $2 > 3^\alpha$. Thus if $\alpha_0 = (\log 2)/(\log 3)$, then for

$0 < \alpha < \alpha_0$ we have

$$H_\alpha^\epsilon(C) = \inf\left\{ \sum_{n=1}^{\infty} [\delta_1(U_n)]^\alpha, \delta_1(U_n) \leqslant \epsilon, C \subset \bigcup_{n=1}^{\infty} U_n \right\}$$

$$\geqslant \epsilon^{\alpha-\alpha_0} \inf\left\{ \sum_{n=1}^{\infty} (\delta_1(U_n))^{\alpha_0}, \delta_1(U_n) \leqslant \epsilon, C \subset \bigcup_{n=1}^{\infty} U_n \right\}$$

$$= \epsilon^{\alpha-\alpha_0} H_{\alpha_0}(C) \to +\infty \qquad \text{as} \quad \epsilon \downarrow 0.$$

Hence

$$\beta(C) = \sup\{\alpha > 0: H_\alpha(C) = \infty\} = \alpha_0.$$

Thus C has Hausdorff dimension α_0, as asserted.

The point of this example is that using the ideas of metric outer measures, one can magnify the structure of "small sets" such as the Cantor set C, which has Lebesgue measure zero. The construction given by Theorem 3 yields a number of distinct classes of metric measures by specializing φ and the basic class $\mathscr{A}$ of sets. Now for $A \in \mathscr{A}$, we replace diam(A) by φ(diam(A)). Here are two examples in addition to the α-dimensional Hausdorff measure of the preceding proposition:

1. Let $\mathscr{A}$ be the family of all closed balls $B_{a,\delta} = \{x \in \Omega: d(a, x) \leqslant \delta\}$, and $\varphi(x) = x^p$, $p > 0$. The resulting construct is the *p-dimensional spherical* measure over Ω, a metric space.
2. Let $\Omega = \mathbb{R}^n$, and $\alpha(\cdot)$ (of Theorem 3) be a p-dimensional spherical measure. If $\mathscr{A}$ is the class of open convex subsets of Ω, the construction of Theorem 3 gives a *p-dimensional Carathéodory measure* on Ω. This can be further extended.

These and many other measures of this type, using the geometry of the underlying metric spaces, are treated in, e.g., Federer (1969). For a special account, with applications of Hausdorff measures, one may consult Rogers (1970). Since we are interested only in the basic theory here, these specializations, which naturally demand much higher preparation, will not be considered further. But they are important is such studies as variational calculus, surface areas, random currents, and stochastic geometry, as well as parts of statistical inference.

EXERCISES

0. Let μ_α be a Hausdorff α-dimensional outer measure on $2^{\mathbb{R}}$, $0 < \alpha < \infty$. If $A \subset \mathbb{R}$, $\mu_\alpha(A) < \infty$, verify that for all $\beta > \alpha$, $\mu_\beta(A) = 0$. Does it follow that for $0 < \delta < \alpha$, $\mu_\delta(A) = +\infty$? [Recall that $2^{\mathbb{R}}$ is the power set of $\mathbb{R}$.]

1. Complete the details of proof of the first part of Theorem 3.

2. Prove that $H_\varphi(\cdot)$ is a metric outer measure, as asserted in Proposition 4.

3. If $A_n \subset A_{n+1} \subset \Omega$, $n \geqslant 1$, is a sequence and $H_\varphi(\cdot)$ is the Hausdorff φ-measure as given in Proposition 4, show that

$$\lim_k \lim_n H_\varphi^{1/k}(A_n) = H_\varphi\Big(\lim_n A_n\Big),$$

where $H_\varphi^\varepsilon(\cdot)$ is given by (15).

4. Using the argument of the proof of Theorem 3, if $A_n \subset A_{n+1} \subset \Omega$ and μ is any metric outer measure on Ω (instead of the one given in Exercise 3), show that

$$\lim_n \mu(A_n) = \mu\Big(\lim_n A_n\Big),$$

whenever each A_{n-1} and A_n^c, $n \geqslant 1$, are a positive distance apart.

5. Show that the p-dimensional spherical measure given above is the metric outer measure and is the same whether the class $\mathscr{A}$ in its definition consists of closed or open (strict inequality between d and δ) balls.

6. If $A \subset \mathbb{R}^n$ and $\beta(A)$ is its Hausdorff dimension, show that $H_p(A) = 0$ for $p > \beta(A)$ and $\beta(A) \leqslant n$. [Thus for the Cantor set $\beta(C) = (\log 2)/(\log 3) < 1$.] Also verify that every countable set in $\mathbb{R}^n$ has Hausdorff dimension zero, and if $n = 1$ then $H_1(\cdot)$ is the same as Lebesgue outer measure μ^*. [If $n \geqslant 2$, then $c_1 H_n(\cdot) \leqslant \mu^*(\cdot) \leqslant c_2 H_n(\cdot)$ for some constants $c_1, c_2 > 0$.]

2.6. LEBESGUE-STIELTJES MEASURES

We now present an important specialization of Radon measures considered in the last part of Section 3. This is the Lebesgue-Stieltjes measure on $\mathbb{R}^n$ induced by monotone functions. There are some particular problems here, since a monotone function can have jump discontinuities on $\mathbb{R}$, and this possibility becomes complicated in $\mathbb{R}^n$, $n > 1$. But these measures are basic in such areas as probability and statistics, where they appear as "distribution functions", and in optimization problems, among others. To understand the necessary normalizations for a trouble-free calculus, we first treat the case on $\mathbb{R}$, and then show how it can be extended to $\mathbb{R}^n$.

Let $f: \mathbb{R} \to \mathbb{R}$ be a nondecreasing function. It can have at most a countable number of jump discontinuities. For, with each jump of f in $\mathbb{R}$ we can associate a nonempty open interval in its range, and the monotonicity property implies that these intervals may be taken disjoint. Since each interval contains a rational number, and the rational numbers form a countable set, our statement on the set of discontinuities follows. However, f need not be left or right continuous at each discontinuity. So we associate f^* and f_*, the *normalized* (or regularized) *functions*, with f as follows:

$$f^*(x) = \lim_{t \downarrow 0} f(x+t), \quad f_*(x) = \lim_{t \downarrow 0} f(x-t), \qquad x \in \mathbb{R}. \tag{1}$$

Then f_* (f^*) is left (right) continuous and coincides with f at all continuity points, which form a dense set. We introduce measures with these three functions and present their properties.

Thus let $\mathscr{I}$ be the class of all open intervals of $\mathbb{R}$. Define a nonnegative set function α_f associated with f on $\mathscr{I}$, and let μ_f be the generated outer measure by the pair $(\alpha_f, \mathscr{I})$ [Theorem 1(iii) below gives alternate definitions, since $\alpha_f(I) = \mu_f(I)$]:

$$\alpha_f(I) = f(b) - f(a), \qquad I = (a, b) \in \mathscr{I},$$

$$\mu_f(A) = \inf\left\{ \sum_{k=1}^{\infty} \alpha_f(I_k) : A \subset \bigcup_{k=1}^{\infty} I_k,\ I_k = (a_k, b_k) \in \mathscr{I} \right\}. \tag{2}$$

It is clear that $\alpha_f(\cdot)$ is additive on disjoint intervals of $\mathscr{I}$, and then by Theorem 2.10, μ_f is a generated outer measure such that $\mathscr{I} \subset \mathscr{M}_{\mu_f}$ and $\mu_f | \mathscr{M}_{\mu_f}$ is a measure which extends α_f. This restriction (to $\mathscr{M}_{\mu_f}$) is called the *Lebesgue-Stieltjes measure* on $\mathbb{R}$, induced by the nondecreasing real function f. It is useful to note that if $f(x) = x$, then μ_f is simply the Lebesgue measure considered before [cf. also Exercise 1.2.6, and Eq. (20) in Section 2]. Another interesting feature is that the μ_f is a metric outer measure on $(\mathbb{R}, d)$, where $d(\cdot) = |\cdot|$ is the usual absolute value metric on $\mathbb{R}$. Indeed, the only point to verify here is $\mu_f(A \cup B) = \mu_f(A) + \mu_f(B)$ for any subsets A, B of $\mathbb{R}$ which are at a positive distance apart. Since μ_f is subadditive, it suffices to show that $\mu_f(A \cup B) \geqslant \mu_f(A) + \mu_f(B)$ whenever $\mu_f(A \cup B) < \infty$. Let $0 < \epsilon < d(A, B)$, the distance between A and B. Choose $I_k = (a_k, b_k) \in \mathscr{I}$ of length less than ϵ such that $A \cup B \subset \bigcup_{k=1}^{\infty} I_k$, and

$$\sum_{k=1}^{\infty} \alpha_f(I_k) < \mu(A \cup B) + \epsilon.$$

In fact, if any $J = (a, b)$ has length $> \epsilon$, then by the compactness of $[a, b]$ it is $\subset \bigcup_{i=1}^{n} J_i$, where $a = \xi_0 < \xi_1 < \cdots < \xi_n = b$, ξ_i are continuity points of f for $1 \leqslant i \leqslant n-1$, $J_i = (\xi_{i-1}, \xi_i)$ has length $< \epsilon$, $1 \leqslant i \leqslant n$, and $\sum_{i=1}^{n} \alpha_f(J_i) < \alpha_f(J) + \epsilon/2^{n+1}$. Replace each such J by this finite subcollection, and the new family satisfies the above requirements. Since continuity points of f are everywhere dense in its domain, this can be done. Hence the covering $\{I_k: k \geqslant 1\}$ splits into two parts, one for A and the other for B. Consequently

$$\mu_f(A) + \mu_f(B) \leqslant \sum_{k=1}^{\infty} \alpha_f(I_k) < \mu(A \cup B) + \epsilon,$$

and $\epsilon > 0$ being arbitrary, our assertion holds.

Denoting by μ_f, μ_{f_*}, and μ_{f^*} the outer measures generated by f, f_*, and f^*, we can now present the properties of these measures and the important fact that all of them induce the same Lebesgue-Stieltjes measure as follows.

Theorem 1. *Let $f: \mathbb{R} \to \mathbb{R}$ be nondecreasing and μ_f, μ_{f_*}, μ_{f^*} be the associated outer measures. If a, b are points of $\mathbb{R}$, $a \leqslant b$, then:*

(i) $\mu_f(\{a\}) = f^*(a) - f_*(a)$, $\{x: \mu_f(\{x\}) > 0\} \subset \mathbb{R}$ *is at most countable, and thus* $\mu_f(\{x_0\}) = 0$ *iff f is continuous at* x_0.

(ii) *All open sets and hence all Borel sets are μ_f-measurable.*

(iii) $\mu_f([a, b]) = f^*(b) - f_*(a)$, $\mu_f((a, b)) = f_*(b) - f^*(a)$, $\mu_f([a, b)) = f_*(b) - f_*(a)$, $\mu_f((a, b]) = f^*(b) - f^*(a)$.

(iv) *If $\mathscr{G}$ is the class of all open sets and $\mathscr{C}$ the class of all compact sets of $\mathbb{R}$, then*

$$\mu_f(A) = \inf\{\mu_f(G): G \supset A, G \in \mathscr{G}\}, \qquad A \subset \mathbb{R},$$

$$\mu_f(B) = \sup\{\mu_f(C): C \subset B, C \in \mathscr{C}\}, \qquad B \in \mathscr{M}_{\mu_f},$$

so that μ_f is outer regular (for $\mathscr{G}$), and also inner regular (for $\mathscr{C}$).

(v) $\mu_f = \mu_{f_*} = \mu_{f^*}$, *i.e., the three outer measures induced by f, f_*, f^* are identical.*

(vi) *Conversely, if μ is any outer measure on $\mathbb{R}$ which has the properties (with the above notation)* (a) $\mathscr{G} \subset \mathscr{M}_\mu$, (b) $\mu(C) < \infty$ *for* $C \in \mathscr{C}$, (c) μ *is outer regular for $\mathscr{G}$, then $\mu = \mu_f$ for some nondecreasing $f: \mathbb{R} \to \mathbb{R}$, and hence it is also inner regular, so that it is automatically a Lebesgue-Stieltjes (hence a Radon) measure.*

Proof. (i): Since $\{a\} \subset (a - 1/n, a + 1/n)$, we have

$$\mu_f(\{a\}) \leqslant \alpha_f\left(\left(a - \frac{1}{n}, a + \frac{1}{n}\right)\right) = f\left(a + \frac{1}{n}\right) - f\left(a - \frac{1}{n}\right).$$

Hence $\mu_f(\{a\}) \leqslant f^*(a) - f_*(a)$, on letting $n \to \infty$. For the opposite inequality, given $\epsilon > 0$, there exists an n_0 such that $n \geqslant n_0 \Rightarrow$

$$\mu_f(\{a\}) + \epsilon > \alpha_f\left(\left(a - \frac{1}{n}, a + \frac{1}{n}\right)\right) \geqslant f\left(a + \frac{1}{n}\right) - f\left(a - \frac{1}{n}\right).$$

Since $\epsilon > 0$ is arbitrary, this gives $\mu_f(\{a\}) \geqslant f^*(a) - f_*(a)$, and hence equality. For the second part, consider $\{x \in (a, b): \mu_f(\{x\}) \geqslant \epsilon\}$. If $a = x_0 < x_1 < \cdots < x_{N+1} = b$ and $\mu_f(\{x_i\}) \geqslant \epsilon$, $i = 1, \ldots, N$, then we can find continuity points a_i, b_i of f such that $a < a_i < x_i < b_i < a_{i+1} < b$ (by density of the continuity set of f) satisfying $\mu_f(\{x_i\}) \leqslant f(b_i) - f(a_i)$, using the first part. Thus

$$f(b) - f(a) \geqslant \sum_{i=1}^{N} [f(b_i) - f(a_i)]$$

$$\geqslant \sum_{i=1}^{N} \mu_f(\{x_i\}) \geqslant \epsilon N.$$

Hence $N \leqslant [f(b) - f(a)]/\epsilon$, so that the set of points in (a, b) having a μ_f-mass of size at least ϵ is finite. But $\mathbb{R}$ can be covered by a countable set of such intervals. So it follows that

$$\{x: \mu_f(\{x\}) > 0\} = \bigcup_{m \geqslant 1} \bigcup_{n \geqslant 1} \left\{x \in (-n, n): \mu(\{x\}) > \frac{1}{m}\right\},$$

is at most countable. Since $f^*(a) = f_*(a)$ iff a is a continuity point of f, $\mu_f(\{a\}) = 0$ iff a is a continuity point of f. So (i) holds.

(ii): We have already seen that μ_f is a metric outer measure and then, by Theorem 5.3, every Borel set and hence every open (and closed) set is μ_f-measurable.

(iii): To establish the formulas, since $[a, b] \subset (a - 1/n, b + 1/n)$, one has

$$\mu_f([a, b]) \leqslant \alpha_f\left(\left(a - \frac{1}{n}, b + \frac{1}{n}\right)\right) = f\left(b + \frac{1}{n}\right) - f\left(a - \frac{1}{n}\right).$$

Since n is arbitrary, we get $\mu_f([a, b]) \leqslant f^*(b) - f_*(a)$. The opposite inequality needs some computation.

Thus let $\{(a_i, b_i), i \geqslant 1\}$ be an open covering of $[a, b]$, a compact interval. So we can extract a finite subcover $\{(a_i, b_i): 1 \leqslant i \leqslant N\}$ (say). It is claimed that

$$f^*(b) - f_*(a) \leqslant \sum_{i=1}^{N} \alpha_f((a_i, b_i)). \tag{3}$$

If $N = 1$, this is true because $(a_1, b_1) \supset [a, b]$ implies $(a - \epsilon, b + \epsilon) \subset (a_1, b_1)$ for a suitable $\epsilon > 0$, so that $f^*(b) - f_*(a) \leqslant f(b_1) - f(a_1)$. To use induction, suppose (3) is true for $N = m$. In $\{(a_i, b_i): 1 \leqslant i \leqslant m + 1\}$, let (a_{m+1}, b_{m+1}) be the last interval which thus contains b. In this case $[a, a_{m+1}]$ is covered by the other m intervals, so that by the inductive hypothesis,

$$f^*(a_{m+1}) - f_*(a) \leqslant \sum_{i=1}^{m} \alpha_f((a_i, b_i)). \tag{4}$$

But $[a_{m+1} + \epsilon', b] \subset (a_{m+1}, b_{m+1})$ for $\epsilon' > 0$ small enough. Then $(a_{m+1} + \epsilon', b + \epsilon') \subset (a_{m+1}, b_{m+1})$, so that on letting $\epsilon' \downarrow 0$,

$$f^*(b) - f^*(a_{m+1}) \leqslant \alpha_f((a_{m+1}, b_{m+1})). \tag{5}$$

Adding (4) and (5), one has (3). Hence

$$f^*(b) - f_*(a) \leqslant \sum_{i \geqslant 1} \alpha_f((a_i, b_i)). \tag{6}$$

Taking infimum of all such coverings, we get $f^*(b) - f_*(a) \leqslant \mu_f([a, b])$, so that with the earlier inequality in the opposite direction, the desired equality follows.

Since $\{a\}$ is a closed (in fact a G_δ-) set, points are μ_f-measurable and by (i) have finite mass. Since $(a, b) = [a, b] - (\{a\} \cup \{b\})$,

$$\begin{aligned} \mu_f((a, b)) &= \mu_f([a, b]) - \mu_f(\{a\}) - \mu_f(\{b\}) \\ &= f^*(b) - f_*(a) - [f^*(a) - f_*(a)] - [f^*(b) - f_*(b)] \\ &= f_*(b) - f^*(a). \end{aligned}$$

The other formulas are similar, and (iii) holds.

(iv): Since μ_f is a generated measure by $(\alpha_f, \mathscr{G})$, by Theorem 2.10 it follows that μ_f is $\mathscr{G}_\sigma(= \mathscr{G})$-outer regular. The inner regularity of μ_f, as shown in Theorem 2.17(ii), depends only on the outer regularity of μ_f, which is now noted to be true. Hence by that proof, the inner regularity of μ_f follows, and so (iv) is true as stated.

(v): Let $g = f^*$. Then $g: \mathbb{R} \to \mathbb{R}$ is increasing and right continuous. By (iii) the generated Lebesgue-Stieltjes measure μ_g can be evaluated for $(a, b]$ to obtain the following (for μ_f and μ_g):

$$\begin{aligned} \mu_f((a, b]) &= f^*(b) - f^*(a) = g(b) - g(a) \\ &= g^*(b) - g^*(a) = \mu_g((a, b]). \end{aligned} \tag{7}$$

This means μ_f and μ_g agree on the semiring $\mathscr{S}$ of such intervals, and hence by the Hahn extension theorem (Theorem 3.1), they agree on $\sigma(\mathscr{S})$, the Borel σ-algebra of $\mathbb{R}$. Since α_f and α_g are σ-finite on $\mathscr{S}$, it also follows from Theorem 3.1 that $\mu_f = \mu_g$ on $\mathscr{M}_{\mu_f}$ $(= \mathscr{M}_{\mu_g})$, so that $\mu_f = \mu_g$. Similarly, considering f and f_*, we conclude that $\mu_f = \mu_{f_*}$ $(= \mu_{f^*})$, and all are the same outer measures.

(vi): Finally, for the converse part, given an arbitrarily fixed $a_0 \in \mathbb{R}$, define $f: \mathbb{R} \to \mathbb{R}$, for μ, by the equation

$$f(x) = \begin{cases} \mu([a_0, x)) & \text{if} \quad x \geqslant a_0, \\ -\mu([x, a_0)) & \text{if} \quad x < a_0. \end{cases}$$

Then $x_1 \leqslant x_2$ implies, after a simple computation about the location of a_0 relative to x_1, x_2,

$$\begin{aligned} 0 \leqslant \mu([x_1, x_2)) &= \mu([a_0, x_2)) + \mu([x_1, a_0)) \\ &= f(x_2) - f(x_1). \end{aligned}$$

So f is nondecreasing, and we have for $x > x_1$

$$\begin{aligned} \infty > f(x) - f(x_1) &= \mu([x_1, x)) \\ &= \lim_{h_n \downarrow 0} \mu([x_1, x - h_n)), \qquad \text{by Proposition 1.1} \\ &= \lim_{h_n \downarrow 0} [f(x - h_n) - f(x_1)]. \end{aligned} \tag{8}$$

Hence, f is left continuous. So if μ_f is the Lebesgue-Stieltjes outer measure generated by $(\alpha_f, \mathscr{G})$, then $\mu_f = \mu$ on $\mathscr{S}$, and so everywhere, as noted above. Consequently μ is automatically inner regular, and hence a Radon measure on $\mathbb{R}$. This proves the theorem completely.

Remark. Because of parts (iii) and (v) of the above theorem, it is convenient to replace a monotone function f by its left (or right) continuous modification. This is the main reason why in such areas as probability and statistics a monotone nondecreasing bounded function is assumed to be left (or right) continuous. The resulting formulas take a simpler form.

How does one proceed in $\mathbb{R}^n$, $n \geqslant 2$? Here a new complication is that $f: \mathbb{R}^n \to \mathbb{R}$ is not necessarily monotone if it is so in each coordinate. Thus a more careful definition is needed. We indicate the necessary modifications and the corresponding result which extends Theorem 1 to $\mathbb{R}^n$.

Recall that a function $f: \mathbb{R} \to \mathbb{R}$ is monotone nondecreasing if its *increment* on each internal is nonnegative, i.e., $f(I) = f(b) - f(a) \geqslant 0$ for $I = (a, b)$. To see how this extends, consider $g: \mathbb{R}^2 \to \mathbb{R}$. If I is a rectangle, $I = \{(x, y): a_1 < x \leqslant b_1, a_2 < y \leqslant b_2\}$, then one sees (by drawing a picture)

$$g(I) = g(b_1, b_2) - g(a_1, b_2) - g(b_1, a_2) + g(a_1, a_2). \tag{9}$$

Thus we say that g is nondecreasing if $g(I) \geqslant 0$ for each rectangle I in $\mathbb{R}^2$. To generalize this to n dimensions, $n > 2$, note that (9) can be viewed alternatively as follows. Let $\tilde{g}(a) = g(a, b_2) - g(a, a_2)$, the increment in the second coordinate. Next treating $\tilde{g}(\cdot)$ as a function of one variable, we get $g(I) = \tilde{g}(b_1) - \tilde{g}(a_1)$. Thus we may consider the increment of $g: \mathbb{R}^n \to \mathbb{R}$ over a rectangle as n distinct *increment functions* differenced step by step using the untreated variables. Thus one gets after an elementary computation, for $I = \{(x_1, \ldots, x_n): a_i < x_i \leqslant b_i,\ i = 1, \ldots, n\}$, the following:

$$g(I) = g(b_1, \ldots, b_n) - \sum_{i=1}^{n} g(b_1, \ldots, b_{i-1}, a_i, b_{i+1}, \ldots, b_n)$$

$$+ \sum_{1 \leqslant i < j \leqslant n} g(b_1, \ldots, b_{i-1}, a_i, b_{i+1}, \ldots, b_{j-1}, a_j, b_{j+1}, \ldots, b_n)$$

$$+ \cdots + (-1)^n g(a_1, \ldots, a_n). \tag{10}$$

One says that $g: \mathbb{R}^n \to \mathbb{R}$ is *nondecreasing* if for each rectangle I of $\mathscr{S}$, $g(I) \geqslant 0$. This is the precise extension for $n > 1$. Next, g is taken to be right (or left) continuous, i.e.,

$$\lim_{h_i \downarrow 0} g(x_1 + h_1, \ldots, x_n + h_n) = g(x_1, \ldots, x_n), \tag{11}$$

with $h_i \uparrow 0$ for left continuity. Note that the limit should exist for all coordinates simultaneously. With these two concepts at hand, one can define the n-dimensional mass functions $\alpha_g(\cdot)$ as

$$\alpha_g(I) = g(I), \qquad I \in \mathscr{S}, \tag{12}$$

where $g(I)$ is defined by (10), $\mathscr{S}$ being the usual semiring of the rectangles of $\mathbb{R}^n$, as in Proposition 1.2.1. Using that proposition, one can show immediately that $\alpha_g: \mathscr{S} \to \mathbb{R}^+$ is an additive function and $\alpha_g(\varnothing) = 0$. If μ_g is the generated outer measure by the pair $(\alpha_g, \mathscr{S})$, then $\mu_g|\mathscr{M}_{\mu_g}$ is called the *n-dimensional Lebesgue-Stieltjes measure*. Since α_g has clearly a unique additive extension to the elementary class $\mathscr{E}$, it can be shown by some modifications of the proof of Theorem 1.2.8, on using the right continuity of g, that α_g is σ-additive on $\mathscr{E}$ so that $\mu_g|\mathscr{E} = \alpha_g$ by Theorem 2.10 (cf. Exercise 2). Taking $g(x_1, \ldots, x_n) = x_1 x_2 \cdots x_n$, one gets the n-dimensional Lebesgue measure from this.

If $g: \mathbb{R}^n \to \mathbb{R}$ is monotone, as defined above, then it is also monotone componentwise. But as noted already the converse is not true. Thus the above definition that $g(I) \geqslant 0$, as given by (10), is a more stringent condition than the mere nonnegativity of each componentwise increment, as easy examples show. Also the set of discontinuities for g is more complicated, and is not necessarily countable, since g can have stretches of, and lower dimensional sets (or planes), as discontinuities. This aspect of the structure is more involved than the one dimensional case. Consequently only a restricted version of Theorem 1 can be formulated here. Thus we have the following

Proposition 2. *If $f: \mathbb{R}^n \to \mathbb{R}$ is a nondecreasing right-continuous function in the sense that $f(I)$ of* (10) *is nonnegative for each rectangle I ($\in \mathcal{S}$) and of* (11), *let μ_f be the outer measure generated by the pair $(\alpha_f, \mathcal{S})$, where α_f is defined by* (12). *Then*:

(i) *μ_f is a Radon measure in the sense of Definition* 3.9.
(ii) *$\mu_f(\{a\}) = 0$ for each a at which f is continuous.*
(iii) *On the other hand, if μ is a bounded Radon measure on $\mathbb{R}^n$, then there is a nonnegative, nondecreasing, right-continuous, and bounded function $f: \mathbb{R}^n \to \mathbb{R}$ such that $\mu = \mu_f$, the generated Lebesgue-Stieltjes measure. In fact, we may define f by the equation*

$$f(a_1, \ldots, a_n) = \mu(\{(x_1, \ldots, x_n): -\infty < x_i \leqslant a_i, i = 1, \ldots, n\}) \quad (13)$$

for $a_i \in \mathbb{R}$, $i = 1, \ldots, n$.

The proof of this result is an extension of that of Theorem 1, and will be left to the reader (Exercise 5).

It should be noted that the Lebesgue-Stieltjes measures afford a considerable flexibility for applications. For instance, if $f: \mathbb{R} \to \mathbb{R}$ is an increasing function which is constant between jumps, one gets a discrete measure. If $a_i > 0$, $i = 1, 2, \ldots$, and $f(x) = \sum_{a_i \leqslant x} a_i$, then the corresponding μ_f is such an example. Note, however, that a characterization of discontinuity points of f in Proposition 2 is somewhat involved.

By now the reader has undoubtedly noted that all the preceding results in this chapter only relate to various classes of sets that can be measured by different types of measure functions. However, we did not include a recipe for checking the measurability of a given set, relative to a measure or a σ-algebra, save for the defining system of equations in the Carathéodory process. This is because there is no simple method except for an approximation as given in Proposition 2.16, or, what is equivalent, to see whether the set can be obtained from known relatively simple ones using a countable number of operations of unions, intersections, and differences. This is always one of the basic problems in utilizing the theory. The point to recognize here is that most familiar sets (or figures) are measurable, and the more complicated ones must be obtained from these by the tedious process described above. Thus *only experience with set operations is a guide to verifying measurability, both here and in future work on such problems.* However, in most cases, this will not cause much difficulty.

With this, we proceed to the next chapter for a study of functions which can be used for various measuring purposes without insisting upon any continuity properties.

EXERCISES

0. Show that a Lebesgue-Stieltjes measure μ, on $\mathbb{R}$ is separable, and that the same holds in $\mathbb{R}^n$ provided μ is diffuse.

1. Let $g: \mathbb{R}^n \to \mathbb{R}$ be a monotone nondecreasing function. If $\alpha_g(\cdot)$ is defined on $\mathscr{S}$, the semiring of $\mathbb{R}^n$, by (12), show that $\alpha_g(\cdot)$ is additive and has a unique extension to the algebra $\mathscr{E}$ of elementary sets of $\mathbb{R}^n$.

2. If g of the above problem is taken to be right continuous, and $\alpha_g(\cdot)$ is given by (12), show that α_g is σ-additive on $\mathscr{E}$ and is σ-finite there.

3. Let μ_g be the Lebesgue-Stieltjes measure defined by g of the preceding problem. If g is continuous, show that μ_g is a diffuse σ-finite measure.

4. Suppose that μ is a Radon outer measure on $\mathbb{R}$, and let f be the associated monotone function of μ. Show that $\lim_{x \to +\infty} f(x)$ exists iff $\mu(\mathbb{R}^+) < \infty$, and $\lim_{x \to -\infty} f(x)$ exists iff $\mu(\mathbb{R}^-) < \infty$. Deduce that if μ is bounded, we may choose f as in (13).

5. Adapting the arguments of Theorem 1 (this is not entirely simple), give a detailed proof of Proposition 2. Show that in (13), we may define (a nonbounded) $f: \mathbb{R}^n \to \mathbb{R}$ if $\mu(A) < \infty$ for some A of the form $A = \{(x_1, \dots, x_n): -\infty < x_i \leqslant a\}$ where $a \leqslant 0$.

6. Consider $f_i: \mathbb{R} \to \mathbb{R}$, $i = 1, 2$, monotone nondecreasing, with $\delta f_1 \leqslant \delta f_2$. Show that $\mu_{f_1} \leqslant \mu_{f_2}$ and then $\mathscr{M}_{\mu_{f_1}} \supset \mathscr{M}_{\mu_{f_2}}$, where μ_{f_i} is the Lebesgue-Stieltjes outer measure generated by $(\alpha_{f_i}, \mathscr{S})$. What are the conclusions if μ_{f_1}, μ_{f_2} are replaced by two Radon outer measures μ_1, μ_2 such that $\mu_1 \leqslant \mu_2$ on $\mathbb{R}^n$, i.e., does $\mathscr{M}_{\mu_1} \supset \mathscr{M}_{\mu_2}$ still hold? (Here $\delta f =$ increment of f.)

7. Let μ be a diffuse Lebesgue-Stieltjes (outer) measure on $\mathbb{R}^n$. Show that a bounded $A \subset \mathbb{R}^n$ is μ-measurable iff it can be approximated by perfect sets from inside. [Use the Cantor-Bendixon lemma (see Exercise 2.17 for this and for a definition of perfectness) after reducing the result to the case of finite measure.]

8. Show that there is a one to one correspondence between the set of all finite Lebesgue-Stieltjes measures on $\mathbb{R}^n$ and the set of all bounded nondecreasing left-continuous functions $f: \mathbb{R}^n \to \mathbb{R}$ such that $\lim_{x_i \to -\infty} f(x_1, \dots, x_n) = 0$ for some $i = 1, \dots, n$.

9. Let $\{G_i: i \in I\}$ be a collection of open subsets of $\mathbb{R}^n$ such that $\mu(G_i) = 0$, $i \in I$, where μ is a Lebesgue-Stieltjes outer measure on $\mathbb{R}^n$. Show that $G = \bigcup_{i \in I} G_i$ is also a μ-null set, i.e., $\mu(G) = 0$. [Note that μ is σ-finite, hence strictly localizable, and $\mu(C) < \infty$ for each compact $C \subset \mathbb{R}^n$, so that if $C \subset G$, compact, then $\mu(C) = 0$. Deduce that if $\mu(G) > 0$, it contradicts the inner regularity of μ.]

10. Let $\{G_i: i \in I\}$ be a class of open sets in $\mathbb{R}^n$, such that $\mathbb{R}^n = \bigcup_{i \in I} G_i$. If for each $i \in I$ there is a Lebesgue-Stieltjes outer measure μ_i on G_i (i.e., μ_i vanishes outside G_i) such that on the common parts $G_i \cap G_j$ the measures μ_i, μ_j agree (i.e., $\mu_i | G_i \cap G_j = \mu_j | G_i \cap G_j$), show that there is a unique minimal Lebesgue-Stieltjes outer measure μ on $\mathbb{R}^n$ such that $\mu | G_i = \mu_i$, $i \in I$. Here minimal means that if $\tilde{\mu}$ is another such measure, then $\mu \leqslant \tilde{\mu}$. [Use Exercise 2.12, and then verify that the resulting measure is inner regular on $\mathbb{R}^n$.]

3

MEASURABLE FUNCTIONS

The basic properties and structure of measurable functions as well as the limit operations on them are considered on measurable and measure spaces. Also the theorems of Egorov, Luzin, and Riesz are proved. Some results on image measures under measurable mappings are included in this chapter.

3.1. DEFINITION AND BASIC PROPERTIES

If Ω is a set and $f: \Omega \to \mathbb{R}$ is a mapping (= a function), then by definition it is a subset of the Cartesian product $\Omega \times \mathbb{R}$ with the property that for each $\omega \in \Omega$, there is a unique point $f(\omega) \in \mathbb{R}$. Then f is a subset of $\Omega \times \mathbb{R}$, and $\{(\omega, f(\omega)): \omega \in \Omega\} \subset \Omega \times \mathbb{R}$ is the *graph* of f. If Σ is a σ-algebra of Ω and $\mathscr{B}$ is the (natural) Borel σ-algebra of $\mathbb{R}$, then it is clear that one should consider the Cartesian product $\Sigma \times \mathscr{B}$ of subsets of $\Omega \times \mathbb{R}$, and for a σ-algebra [or one considers the smallest σ-algebra $\sigma(\Sigma \times \mathscr{B})$ containing the collection $\Sigma \times \mathscr{B}$], one may discuss the measurability of (the set) f relative to $\sigma(\Sigma \times \mathscr{B})$. But this implies that we should first examine properties of such product algebras and then utilize them. Since the graph of f describes a figure geometrically, we need to define the measurability of f in such a way that each of the parts of the figure can be measured. Thus various subsets of the set f in $\Omega \times \mathbb{R}$ should be measurable. There are two equivalent ways of formulating this concept. We first introduce the measurability of f directly without looking into the products of algebras, although the latter will be needed for future work, and then show that the measurability of the graph and our special (simpler but unfortunately less motivated) definition are interrelated. Thus the initial step is as follows.

Definition 1. Let (Ω, Σ) be a measurable space and $f: \Omega \to \mathbb{R}$ be a function. Then f is said to be *measurable* (for Σ and the Borel σ-algebra $\mathscr{B}$ of $\mathbb{R}$) if for each interval $(-\infty, a)$ one has $\{\omega: f(\omega) < a\} = f^{-1}(-\infty, a) \in \Sigma$, $a \in \mathbb{R}$, where f^{-1} is as usual the inverse image of f.

(Note that no particular measure function played any part in this definition.)

This concept may be thought of as an abstraction of the familiar fact that $f: \mathbb{R} \to \mathbb{R}$ is continuous iff $f^{-1}(G)$ is open for each open $G \subset \mathbb{R}$. Some "operational" forms of measurability are given in

Lemma 2. *The following statements are all equivalent for $f: \Omega \to \mathbb{R}$ with (Ω, Σ) as a measurable space*:

(i) *f is measurable*,
(ii) $\{\omega: f(\omega) \leqslant a\} \in \Sigma$, $a \in \mathbb{R}$,
(iii) $\{\omega: f(\omega) < a) \in \Sigma$, $a \in \mathbb{R}$,
(iv) $\{\omega: f(\omega) \geqslant a\} \in \mathbb{R}$, $a \in \mathbb{R}$.

Moreover, $\mathbb{R}$ *can be replaced by one of its countable dense subsets D.*

Proof. (i) $\Rightarrow$ (ii): If f is measurable (for Σ and $\mathscr{B}$—this will be omitted below if there is no ambiguity), then

$$\{\omega: f(\omega) \leqslant a\} = \{\omega: f(\omega) > a\}^c \in \Sigma, \qquad a \in \mathbb{R},$$

since Σ is closed under complements. So (ii) holds.

(ii) $\Rightarrow$ (iii): $\{\omega: f(\omega) < a\} = \bigcup_{n \geqslant 1}\{\omega: f(\omega) \leqslant a - 1/n\} \in \Sigma$, $a \in \mathbb{R}$, since Σ is also closed under countable unions. So (iii) holds.

(iii) $\Rightarrow$ (iv): $\{\omega: f(\omega) \geqslant a\} = \{\omega: f(\omega) < a\}^c \in \Sigma$, $a \in \mathbb{R}$, whence (iv).

(iv) $\Rightarrow$ (i): $\{\omega: f(\omega) > a\} = \bigcup_{n \geqslant 1}\{\omega: f(\omega) \geqslant a + 1/n\} \in \Sigma$. Thus (i) holds.

Finally, if $D \subset \mathbb{R}$ is dense and $\{\omega: f(\omega) > a\} \in \Sigma$ for all $a \in D$, then for each $b \in \mathbb{R}$ there is a sequence $a_n \in D$ such that $a_n \downarrow b$, so that

$$\{\omega: f(\omega) > b\} = \bigcup_{n \geqslant 1} \{\omega: f(\omega) > a_n\} \in \Sigma.$$

This completes the proof.

Remark 1. The same definition works if f is extended real valued, since $\{\omega: f(\omega) = \infty\} = \bigcap_{n \geqslant 1}\{\omega: f(\omega) > n\} \in \Sigma$ and $\{\omega: f(\omega) = -\infty\}$ similarly is in Σ. Note that we have used the fact that Σ is closed under complements. If Σ is only a σ-ring, the definition needs a modification, but this generality will be forgone to avoid an annoying complication. The σ-ring definition will be indicated in Exercises 3–5.

Remark 2. If f is measurable, then $\{\omega: f(\omega) = a\} = \{\omega: f(\omega) \geqslant a\} \cap \{\omega: f(\omega) \leqslant a\} \in \Sigma$. However, $\{\omega: f(\omega) = a\} \in \Sigma$, $a \in \mathbb{R}$, need not imply the measurability of f. (See Exercise 2.)

Another result on elementary operations with measurable functions is given by the following:

Lemma 3. *Let* (Ω, Σ) *be a measurable space and* $f_n: \Omega \to \overline{\mathbb{R}}$ *be measurable functions for each* $n \geqslant 1$. *Then all the following new functions obtained from the* f_n *are measurable for* Σ: (i) $f_1\chi_A$, $A \in \Sigma$; (ii) $f_1^+ = \max(f_1, 0)$; (iii) $f_1^- = \max(-f_1, 0)$; (iv) $af_1 + f_2$, $a \in \mathbb{R}$; (v) $f_1 \cdot f_2$; (vi) $1/f_1$ *if* $f_1 \neq 0$; (vii) $\inf_{n \geqslant 1} f_n$; (viii) $\sup_{n \geqslant 1} f_n$; (ix) $\liminf_n f_n$; (x) $\limsup_n f_n$, (xi) $|f_1|^\beta$, $\beta > 0$, (xii) $g \circ f_1$ *if* $g: \mathbb{R} \to \mathbb{R}$ *is continuous and* $f_1: \Omega \to \mathbb{R}$, *in the above.*

Proof. All the proofs follow from the definition. We discuss a few of these as examples.

(iv): $\{\omega: (af_1 + f_2)(\omega) < t\} = \bigcup_{r \in D} \{\omega: af_1(\omega) < r\} \cap \{\omega: f_2(\omega) < t - r\}$, where $D \subset \mathbb{R}$ is a countable dense set (e.g., rationals) and each set in the union is in Σ for any $a \in \mathbb{R}$. So the left side set is in Σ for each $t \in \mathbb{R}$.

(v): Note that

$$\{\omega: f_1^2(\omega) > a\} = \begin{cases} \Omega & \text{if} \quad a < 0, \\ \{\omega: f(\omega) > \sqrt{a}\} \cup \{\omega: f(\omega) < -\sqrt{a}\} & \text{if} \quad a \geqslant 0, \end{cases}$$

so f_1^2 is measurable. Hence

$$f_1 \cdot f_2 = \tfrac{1}{2}\{(f_1 + f_2)^2 - f_1^2 - f_2^2\}$$

is measurable by (iv) and the preceding.

(viii): Observe that

$$\left\{\omega: \sup_n f_n(\omega) > a\right\} = \bigcup_{n \geqslant 1} \{\omega: f_n(\omega) > a\} \in \Sigma, \qquad a \in \mathbb{R}.$$

Similarly, if each f_n is measurable, then since $\limsup_n f_n = \inf_{k \geqslant 1}\{\sup_{n \geqslant k} f_n\}$, this is also measurable.

The following consequence of the above lemma is useful.

Corollary 4. *Let* $f_1, f_2, \ldots$ *be a sequence of extended real valued measurable functions in* (Ω, Σ). *Then the set* A *of points* ω *at which* $f_n(\omega)$ *converges is measurable.*

Proof. The set A can be expressed as the union of A_1, A_2, A_3 where

$$A_1 = \left\{\omega: \limsup_n f_n(\omega) = +\infty\right\} \cap \left\{\omega: \liminf_n f_n(\omega) = +\infty\right\},$$

$$A_2 = \left\{\omega: \liminf_n f_n(\omega) = -\infty\right\} \cap \left\{\omega: \limsup_n f_n(\omega) = -\infty\right\},$$

$$A_3 = \left\{\omega: \limsup_n f_n(\omega) = \liminf_n f_n(\omega), \text{finite}\right\}.$$

But $A_i \in \Sigma$, $i = 1, 2, 3$, so that $A \in \Sigma$. (For A_3, see Corollary 4′ below.)

Simlarly we get a companion result as

Corollary 4′. *If f, g are extended real measurable functions in (Ω, Σ), then $\{\omega: f(\omega) < g(\omega)\}$, $\{\omega: f(\omega) \leqslant g(\omega)\}$, $\{\omega: f(\omega) = g(\omega)\}$, $\{\omega: f(\omega) > g(\omega)\}$, and $\{\omega: f(\omega) \geqslant g(\omega)\}$ are all measurable.*

To see this, one notes that

$$\{\omega: f(\omega) < g(\omega)\} = \bigcup_{r \in D} \{\omega: f(\omega) < r\} \cap \{\omega: g(\omega) > r\} \in \Sigma, \quad (1)$$

where $D \subset \mathbb{R}$ is a countable dense set. The others follow similarly.

Another property is given by

Lemma 5. *Let (Ω, Σ) be a measurable space, $D \subset \mathbb{R}$ be any countable set, and $\{A_r: r \in D\} \subset \Sigma$ be an increasing family, i.e., $r_1 < r_2 \Rightarrow A_{r_1} \subset A_{r_2}$. Then there is a function $f: \Omega \to \mathbb{R}$, measurable for Σ, satisfying $\omega \in A_r \Rightarrow f(\omega) \leqslant r$, and if $\omega \notin A_r$ then $f(\omega) \geqslant r$.*

Proof. The following method of constructing f is useful. For each $\omega \in \Omega$, let $f(\omega)$ be the first A_r which contains ω, i.e. (the elements of D are labeled r)

$$f(\omega) = \inf\{r: \omega \in A_r\}, \qquad \omega \in \Omega, \quad (2)$$

and, as usual, $\inf(\varnothing) = +\infty$. To see that f is the desired function, since $A_r \subset A_{r'}$ for $r < r'$, if $\omega \notin A_{r'}$ then none of the A_r can contain ω. Hence $f(\omega) \geqslant r'$, and if $\omega \in A_r$ then $f(\omega) \leqslant r$. Now f is clearly a function, and its measurability relative to Σ follows from

$$\{\omega: f(\omega) < a\} = \bigcup_{r<a} \{\omega: f(\omega) \leqslant r\} = \bigcup_{r<a} A_r, \qquad a \in \mathbb{R}, \quad (3)$$

and the right side is a countable union at most. That the left side is in the right side is clear, and if $\omega \in A_r$ for some r, then $f(\omega) \leqslant r < a$ and so it is in the left side, so that (3) is also simple. This finishes the proof.

Remark. If $f: \Omega \to \overline{\mathbb{R}}$ is any function measurable for a σ-algebra Σ of Ω, then we can always construct for f a class $\{A_r: r \in D\} \subset \Sigma$ as required in the above lemma. In fact, set $A_r = \{\omega: f(\omega) \leqslant r\}$, $r \in D$, and this has all the properties there.

We next present a basic structure theorem for measurable functions. Let us introduce a concept for this purpose.

Definition 6. If (Ω, Σ) is a measurable space, then a mapping $\varphi: \Omega \to \mathbb{R}$ is a *simple* (measurable) *function* if it can be expressed as

$$\varphi = \sum_{i=1}^{n} a_i \chi_{A_i}, \quad a_i \in \mathbb{R}, \ A_i \in \Sigma, \text{ disjoint}, \qquad i \geqslant 1. \quad (4)$$

A mapping $\psi\colon \Omega \to \mathbb{R}$ is an *elementary* (measurable) *function* if it can be written as

$$\psi = \sum_{-\infty < n < \infty} a_n \chi_{A_n}, \quad a_n \in \mathbb{R},\ A_n \in \Sigma, \text{ disjoint}, \qquad n \geqslant 1. \tag{5}$$

With this we can present the following, in which all limits are taken as $n \to \infty$.

Theorem 7 (Structure). *If (Ω, Σ) is a measurable space, then $f\colon \Omega \to \overline{\mathbb{R}}$ is measurable (for Σ) iff there exists a sequence of simple functions $\varphi_n\colon \Omega \to \mathbb{R}$ such that $\varphi_n(\omega) \to f(\omega)$. We can have $|\varphi_n(\omega)| \uparrow |f(\omega)|$ for each $\omega \in \Omega$. In case $f \geqslant 0$, then the φ_n may be chosen such that $0 \leqslant \varphi_n(\omega) \uparrow f(\omega)$, $\omega \in \Omega$. Alternatively, f is measurable iff there exists a sequence $\psi_n\colon \Omega \to \mathbb{R}$ of elementary functions such that $\psi_n(\omega) \to f(\omega)$ uniformly in $\omega \in \Omega$.*

Proof. Let $\varphi_n\colon \Omega \to \mathbb{R}$ be a simple (or elementary) function, so that it is measurable. If $\varphi_n \to f$, then f is measurable, since, more generally, the pointwise limit of a sequence of measurable functions is measurable by Lemma 3. (Recall that $f = \lim_n \varphi_n = \limsup_n \varphi_n = \liminf_n \varphi_n$ because the limit exists iff the lim sup and lim inf are equal.)

For the converse, first let $f \geqslant 0$. Define φ_n as

$$\varphi_n(\omega) = \begin{cases} \dfrac{k}{2^n} & \text{if} \quad \omega \in A_{kn} = \left\{\omega\colon \dfrac{k}{2^n} \leqslant f(\omega) < \dfrac{k+1}{2^n}\right\}, \\ n & \text{otherwise}. \end{cases}$$

$$k = 0, 1, \ldots, n2^n - 1,$$

Then $0 \leqslant \varphi_n \leqslant \varphi_{n+1} \leqslant f$, and since f is measurable, $A_{kn} \in \Sigma$, so that $\varphi_n = \sum_{k=0}^{n2^n} (k/2^n) \chi_{A_{kn}} + n\chi_B$, where $B = \Omega - \bigcup_{k=0}^{n2^n} A_{kn} \in \Sigma$, is a simple function. Also $0 \leqslant f(\omega) - \varphi_n(\omega) \leqslant 2^{-n}$ for each ω and large enough n. Hence as $n \to \infty$, this difference tends to zero for each $\omega \in \Omega$. Thus $0 \leqslant \varphi_n \uparrow f$.

If f is not necessarily positive, then $f = f^+ - f^-$ and, by Lemma 3, both f^+, f^- are nonnegative measurable functions. So by the preceding paragraph, there exist simple functions $0 \leqslant \varphi_{1n} \uparrow f^+$ and $0 \leqslant \varphi_{2n} \uparrow f^-$ pointwise, and if we set $\varphi_n = \varphi_{1n} - \varphi_{2n}$, then φ_n is simple and $\varphi_n \to f$, pointwise, as $n \to \infty$. Note that by construction, we automatically have $|\varphi_n| = \varphi_{1n} + \varphi_{2n} \uparrow f^+ + f^- = |f|$ pointwise as $n \to \infty$.

Regarding the last part, we modify the construction of φ_n to get uniformity by allowing countably many values. Thus the simple φ_n will be replaced by the (less simple) elementary ψ_n to get the (stronger) uniform convergence. For this let

$$\psi_n = \sum_{-\infty < k < \infty} \frac{k}{2^n} \chi_{A_{kn}}, \qquad A_{kn} = \left\{\omega\colon \frac{k}{2^n} \leqslant f(\omega) < \frac{k+1}{2^n}\right\}.$$

Then ψ_n is elementary and $|f(\omega) - \psi_n(\omega)| < 2^{-n}$, $\omega \in \Omega$. Thus $\psi_n \to f$ uniformly in ω. Since the converse is evident, the proof is complete.

The next result shows a refined structure of measurable functions if we know that they are already measurable relative to certain σ-subalgebras of Σ. If $f: \Omega \to \mathbb{R}$ is measurable for Σ, then we define $\sigma(f) = f^{-1}(\mathscr{B}) \subset \Sigma$, where $\mathscr{B}$ is the Borel σ-algebra of $\mathbb{R}$. Since f^{-1} preserves all the set operations (verify this), it is clear that $\sigma(f)$ is a σ-algebra. It is called the *σ-algebra generated by f*.

We have the following result, due to J. L. Doob, who proved it for certain applications when $(S, \mathscr{S}) = (\mathbb{R}^n, \mathscr{B}^n)$. It is presented in a general form, due to E. B. Dynkin. Sometimes one terms it the Doob-Dynkin lemma.

Theorem 8. *Let (Ω, Σ) and $(S, \mathscr{S})$ be measurable spaces and $f: \Omega \to S$ be $(\Sigma, \mathscr{S})$-measurable [i.e., $\sigma(f) = f^{-1}(\mathscr{S}) \subset \Sigma$]. A function $g: \Omega \to \mathbb{R}$ is measurable relative to the σ-subalgebra $\sigma(f)$ iff there exists a function $h: S \to \mathbb{R}$ which is measurable for $\mathscr{S}$ (and $\mathscr{B}$), such that $g = h \circ f$ [$= h(f)$].*

Proof. If $g = h \circ f$ with h, f as given, then from the facts that $h^{-1}(\mathscr{B}) \subset \mathscr{S}$, $f^{-1}(\mathscr{S}) \subset \Sigma$, and $g^{-1} = f^{-1} \circ h^{-1}$ we have

$$g^{-1}(\mathscr{B}) = f^{-1}\left(h^{-1}(\mathscr{B})\right) \subset f^{-1}(\mathscr{S}) = \sigma(f).$$

Hence g is measurable for $\sigma(f)$. The converse is not as simple.

Thus let g be $\sigma(f)$-measurable. First suppose that g is a simple function, so that $g = \sum_{i=1}^{n} a_i \chi_{A_i}$, $A_i \in \sigma(f)$, disjoint, $a_i \in \mathbb{R}$. By definition of $\sigma(f)$, there exist $B_i \in \mathscr{S}$ such that $A_i = f^{-1}(B_i)$, $i = 1, \ldots, n$. Define $h = \sum_{i=1}^{h} a_i \chi_{B_i}$. Then $h: S \to \mathbb{R}$ and is an $\mathscr{S}$-measurable function. [In the definition of a simple function, we stipulated that the B_i should be disjoint. Even though A_i are disjoint, B_i need not be. This is not important here, but we may replace them with C_i, where $C_1 = B_1$ and for $i > 1$, $C_i = B_i - \bigcup_{j=1}^{i-1} B_j$, so that $C_j \in \mathscr{S}$, disjoint, and $f^{-1}(C_i) = A_i$. Now instead, consider $\tilde{h} = \sum_{i=1}^{n} a_i \chi_{C_i}$ in the above to conform with Definition 6.] Hence

$$\begin{aligned} h(f(\omega)) &= \sum_{i=1}^{n} a_i \chi_{B_i}(f(\omega)) \\ &= \sum_{i=1}^{n} a_i \chi_{f^{-1}(B_i)} = \sum_{i=1}^{n} a_i \chi_{A_i} = g(\omega), \qquad \omega \in \Omega, \end{aligned}$$

Consequently the result holds if g is simple.

In the general case, by the structure theorem (i.e., Theorem 7) there exists a sequence of simple [for $\sigma(f)$] functions $\varphi_n \to g$ pointwise, as $n \to \infty$. But by the preceding paragraph, there exist $h_n: S \to \mathbb{R}$, simple (for $\mathscr{S}$), such that $\varphi_n = h_n \circ f$, $n \geqslant 1$. Since $\{h_n: n \geqslant 1\}$ need not be a convergent sequence, let

S_1 be the subset of S on which the h_n's converge to some function $\bar{h}$. Then by Corollary 4, $S_1 \in \mathscr{S}$ and $\bar{h}$ is $\mathscr{S}$-measurable. Also, since $\varphi_n \to g$, it follows that $f(\Omega) \subset S_1$ [but $f(\Omega)$ need not be in $\mathscr{S}$], since $h_n(f(\omega)) = \varphi_n(\omega)$ converges. Define $h: S \to \mathbb{R}$ by the equation

$$h(s) = \begin{cases} \bar{h}(s), & s \in S_1, \\ 0, & s \in S_1^c. \end{cases}$$

Then h is $\mathscr{S}$-measurable. Moreover, $h(f(\omega)) = \lim_n h_n(f(\omega)) = \lim_n \varphi_n(\omega) = g(\omega)$, $\omega \in \Omega$. This completes the proof.

Doob's original result is obtained by taking $S = \mathbb{R}^n$, and $\mathscr{S}$ as the Borel σ-algebra of $\mathbb{R}^n$. Then $f: \Omega \to \mathbb{R}^n$ is an n-vector, $f = (f_1, \dots, f_n)$, where $f_i: \Omega \to \mathbb{R}$ is a scalar measurable function for $i = 1, \dots, n$. With this specialization, we get the following result.

Corollary 9. *Let (Ω, Σ) be a measurable space and $f: \Omega \to \mathbb{R}^n$ be a measurable vector function (i.e., $f = (f_1, \dots, f_n)$ and each f_i is measurable). Then a function $g: \Omega \to \mathbb{R}$ is measurable relative to $\sigma(f) \subset \Sigma$ iff there is a Borel function $h: \mathbb{R}^n \to \mathbb{R}$ such that $g = h(f_1, \dots, f_n)$.*

Note that thus far a measure function has not intervened. In the next section we show what other operations are possible when a measure μ on Σ is also available.

EXERCISES

0. Let $(\Omega, \mathscr{A})$ be a measurable space and $\mathscr{B} = \{B \subset \Omega: B \cap A \in \mathscr{A}, A \in \mathscr{A}\}$. Show that $\mathscr{B}$ is a σ-algebra (what if $\mathscr{A}$ is only a σ-ring). If $f_i: \Omega \to \mathbb{R}$, $i = 1, 2$ are functions, f_1 is $\mathscr{A}$-measurable, f_2 is $\mathscr{B}$-measurable, then verify that $f_1 f_2$ is $\mathscr{A}$-measurable. [This is an "ideal" kind of property.] Can you generalize this to a "function of a function" statement?

1. Complete the proof of measurability of the unproved parts of Lemma 3.

2. Show that on a measurable space (Ω, Σ), $f: \Omega \to \mathbb{R}$ and $\{\omega: f(\omega) = a\} \in \Sigma$ for each $a \in \mathbb{R}$ does not necessarily imply that f is measurable for Σ. [Consider $\Omega = (0,1)$, Σ the Lebesgue σ-algebra, and A a nonmeasurable set, and take $f(\omega) = \omega\chi_A(\omega) + (1 + \omega)\chi_{A^c}(\omega)$, $\omega \in \Omega$.] Show also that there exist $g: \Omega \to \mathbb{R}$ such that $|g|$ is measurable, but g is not measurable, for Σ.

3. Let $(\Omega, \mathscr{A})$ be a pair where $\mathscr{A}$ is a σ-ring of Ω. We define a function $f: \Omega \to \mathbb{R}$ to be "$\mathscr{A}$-measurable" if for each interval $I \subset \mathbb{R}$, $f^{-1}(I) \cap \{\omega: f(\omega) \neq 0\} \in \mathscr{A}$. Show that a function $f: \Omega \to \mathbb{R}$ is $\mathscr{A}$-measurable iff $N(f) = \{\omega: f(\omega) \neq 0\} \in \mathscr{A}$ and $f^{-1}(B) \cap A \in \mathscr{A}$ for each $A \in \mathscr{A}$ and

Borel set $B \subset \mathbb{R}$. [The point here is that, since Ω is not necessarily in $\mathscr{A}$, the set $N(f)$ of f has to be treated differently. If we use Definition 6 for $\mathscr{A}$ here, constants will not be measurable.]

4. Show that in the above exercise the class of Borel sets B can be replaced by the class of all open or closed or half-open intervals.

5. Under this $\mathscr{A}$-measurability, one can show that all statements of Lemmas 2 and 3 are true. Prove some of the nontrivial ones here. Hereafter we always use our Definition 6, unless the contrary is stated. In fact, verify that if $\Omega \in \mathscr{A}$, then the two definitions agree.

6. Let $f: \Omega \to \mathbb{R}^n$ and Σ be a σ-algebra of Σ. Writing $f = (f_1, \ldots, f_n)$ and recalling that f is measurable for Σ if $f^{-1}(\mathscr{R}) \subset \Sigma$, where $\mathscr{R}$ is the Borel σ-algebra of $\mathbb{R}^n$, show that f is measurable iff each scalar component $f_i: \Omega \to \mathbb{R}$ is measurable, $i = 1, \ldots, n$. Deduce that a complex function $g: \Omega \to \mathbb{C}$ is measurable (i.e. $g^{-1}(\mathscr{C}) \subset \Sigma$, where $\mathscr{C}$ is the Borel σ-algebra of the complex plane $\mathscr{C}$) if $g_1 = \operatorname{Re} g$ and $g_2 = \operatorname{Im} g$, the real and imaginary parts, so $g_j: \Omega \to \mathbb{R}$, $j = 1, 2$, are measurable. [First verify that for any family $\mathscr{D}$ of sets of $\mathbb{R}$ and $h: \Omega \to \mathbb{R}$, $\sigma(h^{-1}(\mathscr{D})) = h^{-1}(\sigma(\mathscr{D}))$.]

7. Let $\Omega = [0,1]$ and $\Sigma = \mathscr{M}_\mu$, the Lebesgue σ-algebra. Let $\varphi: \Omega \to \mathbb{R}$ and $h: \Omega \to (0,1)$ be a pair of (real) Σ-measurable functions. If $g = \varphi \circ h: \Omega \to \mathbb{R}$, then show by an example that g is not necessarily measurable for Σ. Point out why this does not contradict Corollary 9. [This shows the possible difficulties in the statement "a measurable function of a measurable function is measurable", without explicitly specifying all the σ-algebras involved. To get the negative statement, consider the Cantor set $C \subset \Omega$, i.e., $C = \bigcap_{n=1}^{\infty} F_n$, where F_n is the closed subset of Ω obtained at the nth stage after removing the 2^{n-1} open middle third intervals. We first define a monotone nondecreasing function $f: \Omega \to [0,1]$ that is constant on $\Omega - C$, called the *Cantor function*. In fact let $f_n: \Omega \to [0,1]$ be such that $f_n(0) = 0$, $f_n(1) = 1$, and $f_n(\omega) = k/2^n$ if $\omega \in F_{kn}$, the kth middle third interval removed from F_{n-1}, $k = 1, 2, \ldots, 2^n - 1$, with these points connected by line segments. Thus f_n is continuous and $|f_n - f_{n+1}| \leqslant 2^{-n} \to 0$ uniformly on Ω. Let $f = \lim_n f_n$. It is continuous and nondecreasing, and is the required function. Now define h as an inverse function to f, i.e., $(h \circ f)(x) = x$, $x \in \Omega$. Thus h is measurable, but $h(\omega)$ contains a non-Borel set A. If $\varphi = \chi_A$, then φ is Lebesgue measurable, and $h^{-1}(A) \subset C$. But $g = \varphi \circ h$ is not Lebesgue measurable. Note that this possibility is eliminated if, before the composition, we demand that $\varphi^{-1}(\mathscr{M}_\mu) \subset \mathscr{M}_\mu$. Here $h^{-1}(A) \in \mathscr{M}_\mu$, but it is not a Borel set. The existence of such A follows from Theorem 2.2.19.]

8. If (Ω, Σ) is a measurable space, and $\mathscr{L}_0$ is the set of all real measurable functions on Ω, verify that $\mathscr{L}_0$ is an algebra; its set of nonnegative elements is then closed under pointwise limits with extended real values.

9. If $\Omega = \mathbb{R}$ and $\Sigma = \mathscr{B}$ the Borel σ-algebra of $\mathbb{R}$, then show that each continuous real function on $\mathbb{R}$ is measurable (for $\mathscr{B}$), or *Borel measurable*.

Using the separability of $\mathbb{R}^n$, show that each $f: \mathbb{R}^n \to \mathbb{R}$ which is continuous in each component (is not necessarily continuous on $\mathbb{R}^n$ itself but) is measurable for $\mathscr{R}$, the Borel σ-algebra of $\mathbb{R}^n$. [Consider first $n = 2$ and extend.] (We shall see in the next section that a certain converse of this holds after we prove the Luzin theorem. However, if the Borel σ-algebra is replaced by a smaller subalgebra here, a continuous function need not be measurable relative to it.)

10. Let $f: \mathbb{R} \to \mathbb{R}$ be a function of bounded variation on each compact interval. Show that f is Borel measurable on $\mathbb{R}$. [Note that $f\chi_I$ can be expressed as a difference of two nondecreasing functions for each compact interval I.]

11. If $f_i: \Omega \to \mathbb{R}$, $i \in I$, is a collection of functions and Σ is a σ-algebra, let $\Sigma_0 = \sigma(\bigcup_i f_i^{-1}(\mathscr{B})) = \sigma(f_i, i \in I)$, so that Σ_0 is the smallest σ-algebra relative to which each f_i is measurable, $i \in I$, where $\mathscr{B}$ is the Borel σ-algebra of $\mathbb{R}$. Let $\mathscr{I}$ be the collection of all finite or countable subsets of I, and $\mathscr{S}_J = \sigma(f_j, j \in J)$. Show that Σ_0 is generated by $\mathscr{S}_0 = \bigcup\{\mathscr{S}_J: J \in \mathscr{I}\}$. Deduce that if $g: \Omega \to \mathbb{R}$ is a function, measurable relative to Σ_0, then there exists a Borel function $h: \mathbb{R}^N \to \mathbb{R}$ (N is an integer which may be infinite) such that $g = h \circ (f_{i_1}, f_{i_2}, \ldots, f_{i_n}, \ldots)$. [Use Theorem 8. Compare with Exercise 2.1.5.]

3.2. MEASURABILITY WITH MEASURES AND CONVERGENCE

If (Ω, Σ) is an abstract measurable space and $(\mathbb{R}, \mathscr{B})$ is the real line with its Borel σ-algebra (this is also called a *Borelian space*), then a function $f: \Omega \to \mathbb{R}$ was termed measurable if $f^{-1}(\mathscr{B}) \subset \Sigma$, so that measurability is a property of the σ-algebras. This is entirely natural and proper, since any mapping (or even a relation) is a certain subset of $\Omega \times \mathbb{R}$. However, if a measure function $\mu: \Sigma \to \overline{\mathbb{R}}^+$ is available, one should take advantage of this additional structure and extend the notion of measurability in the sense that if f and g are two functions and $\{\omega: f(\omega) \neq g(\omega)\}$ is a μ-null set, then for measuring purposes f and g should be equally satisfactory. We make this precise in the following manner by incorporating the special features that frequently add flexibility (and sometimes problems) to the subject.

Definition 1. Let (Ω, Σ, μ) be a measure space and $f: \Omega \to \mathbb{R}$ be a function. Then f is called *μ-measurable* if $f^{-1}(B)$ is μ-measurable for each Borel set $B \subset \mathbb{R}$, so that $f^{-1}(B) \in \mathscr{M}_\mu$, the σ-algebra of μ-measurable sets.

Note that from the work of the preceding chapter (Theorem 2.2.10), if μ^* is the outer measure generated by the pair (μ, Σ), then $\mu^*|\mathscr{M}_\mu$ is σ-additive, $\Sigma \subset \mathscr{M}_\mu$, and $\mu^*|\Sigma = \mu$. Typically $\mathscr{M}_\mu$ is a much larger σ-algebra than Σ, and thus the class of μ-measurable sets is considerably larger than the measurable class relative to only Σ. For instance, let $\Omega = [0,1]$, Σ be the σ-algebra of countable sets and those with countable complements (also called "cocountable"), and $\mu(A)$ be the number of points in A. Then (Ω, Σ, μ) is a measure space, and $\mathscr{M}_\mu$ is the power set of Ω, so that it is much bigger than Σ. Here even the completion of Σ is itself, since only the empty set is μ-null. This shows how μ-measurability enlarges the class of measurable functions.

Since only Σ is replaced by $\mathscr{M}_\mu$, the calculus of measurable functions developed in the last section automatically holds if the word "μ-measurable" is substituted for "measurable" there. However, regarding the limits of sequences, we have some new concepts and results due to the availability of a measure. To gain a little more generality, one starts with a possibly noncomplete measure space (Ω, Σ, μ). A proposition $p(\omega)$ on a set $A \subset \Omega$ is said to be true *almost everywhere* (abbreviated a.e.) if there is a set B satisfying $\mu(B) = 0$ and $p(\omega)$ is true for each $\omega \in A - B$. This is also stated as $p(\omega)$ is true for *almost all* ω [for short, a.a.(ω)] and if there are more than one measure, we say that $p(\cdot)$ is *true a.e.*$[\mu]$, or μ-*a.a.*(ω). When Σ is not completed, a trite generality implied here may be seen from a simple example: Let $B \subset \Omega$ be μ-null. If $B_1 \subset B$ is not measurable, then $f_1 = 0$ and $f_2 = \chi_{B_1}$ are two distinct functions such that $f_1(\omega) = f_2(\omega)$ a.a.(ω) (i.e. for every $\omega \in \Omega - B$). Here f_1 is measurable while f_2 is not. If Σ is completed, then both will be measurable and also $f_1 = f_2$ a.e.(μ) (but still not equal everywhere).

Regarding this new relation, we have

Lemma 2. *If $\mathscr{F}$ is the class of all extended real valued functions on* (Ω, Σ, μ), *and if for $f, g \in \mathscr{F}$ we set $f \sim g$ to mean $f = g$* a.e., *then $\sim$ is an equivalence relation on $\mathscr{F}$. Moreover $\sim$ preserves the order* (*or lattice*) *properties.*

Proof. The relation $\sim$ is clearly reflexive and symmetric, since $f = f$ a.e., and $f = g$ a.e. $\Rightarrow$ $g = f$ a.e. Regarding transitivity, let $f_1 = f_2$ a.e. and $f_2 = f_3$ a.e. Then there exist μ-null sets B_1, B_2 such that $f_1(\omega) = f_2(\omega)$, $\omega \in \Omega - B_1$, and $f_2(\omega) = f_3(\omega)$, $\omega \in \Omega - B_2$. If $B = B_1 \cup B_2$ then $\mu(B) = 0$, and for $\omega \in \Omega - B$ we have $f_1(\omega) = f_2(\omega) = f_3(\omega)$, so that $f_1 = f_3$ a.e. Similarly, if $f_1 = f_2$ a.e., $f_3 = f_4$ a.e. and $f_1 \leqslant f_3$, a.e., then there exist μ-null sets B_1, B_2, and B_3 such that $f_1(\omega) = f_2(\omega)$, $\omega \in \Omega - B_1$; $f_3(\omega) = f_4(\omega)$, $\omega \in \Omega - B_2$; and $f_1(\omega) \leqslant f_3(\omega)$, $\omega \in \Omega - B_3$. So if $B = B_1 \cup B_2 \cup B_3$, then B is μ-null and on $\Omega - B$ we have $f_2(\omega) = f_1(\omega) \leqslant f_3(\omega) = f_4(\omega)$, so that $f_2 \leqslant f_4$ a.e., as asserted.

The preceding argument shows that as long as the statement involves at most a countable number of exceptions (i.e., μ-null sets), then we can extend the pointwise assertions to the a.e. context. The facility thus obtained is illuminated by the next result. First we introduce some additional terminology.

Definition 3. Let f_n, f be scalar functions on (Ω, Σ, μ), and μ^* be the outer measure generated by (μ, Σ) (or μ may be an outer measure). Then one says:

(i) f_n *converges a.e.* to f, if there is a μ^*-null set B such that for each $\omega \in \Omega - B$, the scalar sequence $f_n(\omega) \to f(\omega)$. [Similarly, f_n is *fundamental a.e.* if $\{f_n(\omega): n \geqslant 1\}$ is a Cauchy sequence of scalars for each $\omega \in \Omega - B_0$, $\mu^*(B_0) = 0$.]

(ii) f_n *converges in μ-measure* to f if for each $\epsilon > 0$, $\mu^*(\{\omega: |f_n(\omega) - f(\omega)| \geqslant \epsilon\}) \to 0$ as $n \to \infty$. (Here $\{f_n: n \in I\}$ can be a net and the same concept holds without change.) This is often denoted as

$$f_n \xrightarrow{\mu} f.$$

(Note that since subtraction is involved here, we must have all functions at least a.e. finite valued.)

We present a few consequences, leaving some others as exercises, since the arguments will show the necessary modifications from the work of the last section.

Proposition 4. *Let (Ω, Σ, μ) be a measure space and $\mathscr{F}$ be the class of a.e. finite valued functions on Ω. If $f_n, f, f_0 \in \mathscr{F}$ and for $n \geqslant 1$ we have $f_n \leqslant f_0$ a.e. and $f_n \to f$ a.e. as $n \to \infty$, then $f \leqslant f_0$ a.e. Moreover, if the measure space is complete and each f_n is measurable (for Σ), then so is f. Further, if $f = g$ a.e, then $f_n \to g$ a.e.*

Proof. Let B_n and B_0 be μ-null sets such that $f_n(\omega) \leqslant f_0(\omega)$ for $\omega \in \Omega - B_n$ and $f_n(\omega) \to f(\omega)$ for $\omega \in \Omega - B_0$. If $B = \bigcup_{n=0}^{\infty} B_n$, then B is μ-null, and for all $\omega \in \Omega - B$ we have $f_n(\omega) \leqslant f_0(\omega)$ and $f_n(\omega) \to f(\omega) \leqslant f_0(\omega)$. Thus $f \leqslant f_0$ a.e. If each f_n is measurable and $f_n \to f$ a.e., then as noted above, $f_n(\omega) \to f(\omega)$ for all $\omega \in \Omega - B_0$ $[\mu(B_0) = 0]$, and hence for each $a \in \mathbb{R}$,

$$\begin{aligned}\{\omega: f(\omega) > a\} &= [\{\omega: f(\omega) > a\} \cap B_0^c] \cup [\{\omega: f(\omega) > a\} \cap B_0] \\ &= \left[\left\{\omega: \limsup_n f_n(\omega) > a\right\} \cap B_0^c\right] \cup [\{\omega: f(\omega) > a\} \cap B_0].\end{aligned} \tag{1}$$

The first set on the right is in Σ for all a, since $B_0^c \in \Sigma$ and each f_n is measurable. The last term of (1) is a subset of the μ-null set B_0, which is in Σ if Σ is complete. Hence when (Ω, Σ, μ) is complete, f is measurable for Σ; it may not be Σ-measurable in the noncomplete case.

Finally, if $f = g$ a.e., then there is a μ-null set $\tilde{B}$ such that $f(\omega) = g(\omega)$, $\omega \in \Omega - \tilde{B}$. Hence if $\bar{B} = B_0 \cup \tilde{B}$, then $\bar{B}$ is μ-null, and for each $\omega \in \Omega - \bar{B}$,

$f_n(\omega) \to f(\omega) = g(\omega)$, so that $f_n \to g$ a.e. Note that the same argument shows that if $f_n = f_n'$ a.e., $n \geqslant 1$, and $f_n \to f$ a.e., then $f_n' \to f$ a.e. also holds. This completes the proof.

The other concept of convergence (in measure) is not very intuitive. It is a weaker notion than a.e. convergence on finite measure spaces, but it is a very useful one in measure theory, since it holds for nets instead of sequences, and is especially suited if μ is only finitely additive. The relation between these concepts is made precise in the following important result, the second part of which is due to F. Riesz and the last to H. Lebesgue.

Theorem 5. *Let (Ω, Σ, μ) be a measure space, and f_n, f be a.e. finite valued functions on Ω. Then we have:*

(i) *$f_n \xrightarrow{\mu} f$ and $f_n \xrightarrow{\mu} g \Rightarrow f = g$ a.e., and conversely, if $f_n \xrightarrow{\mu} f$, $f = g$ a.e., then $f_n \xrightarrow{\mu} g$.*

(ii) *$f_n \xrightarrow{\mu} f \Rightarrow f_{n_k} \to f$ a.e. for some (cofinal) subsequence $\{n_k: k \geqslant 1\}$.*

(iii) *If $\mu(\Omega) < \infty$, $f_n \to f$ a.e., each f_n is measurable, and Σ is complete then $f_n \xrightarrow{\mu} f$.*

Proof. (i): To show that convergence in measure defines an a.e. unique limit, let $\epsilon > 0$ and consider the limits f, g as given. Then

$$\{\omega: |f(\omega) - g(\omega)| \geqslant \epsilon\} \subset \{\omega: |f_n(\omega) - f(\omega)| \geqslant \epsilon/2\}$$
$$\cup \{\omega: |f_n(\omega) - g(\omega)| \geqslant \epsilon/2\}.$$

Hence, considering the (outer) measures of these sets and using the hypothesis that $f_n \xrightarrow{\mu} f$, $f_n \xrightarrow{\mu} g$ [μ^* is generated by (μ, Σ)],

$$\begin{aligned}
&\mu^*(\{\omega: |f(\omega) - g(\omega)| \geqslant \epsilon\}) \\
&\quad\leqslant \lim_{n\to\infty} \mu^*(\{\omega: |f_n(\omega) - f(\omega)| \geqslant \epsilon/2\}) \\
&\qquad + \lim_{n\to\infty} \mu^*(\{\omega: |f_n(\omega) - g(\omega)| \geqslant \epsilon/2\}) = 0.
\end{aligned} \tag{2}$$

Since $\epsilon > 0$, on taking $\epsilon = 1/k$ and using Theorem 2.2.10, we get

$$\begin{aligned}
&\mu^*(\{\omega: |f(\omega) - g(\omega)| > 0\}) \\
&\quad= \mu^*\Big(\lim_{k\to\infty} \{\omega: |f(\omega) - g(\omega)| > 1/k\}\Big) \\
&\quad= \lim_{k\to\infty} \mu^*(\{\omega: |f(\omega) - g(\omega)| > 1/k\}) = 0,
\end{aligned}$$

by (2). Hence $f = g$ a.e.

The converse is immediate, since

$$\{\omega : |f_n(\omega) - g(\omega)| \geqslant \epsilon\} \subset \{\omega : |f_n(\omega) - f(\omega)| \geqslant \epsilon\}$$
$$\cup \{\omega : |f(\omega) - g(\omega)| > 0\}, \tag{3}$$

and μ^* is subadditive.

(ii): Since $f_n \xrightarrow{\mu} f$, taking $\epsilon = 1, 1/2, 1/2^2, \ldots$ in the definition, we can first choose an integer $n_1 \geqslant 1$ such that

$$\mu^*(\{\omega : |f_{n_1}(\omega) - f(\omega)| \geqslant 1\} < \tfrac{1}{2},$$

and, by induction, if $n_{k-1} > n_1$ is obtained, let $n_k > n_{k-1}$ be chosen such that

$$\mu^*\left(\left\{\omega : |f_{n_k}(\omega) - f(\omega)| > \frac{1}{k}\right\}\right) < \frac{1}{2^k}. \tag{4}$$

To simplify writing, let B_k be the set inside μ^*. To form a monotone sequence out of this collection, let $A_k = \bigcup_{j=k}^{\infty} B_j$. So $A_k \supset A_{k+1}$, and if $A = \bigcap_{n=1}^{\infty} A_n$, then for all n

$$\mu^*(A) \leqslant \mu^*(A_n) \leqslant \sum_{k=n}^{\infty} \mu^*(B_k) \leqslant \sum_{k=n}^{\infty} 2^{-k} = 2^{-n+1}. \tag{5}$$

Since n is arbitrary and the left side does not depend on n, it follows that $\mu^*(A) = 0$. Hence if we set $g_k = f_{n_k}$ and $\omega \in \Omega - A = \bigcup_{k=1}^{\infty} A_k^c$, so that $\omega \in A_{k_0}^c \subset A_{k_0+1}^c$ for some k_0, it follows, for all $k \geqslant k_0$, that

$$|g_k(\omega) - f(\omega)| < \frac{1}{k}. \tag{6}$$

Thus $g_k \to f$ a.e., or $f_{n_k} \to f$ a.e., as $n_k \to \infty$, as asserted. Note that the μ-null set A constructed above depends on the subsequence.

(iii): Finally suppose $\mu(\Omega) < \infty$, and $f_n \to f$ a.e. So there is a μ-null set A such that $f_n(\omega) \to f(\omega)$ for $\omega \in \Omega - A$. Since each f_n is measurable and Σ is complete, then f is measurable by Proposition 4. So given $\epsilon > 0$, we have

$$\{\omega : |f_n(\omega) - f(\omega)| \geqslant \epsilon\} \subset \bigcup_{k=n}^{\infty} \{\omega : |f_k(\omega) - f(\omega)| \geqslant \epsilon\},$$

and then by Proposition 2.1.1,

$$\begin{aligned}\limsup_n \mu(\{\omega: |f_n(\omega) - f(\omega)| \geqslant \epsilon\}) \\ &\leqslant \lim_{n\to\infty} \mu\Big(\bigcup_{k\geqslant n} \{\omega: |f_n(\omega) - f(\omega)| \geqslant \epsilon\}\Big) \\ &= \mu\Big(\limsup_n \{\omega: |f_n(\omega) - f(\omega)| \geqslant \epsilon\}\Big) \\ &= 0.\end{aligned}$$

Hence the limit of the left side exists and is zero, so that $f_n \xrightarrow{\mu} f$. This completes the proof.

Remark. As seen in the last part of the above theorem, if each f_n is assumed measurable (for Σ) and if Σ is complete then f is measurable for Σ, and so also is μ-measurable. Note here how completeness of Σ enters, or else one has to *assume* that each f_n and f are all a.e. finite measurable functions.

We now give an example showing that the converse of (iii) is false, i.e., there exist sequences which converge in μ-measure but which do not converge pointwise at even a single point. In fact, let $\Omega = [0, 1)$, μ be Lebesgue measure, and Σ be the Lebesgue σ-algebra of Ω. Consider

$$A_{nk} = \left\{\omega: \frac{k-1}{n} \leqslant \omega < \frac{k}{n}\right\}, \qquad k = 1, 2, \ldots, n,$$

and define $f_{nk} = \chi_{A_{nk}}$. Then relabel the double sequence

$$f_{11}, f_{21}, f_{22}, f_{31}, f_{32}, f_{33}, \ldots, f_{n1}, \ldots, f_{nn}, \ldots, \text{ as } g_1, g_2, g_3, \ldots.$$

Since there is a one to one correspondence between these two sequences, we have a well-defined measurable (nonnegative) sequence $\{g_n: n \geqslant 1\}$. Given $0 < \epsilon < 1$, one has for a unique $0 \leqslant k \leqslant n$ depending on m

$$\mu(\{\omega: g_m(\omega) > \epsilon\}) \leqslant \mu(A_{nk}) = \frac{1}{n} \to 0,$$

as m and hence $n \to \infty$. Thus $g_n \xrightarrow{\mu} 0$. But for each $\omega \in \Omega$, we have $g_n(\omega) = 1$ or 0 for infinitely many n, so that $\limsup_n g_n(\omega) = 1$ and similarly $\liminf_n g_n(\omega) = 0$. Hence $\{g_n(\omega): n \geqslant 1\}$ converges for no $\omega \in \Omega$.

Even the direct assertion of (iii) is false if $\mu(\Omega) = +\infty$. Indeed, let $\Omega = [0, \infty)$, μ be Lebesgue measure on Ω, Σ be the Lebesgue σ-algebra, and $A_n = [n, \infty)$. Then $A_n \downarrow \varnothing$, $A_n \in \Sigma$, and $\mu(A_n) = +\infty$ for each n. If $f_n = \chi_{A_n}$

for each $\omega \in \Omega$, then $f_n(\omega) \to 0$ as $n \to \infty$. However, if $0 < \epsilon < 1$, then

$$\mu(\{\omega: f_n(\omega) \geqslant \epsilon\}) = \mu(A_n) = +\infty, \qquad n \geqslant 1,$$

so that $f_n \nrightarrow 0$ in μ-measure.

The preceding examples distinguish the a.e. and in-measure convergences, and show at the same time that the latter is not quite intuitive. However it has several advantages in analysis because of its intimate relationship with a given measure. Thus the following result, which is analogous to Proposition 2.4.1, shows how a metric topology can be introduced into the space of (arbitrary) functions also.

Proposition 6. *Let μ be any outer measure on Ω, and $\mathscr{F}$ be the class of all scalar (not necessarily measurable) functions. Then it becomes a linear (semi-)metric space under $d_\varphi(\cdot, \cdot)$: $\mathscr{F} \times \mathscr{F} \to \overline{\mathbb{R}}^+$, where $d_\varphi(f, g) = d_\varphi(f - g, 0)$ and*

$$d_\varphi(f, 0) = \inf\{\varphi(\alpha + \mu(\{\omega: |f(\omega)| > \alpha\})): \alpha > 0\}. \tag{7}$$

Here $\varphi: \overline{\mathbb{R}}^+ \to \mathbb{R}^+$ is a bounded continuous concave function such that $\varphi(x) = 0$ iff $x = 0$. If $\mu(\Omega) < \infty$, then $\varphi(x) = x$ may (and will) be taken. Also $d_\varphi(\cdot, \cdot)$ defines a metric if $\mathscr{F}$ is replaced by $\mathscr{F}/\sim$, the set of a.e. equal functions under μ. Further a sequence $f_n \xrightarrow{\mu} f$ iff $d_\varphi(f_n - f, 0) \to 0$ as $n \to \infty$. In particular, if $\mathscr{G}$ is the class of all a.e. finite μ-measurable scalar functions on Ω, then $\tilde{\mathscr{G}} = \mathscr{G}/\sim$ becomes a complete metric space under the induced (equivalent) metric $\tilde{d}_\varphi(\cdot, \cdot)$ by the convergence in μ-measure.

Proof. As noted in Proposition 2.4.1, $\varphi(x + y) \leqslant \varphi(x) + \varphi(y)$, $x, y \in \mathbb{R}^+$; let $d(f) = d_\varphi(f, 0)$ for simplicity. It is clear that $d(f) \geqslant 0$ and $d(0) = 0$, since $\mu(\varnothing) = 0$ and $\varphi(x) = 0$ iff $x = 0$. To see that $d(\cdot)$ gives a distance on $\mathscr{F}/\sim$, let $d(f) = 0$; we assert that $f = 0$ a.e. In fact by (7), given $\epsilon > 0$, there exists an $\alpha > 0$ such that (since $\varphi(x) < \infty$ and $= 0$ only for $x = 0$)

$$\alpha + \mu(\{\omega: |f(\omega)| > \alpha\}) < \epsilon. \tag{8}$$

So $0 < \alpha < \epsilon$, and since $\{\omega: |f(\omega)| > \epsilon\} \subset \{\omega: |f(\omega)| > \alpha\}$, for $0 < \alpha < \epsilon$ we have by (8), for any $0 < \epsilon < \delta$,

$$\mu(\{\omega: |f(\omega)| > \delta\}) \leqslant \mu(\{\omega: |f(\omega)| > \epsilon\}) < \epsilon.$$

Since $\epsilon > 0$ is arbitrary, $\mu(\{\omega: |f(\omega)| > \delta\}) = 0$. Taking $\delta = 1/n$, we get

$$\mu(\{\omega: |f(\omega)| > 0\}) \leqslant \sum_{n=1}^{\infty} \mu\left(\left\{\omega: |f(\omega)| > \frac{1}{n}\right\}\right) = 0.$$

Hence $f = 0$ a.e. Since clearly $d_\varphi(f, g) = d_\varphi(g, f) = d(f - g)$, we only need to show the validity of the triangle inequality.

Let $f, g \in \mathscr{F}$ and $\alpha > 0$, $\beta > 0$. Then from the fact that $\{\omega: |f + g|(\omega) > \alpha + \beta\} \subset \{\omega: |f(\omega)| > \alpha\} \cup \{\omega: |g(\omega)| > \beta\}$ we have with (7)

$$d(f+g) = \inf\{\varphi(\alpha + \beta + \mu(\{\omega: |f+g|(\omega) > \alpha + \beta\})): \alpha > 0, \beta > 0\}$$

$$\leqslant \inf\{\varphi(\alpha + \mu(\{\omega: |f(\omega)| > \alpha\}))$$

$$+\varphi(\beta + \mu(\{\omega: |g(\omega)| > \beta\})): \alpha > 0, \beta > 0\}$$

(by the subadditivity of μ and φ)

$$\leqslant \inf\{\varphi(\alpha + \mu(\{\omega: |f(\omega)| > \alpha\})): \alpha > 0\}$$

$$+\inf\{\varphi(\beta + \mu(\{\omega: |f(\omega)| > \beta\})): \beta > 0\},$$

$$= d(f) + d(g).$$

Hence $d(\cdot)$ gives a distance on $\mathscr{F}/\sim$. Note that if $\mu(\Omega) < \infty$, then (7) may be replaced by

$$d(f) = \inf\{\mu(\{\omega: |f(\omega)| > \alpha\}): \alpha > 0\},$$

and the same argument shows that $\{\mathscr{F}, d(\cdot)\}$ is a (semi-)metric space.

To see that $f_n \xrightarrow{\mu} f$ is equivalent to $d(f_n - f) \to 0$, suppose that $d(f_n - f) \nrightarrow 0$. Then there exists a $\delta > 0$ such that by (7),

$$\varphi(\alpha + \mu(\{\omega: |f_n - f|(\omega) > \alpha\})) \geqslant d(f_n - f) > \delta > 0 \tag{9}$$

for all $\alpha > 0$ and all large enough n. Since φ is a continuous concave function, it has a unique inverse φ^{-1} such that $\varphi^{-1}(y) = 0$ iff $y = 0$. Thus (9) implies

$$\mu(\{\omega: |f_n - f|(\omega) > \alpha\}) \geqslant \varphi^{-1}(\delta) - \alpha. \tag{10}$$

But $\varphi^{-1}(\delta) > 0$, and $\alpha > 0$ is arbitrary. Choose $0 < \alpha_0 < \frac{1}{2}\varphi^{-1}(\delta)$, so that (10) becomes, for $\alpha = \alpha_0$,

$$\mu(\{\omega: |f_n - f|(\omega) > \alpha_0\}) > \tfrac{1}{2}\varphi^{-1}(\delta) > 0,$$

for all large enough n. Hence $f_n \nrightarrow f$ in μ-measure by Definition 3.

Conversely, if $f_n \nrightarrow f$ in μ-measure, then for each $\epsilon > 0$ there is a $\delta_\epsilon > 0$ such that for all large enough n

$$\mu(\{\omega: |f_n - f|(\omega) > \epsilon\}) > \delta_\epsilon > 0.$$

Hence for $0 < \epsilon' < \epsilon$, we get

$$\varphi(\epsilon' + \mu(\{\omega: |f_n - f|(\omega) > \epsilon'\})) > \varphi(\epsilon' + \delta_\epsilon) \geqslant \varphi(\delta_\epsilon) > 0.$$

Since this is true for all $0 < \epsilon' < \epsilon$, and $\varphi(\delta_\epsilon) > 0$ is independent of ϵ', we conclude that $d(f_n - f) \geqslant \varphi(\delta_\epsilon) > 0$ for all large enough n. Hence $f_n \nrightarrow f$ in the d-metric.

Finally, if $\{f_n: n \geqslant 1\} \subset (\mathscr{G}, d)$ is a Cauchy sequence, let f be its limit. We need to show that $f \in \mathscr{G}$. This follows if f is measurable. But by the above part, the Cauchy condition implies

$$\mu(\{\mu: |f_n - f|(\omega) \geqslant \epsilon\}) \to 0, \qquad n \to \infty,$$

and each f_n is now μ-measurable. Hence by Theorem 5(ii) and Proposition 4, f is also μ-measurable, so that $f \in \mathscr{G}$, and the space is complete. This establishes the proposition.

Next we present an equivalent concept to a.e. convergence in a *finite* measure space, due to D. F. Egorov, which is useful in applications. For this purpose let us introduce a term which explains the restricted nature of the new concept. A sequence $\{f_n: n \geqslant 1\}$ of scalar measurable functions on a measure space (Ω, Σ, μ) *converges* μ*-uniformly* to f if for each $\epsilon > 0$, there exists a set $A_\epsilon \in \Sigma$ such that $\mu(A_\epsilon) < \epsilon$ and on $\Omega - A_\epsilon$, $f_n \to f$ uniformly. In the literature, this is also (unfortunately) called "almost uniform" convergence and abbreviated *a.u. convergence*. Clearly, for this definition it is sufficient if μ is an outer measure, and the measurability is irrelevant. Its importance comes usually when the latter hypothesis as well as $\mu(\Omega) < \infty$ are added.

We can now present the desired result as

Theorem 7 (Egorov). *Let $\{f_n, n \geqslant 1\}$ be a sequence of a.e. finite measurable functions on a **finite** measure space (Ω, Σ, μ). Then f_n converges a.e. to an a.e. finite measurable function f iff $f_n \to f$ μ-uniformly.*

Proof. One direction is immediate. Indeed, if $f_n \to f$ μ-uniformly, then take $\epsilon = 1/n$ such that $f_n(\omega) \to f(\omega)$ uniformly for $\omega \in \Omega - A_m$ where $\mu(A_m) < 1/m$. Let $A = \bigcap_{m=1}^\infty A_m$. Then $A \in \Sigma$, $\mu(A) = 0$, and for ω in $A^c = \Omega - A = \bigcup_{m=1}^\infty(\Omega - A_m)$ we have $f_n(\omega) \to f(\omega)$. Hence $f_n \to f$ a.e. The converse is nontrivial and uses the full hypothesis.

Thus let $f_n \to f$ a.e., so that $f_n(\omega) \to f(\omega)$, $\omega \in \Omega - A$, $\mu(A) = 0$. Then by definition of pointwise convergence, for each $k \geqslant 1$ and $m \geqslant 1$, the set

$$A_{mk} = \left\{\omega: |f_n(\omega) - f(\omega)| < \frac{1}{m} \text{ for all } n \geqslant k\right\} \tag{11}$$

satisfies (i) $A_{mk} \in \Sigma$ (by the hypothesis of measurability of f_n and f), and (ii)

$A_{mk} \subset A_{m(k+1)}$. Let $B_m = \lim_{k \to \infty} A_{mk} = \bigcup_{k=1}^{\infty} A_{mk}$. Since $\mu(\Omega) < \infty$ and $\mu(B_m) = \lim_k \mu(A_{mk})$, given $\epsilon > 0$, there is a $k_0(\epsilon, m)$ such that $k \geqslant k_0(\epsilon, m)$ implies

$$\mu\left(B_m - A_{mk_0(\epsilon, m)}\right) < \epsilon/2^m. \tag{12}$$

Note that $B_m^c \subset A$, since $\omega_0 \in B_m^c$ implies $|f_n(\omega_0) - f(\omega_0)| \geqslant 1/m$ for all large enough n, so that $f_n(\omega_0) \nrightarrow f(\omega_0)$ and $\omega_0 \in A$. Thus $\mu(B_m) \geqslant \mu(A^c) = \mu(\Omega)$, so that Ω and B_m are equal except for a μ-null set for each $m \geqslant 1$. If we let $A_\epsilon = \bigcap_{m=1}^{\infty} A_{mk_0(\epsilon, m)}$, then

$$\begin{aligned}
\mu(A_\epsilon^c) &= \mu\left(\bigcup_{m=1}^{\infty} A_{mk_0(\epsilon, m)}^c\right) \\
&\leqslant \sum_{m=1}^{\infty} \mu\left(A_{mk_0(\epsilon, m)}^c\right) \\
&= \sum_{m=1}^{\infty} \mu\left(B_m - A_{mk_0(\epsilon, m)}\right), \qquad \text{since,} \quad \mu(B_m) = \mu(\Omega) \\
&\leqslant \sum_{m=1}^{\infty} \frac{\epsilon}{2^m} = \epsilon, \qquad \text{by (12).}
\end{aligned}$$

It remains to show that on A_ϵ^c, $f_n \to f$ uniformly. Indeed, if $n \geqslant k_0(\epsilon, m)$, then $|f_n(\omega) - f(\omega)| < 1/m$ for $\omega \in A_{mk_0(\epsilon, m)}$. Hence

$$\sup_{\omega \in A_\epsilon} |f_n(\omega) - f(\omega)| \leqslant \sup_{\omega \in A_{mk_0(\epsilon, m)}} |f_n(\omega) - f(\omega)| < \frac{1}{m},$$

so that for every $m \geqslant 1$, $f_n(\omega) \to f(\omega)$ uniformly if $\omega \in A_\epsilon$ [or $f_n \to f$ a.u.]. This completes the proof.

It should be noted that not only (the converse part of) the proof depends on the hypothesis that $\mu(\Omega) < \infty$; the result itself is false if $\mu(\Omega) = +\infty$. Indeed, if $\Omega = [0, \infty)$, μ is Lebesgue measure, and Σ is the Lebesgue σ-algebra, let $f_n = \chi_{A_n}$, where $A_n = [n, \infty)$. Then $f_n \to 0$ everywhere. But no matter what $\epsilon > 0$ is given, there can be no A_ϵ such that $\mu(A_\epsilon^c) < \epsilon$ and $f_n \to 0$ on A_ϵ uniformly: for instance, if $\epsilon < 1$, then choose $x_n = n + 2$, so that $f_n(x_n) = 1$, $f(x_n) = 0$, and $f_n(x_n) \nrightarrow 0$, i.e., $f_n \nrightarrow f = 0$ uniformly.

A feeble extension to σ-finite measures can be given, which is an immediate consequence of the above result. We state this as

Corollary 8. *Let* (Ω, Σ, μ) *be a* σ*-finite measure space. If* f_n *is an a.e. finite sequence of measurable functions converging a.e. to* f*, which is an a.e. finite*

measurable function, then there exist a disjoint sequence $A_n \in \Sigma$, $n \geq 1$, $\mu(A_n) < \infty$, *and a* μ*-null set* N *such that* $\Omega = N \cup \bigcup_{k=1}^{\infty} A_k$ *and* $f_n \to f$ *uniformly on each* A_k, $k \geq 1$.

Since $\Omega = \bigcup_{n=1}^{\infty} \Omega_n$, $\mu(\Omega_n) < \infty$, by σ-finiteness, we can apply the theorem to each Ω_n, and taking $\epsilon = 1/m$, one can find A_{mn} such that $\mu(\Omega_n - A_{kn}) < 1/k$ and $f_n \to f$ uniformly on A_{kn}. But $\mu(\Omega_n - \bigcup_{k=1}^{\infty} A_{kn}) = 0$, and the uniform convergence holds on each A_{kn}, $k \geq 1$, $n \geq 1$. Renumber and disjunctify the sequence. We may now leave the details to the reader.

Finally, if there is topology, one can relate the measurability and continuity of functions with the following important result (for integration theory) due to N. N. Luzin. The significance of the hypotheses to be imposed is seen from the remark that on a measure space $(\mathbb{R}, \Sigma)$ a continuous function $f: \mathbb{R} \to \mathbb{R}$ need not be measurable if Σ is not rich enough; i.e., if Σ does not contain all the open sets (e.g. $\Sigma = \{\emptyset, A, A^c, \mathbb{R}\}$), then this situation prevails. Thus we have to restrict μ so that $\Sigma = \mathscr{M}_\mu$ is rich enough. The desired result is as follows.

Theorem 9 (Luzin). *Let* Ω *be a (Hausdorff) topological space, and* μ *be an outer measure with* $\mathscr{M}_\mu$ *as its* σ*-algebra of* μ*-measurable sets. Suppose that if* $\mathscr{G}$ *is the collection of the open sets of* Ω*, we have* (i) $\mathscr{G} \subset \mathscr{M}_\mu$, *and* (ii) *each* $A \in \mathscr{M}_\mu$ *can be approximated by open sets from above, i.e.,*

$$\inf\{\mu(G - A): A \subset G \in \mathscr{G}\} = 0, \qquad A \in \mathscr{M}_\mu. \tag{13}$$

Then $f: \Omega \to \overline{\mathbb{R}}$ *is* μ*-measurable iff for each* $\epsilon > 0$ *there exists a closed set* C_ϵ *(hence* $C_\epsilon \in \mathscr{M}_\mu$*) such that* $\mu(C_\epsilon^c) < \epsilon$ *and* $f|C_\epsilon$ *is continuous. Moreover, if* μ *is a Radon measure (Definition* 2.3.9*) with* $\mu(\Omega) < \infty$*, then* C_ϵ *can be chosen to be compact.*

Proof. First let f be μ-measurable, and $\{I_n: n \geq 1\}$ be a base of open sets of $\mathbb{R}$. Here the I_n may be taken to be open intervals with rational end points. Then $A_n = f^{-1}(I_n) \in \mathscr{M}_\mu$, by hypothesis. By (i) and (ii), for each $\epsilon > 0$ there exists a $G_n \in \mathscr{G} \subset \mathscr{M}_\mu$ such that $G_n \supset A_n$ and

$$\mu(G_n - A_n) < \epsilon/2^{n+1}. \tag{14}$$

If $B_\epsilon = \bigcup_{n=1}^{\infty}(G_n - A_n)$ $(\in \mathscr{M}_\mu)$, then clearly

$$\mu(B_\epsilon) \leq \sum_{n=1}^{\infty} \mu(G_n - A_n) < \frac{\epsilon}{2}, \qquad \text{by (14).}$$

We claim that on $\Omega - B_\epsilon$, f is continuous. In fact, if $g = f|B_\epsilon^c$, then

$$g^{-1}(I_n) = B_\epsilon^c \cap f^{-1}(I_n) = B_\epsilon^c \cap A_n, \qquad \text{by definition.}$$

Also $B_\epsilon^c \cap A_n = B_\epsilon^c \cap G_n$, and since $G_n \supset A_n$, (expand to see) it is relatively open. Hence g is continuous. But we need to replace B_ϵ by a closed set. So using (13) again, one can find a set $(B_\epsilon \subset)\, G_\epsilon \in \mathcal{G}$ such that $\mu(G_\epsilon - B_\epsilon) < \epsilon/2$. Consequently

$$\mu(G_\epsilon) = \mu((G_\epsilon - B_\epsilon) \cup B_\epsilon) < \frac{\epsilon}{2} + \frac{\epsilon}{2} = \epsilon.$$

Let $C_\epsilon = \Omega - G_\epsilon \subset \Omega - B_\epsilon = B_\epsilon^c$. Then C_ϵ is closed, $\mu(C_\epsilon^c) < \epsilon$, and $f|C_\epsilon = g|C_\epsilon$ is continuous.

For the converse, suppose conditions (i) and (ii) and the continuity of $f|C_\epsilon$ are satisfied. Then for any closed set $J \subset \mathbb{R}$, we show that $f^{-1}(J) \in \mathcal{M}_\mu$. Let $\epsilon = 1/n$; then by (13) there is a closed $C_n \subset \Omega$ such that $\mu(C_n^c) < 1/n$ and $f|C_n$ is continuous. Hence $\mu(\bigcap_{n=1}^\infty C_n^c) = 0$, $C_n \cap f^{-1}(J)$ is relatively closed, and

$$f^{-1}(J) = \bigcup_{n=1}^{\infty} [C_n \cap f^{-1}(J)] \cup \left[\left(\bigcap_{n=1}^{\infty} C_n^c\right) \cap f^{-1}(J)\right]. \tag{15}$$

The second set on the right side is μ-null, and the first one is a countable union of relatively closed sets each of which is in $\mathcal{M}_\mu$. Thus $f^{-1}(J) \in \mathcal{M}_\mu$, and since $J \subset \mathbb{R}$ is an arbitrary closed set, f is μ-measurable.

If μ is a finite Radon measure, then $\mu(C_\epsilon) < \infty$, and by the inner regularity [see Definition 2.3.9(ii)], there exists a compact set K_ϵ such that $K_\epsilon \subset C_\epsilon$ and $\mu(C_\epsilon - K_\epsilon) < \epsilon$. Thus $f|K_\epsilon$ is continuous, and $\mu(K_\epsilon^c) \leqslant \mu(C_\epsilon^c) + \mu(C_\epsilon - K_\epsilon) < 2\epsilon$. This completes the proof of the theorem.

Since the Lebesgue measure satisfies all the hypotheses of this theorem, we have as a consequence

Corollary 10. *Let $\Omega = \mathbb{R}^n$, and (Ω, Σ, μ) be the Lebesgue measure space. Then $f: \Omega \to \mathbb{R}$ is μ-measurable iff for each $\epsilon > 0$ there is a closed set C_ϵ such that $\mu(C_\epsilon^c) < \epsilon$ and $f|C_\epsilon$ is continuous. In particular, if f vanishes outside a set of finite μ-measure, then C_ϵ can be chosen compact.*

As another immediate application of this corollary and Egorov's theorem, we have the following result.

Corollary 11. *Let (Ω, Σ, μ) be the Lebesgue measure space, and $\Omega = \mathbb{R}^n$. If $\Omega_0 \in \Sigma$, $\mu(\Omega_0) < \infty$, and $f_n: \Omega_0 \to \overline{\mathbb{R}}$, $n \geqslant 1$, is a sequence of a.e. finite measurable functions which converge a.e. to a finite measurable function $f: \Omega_0 \to \mathbb{R}$, then for each $\epsilon > 0$ there exists a measurable set $A_\epsilon \subset \Omega_0$ (an $F_{\sigma\delta}$-set) such that $\mu(\Omega_0 - A_\epsilon) < \epsilon$ and $f_n \to f$ uniformly on A_ϵ.*

This result may be regarded as an extension of the following classical result due to G. Pólya without using measure theory, but strengthening the hypothesis on the sequence: If $f_n: [a, b] \to \mathbb{R}$, $n \geqslant 1$, is a monotone increasing sequence of (not necessarily continuous) functions which converge pointwise to a continuous function $f: [a, b] \to \mathbb{R}$, then the convergence is uniform. Here both the hypothesis and the conclusion are stronger. Note that f_n being monotone implies it is Borel (hence Lebesgue) measurable on $[a, b]$, an interval of finite measure.

In the next section, we consider some consequences and complements with images of a measure under measurable mappings into another space.

EXERCISES

0. Let f_n, g_n be measurable functions on a measure space (Ω, Σ, μ).
   (a) If $f_n = g_n$ a.e., and $f_n \to f$ a.e. or in measure, show that $g_n \to f$ in the same sense.
   (b) If $f_n = g_n$ a.e., $f_n \to f$ and $g_n \to g$ in measure or a.e., then show that $f = g$ a.e.
   (c) If $f_n \to f$ and $g_n \to g$ in one of the two senses (a.e. or in measure), when can we conclude $f_n g_n \to fg$ in the same sense? [Answer this without looking at a more detailed version in a later problem.]

1. Let (Ω, Σ, μ) be a finite measure space. Show that there is a sequence f_n, $n \geqslant 1$, of real functions on Ω such that $f_n \xrightarrow{\mu} f$, having at least one subsequence $f_{n_k} \to g$ a.e., and such that $f \neq g$ a.e. [Examine the first example after Theorem 5.]

2. Let (Ω, Σ, μ) be a finite measure space, and f_n, f be extended real valued measurable functions on Ω such that $f_n \to f$ a.e. Show that for each $\epsilon > 0$, there exists an $A_\epsilon \in \Sigma$ such that $\mu(A_\epsilon^c) < \epsilon$ and $f_n(\omega) \to f(\omega)$ uniformly on A_ϵ even though both f_n and f may take infinite values on sets of positive measure. (Consider $\varphi \circ f_n$ and $\varphi \circ f$, where φ is as in Proposition 6.)

3. Let (Ω, Σ, μ) be a topological measure space satisfying the hypothesis of Theorem 9. If $f: \Omega \to \mathbb{R}$ is a μ-measurable function, show that there exists a $g: \Omega \to \mathbb{R}$ such that g is Borel measurable [i.e. relative to $\sigma(\mathscr{G}) \subset \mathscr{M}_\mu$ of Theorem 9] and that $f = g$ a.e. This is applicable in particular for any Radon measure space.

4. Complete the proof of (a) Corollary 8 and (b) Corollary 11.

5. Let $f_n: \Omega \to \mathbb{R}$, $n \geqslant 1$, be a sequence which is almost μ-uniformly Cauchy, i.e.,, for each $\epsilon > 0$, there is an n_0 such that $\mu(A_\epsilon) < \epsilon$, and the condition holds on A_ϵ^c (μ = outer measure). Verify that there exists an $f: \Omega \to \mathbb{R}$ such that $f_n \to f$ a.u., and if each f_n is μ-measurable then f can be chosen to be μ-measurable.

6. Suppose that (Ω, Σ, μ) is a measure space and $f_n: \Omega \to \mathbb{R}$, $n \geqslant 1$, is a sequence such that $f_n \xrightarrow{\mu} f$. Show that there is a subsequence $f_{n_k} \to f$ a.u.

Deduce that, conversely, if each subsequence of $\{f_n: n \geq 1\}$ has a further subsequence which converges a.u. to f, then $f_n \xrightarrow{\mu} f$. Give an example to show that in the last statement a.u. cannot be replaced by a.e., unless $\mu(\Omega) < \infty$.

7. The calculus of convergent sequences for a.e. and in-measure is really different in some cases. Verify the following statements for $f_n, g_n: \Omega \to \mathbb{R}$, where (Ω, Σ, μ) is a measure space:

(a) If $f_n \to f$, $g_n \to g$, a.e. or in measure, then $af_n + bg_n \to af + bg$, in the same sense, as $n \to \infty$.

(b) If f_n, f are as in (a), then $f_n^{\pm} \to f^{\pm}$ in the same sense.

(c) If f_n, g_n are as in (a), then $f_n g_n \to fg$ for a.e. convergence, but not for convergence in measure. Give an example to illustrate the latter. Show, however, that the result holds if $\mu(\Omega) < \infty$. In this case, if $g \neq 0$ a.e., then $f_n/g_n \to f/g$ both a.e. and in measure. [Again the second statement is false if $\mu(\Omega) = \infty$.]

(d) If A_n, A are sets in Ω and $f_n = \chi_{A_n}$, $f = \chi_A$, verify that, in the notation of Proposition 6, $d(f_n - f) \to 0$ iff $\mu(A_n \triangle A) \to 0$, where μ stands for an outer measure in case the sets are not measurable.

8. If $f: \Omega \to \overline{\mathbb{R}}$ and μ is an outer measure, define $\|f\|_\infty$ as the essential supremum of $|f|$ [$= \inf\{\alpha > 0: \mu(\{\omega: |f(\omega)| > \alpha\}) = 0\}$]. Let $\mathscr{X} = \{f: \Omega \to \mathbb{R}, \|f\|_\infty < \infty\}$. Show that $\{\mathscr{X}, \|\cdot\|_\infty\}$ is a linear complete semi-metric space. Since for any scalar a we have $\|af\|_\infty = |a| \|f\|_\infty$, the (semi-)distance function $\|\cdot\|_\infty$ on $\mathscr{X}$ is called a (semi-)*norm*. If equivalence classes are identified, ($\tilde{\mathscr{X}} = \mathscr{X}/\sim$, where $f \sim g$ iff $f = g$ a.e.), then $(\tilde{\mathscr{X}}, \|\cdot\|_\infty)$ is a complete normed linear space, called a *Banach space*, where $\tilde{f} \in \tilde{\mathscr{X}} \Rightarrow \|\tilde{f}\|_\infty = \|f\|_\infty$, $f \in \tilde{f} = \{g: f \sim g\}$.

9. If the measure space (Ω, Σ, μ) is σ-finite and complete (or more generally localizable), we can assert the existence of suprema of (not necessarily countable) collections of μ-measurable functions. Establish this by using the localizability of these measures. If $\mathscr{F}$ is a collection of μ-measurable functions on Ω into $\overline{\mathbb{R}}$, one says $g: \Omega \to \overline{\mathbb{R}}$ is a *supremum* of $\mathscr{F}$ if (i) g is μ-measurable, (ii) $g \geq f$ a.e. for each $f \in \mathscr{F}$; and if $\tilde{g}$ is another function with properties (i) and (ii), then $\tilde{g} \geq g$ a.e., so that g is the smallest μ-measurable function which majorizes $\mathscr{F}$. Show that for every such collection $\mathscr{F}$, there exists a supremum. [Let $A_{f,r} = \{\omega: f(\omega) \geq r\}$, where $f \in \mathscr{F}$ and r is a rational, and let B_r be the supremum of $\{A_{f,r}: f \in \mathscr{F}\} \subset \Sigma$ (cf. Definition 2.3.4). If $r_1 < r_2$ then $B_{r_1} \supset B_{r_2}$ a.e. and if $C_r = \bigcup_{r_1 > r} B_{r_1}$, we get $C_r \in \Sigma$, $C_r \subset C_{r'}$ for $r > r'$. Define $g_r(\omega) = r$ for $\omega \in C_r$, and $= -\infty$ otherwise. If $g = \sup\{g_r: r \text{ rational}\}$, then g is μ-measurable. Verify that it also satisfies conditions (ii) and (iii), so that it is a supremum. Observe that if $\mathscr{F}$ is countable, this procedure simplifies to yield our previous Lemma 1.3.]

10. Let Ω be a Hausdorff topological space, and (Ω, Σ, μ) be a Radon measure space (Definition 2.3.9). If $f: \Omega \to \overline{\mathbb{R}}$ is an a.e. finite measurable function, verify that there exists a sequence of continuous functions $f_n: \Omega \to \mathbb{R}$ such that $f_n \xrightarrow{\mu} f$, and moreover, if $\mu(\Omega) < \infty$, then each f_n can be chosen to have a compact support.

11. Let Ω be a set, and μ be an outer measure on it. If $\mathscr{F}$ is the class of all real μ-measurable functions on Ω, let $\mathscr{N}$ be the subset of all μ-null functions such that $f \in \mathscr{N}$ is an ideal (in the ordinary sense of linear algebra). Show that $\mathscr{F}$ is a vector space, and is closed under a.e. pointwise multiplication, and $\mathscr{N} \subset \mathscr{F}$ has the same properties plus $g \in \mathscr{F}$, $f \in \mathscr{N}$ $\Rightarrow fg \in \mathscr{N}$.

3.3. IMAGE MEASURES AND VAGUE CONVERGENCE

In this section we present a few results on the behavior of measures and their images under measurable mappings. These image measures play an important role in certain applications, especially in probability and statistics, but also in classical harmonic analysis among others.

The concept of an image measure appeared briefly in Exercise 2.2.7. However, to obtain some interesting results, as the following work reveals, our measures must be finite. Recall that if Ω, $\tilde{\Omega}$ are two point sets and $f: \Omega \to \tilde{\Omega}$ is a mapping, then the *inverse image* f^{-1} of f is the set function on the power set of $\tilde{\Omega}$ into that of Ω where $f^{-1}(A) = \{\omega: f(\omega) \in A\}$ for each $A \subset \tilde{\Omega}$. In case $A \subset \tilde{\Omega} - f(\Omega)$, then $f^{-1}(A)$ is defined to be the empty set (and then f^{-1} is sometimes called a "complete inverse image").

We first record some elementary operations with f and f^{-1} for ready reference, even though we have used some of them already.

Lemma 1. *Let $f: \Omega \to \tilde{\Omega}$ be a mapping and $A_i \subset \Omega$, $i \in I$, be a family of sets. Also let $B_j \subset \tilde{\Omega}$, $j \in J$, be a similar family. Then*

(i) $f(\bigcup_{i \in I} A_i) = \bigcup_{i \in I} f(A_i)$;

(ii) $f(A_1 \cap A_2) \subset f(A_1) \cap f(A_2)$; *and* $f(\bigcap_{i \in I} A_i) = \bigcap_{i \in I} f(A_i)$ *for every such family iff f is one to one*;

(iii) $f(A_1^c) = (f(A_1))^c$ *for every* $A_1 \subset \Omega$ *iff f is one to one and onto*;

(iv) $f^{-1}(\bigcap_{j \in J} B_j) = \bigcap_{j \in J} f^{-1}(B_j)$, $f^{-1}(\bigcup_{j \in J} B_j) = \bigcup_{j \in J} f^{-1}(B_j)$, $f^{-1}(B_1 - B_2) = f^{-1}(B_1) - f^{-1}(B_2)$ *for any pair* B_1, B_2;

(v) $f(f^{-1}(B)) \subset B$, *and there is equality for all* $B \subset \tilde{\Omega}$ *iff f is onto*;

(vi) $f^{-1}(f(A)) \supset A$, *and there is equality for all* $A \subset \Omega$ *iff f is one to one.*

Proof. All these relations follow from the definition. As an example let us establish (v). Let $y \in f(f^{-1}(B))$. So there is an $x \in f^{-1}(B)$ such that $y = f(x)$. Also there must be a $y' \in B$ such that $f(x) = y'$. Since f is a function, we must have $y = f(x) = y' \in B$, so that $f(f^{-1}(B)) \subset B$. If f is onto and $B \subset \tilde{\Omega}$ is arbitrary, we need to verify the opposite inclusion. Thus let $y \in B$. Since

$f(\Omega) = \tilde{\Omega}$, there is an $x \in \Omega$ with $y = f(x)$, so that $x \in f^{-1}(B)$, i.e., $y = f(x) \in f(f^{-1}(B))$. Consequently $B \subset f(f^{-1}(B))$, and thus equality holds.

Conversely, let $f(f^{-1}(B)) = B$ for each $B \subset \tilde{\Omega}$. If $f(\Omega) \neq \tilde{\Omega}$, then there exists a $y_1 \in \tilde{\Omega} - f(\Omega)$. If $B = \{y_1\}$, then $f^{-1}(B) = \varnothing$, and $f(f^{-1}(B)) = \varnothing \neq B$, a contradiction. So $f(\Omega) = \tilde{\Omega}$ must be true. The other relations are entirely analogous.

Let (Ω, Σ, μ) be a measure space and $(\tilde{\Omega}, \tilde{\Sigma})$ be a (possibly distinct) measurable space. If $f: \Omega \to \tilde{\Omega}$ is a measurable function [i.e., $f^{-1}(\tilde{\Sigma}) \subset \Sigma$], then the image of μ under f on $\tilde{\Sigma}$, denoted $\mu \circ f^{-1}$, μ_f, or ν, is the map:

$$\nu: A \mapsto \nu(A) = \mu_f(A) = \mu(f^{-1}(A)), \qquad A \in \tilde{\Sigma}. \tag{1}$$

It is clear that ν is σ-additive and nonnegative on $\tilde{\Sigma}$, and $\nu(\varnothing) = 0$, so that it is a measure. Since, as noted in Exercise 2.2.7, the class $\mathscr{A} = \{A \subset \tilde{\Omega}: f^{-1}(A) \in \Sigma\}$ is a σ-algebra containing $\tilde{\Sigma}$ when f is measurable relative to Σ and $\tilde{\Sigma}$, it follows that ν is σ-additive on $\mathscr{A}$ also. So $\tilde{\Sigma} \subset \mathscr{A} \subset \mathscr{M}_\nu$, the largest class of ν-measurable sets. The last inclusion follows from the fact that one may define the generated outer measure ν^* by the pair $(\nu, \mathscr{A})$, and then $\mathscr{M}_{\nu^*}$ ($= \mathscr{M}_\nu$ by definition here) is the largest class on which ν^* is σ-additive, and $\mathscr{A} \subset \mathscr{M}_\nu$ with $\nu^*|\mathscr{A} = \nu$ by Theorem 2.2.10. However, there may be other extensions of ν from $\tilde{\Sigma}$ to $\mathscr{A}$ and beyond, so that unpleasant phenomena including nonuniqueness can result.

An important reason for a study of the image measure by a function from an abstract (Ω, Σ, μ) to another (usually topological) measurable space is to take advantage of a finer structure of the range space. If $(\tilde{\Omega}, \tilde{\Sigma})$ is $(\mathbb{R}, \mathscr{B})$, where $\mathscr{B}$ is the Borel σ-algebra of $\mathbb{R}$, then there is both a topology and a linear order structure for $\mathbb{R}$. If $f = (f_1, \ldots, f_n)$ is a vector of real valued components, we take $\tilde{\Omega} = \mathbb{R}^n$. Let us now specialize this aspect. Taking $A = (-\infty, x)$ in (1), $x \in \mathbb{R}$, we may associate a *point function* F with the set function ν as

$$F_f(x) = \nu((-\infty, x)) = \mu(\{\omega: f(\omega) < x\}). \tag{2}$$

If $\epsilon_n \downarrow 0$, then

$$\lim_{n \to \infty} F_n(x - \epsilon_n) = \lim_{n \to \infty} \mu(\{\omega: f(\omega) < x - \epsilon_n\})$$

$$= \mu\left(\bigcup_{n=1}^{\infty} \{\omega: f(\omega) < x - \epsilon_n\}\right)$$

(by Proposition 2.1.1, since the sets are increasing)

$$= \mu(\{\omega: f(\omega) < x\}) = F_f(x).$$

Thus $F_f(\cdot)$ is left continuous and is clearly nondecreasing. However, if μ is the trivial two valued $\{0, \infty\}$-measure or is infinite on proper subsets of Ω, even with the finite subset property, the monotone function F_f will not be very useful. Further restrictions on μ are necessary. The finiteness of μ is a natural assumption but others such as σ-finiteness can be made. In case $\mu(\Omega) < \infty$, one can also define

$$\tilde{F}_f(x) = \mu(\{\omega: f(\omega) \leqslant x\}), \qquad x \in \mathbb{R}, \tag{3}$$

and easily verify that $\tilde{F}_f$ is right continuous. Moreover, from (2) and (3), $\lim_{x\to-\infty} F_f(x) = 0 = \lim_{x\to-\infty} \tilde{F}_f(x)$. When $0 < \mu(\Omega) < \infty$, one normalizes the space so that $\mu(\Omega) = 1$ can be taken for convenience (and then μ is called a *probability measure*). With such a normalization the image ν of μ under f is termed an induced probability, and F_f (or $\tilde{F}_f$) the (probability) *distribution function* of f. But now, convergence in measure (or probability) gives rise to a new concept as follows.

To motivate the definition, let (Ω, Σ, μ) be a measure space with $0 < \mu(\Omega) < \infty$, and $f_n(\omega) = (-1)^n/n$, $n \geqslant 1$, $\omega \in \Omega$. Then the f_n are measurable and $f_n \to f = 0$ pointwise. Let F_n, F be their distribution functions, so that

$$F_n(x) = \begin{cases} 0, & \text{if } x < (-1)^n/n, \\ \mu(\Omega) & \text{if } x \geqslant (-1)^n/n \end{cases}$$

$$F(x) = \begin{cases} 0 & \text{if } x < 0, \\ \mu(\Omega) & \text{if } x \geqslant 0. \end{cases}$$

However, $F_n(0) = \mu(\Omega)$ if n is even, $= 0$ if n is odd, so that $\{F_n(0): n \geqslant 1\}$ does not converge to $F(0)$, even though $F_n(x) \to F(x)$ if $x \in \mathbb{R} - \{0\}$. The trouble is that 0 is not a continuity point of F. Since F is monotone nondecreasing and left continuous, and since there can be at most a countable set of discontinuities for a monotone function, we may exclude these points from consideration. Bypassing these problems leads to a weaker concept of convergence, called "vague" convergence by N. Bourbaki, and stated precisely as follows.

Definition 2. Let (Ω, Σ, μ) be a finite measure space, and $f, f_n: \Omega \to \mathbb{R}$, $n \geqslant 1$, be a sequence of measurable functions. If F, F_n are the distribution functions of f, f_n respectively, then f_n is said to *converge in distribution* to f, written $f_n \xrightarrow{D} f$, if $F_n(x) \to F(x)$ at all the continuity points x of F.

Since $\nu_n = \mu_{f_n}$ is a finite measure on $(\mathbb{R}, \mathscr{B})$, and $\nu_n([a, b)) = F_n(b) - F_n(a) = \mu(\{\omega: a \leqslant f_n(\omega) < b\})$ for all $a < b$ in $\mathbb{R}$, the above definition can be interpreted as saying that a sequence of finite measures ν_n on $(\mathbb{R}, \mathscr{B})$ such that $\nu_n(\mathbb{R}) \leqslant M < \infty$, $n \geqslant 1$, *converges vaguely* to a finite measure ν on $\mathscr{B}$ if there exists a dense set of points $B \subset \mathbb{R}$ such that

$$\nu_n([a, b)) \to \nu([a, b)), \qquad a < b, \quad a, b \in B. \tag{4}$$

In this case we may use the notation $\nu_n \xrightarrow{v} \nu$. Note that this is defined for all such uniformly bounded measures on $(\mathbb{R}, \mathscr{B})$, even if they are not related to distribution functions. This generality is however illusory, since it is always possible to manufacture a finite measure space and a sequence of measurable mappings on it such that each ν_n is an image measure of that measure space under these mappings. A proof of this statement follows from Theorem 5 below (cf. Exercise 6).

The relations between convergences a.e., in-measure, and in-distribution are given by the following:

Proposition 3. *Let (Ω, Σ, μ), $\mu(\Omega) < \infty$, be a measure space, and $f_n, f, n \geqslant 1$, be a.e. finite measurable functions on Ω. Then $f_n \to f$ a.e. $\Rightarrow$ $f_n \xrightarrow{\mu} f$, and $f_n \xrightarrow{\mu} f$ $\Rightarrow$ $f_n \xrightarrow{D} f$. If $f = c$ a.e., a constant, then $f_n \xrightarrow{D} c$ $\Rightarrow$ $f_n \xrightarrow{\mu} c$ also, and that the last two are equivalent iff f is a constant.*

Proof. The first implication has already been shown in Theorem 2.5(iii). So for the second one let $f_n \xrightarrow{\mu} f$, and F_n, F be distribution functions of f_n, f. If $x < y$, then from the inclusions

$$\begin{aligned}\{\omega: f(\omega) < x\} &= (\{\omega: f(\omega) < x\} \cap \{\omega: f_n(\omega) < y\}) \\ &\quad \cup(\{\omega: f(\omega) < x\} \cap \{\omega: f_n(\omega) \geqslant y\}) \\ &\subset (\{\omega: f_n(\omega) < y\}) \cup (\{\omega: |f_n - f|(\omega) \geqslant y - x\})\end{aligned}$$

one gets

$$\begin{aligned}F(x) = \mu(\{\omega: f(\omega) < x\}) &\leqslant \mu(\{\omega: f_n(\omega) < y\}) \\ &\quad + \mu(\{\omega: |f_n - f|(\omega) \geqslant y - x\}). \qquad (5)\end{aligned}$$

Taking $\epsilon = y - x > 0$, and letting $n \to \infty$, (5) implies

$$F(x) \leqslant \liminf_n F_n(y), \qquad \text{since} \quad f_n \xrightarrow{\mu} f. \qquad (6)$$

Interchanging f_n and f in (5), and replacing x, y by $y < z$, we get

$$F_n(y) \leqslant F(z) + \mu(\{\omega: |f - f_n|(\omega) \geqslant (z - y)\}).$$

Letting $n \to \infty$, this gives similarly

$$\limsup_n F_n(y) \leqslant F(z), \qquad \text{since} \quad f_n \xrightarrow{\mu} f. \qquad (7)$$

Thus for any $x < y < z$, (6) and (7) yield

$$F(x) \leqslant \liminf_n F_n(y) \leqslant \limsup_n F_n(y) \leqslant F(z). \qquad (8)$$

If y is a continuity point of F, letting $x \uparrow y$ and $z \downarrow y$, (8) implies that the extreme terms, which are finite since $\mu(\Omega) < \infty$, tend to $F(y)$. Hence $F_n(y) \to F(y)$. Since y is any continuity point, and such points are everywhere dense in $\mathbb{R}$, we conclude that $f_n \xrightarrow{D} f$.

For the last statement, let $f = c$ a.e. Then $F(x) = 0$ if $x < c$, and $= \mu(\Omega)$ if $x \geqslant c$. Then for each $\epsilon > 0$

$$\begin{aligned}\mu(\{\omega: |f_n - c|(\omega) \geqslant \epsilon\}) &= \mu(\{\omega: f_n(\omega) \geqslant c + \epsilon\}) + \mu(\{\omega: f_n(\omega) \leqslant c - \epsilon\}) \\ &= \mu(\Omega) - F_n(c+\epsilon) + F_n(c-\epsilon) \\ &\to \mu(\Omega) - F(c+\epsilon) + F(c-\epsilon) \\ &= \mu(\Omega) - \mu(\Omega) = 0,\end{aligned}$$

since $c \pm \epsilon$ are continuity points of F and $F(c+\epsilon) = \mu(\Omega)$, $F(c-\epsilon) = 0$. Thus $F_n \xrightarrow{\mu} c$, and the proof is complete, because of the Example A below.

If the limit f of the f_n is at least a three valued (or even a two valued) function, then the last statement does not hold, as seen from the following (the two valued case is even simpler and is left to the reader).

Example 4. Let $\Omega = \{a, b, c\}$, Σ = power set of Ω, $\mu(\{a\}) = \mu(\{b\}) = \mu(\{c\}) = \frac{1}{3}$. If $f, f_n: \Omega \to \mathbb{R}$ are defined by

$$f_n(a) = 1, \quad f_n(b) = 2, \quad f_n(c) = 3, \quad n \geqslant 1,$$

and

$$f(a) = 3, \quad f(b) = 1, \qquad f(c) = 2,$$

then $|f_n - f|(\omega) = 1$ for all $\omega \in \Omega$. But $F_n(x) = F(x)$, $x \in \mathbb{R}$, where

$$F(x) = \begin{cases} 0, & x < 1, \\ \frac{1}{3}, & 1 \leqslant x < 2, \\ \frac{2}{3}, & 2 \leqslant x < 3, \\ 1, & x \geqslant 3. \end{cases}$$

Thus $f_n \xrightarrow{D} f$ trivially, but $\mu(\{\omega: |f_n - f|(\omega) \geqslant \epsilon\}) = 1$ for $0 < \epsilon < 1$.

In view of the above example, one may want to know whether $F_n \to F$ and $F_n \to G$ imply $F = G$. This means $F_n(x) \to F(x)$ and $F_n(x) \to G(x)$ at all continuity points of F and G. Since these are distributions, it follows that each of the discontinuity sets D_1, D_2 of these functions, and hence $D_1 \cup D_2$, is countable. So $\mathbb{R} - (D_1 \cup D_2) = A$ is dense in $\mathbb{R}$, and $F(x) = G(x)$ for all

$x \in A$. If $a < b$, a, b in A, then $\nu_1([a, b)) = F(b) - F(a) = G(b) - G(a) = \nu_2([a, b))$, and the additive set functions ν_1, ν_2 agree on the class of elementary sets $\mathscr{E}$ of $\mathbb{R}$. The left continuity of F, G implies that ν_1, ν_2 are σ-additive, and by Theorem 2.3.1, $\nu_1 = \nu_2$ on $\mathscr{B}$, so that $F(x) = \nu_1((-\infty, x)) = \nu_2((-\infty, x)) = G(x)$ for all $x \in \mathbb{R}$. Thus the limits are unique. (However, $f_n \xrightarrow{D} f$, $f_n \xrightarrow{D} g \nRightarrow f = g$, a.e. as the above example shows.)

In spite of the example and proposition, the following result, due to A. V. Skorokhod, shows that there is a hidden equivalence between a.e. convergence on one space and convergence in distribution in *another* space. More precisely:

Theorem 5 (Skorokhod). *Let (Ω, Σ, μ) be a finite measure space and f_n, $f: \Omega \to \mathbb{R}$ be measurable functions such that $f_n \xrightarrow{D} f$. Then on the Lebesgue measure space $(I, \mathscr{B}, \lambda)$, where $I = (0, \mu(\Omega))$, $\mathscr{B}$ is the Borel σ-algebra of I, and λ is the Lebesgue measure, there exist measurable functions g_n, $g: I \to \mathbb{R}$ such that $g_n(t) \to g(t)$ a.e.$[\lambda]$, and*

$$\mu(\{\omega: f_n(\omega) < x\}) = \lambda(\{t: g_n(t) < x\})$$

$$\mu(\{\omega: f(\omega) < x\}) = \lambda(\{t: g(t) < x\}), \tag{9}$$

$x \in \mathbb{R}$, $n \geqslant 1$.

Proof. Excluding the trivial case that $\mu(\Omega) = 0$, we may assume (by normalization) that $\mu(\Omega) = 1$, for convenience. Then $(I, \mathscr{B}, \lambda)$ is the Lebesgue unit interval. Let F_n and F be the distribution functions of f_n and f, so that $F_n(x) \to F(x)$ at all continuity points x of F. We can now define g_n and g as the (generalized) inverses of F_n and F. In fact, if $0 < t < 1$, let

$$g_n(t) = \inf\{x: F_n(x) > t\}, \qquad g(t) = \inf\{x: F(x) > t\}. \tag{10}$$

If F_n and F are strictly increasing, then g_n, g are the usual inverse functions of F_n, F. If F_n is not strictly increasing or has discontinuities, then (10) gives g_n uniquely, and $\{x: F_n(x) > t\}$ is a left open infinite interval $(g_n(t), \infty)$ by the left continuity of F_n. Thus $g_n(b) < x$ holds iff $F_n(x) > t$. Hence g_n on I is a (Borel) measurable function. Likewise g is a (Borel) measurable function. Also

$$\lambda(\{t: g_n(t) < x\}) = \lambda(\{t: t < F_n(x)\}) = F_n(x), \qquad x \in \mathbb{R}. \tag{11}$$

Similarly the second equation holds and (9) follows.

Next, using the fact that $F_n \to F$, we deduce that $g_n \to g$ a.e.$[\lambda]$. Let $0 < t < 1$ and $0 < \epsilon < 1$ be given. Then there exists a continuity point x of F in the open interval $(g(t) - \epsilon, g(t))$, and $F(x) < t$. But $F_n(x) \to F(x)$ for this x, so that there is an $n_0 = n_0(\epsilon, t)$ such that $n \geqslant n_0 \Rightarrow F_n(x) < t$; whence $g(t) - \epsilon < x < g_n(t)$. Thus $\liminf_n g_n(t) \geqslant g(t)$ [corresponding to

(6)]. Similarly, for $s > t$ let y be a continuity point of F in the open interval $(g(s), g(s)+\epsilon)$. Hence $t < s \leqslant F(g(s)) \leqslant F(y)$, by choice of y. So there exists n_1 $(= n_1(s, \epsilon))$ such that $n \geqslant n_1 \Rightarrow F_n(y) > t$, and then $g_n(t) < y < g(s) + \epsilon$. Consequently $\limsup_n g_n(t) \leqslant g(s)$ if $s > t$. Hence

$$g(t) \leqslant \liminf_n g_n(t) \leqslant \limsup_n g_n(t) \leqslant g(s). \tag{12}$$

Since g (and g_n) is nondecreasing, let t be in its continuity set which is everywhere dense in I. Letting $s \downarrow t$, (12) implies that $g_n(t) \to g(t)$ on that set. But the set of discontinuities of g is at most countable and so has λ-measure zero. So $g_n \to g$ a.e.[λ]. Note that if we now redefine both g_n and g to be of any fixed value (= 1, say, so that g_n, g are not necessarily monotone any more —but this is irrelevant here), then $g_n(t) \to g(t)$ for all $t \in I$, and again (9) holds. This completes the proof.

This result shows how image measure spaces bring several new ideas into measure theory, which usually studies structural properties of functions and measures on one given measure space. The above result shows, on the other hand, how new and surprising things can happen if the distribution functions (including the monotone functions associated with the Lebesgue-Stieltjes measures on $\mathbb{R}$ or $\mathbb{R}^n$) are fixed. We shall indicate later how the integration theory can be enriched with these ideas, using results of the above type. Further image measures, in the form of distribution functions, play a central role in probability and statistics, and thus their theory has branched out and has taken a distinct character of its own.

There is a closely related concept called "equimeasurability" which is useful in the study of function spaces and Fourier analysis. Thus let (Ω, Σ, μ) be a finite measure space and $f: \Omega \to \mathbb{R}$ be a measurable function. Let $d_f(t) = \mu(\{\omega: f(\omega) \geqslant t\})$, $t \in \mathbb{R}$. Since $\mu(\Omega) < \infty$, it is clear that $d_f(t) = \mu(\Omega) - F_f(t)$, where F_f is the distribution function of f as in (2). Then $d_f(\cdot)$ is a left-continuous decreasing function, $d_f: \mathbb{R} \to \mathbb{R}^+$, and is Borel measurable. If $f, g: \Omega \to \mathbb{R}$ are a pair of measurable functions, then they are called *equimeasurable* relative to μ if $d_f(t) = d_g(t)$, $t \in \mathbb{R}$. Defining the (generalized) inverse f^* of the monotone function $d_f(\cdot)$ as in (10), we say that f^* is a *decreasing rearrangement* of f relative to μ. Thus for $0 \leqslant x \leqslant \mu(\Omega)$,

$$f^*(x) = \inf\{t: d_f(t) < x\} = \sup\{t: d_f(t) > x\}, \tag{13}$$

with, as usual, $\inf\{\varnothing\} = +\infty$ and $\sup\{\varnothing\} = -\infty$. Moreover, f^* and d_f are (generalized) inverses of each other. Note that for each measurable $f: \Omega \to \mathbb{R}$, we have that $f^*(\cdot)$ on $(\tilde{\Omega}, \mathscr{B}, \lambda)$ is measurable, where $\tilde{\Omega} = [0, \mu(\Omega))$, and since $d_{f^*}(t) = d_f(t)$, $t \in \mathbb{R}$ (which follows from definition), f and f^* *are* (by definition) *equimeasurable*, since they have the same $d(\cdot)$-functions on $\mathbb{R}$, even though they are defined on different finite measure spaces of equal total measure. (See Exercise 9.) This also implies that if $f, g: \Omega \to \mathbb{R}$ are equimeasurable relative to μ, and f^*, g^* are their decreasing rearrangements, then $f^* = g^*$ a.e.[λ].

The following result collects some elementary properties, which are of interest in applications, particularly after the theory of integration is presented.

Proposition 6. *Let* (Ω, Σ, μ) *be a finite measure space and* $f, g: \Omega \to \mathbb{R}$ *be measurable. Then we have for the decreasing rearrangements on* $([0, \mu(\Omega)), \mathscr{B}, \lambda)$:

(i) $f \leqslant g$ *a.e.*$[\mu]$ $\Rightarrow$ $f^* \leqslant g^*$ *a.e.*$[\lambda]$, *and* $d_f = d_g$ *iff* $d_{f^\pm} = d_{g^\pm}$, *so that* f, g *are equimeasurable iff* $f^\pm$ *and* $g^\pm$ *have the same property, where* $f^+ = \max(f, 0)$ *and* $f^- = f^+ - f$.

(ii) $(af^*) = af^*$ *for* $a \geqslant 0$, *and* $d_{af}(t) = d_f(t/a)$ *for* $a > 0$, $t \in \mathbb{R}$.

(iii) *If* $h: \mathbb{R} \to \mathbb{R}$ *is continuous and increasing, then* $[h(f)]^* = h(f^*)$ *a.e.*$[\lambda]$.

(iv) *If* $f, f_n: \Omega \to \mathbb{R}$ *are measurable, and if* $g, g_n: \Omega \to \mathbb{R}$ *are measurable such that* $d_{f_n} = d_{g_n}$, $n \geqslant 1$ *and* $f_n(\omega) \to f(\omega)$, $g_n(\omega) \to g(\omega)$, $\omega \in \Omega$, *then* $d_f = d_g$ *a.e.*$[\lambda]$.

(v) *If* $f_n: \Omega \to \mathbb{R}$ *is measurable,* $n \geqslant 1$, $f_n \xrightarrow{\mu} 0$, *then* $f_n^* \to 0$ *uniformly on compact subsets of* $[0, \mu(\Omega))$.

(vi) *If* $f: \Omega \to \mathbb{R}$ *is measurable, then* $f^*(t) = (f^+)^*(t) - (f^-)^*(\mu(\Omega) - t)$, *and* $|f|^*(t) = (f^+)^*(t) + (f^-)^*(\mu(\Omega) - t) \geqslant \frac{1}{2}[(f^+)^*(t) + (f^-)^*(t)]$ *for a.a.* t *in* $[0, \mu(\Omega))$.

Proof. All these statements follow from definitions. For instance we prove (vi). Consider $d_f(\cdot)$, i.e.,

$$d_f(t) = \mu(\{\omega: f^+(\omega) - f^-(\omega) > t\})$$

$$= \begin{cases} \mu(\{\omega: f^+(\omega) > t\}) = d_{f^+}(t) & \text{if } t \geqslant 0, \\ \mu(\Omega) - \mu(\{\omega: -f(\omega) < t\}) = \mu(\Omega) - d_{f^-}(t) & \text{if } t < 0. \end{cases}$$

Hence taking the inverses we get that if $d_f(x) < t \Rightarrow x \geqslant 0$, then $f^* = (f^+)^*$ a.e., and if $x < 0$, then

$$f^*(t) = \inf\{x: \mu(\Omega) - d_{f^-}(-x) < t\}$$

$$= \inf\{x: d_{f^-}(-x) > \mu(\Omega) - t\} = -(f^-)^*(\mu(\Omega) - t).$$

The other parts need similar reasoning.

In the next chapter we develop the basic theory of abstract integration of real functions on general measure spaces. All the above work will be utilized. Then new and fundamental results are established and used in the rest of the book.

EXERCISES

0. Let (Ω, Σ, μ) be a finite measure space, and in fact let $\mu(\Omega) = 1$. If $f: \Omega \to \mathbb{R}$ is a measurable function and F_f is its distribution function, and $\{x_i: i \in I\}$ is the set of the discontinuities of F_f, calculate

$$\lim_{\epsilon \downarrow 0} \sum_{x_i \in (t-\epsilon, t)} (F(x_i) - F(x_i - 0)).$$

If F_f has no discontinuities, find the distribution of the new function $G_f = F_f \circ f (= F_f(f): \Omega \to [0,1])$.

1. Complete the proof of all the assertions given in Lemma 1.

2. Let (Ω, Σ, μ) be a finite measure space. If $\{f, f_n: n \geqslant 1\}$ and $\{g, g_n: n \geqslant 1\}$ are two sequences of real measurable functions such that $f_n \xrightarrow{\mu} f$, $g_n \xrightarrow{\mu} g$, show that $h(f_n, g_n) \xrightarrow{\mu} h(f, g)$ for any continuous function $h: \mathbb{R}^2 \to \mathbb{R}$. [Continuity of h on a metric space implies its sequential continuity.]

3. Let the hypothesis of the above exercise hold. If $h: \mathbb{R}^2 \to \mathbb{R}$ is a Borel function whose discontinuity points form a set of ν-measure zero, where $\nu(A) = \mu(\{\omega: (f(\omega), g(\omega)) \in A\})$, if $A \subset \mathbb{R}^2$ is Borel, and if $F_{f,g}(x, y) = \nu((-\infty, x) \times (-\infty, y))$, $x, y \in \mathbb{R}$ (here $F_{f,g}$ is termed the *joint distribution* of f, g), show that $h(f_n, g_n) \xrightarrow{D} h(f, g)$. Show also that the same statement holds if $\alpha f_n + \beta g_n \xrightarrow{D} \alpha f + \beta g$ for each pair of reals α, β is substituted for the hypothesis of convergence in measure. [Note that the discontinuity points of h form a Borel set.]

4. If (Ω, Σ, μ) and $\{f, f_n: n \geqslant 1\}$ and $\{g_n: n \geqslant 1\}$ are as in Exercise 2, establish the following (Slutsky's theorem):

(a) Let $f_n \xrightarrow{D} f$ and $g_n \xrightarrow{D} \alpha \neq 0$, where α is a constant. Then $f_n/g_n \xrightarrow{D} f/\alpha$.

(b) If the second part of (a) is replaced by $(f_n - g_n) \xrightarrow{\mu} 0$, conclude that $g_n \xrightarrow{D} f$.

5. Expanding the discussion following the definition of vague convergence, show that a uniformly bounded sequence $\{\nu, \nu_n: n \geqslant 1\}$ of measures on $(\mathbb{R}, \mathscr{B})$ converges vaguely iff $\nu_n((a, b]) \to \nu((a, b])$ for all left open intervals $(a, b]$ such that neither $\{a\}$ nor $\{b\}$ is an atom of ν, so that $\nu((a, b)) = \nu([a, b]) = \nu([a, b)) = \nu((a, b])$. Such a set is called a *continuity interval* of ν.

6. Let $\{\nu, \nu_n: n \geqslant 1\}$ be probability measures on $(\mathbb{R}, \mathscr{B})$, and $f: \mathbb{R} \to \mathbb{R}$ be measurable. If $\nu_n \xrightarrow{v} \nu$ and the discontinuity points of f form a set of

ν-measure zero, then deduce from Theorem 2.6.1(vi) and Skorokhod's theorem that $\tilde{\nu}_n \xrightarrow{v} \tilde{\nu}$, where $\tilde{\nu}_n = \nu_n \circ f^{-1}$ and $\tilde{\nu} = \nu \circ f^{-1}$. (A similar statement in $(\mathbb{R}^n, \mathscr{B})$ is true, but first one has to prove the $\mathbb{R}^n$-version of Theorem 5.)

7. Prove Pólya's theorem, noted at the end of the last section, in the following form: Let (Ω, Σ, μ) be a finite measure space with $\{f, f_n: n \geqslant 1\}$ as measurable real functions on Ω such that $f_n \xrightarrow{D} f$ and $\mu \circ f^{-1}$ is diffuse. Then the distribution functions F_n, F of f_n, f have the property that $F_n(x) \to F(x)$ uniformly in $x \in \mathbb{R}$.

8. Complete the proof of Proposition 6 for the unproved parts.

9. If $f: \Omega \to \mathbb{R}$ is measurable on a finite measure space (Ω, Σ, μ), show that f and its decreasing rearrangement f^* are equimeasurable. What happens if $\mu(\Omega) = +\infty$ is allowed? Where do the arguments break down, if we want to use the same proof as in the case of $\mu(\Omega) < \infty$?

4

INTEGRATION

The general outline and fundamental properties of the Lebesgue integral, including the basic limit theorems of the subject, are presented in this chapter. As applications, Lebesgue spaces L^p are discussed. A characterization of Riemann integration and the Vitali-Hahn-Saks theorem are also included.

4.1. THE ABSTRACT LEBESGUE INTEGRAL

We start with the familiar idea of the ancients that the integral of a (continuous) nonnegative function is the area (or volume) between the curve (or surface) and the basic (domain) set. In using the geometry of these objects, we observe that the integral measures the "volume" under the graph of such functions. This can be abstracted and the concept made precise as follows.

Let $f: \Omega \to \overline{\mathbb{R}}^+$ be a function. The *region* R_f under f is the subset of $\Omega \times \overline{\mathbb{R}}^+$ defined as

$$R_f = \{(\omega, y): 0 < y \leqslant f(\omega), \omega \in \Omega\},$$

with strict inequality on the right whenever $f(\omega) = +\infty$. Suppose there is given an outer measure μ on Ω. Then to measure R_f it is natural to approximate it by "rectangles", i.e., by sets of the form $A \times [0, t]$ for $0 \leqslant t \leqslant \infty$. Thus let $\mathscr{R}_0$ be a class of such rectangles $\{A \times [0, t]: t \geqslant 0, A \subset \Omega\}$, $\varnothing \in \mathscr{R}_0$. The idea now is to associate a value to each of these rectangles and then produce a measure for R_f using the Carathéodory process as in Definition 2.2.7. Thus we let $\alpha(A \times [0, t]) = \mu(A)m([0, t]) = t\mu(A)$ for each member of $\mathscr{R}_0$, where $m(\cdot)$ is the Lebesgue length function. Then we have:

Definition 1. For any set $E \subset \Omega \times \overline{\mathbb{R}}^+$, its volume is defined as

$$\nu(E) = \inf\left\{ \sum_{i=1}^{\infty} \alpha(B_i): E \subset \bigcup_{i=1}^{\infty} B_i,\ B_i \in \mathscr{R}_0 \right\}, \tag{1}$$

where as usual $\inf\{\varnothing\} = +\infty$.

The measuring function ν is defined for all subsets of $\Omega \times \overline{\mathbb{R}}^+$ and by Theorem 2.2.10, $\nu(\cdot)$ is a generated outer measure by the pair $(\alpha, \mathscr{R}_0)$. In particular we have $0 \leqslant \nu(R_f) \leqslant \infty$. Following the classical procedure, we may introduce:

Definition 2. Let μ be an outer measure on a set Ω, and $f: \Omega \to \overline{\mathbb{R}}^+$ be a (not necessarily μ-measurable) function. Then the (*outer*) *integral* of f relative to μ, denoted $\int_\Omega^* f\,d\mu$ or $\int_\Omega^* f(\omega)\mu(d\omega)$, is given by

$$\int_\Omega^* f\,d\mu = \nu(R_f), \tag{2}$$

where $\nu(\cdot)$ is the generated outer measure by $(\alpha, \mathscr{R}_0)$. More generally let $f: \Omega \to \overline{\mathbb{R}}$ be any function. If $f = f^+ - f^-$, where $f^+ = \max(f, 0)$ and $f^- = \max(-f, 0)$, then we set

$$\int_\Omega^* f\,d\mu = \nu(R_{f^+}) - \nu(R_{f^-}) \tag{3}$$

whenever at least one of the right side terms is finite. Finally for any set $A \subset \Omega$, define

$$\int_A^* f\,d\mu = \int_\Omega^* (f\chi_A)\,d\mu \tag{4}$$

if the right side symbol exists, as given by (3).

The point of the above integral is that it is defined for all functions, and the concept is a natural extension of the area of the region under the curve (or surface). It is now necessary to examine the properties of the symbol used in (2)–(4) and justify the appellation "integral". Recalling our convention that $a \cdot \infty = \pm\infty$ according as $a \gtrless 0$ and $= 0$ if $a = 0$, one has the following properties of the thus defined integral.

Theorem 3. *Let μ be an outer measure on Ω, and $f_i: \Omega \to \overline{\mathbb{R}}^+$, $i = 1, 2$, be a pair of functions. Then we have:*

(i) $0 \leqslant f_1 \leqslant f_2$ *a.e.* $\Rightarrow$ $0 \leqslant \int_\Omega^* f_1\,d\mu \leqslant \int_\Omega^* f_2\,d\mu \leqslant \infty$, *and* $0 \leqslant f_1 = c \cdot f_2$ *a.e.* $\Rightarrow$ $\int_\Omega^* f_1\,d\mu = c \cdot \int_\Omega^* f_2\,d\mu$ $(0 \leqslant c < \infty)$.

(ii) $f = \chi_A$, $A \subset \Omega$ $\Rightarrow$ $\int_\Omega^* f\,d\mu = \mu(A)$; *for* $f \geqslant 0$, $\int_\Omega^* f\,d\mu = 0$ *iff* $f = 0$ *a.e.*

(iii) *If* $0 \leqslant f$ *and* $\int_\Omega^* f\,d\mu < \infty$, *then* $\{\omega: f(\omega) > 0\} \subset \bigcup_{n=1}^\infty A_n$ *with* $\mu(A_n) < \infty$ *for each* $n \geqslant 1$. *If moreover either f is μ-measurable or μ is Carathéodory regular, then each A_n can be chosen μ-measurable so that the support of f is contained in a σ-finite μ-measurable set.*

(iv) *If* $f = f^+ - f^-: \Omega \to \overline{\mathbb{R}}$ *is a function and* $\int_\Omega^* f\,d\mu$ *exists, then* $\int_\Omega^* f\,d\mu = \int_\Omega^* f^+\,d\mu - \int_\Omega^* f^-\,d\mu$, *and*

$$\left|\int_\Omega^* f\,d\mu\right| \leqslant \int_\Omega^* |f|\,d\mu \leqslant \int_\Omega^* f^+\,d\mu + \int_\Omega^* f^-\,d\mu. \tag{5}$$

(v) *For any* $f: \Omega \to \overline{\mathbb{R}}$, $|\int_\Omega^* f\,d\mu| < \infty$ *iff* $\int_\Omega^* |f|\,d\mu < \infty$, *and in this case* $\int_A^* f\,d\mu$ *exists for any* $A \subset \Omega$.

(vi) *In general, the integral is only subadditive: for* $f_i: \Omega \to \mathbb{R}^+$, $i = 1, 2$,

$$\int_\Omega^* (f_1 + f_2)\,d\mu \leqslant \int_\Omega^* f_1\,d\mu + \int_\Omega^* f_2\,d\mu. \tag{6}$$

Remark. Integration of functions without measurability conditions has been a recurring theme from time to time. In the 1920s it was discussed by T. H. Hildebrandt and L. M. Graves. It was considered again by E. Zakon (1966). It is of interest to us because several properties can be studied without extra effort. The work was streamlined and clarified with the Carathéodory process by M. Sion (1968), thereby simplifying the treatment and exhibiting its structure more clearly, and our account is essentially influenced by that work.

Proof. (i): Since by (1) we have $0 \leqslant f_1 \leqslant f_2$ a.e., $\Rightarrow$ $R_{f_1} \subset R_{f_2}$ a.e., so that $\nu(R_{f_1}) \leqslant \nu(R_{f_2})$, the integral inequality follows from (2). If $c = 0$, then $f_1 = 0$, so that $R_{f_1} = \varnothing$. Hence the equality between integrals holds. If $c > 0$, then $R_{f_1} \subset \bigcup_{i=1}^\infty (A_i \times [0, t_i])$ iff $R_{cf_2} \subset \bigcup_{i=1}^\infty (A_i \times [0, ct_i])$. But by definition $\alpha(A_i \times [0, ct_i]) = ct_i\mu(A_i) = c\alpha(A_i \times [0, t_i])$, so that $\nu(R_{f_1}) = c\nu(R_{f_2})$, and thus the equality holds in all cases.

(ii): If $f = \chi_A$, then $R_f = A \times [0,1]$, so that $\nu(R_f) \leqslant \alpha(A \times [0,1]) = \mu(A)$. For the opposite inequality, let $R_f \subset \bigcup_{i=1}^\infty B_i$, $B_i \in \mathscr{R}_0$, where $B_i = A_i \times [0, t]$. Consider the (at most countable) set $J = \{i: t_i \geqslant 1\}$. Then for each $\omega \in A$, $(\omega, 1) \in R_f \Rightarrow (\omega, 1) \in B_i = A_i \times [0, t_i]$ for some $i \in J$. Thus $\omega \in A_i$, so that $A \subset \bigcup_{i \in J} A_i$. Hence

$$\mu(A) \leqslant \sum_{i \in J} \mu(A_i) \leqslant \sum_{i \in J} t_i \mu(A_i)$$

$$= \sum_{i \in J} \alpha(A_i \times [0, t_i]) \leqslant \sum_{i=1}^\infty \alpha(B_i).$$

Taking infimum of all such covers, we get $\mu(A) \leqslant \nu(R_f)$. This and the earlier inequality show $\int_\Omega^* f\,d\mu = \mu(A)$.

Let $A_0 = \{\omega: f(\omega) > 0\} = \bigcup_{n=1}^\infty A_n$, with $A_n = \{\omega: f(\omega) > 1/n\}$. Then $(1/n)\chi_{A_n} \leqslant f$, so that by (i) and the above result,

$$0 \leqslant \frac{1}{n}\mu(A_n) \leqslant \int_\Omega^* f\,d\mu = 0,$$

the last equality by hypothesis. So $\mu(A_n) = 0$, and hence $\mu(A_0) \leqslant \sum_{n=1}^{\infty}\mu(A_n) = 0$, since μ is an outer measure. Thus $f = 0$ a.e. Conversely, if $f = 0$ a.e., then $\mu(A_0) = 0$ and $R_f \subset A_0 \times [0, \infty)$, $\nu(R_f) \leqslant \mu(A_0) \cdot \infty = 0 \cdot \infty = 0$. Hence $\int_\Omega^* f\, d\mu = 0$, and (ii) follows.

(iii): Let A_0, A_n be as above and $\int_\Omega^* f\, d\mu < \infty$. Then

$$\frac{1}{n}\mu(A_n) \leqslant \int_\Omega^* f\, d\mu < \infty, \qquad n \geqslant 1.$$

So $\mu(A_n) < \infty$ and $A_0 \subset \bigcup_{n=1}^{\infty} A_n$. (There is even equality.)

If f is μ-measurable, then A_0 and each A_n are both μ-measurable. Thus A_0 is σ-finite. On the other hand, if μ is Carathéodory regular, then each A_n has a μ-measurable cover, i.e., there exist $C_n \in \mathcal{M}_\mu$, $C_n \supset A_n$, and $\mu(C_n) = \mu(A_n)$ such that $A_0 \subset \bigcup_{n=1}^{\infty} C_n$, and the last union is a μ-measurable σ-finite set, as asserted.

(iv): If $f\colon \Omega \to \overline{\mathbb{R}}$ and $\int_\Omega^* f\, d\mu$ exists, then $\nu(R_{f^+}) < \infty$ or $\nu(R_{f^-}) < \infty$. But then by (3) one has

$$\int_\Omega^* f\, d\mu = \nu(R_{f^+}) - \nu(R_{f^-}) = \int_\Omega^* f^+\, d\mu - \int_\Omega^* f^-\, d\mu \qquad [\text{cf. } (2)].$$

Also since f^+ and f^- have disjoint supports and $|f| = f^+ + f^-$, it follows that $R_{|f|} \subset R_{f^+} \cup R_{f^-}$ and hence

$$\int_\Omega^* |f|\, d\mu = \nu(R_{|f|}) \leqslant \nu(R_{f^+}) + \nu(R_{f^-}) \text{ (by the subadditivity of } \nu)$$

$$= \int_\Omega^* f^+\, d\mu + \int_\Omega^* f^-\, d\mu.$$

This gives the last inequality of (5). For the first one, since $f^{\pm} \leqslant |f|$, we have by (i) $\int_\Omega^* f^{\pm}\, d\mu \leqslant \int_\Omega^* |f|\, d\mu$. But then the triangle inequality gives

$$\left|\int_\Omega^* f\, d\mu\right| = \left|\int_\Omega^* f^+\, d\mu - \int_\Omega^* f^-\, d\mu\right|$$

$$\leqslant \max\left\{\int_\Omega^* f^+\, d\mu, \int_\Omega^* f^-\, d\mu\right\} \leqslant \int_\Omega^* |f|\, d\mu < \infty.$$

This proves (5).

(v): If $|\int_\Omega^* f\, d\mu| < \infty$, then by (iv) $\int_\Omega^* f^{\pm}\, d\mu < \infty$, so that

$$\int_\Omega^* |f|\, d\mu < \infty.$$

The converse is immediate.

If the integral exists, then for any $A \subset \Omega$

$$\int_A^* f^{\pm}\, d\mu = \int_\Omega^* (f^{\pm}\chi_A)\, d\mu \leqslant \int_\Omega^* f^{\pm}\, d\mu < \infty,$$

where we have used $f^{\pm}\chi_A \leqslant f^{\pm}$ and (i). This proves (v).

(vi): We establish this part by an example. Let $\Omega = [0,1]$, μ be Lebesgue outer measure, and $A \subset \Omega$ be a nonmeasurable set such that $\mu(A) = 1 = \mu(A^c)$. The existence of such an A was shown in Proposition 2.2.20. Let $f_1 = \chi_A$, $f_2 = \chi_{A^c}$, so that $f_1 + f_2 = 1$, which is μ-measurable. Also

$$1 = \mu(\Omega) = \int_\Omega^* (f_1 + f_2)\, d\mu$$

$$< \int_\Omega^* f_1\, d\mu + \int_\Omega^* f_2\, d\mu = \mu(A) + \mu(A^c) = 2.$$

Thus only subadditivity holds for the outer integral in general. This completes the proof.

Remark. If $\Omega = [0,1]$, and $f\colon \Omega \to \mathbb{R}$ is any continuous function or any bounded Riemann integrable function, then by decomposing it into its positive and negative parts, we deduce that $\int_\Omega^* f\, d\mu = \int_0^1 f(x)\, dx$ if μ is the Lebesgue length function. Thus the definition here generalizes the classical Riemann case for bounded functions.

In view of the known fact that for the general Riemann integrable functions (e.g., $\int_{\mathbb{R}} (\sin x)/x\, dx$ exists, but $\int_{\mathbb{R}} |\sin x|/|x|\, dx = +\infty$), our Definition 1 must contain something different from this case, in view of Theorem 2(v). That is, we have absolute integrability built into the definition. As we shall find later, this is a consequence of using countable covers in calculating the "volumes" of regions under the surfaces (or curves) in (1).

The definition of the integral (2) or (3) may not be easy for applications, although it is quite general. For μ-measurable functions (i.e., relative to $\mathcal{M}_\mu$) it is equivalent to other classical definitions, as shown in the next result. (For measurable functions $\int_\Omega^* f\, d\mu$ is written simply as $\int_\Omega f\, d\mu$.)

Theorem 4. *Let μ be an outer measure on a set Ω, and $f\colon \Omega \to \overline{\mathbb{R}}^+$ be a μ-measurable elementary function, so that $f = \sum_{i=1}^\infty a_i \chi_{A_i}$, A_i disjoint μ-measurable sets, $a_i \in \overline{\mathbb{R}}^+$. Then the following statements hold*:

(i) *We have*

$$\int_\Omega^* f\, d\mu = \int_\Omega f\, d\mu = \sum_{i=1}^\infty a_i \mu(A_i). \tag{7}$$

(ii) *If* $g: \Omega \to \overline{\mathbb{R}}^+$ *is any* μ*-measurable function, then* (2) *is equivalent to each of the following alternative definitions*:

(a) *Structural*:

$$\int_\Omega g\,d\mu = \lim_{r\downarrow 1} \sum_{n=-\infty}^{\infty} r^n\mu\big(g^{-1}(r^n, r^{n+1})\big) + \infty\cdot\mu\big(g^{-1}(\{\infty\})\big). \quad (8)$$

(b) *Lebesgue*:

$$\int_\Omega g\,d\mu = \begin{cases} \inf\left\{\int_\Omega h\,d\mu\colon h \geqslant g,\ h \text{ is } \mu\text{-elementary as in}(7)\right\}, \\ \sup\left\{\int_\Omega h'\,d\mu\colon h' \leqslant g,\ h' \text{ is } \mu\text{-elementary}\right\}. \end{cases} \quad (9)$$

(c) *Saks*:

$$\int_\Omega g\,d\mu = \sup\left\{\sum_{i=1}^{\infty} \inf g(A_i)\mu(A_i)\colon A_i \in \mathscr{M}_\mu,\ \text{disjoint},\ \bigcup_{i=1}^{\infty} A_i = \Omega\right\}. \quad (10)$$

(d) *Darboux*:

$$\int_\Omega g\,d\mu = \inf\left\{\sum_{i=1}^{\infty} \sup g(A_i)\mu(A_i)\colon A_i \in \mathscr{M}_\mu,\ \text{as above}\right\}. \quad (11)$$

Thus for μ*-measurable functions, any one of these equivalent forms can* (*and will*) *be used.*

Proof. (i): Let f be a μ-measurable elementary function as given. Then $R_f \subset \bigcup_{i=1}^{\infty}(A_i \times [0, a_i])$, so that

$$\int_\Omega f\,d\mu = \nu(R_f) \leqslant \sum_{i=1}^{\infty} a_i\mu(A_i). \quad (12)$$

For the opposite inequality, consider an arbitrary covering from $\mathscr{R}_0$ of R_f, namely $R_f \subset \bigcup_{i=1}^{\infty}(C_i \times [0, t_i])$. Since $A_i \times [0, a_i] \subset R_f$, let $J_i = \{j\colon A_i \times [0, a_i] \subset \bigcup_j C_j \times [0, t_j]\}$. Then using the same argument as in the proof of (ii) of the preceding theorem, we conclude that $t_j \geqslant a_i$, $j \in J_i$, and $\bigcup_{j\in J_i} C_j \supset A_i$.

Thus $A_i = \bigcup_{j \in J_i} C_j \cap A_i \in \mathscr{M}_\mu$, disjoint, and hence

$$\sum_{i=1}^{\infty} a_i \mu(A_i) = \sum_{i=1}^{\infty} a_i \mu\Big(A_i \cap \bigcup_{j \in J_i} C_j\Big)$$

$$\leqslant \sum_{i=1}^{\infty} a_i \sum_{j \in J_i} \mu(A_i \cap C_j), \qquad \text{by the } \sigma\text{-subadditivity of } \mu$$

$$\leqslant \sum_{i=1}^{\infty} \sum_{j \in J_i} t_j \mu(A_i \cap C_j), \qquad \text{since } t_j \geqslant a_i \text{ for } j \in J_i$$

$$\leqslant \sum_{i=1}^{\infty} \sum_{j=1}^{\infty} t_j \mu(A_i \cap C_j)$$

$$= \sum_{j=1}^{\infty} t_j \sum_{i=1}^{\infty} \mu(A_i \cap C_j)$$

$$= \sum_{j=1}^{\infty} t_j \mu\Big(\bigcup_{i=1}^{\infty} A_i \cap C_j\Big)$$

[since A_j are disjoint and $\mu|\mathscr{M}_\mu(C_j)$ is σ-additive by Theorem 2.2.9(ii)]

$$\leqslant \sum_{j=1}^{\infty} t_j \mu(C_j) = \sum_{j=1}^{\infty} \alpha\big(C_j \times [0, t_j]\big).$$

Hence taking infimum over all such coverings of R_f, the right side becomes $\int_\Omega f\, d\mu$ by (2). This and (12) imply (7).

(ii): We now use (7) and obtain the structural formula (8) for any nonnegative μ-measurable function g on Ω. If $r > 1$ is a real number, for each $-\infty < n < \infty$, let $A_n = g^{-1}((r^n, r^{n+1}))$ $(\in \mathscr{M}_\mu)$ and put $A_\infty = g^{-1}(\{\infty\})$ $(\in \mathscr{M}_\mu)$. Now we associate a μ-measurable elementary function g_r with these sets as

$$g_r = \sum_{n=-\infty}^{\infty} r^n \chi_{A_n} + \infty \cdot \chi_{A_\infty}. \tag{13}$$

Since $r > 1$, it follows from definition that $g_r \leqslant g \leqslant r g_r$. So by Theorem 3(i), we have $\int_\Omega g_r\, d\mu \leqslant \int_\Omega g\, d\mu \leqslant r \int_\Omega g_r\, d\mu$. Thus taking the lim sup and lim inf, we get from these inequalities

$$\limsup_{r \downarrow 1} \int_\Omega g_r\, d\mu \leqslant \int_\Omega g\, d\mu \leqslant \liminf_{r \downarrow 1} r \int_\Omega g_r\, d\mu = \liminf_{r \downarrow 1} \int_\Omega g_r\, d\mu, \tag{14}$$

so that $\lim_{r\downarrow 1}\int_\Omega g_r\,d\mu$ exists and $= \int_\Omega g\,d\mu$. Using (7) for the former integral, we conclude that (8) holds as stated.

The other equivalences will now be deduced from (8). Since by definition $g_r \leqslant g$, and g_r is μ-elementary, we have by (7) and (8)

$$\begin{aligned}\int_\Omega g\,d\mu &= \lim_{r\downarrow 1}\int_\Omega g_r\,d\mu \\ &\leqslant \sup\left\{\int_\Omega h'\,d\mu \colon h' \leqslant g,\ h' \text{ is } \mu\text{-elementary}\right\} \\ &\leqslant \int_\Omega g\,d\mu, \qquad \text{by Theorem 3(i).}\end{aligned} \tag{15}$$

Similarly, considering $g \leqslant rg_r$ (μ-elementary), we get

$$\begin{aligned}\int_\Omega g\,d\mu &= \lim_{r\downarrow 1}\int_\Omega rg_r\,d\mu \\ &\geqslant \inf\left\{\int_\Omega h\,d\mu \colon h \geqslant g,\ h \text{ is } \mu\text{-elementary}\right\} \\ &\geqslant \int_\Omega g\,d\mu, \qquad \text{by Theorem 3(i).}\end{aligned} \tag{16}$$

Hence there is equality in (15) and (16), and together they prove (9).

To obtain (10) and (11), note that on A_n, $r^n \leqslant \inf\{g(\omega)\colon \omega \in A_n\}$ and $r^{n+1} \geqslant \sup\{g(\omega)\colon \omega \in A_n\}$. Also $\Omega = (\bigcup_n A_n) \cup A_\infty \cup g^{-1}(\{0\})$. Consequently (13) implies

$$g_r \leqslant \sum_{n=-\infty}^{\infty} \inf g(A_n)\cdot\chi_{A_n} + \infty\cdot\chi_{A_\infty} \leqslant g \leqslant \sum_{n=-\infty}^{\infty} r^{n+1}\chi_{A_n} + \infty\cdot\chi_{A_\infty}. \tag{17}$$

Hence by (14) and (17),

$$\begin{aligned}\int_\Omega g\,d\mu &= \limsup_{r\downarrow 1}\int_\Omega g_r\,d\mu \\ &\leqslant \sup\left\{\sum_{n=-\infty}^{\infty} \inf g(A_n)\cdot\mu(A_n) + \infty\cdot\mu(A_\infty)\right\} \\ &\leqslant \int_\Omega g\,d\mu.\end{aligned}$$

This is (10). Interchanging the sup and inf here and using the second part of (17), we get (11). The equivalences of the definitions are thus established. This completes the proof of the theorem.

Now that the equivalence of our definition of the integral by (2) with (8)–(11) is at hand, we should compare this and the traditional Riemann-Darboux method. If g is bounded, then (7) and (8) imply that the sums on the right have only a finite number of nonzero terms. Moreover, (8) and (9) then imply that the integral of a measurable function is obtained by approximating it by upper or lower envelopes of *μ-simple* functions (and elementary functions are needed only in the unbounded case). An essential difference between the (abstract) Lebesgue and Riemann-Darboux definitions is that in the latter, one partitions the domain of the function and then approximates the values of the function accordingly. On the other hand, for the Lebesgue case [see (8)] it is the range of the function that is partitioned, and then the domain is decomposed accordingly in the approximation. This is described picturesquely by saying that "the integral is tailored to order" here, in contrast to the Riemann-Darboux "ready-to-wear" approach. Thus a flexibility afforded by Lebesgue's method compared to Riemann's must be noted.

As a simple illustration of the difference between these two approaches, consider the classical Dirichlet function f on $\Omega = [0,1]$ with $\mu =$ Lebesgue measure, where $f(\omega) = 1$ if ω in Ω is rational, and $= 0$ if ω is irrational. Then it is well known and is also immediate to verify that f is not Riemann integrable. However, $f = 0$ a.e., and by (7) and Theorem 3(i), since f is clearly measurable (being an elementary function), f is Lebesgue integrable with value zero.

We shall show later that each bounded Riemann-integrable function is Lebesgue integrable, with the same value. Thus for this class of functions the Lebesgue integral is a generalization of the classical one. (This comment should be contrasted with that on unbounded functions made after the remark following Theorem 3.)

If we consider only measurable functions, then Theorem 3(vi) can be improved to additivity, as seen from

Proposition 5. *If* $f_i\colon \Omega \to \overline{\mathbb{R}}$, $i = 1, 2$, *are μ-measurable where μ is an outer measure on* Ω, *and* f_1, f_2 *are μ-integrable* [*i.e.*, (3) *holds*], *then*

$$\int_\Omega (f_1 + f_2)\, d\mu = \int_\Omega f_1\, d\mu + \int_\Omega f_2\, d\mu, \tag{18}$$

is true whenever the right side is defined (*i.e. is not of the form* $\infty - \infty$).

Proof. First we establish (18) for nonnegative elementary functions, then for all nonnegative μ-measurable ones, and finally for the general case. This will become a standard procedure in the Lebesgue theory, and the reader should keep it in mind.

Thus let $f_i = \sum_{j=1}^{\infty} a_j^i \chi_{A_j^i}$, $i = 1, 2$, $\{A_j^i: j \geqslant 1\}$ be disjoint partitions of Ω in $\mathscr{M}_\mu$. Then $f_1 + f_2$ is also μ-elementary, and since

$$f_1 + f_2 = \sum_{j=1}^{\infty} \sum_{k=1}^{\infty} \left(a_j^1 + a_k^2\right) \chi_{A_j^1 \cap A_k^2}, \qquad a_j^i \geqslant 0,$$

we have by (7)

$$\int_\Omega (f_1 + f_2)\, d\mu = \sum_{j=1}^{\infty} \sum_{k=1}^{\infty} \left(a_j^1 + a_k^2\right) \mu\left(A_j^1 \cap A_k^2\right)$$

$$= \sum_{j=1}^{\infty} \sum_{k=1}^{\infty} a_j^1 \mu\left(A_j^1 \cap A_k^2\right) + \sum_{j=1}^{\infty} \sum_{k=1}^{\infty} a_k^2 \mu\left(A_j^1 \cap A_k^2\right)$$

$$= \sum_{j=1}^{\infty} a_j^1 \mu\left(\bigcup_{k=1}^{\infty} A_k^2 \cap A_j^1\right) + \sum_{k=1}^{\infty} a_k^2 \mu\left(A_k^2 \cap \bigcup_{j=1}^{\infty} A_j^1\right)$$

(since $\mu | \mathscr{M}_\mu$ is σ-additive)

$$= \sum_{j=1}^{\infty} a_j^1 \mu\left(A_j^1\right) + \sum_{k=1}^{\infty} a_k^2 \mu\left(A_k^2\right)$$

$$= \int_\Omega f_1\, d\mu + \int_\Omega f_2\, d\mu.$$

Thus (18) holds in this case. Next, if $f_1, f_2 \geqslant 0$ are μ-measurable, let h_{ij} be μ-elementary and $h_{i1} \leqslant f_i \leqslant h_{i2}$, $i = 1, 2$. Then by the preceding special case and Theorem 3(i), we have

$$\int_\Omega h_{11}\, d\mu + \int_\Omega h_{21}\, d\mu = \int_\Omega (h_{11} + h_{21})\, d\mu \leqslant \int_\Omega (f_1 + f_2)\, d\mu$$

$$\leqslant \int_\Omega (h_{12} + h_{22})\, d\mu = \int_\Omega h_{12}\, d\mu + \int_\Omega h_{22}\, d\mu.$$

Taking infimum on the right and supremum on the left as the h_{ij} vary, one gets by (9) that

$$\int_\Omega f_1\, d\mu + \int_\Omega f_2\, d\mu = \int_\Omega (f_1 + f_2)\, d\mu,$$

and so (18) holds for all nonnegative μ-measurable f_1, f_2.

For the general case, let f_1, f_2 be μ-measurable. So we have, on separating $f_1 + f_2$ into the positive and negative parts written in two ways,

$$f_1 + f_2 = (f_1 + f_2)^+ - (f_1 + f_2)^- = f_1^+ - f_1^- + f_2^+ - f_2^-.$$

Now transposition gives

$$(f_1 + f_2)^+ + f_1^- + f_2^- = (f_1 + f_2)^- + f_1^+ + f_2^+. \tag{19}$$

By hypothesis, for f_1, f_2 the μ-integrals are defined and $\int_\Omega f_1\, d\mu + \int_\Omega f_2\, d\mu$ is also meaningful. Let this be finite. Then by (3) $\int_\Omega f_1^+\, d\mu + \int_\Omega f_2^+\, d\mu < \infty$, so that from $(f_1 + f_2)^+ \leqslant f_1^+ + f_2^+$ we have $\int_\Omega (f_1 + f_2)^+\, d\mu < \infty$. Consequently, integrating the expressions in (19) and using the preceding step and subtracting these finite terms, it follows that

$$-\int_\Omega (f_1 + f_2)\, d\mu = \int_\Omega (f_1^- + f_2^-)\, d\mu - \int_\Omega (f_1 + f_2)^+\, d\mu$$

$$= \int_\Omega f_1^-\, d\mu + \int_\Omega f_2^-\, d\mu - \int_\Omega f_1^+\, d\mu - \int_\Omega f_2^+\, d\mu$$

[(19) is used here],

$$= -\int_\Omega (f_1^+ - f_1^-)\, d\mu - \int_\Omega (f_2^+ - f_2^-)\, d\mu$$

$$= -\int_\Omega f_1\, d\mu - \int_\Omega f_2\, d\mu.$$

Multiplying by -1 on both sides, we get (18). Similarly, if the right side of (18) is $> -\infty$, one gets the same result from $\int_\Omega (f_1 + f_2)\, d\mu$. Thus (18) holds as stated, and the proof is complete.

On a measure space (Ω, Σ, μ), a measurable function $f\colon \Omega \to \overline{\mathbb{R}}$ is called *integrable* if $\int_\Omega f^\pm\, d\mu < \infty$, so that $\int_\Omega |f|\, d\mu < \infty$, or equivalently $|\int_\Omega f\, d\mu| < \infty$, by Theorem 3(v). [Sometimes the word "summable" is used, and if (3) holds with a possibly infinite value, so $\int_\Omega f^+\, d\mu < \infty$ *or* $\int_\Omega f^-\, d\mu < \infty$, but not necessarily both, then f is called "integrable". Instead of using these extra distinctions, we shall use "integrable" as defined above, since that concept occurs often, and on the rare occasion when the more general one is desired, we say that f is integrable in the *extended sense* or that $\int_\Omega f\, d\mu$ is defined.]

With this we may summarize the definition and properties for reference in the following form.

Theorem 6. *Let* (Ω, Σ, μ) *be an arbitrary measure space and* $f_i\colon \Omega \to \overline{\mathbb{R}}$, $i = 1, 2$, *be* μ*-measurable. Then we have*:

(i) *for* $f_1 \geqslant 0$, $f_1 = \sum_{i=1}^{\infty} a_i \chi_{A_i}$, *implies*

$$\int_\Omega f_1 \, d\mu = \sum_{i=1}^{\infty} a_i \mu(A_i);$$

(ii) *for* $f_1 \geqslant 0$ *but not necessarily elementary,*

$$\int_\Omega f_1 \, d\mu = \inf\left\{ \sum_{i=1}^{\infty} a_i \mu(A_i) : f_1 \leqslant h = \sum_{i=1}^{\infty} a_i \chi_{A_i},\ A_i \in \Sigma, \text{disjoint} \right\}$$

$$= \sup\left\{ \sum_{i=1}^{\infty} b_i \mu(B_i) : f_1 \geqslant h' = \sum_{i=1}^{\infty} b_i \chi_{B_i},\ B_i \in \Sigma, \text{disjoint} \right\}$$

(iii) *if* $f_1 = f_2$ *a.e. and* f_1 *is integrable, then* f_2 *is integrable and both have the same integrals*;

(iv) *the mapping* $f \mapsto \int_\Omega f \, d\mu$ *is linear and positivity preserving, and the integral is absolute in the sense that* f *is integrable iff* $|f|$ *is*;

(v) $\int_A f \, d\mu = \int_\Omega (f \chi_A) \, d\mu$, *and the integrability of* f *implies that the set* $\{\omega : f(\omega) \neq 0\}$ *is* σ*-finite for* μ.

In the next section we record a few properties of integrals of nonmeasurable functions before moving on to the key limit theorems. The usefulness of the results of such (general) functions will become clear when we study product measures and integrals in Chapter 6. Otherwise they can be skipped on a first reading.

EXERCISES

0. Let Ω be a set and μ be an outer measure on it. If $f\colon \Omega \to \overline{\mathbb{R}}^+$ is μ-measurable, show, with the assumption of the finite subset property for μ, that $\int_\Omega f \, d\mu < \infty$ iff $\sup\{\int_A f \, d\mu : A \in \Sigma_0\} < \infty$, where $\Sigma_0 = \{A \subset \Omega;\ \mu(A) < \infty\}$. Show also that there is then equality between these two numbers. [Consider $\nu\colon B \mapsto \int_B f \, d\mu$.]

1. Prove the comment made after Theorem 4 in the form: If $f\colon \Omega \to \mathbb{R}$ is a bounded measurable function where (Ω, Σ, μ) is a finite measure space, then

$$\int_{\Omega} f\, d\mu = \inf\left\{\int_{\Omega} g\, d\mu : g \geqslant f,\ g \text{ is } (\Sigma\text{-})\text{simple}\right\}$$

$$= \sup\left\{\int_{\Omega} h\, d\mu : h \leqslant f,\ h \text{ is } (\Sigma\text{-})\text{simple}\right\}.$$

2. Let μ be an outer measure on Ω, and $f: \Omega \to \overline{\mathbb{R}}^+$ be a function. Define $\mu_1(A) = \int_A^* f\, d\mu$, $A \subset \Omega$. Show that $\mu_1(\cdot)$ is σ-subadditive and that if f is μ-measurable then $\mathscr{M}_{\mu_1} \supset \mathscr{M}_{\mu}$. Is the inclusion relation strict? [Observe that $\mu_1(A) = \nu(R_{f\chi_A})$ in the notation of (2) and that $R_{f_1+f_2} \subset \bigcup_i R_{f_i}$.]

3. Let Ω, μ be as above and $f: \Omega \to \overline{\mathbb{R}}$ be a function such that $|\int_\Omega^* f\, d\mu| < \infty$. If $\int_A^* f\, d\mu = 0$ for all $A \subset \Omega$, show that $f = 0$ a.e. If f is μ-measurable, show that the same conclusion holds when we restrict A to μ-measurable sets.

4. If (Ω, Σ, μ) is a measure space and $f: \Omega \to \mathbb{R}$ is a bounded function and $\mu(\Omega) = 0$, verify that $\int_\Omega^* f\, d\mu = 0$.

5. Let μ be an outer measure on Ω and $f: \Omega \to \overline{\mathbb{R}}$ be any function such that $|\int_\Omega^* f\, d\mu| < \infty$. Show that for any $\beta > 0$ we have the inequality

$$\mu(\{\omega : |f(\omega)| \geqslant \beta\}) \leqslant \frac{1}{\beta}\int_\Omega^* |f|\, d\mu.$$

Deduce that $\{\omega: |f(\omega)| > 0\} \subset \bigcup_{n=1}^\infty A_n$, $\mu(A_n) < \infty$, $n \geqslant 1$. If f is μ-measurable, then the above result is called the *Markov inequality*. If $\varphi: \mathbb{R}^+ \to \mathbb{R}^+$ is an increasing function such that $\varphi(x) = 0$ implies $x = 0$, then verify that the following extension of the above [*Čebyšev's inequality* if $\varphi(x) = x^2$] is valid:

$$\mu(\{\omega : |f(\omega)| > \beta\}) \leqslant \frac{1}{\varphi(\beta)}\int_\Omega^* \varphi(|f|)\, d\mu.$$

6. Let μ be a finite outer measure on Ω, and $\mathscr{F}$ be the collection of a.e. finite functions $f: \Omega \to \overline{\mathbb{R}}$. Define $d: \mathscr{F} \times \mathscr{F} \to \mathbb{R}^+$ by

$$d(f, g) = \int_\Omega^* \frac{|f - g|}{1 + |f - g|}\, d\mu.$$

Show that $d(\cdot,\cdot)$ is a (semi-)metric on $\mathscr{F}$ and that $f_n \xrightarrow{\mu} f$ iff $d(f_n, f) \to 0$. Deduce that if $\mathscr{F}$ is replaced by $\tilde{\mathscr{F}} = \mathscr{F}/\sim$, the set of a.e. equal functions, if $\tilde{d}(\tilde{f}, \tilde{g}) = d(f, g)$, and if $\mathscr{F}$ consists of all μ-measurable functions, then $(\tilde{\mathscr{F}}, \tilde{d})$ is a complete metric space. [Observe that if $\varphi(x) = x/(1 + x)$, then $\varphi\colon \mathbb{R}^+ \to \mathbb{R}^+$ is a strictly increasing (bounded) continuous concave function, so that it is subadditive.]

7. The integral of a measurable function f on (Ω, Σ, μ) can also be defined in another way, starting with the first result of Theorem 4. Thus if $f\colon \Omega \to \mathbb{R}$ is a simple function, i.e., $f = \sum_{i=1}^{n} a_i \chi_{A_i}$, with $A_i \in \Sigma$, disjoint, *and* $\mu(A_i) < \infty$, then define

$$\int_B f\,d\mu = \sum_{i=1}^{n} a_i \mu(A_i \cap B), \qquad B \in \Sigma.$$

(a) Show that this definition of the integral does not depend on the representation of the simple function f (i.e., $f = \sum_{j=1}^{k} b_j \chi_{B_j} = \sum_{i=1}^{n} a_i \chi_{A_i}$ gives the same result).

(b) Let $f\colon \Omega \to \mathbb{R}$ be any measurable function. By the structure theorem there exist simple μ-measurable functions $f_n\colon \Omega \to \mathbb{R}$ such that $f_n \to f$ pointwise. Suppose either that $\mu(\Omega) < \infty$ or else that the support of each f_n has finite μ-measure. If $\{\int_B f_n\colon n \geqslant 1\}$ is a Cauchy sequence with limit f_B, then one defines

$$f_B = \int_B f\,d\mu = \lim_{n \to \infty} \int_B f_n\,d\mu, \qquad B \in \Sigma.$$

Show that the integral is uniquely defined and does not depend on the particular sequence $\{f_n\colon n \geqslant 1\}$ used. Moreover, $f \mapsto \int_\Omega f\,d\mu$ is a linear mapping which preserves order, $|\int_\Omega f\,d\mu| \leqslant \int_\Omega |f|\,d\mu$, and f is integrable iff $|f|$ is. Conclude that this and Definition 2 agree for all such f.

8. Let $f_i\colon \Omega \to \overline{\mathbb{R}}$, $i = 1, 2$, be a pair of functions, and μ be an outer measure on Ω. If $|\int_\Omega^* f_i\,d\mu| < \infty$, $i = 1, 2$, verify that

$$\max(f_1, f_2) = \frac{f_1 + f_2 + |f_1 - f_2|}{2}$$

and

$$\min(f_1, f_2) = \frac{f_1 + f_2 - |f_1 - f_2|}{2}, \qquad \text{a.e.,}$$

and deduce that

$$\left|\int_\Omega^* \max(f_1, f_2)\,d\mu\right| < \infty \quad \text{and} \quad \left|\int_\Omega^* \min(f_1, f_2)\,d\mu\right| < \infty.$$

[The condition implies $\mu(\{\omega\colon |f_i(\omega)| = \infty\}) = 0$, by Exercise 5; then invoke Theorem 3(iv).]

9. Let $f_i\colon \Omega \to \overline{\mathbb{R}}$, $i = 1, 2$, satisfy the hypothesis of Exercise 8 for some outer measure μ on Ω (or just be real valued). Then $f(\omega) = (f_1 + if_2)(\omega)$ is a complex-valued function for a.a. ω (or for all ω), and we define its upper integral as

$$\int_\Omega^* f\,d\mu = \int_\Omega^* f_1\,d\mu + i\int_\Omega^* f_2\,d\mu$$

(whenever this is meaningful). Show that $|\int_\Omega^* f\,d\mu| < \infty$ iff $\int_\Omega^* |f|\,d\mu$ is, and we have $|\int_\Omega^* f\,d\mu| \leqslant \int_\Omega^* |f|\,d\mu$, $|\cdot|$ being the absolute value. [Note that $|\int_\Omega^* f\,d\mu| = e^{i\theta}\int_\Omega^* f\,d\mu$ for some $0 \leqslant \theta \leqslant 2\pi$, and then verify that this is $= \int_\Omega^* e^{i\theta} f\,d\mu = \int_\Omega^* \mathrm{Re}(e^{i\theta} f)\,d\mu \leqslant \int_\Omega^* |f|\,d\mu$.]

10. Using the notation of the preceding exercise, let f_1, f_2 be a pair of complex functions on Ω such that $|\int_\Omega^* f_j\,d\mu| < \infty$, $j = 1, 2$. Show that $f_1 = f_2$ a.e. iff $\int_A^* f_1\,d\mu = \int_A^* f_2\,d\mu$ for all $A \subset \Omega$. (Compare with Exercise 3 above.)

4.2. INTEGRATION OF NONMEASURABLE FUNCTIONS

The work of the preceding section raises the question of developing integration theory for all functions without regard to measurability. As noted in the above exercises, some of it can be done. In fact, as pointed out there, E. Zakon (1966) has presented such an exposition, based on a "measurable partition", generalizing in some respects the early work of S. C. Fan (1941) and others. This means that if (Ω, Σ, μ) is a measure space and μ^* is the generated outer measure by (μ, Σ), so that (see Theorem 2.2.10) μ^* is Carathéodory regular and each set $A \subset \Omega$ has a measurable cover B [i.e., $A \subset B \in \mathscr{M}_{\mu^*}$ and $\mu^*(A) = \mu^*(B)$], then a sequence $\{A_n : n \geqslant 1\}$ is a "measurable partition" if $\Omega = \bigcup_{n=1}^\infty A_n$, A_n disjoint, and the measurable covers B_n, $n \geqslant 1$, of this family satisfy $\mu^*(B_j \cap B_k) = 0$ if $j \neq k$. However, the existence of such a partition is a nontrivial restriction in applying this theory for genuinely nonmeasurable functions.

In what follows, we consider a modification of these authors' computations and remove the above-noted restriction. In exchange, a form of the monotone convergence statement is here *assumed* for measurable functions. This is a basic limit theorem in Lebesgue's theory, and it is proved in the following section (see Theorem 3.1). Thus the work of this section can be regarded as a motivation for and an extension of the standard theory. However, it will also be of interest in later applications.

Consequently, we shall present here some results with the altered procedure applied to nonmeasurable functions, which contain essentially the results exposed by Fan, Zakon, Bourbaki (1965) and others. As noted earlier, a study of nonmeasurable functions illuminates the notion of measurability and will also be useful in Section 6.2 on product measures.

The first result is an alternative condition for expressing the outer integral of nonmeasurable functions by measurable approximations.

Proposition 1. *Let* Ω *be a set and* μ *be a Carathéodory regular outer measure on* Ω, *so that* $\mu(A) = \inf\{\mu(B): B \supset A,\ B \in \mathscr{M}_\mu\}$ *for each* $A \subset \Omega$. *If* $f: \Omega \to \overline{\mathbb{R}}^+$ *is any function, then its outer integral is given by*

$$\int_\Omega^* f\,d\mu = \inf\left\{\int_\Omega h\,d\mu: h \geqslant f,\ h \text{ is } \mu\text{-elementary}\right\}, \tag{1}$$

provided that for μ*-simple* $0 \leqslant h_n \uparrow$ *one has* $\lim_n \int_\Omega h_n\,d\mu = \int_\Omega \lim_n h_n\,d\mu$ *and* $\int_{(\cdot)} f\,d\mu$ *is* σ*-additive on* $\mathscr{M}_\mu$.

Remark. If we replace inf by sup and the inequality is reversed between h and f, then one obtains the "lower integral" $\int_{*\Omega} f\,d\mu$, which is, in general, strictly smaller than the above number. However, if h is allowed to be based on "measurable partitions", then it is possible that for some nonmeasurable $f: \Omega \to \overline{\mathbb{R}}^+$ the lower and upper integrals are equal, and one can say that such f is integrable in a generalized sense. ($\int_\Omega^* f\,d\mu = \int_{*\Omega} f\,d\mu = \int_\Omega f\,d\mu$ for measurable f by Theorem 1.4.)

Proof. If $f: \Omega \to \overline{\mathbb{R}}^+$, then by definition of the outer integral (Definition 1.2), we get (1) by the computation

$$\int_\Omega^* f\,d\mu = \nu(R_f) = \inf\left\{\sum_{i=1}^\infty \alpha(B_i): \bigcup_{i=1}^\infty B_i \supset R_f,\ B_i \in \mathscr{R}_0\right\}$$

$$= \inf\left\{\sum_{i=1}^\infty a_i\mu(A_i): \{B_i\} = A_i \times [0, a_i],\ \{B_i\} \text{ as above}\right\}$$

$$= \inf\left\{\sum_{i=1}^\infty a_i\mu(C_i): \{C_i\} \text{ is a } \mu\text{-measurable cover of } A_i,\ i \geqslant 1\right\}$$

$$= \inf\left\{\lim_n \int_\Omega \left(\sum_{i=1}^n a_i\chi_{C_i}\right) d\mu: C_i \supset A_i,\ i \geqslant 1\ \{C_i\} \text{ as above}\right\}$$

(by Proposition 1.5)

$$= \inf\left\{\int_\Omega g\,d\mu: g = \sum_{i=1}^\infty a_i\chi_{C_i} \geqslant f\right\},$$

(by the hypothesis on monotone limits)

$$= \inf\left\{\int_\Omega h\,d\mu: h \geqslant f,\ h \text{ is } \mu\text{-elementary}\right\}$$

(h is obtained from g with disjunctified C_i's).

For the second part, ν is an outer measure, and if we let $\tilde{\nu}(A) = \int_A^* f\,d\mu = \int_\Omega^* f\chi_A\,d\mu$, then $\tilde{\nu}(A) = \nu(R_{f\chi_A})$ and $\tilde{\nu}(\varnothing) = 0$. If $A_1, A_2, \dots$ is any sequence of sets, and $A = \bigcup_n A_n \subset \Omega$, then it is clear that $R_{f\chi_A} \subset \bigcup_{i=1}^\infty R_{f\chi_{A_i}}$. Since the σ-subadditivity of ν thus implies the same for $\tilde{\nu}$ on $\mathscr{M}_\mu$, it suffices to establish its finite additivity by Lemma 1.3.5. So let A_1, A_2 be disjoint sets in $\mathscr{M}_\mu$, and $A = A_1 \cup A_2$. By the subadditivity, we may assume that $\int_A^* f\,d\mu < \infty$, and show that it is superadditive.

Let $\epsilon > 0$ be given. Then by (1) there exists a μ-elementary function h_ϵ that vanishes outside A such that $h_\epsilon \geqslant f\chi_A$ and

$$\int_A^* f\,d\mu = \int_\Omega^* f\chi_A\,d\mu > \int_A h_\epsilon\,d\mu - \epsilon. \tag{2}$$

Also note that $h_\epsilon\chi_{A_i}$ is μ-measurable and $h_\epsilon\chi_{A_i} \geqslant f\chi_{A_i}$, $i = 1, 2$. Hence using the μ-measurability of A_i and h_ϵ, together with Proposition 1.5 we can express (2) as

$$\begin{aligned}\int_A^* f\,d\mu + \epsilon &> \int_{A_1} h_\epsilon\,d\mu + \int_{A_2} h_\epsilon\,d\mu \\ &= \int_{A_1} h_\epsilon\chi_{A_1}\,d\mu + \int_{A_2} h_\epsilon\chi_{A_2}\,d\mu \\ &\geqslant \int_{A_1}^* f\,d\mu + \int_{A_2}^* f\,d\mu, \qquad \text{by (1)}.\end{aligned}$$

Since $\epsilon > 0$ is arbitrary, this establishes the additivity of $\tilde{\nu}$, and hence its σ-additivity on $\mathscr{M}_\mu$. This completes the proof.

Decomposing $f: \Omega \to \overline{\mathbb{R}}$ into its positive and negative parts and remembering Theorem 1.3(iv) regarding the upper integral, we get from the above result the following immediate extension.

Corollary 2. *Let $f: \Omega \to \overline{\mathbb{R}}$ and μ be a Carathéodory regular outer measure on Ω. Let $\int_\Omega^* f\,d\mu$ exist. Then the function $A \mapsto \int_A^* f\,d\mu$ is σ-additive on $\mathscr{M}_\mu$ provided the monotone limit result holds for μ-simple functions.*

The following result is a slight refinement of Proposition 1 and will be useful as a technical tool in computations. It shows how in many cases work with nonmeasurable functions may be reduced to measurable ones. We assume the result on monotone limits again.

Proposition 3. *If μ is a Carathéodory-regular outer measure on Ω, and $f: \Omega \to \overline{\mathbb{R}}$ is a function for which $\int_A^* f\,d\mu$ is defined, where A is in $\mathscr{M}_\mu$, then there exists a* *μ-measurable function h such that* $h \geqslant f$ *and* $\int_A h\,d\mu = \int_A^* f\,d\mu$, *provided the monotone limit result holds.*

Proof. By replacing f with $f\chi_A$, we may take $A = \Omega$. Also, because of Theorem 1.3(iv) it suffices to prove the result for f^+, i.e., for $f \geqslant 0$. In case $\int_\Omega^* f\,d\mu = \infty$, then we may take $h = +\infty$ and the result holds. So let $\int_\Omega^* f\,d\mu < \infty$. Then by Proposition 1, given $\epsilon > 0$, there exists a μ-elementary function $h_\epsilon \geqslant f$ such that

$$\int_\Omega^* f\,d\mu + \epsilon > \int_\Omega h_\epsilon\,d\mu \geqslant \int_\Omega^* f\,d\mu. \tag{3}$$

Let $\epsilon = 1/n$, and write h_n for h_ϵ. Then $h = \inf_n h_n \geqslant f$ and h is μ-measurable. Consequently (3) gives

$$\int_\Omega^* f\,d\mu + \frac{1}{n} \geqslant \int_\Omega h\,d\mu \geqslant \int_\Omega^* f\,d\mu. \tag{4}$$

Since n is arbitrary, and because of the initial simplification, this finishes the proof.

The following conclusion resulting from the preceding two propositions is noteworthy. If the outer measure under consideration is Carathéodory regular, then for the integration theory the measurability of functions is of minimal concern. But by Proposition 2.2.14, an arbitrary outer measure on Ω can be replaced, with just one extension, by a Carathéodory regular one, so that the Carathéodory process does present us with a very general and natural framework for integration. Our assumption on monotone limits for μ is illuminated by the following:

Proposition 4. *Let* μ *be a Carathéodory regular outer measure on* Ω, *and* $f_n\colon \Omega \to \overline{\mathbb{R}}^+$ *be such that* $f_n \leqslant f_{n+1} \to f$ *pointwise. Then the monotone convergence theorem holds for all such* f_n *iff it holds for the* μ*-measurable increasing sequences, i.e.,*

$$\lim_n \int_\Omega^* f_n\,d\mu = \int_\Omega^* f\,d\mu, \tag{5}$$

holds for all sequences $0 \leqslant f_n \uparrow f$ *iff it holds for the* μ*-measurable monotone sequences.*

Proof. If (5) holds for all sequences, then it certainly holds for the μ-measurable ones. Conversely, let (5) be true for the μ-measurable sequences, and let $0 \leqslant f_n \uparrow f$ be an arbitrary sequence. Then by Proposition 3, there exist

μ-measurable g_n and g such that $g_n \geqslant f_n$, $g \geqslant f$, and

$$\int_{\Omega}^{*} f_n \, d\mu = \int_{\Omega} g_n \, d\mu, \quad \int_{\Omega}^{*} f \, d\mu = \int_{\Omega} g \, d\mu, \qquad n \geqslant 1. \tag{6}$$

For each n, $g_n \geqslant f_n \geqslant f_k$, $1 \leqslant k \leqslant n$, and similarly $g \geqslant f_k$, $k \geqslant 1$. But the g_n's need not be monotone. So let us form another sequence by letting $h_n = \inf\{g, g_k \colon k \geqslant n\}$. Then $h_n \leqslant h_{n+1}$ and each h_n is μ-measurable. If $h = \lim_n h_n$, then since $h_n \leqslant \min(g_n, g)$, we get $h \leqslant g$. Also $f_n \leqslant f_{n+1}$, so that $f_n \leqslant g_k, g$ for $k \geqslant n$. Hence $f_n \leqslant \inf\{g, g_k \colon k \geqslant n\} = h_n \leqslant h \leqslant g$, and so $f = \lim_n f_n \leqslant h \leqslant g$. Now using the fact that the outer integral is order preserving for positive functions [Theorem 1.3(i)], we get

$$\begin{aligned} \int_{\Omega}^{*} f_n \, d\mu &\leqslant \int_{\Omega} h_n \, d\mu \leqslant \int_{\Omega} g_n \, d\mu = \int_{\Omega}^{*} f_n \, d\mu, \\ \int_{\Omega}^{*} f \, d\mu &\leqslant \int_{\Omega} h \, d\mu \leqslant \int_{\Omega} g \, d\mu = \int_{\Omega}^{*} f \, d\mu. \end{aligned} \tag{7}$$

It follows from (7) that we may replace g_n and g of (6) by the monotone set h_n and h, $n \geqslant 1$, and if (5) is true for μ-measurable sequences (here the h_n-sequence), then we also have the result for the general case of the f_n-sequence, completing the demonstration.

The following result for any sequence of nonnegative functions is an easy consequence of the above proposition and is of interest.

Corollary 5. *Let* (Ω, μ) *be as in the proposition. If* $f_n \colon \Omega \to \overline{\mathbb{R}}^{+}$, $n \geqslant 1$, *is arbitrary, then we have*

$$\int_{\Omega}^{*} \liminf_n f_n \, d\mu \leqslant \liminf_n \int_{\Omega}^{*} f_n \, d\mu \tag{8}$$

whenever the monotone convergence theorem is valid for the measurable case.

Proof. Let $g_n = \inf\{f_k \colon k \geqslant n\}$. Then $0 \leqslant g_n \leqslant g_{n+1} \leqslant f_{n+1}$, and by definition $g = \lim_n g_n = \liminf_n f_n$. Since the validity of monotone convergence for measurable functions implies the truth of (5), we have

$$\begin{aligned} \int_{\Omega}^{*} g \, d\mu = \lim_n \int_{\Omega}^{*} g_n \, d\mu &= \liminf_n \int_{\Omega}^{*} g_n \, d\mu \\ &\leqslant \liminf_n \int_{\Omega}^{*} f_n \, d\mu. \end{aligned}$$

This proves (8).

The preceding proposition and corollary show that for a deeper analysis of integrable functions one has to go to the class of measurable functions. The study of general functions by itself will not be strong enough to be used in future work. However, before turning to this class, we present the following *important result* connecting the integral relative to a measure and that relative to an image measure by an arbitrary function.

Theorem 6. *Let Ω and Ω' be two sets, and μ be a Carathéodory regular outer measure on Ω. Let $\pi: \Omega \to \Omega'$ be a mapping, and $\mu' = \mu \circ \pi^{-1}$ be the image of μ. Then μ' is an outer measure on Ω', and we have the following assertions*:

(i) *π is $(\mathscr{M}_{\mu}, \mathscr{M}_{\mu'})$-measurable, so that $\pi^{-1}(\mathscr{M}_{\mu'}) \subset \mathscr{M}_{\mu}$,*
(ii) *for each μ'-measurable $f: \Omega' \to \overline{\mathbb{R}}$,*

$$\int_{\Omega'} f \, d\mu' = \int_{\Omega} f \circ \pi \, d\mu \tag{9}$$

in the sense that the existence of the integral on either side implies the other exists and equality holds.

Proof. Since π^{-1} preserves all set operations and μ is an outer measure, we deduce that $\mu'(\varnothing) = 0$ and μ' is σ-subadditive, because

$$\mu'\left(\bigcup_{n=1}^{\infty} A_n\right) = \mu\left(\bigcup_{n=1}^{\infty} \pi^{-1}(A_n)\right) \leqslant \sum_{n=1}^{\infty} \mu\left(\pi^{-1}(A_n)\right) = \sum_{n=1}^{\infty} \mu'(A_n).$$

Also $\mathscr{A} = \pi^{-1}(\mathscr{M}_{\mu'})$ is a σ-algebra on Ω. We first assert that μ is σ-additive on $\mathscr{A}$.

In fact, let A_1, A_2 be disjoint sets in $\mathscr{A}$. Then there exist B_1, B_2 in $\mathscr{M}_{\mu'}$, $A_i = \pi^{-1}(B_i)$, $i = 1, 2$. Let $C_1 = B_1$, $C_2 = B_2 - B_1$. Then C_1, C_2 are disjoint, $C_i \in \mathscr{M}_{\mu'}$, and $A_i = \pi^{-1}(C_i)$, since $\mathscr{A}$ is the smallest σ-algebra relative to which π is measurable [i.e., π is $(\mathscr{A}, \mathscr{M}_{\mu'})$-measurable]. Thus

$$\begin{aligned} \mu(A_1 \cup A_2) &= \mu\left(\pi^{-1}(C_1) \cup \pi^{-1}(C_2)\right) \\ &= \mu\left(\pi^{-1}(C_1 \cup C_2)\right) = \mu'(C_1 \cup C_2) \\ &= \mu'(C_1) + \mu'(C_2), \qquad \text{since } \mu' | \mathscr{M}_{\mu'} \text{ is a measure} \\ &= \mu(A_1) + \mu(A_2). \end{aligned}$$

But μ is σ-subadditive and additive on $\mathscr{A}$. So it follows that μ is σ-additive on (the σ-algebra) $\mathscr{A}$. But by Theorem 2.2.9 and Proposition 2.2.14, the outer regularity of μ implies that $\mathscr{M}_\mu$ is the largest class of sets on which μ is σ-additive. Hence $\mathscr{A} \subset \mathscr{M}_\mu$. This is (i) in another notation. Using this, we now prove (9).

If $f: \Omega' \to \overline{\mathbb{R}}$ is μ'-measurable, then for any Borel set $B \subset \overline{\mathbb{R}}$, $f^{-1}(B) \in \mathscr{M}_{\mu'}$. Hence $(f \circ \pi)^{-1}(B) = \pi^{-1}(f^{-1}(B)) \in \mathscr{A} \subset \mathscr{M}_\mu$ by (i). So $f \circ \pi: \Omega \to \overline{\mathbb{R}}$ is μ-measurable. Thus if $f = \chi_A$, $A \in \mathscr{M}_{\mu'}$, then

$$\int_{\Omega'} f \, d\mu' = \mu'(A) = \mu(\pi^{-1}(A)) = \int_\Omega \chi_{\pi^{-1}(A)} \, d\mu = \int_\Omega f \circ \pi \, d\mu.$$

If $f = \sum_{i=1}^n a_i \chi_{A_i}$ is a simple μ'-measurable function, then by the linearity of the integral on measurable functions (Proposition 1.5), we deduce that (9) holds for all simple functions. If $f = f^+ - f^-$ is any μ'-measurable function, then considering $f^\pm$ separately and using the additivity of the integral on measurable functions whenever the sum is defined, it suffices to establish (9) for all $f \geqslant 0$ which are μ-measurable. But by the structure theorem (Theorem 3.1.7) there exists a sequence of μ'-simple functions $0 \leqslant \varphi_n \uparrow f$ pointwise, and then $\varphi_n \circ \pi \uparrow f \circ \pi$. So, from the already established result above, we get

$$\int_{\Omega'} \varphi_n \, d\mu' = \int_\Omega \varphi_n \circ \pi \, d\mu, \qquad n \geqslant 1.$$

Let $n \to \infty$ on both sides, and interchange the limit and integral by the monotone convergence theorem, which we shall establish next. That gives (9) as stated, and the result will follow.

However, we can establish (9) *without* using the monotone convergence theorem, but with Theorem 1.4(ii) directly. Thus for any μ'-measurable $f \geqslant 0$,

$$\begin{aligned}
\int_{\Omega'} f \, d\mu' &= \lim_{r \downarrow 1} \sum_{n=-\infty}^{\infty} r^n \mu'\big(f^{-1}(r^n, r^{n+1})\big) + \infty \cdot \mu'\big(f^{-1}(\{\infty\})\big) \\
&= \lim_{r \downarrow 1} \sum_{n=-\infty}^{\infty} r^n \mu\big((f \circ \pi)^{-1}(r^n, r^{n+1})\big) + \infty \cdot \mu\big((f \circ \pi)^{-1}(\{\infty\})\big) \\
&= \int_\Omega f \circ \pi \, d\mu, \qquad \text{by Theorem 1.4.}
\end{aligned}$$

This gives (9), and the theorem is completely proved.

As an immediate consequence we obtain the useful result with distribution functions, introduced in Section 3.3, as:

Corollary 7 (Fundamental theorem of probability). *Let* (Ω, Σ, μ) *be a finite measure space* $[\mu(\Omega) = 1$ *for the probability case*]. *If* $f: \Omega \to \mathbb{R}$ *is a measurable* *function (for Σ), and F_f is its distribution on $\mathbb{R}$, then for any Borel function $g: \mathbb{R} \to \mathbb{R}$ we have*

$$\int_\Omega g(f(\omega))\mu(d\omega) = \int_{\mathbb{R}} g(x) F_f(dx), \tag{10}$$

in the sense that if either side exists, so does the other and equality holds.

Note that the right side integral, by definition, is the same as $\int_{\mathbb{R}} g\,d\nu$, where $\nu([a, b)) = F_f(b) - F_f(a)$, so that ν is a Lebesgue-Stieltjes measure induced by F_f.

In all the preceding results, the monotone convergence theorem for measurable functions is in the background. We now take up this and other limit theorems in the next section. They are the basic results of the abstract Lebesgue integration.

EXERCISES

(It is assumed that the outer measure μ in *each problem is Carathéodory regular and that the monotone convergence for μ-simple functions is valid.*)

0. Let $f: \Omega \to \mathbb{R}^+$ be a function and μ be an outer measure such that $\nu: A \mapsto \int_A^* f\,d\mu$ has the finite subset property. Show that $\nu(\Omega) < \infty$ if $\alpha = \sup\{\nu(A): A \in \Sigma_0\} < \infty$ where Σ_0 is the class of subsets of Ω of ν–finite measure. Is it then true that $\alpha = \nu(\Omega)$?

1. If $f: \Omega \to \mathbb{R}^+$ is any bounded function and μ is an outer measure, then verify that in Proposition 1, the integral can be approximated from above with μ-simple functions that dominate f.

2. Let $f: \Omega \to \overline{\mathbb{R}}$ be a function and μ an outer measure. If $|\int_\Omega^* f\,d\mu| < \infty$, show that for each $\epsilon > 0$ there exists a μ-elementary function h_ϵ such that $h_\epsilon \geqslant f$ and $\int_\Omega^* f\,d\mu \leqslant \int_\Omega^* h_\epsilon\,d\mu \leqslant \int_\Omega^* f\,d\mu + \epsilon$.

3. If $f, g: \Omega \to \overline{\mathbb{R}}$ satisfy the hypothesis of Exercise 2, show that

$$\int_\Omega^* (f+g)\,d\mu \leqslant \int_\Omega^* f\,d\mu + \int_\Omega^* g\,d\mu.$$

[Note that $f_1 \leqslant g_1 \Rightarrow f_1^+ \leqslant g_1^+$ and $f_1^- \geqslant g_1^-$; then use Definition 1.2 to deduce that the outer integral is order preserving. Next, assuming that the result is false, get a contradiction on applying the result of Exercise 2.]

4. Complete the proof of Corollary 7.

5. Let $\Omega = \mathbb{R}$, and $g: \mathbb{R} \to \mathbb{R}$ be a nondecreasing differentiable function with derivative g'. If μ_g^* is the Lebesgue-Stieltjes outer measure generated by g, and $f\colon \Omega \to \overline{\mathbb{R}}$ is an arbitrary function, show that the following equation holds if either outer integral exists:

$$\int_{\mathbb{R}}^{*} f(x)\mu_g^*(dx) = \int_{\mathbb{R}}^{*} f(x)g'(x)\lambda^*(dx),$$

where $\lambda^*(\cdot)$ is the Lebesgue outer measure on $\mathbb{R}$. If f is a measurable function, then the right side quantity is the *Lebesgue integral*, and the left side one is termed the *Lebesgue-Stieltjes integral*.

6. Let (Ω, μ) be as in Exercise 1, and $f\colon \Omega \to \overline{\mathbb{R}}$ be a function such that $\int_A^* f\,d\mu \in \mathbb{R}$ for each $A \subset \Omega$. Show that $\int_\Omega^* |f|\,d\mu < \infty$.

4.3. THE LEBESGUE LIMIT THEOREMS

The fundamental limit theorems that are often used in applications of the abstract integral are the Lebesgue monotone and dominated convergences. One of these has already come up in the preceding section. Here we establish both these theorems and some related ones for the crucial classes of measurable functions.

Theorem 1 (Monotone convergence). *Let (Ω, Σ, μ) be a measure space, and $f, f_n\colon \Omega \to \overline{\mathbb{R}}^+$, $n \geqslant 1$, be a sequence of measurable (for Σ) functions such that $f_n \leqslant f_{n+1}$ a.e. and $f_n \to f$. Then*

$$\lim_n \int_\Omega f_n\,d\mu = \int_\Omega f\,d\mu. \tag{1}$$

Proof. By hypothesis $0 \leqslant f_n \uparrow f$, and these are measurable (for Σ). Since $\int_\Omega f_n\,d\mu \leqslant \int_\Omega f_{n+1}\,d\mu$ and $f \geqslant f_n$, it is clear that

$$\lim_n \int_\Omega f_n\,d\mu \leqslant \int_\Omega f\,d\mu. \tag{2}$$

But (1) is true if the left side in (2) is infinite. So let $t < \int_\Omega f\,d\mu$ be an arbitrary real number. Then by using Lebesgue's definition of the integral [cf. Theorem 1.4(ii)], there exists an elementary function h_t such that $h_t \leqslant f$, measurable for Σ, and

$$t < \int_\Omega h_t\,d\mu \leqslant \int_\Omega f\,d\mu. \tag{3}$$

Let $\Omega_n = \{\omega\colon h_t(\omega) \leqslant f_n(\omega)\}$. Since h_t and f_n are measurable for Σ, $\Omega_n \in \Sigma$ (cf. Corollary 3.1.4), and since $f_n \leqslant f_{n+1}$, we get $\Omega_n \subset \Omega_{n+1}$ and $\lim_n \Omega_n = \Omega$.

Also, if $\nu(A) = \int_A h_t \, d\mu$, then we have already seen that $\nu(\cdot)$ is an outer measure on Ω and by Proposition 1.5 it is additive on Σ. Hence $\nu(\cdot)$ is σ-additive on the σ-algebra Σ (see Lemma 1.3.5). Consequently,

$$\lim_n \int_{\Omega_n} h_t \, d\mu = \lim_n \nu(\Omega_n) = \nu\Big(\lim_n \Omega_n\Big)$$

$$= \nu(\Omega) = \int_\Omega h_t \, d\mu, \text{ by Proposition 2.1.1(i).} \tag{4}$$

It follows from (3) and (4) that

$$t < \int_\Omega h_t \, d\mu = \lim_n \int_{\Omega_n} h_t \, d\mu$$

$$\leqslant \lim_n \int_{\Omega_n} f_n \, d\mu, \qquad \text{since} \quad h_t \leqslant f_n \text{ on } \Omega_n$$

$$\leqslant \lim_n \int_\Omega f_n \, d\mu, \qquad \text{since} \quad f_n \geqslant 0 \text{ and } \Omega_n \subset \Omega$$

$$\leqslant \int_\Omega f \, d\mu, \qquad \text{by (2).} \tag{5}$$

Letting $t \uparrow \int_\Omega f \, d\mu$, we see that there is equality in (5), and (1) follows. This completes the proof.

From this result a number of consequences will be recorded, since they often appear in applications. The first one is deduced from the above *and* Proposition 2.4.

Corollary 2. *Let* μ *be a Carathéodory regular outer measure on* Ω, *and* $f_n\colon \Omega \to \overline{\mathbb{R}}^+$, $n \geqslant 1$, *be any increasing sequence of functions. Then*

$$\lim_n \int_\Omega^* f_n \, d\mu = \int_\Omega^* \lim_n f_n \, d\mu. \tag{6}$$

The next result is a different form of the above theorem, and is given for ready reference.

Corollary 3. *Let* (Ω, Σ, μ) *be a measure space, and* $f_n\colon \Omega \to \overline{\mathbb{R}}$, $n \geqslant 1$, *be measurable*, $f_n \leqslant f_{n+1}$. *If* $|\int_\Omega f_1 \, d\mu| < \infty$, *then*

(a) *we have*

$$\lim_n \int_\Omega f_n \, d\mu = \int_\Omega \lim_n f_n \, d\mu. \tag{7}$$

(b) *If* $A_n \in \Sigma$, $A_n \subset A_{n+1}$, *then*

$$\lim_n \int_{A_n} f_1 \, d\mu = \int_{\bigcup_n A_n} f_1 \, d\mu. \tag{8}$$

(c) *If* $f_n \geqslant 0$, $n \geqslant 1$, *but only measurable, then we have*

$$\int_\Omega \sum_{n=1}^{\infty} f_n \, d\mu = \sum_{n=1}^{\infty} \int_\Omega f_n \, d\mu. \tag{9}$$

Proof. (a): Let $h_n = f_n - f_1$, $n \geqslant 2$. Then h_n satisfies the hypothesis, so that, since $|\int_\Omega f_1 \, d\mu| < \infty$ and $\lim_n h_n = f - f_1$,

$$\lim_n \int_\Omega f_n \, d\mu - \int_\Omega f_1 \, d\mu = \lim_n \int_\Omega h_n \, d\mu = \int_\Omega \lim_n h_n \, d\mu, \text{ by Theorem 1}$$

$$= \int_\Omega (f - f_1) \, d\mu = \int_\Omega f \, d\mu - \int_\Omega f_1 \, d\mu.$$

Canceling the finite number $\int_\Omega f_1 \, d\mu$ from both sides, we get (7).

(b): This has already been noted [see (4) above].

(c): Let $g_n = \sum_{k=1}^{n} f_k$. Then $\{g_n: n \geqslant 1\}$ satisfies the hypothesis of the theorem, so that

$$\int_\Omega \left(\sum_{n=1}^{\infty} f_n \right) d\mu = \int_\Omega \lim_n g_n \, d\mu = \lim_n \int_\Omega g_n \, d\mu$$

$$= \lim_n \sum_{k=1}^{n} \int_\Omega f_k \, d\mu, \text{ by Proposition 1.5}$$

$$= \sum_{k=1}^{\infty} \int_\Omega f_k \, d\mu.$$

This proves (9).

This next result becomes Corollary 2.5 for measurable functions. It is used when there is no order relation between the f_n.

Theorem 4 (Fatou's lemma). *Let* (Ω, Σ, μ) *be a measure space, and* f_0, f_n, g: $\Omega \to \overline{\mathbb{R}}$ *be measurable for* Σ. *If*

(i) $f_0 \leqslant f_n$, $n \geqslant 1$, *and* $\int_\Omega f_0 \, d\mu > -\infty$, *then*

$$\int_\Omega \liminf_n f_n \, d\mu \leqslant \liminf_n \int_\Omega f_n \, d\mu; \tag{10}$$

(ii) $f_n \leqslant g$, $n \geqslant 1$, *and* $\int_\Omega g\,d\mu < \infty$, *then*

$$\int_\Omega \limsup_n f_n\,d\mu \geqslant \limsup_n \int_\Omega f_n\,d\mu. \tag{11}$$

Proof. (i): Since $\int_\Omega f_0\,d\mu > -\infty$, the integral exists and equals $\int_\Omega f_0^+\,d\mu - \int_\Omega f_0^-\,d\mu$. So $\int_\Omega f_0^-\,d\mu < \infty$ and $0 \leqslant f_0^- < \infty$ a.e. Consequently $f_n \geqslant f_0 = f_0^+ - f_0^- \Rightarrow f_n + f_0^- \geqslant f_0^+ \geqslant 0$. Using the same argument as in the proof of Corollary 3, it may be assumed that $f_n \geqslant 0$ holds for all $n \geqslant 1$. Letting $g_n = \inf\{f_k: k \geqslant n\}$, we note that $0 \leqslant g_n \leqslant g_{n+1}$, each g_n is measurable for Σ, and $\lim_n g_n = \liminf_n f_n$ is measurable for Σ (cf. Lemma 3.1.3). Hence by Theorem 1 we have (10), because

$$\int_\Omega \liminf_n f_n\,d\mu = \int_\Omega \lim_n g_n\,d\mu = \lim_n \int_\Omega g_n\,d\mu$$

$$= \liminf_n \int_\Omega g_n\,d\mu \leqslant \liminf_n \int_\Omega f_n\,d\mu.$$

(ii): We may reduce this result to part (i) by considering $-f_n \geqslant -g$ and $\int_\Omega -g\,d\mu = -\int_\Omega g\,d\mu > -\infty$. Thus (10), for the sequence $\{-f_n: n \geqslant 1\}$, gives

$$-\int_\Omega \limsup_n f_n\,d\mu = \int_\Omega \liminf_n(-f_n)\,d\mu$$

$$\leqslant \liminf_n \int_\Omega (-f_n)\,d\mu = \liminf_n\left(-\int_\Omega f_n\,d\mu\right)$$

$$= -\limsup_n \int_\Omega f_n\,d\mu.$$

Multiplying through by -1, we get (11) and the proof is complete.

It is to be noted that the monotone convergence theorem has played a fundamental role in this work. With these results we can obtain the dominated convergence theorem in two forms.

Theorem 5 (Dominated convergence, first form). *Let* (Ω, Σ, μ) *be any measure space, and* $f_n, f, g: \Omega \to \overline{\mathbb{R}}$ *be measurable functions such that* (i) $|f_n| \leqslant g$, *a.e.,* (ii) $\int_\Omega g\,d\mu < \infty$, *and* (iii) $f_n \to f$ *a.e. Then* $\int_\Omega |f|\,d\mu < \infty$, *and*

$$\lim_n \int_\Omega f_n\,d\mu = \int_\Omega \lim_n f_n\,d\mu = \int_\Omega f\,d\mu. \tag{12}$$

Proof. By hypothesis $|f_n| \leqslant g$ and $f_n \to f$, so that $|f| \leqslant g$ a.e. also. Since our integral is order preserving, we get $\int_\Omega |f|\,d\mu \leqslant \int_\Omega g\,d\mu < \infty$. To establish (12), note that $-g \leqslant f_n \leqslant g$ and $\limsup_n f_n = f$, so that by Fatou's lemma (both parts can be used, since g is integrable) we get

$$\begin{aligned}\int_\Omega f\,d\mu &= \int_\Omega \liminf_n f_n\,d\mu \leqslant \liminf_n \int_\Omega f_n\,d\mu \\ &\leqslant \limsup_n \int_\Omega f_n\,d\mu \\ &\leqslant \int_\Omega \limsup_n f_n\,d\mu = \int_\Omega f\,d\mu.\end{aligned}$$

Hence there is equality throughout, and this gives (12), and the theorem follows.

If $\mu(\Omega) < \infty$, then every constant has a finite integral, so that taking $g = k_0$ in the theorem as a common bound for all functions, we get:

Corollary 6 (Bounded convergence). *Let* (Ω, Σ, μ) *be a finite measure space and* $f, f_n\colon \Omega \to \mathbb{R}$, $n \geqslant 1$, *be a uniformly bounded sequence of measurable functions for* Σ. *If* $f_n \to f$, *pointwise, then* f *is bounded and one has*

$$\lim_n \int_\Omega f_n\,d\mu = \int_\Omega f\,d\mu. \tag{13}$$

The results described above are the best possible in the sense that relaxing the hypotheses renders them false or [for (10) and (11)] the inequalities cannot be strengthened. The following examples explain this.

Example A. Let (Ω, Σ, μ) be the Lebesgue unit interval, and $f_n\colon \Omega \to \mathbb{R}^+$ be defined by $f_{2n}(\omega) = 0$ if $\omega \in [0, \frac{1}{2}]$, $= 1$ if $\omega \in (\frac{1}{2}, 1]$, and $f_{2n+1}(\omega) = 1$ if $\omega \in [0, \frac{1}{2}]$, $= 0$ if $\omega \in (\frac{1}{2}, 1]$. Thus $\limsup_n f_n = 1$ and $\liminf_n f_n = 0$. Now for (10) and (11)

$$\begin{aligned}0 &= \int_\Omega \liminf_n f_n\,d\mu \\ &< \liminf_n \int_\Omega f_n\,d\mu = \tfrac{1}{2} = \limsup_n \int_\Omega f_n\,d\mu \\ &< \int_\Omega \limsup_n f_n\,d\mu = 1.\end{aligned}$$

Example B. On the same measure space, let $g_n(\omega) = n$ if $\omega \in (0, 1/n]$, $= 0$ if $\omega \in (1/n, 1]$. Then $g_n \to g = 0$. But the g_n-sequence is not dominated, and

$$0 = \int_\Omega g\, d\mu < \lim_n \int_\Omega g_n\, d\mu = 1.$$

Since $\limsup_n g_n = \liminf_n g_n = \lim_n g_n = 0$, for this sequence we also have

$$0 = \int_\Omega \limsup_n g_n\, d\mu < \limsup_n \int_\Omega g_n\, d\mu = 1,$$

so (11) is false in this case. Considering $\{-g_n: n \geqslant 1\}$, (10) is also false for this sequence. Thus the hypotheses cannot be relaxed.

Example C. Again Theorem 1 (and hence the others based on it) will not be true if the sequences are replaced with generalized (uncountable) sequences. Indeed, consider the same measure space, and let $\mathscr{F}$ be the collection of all finite subsets of Ω ordered by inclusion. For each $\alpha \in \mathscr{F}$, consider F_α, a finite set. If $f_\alpha = \chi_{F_\alpha}$, then $\alpha \leqslant \beta \Rightarrow f_\alpha \leqslant f_\beta$ and $\sup_\alpha f_\alpha = 1$ (i.e., $f_\alpha \uparrow 1$). However, for the Lebesgue measure μ, $\mu(F_\alpha) = 0$ for each $\alpha \in \mathscr{F}$, so that

$$1 = \int_\Omega \sup_\alpha f_\alpha\, d\mu > \sup_\alpha \int_\Omega f_\alpha\, d\mu = 0.$$

Remark. We can salvage a small part of the generalization of monotone convergence, namely, if the generalized sequence has a countable *cofinal* sequence (i.e., there is a sequence $\alpha_1 < \alpha_2 < \cdots$ such that for each β we can find an α_i such that $\beta < \alpha_i$). (See Exercise 13 for a slightly modified result.)

A precise statement is as follows:

Corollary 7. *Let* (Ω, Σ, μ) *be a measure space;* $f_i, f: \Omega \to \overline{\mathbb{R}}^+$, $i \in I$, *be measurable functions; and* I *be an increasingly ordered set having a countable cofinal set* $J \subset I$. *If* $f_i \leqslant f_{i'}$, $i < i'$, *and* $\sup_i f_i = f$, *then* f *is* μ*-measurable and*

$$\int_\Omega f\, d\mu = \lim_i \int_\Omega f_i\, d\mu = \sup_i \int_\Omega f_i\, d\mu. \tag{14}$$

The proof is easily obtained, since $\sup_{i \in I} f_i = \sup_{j \in J} f_j = f$ by cofinality, and is left to the reader. Actually this result holds if all the f_i are defined on a Radon space and each f_i is lower semicontinuous, so that $\sup_i f_i$ is also measurable (and then no additional cofinality of I is needed).

In order to use in the next result, we establish the following property of the Lebesgue integral, which is of independent interest.

Proposition 8. *Let* (Ω, Σ, μ) *be a measure space, and* $f: \Omega \to \overline{\mathbb{R}}$ *be measurable (for* Σ*) and* μ*-integrable. Then the integral is* μ*-continuous in the sense that*

$$\lim_{\mu(A)\to 0} \int_A |f|\, d\mu = 0. \tag{15}$$

Proof. Since by hypothesis $\int_\Omega |f|\, d\mu < \infty$ [recall that $|\int_\Omega f\, d\mu| < \infty$ iff the former is: see Theorem 1.3(iv)], the function $\nu: A \mapsto \int_A |f|\, d\mu$, $A \in \Sigma$, is a finite measure. If (15) is false, then there exists a $\delta > 0$ such that for each $\epsilon > 0$ and each $A \in \Sigma$ with $\mu(A) < \epsilon$ we have $\nu(A) \geqslant \delta$. Choose $\epsilon_n > 0$ such that $\sum_{n=1}^{\infty} \epsilon_n < \infty$ and $A_n \in \Sigma$ satisfy $\mu(A_n) < \epsilon_n$. If $B_n = \bigcup_{k=n}^{\infty} A_k (\in \Sigma)$, then $B_n \supset B_{n+1}$ and

$$\mu(B_n) \leqslant \sum_{k=n}^{\infty} \mu(A_k) \leqslant \sum_{k=n}^{\infty} \epsilon_k \to 0 \quad \text{as} \quad n \to \infty.$$

Hence $\mu(\lim_n B_n) = \lim_n \mu(B_n) = 0$, by Proposition 2.1.1, since $\mu(B_1) < \infty$. Consequently, letting $B = \lim_n B_n$, we get

$$\nu(B) = \int_B |f|\, d\mu = \int_\Omega \chi_B |f|\, d\mu = 0,$$

since $|f|\chi_B = 0$ a.e. But then (taking $\epsilon = \epsilon_1$)

$$0 < \delta \leqslant \nu(A_n) \leqslant \nu(B_n) \to \nu(B) = 0,$$

by Proposition 2.1.1 again, since ν is a finite measure. This contradiction shows that our supposition is false, and the proposition holds as stated.

A consequence of this result which we use below is given by:

Corollary 9. *Let* (Ω, Σ, μ) *be a measure space; let* $0 \leqslant f_n \leqslant g$*, where* $f_n, g: \Omega \to \overline{\mathbb{R}}^+$ *are measurable (for* Σ*) and* $f_n \xrightarrow{\mu} 0$*. Then* $\int_\Omega g\, d\mu < \infty \Rightarrow \int_\Omega f_n\, d\mu \to 0$ *as* $n \to \infty$.

Proof. If $A_n = \{\omega: f_n(\omega) \geqslant \delta\}$ for any prescribed $\delta > 0$, then $\mu(A_n) \to 0$ by hypothesis. Also, if $F_k = \{\omega: 1/k \leqslant g(\omega) \leqslant k\}$, then $F_k \subset F_{k+1}$ and

$$\mu(F_k) \leqslant \mu\left(\left\{\omega: g(\omega) \geqslant \frac{1}{k}\right\}\right) \leqslant k \int_\Omega g\, d\mu < \infty, \qquad k \geqslant 1.$$

But if $F = \{\omega: g(\omega) > 0\}$, then $F_m \uparrow F$, so that, by Theorem 5, $\nu(F - F_m) = \int_{F-F_m} g\, d\mu \to 0$, since $F - F_m \downarrow \varnothing$ and $g\chi_{F_m} \uparrow g$. Given $\epsilon > 0$, choose m_0 such that $m \geqslant m_0 \Rightarrow \nu(F - F_m) < \epsilon/2$. Take $\delta > 0$ such that $\mu(F_{m_0}) \leqslant \epsilon/2\delta$.

(Note that F_m does not depend on A_n, and so we use this δ for A_n.) Then $0 \leqslant f_n \chi_{F_{m_0}} \leqslant g \chi_{F_{m_0}} \leqslant m_0 \chi_{F_{m_0}}$, and

$$\begin{aligned}\int_\Omega f_n \, d\mu &= \int_{A_n} f_n \, d\mu + \int_{A_n^c \cap F_{m_0}} f_n \, d\mu + \int_{A_n^c \cap F \cap F_{m_0}^c} f_n \, d\mu \\ &\leqslant \int_{A_n} g \, d\mu + \delta \cdot \mu\left(F_{m_0} \cap A_n^c\right) + \int_{F \cap F_{m_0}^c} g \, d\mu \\ &\leqslant \nu(A_n) + \delta \cdot \mu\left(F_{m_0}\right) + \nu\left(F - F_{m_0}\right) \\ &< \nu(A_n) + \delta \cdot \frac{\epsilon}{2\delta} + \frac{\epsilon}{2} = \epsilon + \nu(A_n).\end{aligned}$$

But by Proposition 8, $\lim_{\mu(A_n) \to 0} \nu(A_n) = 0$, and $f_n \xrightarrow{\mu} 0 \Rightarrow \mu(A_n) \to 0$, so

$$\lim_{n \to \infty} \int_\Omega f_n \, d\mu < \epsilon.$$

Since $\epsilon > 0$ is arbitrary, the assertion follows.

With this result, we are now ready to present

Theorem 10 (Dominated convergence, second form). *If* (Ω, Σ, μ) *is a measure space,* $f_n, f, g\colon \Omega \to \overline{\mathbb{R}}$ *are measurable,* $|f_n| \leqslant g$, $f_n \xrightarrow{\mu} f$, *and* $\int_\Omega g \, d\mu < \infty$, *then* $\int_\Omega f_n \, d\mu \to \int_\Omega f \, d\mu$, *as* $n \to \infty$.

Proof. Clearly $\int_\Omega |f_n| \, d\mu \leqslant \int_\Omega g \, d\mu < \infty$, and by F. Riesz's theorem, $f_{n_i} \to f$ a.e., for a subsequence and $|f| \leqslant g$. So f is also integrable. Moreover, $|f_n - f| \leqslant 2g$ and $f_n - f \xrightarrow{\mu} 0$. Since $2g$ is integrable, by Corollary 9 we have

$$\left| \int_\Omega f_n \, d\mu - \int_\Omega f \, d\mu \right| \leqslant \int_\Omega |f_n - f| \, d\mu \to 0 \quad \text{as} \quad n \to \infty.$$

This proves the result in a stronger form than the assertion.

Both forms of dominated convergence are extremely useful. Moreover, the same type of argument shows that the bounded as well as the monotone convergence theorems are valid if a.e. convergence is replaced by convergence in measure. Note also that if the dominated convergence theorem is given, then the monotone convergence follows from it.

Suppose a dominating function is not available. Is there a condition that can be used to save the preceding convergence theorems? By Example 2 above, the result will be false if no additional condition is imposed. A characterization of the limit relation, due to G. Vitali, will now be given, since it also turns out to be used frequently in applications; it is in fact another major result of this section.

Theorem 11 (Vitali). *Let* (Ω, Σ, μ) *be a measure space, and* $f_n, f\colon \Omega \to \overline{\mathbb{R}}$ *be a sequence of measurable functions (for* Σ*) such that* (i) $\int_\Omega |f_n|\,d\mu < \infty$, $n \geqslant 1$, *and* (ii) $f_n \to f$ *a.e. Then* f *is integrable and* $\int_\Omega |f_n - f|\,d\mu \to 0$, *iff* (a) $\lim_{\mu(A)\to 0} \int_A |f_n|\,d\mu = 0$ *uniformly in n and* (b) *for each* $\epsilon > 0$, *there exists a set* $A_\epsilon \in \Sigma$, $\mu(A_\epsilon) < \infty$, *satisfying* $\int_{A_\epsilon^c} |f_n|\,d\mu < \epsilon$, $n \geqslant 1$.

Remark. In view of Proposition 8, if $|f_n| \leqslant g$, $\int_\Omega g\,d\mu < \infty$, then (a) and (b) are satisfied, so that Theorem 5 is a consequence of this result, whose proof, given below, is independent of the monotone convergence theorem. Here again $f_n \to f$ a.e. in (ii) can be replaced by $f_n \xrightarrow{\mu} f$, and we omit this case. [If $\mu(\Omega) < \infty$, even vague convergence suffices. See Exercises 2 and 19.]

Proof. We first prove the sufficiency. So let (a) and (b) hold. The idea of proof is to show that $\{f_n\colon n \geqslant 1\}$ is Cauchy in the sense that $\|f_n - f_m\|_1 = \int_\Omega |f_n - f_m|\,d\mu \to 0$ as $n, m \to \infty$, and then deduce that f is integrable as well as $\|f_n - f\|_1 \to 0$. Thus consider for $\epsilon > 0$ with the A_ϵ of (b):

$$\begin{aligned}
\|f_m - f_n\|_1 &= \int_{A_\epsilon} |f_m - f_n|\,d\mu + \int_{A_\epsilon^c} |f_m - f_n|\,d\mu \\
&\leqslant \int_{A_\epsilon} |f_m - f_n|\,d\mu + \int_{A_\epsilon^c} |f_m|\,d\mu + \int_{A_\epsilon^c} |f_n|\,d\mu \\
&< 2\epsilon + \int_{A_\epsilon} |f_m - f_n|\,d\mu. \qquad (16)
\end{aligned}$$

Hence it suffices to show that the second term tends to zero as $m, n \to \infty$ in (16). Since $\mu(A_\epsilon) < \infty$, and $f_n \to f$ a.e. on A_ϵ also, it converges in measure on this set by Theorem 3.2.5(iii). So given $\epsilon' > 0$, there is a $\delta_{\epsilon'} > 0$ and n_0 $[= n_0(\epsilon')]$ such that $m, n \geqslant n_0$ implies that if $A_{mn} = \{\omega \in A_\epsilon\colon |f_m - f_n|(\omega) > \epsilon'\}$ then $\mu(A_{mn}) < \delta_{\epsilon'}$. This gives

$$\begin{aligned}
\int_{A_\epsilon} |f_m - f_n|\,d\mu &= \int_{A_\epsilon - A_{mn}} |f_m - f_n|\,d\mu + \int_{A_{mn}} |f_m - f_n|\,d\mu \\
&\leqslant \epsilon'\mu(A_\epsilon - A_{mn}) + \int_{A_{mn}} |f_m|\,d\mu + \int_{A_{mn}} |f_n|\,d\mu \\
&\leqslant \epsilon'\mu(A_\epsilon) + 2\epsilon', \quad \text{by (a)}, \qquad \text{if} \quad m, n \geqslant n_1 \geqslant n_0. \quad (17)
\end{aligned}$$

Since $\epsilon' > 0$ is arbitrary, (16) and (17) imply that $\|f_m - f_n\|_1 \to 0$ as $m, n \to \infty$. Hence $|\|f_m\|_1 - \|f_n\|_1| \leqslant \|f_m - f_n\|_1 \to 0$, so that $\{\|f_n\|_1, n \geqslant 1\}$ is bounded.

Now $f_n \to f$ a.e. implies $|f_n| \to |f|$ a.e., so that by Fatou's lemma

$$\int_\Omega |f|\,d\mu \leqslant \liminf_n \int_\Omega |f_n|\,d\mu = \liminf_n \|f_n\|_1 < \infty,$$

since the right side is bounded, by the preceding sentence. Thus f is integrable. Next we assert that $\|f_n - f\|_1 \to 0$ as $n \to \infty$.

By Proposition 8, f also satisfies (a) and if $A_n = \{\omega: 1/n \leqslant |f(\omega)| < n\}$, then $A_n \uparrow A_0 = \{\omega: |f(\omega)| > 0\}$. Consequently

$$\int_{A_n} |f|\,d\mu = \int_\Omega |f| \chi_{A_n}\,d\mu \to \int_{A_0} |f|\,d\mu,$$

by Proposition 2.1.1, and let $\nu: A \mapsto \int_A |f|\,d\mu$. Being a finite measure, $\nu(A_0 - A_n) \to 0$ as $n \to \infty$. Hence, given $\epsilon_1 > 0$, there is an n_0 such that $\mu(A_{n_0}) < \infty$ and $\nu(A - A_{n_0}) < \epsilon_1$. If we take $\tilde{A}_\epsilon = A_\epsilon \cup A_{n_0}$, then $\mu(\tilde{A}_\epsilon) < \infty$ and writing $f = f_\infty$ for symmetry (temporarily),

$$\int_{\Omega - \tilde{A}_\epsilon} |f_n|\,d\mu < \epsilon_1, \qquad 1 \leqslant n \leqslant \infty. \tag{18}$$

Thus f_n, $1 \leqslant n \leqslant \infty$, also satisfies (b), i.e., both (a) and (b) hold for this set. Hence by (16) and (17) we can conclude that $\|f_n - f_\infty\|_1 \to 0$ if we show that

$$\int_{\tilde{A}_\epsilon} |f_n - f_\infty|\,d\mu \to 0 \qquad \text{as} \quad n \to \infty. \tag{19}$$

To prove (19), by hypothesis $f_n \to f_\infty$ a.e. on $\tilde{A}_\epsilon$, and so by the Egorov theorem, for each $\delta > 0$, there is a measurable set $B_\delta \subset \tilde{A}_\epsilon$ such that $\mu(B_\delta) < \delta$ $\Rightarrow$ $f_n \to f_\infty$ uniformly on $\tilde{A}_\epsilon - B_\delta$. Thus, given $\eta > 0$, there is an n_2 such that $n \geqslant n_2 \Rightarrow |f_n(\omega) - f_\infty(\omega)| < \eta$ for all $\omega \in \tilde{A}_\epsilon - B_\delta$. This implies that, for any $\eta' > 0$, there is a $\delta' > 0$ such that $\int_{B_{\delta'}} |f_n|\,d\mu < \eta'$ for all $n_2 \leqslant n \leqslant \infty$, if $\mu(B_{\delta'}) < \delta'$. Hence

$$\begin{aligned}
\int_{\tilde{A}_\epsilon} |f_n - f_\infty|\,d\mu &= \int_{\tilde{A}_\epsilon - B_{\delta'}} |f_n - f_\infty|\,d\mu + \int_{B_{\delta'}} |f_n - f_\infty|\,d\mu \\
&\leqslant \eta\mu(\tilde{A}_\epsilon - B_{\delta'}) + \int_{B_{\delta'}} |f_n|\,d\mu + \int_{B_{\delta'}} |f_\infty|\,d\mu, \qquad n \geqslant n_2 \\
&\leqslant \eta\mu(\tilde{A}_\epsilon) + 2\eta'.
\end{aligned} \tag{20}$$

Consequently, letting $n \to \infty$ and then $\eta \to 0$ in (20), we get the right side $< 2\eta'$. This gives (19), since η' is arbitrary.

For the converse, let $f_n \to f$ a.e., $\|f\|_1 < \infty$, and $\|f_n - f\|_1 \to 0$. Given $\epsilon > 0$, let n_0 [$= n_0(\epsilon)$] be chosen such that $\|f_n - f\|_1 < \epsilon$ for $n \geqslant n_0$. Then as seen in (18), there exists an A_ϵ and a B_ϵ in Σ such that $\mu(A_\epsilon) < \infty$, $\mu(B_\epsilon) < \infty$, and

$$\int_{A_\epsilon^c} |f| \, d\mu < \epsilon, \qquad \int_{B_\epsilon^c} |f_n| \, d\mu < \infty, \quad \text{for } 1 \leqslant n \leqslant n_0. \tag{21}$$

If we set $\tilde{A}_\epsilon = A_\epsilon \cup B_\epsilon$, then $\mu(\tilde{A}_\epsilon) < \infty$, and by (21), (b) holds for $1 \leqslant n \leqslant n_0$. Now for $n > n_0$,

$$\begin{aligned} \int_{\tilde{A}_\epsilon^c} |f_n| \, d\mu &\leqslant \int_{\tilde{A}_\epsilon^c} |f_n - f| \, d\mu + \int_{\tilde{A}_\epsilon^c} |f| \, d\mu \\ &\leqslant \|f_n - f\|_1 + \epsilon < 2\epsilon, \end{aligned} \tag{22}$$

by the choice of n_0 and (21).

Regarding (a), by Proposition 8, $\lim_{\mu(A)\to 0} \int_A |f| \, d\mu = 0$, and similarly for the finite collection f_n, $1 \leqslant n \leqslant n_0$, we get $\lim_{\mu(A)\to 0} \int_A |f_n| \, d\mu = 0$. If $n \geqslant n_0$, one has

$$\begin{aligned} \int_A |f_n| \, d\mu &\leqslant \|f_n - f\|_1 + \int_A |f| \, d\mu, \qquad \text{as in (22)} \\ &\leqslant \epsilon + \int_A |f| \, d\mu. \end{aligned}$$

Hence,

$$\lim_{\mu(A)\to 0} \int_A |f_n| \, d\mu \leqslant \epsilon \qquad \text{uniformly in } n.$$

Since $\epsilon > 0$ is arbitrary, this shows (a) also is true. Thus the theorem is completely proved.

Conditions (a) and (b) play an important role in the linear analysis of functions on measure spaces. If $\mu(\Omega) < \infty$, then in (b) we may take $A_\epsilon = \Omega$, so that it is automatically satisfied. In this case we use the following concept.

Definition 12. Let (Ω, Σ, μ) be a finite measure space, and $\mathscr{F}$ be a class of integrable functions on Ω. Then the set $\mathscr{F}$ is said to be *uniformly integrable* if (i) $\sup\{\int_\Omega |f| \, d\mu : f \in \mathscr{F}\} < \infty$, and (ii) $\lim_{\mu(A)\to 0} \int_A |f| \, d\mu = 0$ uniformly in $f \in \mathscr{F}$.

Note that if $\mathscr{F} = \{f_n : n \geqslant 1\}$ and $f_n \to f$ a.e. (or in measure), then the sufficiency part of the theorem shows that (i) is automatic, and only (ii) is crucial for sequences when $\mu(\Omega) < \infty$. Thus we have the following consequence, which is the usual form of the above result.

Corollary 13 (Vitali's theorem, original case). *Let (Ω, Σ, μ) be a finite measure space, and $f_n\colon \Omega \to \mathbb{R}$, $n \geqslant 1$, be a sequence of measurable functions converging a.e. (or in measure) to a measurable f. Then $\|f_n - f\|_1 \to 0$ as $n \to \infty$ iff $\{f_n\colon n \geqslant 1\}$ is uniformly integrable. When the condition is satisfied, we have*

$$\lim_{n\to\infty} \int_\Omega f_n \, d\mu = \int_\Omega f \, d\mu. \tag{23}$$

Now that the essentials of abstract Lebesgue integration are established, we can present a characterization of bounded Riemann integrable functions. Ironically, it is not a result in the classical Riemann theory itself.

Theorem 14. *Let $\Omega = [a, b]$, and $f\colon \Omega \to \mathbb{R}$ be a bounded (i.e., classical) Riemann integrable function. Then f is also Lebesgue integrable, and both integrals have the same value. In symbols,*

$$(\mathrm{R})\int_a^b f(x)\,dx = (\mathrm{L})\int_\Omega f(x)\lambda(dx), \tag{24}$$

where λ is the Lebesgue measure on Ω. Further, a bounded function $f\colon \Omega \to \mathbb{R}$ is Riemann integrable iff it is continuous outside of a set of Lebesgue measure zero (so that f is "a.e. continuous").

Proof. Consider the lower and upper Darboux sums of f over $[a, b]$. If $P_n\colon a = x_0 < x_1 < \cdots < x_n = b$ is a partition, let P_{k+1} be a refinement of P_k, and define for $\Delta_k^n = (x_{k-1}, x_k)$ in P_n and $\delta_k^n = x_k - x_{k-1}$:

$$S_n = \sum_{k=1}^n M_{nk}\delta_k^n, \qquad s_n = \sum_{k=1}^n m_{nk}\delta_k^n, \tag{25}$$

where $M_{nk} = \sup\{f(x)\colon x \in \Delta_k^n\}$, $m_{nk} = \inf\{f(x)\colon x \in \Delta_k^n\}$. Since f is bounded, the M_{nk}, m_{nk} are finite, and by the classical Riemann criterion for integrability [Bartle (1976, p. 228)], f is Riemann integrable iff $\lim_n S_n = \lim_n s_n$, the limits existing, as $\max_{1\leqslant k\leqslant n}(x_k - x_{k-1}) \to 0$. Then the common value is denoted by $(\mathrm{R})\int_a^b f(x)\,dx$. We convert this into the Lebesgue formalism. Define

$$\varphi_n = \sum_{k=1}^n M_{nk}\chi_{\Delta_k^n}, \qquad \psi_n = \sum_{k=1}^n m_{nk}\chi_{\Delta_k^n}. \tag{26}$$

The simple functions φ_n, ψ_n, $n \geqslant 1$, are Lebesgue measurable, and since P_n is refined by P_{n+1}, we get $\varphi_n \geqslant \varphi_{n+1} \geqslant f \geqslant \psi_{n+1} \geqslant \psi_n$, $n \geqslant 1$. If $\varphi = \lim_n \varphi_n$ and

$\psi = \lim_n \psi_n$, then $\varphi \geqslant f \geqslant \psi$. Further

$$S_n = \int_\Omega \varphi_n \, d\lambda, \qquad s_n = \int_\Omega \psi_n \, d\lambda, \tag{27}$$

and we have

$$\begin{aligned}
(\mathrm{R})\int_a^b f(x)\,dx &= \lim_n S_n = \lim_n \int_\Omega \varphi_n \, d\lambda \\
&= \int_\Omega \varphi \, d\lambda, \qquad \text{by Corollary 6} \\
&= \lim_n s_n = \lim_n \int_\Omega \psi_n \, d\lambda \\
&= \int_\Omega \psi \, d\lambda, \quad \text{by Theorem 1.}
\end{aligned} \tag{28}$$

Hence

$$\int_\Omega (\varphi - \psi)\, d\lambda = (\mathrm{R})\int_a^b f(x)dx - (\mathrm{R})\int_a^b f(x)\,dx = 0.$$

Since $\varphi - \psi \geqslant 0$, this implies $\varphi = \psi$ a.e., by Theorem 1.3(ii). Thus $\varphi = \psi = f$ a.e., and f is Lebesgue integrable. Thus (28) implies (24).

For the second statement of the theorem, let A_1 be the Lebesgue null set such that $\varphi(\omega) = \psi(\omega) = f(\omega)$ for $\omega \in \Omega - A_1$. Also let A_{2k} be the partition points of P_k. Since, for each k, A_{2k} is a finite set, $A_2 \equiv \bigcup_k A_{2k}$ is countable, using P_k, $k \geqslant 1$. Let $A = A_1 \cup A_2$, so that $\lambda(A) = 0$. If $\omega \in \Omega - A$, then f is continuous at ω. For otherwise, there is a neighborhood $U(\omega)$ of ω at which f is discontinuous, i.e., there is an $\epsilon > 0$ such that $\varphi_k(\omega) - \psi_k(\omega) \geqslant \epsilon$ for all k, so that $\varphi(\omega) \geqslant \psi(\omega) + \epsilon$. But then $\omega \in A_1 \subset A$, contradicting the fact that $\omega \in \Omega - A$. Thus f is continuous on $\Omega - A$.

Conversely, if f is continuous on $\Omega - A$, where $\lambda(A) = 0$ and f is bounded on Ω, consider the partitions Q_n, $n \geqslant 1$, of Ω such that the maximum subdivision tends to zero as $n \to \infty$ whether or not Q_{n+1} is a refinement of Q_n. The corresponding $\tilde{\varphi}_n$, $\tilde{\psi}_n$ of (26) are not necessarily monotone, but they converge to f, boundedly (since f is bounded) outside of A, by the continuity of f. Hence by Corollary 6, the integrals converge to $\int_\Omega f \, d\lambda$. However, $\int_\Omega \tilde{\varphi}_n \, d\lambda = \tilde{S}_n$ and $\int_\Omega \tilde{\psi}_n \, d\lambda = \tilde{s}_n$, the Riemann-Darboux upper and lower sums. By the preceding result, $\lim_n \tilde{S}_n$ and $\lim_n \tilde{s}_n$ both exist and equal $\int_\Omega f(x)\, d\lambda$. Hence by the Riemann criterion f is Riemann integrable, which proves the theorem.

Remark. For the Dirichlet function g on $[0, 1]$, we have seen that it is not Riemann integrable, since it is discontinuous a.e. But g is a.e. equal to a Lebesgue-integrable function, namely zero, and has its Lebesgue integral zero. Hence by the structure theorem, there exist simple functions $0 < h_n \uparrow g$ pointwise. Then $\int_0^1 h_n \, d\lambda = (\mathrm{R})\int_0^1 h_n(x)\, dx = 0$, $n \geqslant 1$, but the limit and integral cannot be interchanged for the Riemann integral. Thus the Lebesgue limit theorems are not valid for the Riemann case without further stringent restrictions. This facility of the Lebesgue limit theory is one of the main reasons for employing it in modern mathematical analysis.

In the next section we present another important extension of Vitali's theorem, but first we provide several applications of the preceding results as exercises. More theoretical applications will be discussed in the last section of this chapter.

EXERCISES

0. Let (Ω, Σ, μ) be a measurable space and $f_n \colon \Omega \to \overline{\mathbb{R}}$ be a sequence of measurable functions. If f_n, $n \geqslant 1$, is monotone, f_{n_0} is μ-integrable for some n_0, and each f_n^+, $n \geqslant 1$ (or each f_n^-, $n \geqslant 1$) is μ-integrable, then show that

$$\lim_n \int_\Omega f_n \, d\mu = \int_\Omega \lim_n f_n \, d\mu$$

in the sense that both sides are finite and equality holds or both sides are $\pm\infty$. [This is a form of Beppo Levi's theorem.]

1. Complete the details of proof of the monotone convergence theorem for the form given in Corollary 7.

2. Let (Ω, Σ, μ) be a finite measure space, and $f_n, f \colon \Omega \to \mathbb{R}$ be measurable functions such that $f_n \xrightarrow{D} f$. Suppose that $\{f_n \colon n \geqslant 1\}$ is uniformly integrable in the sense that only Definition 12(ii) holds. Show that f is integrable and

$$\lim_n \int_\Omega f_n \, d\mu = \int_\Omega f \, d\mu.$$

[Use Theorem 3.3.5 and reduce the result to Corollary 13.] (Note how the image measure theory gives a significant extension of Vitali's theorem. For further extensions see Exercise 19 below.)

3. Prove Theorem 11 if "$f_n \to f$ a.e." there is replaced by "$f_n \xrightarrow{\mu} f$".

4. Let (Ω, Σ, μ) be a measure space and $f_n \colon \Omega \to \overline{\mathbb{R}}$, $n \geqslant 1$, be a sequence of a.e. finite measurable functions such that $\|f_n - f_m\|_1 = \int_\Omega |f_n - f_m| \, d\mu \to 0$ as $n, m \to \infty$. Show that $\{f_n \colon n \geqslant 1\}$ is Cauchy in measure and hence converges in measure to a measurable function f. Conclude that f is integrable. [Use Fatou's lemma.] Deduce that $\|f_n - f\|_1 \to 0$ as $n \to \infty$. [Use Exercise 3.]

5. If $\Omega = \mathbb{R}$, Σ is the Lebesgue σ-algebra, and μ is Lebesgue measure, let $f_n = -\chi_{A_n}$, where $A_n = (n, \infty)$. Show that $f_n \leqslant f_{n+1}$ but that Theorem 1 is false for this monotone sequence. Explain why.

6. Let $\Omega = (0, 1)$, Σ be the σ-algebra of subsets of Ω which are at most countable or which have countable complements, and $\mu(A)$ be the number of points in A. Let $f(x) = x$, and $g\colon \Omega \to \mathbb{R}$ be measurable (for Σ). Show that fg is measurable and integrable on (Ω, Σ, μ) if g is integrable even though f is not measurable (Σ). Show however if (Ω, Σ, μ) is any σ-finite (or more generally localizable) measure space, then the integrability of fg for each integrable g implies that f is measurable and essentially bounded, i.e., $\mu(\{\omega\colon |f(\omega)| > N\}) = 0$ for some $N > 0$. [Start with simple f, and then, by the argument of the structure theorem (Theorem 3.1.7), establish that f is measurable in general, and finally that it must be essentially bounded, by deriving a contradiction in the opposite case. The problem is not simple.]

7. The relation between the Riemann and Lebesgue integrals given in Theorem 14 is elaborated as follows: Let $(\Omega, \Sigma, \lambda)$ be the Lebesgue unit interval, where $\Omega = (0, 1]$. Let $f(x) = x^{-\alpha}$, $0 < \alpha < 1$. Then $(\mathrm{R})\int_0^1 f(x)\,dx = \lim_{\epsilon \downarrow 0}(\mathrm{R})\int_\epsilon^1 f(x)\,dx = (1-\alpha)^{-1}$. This is an improper Riemann integral. Show that for this continuous f on Ω, $(\mathrm{L})\int_\Omega f\,d\lambda$ does not exist. (Use the Saks definition in Theorem 1.4.)

8. Let (Ω, Σ, μ) be a finite measure space, and $f\colon \Omega \to \mathbb{R}$ be measurable. If $g\colon \mathbb{R} \to \mathbb{R}^+$ is a symmetric increasing differentiable function such that $g(0) = 0$, show that we then have (with g' for the derivative of g)

$$\int_\Omega g(f)\,d\mu = \int_0^\infty g'(t)\mu(\{\omega\colon |f(\omega)| \geqslant t\})\,dt,$$

in that both sides are finite and equal or both sides are infinite. If in particular $g(x) = |x|^p$, deduce that

$$\int_\Omega |f|^p\,d\mu = p\int_0^\infty t^{p-1}\mu(\{\omega\colon |f(\omega)| \geqslant t\})\,dt.$$

[Let $F_{|f|}$, F_f be the distributions of $|f|$ and f. Use Corollary 2.7 and the fact that $F_{|f|}(x) = F_f(x) - F_f(-x+0)$ to deduce

$$\begin{aligned}\int_\Omega |f|^p\,d\mu &= \lim_{M\to\infty}\left[\int_0^M g(x)\,dF_f(x) - \int_0^M f(x)\,dF_f(-x+0)\right]\\ &= \lim_{M\to\infty}\int_0^M g'(t)\big[\mu(\{\omega\colon |f(\omega)| \geqslant t\})\\ &\qquad -\mu(\{\omega\colon |f(\omega)| < t\})\big]\,dt,\end{aligned}$$

with integration by parts. Simplify the right side, and use Theorem 1 after noting that g' is nonnegative. This gives the main result.]

9. Let (Ω, Σ, μ) be the Lebesgue unit interval and $h = 2 + f$, where f is the Dirichlet function on Ω, so that h is Lebesgue (but not Riemann) integrable. Conclude that there exists no (bounded or not) Riemann-integrable function $g: \Omega \to \mathbb{R}$ such that $\int_\Omega |h - g|\, d\lambda = 0$.

10. Let (Ω, Σ, μ) be an arbitrary measure space and $g, f, f_i: \Omega \to \overline{\mathbb{R}}$, $i \in I$, be measurable functions such that $|f_i| \leqslant g$ a.e. and g is integrable. Prove the following extension of Theorem 10: If I is ordered and $f_i \xrightarrow{\mu} f$ in the sense that $\lim_i \mu(\{\omega: |f_i - f|(\omega) > \epsilon\}) = 0$ for each $\epsilon > 0$, then

$$\lim_i \int_\Omega f_i\, d\mu = \int_\Omega f\, d\mu,$$

and even

$$\int_\Omega |f_i - f|\, d\mu \to 0 \qquad \text{as "} i \to \infty \text{".}$$

Conversely, if the second expression above goes to zero, then $f_i \xrightarrow{\mu} f$ also holds. [If I is the positive integers or an ordered countable set, this result is the usual dominated convergence theorem. In the general case let $d(\cdot, \cdot)$ be the metric associated with convergence in measure, as in Proposition 3.2.6. If $\|f_i - f\|_1 \nrightarrow 0$, then find a subsequence $\{f_{i_j}: j \geqslant 1\}$ such that $\|f_{i_j} - f\|_1 \geqslant \alpha > 0$. Since in the d-metric $\{f_i: i \in I\}$ is Cauchy with limit f, the same must be true of $d(f_{i_j}, f)$, since a Cauchy sequence in a metric space has a unique limit. Choose $\{f_{i_{j_n}}: n \geqslant 1\}$ such that $d(f_{i_{j_n}}, f) \to 0$; then for this ordinary sequence, since $f_{i_{j_n}} \xrightarrow{\mu} f$, again by Theorem 10, we have $\|f_{i_{j_n}} - f\|_1 \to 0$, contradicting the assumption. For the converse, use the Markov inequality and deduce the result.]

11. Let f be an integrable function on a measure space (Ω, Σ, μ). Using Theorems 5 and 3.7, show that for each $\epsilon > 0$, there exists a simple function g_ϵ vanishing outside a set of finite μ-measure such that $\|f - g_\epsilon\|_1 = \int_\Omega |f - g_\epsilon|\, d\mu < \epsilon$. [Reduce the result to the case of finite measures.]

12. We have seen that on a measurable space (Ω, Σ, μ), a measurable function f vanishes a.e. if $\int_A f\, d\mu = 0$ holds for all $A \in \Sigma$. Even less is needed if we specialize the measure space. Let $(\Omega, \Sigma, \lambda)$ be the Lebesgue unit interval $\Omega = [0, 1]$, and $f: \Omega \to \mathbb{R}$ be measurable. If $\int_A f\, d\lambda = 0$ for all A of the form $[0, x)$, $0 \leqslant x \leqslant 1$, then $f = 0$ a.e. [Let $\nu(A) = \int_A f\, d\lambda$, and note that $\nu(A) = \int_A f^+\, d\lambda - \int_A f^-\, d\lambda = \nu_1(A) - \nu_2(A)$ and that ν_i, $i = 1, 2$, agree on the semiring of left closed intervals of $[0, 1]$. Hence they agree on Σ by Theorem 2.3.1, so that $\nu(A) = 0$ for all $A \in \Sigma$.]

13. This exercise strengthens Corollary 7 in some respects. Let (Ω, Σ, μ) be a localizable measure space (Definition 2.3.4). If $0 \leqslant f_\alpha: \Omega \to \overline{\mathbb{R}}^+$, $\alpha \in I$, is a generalized monotone net of measurable functions, I ordered upwards, (so $\alpha, \alpha' \in I$ and $\alpha \leqslant \alpha'$ implies $f_\alpha \leqslant f_{\alpha'}$ a.e.), then $f = \sup_\alpha f_\alpha$ a.e. is the upper envelope of the set, f exists as a measurable function, and $\int_\Omega f\, d\mu = \sup_\alpha \int_\Omega f_\alpha\, d\mu$. [The point here is that I is not assumed to have a *countable* cofinal sequence, but the measurability of f is guaranteed by the localizability of μ; cf. Exercise 3.2.9. The equation has to be established only if the right side is $\beta < \infty$. Then there exists a sequence $\alpha_1 < \alpha_2 < \dots$ such that $\lim_n \int_\Omega f_{\alpha_n}\, d\mu = \beta$. One shows that $\sup_n f_{\alpha_n}$ $(= \tilde{f})$ and f differ on at most a μ-null set, so that the equation must hold. In Example 3 following Corollary 6, the set $\{f_\alpha : \alpha \in \mathscr{F}\}$ has an upper bound, treating them as individual functions, that is different from their *a.e. supremum*, which is 0. This shows the clear difference between equivalence classes and their pointwise ordering and analysis. A more general limit theorem is in Tulcea (1969, p. 40). In the case that Ω is a locally compact space, μ is a Radon measure, and the f_α are lower semicontinuous, the result was given by Bourbaki (1965) for the case of the *pointwise* upper bound.]

14. Let $f_n, f: \Omega \to \overline{\mathbb{R}}^+$ be a sequence of measurable functions on a measure space (Ω, Σ, μ) such that $f_n \to f$ a.e. (or only in measure), and each f_n is integrable, $n \geqslant 1$. Then $\int_\Omega f_n\, d\mu \to \int_\Omega f\, d\mu < \infty$ iff $\{f_n : n \geqslant 1\}$ is uniformly integrable in the sense of Definition 12. (Hence it holds iff $\|f_n - f\|_1 \to 0$, By Theorem 11.) [The only direction needing a new proof is that $\int_\Omega f_n\, d\mu \to \int_\Omega f\, d\mu < \infty$ implies that $\{f_n : n \geqslant 1\}$ is uniformly integrable. For this, note that $f_n + f = \min(f_n, f) + \max(f_n, f)$ and $|f_n - f| = \max(f_n, f) - \min(f_n, f)$. By the dominated convergence theorem $\int_\Omega \min(f_n, f)\, d\mu \to \int_\Omega f\, d\mu$, and then $\int_\Omega \max(f_n, f)\, d\mu \to \int_\Omega f\, d\mu$. Thus $\|f_n - f\|_1 \to 0$.] (A form of this useful lemma is due to H. Scheffé, who proved it for statistical applications.)

15. Let f_n, g_n, h_n and f, g, h be measurable functions on a measure space (Ω, Σ, μ), $n \geqslant 1$. Suppose that $f_n \to f$ a.e., $g_n \to g$ a.e., $h_n \to h$ a.e., and $f_n \leqslant g_n \leqslant h_n$, $n \geqslant 1$. If $\int_\Omega f_n\, d\mu \to \int_\Omega f\, d\mu \in \mathbb{R}$ and $\int_\Omega h_n\, d\mu \to \int_\Omega h\, d\mu \in \mathbb{R}$, show that $\lim_n \int_\Omega g_n\, d\mu = \int_\Omega g\, d\mu \in \mathbb{R}$, which is another form of Theorem 5. [We may assume for this work that f_n, h_n, $n \geqslant 1$, are integrable. Then $h_n - f_n \geqslant 0$ and $\int_\Omega (h_n - f_n)\, d\mu \to \int_\Omega (h - f)\, d\mu$, and by Scheffé's lemma above, $\{h_n - f_n : n \geqslant 1\}$ is uniformly integrable. Since $0 \leqslant g_n - f_n \leqslant h_n - f_n$, $n \geqslant 1$, $\{g_n - f_n : n \geqslant 1\}$ is also uniformly integrable. Apply Exercise 14.]

16. Let $f_1, \dots, f_n$ be integrable functions on a measure space (Ω, Σ, μ). Show that $(\sum_{i=1}^n |f_i|^p)^{1/p}$ and $\sum_{1 \leqslant i, j \leqslant n} |f_i f_j|^{1/2}$ are integrable, where $p \in [1, \infty)$. [For the first observe that if $\varphi(x) = x^{1/p}$, $p \geqslant 1$, then $\varphi: \mathbb{R}^+ \to \mathbb{R}^+$ is a continuous concave function (cf. Remarks after Proposition 2.4.1). For the second use the inequality between the arithmetic and geometric means.]

17. Let $(\Omega, \Sigma, \lambda)$ be the Lebesgue unit interval, and $f: \mathbb{R} \to \mathbb{R}$ be a Lebesgue-integrable periodic function of period 1, i.e., $f(x + 1) = f(x)$, $x \in \mathbb{R}$. For each $0 < x < 1$ and $n \geqslant 1$, if $P_n: 0 = x_0 < x_1 < \cdots < x_n$, $x_i - x_{i-1} =$

n^{-1}, then define $f_n(x) = (1/n)\Sigma_{i=0}^{n-1} f(x + x_i)$. Show that $f_n(x) \to \alpha$, a constant, for a.a. x and that $\alpha = \int_\Omega f\,d\lambda$. [Note that λ is translation invariant and follow the proof of Theorem 14.]

18. Let $\Omega = \{x: x = (x_1, x_2, \dots),\ x_i = 0 \text{ or } 1,\ i \geqslant 1\}$, the coin tossing space. Let $\mathscr{S} = \{I_n \subset \Omega: I_n$ consists of those x whose first n coordinates have a prescribed pattern$\}$. Thus, for instance, I_2 can be all x's whose first two coordinates are zeros, etc. Show that $\mathscr{S}$ is a semialgebra. Let $\Sigma = \sigma(\mathscr{S})$, and define $\mu: \mathscr{S} \to \mathbb{R}^+$ by the equation $\mu(I_n) = p^k q^{n-k}$, where I_n has k among its first n prescribed coordinates 0's and the other $n - k$ of them 1's, $0 \leqslant k \leqslant n$. Verify that μ is σ-additive on $\mathscr{S}$ and that $\mu(\Omega) = 1$ when $0 < p = 1 - q < 1$. Let μ also denote its extension to Σ. If $f: \Omega \to \overline{\mathbb{R}}^+$ is defined as $f(x) =$ number of 0's that precede the first 1 in x, show that f is measurable for Σ, and $\int_\Omega f\,d\mu = p/q$, as well as $\int_\Omega [f - (p/q)]^2\,d\mu = p/q^2$. [Note that for $|t| < 1$ the series $\Sigma_{n=1}^\infty nt^{n-1}$ is uniformly and absolutely convergent and the sum equals $(d/dt)(\Sigma_{n=1}^\infty t^n) = (1 - t)^{-2}$. In probability language, this says that the expected value of f is p/q and the variance of f is p/q^2, which result has interest in such studies.]

19. This exercise again gives extensions of the limit theorems when the convergence hypothesis is weakened, as in Exercise 2. Thus let (Ω, Σ, μ) be a finite measure space, and $f_n, f: \Omega \to \mathbb{R}$ be measurable functions. Establish the following statements if only $f_n \xrightarrow{D} f$:

(a) (Fatou's lemma) If $f_n, f \geqslant 0$, μ-integrable, then $\int_\Omega f\,d\mu \leqslant \liminf_n \int_\Omega f_n\,d\mu$.

(b) (Dominated convergence) If $|f_n| \leqslant g$, g μ-integrable, then $\int_\Omega f\,d\mu = \lim_n \int_\Omega f_n\,d\mu$. [Use Theorem 3.3.5, and reduce these to the standard case.]

(c) Formulate the other form of Fatou's lemma (with "lim sup"), and the monotone convergence statement.

4.4. THE VITALI-HAHN-SAKS THEOREM AND SIGNED MEASURES

In the last section, we have established the important Vitali convergence theorem on a measure space (Ω, Σ, μ) for a sequence of integrable functions $f_n \to f$ a.e. (or $f_n \xrightarrow{D} f$), that $\int_\Omega f_n\,d\mu \to \int_\Omega f\,d\mu$ under some (precise) conditions. If μ is replaced by $\mu_A(\cdot) = \mu(A \cap \cdot)$, $A \in \Sigma$, then the same result holds. But this can be written as $\int_A f_n\,d\mu \to \int_A f\,d\mu$ for each $A \in \Sigma$. Following a decade and a half of G. Vitali's work, H. Hahn extended this result in 1922 by proving that if $\int_A f_n\,d\mu \to \int_A f\,d\mu$ for each $A \in \Sigma$, then the set functions $\nu_n(A) = \int_A f_n\,d\mu$ and $\nu(A) = \int_A f\,d\mu$, $A \in \Sigma$, satisfy the condition that $\lim_{\mu(A)\to 0} \nu_n(A) = 0$ uniformly in n. In both cases (Ω, Σ, μ) is the Lebesgue measure space of the interval $(0, 1)$. If the functions $\nu_n: \Sigma \to \mathbb{R}$, $n \geqslant 1$, are not necessarily defined as indefinite integrals, but are σ-additive such that $\nu_n(A) \to \nu(A)$ for each $A \in \Sigma$, then one wants to know the corresponding properties of these functions. This question was solved abstractly by S. Saks, after another decade. And the final version, which turned out to be very powerful for applications, will be proved here with Saks's method. For this we need a topological property of metric spaces due to R. Baire, which is of independent interest. So we first prove the Baire category theorem and then use it in the main result.

Let us restate the notions of "nowhere dense" and "category" mentioned in Exercise 2.2.18. Thus in a topological space T, a set $A \subset T$ is *nowhere dense* if its closure $\bar{A}$ has an empty interior. For example, the set of integers in $\mathbb{R}$ is nowhere dense, and the Cantor set in $[0, 1]$ is nowhere dense. A set $B \subset T$ is said to be of the *first category* or *meager* if B is a countable union of nowhere dense sets. Any set which is not of first category is said to be of the *second category* or *nonmeager*. Thus the set of all rationals in $\mathbb{R}$ is meager, and hence the set of all irrationals is of the second category in $\mathbb{R}$ (with its usual metric topology).

Proposition 1 (Baire category theorem). *A complete metric space (M, d) is of the second category as a subset of itself. Thus, if $M = \bigcup_{n=1}^{\infty} A_n$, where each A_n is a closed set, then at least one of the A_n's contains a nonempty open set.*

Proof. A short indirect proof will be given. Thus suppose $M = \bigcup_{n=1}^{\infty} A_n$, where each A_n is closed and nowhere dense, so that each A_n has an empty interior. Since a complete metric space has nonempty open sets (e.g., open balls), we have $A_1^c \subsetneqq M$, and if $x_1 \in A_1^c$, then there is a positive r_1 $(0 < r_1 < 1)$ and an open ball $B_1(x_1, r_1) = \{x: d(x, x_1) < r_1\} \subset A_1^c$. Since A_2 is nowhere dense, A_2^c is open and $A_2^c \cap B_1(x_1, r_1/2)$ is open and nonvoid. So it contains an open ball $B_2(x_2, r_2)$ with $0 < r_2 < r_1/2$. Continuing by induction, we find an open ball $B_n(x_n, r_n) \subset A_n^c \cap B_{n-1}(x_{n-1}, r_{n-1}/2)$. Thus we have

$$B_1(x_1, r_1) \supset B_2(x_2, r_2) \supset \cdots, \qquad 0 < r_n < 2^{-n+1}, \quad n \geqslant 1. \tag{1}$$

Since $B_n(x_n, r_n) \subset A_n^c$, $n \geqslant 1$, for the sequence $\{x_n: n \geqslant 1\}$ of the centers, we have

$$\begin{aligned} d(x_m, x_n) &\leqslant \sum_{i=1}^{n-m} d(x_{m+i-1}, x_{m+i}) \\ &< \sum_{i=1}^{\infty} \frac{1}{2^{m+i}} = \frac{1}{2^{m-1}}, \qquad m < n. \end{aligned}$$

So $\{x_n: n \geqslant 1\}$ forms a Cauchy sequence, and by the completeness of M, $x_n \to x \in M$. (The proof breaks down here if M is incomplete.) Also

$$d(x_n, x) \leqslant d(x_n, x_{n+k}) + d(x_{n+k}, x) < 2^{-n-k} + d(x_{n+k}, x). \tag{2}$$

Letting $k \to \infty$, we see that $x \in B_n(x_n, r_n)$ for all n and so $x \in A_n^c$ for all n.

Hence $x \notin A_n$ for all n, or $x \notin \bigcup_{n=1}^{\infty} A_n = M$, which is a contradiction. This proves the result.

As an interesting application we get the following fact which illuminates the sets in Example 1 near the beginning of Section 2.2, and explains further the difference between the Jordan and Lebesgue definitions of the outer measure there.

Corollary 2. *The set of rationals is not a G_δ-set in $\mathbb{R}$.*

Proof. Let $Q \subset \mathbb{R}$ be the set of rationals. If it is a G_δ-set, then $Q = \bigcap_{n=1}^{\infty} G_n$, where G_n is open in $\mathbb{R}$ for each $n \geqslant 1$. Since G_n^c is closed and $G_n^c \subset Q^c$, it does not contain any rationals. The density of the latter in $\mathbb{R}$ implies G_n^c must be nowhere dense. But if we enumerate Q as $\{r_n: n \geqslant 1\}$, then $\mathbb{R} = \bigcup_{n=1}^{\infty}(G_n^c \cup \{r_n\})$. Since the metric space $\mathbb{R}$ is complete and since each set $G_n^c \cup \{r_n\}$ is still nowhere dense, this equation leads to a contradiction to the proposition. Hence Q cannot be a G_δ-set, proving the assertion.

With these preliminaries out of the way, we are ready to establish the main result of this section.

Theorem 3 (Vitali, Hahn, and Saks). *Let (Ω, Σ, μ) be a measure space, and $\nu_n: \Sigma \to \mathbb{R}$ be a σ-additive function such that $\lim_{\mu(A) \to 0} \nu_n(A) = 0$ for each $n \geqslant 1$. If $\nu(A) = \lim_n \nu_n(A)$ exists for each A in the σ-algebra Σ, then $\lim_{\mu(A) \to 0} \nu_n(A) = 0$ uniformly in $n \geqslant 1$. Further, if $\mu(\Omega) < \infty$, then ν is σ-additive.*

Proof. If $d: \Sigma \times \Sigma \to \mathbb{R}^+$ is the semimetric defined as in Proposition 2.4.1 $[d(A, B) = \varphi(\mu(A \triangle B))$ for $A, B \in \Sigma]$, then (Σ, d) is a complete (semi-)metric space, as shown there. Also, if d is replaced by $\tilde{d}$, where

$$\tilde{d}(\tilde{A}, \tilde{B}) = d(A, B)$$

with $\tilde{A}$ $(\tilde{B})$ as the class of all sets that differ from A (B) by μ-null sets, then $(\tilde{\Sigma}, \tilde{d})$ is a complete metric space, $(\tilde{\Sigma} = \Sigma/\sim\,,\ A \in \tilde{A})$. Further

$$|\nu_k(A_n) - \nu_k(B)| \leqslant |\nu_k(A_n - A_n \cap B)| + |\nu_k(B - B \cap A_n)|, \tag{3}$$

since ν_k is additive on Σ. Now let $d(A_n, B) \to 0$. Then $\varphi \circ \mu(A_n \triangle B) \to 0$, so that $\varphi \circ \mu(A_n - A \cap B) \to 0$ and $\varphi \circ \mu(B - A_n \cap B) \to 0$. Since ν_k is μ-continuous by hypothesis, we get from (3) that the right side approaches 0 as $d(A_n, B) \to 0$. Thus $\nu_k(A_n) \to \nu_k(B)$, and each ν_k is a continuous function on (Σ, d). Consider, for each $\epsilon > 0$, the sets

$$\begin{aligned}\tilde{\Sigma}_k &= \left\{\tilde{A} \in \tilde{\Sigma}: |\nu_n(A) - \nu_m(A)| \leqslant \epsilon/3, \text{ all } m, n \geqslant k,\ A \in \tilde{A}\right\} \\ &= \bigcap_{m, n \geqslant k} \left\{\tilde{A} \in \tilde{\Sigma}: |\nu_n(A) - \nu_m(A)| \leqslant \epsilon/3\right\}.\end{aligned}$$

By the continuity of ν_n, $n \geqslant 1$, the $\tilde{\Sigma}_k$ is a closed set in $(\tilde{\Sigma}, \tilde{d})$. Now using the additional hypothesis that $\lim_n \nu_n(A) = \nu(A)$ exists for each $A \in \tilde{A} \in \tilde{\Sigma}$, we note that any $\tilde{A}$ in $\tilde{\Sigma}$ belongs to $\tilde{\Sigma}_k$ for some $k \geqslant 1$, so that $\tilde{\Sigma} = \bigcup_{k=1}^{\infty} \tilde{\Sigma}_k$. But $(\tilde{\Sigma}, \tilde{d})$ is a complete metric space which is a countable union of closed sets. Hence by Proposition 1, at least one of the $\tilde{\Sigma}_k$ must have a nonempty interior. So there exists an $r > 0$, an integer $k_0 \geqslant 1$, and a point $\tilde{A} \in \tilde{\Sigma}$ such that the ball

$$B(\tilde{A}, r) = \{\tilde{B} \in \tilde{\Sigma}: \tilde{d}(\tilde{B}, \tilde{A}) < r\} \subset \tilde{\Sigma}_{k_0}. \tag{4}$$

This implies

$$|\nu_m(D) - \nu_n(D)| < \epsilon/3 \qquad \text{for all} \quad D \in \tilde{D} \in B(\tilde{A}, r), \quad m, n \geqslant k_0. \tag{5}$$

Consider the finite set of ν_n, $1 \leqslant n \leqslant k_0$. By hypothesis, if $0 < \delta < \varphi^{-1}(r)$ is chosen so that $\mu(C) < \delta$, $C \in \Sigma$, then $|\nu_n(C)| < \epsilon/3$, $1 \leqslant n \leqslant k_0$. But for each such $C \in \Sigma$, $d(A \cup C, A) \leqslant \varphi(\mu(C)) < \varphi(\delta) < r$, and similarly

$$d(A, A - C) \leqslant \varphi(\mu(C)) < r, \quad \text{so that} \quad \{\tilde{A} \cup \tilde{C}, \tilde{A} - \tilde{C}\} \subset B(\tilde{A}, r).$$

Hence for any $n \geqslant 1$, since $C, C' \in \tilde{C} \Rightarrow \nu_n(C) = \nu_n(C')$, we have

$$\begin{aligned}
|\nu_n(C)| &= \left|\nu_{k_0}(C) + \left(\nu_n(C) - \nu_{k_0}(C)\right)\right| \\
&\leqslant \left|\nu_{k_0}(C)\right| + \left|\left(\nu_n(A \cup C) - \nu_n(A - C)\right)\right. \\
&\qquad \left. - \left(\nu_{k_0}(A \cup C) - \nu_{k_0}(A - C)\right)\right| \\
&\leqslant \left|\nu_{k_0}(C)\right| + \left|\nu_n(A \cup C) - \nu_{k_0}(A \cup C)\right| \\
&\quad + \left|\nu_n(A - C) - \nu_{k_0}(A - C)\right| \\
&< \frac{\epsilon}{3} + \frac{\epsilon}{3} + \frac{\epsilon}{3} = \epsilon,
\end{aligned}$$

by (5) and the choice of $\delta > 0$. Since $\epsilon > 0$ is arbitrary, we have

$$\lim_{\mu(C) \to 0} \nu_n(C) = 0 \qquad \text{uniformly in} \quad n \geqslant 1. \tag{6}$$

Now let $\mu(\Omega) < \infty$ [so $\varphi(x) = x$ in the above]. Since $\nu(A) = \lim_n \nu_n(A)$, $A \in \Sigma$ by hypothesis, and since each ν_n is (σ-)additive, it follows that ν is additive. If $\{A_n: n \geqslant 1\} \subset \Sigma$ is a disjoint collection, then

$$\nu\left(\bigcup_{n=1}^{\infty} A_n\right) = \sum_{i=1}^{n} \nu(A_i) + \nu\left(\bigcup_{k>n} A_k\right), \qquad \text{by additivity of} \quad \nu. \tag{7}$$

It suffices to show that the second term on the right of (7) tends to zero as $n \to \infty$. This is equivalent to showing that for any $B_n \in \Sigma$, $B \downarrow \varnothing \Rightarrow |\nu(B_n)| \to 0$. But since μ is σ-additive and finite, by Proposition 2.1.1 we have $\mu(B_n) \to 0$. Hence given $\epsilon > 0$, there is an n_0 such that $n \geqslant n_0 \Rightarrow \mu(B_n) < \epsilon$. Consequently by the first part for each $\eta > 0$, there is a $0 < \delta_\eta < \epsilon$ and an $n_1 \geqslant n_0$ such that $n \geqslant n_1 \Rightarrow \mu(B_n) < \delta_\eta$ and so $|\nu_k(B_n)| < \eta$ for all $k \geqslant 1$. Letting $k \to \infty$, we get $|\nu(B_n)| < \eta$ for $n \geqslant n_1$. So ν is σ-additive. This completes the proof of the theorem.

Remark (i). It should be noted that in the above proof we have not used any special properties of the ν_n except that they are σ-additive and that subtraction between $\nu_n(A)$, $\nu_m(B)$ is defined for all n, m, A, and B. Only the (extended) positive valued nature of μ played a key role. Thus the result holds for all set functions where these concepts are well defined. In particular, if $\nu_n: \Sigma \to \mathscr{X}$, where $\mathscr{X}$ is a complete normed vector space and each ν_n is σ-additive, the result holds without any change in proof or statement.

Remark (ii). The fact that Σ is a σ-algebra, not merely an algebra, is also important here. This is because (Σ, d) will not necessarily be a complete (semi-)metric space unless Σ is a σ-algebra. For the Baire category theorem the completeness is essential. In fact there are simple counterexamples to the theorem in the case that Σ is merely an algebra. The following is one such.

Example 4. Let $(\Omega, \Sigma, \lambda)$ be the Lebesgue unit interval. For each n let $\Sigma_n \subset \Sigma$ be the σ-algebra generated by the intervals $[0, 2^{-n-1}], (i\,2^{-n-1}, (i+1)2^{-n-1})$, $i = 1, \ldots, 2^n - 2, 2^n, 2^n + 1, \ldots, 2^{n+1} - 1$, $I_n = ((2^n - 1)/2^{n+1}, \frac{1}{2})$. Define $\nu_n(A) = 2^{n+1}\lambda(A \cap I_n)$, $A \in \Sigma_n$. If $\Sigma_0 = \bigcup_{n=1}^{\infty}\Sigma_n$ and $A \in \Sigma_0$, then $A \in \Sigma_n$ for some n, so $\nu(A) = \nu_n(A)$. It is easily seen that $\nu(A) = \lim_n \nu_n(A)$ for each $A \in \Sigma_0$ which is an algebra, but not a σ-algebra, ν is additive. However ν is not σ-additive, even though $\nu_n = \nu|\Sigma_n$ is σ-additive for $n \geqslant 1$.

Before we present a couple of important applications of Theorem 3, it will be convenient to show how many of the results on measures can be generalized to real σ-additive set functions, called *real* or *signed measures*, since both positive and negative signs are allowed. In some respects the lattice structure of these measures is analogous to that of real valued functions, although the proofs are, of necessity, different. Naturally we define a partial order between a pair of measures μ_1, μ_2 as follows: $\mu_1 \geqslant \mu_2$ iff $\mu_1 - \mu_2 \geqslant 0$, i.e., a measure $[(\mu_1 - \mu_2)(A) \geqslant 0$ for all $A \in \Sigma]$. Defining $(a\mu)(A) = a\mu(A)$, $a \in \mathbb{R}$, and $(\mu_1 + \mu_2)(A) = \mu_1(A) + \mu_2(A)$, $A \in \Sigma$, we see that the set of all finite linear combinations of finite measures, defines a vector space of real (σ-)additive set functions. Since there is a partial order, as noted above, we can make it a vector lattice by saying that $\inf(\mu_1, \mu_2)$ is the largest (σ-)additive set function μ such that $\mu(A) \leqslant \mu_i(A)$, $i = 1, 2$, for a given pair of (σ-)additive set functions. Similarly, $\nu = \sup(\mu_1, \mu_2)$ is defined as the smallest such function satisfying $\nu(A) \geqslant \mu_i(A)$, $i = 1, 2$, $A \in \Sigma$.

For computational facility, it will be desirable to get alternative definitions of inf and sup, to the ones given there. In the process we shall also show the existence of μ, ν. Thus let

$$\mu(A) = \inf\{\mu_1(A_1) + \mu_2(A_2): A = A_1 \cup A_2,\ A_i \in \Sigma, \text{disjoint}\}. \quad (8)$$

Since $\sup(\mu_1, \mu_2)(A) = -\inf(-\mu_1, -\mu_2)(A)$ by definition, there is no need to consider sup separately. That (8) is an alternative definition is seen from:

Proposition 5. *Let $\mathscr{A}$ be an algebra on a set Ω. If $\mu_1, \mu_2: \mathscr{A} \to \mathbb{R}$ are bounded additive functions, then μ of (8) exists and is the $\inf(\mu_1, \mu_2)$. Similarly $\nu = \sup(\mu_1, \mu_2)$ exists and is given by the corresponding formula. Further, setting $\mu^+ = \sup(\mu, 0)$ and $\mu^- = -\inf(\mu, 0)$, we have the* ***Jordan decomposition***

$$\mu = \mu^+ - \mu^-. \quad (9)$$

If μ_1, μ_2 are σ-additive, then so are μ, ν and $\mu^\pm$.

Proof. Since the μ_i are bounded, the definition (8) implies that given $\epsilon > 0$, there exist disjoint sets A_1, A_2 in $\mathscr{A}$, $A = A_1 \cup A_2$, such that

$$\mu(A) + \epsilon > \mu_1(A_1) + \mu_2(A_2) \geqslant \mu(A_1) + \mu(A_2).$$

Since $\epsilon > 0$ is arbitrary, this shows $\mu(A_1 \cup A_2) \geqslant \mu(A_1) + \mu(A_2)$. For the opposite inequality, let $A, B \in \mathscr{A}$ be disjoint. If $A = A_1 \cup A_2$, $B = B_1 \cup B_2$ is a partition from $\mathscr{A}$, such that

$$\mu(A) + \frac{\epsilon}{2} > \mu_1(A_1) + \mu_2(A_2), \qquad \mu(B) + \frac{\epsilon}{2} > \mu_1(B_1) + \mu_2(B_2).$$

Then adding these and using (8), we get

$$\mu(A) + \mu(B) + \epsilon > (\mu_1(A_1) + \mu_1(B_1)) + (\mu_2(A_2) + \mu_2(B_2))$$

$$= \mu_1(A_1 \cup B_1) + \mu_2(A_2 \cup B_2)$$

(since the μ_i are additive)

$$\geqslant \mu(A \cup B),$$

since $\{A_i \cup B_i\}_1^2$ is a partition of $A \cup B$. The arbitrariness of $\epsilon > 0$ shows that $\mu(A) + \mu(B) \geqslant \mu(A \cup B)$ and, with the previous inequality, gives the additivity of μ.

To see that μ is the largest such function, let $\tilde{\mu} \leqslant \mu_i$, $i = 1, 2$, be any other additive function on $\mathscr{A}$. If $A \in \mathscr{A}$, $A = A_1 \cup A_2$ is again any partition from

$\mathscr{A}$, then $\tilde{\mu}(A_j) \leqslant \mu_i(A_j)$, $j = 1, 2$ and $i = 1, 2$. Thus

$$\tilde{\mu}(A) = \tilde{\mu}(A_1) + \tilde{\mu}(A_2) \leqslant \mu_1(A_1) + \mu_2(A_2).$$

Taking infimum on the right over all such partitions of A, we get $\tilde{\mu}(A) \leqslant \mu(A)$, so that $\mu = \inf(\mu_1, \mu_2)$ as asserted. Since $\sup(\mu_1, \mu_2) = -\inf(-\mu_1, -\mu_2)$, the corresponding formula is obtained from what is already established.

Let $\mu^+ = \sup(\mu, 0) \geqslant 0$ and $\mu^- = -\inf(\mu, 0) \geqslant 0$. Then by (8), since $\mu^+ \geqslant \mu$ (and $-\mu^- \leqslant \mu$), we have for all $A \in \mathscr{A}$

$$\begin{aligned} 0 \leqslant \mu^+(A) - \mu(A) &= \sup\{\mu(B): B \subset A,\, B \in \mathscr{A}\} - \mu(A) \\ &= \sup\{\mu(B) - \mu(A): B \subset A,\, B \in \mathscr{A}\} \\ &= \sup\{-\mu(A - B): B \subset A,\, B \in \mathscr{A}\} \\ &= -\inf\{\mu(G): G \subset A,\, G \in \mathscr{A}\} \\ &= \mu^-(A). \end{aligned}$$

Hence $\mu(A) = \mu^+(A) - \mu^-(A)$, $A \in \mathscr{A}$, which is (9).

Suppose finally that μ_1, μ_2 are σ-additive. Since $\mu = \inf(\mu_1, \mu_2)$ is already shown to be additive, we need to verify its σ-additivity. But this follows from the fact that for $A_n \in \mathscr{A}_n$, $A_n \downarrow \varnothing$, μ_i bounded, and

$$|\mu(A_n)| \leqslant |\mu_1(A_n)| + |\mu_2(A_n)| \to 0 \qquad \text{as} \quad n \to \infty,$$

since μ_1, μ_2 are σ-additive. In particular, if μ is σ-additive, then $\mu^{\pm}$ are also σ-additive. This completes the proof.

The first part extends immediately, by induction, to finite collections. The proof of the second part shows that (9) is the most efficient decomposition in that if $\mu = \mu_1 - \mu_2$, $\mu_i \geqslant 0$, is another representation then $\mu^+ \leqslant \mu_1$ and $\mu^- \leqslant \mu_2$. In fact $\mu \leqslant \mu_1$, so that $\mu^+ \leqslant \mu_1$ and similarly $\mu^- = (-\mu)^+ \leqslant \mu_2$. Let us state these and another consequence for reference as:

Corollary 6. *Let $\mu_1, \ldots, \mu_n$ be a finite collection of (σ-) additive bounded functions on an algebra $\mathscr{A}$ of Ω. Then $\mu = \inf_{i \leqslant n} \mu_i$, $\nu = \sup_{i \leqslant n} \mu_i$ exist and are (σ-) additive. Moreover, every $\mu: \mathscr{A} \to \mathbb{R}$ of the proposition has the most efficient decomposition* (9). *If μ is σ-additive, then μ has a unique σ-additive extension $\hat{\mu}$ to $\sigma(\mathscr{A})$, and it is automatically bounded.*

Proof. We have to establish only the last part. By (9), $\mu(\mathscr{A}) \subset \mathbb{R}$, $\mu = \mu^+ - \mu^-$, and $\mu^{\pm}$ are σ-additive on $\mathscr{A}$. So $\mu^{\pm}(\Omega) < \infty$, and hence by Theorem 2.3.1, the $\mu^{\pm}$ have unique σ-additive extensions $\hat{\mu}^{\pm}: \sigma(\mathscr{A}) \to \mathbb{R}^+$. Let $\hat{\mu} = \hat{\mu}^+ - \hat{\mu}^-$. Then $\hat{\mu}: \sigma(\mathscr{A}) \to \mathbb{R}$ is σ-additive and bounded, and $\hat{\mu}|\mathscr{A} =$

$\mu^+ - \mu^- = \mu$. To see that $\hat{\mu}$ is unique, let $\tilde{\mu}$ be any other σ-additive extension of μ to $\sigma(\mathscr{A})$. Then by (9), $\tilde{\mu} = \tilde{\mu}^+ - \tilde{\mu}^-$, $\tilde{\mu}|\mathscr{A} = \mu$. Let $\xi = \hat{\mu}^+ + \hat{\mu}^- + \tilde{\mu}^+ + \tilde{\mu}^-$. Then $\xi: \sigma(\mathscr{A}) \to \mathbb{R}^+$ is a finite measure. If $d(A, B) = \xi(A \triangle B)$, then $(\sigma(\mathscr{A}), d)$ is a complete (semi-)metric space on which $\hat{\mu}$ and $\tilde{\mu}$ are both uniformly continuous. They agree on $\mathscr{A}$, a dense subspace of $(\sigma(\mathscr{A}), d)$. Hence $\hat{\mu} = \tilde{\mu}$ on $\sigma(\mathscr{A})$ must hold. This completes the proof.

An important consequence of Theorem 3 and the above corollary is the following result, established differently by O. M. Nykodým.

Corollary 7 (Nikodým). *Let* (Ω, Σ) *be a measurable space and* $\nu_n: \Sigma \to \mathbb{R}$ *be* σ*-additive,* $n \geqslant 1$. *If* $\lim_n \nu_n(A) = \nu(A)$ *exists for each* A *in the* σ*-algebra* Σ, *then* ν *is* σ*-additive, and for any sequence* $B_n \in \Sigma$, $B_n \downarrow \varnothing$, *we have* $\lim_n \nu_k(B_n) = 0$ *uniformly in* $k \geqslant 1$, *i.e.,* ν_k *is uniformly* σ*-additive on* Σ.

Proof. The result will follow from Theorem 3 if we can find a finite measure μ on (Ω, Σ) such that $\lim_{\mu(A) \to 0} \nu_n(A) = 0$ for each n. Since by Corollary 6 each ν_n is bounded, let $\tilde{\nu}_n = \nu_n^+ + \nu_n^-$, using (9). Then $0 \leqslant \tilde{\nu}_n(B) \leqslant \tilde{\nu}_n(\Omega) < \infty$, $n \geqslant 1$, $B \in \Sigma$. Define μ on Σ as

$$\mu(A) = \sum_{n=1}^{\infty} \frac{2^{-n}\tilde{\nu}_n(A)}{1 + \tilde{\nu}_n(\Omega)}, \qquad A \in \Sigma. \tag{10}$$

Then $\mu(\Omega) \leqslant 1$ and μ is a measure. It is clear that $\mu(A) \to 0$ implies $\tilde{\nu}_n(A) \to 0$ for each $n \geqslant 1$, as desired.

Some important applications of Theorem 3 and the above result will appear below and in the next section. First note that it is not always easy to verify the μ-continuity of ν. At least for the σ-additive case, this requirement can be simplified, as seen from the following:

Proposition 8. *Let* μ, ν *be finite measures on a measurable space* (Ω, Σ). *Then* $\lim_{\mu(A) \to 0} \nu(A) = 0$ *iff* $\mu(A) = 0 \Rightarrow \nu(A) = 0$.

Proof. Since it is evident that μ-continuity implies $\nu(A) = 0$ for $\mu(A) = 0$, only the converse is nontrivial.

Now if the assertion is false, then there exists an $\epsilon > 0$ and a sequence $A_n \in \Sigma$, $n \geqslant 1$, such that $\mu(A_n) < n^{-2}$, but $\nu(A_n) \geqslant \epsilon$. If $\tilde{A} = \limsup_n A_n = \bigcap_{k \geqslant 1} \bigcup_{n \geqslant k} A_n$ $(\in \Sigma)$, we have

$$\mu(\tilde{A}) \leqslant \mu\left(\bigcup_{n \geqslant k} A_n\right) \leqslant \sum_{n \geqslant k} \mu(A_n) < \sum_{n \geqslant k} n^{-2} \to 0$$

as $k \to \infty$. So $\mu(\tilde{A}) = 0$, and by hypothesis $\nu(\tilde{A}) = 0$. But by Proposition 2.1.1,

$$0 = \nu(\tilde{A}) = \nu\left(\lim_k \bigcup_{n \geqslant k} A_n\right) = \lim_k \nu\left(\bigcup_{n \geqslant k} A_n\right)$$

$$\geqslant \limsup_n \nu(A_n) \geqslant \epsilon > 0.$$

This contradiction proves the result.

For an application, we introduce a concept. A nonnegative additive function μ on an algebra $\mathscr{A}$ is called *purely finitely additive* if it dominates no nontrivial measure, i.e., if for $0 \leqslant \nu \leqslant \mu$ with ν as a measure we have $\nu = 0$.

The desired result, due to K. Yosida and E. Hewitt, is that every finite additive positive set function is composed of a measure and a purely finitely additive function. More precisely,

Theorem 9. *Let $\mathscr{A}$ be an algebra of subsets of Ω, and $\mu: \mathscr{A} \to \mathbb{R}^+$ be additive. Then there exist $\mu_i: \mathscr{A} \to \mathbb{R}^+$, $i = 1,2$, such that μ_1 is σ-additive, μ_2 is purely finitely additive, and $\mu = \mu_1 + \mu_2$. Moreover, this decomposition of μ is unique.*

Proof.† Consider the set $M = \{\nu: 0 \leqslant \nu \leqslant \mu,\ \nu \text{ is } \sigma\text{-additive on } \mathscr{A}\}$. Let $\alpha = \sup\{\nu(\Omega): \nu \in M\} \leqslant \mu(\Omega) < \infty$. By definition of sup we can find a sequence $\{\nu_n: n \geqslant 1\} \subset M$ such that $\nu_n(\Omega) \to \alpha$. Since this sequence need not be monotone, let $\tilde{\nu}_n = \sup(\nu_1, \dots, \nu_n)$, $n \geqslant 1$, which exists by Corollary 6. Thus $\tilde{\nu}_1 \leqslant \tilde{\nu}_2 \leqslant \cdots \leqslant \mu$. If $\Sigma = \sigma(\mathscr{A})$, then again by Corollary 6, each $\tilde{\nu}_i$ has a unique σ-additive extension $\bar{\nu}_i$ to Σ, and then $\bar{\nu}_1 \leqslant \bar{\nu}_2 \leqslant \cdots$, with $\bar{\nu}_i(\Omega) \leqslant \mu(\Omega) < \infty$, $i \geqslant 1$. If $\mu_1(A) = \lim_n \bar{\nu}_n(A)$, $A \in \Sigma$, then the limit exists for each $A \in \Sigma$, since $\{\bar{\nu}_n(A): n \geqslant 1\}$ is a bounded monotone sequence of reals. But by Corollary 7, $\mu_1: \Sigma \to \mathbb{R}^+$ is a measure, so that $\mu_1|\mathscr{A}$ is also σ-additive. Further, $\alpha \geqslant \mu_1(\Omega) = \lim_n \bar{\nu}_n(\Omega) = \lim_n \tilde{\nu}_n(\Omega) = \sup_n \nu_n(\Omega) = \alpha$. Let $\mu_2: \mathscr{A} \to \mathbb{R}^+$ be defined by $\mu_2(A) = \mu(A) - \mu_1(A) \geqslant 0$, $A \in \mathscr{A}$. We now claim that $\mu = \mu_1 + \mu_2$ is the desired decomposition.

To see that μ_2 is purely finitely additive, let $0 \leqslant \xi \leqslant \mu_2$ be a measure. Then $\xi + \mu_1 \leqslant \mu_2 + \mu_1 = \mu$, and we have

$$\begin{aligned} \alpha &= \lim_n \tilde{\nu}_n(\Omega) = \sup\{\nu(\Omega): \nu \in M\} \\ &= \mu_1(\Omega) < \mu_1(\Omega) + \xi(\Omega) \\ &\leqslant \sup\{\nu(\Omega): \nu \in M\} = \alpha < \infty. \end{aligned}$$

This is a contradiction unless $\xi \equiv 0$. Thus μ_2 is purely finitely additive.

† This argument is classical and is essentially due to the authors. There are other (even shorter) proofs, needing a somewhat different preparation. One such method, valid for both scalar *and* vector cases at the same time, can be found in R. E. Huff (1973).

For the uniqueness, let $\mu_1 + \mu_2 = \mu = \bar{\mu}_1 + \bar{\mu}_2$ be two such representations. Then $\mu_1 - \bar{\mu}_1 = \bar{\mu}_2 - \mu_2 \leqslant \bar{\mu}_2$. Since $\mu_1 - \bar{\mu}_1$ is a signed measure on $\mathscr{A}$, we can apply Corollary 6 and deduce that $\sup(\mu_1 - \bar{\mu}_1, 0)$ is a (nonnegative) measure dominated by $\bar{\mu}_2$. But $\bar{\mu}_2$ is purely finitely additive. So $\sup(\mu_1 - \bar{\mu}_1, 0) = 0$ and $\mu_1 \leqslant \bar{\mu}_1$. By a similar argument applied to $\bar{\mu}_1 - \mu_1 = \mu_2 - \bar{\mu}_2 \leqslant \mu_2$, we get $\bar{\mu}_1 \leqslant \mu_1$, so that $\mu_1 = \bar{\mu}_1$ and then $\mu_2 = \bar{\mu}_2$. This completes the proof.

EXERCISES

0. Let (Ω, Σ) be a measurable space, and $\nu: \Sigma \to \mathbb{R}$ be a measure or a signed measure. Show that ν is bounded. If $\mathscr{N} = \{A \in \Sigma: \nu(A) = 0\}$, verify that $\mathscr{N}$ is a σ-ideal (i.e., $A \in \Sigma$, $B \in \mathscr{N} \Rightarrow A \cap B \in \mathscr{N}$, and it is a σ-ring) for measures ν, but it is not even a ring for signed measures.

1. Complete the details of the assertions of Example 4.

2. Let $\Omega = \{1, 2, \dots\}$, the natural numbers, and Σ be the algebra of all finite subsets of Ω and of the sets with finite complements. Define $\mu(A) = \sum_{n \in A} (-1)^n/n$. Show that $\mu: \Sigma \to \mathbb{R}$ is an additive function which is not bounded. (This illustrates one of the conditions in Corollary 6.)

3. If μ_1, μ_2 are two bounded real additive functions on an algebra $\mathscr{A}$ of a set Ω, show that $\inf(\mu_1, \mu_2), \sup(\mu_1, \mu_2)$ of Proposition 5 are the same as the following:

$$\inf(\mu_1, \mu_2) = \tfrac{1}{2}(\mu_1 + \mu_2 - |\mu_1 - \mu_2|),$$

$$\sup(\mu_1, \mu_2) = \tfrac{1}{2}(\mu_1 + \mu_2 + |\mu_1 - \mu_2|).$$

4. If $\mu: \mathscr{A} \to \mathbb{R}$ is a bounded additive function, and $\mu = \mu^+ - \mu^-$ is the Jordan decomposition of (9), show that the additive set function $\nu = \inf(\mu^+, \mu^-)$ is identically zero. [Examine the defining equation (8), and show $\mu^{\pm}(A) > 0 \Rightarrow \mu^{\mp}(A) = 0$.]

5. Let $ba(\Omega, \Sigma)$ be the space of all real bounded additive set functions on an algebra Σ of Ω. For each $\mu \in ba(\Omega, \Sigma)$, define the functional $\|\cdot\|: \mu \mapsto \|\mu\| = \mu^+(\Omega) + \mu^-(\Omega)$, with the notation of (9). Show that $\|\cdot\|$ is a norm and $(ba(\Omega, \Sigma), \|\cdot\|)$ is a real Banach space, i.e., a complete normed real linear space. Also verify that its subset of σ-additive elements is a closed subspace, so that it is also a Banach space, denoted by $ca(\Omega, \Sigma)$.

6. Let (Ω, Σ, μ) be a measure space, and $f_n: \Omega \to \overline{\mathbb{R}}$ be μ-integrable, $n \geqslant 1$. Suppose that $\nu_n: A \mapsto \int_A f_n \, d\mu$, $A \in \Sigma$, and that $\lim_n \nu_n(A) = \nu(A)$ exists for each A in the σ-algebra Σ. Show that ν is σ-additive and μ-continuous. Also verify that

$$\lim_{k \to \infty} \int_{[|f_n| > k]} |f_n| \, d\mu = 0$$

uniformly in $n \geqslant 1$. If $f_n - f_m \xrightarrow{\mu} 0$ as $m, n \to \infty$, show that there exists an integrable f such that $\nu(A) = \int_A f\,d\mu$, $A \in \Sigma$, and $\|f_n - f\|_1 \to 0$. [μ can be taken σ-finite; apply Corollary 7 and Exercise 3.3.]

7. Let $\mathscr{A}$ be an algebra on a set Ω, and $\mu: \mathscr{A} \to \mathbb{C}$ an additive (complex) function. Define $\mu_1(A) = \operatorname{Re}\mu(A)$, $\mu_2(A) = \operatorname{Im}\mu(A)$, $A \in \mathscr{A}$, and show that $\mu_j: \mathscr{A} \to \mathbb{R}$, $j = 1, 2$, are additive, and that $\mu = \mu_1^+ - \mu_1^- + i(\mu_2^+ - \mu_2^-)$ is the most efficient decomposition of μ into $\mu_j^\pm: \mathscr{A} \to \mathbb{R}^+$, $j = 1, 2$, additive. (This is the *Jordan decomposition for complex additive set functions*.) Deduce that μ has a unique σ-additive extension to $\sigma(\mathscr{A})$ if μ is σ-additive on $\mathscr{A}$.

8. A complex additive set function μ on $\mathscr{A}$ is called purely finitely additive if its components $\mu_j^\pm$, $j = 1, 2$, given in Exercise 7 are purely finitely additive, as defined before. Show that each complex additive set function on $\mathscr{A}$ admits the Yosida-Hewitt decomposition.

4.5. THE L^p-SPACES

One of the major goals of integration theory is to study the structure of spaces of various types of integrable functions on a measure space, called the Lebesgue spaces and denoted L^p, as well as some related extensions. These function spaces appear in many different contexts in analysis, including differential equations, harmonic analysis, probability, and statistics. Here we present an introduction to these spaces together with important complements in exercises, and consider this from time to time in various contexts both illustrating some work and extending it on occasion. Also, these spaces serve as a concrete illustration for abstract analysis in the same way that the Lebesgue measure helps in a study of the corresponding general case.

Definition 1. Let (Ω, Σ, μ) be a measure space, and $p \geqslant 0$ be a fixed number. We denote by $\mathscr{L}^p(\Omega, \Sigma, \mu)$, or $\mathscr{L}^p(\mu)$ for short, the set of functions on $\Omega \to \mathbb{R}$, as follows:

$$\mathscr{L}^0(\mu) = \{f: f \text{ is real measurable for } \Sigma\}, \tag{1}$$

and

$$\mathscr{L}^p(\mu) = \left\{f \in \mathscr{L}^0(\mu): \int_\Omega |f|^p\,d\mu < \infty\right\} \quad \text{if} \quad 0 < p < \infty. \tag{2}$$

Also

$$\mathscr{L}^\infty(\mu) = \{f \in \mathscr{L}^0(\mu): \mu(\{\omega: |f(\omega)| > k\}) = 0 \text{ for some } k = k_f > 0\}. \tag{3}$$

It is useful to make them topological spaces by introducing a (semi-)metric on them. Some of these already appeared, and the key triangle inequality is proved in Theorem 4 below. But we define it precisely for reference purposes.

Definition 2. If $f, g \in \mathscr{L}^0(\mu)$, let $d(f, g) = d(f - g, 0) = d(f - g)$, with the (semi-)metric

$$\|f\|_0 = d(f) = \inf\{\varphi(\alpha + \mu(\{\omega : |f(\omega)| > \alpha\})) : \alpha > 0\}, \tag{4}$$

where $\varphi: \overline{\mathbb{R}}^+ \to \mathbb{R}^+$ is a bounded increasing continuous concave function, $\varphi(x) = x$ if $\mu(\Omega) < \infty$,

$$\|f\|_p = \begin{cases} \int_\Omega |f|^p \, d\mu, & 0 < p < 1, \\ \left[\int_\Omega |f|^p \, d\mu\right]^{1/p}, & 1 \leqslant p < \infty, \end{cases} \tag{5}$$

and

$$\|f\|_\infty = \inf\{k > 0 : \mu(\{\omega : |f(\omega)| > k\}) = 0\}. \tag{6}$$

These $\|\cdot\|_p$ become metrics if a.e. equal functions are identified. Then we denote the new metric by the same symbol, and $L^p(\mu) = \mathscr{L}^p(\mu)/\sim$, the spaces of equivalence classes $(L^p(\mu), \|\cdot\|_p)$, $0 \leqslant p \leqslant \infty$, are the *Lebesgue spaces*.

The first step in the study of these spaces is their linear structure and the basic topological properties induced by the $\|\cdot\|_p$ of (4)–(6). We have already seen in Proposition 3.2.6 that $(L^0(\mu), \|\cdot\|_0)$ is a complete linear metric space. This property is also true for $L^p(\mu)$, $p > 0$, but it needs a separate proof. The fact that the space $\mathscr{L}^p(\mu)$, $0 < p \leqslant \infty$, is linear is immediate. Indeed let $f \in \mathscr{L}^p(\mu)$. Then $a \cdot f \in \mathscr{L}^p(\mu)$, $a \in \mathbb{R}$, by (5) and (6). For the linearity, if $f, g \in \mathscr{L}^p(\mu)$, then

$$\begin{aligned} |f + g|^p &\leqslant (|f| + |g|)^p \leqslant 2^p[\max(|f|, |g|)]^p \\ &= 2^p \max(|f|^p, |g|^p) \leqslant 2^p(|f|^p + |g|^p). \end{aligned}$$

Hence

$$\int_\Omega |f + g|^p \, d\mu \leqslant 2^p\left(\int_\Omega |f|^p \, d\mu + \int_\Omega |g|^p \, d\mu\right) < \infty. \tag{7}$$

So $f + g \in \mathscr{L}^p$, $0 < p < \infty$. The case of $p = \infty$ is trivial, since f and g

essentially bounded implies the same for their sum. Thus it is the completeness property with the topological structure that requires additional argument.

The functional $\|\cdot\|_\infty$ has the following relation with $\|\cdot\|_p$, $0 < p < \infty$.

Proposition 3. *Let* $f \in \mathscr{L}^p(\mu)$, $0 < p < \infty$. *Then* $\lim_r \|f\|_r$ *exists (may be* $= +\infty$) *and* $= \|f\|_\infty$. *On the other hand, if* $\mu(\Omega) < \infty$, *then* $\mathscr{L}^p(\mu) \supset \mathscr{L}^r(\mu)$ *for* $0 \leqslant p < r \leqslant \infty$.

Proof. If $\|f\|_\infty = 0$, then by (6), $f = 0$, a.e.. So $f \in \mathscr{L}^r(\mu)$ for all $r > 0$ and $\|f\|_r = 0$. Thus the result is true and trivial. Hence let $0 < \|f\|_\infty$, but $\|f\|_p < \infty$ by hypothesis. Then by Markov's inequality and (6), for each $0 < \alpha < \|f\|_\infty$, if we let $A_\alpha = \{\omega\colon |f(\omega)| > \alpha\}$,

$$0 < \mu(A_\alpha) \leqslant \alpha^{-p} \int_\Omega |f|^p \, d\mu < \infty. \tag{8}$$

Hence for each $r > 1$, consider

$$\|f\|_r^r = \int_\Omega |f|^r \, d\mu \geqslant \int_{A_\alpha} |f|^r \, d\mu \geqslant \alpha^r \mu(A_\alpha).$$

Since $0 < \mu(A_\alpha) < \infty$, we have

$$\liminf_{r\to\infty} \|f\|_r \geqslant \liminf_{r\to\infty} \alpha[\mu(A_\alpha)]^{1/r} = \alpha.$$

Letting $\alpha \uparrow \|f\|_\infty$, we get $\liminf_{r\to\infty} \|f\|_r \geqslant \|f\|_\infty$.

For the opposite inequality we need only consider if $\beta = \|f\|_\infty < \infty$. Then for $r > p$,

$$\|f\|_r^r = \int_\Omega |f|^r \, d\mu \leqslant \beta^{r-p} \int_\Omega |f|^p \, d\mu,$$

so that

$$\|f\|_r \leqslant \beta^{1-(p/r)} \left(\int_\Omega |f|^p \, d\mu \right)^{1/r}.$$

Since $f \in \mathscr{L}^p(\mu)$, the integral on the right is a fixed positive number and hence

$$\limsup_{r\to\infty} \|f\|_r \leqslant \beta = \|f\|_\infty.$$

This and the previous inequality together establish the first assertion.

Next let $\mu(\Omega) < \infty$. Since for any $p > 0$, $\int_\Omega |f|^p \, d\mu < \infty$ implies f is a.e. finite, it is clear that $f \in \mathscr{L}^0$, so that $\mathscr{L}^p(\mu) \subset \mathscr{L}^0(\mu)$ for $0 < p < \infty$ even if

μ is not finite. If $p = \infty$, then $|f| < \infty$ a.e. (indeed, a.e. bounded), so $\mathscr{L}^\infty(\mu) \subset \mathscr{L}^0(\mu)$ again. In fact (2) and (3) imply this at once. Now let $0 < p < r < \infty$ and $f \in \mathscr{L}^r(\mu)$. Then

$$\int_\Omega |f|^p \, d\mu = \int_{\{\omega: |f|(\omega) \leqslant 1\}} |f|^p \, d\mu + \int_{\{\omega: |f|(\omega) > 1\}} |f|^p \, d\mu$$

$$\leqslant \mu(\Omega) + \int_\Omega |f|^r \, d\mu < \infty, \qquad \text{since} \quad \mu(\Omega) < \infty. \tag{9}$$

Hence $f \in \mathscr{L}^p(\mu)$, so that $\mathscr{L}^p(\mu) \supset \mathscr{L}^r(\mu)$ for $0 < p < r < \infty$. If $r = \infty$, the result is simple, since $\|f\|_\infty = \alpha < \infty \Rightarrow$

$$\int_\Omega |f|^p \, d\mu \leqslant \alpha^p \mu(\Omega) < \infty. \tag{10}$$

This completes the proof.

To show that the functional $\|\cdot\|_p$ defined in (5) is a (semi-)metric, one has to distinguish between the cases that $0 < p < 1$ and $p \geqslant 1$. This corresponds to the fact that $\varphi_p: x \mapsto x^p$ on $\mathbb{R}^+$ is increasing but concave for $0 < p < 1$, and convex for $p \geqslant 1$, so that all the key inequalities reverse in these two cases. As noted in the remarks after Proposition 2.4.1, φ is subadditive for concave φ and hence is superadditive for convex φ. So we need different techniques in each of these cases.

Recall that a twice differentiable function $\varphi: \mathbb{R}^+ \to \mathbb{R}^+$ is convex iff $\varphi''(x) > 0$ for $x \in \mathbb{R}^+$. Since ψ is concave iff $-\psi$ is convex, some properties of concave functions can be obtained directly from those of convex functions. Let $\varphi(x) = -\log x$, $x > 0$. Then $\varphi''(x) = x^{-2} > 0$, so that it is convex. But also, by definition, φ is convex iff for any $\alpha \geqslant 0$, $\beta \geqslant 0$ with $\alpha + \beta = 1$ we have

$$\varphi(\alpha x + \beta y) \leqslant \alpha\varphi(x) + \beta\varphi(y).$$

Geometrically, this means the chord joining $(x, \varphi(x))$ and $(y, \varphi(y))$ lies above the arc. Now taking in particular $\varphi(x) = -\log x$, here we get

$$\log(\alpha x + \beta y) \geqslant \alpha \log x + \beta \log y = \log x^\alpha y^\beta.$$

Since $\log(\cdot)$ is strictly increasing, this implies (see Exercise 25 for more general φ-functions)

$$x^\alpha y^\beta \leqslant \alpha x + \beta y, \qquad x > 0, \quad y > 0, \quad 1 > \alpha > 0, \quad \alpha + \beta = 1. \tag{11}$$

One has the following useful inequalities:

Theorem 4. *Let* (Ω, Σ, μ) *be a measure space,* $f, g \in \mathscr{L}^p(\mu)$, $h \in \mathscr{L}^q(\mu)$, *where* $p \geqslant 1$, $q = p/(p-1)$ (*if* $p = 1$ *we take* $q = +\infty$). *Then*

(i) (*Hölder's inequality*) $fh \in \mathscr{L}^1(\mu)$ *and*

$$\|fh\|_1 \leqslant \|f\|_p \|h\|_q. \tag{12}$$

(ii) (*Minkowski's inequality*) $f + g \in \mathscr{L}^p(\mu)$ *and*

$$\|f + g\|_p \leqslant \|f\|_p + \|g\|_p. \tag{13}$$

(iii) *If* $f, g \in \mathscr{L}^p(\mu)$, $0 < p < 1$, *then* $f + g \in \mathscr{L}^p(\mu)$ *and*

$$\|f + g\|_p \leqslant \|f\|_p + \|g\|_p. \tag{14}$$

There is no such analog for Hölder's inequality here (*but see Proposition* 6 *below*). *Also the difference in definition of* $\|\cdot\|_p$ *in* (5) *should be remembered for* $0 < p < 1$ *and for* $p \geqslant 1$.

Proof. (i): Let $\alpha = 1/p$, $\beta = 1/q$ and $x = |f|^p/\|f\|_p^p$, $y = |h|^q/\|h\|_q^q$ provided both $\|f\|_p > 0$ and $\|h\|_q > 0$. Since (12) is clearly true if either $\|h\|_q = 0$ or $\|f\|_p = 0$, we only need to consider when both of these are positive. Then by (11) we get, on integration and by (5),

$$\int_\Omega \frac{|fh|}{\|f\|_p \|h\|_q} \, d\mu \leqslant \frac{1}{p} + \frac{1}{q} = 1.$$

This proves (12) on multiplying by $\|f\|_p \|h\|_q$ when $1 < p < \infty$, so $q > 1$. If $p = 1$ (so $q = +\infty$), we have $\|h\|_\infty < \infty$ and $|h| \leqslant \|h\|_\infty$ a.e. Hence

$$\int_\Omega |fh| \, d\mu \leqslant \|h\|_\infty \int_\Omega |f| \, d\mu = \|h\|_\infty \|f\|_1.$$

Hence (i) is true as stated.

(ii): If $p = 1$, then [by Proposition 3, $f + g \in \mathscr{L}^p(\mu)$ already]

$$\|f + g\|_1 = \int_\Omega |f + g| \, d\mu \leqslant \int_\Omega (|f| + |g|) \, d\mu = \|f\|_1 + \|g\|_1.$$

Note that there is equality if $f \geqslant 0$ and $g \geqslant 0$ a.e. here.

If $p = +\infty$, then $|f| \leqslant \|f\|_\infty$ a.e. and $|g| \leqslant \|g\|_\infty$ a.e., so that

$$|f + g| \leqslant |f| + |g| \leqslant \|f\|_\infty + \|g\|_\infty \quad \text{a.e.}$$

Hence

$$\|f + g\|_\infty \leqslant \|f\|_\infty + \|g\|_\infty.$$

If $1 < p < \infty$, then we assume $\|f + g\|_p > 0$, since the result is clearly true otherwise. Hence on using the fact that $p - 1 = p/q$ and (i),

$$\|f + g\|_p^p = \int_\Omega |f + g|^p \, d\mu \leqslant \int_\Omega |f + g|^{p-1}(|f| + |g|) \, d\mu$$

$$\leqslant \|f + g\|_p^{p-1}\|f\|_p + \|f + g\|_p^{p-1}\|g\|_p.$$

Since $\|f + g\|_p^{p-1} > 0$, we get, on canceling this factor,

$$\|f + g\|_p \leqslant \|f\|_p + \|g\|_p,$$

which is (13).

(iii): If 0 < p < 1, then again $f + g \in \mathscr{L}^p(\mu)$ by Proposition 3, and

$$\|f + g\|_p = \int_\Omega |f + g|^p \, d\mu$$

$$\leqslant \int_\Omega (|f|^p + |g|^p) \, d\mu$$

(by subadditivity of the concave function, noted before)

$$= \|f\|_p + \|g\|_p.$$

This completes the proof of the theorem.

In view of (13) and (14), we deduce that $\|\cdot\|_p$ is a metric when a.e. equal functions are identified. Hence writing $L^p(\mu) = \mathscr{L}^p(\mu)/\sim$, one has:

Corollary 5. *The space* $(L^p(\mu), \|\cdot\|)$, $0 \leqslant p \leqslant \infty$, *is a linear metric space. If* $p \geqslant 1$, *then* $\|\cdot\|_p$ *is also a norm, meaning* $\|af\|_p = |a|\,\|f\|_p$, *in addition to the property of being a metric.*

A natural question here is to ask for the consequences if we define $\|f\|_p$ by the same formula as in (5) for all $0 < p < \infty$. If we define $\|f\|_p' = [\int_\Omega |f|^p \, d\mu]^{1/p}$, then for $0 < p < 1$ both the Hölder and Minkowski inequalities reverse from those of (12) and (13), and they cease to have much interest here, since then $\|\cdot\|_p'$ cannot be used as a distance function. Let us state this precisely as follows.

Proposition 6. *Let* (Ω, Σ, μ) *be a measure space, and define* $\|\cdot\|'_p$ *for* $f \in \mathscr{L}^p(\mu)$ *by the formula above, i.e., with* $0 < p < 1$, $\|f\|'_p = (\int_\Omega |f|^p \, d\mu)^{1/p}$. *Let* $q = p/(p-1)$, *so that* $1/p + 1/q = 1$ *(but* $q < 0$ *here). If* $\|f\|'_p < \infty$, $\|g\|'_q = [\int_\Omega |g|^q \, d\mu]^{1/q} < \infty$, *and* $\|h\|'_p < \infty$, *then we have*

$$\int_\Omega |fg| \, d\mu \geqslant \|f\|'_p \|g\|'_q, \tag{15}$$

and

$$\big\| |f| + |h| \big\|'_p \geqslant \|f\|'_p + \|h\|'_p. \tag{16}$$

Sketch of Proof. We may assume that the right side of (15) is positive and the left side is finite. Since $q < 0$, this implies $0 < |g| < \infty$ a.e. To use (12), let $u = |g|^{-p}$, $v = |fg|^p$. So $u^{1/(1-p)} = |g|^{p/(p-1)}$, and $u \in \mathscr{L}^{1/(1-p)}(\mu)$, $v \in \mathscr{L}^{1/p}(\mu)$. By (12) then

$$\begin{aligned} \int_\Omega |f|^p \, d\mu &= \int_\Omega vu \, d\mu \\ &\leqslant \|v\|_{1/p} \|u\|_{1/(1-p)} \\ &= \left(\int_\Omega |fg| \, d\mu \right)^p \left(\int_\Omega |g|^q \, d\mu \right)^{1-p}. \end{aligned}$$

This yields (15), and then the same computation as in (13), using (15) in place of (12) there, gives (16). The details will be left to the reader.

Since $\|\cdot\|'_p$ does not satisfy the triangle inequality and hence is not a distance function, we cannot use (15) and (16) in our analysis. On the other hand, if we use $\|f\|''_p = \int_\Omega |f|^p \, d\mu$ for all $p > 0$, then with $\varphi(x) = |x|^p$, $p \geqslant 1$, we get

$$\begin{aligned} \varphi(x+y) &= \int_0^{x+y} \varphi'(t) \, dt = \int_0^x \varphi'(t) \, dt + \int_x^{x+y} \varphi'(t) \, dt \\ &= \varphi(x) + \int_0^y \varphi'(x+t) \, dt \\ &\geqslant \varphi(x) + \int_0^y \varphi'(t) \, dt = \varphi(x) + \varphi(y), \end{aligned}$$

since the derivative $\varphi'(\cdot)$ is increasing. Thus the convex φ is superadditive, and so (13) becomes

$$
\begin{aligned}
\| |f| + |g| \|_p'' &= \int_\Omega (|f| + |g|)^p \, d\mu \\
&\geqslant \int_\Omega |f|^p \, d\mu + \int_\Omega |g|^p \, d\mu \\
&= \|f\|_p'' + \|g\|_p'', \qquad p \geqslant 1.
\end{aligned} \tag{17}
$$

However, since $\psi: x \mapsto |x|^{1/p}$ is convex, for $0 < p < 1$, we have

$$
\begin{aligned}
\| |f| + |g| \|_p' &= \left(\int_\Omega (|f| + |g|)^p \, d\mu \right)^{1/p} \\
&\leqslant \left(\int_\Omega |f|^p \, d\mu + \int_\Omega |g|^p \, d\mu \right)^{1/p} \\
&= 2^{1/p} \left[\frac{1}{2} \left(\int_\Omega |f|^p \, d\mu + \int_\Omega |g|^p \, d\mu \right) \right]^{1/p} \\
&\leqslant 2^{(1-p)/p} \left(\|f\|_p' + \|g\|_p' \right).
\end{aligned} \tag{18}
$$

But $2^{(1-p)/p} \to 1$ as $p \to 1$, and > 1 for $0 < p < 1$, so (18) is not a usable triangle inequality. For these reasons, *we always use* $\| \cdot \|_p$, *of Definition* 2, *unless the contrary is stated.*

Next let us establish another key property of the $\mathscr{L}^p(\mu)$-spaces:

Theorem 7. *Let* (Ω, Σ, μ) *be a measure space, and* $(L^p(\mu), \| \cdot \|_p)$ *the linear metric space defined above (cf. Corollary 5). Then for* $0 \leqslant p \leqslant \infty$, $L^p(\mu)$ *is complete, and if* $0 < p < \infty$, *simple functions are dense in the metric* $\| \cdot \|_p$ *in the sense that for each* $\epsilon > 0$, $f \in \mathscr{L}^p(\mu)$, *there exists a simple function* $h_\epsilon = \sum_{i=1}^n a_i \chi_{A_i} \in \mathscr{L}^p(\mu)$ *such that*

$$
\|f - h_\epsilon\|_p < \epsilon. \tag{19}
$$

Proof. The completeness was already shown if $p = 0$, in Proposition 3.2.6. The idea of the proof for $p > 0$ is essentially the same, and we fill in the details. Thus let $\{f_n: n \geqslant 1\} \subset \mathscr{L}^p(\mu)$ be a Cauchy sequence. First choose an $n_1 \geqslant 1$ so that $n \geqslant n_1 \Rightarrow \|f_{n_1} - f_n\|_p < 1$. If n_k is chosen, let $n_{k+1} > n_k$ be selected such that $\|f_{n_{k+1}} - f_n\|_p < 2^{-k}$ for $n \geqslant n_{k+1}$. This is all possible because by hypothesis $\|f_n - f_m\|_p \to 0$ as $n, m \to \infty$. Consequently

$$
g_k = |f_{n_1}| + \sum_{i=1}^k |f_{n_{i+1}} - f_{n_i}| \in \mathscr{L}^p(\mu), \qquad \text{by linearity.}
$$

And now, by (13) and (14), for all $k > 1$,

$$\|g_k\|_p \leqslant \|f_{n_1}\|_p + \sum_{i=1}^{k} \|f_{n_{i+1}} - f_{n_i}\|_p$$

$$< \|f_{n_1}\|_p + \sum_{i=1}^{k} 2^{-i} < \|f_{n_1}\|_p + 1.$$

Thus $g_k \in \mathscr{L}^p(\mu)$ and so $g_k^p \in \mathscr{L}^1(\mu)$. But $0 \leqslant g_k^p \leqslant g_{k+1}^p \to g^p$ a.e. Hence the monotone convergence theorem implies

$$\int_\Omega g^p \, d\mu = \lim_k \int_\Omega g_k^p \, d\mu \leqslant 1 + \|f_{n_1}\|_p < \infty. \tag{20}$$

But by definition $g = |f_{n_1}| + \sum_{i=1}^{\infty} |f_{n_{i+1}} - f_{n_i}|$, and (20) implies (by Markov's inequality) $0 \leqslant g < \infty$ a.e. Hence

$$\left| f_{n_1} + \sum_{i=1}^{\infty} \left(f_{n_{i+1}} - f_{n_i} \right) \right| = \lim_k \left| f_{n_1} + \sum_{i=1}^{k} \left(f_{n_{i+1}} - f_{n_i} \right) \right|$$

$$\leqslant g < \infty \qquad \text{a.e.},$$

so that the left side sum, which is $|\lim_{k\to\infty} f_{n_k}|$, exists a.e. and $\leqslant g$ a.e. If $f = \lim_{k\to\infty} f_{n_k}$, then $|f|^p \leqslant g^p \in \mathscr{L}^1(\mu)$ and $f \in \mathscr{L}^p(\mu)$. [We define $f(\omega) = 0$ where this limit does not exist, since those points constitute a μ-null set.]

Now $\|f_{n_k} - f_n\|_p < 2^{-k+1}$ and $|f_{n_k} - f_n|^p \to |f - f_n|^p$, a.e., as $k \to \infty$ for each n. Hence, by Fatou's lemma, we conclude that

$$\int_\Omega |f - f_n|^p \, d\mu \leqslant \varliminf_{k\to\infty} \int_\Omega |f_{n_k} - f_n|^p \, d\mu < \infty, \tag{21}$$

on using the Cauchy condition. But the last hypothesis also implies that $\|f_n - f_m\|_p^p \to 0$ as $n, m \to \infty$ (here $0 < p < \infty$). Hence (21) tends to zero if we let $k \to \infty$ and then $n \to \infty$. This shows $\|f_n - f\|_p \to 0$, and since we have already noted that $f \in \mathscr{L}^p(\mu)$, the completeness follows.

If $p = +\infty$, then $|f_n - f_m| \leqslant \|f_n - f_m\|_\infty$ a.e. If $A_{m,n}$ is the exceptional μ-null set here, let $A = \bigcup_{m,n} A_{m,n}$. Thus $\mu(A) = 0$, and for $\omega \in A^c$ we have $|f_n(\omega) - f_m(\omega)| \leqslant \|f_n - f_m\|_\infty \to 0$ as $n, m \to \infty$, uniformly in ω. So $f_n(\omega) \to f(\omega)$, $\omega \in A^c$, and f is bounded. Next, defining $f(\omega) = 0$ on A, we get $f_n \to f$ a.e., so that $f \in \mathscr{L}^\infty(\mu)$. Also the uniform convergence implies $\|f_n - f\|_\infty \to 0$. Thus again the completeness follows.

For the last part on density, since $f \in \mathscr{L}^p(\mu)$ iff f^+ and f^- are in $\mathscr{L}^p(\mu)$, consider $f \geqslant 0$ a.e. Then by the structure theorem for measurable functions (Theorem 3.1.7), there is a sequence of simple functions $0 \leqslant \varphi_n \uparrow f$. Hence

$(f - \varphi_n)^p \to 0$ a.e., and $0 \leqslant (f - \varphi_n)^p \leqslant f^p \in \mathscr{L}^1(\mu)$. So by dominated convergence (Theorem 3.4)

$$\int_\Omega (f - \varphi_n)^p \, d\mu \to 0 \qquad \text{as} \quad n \to \infty.$$

Given $\epsilon > 0$, choose n_0 [$= n_0(\epsilon)$] such that $n \geqslant n_0 \Rightarrow$

$$\int_\Omega (f - \varphi_n)^p \, d\mu < \begin{cases} \epsilon & \text{if} \quad 0 < p < 1, \\ \epsilon^p & \text{if} \quad 1 \leqslant p < \infty. \end{cases} \tag{22}$$

Hence $\|f - \varphi_n\|_p < \epsilon$. In the general case, let $f = f^+ - f^-$, and by (22) choose φ_{in}, $i = 1, 2$, simple functions, such that $\|f^+ - \varphi_{1n}\|_p < \epsilon/2$ and $\|f^- - \varphi_{2n}\|_p < \epsilon/2$ if $n \geqslant n_i$, $i = 1, 2$. Let $n' = \max(n_1, n_2)$ and $h_\epsilon = \varphi_{1n'} + \varphi_{2n'}$. Then (13) implies

$$\|f - h_\epsilon\|_p \leqslant \|f^+ - \varphi_{1n'}\|_p + \|f^- - \varphi_{2n'}\|_p < \frac{\epsilon}{2} + \frac{\epsilon}{2} = \epsilon.$$

This gives (19), and the theorem is proved.

Hereafter, identifying functions that are equal a.e., we say, following custom, that $L^p(\mu)$ is the Lebesgue space of pth order integrable functions when $\mathscr{L}^p(\mu)$ is meant. The reason is that $\|\cdot\|_p$ is a metric on $L^p(\mu)$ and only a semimetric on $\mathscr{L}^p(\mu)$. Unless this distinction is essential in a particular context, we do not need to point it out. Of course, if we are dealing with noncountable subsets of $L^p(\mu)$, then there is a problem as to how functions can be chosen from equivalence classes to satisfy other relations. We deal with that (nontrivial) problem in Chapter 8. Until then this question is not essential and will be ignored. Note that if $L^p(\mu)$ itself has a dense denumerable subset in it, then the above-mentioned problem is not important. We shall present a condition for this to happen, in which case the space [$\mathscr{L}^p(\mu)$ or] $L^p(\mu)$ is termed *separable* in its metric topology. (See Exercise 5.)

In the last part of the preceding theorem the spaces $L^0(\mu)$ and $L^\infty(\mu)$ are excluded. This is because if $\mu(\Omega) = +\infty$ then $\chi_\Omega \in L^\infty(\mu)$ [as well as $L^0(\mu)$], and since by definition a step function should vanish outside a set of finite μ-measure, it is clear that this will be impossible in these two spaces. However, the following weaker result holds.

Proposition 8. *If* (Ω, Σ, μ) *is a measure space, and* $L^0(\mu)$ *and* $L^\infty(\mu)$ *are respectively the spaces of* (*the equivalence classes of*) *real measurable functions and essentially bounded measurable functions on* Ω, *then, in their metrics, the sets of functions of the form* $f_n = \sum_{i=1}^n a_i \chi_{A_i}$, $A_i \in \Sigma$ [*but* $\mu(A_i) = +\infty$ *is allowed*] *for* $n = \infty$ *and for* $n < \infty$ *are dense. Thus, given* f, *and* $\epsilon > 0$, *there exists an* $f_{n_\epsilon} \in L^p(\mu)$, $p = 0$ *or* ∞, (*elementary or finite step respectively*) *such* *that*

$$\|f - f_{n_\epsilon}\|_p < \epsilon. \tag{23}$$

Proof. If $p = 0$ and $f \in L^0(\mu)$, then by the structure theorem there exist elementary functions f_n ($f_n = \sum_{i=-\infty}^{\infty} a_i^n \chi_{A_i^n}$, $\mu(A_i^n) = \infty$ is possible) such that $f_n \to f$ uniformly, and hence in μ-measure. (Indeed, given $\epsilon > 0$, there is an n_0 [$= n_0(\epsilon)$] such that $n \geqslant n_0 \Rightarrow |f(\omega) - f_{n_0}(\omega)| < \epsilon$ uniformly in ω, so that $\mu(\{\omega: |f - f_n| > \epsilon\}) \to 0$ as $n \to \infty$.) But convergence in measure is the same as convergence in the metric of $L^0(\mu)$, by Proposition 3.2.6. Consequently (23) follows. Here f_n is elementary, and it can be chosen simple when $\mu(\Omega) < \infty$, since then a.e. convergence and a.u. convergence are equivalent and both imply convergence in measure. This implication is not true if $\mu(\Omega) = \infty$.

Next consider $p = +\infty$. If $f \in L^\infty(\mu)$, then it is essentially bounded. So there exists a μ-null set A such that $f(\Omega - A) \subset \mathbb{R}$ is bounded. Thus given $\epsilon > 0$, there exist a finite collection of disjoint Borel sets $B_1, \ldots, B_n$ of $\mathbb{R}$ such that each B_n has its diameter not exceeding ϵ. Let $A_i = f^{-1}(B_i)$, $i = 1, \ldots, n$. So $A_i \in \Sigma$. If $a_i \in B_i$, let $f_{n_\epsilon} = \sum_{i=1}^n a_i \chi_{A_i}$. Then f_{n_ϵ} has a finite number of steps [$\mu(A_i) = +\infty$ is possible again], and

$$\left|f(\omega) - f_{n_\epsilon}(\omega)\right| < \epsilon, \qquad \omega \in \Omega - A,$$

by construction. But this is (23) in a different notation, and hence the proof is complete.

Since the metric $\|\cdot\|_p$, $1 \leqslant p \leqslant \infty$, is actually a norm, and since the form of this functional in (5) is determined by $p \geqslant 1$, a great many specialized results are possible for these L^p-spaces. They utilize the integration theory fully and are employed to discover new results in integration.

Note that for each $f \in L^p(\mu)$, and $A_n \in \Sigma$, $A_n \downarrow \varnothing$, we have by the dominated convergence theorem

$$\|f\chi_{A_n}\|_p^p = \int_\Omega |f|^p \chi_{A_n}\, d\mu \to 0 \quad \text{as } n \to \infty, \qquad 0 < p < \infty. \tag{24}$$

This is called the *absolute continuity* property of the metric; it is false if $p = 0$ or $+\infty$. Specializing the abstract measure space to be topological, we can find stronger properties of $\|\cdot\|_p$. The next result is an analog of Theorem 7 for Radon measures, and presents another property.

Theorem 9. *Let* (Ω, Σ, μ) *be a Radon measure space, and* $L^p(\mu)$ *be the Lebesgue space on it. Then for* $0 < p < \infty$, *continuous functions with compact supports are dense in* $L^p(\mu)$. *In particular, if* $\Omega = \mathbb{R}^n$, $\Sigma = \mathscr{M}_\lambda$, *with* $\lambda =$ *Lebesgue measure, then* $\lim_{h \to 0}\|\tau_h f - f\|_p = 0$, $f \in L^p(\lambda)$, *where* τ_h *is the translation operator, i.e.,* $(\tau_h f)(x) = f(x + h)$, $x \in \mathbb{R}^n$, *for each fixed* $h \in \mathbb{R}^n$.

Proof. First note that if φ is a continuous function on Ω with compact support $K \subset \Omega$, then φ is bounded $[\|\varphi\| = \sup_{\omega \in \Omega}|\varphi(\omega)| < \infty]$ and for $0 < p < \infty$ [recall that $\mu(K) < \infty$, since μ is a Radon measure]

$$\int_\Omega |\varphi|^p \, d\mu = \int_K |\varphi|^p \, d\mu \leqslant \|\varphi\|^p \mu(K) < \infty.$$

Hence $\varphi \in L^p(\mu)$. If now $f \in L^p(\mu)$ is any element, then by Theorem 7, for each $\epsilon > 0$ there exists a simple function $h_\epsilon \in L^p(\mu)$ such that $\|f - h_\epsilon\|_p < \epsilon/2$ and h_ϵ has a support Ω_0 such that $\mu(\Omega_0) < \infty$. Being a simple function, h_ϵ is bounded, so that $|h_\epsilon| \leqslant \alpha < \infty$. Hence by Corollary 3.2.10, there exists a compact set $C_\epsilon \subset \Omega$ such that $h_\epsilon | C_\epsilon$ is continuous and $\mu(\Omega_0 - C_\epsilon) < (\epsilon/2\alpha)^p$. Let $\varphi = h_\epsilon | C_\epsilon$. Then φ is continuous with compact support, and if $p \geqslant 1$,

$$\|h_\epsilon - \varphi\|_p = \|h_\epsilon \chi_{\Omega_0 - C_\epsilon}\|_p \leqslant \alpha \cdot \frac{\epsilon}{2\alpha} = \frac{\epsilon}{2}. \tag{25}$$

If $0 < p < 1$, then we replace $(\epsilon/2\alpha)^p$ by $\epsilon/2\alpha^p$ and get the same estimate on the distance in (25). Hence

$$\|f - \varphi\|_p \leqslant \|f - h_\epsilon\|_p + \|h_\epsilon - \varphi\|_p < \frac{\epsilon}{2} + \frac{\epsilon}{2} = \epsilon. \tag{26}$$

Thus continuous functions with compact supports are dense in $L^p(\mu)$.

Let (Ω, Σ, μ) be specialized to the Lebesgue space $(\mathbb{R}^n, \mathcal{M}_\lambda, \lambda)$. If $f \in L^p(\lambda)$, then $\tau_h f \in L^p(\lambda)$ and $\|f\|_p = \|\tau_h f\|_p$. To see this we use the translation invariance of the Lebesgue measure λ [Theorem 2.2.17(v)]. Thus

$$\int_\Omega |f|^p \, d\lambda = \int_\Omega |\tau_h f|^p \, d\lambda, \qquad f = \chi_A, \tag{27}$$

holds if $A \in \mathcal{M}_\lambda$, $\lambda(A) < \infty$. If $f = \sum_{i=1}^n a_i \chi_{A_i}$ [A_i disjoint, $A_i \in \mathcal{M}_\lambda$, $\lambda(A_i) < \infty$], then $|f|^p = \sum_{i=1}^n |a_i|^p \chi_{A_i}$, so that by linearity of the integral, (27) holds if f is a simple function. The general case follows by the monotone convergence theorem, since by the structure of measurable functions, there exist simple functions $\psi_n \to f$ such that $|\psi_n| \uparrow |f|$. Thus (27) holds for all $f \in L^p(\lambda)$.

To prove the last part, if $f \in L^p(\mu)$ is arbitrary, then there exists a continuous φ with compact support such that (26) holds. Hence φ is uniformly continuous. So, with $\mu = \lambda$, for the given $\epsilon > 0$, there exists a $\delta_\epsilon > 0$ such that $|h| < \delta_\epsilon \Rightarrow |\varphi - \tau_h \varphi| < \epsilon/[\lambda(K)]^{1/p}$, $p \geqslant 1$, and $K = \operatorname{spt} \varphi$. Thus $\|\varphi - \tau_h \varphi\|_p < \epsilon$. (If $0 < p < 1$, we replace $\epsilon/[\lambda(K)]^{1/p}$ by $[\epsilon/\lambda(K)]^{1/p}$ to get the same estimate.) Hence

$$\begin{aligned} \|f - \tau_h f\|_p &\leqslant \|f - \varphi\|_p + \|\varphi - \tau_h \varphi\|_p + \|\tau_h \varphi - \tau_h f\|_p \\ &< \epsilon + \epsilon + \epsilon = 3\epsilon, \qquad \text{by (26) and (27).} \end{aligned}$$

Since $\epsilon > 0$ is arbitrary, we get $\lim_{|h| \to 0} \|f - \tau_h f\|_p = 0$. Note that here $|h|$ is the Euclidean length of a vector in $\mathbb{R}^n$. This completes the proof of the theorem.

Specializing the measure space (Ω, Σ, μ), we can recover the familiar spaces of real analysis. Let $\Omega = N$, the natural numbers, $\Sigma =$ the power set, and $\mu =$ the counting measure. Then every $f: \Omega \to \mathbb{R}$ is measurable and is just a sequence, $f = (f_1, f_2, \dots)$. The spaces $L^p(\mu)$ in this case are denoted by l^p. [If $\mu(\{i\}) = a_i > 0$, then we denote it by $l^p(a)$, where $a = (a_1, a_2, \dots)$ are weights, and hence it is l^p when $a_i = 1$, all i.] In case $\Omega = \{1, 2, \dots, n\}$ and μ is the counting measure as above, then the space l^p consists of the same vectors $f = (f_1, \dots, f_n)$ for all $p \geqslant 0$ (since no convergence intervenes here), and is denoted l_n^p. If $p = 2$, with metric $\|f\|_2 = \sqrt{f_1^2 + \cdots + f_n^2}$, then l_n^2 is simply $\mathbb{R}^n$ with its inner product metric. However, in the finite dimensional case all topologies are equivalent in l_n^p, in the sense that there exist positive constants $a_p > 0$, $b_p > 0$ such that

$$a_p \|f\|_p \leqslant \|f\|_2 \leqslant b_p \|f\|_p, \qquad f \in l_n^p.$$

So we only consider the infinite dimensional spaces l^p hereafter.

The simpler space $L^2(\mu)$ with $p = 2$ has many special properties, and we call it the *Hilbert space*. It is a complete normed linear space whose norm is derived from the *inner product*

$$(f, g) = \int_\Omega fg \, d\mu,$$

or (overbar denoting complex conjugate)

$$(f, g) = \int_\Omega f\bar{g} \, d\mu$$

in the complex case, where by definition $\int_\Omega f \, d\mu = \int_\Omega \operatorname{Re} f \, d\mu + i \int_\Omega \operatorname{Im} f \, d\mu$ (Re = real, Im = imaginary part)—cf. Exercise 1.9—though mostly we treat only real functions. Thus

(i) $(f_1 + f_2, g) = (f_1, g) + (f_2, g)$,
(ii) $(af, g) = a(f, g)$, $(f, ag) = \bar{a}(f, g)$, $a \in \mathbb{C}$,
(iii) $(f, g) = \overline{(g, f)}$ and $(f, f) \geqslant 0$, $= 0$ iff $f = 0$ a.e.

If we let $\|f\|_2 = \sqrt{(f, f)}$, then from our Definition 2 for $\|\cdot\|_2$, this is the same thing as (5). Here (i) and (ii) together are called the *sesquilinearity* of $(\cdot, \cdot)$ and (iii) is its *Hermitian positive definiteness*. Thus $(L^2(\mu), (\cdot, \cdot))$ is a complete inner product (or Hilbert) space. Note that the second half of (ii) follows from its first half and (iii).

The space l^2 was first studied in detail by D. Hilbert at the turn of the century and analyzed before the spaces $L^p(\mu)$, $p \neq 2$. The corresponding inequality (12) with $p = 2$ $(= q)$ is called the *Cauchy-Buniakowski-Schwarz* (or CBS) inequality. Thus one has

Corollary 10 (CBS inequality). *Let $f, g \in L^2(\mu)$. Then fg is integrable and*

$$\int_\Omega |f\bar{g}|\, d\mu \leqslant \left(\int_\Omega |f|^2\, d\mu \cdot \int_\Omega |g|^2\, d\mu \right)^{1/2}. \tag{28}$$

Proof. Since this result is important and it historically predates (12), we present a short independent proof. By the linearity of $L^2(\mu)$, $tf + g \in L^2(\mu)$ for any t real or complex. Then using the inner product notation,

$$\begin{aligned} 0 &\leqslant (tf + g, tf + g) \\ &= |t|^2(f, f) + t(f, g) + \bar{t}(g, f) + (g, g) \\ &= |t|^2(f, f) + (g, g) + 2\,\mathrm{Re}(t(f, g)). \end{aligned} \tag{29}$$

Since t is arbitrary here, we may choose any value. Because (28) is clear if the left side is zero, suppose $(f, g) \neq 0$. Let us take $t = -(g, g)/(f, g)$. Then (29) becomes

$$0 \leqslant \frac{(f, f)(g, g)^2}{|(f, g)|^2} + (g, g) - 2(g, g).$$

Thus

$$|(f, g)|^2 \leqslant (f, f)(g, g). \tag{30}$$

Since this is true for all f, g in $L^2(\mu)$, it is also true if we consider $|f|, |g|$ which belong to $L^2(\mu)$, as indicated in Exercise 1.9. But then (30) implies (28), and the result follows.

It is interesting to point out that (28) was proved by A.-L. Cauchy in 1830 if Ω is a finite set with μ as the counting measure (in the current terminology), and it was extended by H. A. Schwarz in 1890 to the case that $\Omega = \mathbb{R}^n$ and the integral is Riemann's, without knowing that V. Buniakowski already had got (28) for the Riemann-Stieltjes integral in 1859. This is why it is called the CBS inequality.

The Hilbert space being an infinite dimensional version of the familiar Euclidean n-dimensional space, it extends many known finite dimensional facts. For instance $f, g \in L^2(\mu)$, $(f, g) = 0, \Rightarrow (g, f) = 0$ also, and f is said to be *orthogonal* to g, denoted $f \perp g$; and they are *orthonormal* if $f \perp g$

and $\|f\|_2 = 1 = \|g\|_2$. The *parallelogram law* holds: if $f_1, f_2 \in L^2(\mu)$ then

$$\|f_1 + f_2\|_2^2 + \|f_1 - f_2\|_2^2 = 2\|f_1\|_2^2 + 2\|f_2\|_2^2. \tag{31}$$

In fact, expanding (31) on the left and using inner products, we get

$$\begin{aligned}(f_1 + f_2, f_1 + f_2) &+ (f_1 - f_2, f_1 - f_2)\\ &= 2(f_1, f_1) + 2(f_2, f_2) + (f_1, f_2) + (f_2, f_1) - (f_1, f_2) - (f_2, f_1)\\ &= 2(f_1, f_1) + 2(f_2, f_2),\end{aligned}$$

which is the right side of (31). Another property is given by:

Proposition 11. *Let* $\{\varphi_n : n \geqslant 1\} \subset L^2(\mu)$ *be an orthonormal sequence and* $f \in L^2(\mu)$. *Then letting* $a_n = (f, \varphi_n)$, *we have* the ***Bessel inequality***:

$$\sum_{n=1}^{\infty} |a_n|^2 \leqslant (f, f). \tag{32}$$

Proof. Consider $f - \sum_{k=1}^{n} a_k\varphi_k \in L^2(\mu)$. Then

$$\begin{aligned}0 &\leqslant \left(f - \sum_{k=1}^{n} a_k\varphi_k, f - \sum_{k=1}^{n} a_k\varphi_k\right)\\ &= (f, f) - \sum_{k=1}^{n} (f, a_k\varphi_k) - \sum_{k=1}^{n} (a_k\varphi_k, f)\\ &\quad + \sum_{k=1}^{n}\sum_{j=1}^{n} (a_k\varphi_k, a_j\varphi_j)\end{aligned}$$

[by using the sesquilinearity of $(\cdot, \cdot)$ etc.]

$$\begin{aligned}&= (f, f) - \sum_{k=1}^{n} \bar{a}_k a_k - \sum_{k=1}^{n} a_k\bar{a}_k + \sum_{k=1}^{n}\sum_{j=1}^{n} a_k\bar{a}_j(\varphi_k, \varphi_j)\\ &= (f, f) - 2\sum_{k=1}^{n} |a_k|^2 + \sum_{k=1}^{n} |a_k|^2,\end{aligned}$$

since $\varphi_k \perp \varphi_j$ if $i \neq j$ and $\|\varphi_k\|_2 = 1$, all k. This implies

$$\sum_{k=1}^{n} |a_k|^2 \leqslant \|f\|_2^2 < \infty.$$

Since n is arbitrary and the right side is a fixed number, we get (32), as asserted. In particular $a_n \to 0$. Orthonormal sequences always exist. (See Exercises 14 and 15). Also observe that the above proof is the same in any dimension.)

Example. Let $\Omega = (0, 2\pi]$; $d\mu = (1/2\pi)\,dx$, the normalized Lebesgue measure; and $f = \chi_A$, where A is a Borel set in Ω. If $\varphi_n(x) = \sqrt{2}\sin nx$, $n \geqslant 1$, then elementary integration shows that $\{\varphi_n: n \geqslant 1\}$ is an orthonormal sequence in $L^2(\mu)$, and if $a_n = (\sqrt{2}/2\pi)\int_0^{2\pi}\chi_A \sin nx\,dx = (\sqrt{2}/2\pi)\int_A \sin nx\,dx$, then $a_n \to 0$ as $n \to \infty$, by the above result, for any Borel set A. In fact this is true for any $f \in L^2(\mu)$.

Motivated by this example, we introduce two convergence concepts in the spaces $L^p(\mu)$, $p \geqslant 1$.

Definition 12. Let $\{f, f_n: n \geqslant 1\} \subset L^p(\mu)$ be a sequence. Then f_n is said to *converge weakly* to f if for each $g \in L^q(\mu)$,

$$\lim_n \int_\Omega (f_n - f)g\,d\mu = 0, \tag{33}$$

where $p \geqslant 1$ and $q = p/(p-1)$ (with $q = +\infty$ when $p = 1$). The sequence is said to converge in *p-mean* or *strongly* if $\|f_n - f\|_p \to 0$ as $n \to \infty$.

The relation between these two concepts may be seen from the following

Proposition 13. *Let (Ω, Σ, μ) be a measure space, and $\{f, f_n: n \geqslant 1\}$ be a sequence in $L^p(\mu)$, $1 \leqslant p < \infty$. If $f_n \to f$ strongly, then it converges weakly to the same limit, and also $f_n \xrightarrow{\mu} f$. But if $f_n \to f$ weakly, then $f_n \nrightarrow f$ strongly or in measure, in general.*

Proof. Let $f_n \to f$ strongly. If $g \in L^q(\mu)$, $q = p/(p-1)$, then

$$\left|\int_\Omega (f_n - f)g\,d\mu\right| \leqslant \|f_n - f\|_p \|g\|_q, \qquad \text{by (12)}$$
$$\to 0 \quad \text{as } n \to \infty, \qquad \text{by hypothesis.}$$

Thus $f_n \to f$ weakly. Also, if $\epsilon > 0$,

$$\mu(\{\omega: |f_n - f|(\omega) > \epsilon\}) = \mu(\{\omega: |f_n - f|^p(\omega) > \epsilon^p)$$
$$\leqslant \frac{1}{\epsilon^p}\int_\Omega |f_n - f|^p\,d\mu, \text{ (Markov's inequality)}$$
$$= \frac{1}{\epsilon^p}\|f_n - f\|_p^p \to 0$$

for $1 \leqslant p < \infty$, by hypothesis. So $f_n \xrightarrow{\mu} f$.

For the negative statements, let $\Omega = (0, 2\pi]$, $\mu = (1/2\pi)\lambda$, the normalized Lebesgue measure, and $f = \chi_A$, $A \subset \Omega$ any Borel set. If $f_n = \sqrt{2}\sin nx$, then for $p = 2$, $f_n \to 0$ weakly, since by the above example

$$\frac{1}{2\pi}\int_A f_n\, d\lambda = \int_\Omega \chi_A \sin nx \frac{dx}{\sqrt{2}\pi} \to 0 \quad \text{as} \quad n \to \infty$$

for all Borel sets. Hence $g = \sum_{i=1}^m a_i \chi_{A_i} \in L^2(\mu)$, and

$$\frac{1}{2\pi}\int_\Omega f_n g\, d\lambda = \sum_{i=1}^m a_i \int_\Omega f_n \chi_{A_i}\, d\lambda \to 0 \quad \text{as} \quad n \to \infty. \tag{34}$$

If $g \in L^2(\mu)$ is arbitrary, then for each $\epsilon > 0$ there is a simple function g_ϵ such that $\|g - g_\epsilon\|_2 < \epsilon$. Hence

$$\begin{aligned}\left|\int_\Omega f_n g\, d\mu\right| &\leqslant \left|\int_\Omega f_n (g - g_\epsilon)\, d\mu\right| + \left|\int_\Omega f_n g_\epsilon\, d\mu\right| \\ &\leqslant \|f_n\|_2 \|g - g_\epsilon\|_2 + \left|\int_\Omega f_n g_\epsilon\, d\mu\right|.\end{aligned}$$

Thus

$$\limsup_n \left|\int_\Omega f_n g\, d\mu\right| \leqslant \epsilon, \qquad \text{by (34)}.$$

Since $\epsilon > 0$ is arbitrary, this shows $\lim_n \int_\Omega f_n g\, d\mu = 0$. However, $\|f_n\|_2 = 1 \nrightarrow 0$, and so $f_n \nrightarrow 0$ strongly. The same example shows that $f_n \nrightarrow 0$ in measure, since otherwise there will exist a subsequence such that $f_{n_k} \to 0$ a.e. Clearly the periodic functions $x \mapsto \sqrt{2}\sin n_k x$ do not converge for any $0 < x < 2\pi$. This completes the proof.

There is however a nontrivial case where the strong and weak convergences coincide. That is for the space l^1. We deduce this, a little more generally, from Theorem 15 below. The next one has some independent interest, and its ideas of proof are used in the following result. It is due to F. Riesz.

Theorem 14. *Let $\{f, f_n : n \geqslant 1\} \subset L^p(\mu)$ be a sequence such that $f_n \to f$, a.e. or only in μ-measure. Then $\|f_n\|_p \to \|f\|_p$ iff $f_n \to f$ strongly in $L^p(\mu)$, $1 \leqslant p < \infty$. On the other hand, if $\{\|f_n\|_p : n \geqslant 1\}$ is only bounded, but $1 < p < \infty$, then $f_n \to f$ weakly, and this result fails for $p = 1$.*

Proof. If $\|f_n - f\|_p \to 0$ as $n \to \infty$, then we have by the triangle inequality

$$\big|\|f_n\|_p - \|f\|_p\big| \leqslant \|f_n - f\|_p \to 0.$$

Thus only the converse is nontrivial.

Suppose $\|f_n\|_p \to \|f\|_p$, so that $\|f_n\|_p^p \to \|f\|_p^p$ also. Then

$$\int_\Omega |f_n|^p \, d\mu \to \int_\Omega |f|^p \, d\mu.$$

Since $\{|f_n|^p: n \geqslant 1\}$ is a nonnegative integrable sequence which converges a.e. to $|f|^p$, by Scheffé's lemma (cf. Exercise 3.14) the set is uniformly integrable and $\int_\Omega ||f_n|^p - |f|^p| \, d\mu \to 0$. By the necessity part of Theorem 3.11, we have

$$\lim_{\mu(A)\to 0} \int_A |f_n|^p \, d\mu = 0 \qquad \text{uniformly in} \quad n \geqslant 1, \tag{35}$$

and given $\epsilon > 0$, there exists an $A_\epsilon \in \Sigma$, $\mu(A_\epsilon) < \infty$, such that

$$\int_{A_\epsilon^c} |f_n|^p \, d\mu < \epsilon, \qquad n \geqslant 1. \tag{36}$$

We now *assert* that (35) and (36) together imply

$$\lim_n \int_{A_\epsilon} |f_n - f|^p \, d\mu = 0. \tag{37}$$

Suppose (37) is true. Then we can complete the strong convergence assertion quickly as follows. We first show that $\{f_n: n \geqslant 1\}$ is Cauchy in $L^p(\mu)$. In fact,

$$\begin{aligned}
\limsup_{m,n} \|f_n - f_m\|_p &= \limsup_{m,n} \left(\int_{A_\epsilon} |f_n - f_m|^p \, d\mu + \int_{A_\epsilon^c} |f_m - f_n|^p \, d\mu \right)^{1/p} \\
&\leqslant \limsup_{m,n} \left(\int_{A_\epsilon} |f_n - f_m|^p \, d\mu \right)^{1/p} + 2\epsilon^{1/p}, \qquad \text{by (36)} \\
&\leqslant \limsup_{m,n} \left[\|(f_n - f)\chi_{A_\epsilon}\|_p + \|(f_m - f)\chi_{A_\epsilon}\|_p \right] + 2\epsilon^{1/p} \\
&\leqslant 2\epsilon^{1/p}, \qquad \text{by (37)}.
\end{aligned}$$

Since $\epsilon > 0$ is arbitrary, $\{f_n: n \geqslant 1\}$ is Cauchy in $L^p(\mu)$. By the completeness of this space (cf. Theorem 7), $f_n \to f'$ strongly, so that $f_n \xrightarrow{\mu} f'$, and for a subsequence $f_{n_k} \to f'$ a.e. But by hypothesis $f_n \to f$ a.e. This implies $f = f'$ a.e., and we have $\|f_n - f\|_p \to 0$.

It remains to prove (37). The proof is essentially identical with the sufficiency of Vitali's theorem, but we sketch the argument for completeness. Indeed, letting $f_\infty = f$, then (36) and (37) hold for f_n, $1 \leqslant n \leqslant \infty$, uniformly in n. Also

$\mu(A_\epsilon) < \infty$ and $f_n \to f_\infty$ a.e. on A_ϵ, and hence if $A_{mn} = \{\omega: |f_n - f_m|(\omega) \geqslant \epsilon\}$ then $\mu(A_{mn}) \to 0$ as $m, n \to \infty$. So for any $\epsilon' > 0$ there is a $\delta_{\epsilon'} > 0$ and n_0 [$= n_0(\epsilon')$] such that $m, n \geqslant n_0 \Rightarrow \mu(A_{mn}) < \delta_{\epsilon'}$, so that (36) holds for this A_{mn} and ϵ', for all f_k, $1 \leqslant k \leqslant \infty$, uniformly. Thus

$$\begin{aligned}\left(\int_{A_\epsilon} |f_n - f_m|^p \, d\mu\right)^{1/p} &\leqslant \left(\int_{A_\epsilon - A_{mn}} |f_n - f_m|^p \, d\mu\right)^{1/p} + \left(\int_{A_{mn}} |f_n - f_m|^p \, d\mu\right)^{1/p} \\ &\leqslant [\epsilon' \mu(A_\epsilon)]^{1/p} \\ &\quad + \left(\int_{A_{mn}} |f_n|^p \, d\mu\right)^{1/p} + \left(\int_{A_{mn}} |f_m|^p \, d\mu\right)^{1/p} \\ &\leqslant \left\{[\mu(A_\epsilon)]^{1/p} + 2\right\}(\epsilon')^{1/p}.\end{aligned}$$

Since $\epsilon' > 0$ is arbitrary and $1 \leqslant p < \infty$, this proves (37), and hence the first part of the theorem follows. Note that the alternate hypothesis $f_n \xrightarrow{\mu} f$ is a consequence of this. Indeed, if the result is not true, then by Theorem 3.2.5 (Riesz theorem) $f_{n_k} \to f$ a.e. for a subsequence, and the result must be false for this too. But this contradicts the preceding proof. (For an alternate proof and a generalization of this result, see Exercise 7.)

Regarding the second part, it suffices to consider a.e. convergence again. Then $|f_n|^p \to |f|^p$ a.e., and since $\alpha = \sup_n \|f_n\|_p < \infty$ by hypothesis, we conclude by Fatou's lemma that $\|f\|_p \leqslant \alpha < \infty$. Next use the fact that $1 < p < \infty$, so that $q = p/(p-1)$ also satisfies $1 < q < \infty$, and hence for any $g \in L^q(\mu)$, $\lim_{\mu(A) \to 0} \|g\chi_A\|_q = 0$ by Proposition 3.8. (If $p = 1$, then $q = +\infty$; this result on the absolute continuity of the norm fails, and the subsequent argument breaks down at this point. Since $\|\chi_A\|_\infty = 1$ all $A \in \Sigma$, $\lim_{\mu(A) \to 0} \|\chi_A\|_\infty = 1$ also.) Now if $A_n = \{\omega: 1/n \leqslant |g(\omega)|^q \leqslant n\}$, then $\mu(A_n) < \infty$, $A_n \uparrow \tilde{A} = \{\omega: |g(\omega)|^q > 0\}$, and hence $\|g\chi_{A_n^c}\|_q \to 0$. So given $\epsilon > 0$, there is an n_0 [$= n_0(\epsilon)$] such that $n \geqslant n_0 \Rightarrow \|g\chi_{A_n^c}\|_q < \epsilon$ and $\mu(A_n) < \infty$. But $f_k \to f$ a.e. on A_n. Then by Egorov's theorem, for each $\delta > 0$ there is a $B \subset A_n$ such that $\mu(A_n - B) < \delta$ and $f_k \to f$ uniformly on B. Thus for the above fixed but arbitrary $g \in L^q(\mu)$, we have

$$\begin{aligned}\left|\int_\Omega (f_k - f) g \, d\mu\right| &\leqslant \int_{A_n} |f_k - f| \, |g| \, d\mu + \int_{A_n^c} |f_k - f| \, |g| \, d\mu \\ &\leqslant \int_B |f_k - f| \, |g| \, d\mu + \int_{A_n - B} |f_k - f| \, |g| \, d\mu \\ &\quad + \|f_k - f\|_p \|g\chi_{A_n^c}\|_q \\ &\leqslant \int_B |f_k - f| \, |g| \, d\mu + \|f_k - f\|_p \left(\|g\chi_{A_n - B}\|_q + \|g\chi_{A_n^c}\|_q\right) \\ &< \int_B |f_k - f| \, |g| \, d\mu + 2\alpha(\epsilon + \epsilon) \qquad \text{if} \quad n \geqslant n_0.\end{aligned}$$

Hence letting $k \to \infty$ and using Corollary 3.13, the integral on the right above goes to zero, since $\mu(B) < \infty$. But because $\epsilon > 0$ is arbitrary, we may infer that $\int_\Omega f_n g\, d\mu \to \int_\Omega fg\, d\mu$, $g \in L^q(\mu)$. Thus $f_n \to f$ weakly.

Finally consider $\Omega = [0, 1]$, μ = Lebesgue measure, $\Sigma = \mathscr{M}_\mu$, and $f_n = n\chi_{[0, 1/n]}$. Then $\|f_n\|_1 = 1$, $f_n \to 0$ a.e., but $\int_A f_n\, d\mu \to 1$ for each closed interval A such that $0 \in A$, and the limit is zero if $0 \notin A$. So $f_n \nrightarrow 0$ weakly. Thus the result fails for $p = 1$. (For a positive result with a strengthened hypothesis, see Exercise 13.) This completes the proof.

As a companion to the above result we now establish the previously announced equivalence of weak and strong convergences for l^1. This exhibits the peculiarities of the spaces $L^1(\mu)$ and l^1. (Some other differences are indicated in exercises.)

Theorem 15. *Let (Ω, Σ, μ) be a measure space such that each $\omega \in \Omega$ satisfies $\{\omega\} \in \Sigma$, $\mu(\{\omega\}) > 0$. Let $\{f, f_n: n \geqslant 1\} \subset L^1(\mu)$. Then $f_n \to f$ weakly iff $f_n \to f$ strongly. And this is generally false in $L^p(\mu)$, $p > 1$.*

Proof. Let $A_n = \{\omega: |f_n(\omega)| > 0\}$. Since $\mu(\{\omega\}) > 0$ and each A_n is σ-finite (so μ has the finite subset property), we must have $0 < \mu(\{\omega\}) < \infty$. Hence A_n is at most countable. Similarly $A_0 = \{\omega: |f(\omega)| > 0\}$ is countable. Here we have used the fact that $\{f, f_n: n \geqslant 1\} \subset L^1(\mu)$. So $\Omega_0 = \bigcup_{k=1}^{\infty} A_k$ is countable, and for this proof we may and do replace Ω by Ω_0. With this each $f_n = (f_{n1}, f_{n2}, \ldots)$ is a sequence. Let $f = (f_{01}, f_{02}, \ldots)$, and g be any bounded function on Ω_0. By hypothesis $f_n \to f$ weakly. So

$$\int_{\Omega_0} f_n g\, d\mu \to \int_{\Omega_0} fg\, d\mu, \qquad \text{i.e.,} \quad \sum_{i=1}^{\infty} f_{ni} g_i \to \sum_{i=1}^{\infty} f_{0i} g_i, \tag{38}$$

where $g = (g_1, g_2, \ldots) \in L^\infty(\mu)$. In particular, taking $g = (0, 0, \ldots, 1, 0, 0, \ldots)$ with 1 in the ith place and 0's elsewhere, (38) gives $f_{ni} \to f_{0i}$, $i \geqslant 1$, so that $f_n(\omega) \to f(\omega)$, $\omega \in \Omega_0$.

It is necessary to show next that $\|f_n\|_1 \to \|f\|_1$ to apply Theorem 14. The preceding conclusion only yields $\sum_{i=1}^m |f_{ni}| \to \sum_{i=1}^m |f_{0i}|$ for each $1 \leqslant m < \infty$, as $n \to \infty$. Thus we need some uniformity to let $m \to \infty$ and interchange the limits. This is not available (and is not possible in general). So we go back to (38) again. Taking $g = \chi_A$, $A \in \Sigma$, and letting $\nu_n(A) = \int_A f_n\, d\mu$, we see that $\nu_n: \Sigma \to \mathbb{R}$ is σ-additive and individually bounded, and that $\nu_n(A) \to \nu(A) = \int_A f\, d\mu$, by (38), for each $A \in \Sigma$. Hence,

(i) $\lim_{\mu(A) \to 0} \nu_n(A) = 0$ uniformly in $n \geqslant 1$, by the Vitali-Hahn-Saks theorem, and

(ii) by (the Nikodým) Corollary 4.7 $\lim_{k \to \infty} \nu_n(A_k) = 0$ uniformly in $n \geqslant 1$ for each $A_k \downarrow \varnothing$.

Now by (i), given $\epsilon > 0$, there is a $\delta_\epsilon > 0$ such that $\mu(A) < \delta_\epsilon \Rightarrow |\nu_n(A)| < \epsilon/2$, uniformly in $n \geqslant 1$. If $B = \{\omega: f_n(\omega) \geqslant 0\}$ and $\tilde{B} = A \cap B \subset A$, so that also $\mu(\tilde{B}) < \delta_\epsilon$, one has

$$\nu_n(\tilde{B}) = \left|\int_{\tilde{B}} f_n \, d\mu\right| = \int_{\tilde{B}} f_n^+ \, d\mu < \frac{\epsilon}{2},$$

and similarly if $B_1 = \{\omega: f_n(\omega) < 0\}$ and $\tilde{B}_1 = A \cap B_1 \subset A$, then

$$\nu_n(\tilde{B}_1) = \left|\int_{\tilde{B}} f_n^- \, d\mu\right| < \frac{\epsilon}{2}.$$

Adding these two gives, for each $A \in \Sigma$, $\mu(A) < \delta_\epsilon$,

$$\int_A |f_n| \, d\mu < \epsilon \qquad \text{uniformly in} \quad n \geqslant 1. \tag{39}$$

Next, to use (ii), note that Ω_0 is σ-finite, so that there exists $C_n \in \Sigma$ such that $\mu(C_n) < \infty$ and $C_n \uparrow \Omega_0$. If $A_k = \Omega_0 - C_k$, then $A_k \downarrow \varnothing$, and by (ii), for each $\epsilon > 0$ there exists a k_0 $[= k_0(\epsilon)]$ such that $k \geqslant k_0 \Rightarrow |\nu_n(A_k)| < \epsilon/2$ uniformly in $n \geqslant 1$. Thus $|\int_{A_k} f_n \, d\mu| < \epsilon/2$. We claim that this implies $\int_{A_k} |f_n| \, d\mu < \epsilon$, $n \geqslant 1$. If not, there exists a subsequence f_{n_k}, $k \geqslant 1$, from our sequence, such that

$$\int_{A_k} f_{n_k}^+ \, d\mu + \int_{A_k} f_{n_k}^- \, d\mu = \int_{A_k} |f_{n_k}| \, d\mu \geqslant \epsilon, \qquad k \geqslant k_0.$$

Hence one of these integrals is at least $\epsilon/2$. Suppose it is the first one. Let $B_k = A_k \cap \{\omega: f_{n_k}^+(\omega) > 0\}$. Then $B_k \subset A_k$, $B_k \in \Sigma$, and since $f_{n_k}^- = 0$ on B_k, we have

$$\left|\int_{B_k} f_{n_k} \, d\mu\right| \geqslant \frac{\epsilon}{2}. \tag{40}$$

But $f_{n_k} \to f$ weakly, so the ν_{n_k} are uniformly σ-additive. Since $B_k \subset A_k \downarrow \varnothing$, we must have

$$\lim_k \left|\int_{B_k} f_{n_k} \, d\mu\right| = 0, \tag{41}$$

and this is a direct contradiction to (40). Hence our supposition is false, and $\int_{A_k} |f_n| \, d\mu < \epsilon$ for $k \geqslant k_0$, uniformly in n. Since $\mu(A_k^c) < \infty$, this and (39), together with the first paragraph showing that $f_n \to f$ pointwise, satisfy the hypothesis of Vitali's theorem (Theorem 3.11). Hence $\|f_n - f\|_1 \to 0$, and also $\|f_n\|_1 \to \|f\|_1$. Thus $f_n \to f$ strongly, as asserted.

If $p > 1$, consider, for a counterexample, l^p. Let $f_n = (0, 0, \ldots 1, 0, \ldots)$ with 1 at the nth place and 0's elsewhere. Then for each $g \in l^q$, $1 < q = p/(p-1) < \infty$, we have $(f_n, g) = \sum_{i=1}^{\infty} f_{n_i} g_i = g_n \to 0$ as $n \to \infty$, since $\sum_{i=1}^{\infty} |g_i|^q < \infty$. Thus $f_n \to 0$ weakly. But $\|f_n\| = 1$, $n \geqslant 1$, and $f_n \nrightarrow 0$ strongly. This proves the theorem.

An immediate consequence is the following

Corollary 16. *In l^1, the weak and strong convergences are the same for any sequence.*

In $L^p(\mu)$, the norm of an element can be calculated by the following method, which leads to generalizations of these spaces to some more inclusive Banach function spaces of interest in applications.

Proposition 17. *Let $L^p(\mu)$, $1 \leqslant p < \infty$, be the Lebesgue space, and $f \in L^p(\mu)$. Then we have, with $q = p/(p-1)$ as usual,*

$$\|f\|_p = \sup\left\{\left|\int_\Omega fg\, d\mu\right| : \|g\|_q \leqslant 1\right\}, \tag{42}$$

and the result also holds for $p = +\infty$ if μ has the finite subset property.

Proof. It is clear that the right side of (42) is never larger than the left side, because of the Hölder inequality. To prove the equality, we consider the cases $p = 1$, $1 < p < \infty$, and $p = \infty$ separately.

If $p = 1$, we produce a function g for which there is equality in (42). Indeed, consider $g = \operatorname{sgn} f$, where sgn is the signum function, i.e., $\operatorname{sgn} x = +1$ if $x > 0$, 0 if $x = 0$, and -1 if $x < 0$. Clearly $\operatorname{sgn}(\cdot)$ is a Borel function on $\mathbb{R}$, so that $g: \Omega \to \mathbb{R}$ is bounded and measurable. In fact $|g(\omega)| \leqslant 1$, and so $\|g\|_\infty \leqslant 1$. But $(fg)(\omega) = |f(\omega)|$, $\omega \in \Omega$, and $\int_\Omega fg\, d\mu = \int_\Omega |f|\, d\mu = \|f\|_1$. Thus for this particular g, there is equality in (42), and hence that the result is true for $p = 1$.

Next, let $1 < p < \infty$, so that $1 < q = p/(p-1) < \infty$. Define $g = |f|^{p-1} \operatorname{sgn} f$. Then again g is measurable, since, by Lemma 3.1.2, $|f|^{p-1}$ is measurable and $\operatorname{sgn} f$ is already seen to be measurable. Also $|g|^q = |f|^{(p-1)q} = |f|^p$, and so $g \in L^q(\mu)$. Define $\tilde{g} = g/\|f\|_p^{p-1}$ if $f \neq 0$, and $= 0$ if $f = 0$ a.e.; then $\|\tilde{g}\|_q = 1$. Hence we have

$$\left|\int_\Omega f\tilde{g}\, d\mu\right| = \int_\Omega \frac{|f|^p\, d\mu}{\|f\|_p^{p-1}} = \|f\|_p.$$

Thus there is again equality in (42) in this case. If $f = 0$ a.e., then (42) is trivial.

Finally, let $p = +\infty$; we may assume that $\alpha = \|f\|_\infty > 0$. Let $0 < a < \alpha$. Now by the finite subset property of μ, there exists an $A \in \Sigma$, $0 < \mu(A) < \infty$, and $A \subset \{\omega: |f(\omega)| > a\}$. Let $g = (\chi_A \operatorname{sgn} f)/\mu(A)$. Then $\|g\|_1 = 1$, and

$$\|f\|_\infty \geqslant \left|\int_\Omega fg\,d\mu\right| = \frac{1}{\mu(A)}\int_A |f|\,d\mu \geqslant a.$$

Letting $a \uparrow \alpha$, we get (42) in this case also. This terminates the proof.

The preceding work clearly shows how much of the integration theory is extended, using only the basic convergence theorems of the last two sections. It is natural to consider compactness criteria in the topological vector spaces $L^p(\mu)$, and based on this experience one may introduce more general spaces. All these are possible avenues, and indeed the results are being extended along these paths. But we shall only consider some important complements in exercises, and proceed to the other aspects of measure and integration theory in the following chapters.

EXERCISES

0. Let (Ω, Σ, μ) be a measure space and $L^p(\mu)$ be the corresponding Lebesgue space on it, where $0 \leqslant p \leqslant \infty$.

(a) Show that $L^1(\mu) \cap L^p(\mu)$ is dense in $L^p(\mu)$ for each $1 < p < \infty$.

(b) Show that $L^\infty(\mu) \cap L^p(\mu)$ is dense in $L^p(\mu)$ for $0 < p < \infty$.

(c) If (Ω, Σ, μ) is a finite measure space in the above, show that $L^\infty(\mu) \cdot L^p(\mu) = \{f \cdot g: f \in L^\infty(\mu), g \in L^p(\mu)\} = L^p(\mu)$, $0 \leqslant p < \infty$.

(d) If (Ω, Σ, μ) is the Lebesgue real line in the first statement, and $f_t: \mathbb{R} \to \mathbb{C}$ is defined as $f_t(x) = (e^{itx} - 1)/x$, show that $f_t \in L^2(\mu)$ for each $t \in \mathbb{R}$. (Integrals for complex f are defined in terms of the real and imaginary parts.) In fact, verify that $\int_{\mathbb{R}} f_t(x)\overline{f_s(x)}\,d\mu(x) = \min(|t|, |s|)$ if $ts \geqslant 0$, and is equal to 0 otherwise.

1. Complete the proof of Proposition 6 on the inequalities for $L^p(\mu)$, $0 < p < 1$.

2. Show that equality holds in the Hölder inequality (12) iff for each $1 < p < \infty$, $|f|^p = A|h|^q$ a.e., where $A > 0$ is a constant, and similarly there is equality in (13) iff $f = A'g$ a.e., $A' > 0$, is a constant. Discuss the cases $p = 1$ and $p = +\infty$ separately.

3. Let $f_i \in L^{p_i}(\mu)$, $i = 1, \ldots, n$, where (Ω, Σ, μ) is a fixed measure space. Suppose that there is an $r \geqslant 1$ such that $\sum_{i=1}^n p_i^{-1} = r^{-1}$, where $p_i \geqslant 1$. Let $g = f_1 f_2 \cdots f_n$. Show that $g \in L^r(\mu)$ and moreover we have the generalized Hölder inequality

$$\|g\|_r \leqslant \prod_{i=1}^n \|f_i\|_{p_i}.$$

[To use induction, consider the case $n = 2$. Let $\alpha = p_1(p_1 + p_2)^{-1}$, so that $0 \leqslant \alpha \leqslant 1$. Set $g = f_1 f_2$,

$$\|g\|_r = \left(\int_\Omega (|f_1|^{p_1})^{1-\alpha} (|f_2|^{p_2})^{\alpha}\, d\mu \right)^{1/r},$$

apply Hölder's inequality, and simplify.]

4. Consider the space $L^p(\mu)$, $1 \leqslant p < \infty$, on a measure space (Ω, Σ, μ), and let $L^p(\tilde{\mu})$ denote the corresponding class on the completed space $(\Omega, \tilde{\Sigma}, \tilde{\mu})$. Show that the $L^p(\mu)$, $L^p(\tilde{\mu})$ are identifiable in the sense that for each $f \in L^p(\mu)$ there is an $\tilde{f} \in L^p(\tilde{\mu})$ satisfying $f = \tilde{f}$ a.e., and $\|f\|_p = \|\tilde{f}\|_p$. [First note that this holds for simple functions.]

5. For each $0 < p < \infty$, let $L^p(\mu)$ be as in Definition 1. Show that the space is separable (i.e., has a dense denumerable subset) iff the measure μ is separable in the sense of Definition 2.4.2. [Consider the set of simple functions with rational coefficients, and show that the above condition is satisfied iff μ is separable for the δ-ring of sets of finite μ-measure.] Deduce that l^p, $0 < p < \infty$, is separable.

6. We have seen that $f_n \xrightarrow{\mu} f$, $g_n \xrightarrow{\mu} g$ does not always imply that $f_n g_n \xrightarrow{\mu} fg$. Show however that if $\{f, f_n : n \geqslant 1\} \subset L^p(\mu)$, $\{g, g_n : n \geqslant 1\} \subset L^q(\mu)$, where $p \geqslant 1$ and $q = p/(p-1)$ and where $f_n \to f$, $g_n \to g$ strongly, then $f_n g_n \xrightarrow{\mu} fg$; and in fact also in $L^1(\mu)$.

7. Theorem 14 admits the following extension. Let $\varphi : \mathbb{R}^+ \to \mathbb{R}^+$ be a continuous increasing concave function satisfying $\varphi(x) = 0$ iff $x = 0$ or a convex function satisfying $\varphi(2x) \leqslant C\varphi(x)$, $x \in \mathbb{R}^+$, $0 < C < \infty$, and $\varphi(x) = 0$ iff $x = 0$. For instance, let $\varphi(x) = x^p$. Then these conditions are satisfied if $0 < p < 1$ or if $p \geqslant 1$, respectively. Let $\{f, f_n : n \geqslant 1\}$ be a sequence of measurable functions on (Ω, Σ, μ) such that (i) $\{\varphi(|f_n|), \varphi(|f|) : n \geqslant 1\} \subset L^1(\mu)$, (ii) $f_n \xrightarrow{\mu} f$, and (iii) $\int_\Omega \varphi(|f_n|)\, d\mu \to \int_\Omega \varphi(|f|)\, d\mu$. Show that, under these conditions, $\int_\Omega \varphi(|f_n - f|)\, d\mu \to 0$ as $n \to \infty$, and that $\{\varphi(|f_n|) : n \geqslant 1\}$ is uniformly integrable. [First note that in both cases $\varphi(x + y) \leqslant \tilde{C}(\varphi(x) + \varphi(y))$, $x, y \in \mathbb{R}^+$, where $\tilde{C} = \max(1, C/2)$. Hence

$$\tilde{C}[\varphi(|f_n|) + \varphi(|f|)] - \varphi(|f_n - f|) \geqslant 0 \qquad \text{a.e.}$$

Integrating and using (iii), (ii) with a.e. convergence and Fatou's inequality, we get

$$2\tilde{C}\int_\Omega \varphi(|f|)\, d\mu \leqslant \lim_n \tilde{C}\int_\Omega \varphi(|f_n|)\, d\mu + \tilde{C}\int_\Omega \varphi(|f|)\, d\mu$$

$$- \limsup_n \int_\Omega \varphi(|f_n - f|)\, d\mu.$$

This gives the result with (ii) replaced by a.e. convergence. If (ii) is taken as given and if the conclusion is false, then it must be so for a subsequence and hence for an a.e. convergent subsequence, contradicting the above.]

8. Give an example showing that the formula (42) of Proposition 17 is no longer valid for $p = +\infty$ if μ does not have the finite subset property.

9. Establish the following generalization of the example given after Proposition 11. Let $\Omega = (0, 2\pi]$, $\mu = (1/2\pi) \times$ (Lebesgue measure), and $f \in L^p(\mu)$, $1 \leqslant p < \infty$. Then $a_n = (1/2\pi)\int_0^{2\pi} f(x) \sin nx \, dx$ is defined, and $a_n \to 0$ as $n \to \pm\infty$. [For $p \geqslant 2$ the result is as in the example, and if $1 \leqslant p < 2$, then for each $\epsilon > 0$ there is a simple function h_ϵ such that $\|f - h_\epsilon\|_p < \epsilon$ and $h_\epsilon \in L^2(\mu)$. Deduce the result from the $L^2(\mu)$ case. If $p = 1$, the most important one, this result is a form of the *Riemann-Lebesgue lemma*. Note that here $\sin nx$ can be replaced by $\cos nx$, and the conclusions remain unchanged.]

10. Another (and perhaps more common) form of the Riemann-Lebesgue lemma is the following. Let $\Omega = \mathbb{R}$, $\mu =$ Lebesgue measure on $\mathbb{R}$, and $\Sigma = \mathcal{M}_\mu$. If $f \in L^1(\mu)$, $\hat{f}(t) = \int_{\mathbb{R}} e^{itx} f(x)\, dx$, $t \in \mathbb{R}$, show that $\hat{f}$ is uniformly continuous on $\mathbb{R}$ and $\lim_{|t| \to \infty} \hat{f}(t) = 0$. [The continuity is a consequence of the dominated convergence theorem, and for the main conclusion, given an $\epsilon > 0$, there is a simple function h_ϵ (which therefore vanishes outside a bounded interval I) such that $\|f - h_\epsilon\|_1 < \epsilon$. Deduce that $\int_{\mathbb{R}} e^{itx} h_\epsilon(x)\, dx = \int_I h_\epsilon(x) \cos tx \, dx + i\int_I h_\epsilon(x) \sin tx \, dx \to 0$ as $|t| \to \infty$ by the result of Exercise 9. Note that if $f \in L^p(\mu)$, $p > 1$, then $\hat{f}(\cdot)$ is not defined as a Lebesgue integral, and $p = 1$ is natural for the case of infinite measures, in contrast to the case that $\mu(\Omega) < \infty$. It may be remarked that the same result holds if $\mathbb{R}$ is replaced by $\mathbb{R}^n$ and tx is interpreted as $(t, x) = \sum_{j=1}^n t_j x_j$ with $|t| = \sqrt{\sum_{j=1}^n t_j^2}$. This result is important in Fourier analysis. The above $\hat{f}$, called the *Fourier transform* of f, is defined for all $f \in L^1(\mathbb{R}, \mathcal{M}_\mu, \mu)$.]

11. Let (Ω, Σ, μ) be a measure space, and $f: \Omega \to \overline{\mathbb{R}}$ be an a.e. finite measurable function. Show that for each $1 \leqslant p < \infty$, $q = p/(p-1)$, we have

$$\|f\|_p = \sup\left\{\left|\int_\Omega fg \, d\mu\right| : \|g\|_q \leqslant 1\right\},$$

even if $\|f\|_p = +\infty$. [The case $\|f\|_p < \infty$ is Proposition 17, so consider only the infinite case.]

12. If $\{f, f_n : n \geqslant 1\} \subset L^1(\mu)$ is a sequence such that $f_n \to f$ weakly, show that by an example this does not necessarily imply that $|f_n| \to |f|$ weakly. [Consider the space of Exercise 9, and let $f_n(x) = e^{inx}$.] Show however that the result holds (even the strong convergence holds) if also $f_n \xrightarrow{\mu_A} F$, where $\mu_A = \mu|\Sigma(A)$ for each $A \in \Sigma$ with $\mu(A) < \infty$. [Note that we may assume that μ is σ-finite, and examine the proof of Theorem 15.]

13. We now present a positive solution to the excluded case of $p = 1$ in Theorem 14, on weak convergence, and also elaborate the last part of the above exercise. Let (Ω, Σ, μ) be a measure space, and $L^1(\mu)$ be the Lebesgue space on it. If $\{f_n: n \geq 1\} \subset L^1(\mu)$ is a sequence such that $\lim_{n\to\infty}\int_A f_n\,d\mu$ exists for each $A \in \Sigma$ of σ-finite measure, then show that (i) $\sup_n \int_\Omega |f_n|\,d\mu < \infty$, and (ii) $(f_n - f_m) \to 0$ weakly as $n, m \to \infty$, i.e. $\{f_n: n \geq 1\}$ is a weak Cauchy sequence. [It will be seen in Chapter 5 (with the Radon-Nikodým theorem) that this implies the existence of an $f \in L^1(\mu)$ such that $f_n \to f$ weakly.] If, moreover, $\{f_n: n \geq 1\}$ is Cauchy in measure on each set of finite μ-measure, show that it converges strongly to an f in $L^1(\mu)$. [First note that $f_n \in L^1(\mu)$ implies $A_n = \{\omega: f_n(\omega) \neq 0\}$ is σ-finite and that if $\Omega_0 = \bigcup_{n=1}^\infty A_n$, then Ω_0 is σ-finite and $\{f_n: n \geq 1\} \subset L^1(\Omega_0, \Sigma(\Omega_0), \mu)$, where μ is σ-finite on $\Sigma(\Omega_0)$. If $\nu_n(A) = \int_A f_n\,d\mu$, then $\nu_n(A) \to \nu(A)$, $A \in \Sigma(\Omega_0)$, and hence by Theorem 4.3 and Corollary 4.7, $\lim_{\mu(A)\to 0}\nu_n(A) = 0$ uniformly in $n \geq 1$, and for each $\epsilon > 0$ there is an $A_\epsilon \in \Sigma(\Omega_0)$ such that $\mu(A_\epsilon) < \infty$ and $\int_{\Omega_0 - A_\epsilon}|f_n|\,d\mu < \epsilon$, $n \geq 1$. (See the proof of Theorem 15.) Hence (i) follows if $\sup_n \int_{A_\epsilon}|f_n|\,d\mu < \infty$. But there is a $\delta_\epsilon > 0$ with $A \in \Sigma(A_\epsilon)$, $\mu(A) < \delta_\epsilon \Rightarrow \int_A |f_n|\,d\mu < \epsilon$, $n \geq 1$. So consider a measurable partition of A_ϵ into sets B_i and A_i with $\mu(A_i) \leq \delta_\epsilon$ and $\mu(B_i) > \delta_\epsilon$, all the sets being disjoint. Since $\mu(A_\epsilon) < \infty$, there can be only a *finite* collection of each of them, because they may be taken as the following specialized kind. There can be at most a countable number of μ-atoms in $\Sigma(A_\epsilon)$, where C is a *μ-atom* if $C \in \Sigma(A_\epsilon)$, $\mu(C) \neq 0$, and if $B \subset C$ and $B \in \Sigma(A_\epsilon)$, then $\mu(B) = \mu(C)$ or $\mu(B) = 0$. This follows from the fact that for each $n \geq 1$, the D_k with $\mu(D)_k \geq 1/n$, disjoint, is a finite sequence, since $\mu(A_\epsilon) < \infty$. Let $\tilde{A}_\epsilon = A_\epsilon - \bigcup_{j=1}^\infty C_j$, C_j a μ-atom. Then $\mu|\Sigma(\tilde{A}_\epsilon)$ is diffuse, and for each $0 < \alpha < \mu(\tilde{A}_\epsilon)$, the collection $\{E \in \Sigma(\tilde{A}_\epsilon): \mu(E) < \alpha\}$ has a supremum, $\tilde{E} \in \Sigma(\tilde{A}_\epsilon)$. Now the number of atoms $\{C_j: \mu(C_j) \geq \delta_\epsilon\}$ is thus finite, and $\tilde{A}_\epsilon$ can be decomposed into a finite number of measurable sets of measure $\leq \delta_\epsilon$. Hence A_ϵ can be split into a finite number of measurable sets of measure $\leq \delta_\epsilon$ and a finite number of atoms of measure $> \delta_\epsilon$. This is the partition we wanted. Then $\sup_n \int_A |f_n|\,d\mu \leq \epsilon$ for all A with $\mu(A) \leq \delta_\epsilon$, and on each atom C_j, each f_n is a constant, and the hypothesis that $\int_{C_j} f_n\,d\mu$ converges implies that $\{f_n(\omega)\mu(C_j): n \geq 1\}$ is bounded. Hence $\sup_n \int_{\bigcup_j C_j}|f_n|\,d\mu < \infty$, and finally $\sup_n \int_{A_\epsilon}|f_n|\,d\mu < \infty$. This gives (i). Regarding (ii), $\int_{\Omega_0}(f_n - f_m)g\,d\mu \to 0$ if $g = \chi_A$, $A \in \Sigma(\Omega_0)$; then apply Proposition 8. The last part is obtained as noted in Exercise 12.]

14. In Proposition 11, we assumed the existence of orthogonal functions in $L^2(\mu)$. Here we construct such sequences. Recall that if $f_1, \ldots, f_n$ are n vectors (or functions), then they are linearly independent if $\sum_{i=1}^n a_i f_i = 0$ a.e. implies $a_1 = a_2 = \cdots = a_n = 0$.

(a) If $\{\varphi_1, \ldots, \varphi_n\}$ in $L^2(\mu)$ is a set of n orthonormal functions, show that they are linearly independent.

(b) If $\{f_1, \ldots, f_n\}$ in $L^2(\mu)$ is any set of n linearly independent elements, consider the functions $\varphi_1, \ldots, \varphi_n$ defined as follows: $\varphi_1 = f_1$, $\varphi_2 = f_1(f_2, f_1) - f_2(f_1, f_1)$, and in general, for $1 < m \leqslant n$, φ_m is given by the determinant

$$\varphi_m = \begin{vmatrix} f_1 & f_2 & \cdots & f_m \\ (f_1, f_1) & (f_2, f_1) & \cdots & (f_m, f_1) \\ \vdots & \vdots & & \vdots \\ (f_1, f_{m-1}) & (f_2, f_{m-1}) & \cdots & (f_m, f_{m-1}) \end{vmatrix}$$

Verify that the φ_m are orthogonal functions in $L^2(\mu)$, where $(f, g) = \int_\Omega f\bar{g}\, d\mu$. Hence the sequence $\psi_1 = \varphi_1/\|\varphi_1\|_2$, $\psi_2 = \varphi_2/\|\varphi_2\|_2, \ldots, \psi_n = \varphi_n/\|\varphi_n\|_2$ is an orthonormal set. (This method of generating an orthonormal set from a linearly independent one is called the *Gram-Schmidt* process.)

15. An orthonormal set $\{\varphi_i : i \in I\} \subset L^2(\mu)$ is said to be *complete* if $(f, \varphi_i) = 0$ for all $i \in I$ implies $f = 0$ a.e. Show that every $L^2(\mu)$ with μ having the finite subset property contains a complete orthonormal set. [Consider the family $\mathcal{O}$ of orthonormal sets in $L^2(\mu)$. This is nonempty; order it by inclusion. Observe that each chain in $\mathcal{O}$ has an upper bound, namely their union. Then by Zorn's lemma (see Appendix: we also used it in the proof of Theorem 2.3.10) there is at least one maximal set $\mathcal{O}_1 \in \mathcal{O}$. Let $\mathcal{O}_1 = \{\varphi_i : i \in I\}$. This is the desired one, since if there is an $f \in L^2(\mu)$ such that $(f, \varphi_i) = 0$, $i \in I$, but $f \neq 0$ a.e., then we can enlarge $\mathcal{O}_1$ by adding $f/\|f\|_2$, and this will contradict the maximality of $\mathcal{O}_1$.]

16. Recall that $L^p(\mu)$ is separable if it has a countable collection of elements, the closure of whose linear span is $L^p(\mu)$. Show that $L^2(\mu)$ is separable iff it has a complete orthonormal set which is countable. [Use the Gram-Schmidt process in one direction.] Let $\{\varphi_n : n \geqslant 1\}$ be such a set in the separable space $L^2(\mu)$. For each $f \in L^2(\mu)$, let $a_n = (f, \varphi_n)$ and $a_f = (a_1, a_2, \ldots)$, and similarly, if $g \in L^2(\mu)$, define $b_g = (b_1, b_2, \ldots)$, $b_n = (g, \varphi_n)$. Let $T: f \mapsto a_f$ be a mapping from $L^2(\mu)$ into l^2, the sequence space, where $a, b \in l^2$ implies $(a, b) = \sum_{n=1}^{\infty} a_n \bar{b}_n$, $(a, a) = \|a\|_2^2 < \infty$. Show that $T: L^2(\mu) \to l^2$ is linear, satisfies $\|f\|_2 = \|Tf\|_2$, and is one to one and onto. [Note that $T\varphi_n = e_n = (0, \ldots 0, 1, 0, \ldots)$ with 1 in the nth place and zeros elsewhere, and $\operatorname{sp}\{e_n : n \geqslant 1\}$ is dense in l^2.] Because of this, $L^2(\mu)$ and l^2 are termed *isometrically isomorphic*. Deduce that any two complete orthonormal sets in $L^2(\mu)$ are isomorphic to each other, and all separable infinite dimensional Hilbert spaces are isomorphic to l^2.

17. Verify that each orthogonal sequence in $L^2(\mu)$ is weakly convergent, and if the space is infinite dimensional, it does not converge strongly.

18. Recall that $\varphi: J \to \mathbb{R}$ is (strictly) convex if it is continuous and for any x, y in the interval J ($\subset R$), $\varphi([x+y]/2) \leqslant [\varphi(x)+\varphi(y)]/2$ (with strict inequality unless $x = y$). Note that if $n = 2^k$, then by induction this implies, for $x_1, \ldots, x_n$ in J, the inequality

$$\varphi\left(\frac{1}{n}\sum_{i=1}^{n} x_i\right) \leqslant \frac{1}{n}\sum_{i=1}^{n}\varphi(x_i), \qquad (*)$$

and if $1 < n < 2^m$ for some integer $m > 1$, we may let $\bar{x} = \sum_{i=1}^{n} x_i/n$ and $x_j = \bar{x}$ for $j = n+1, \ldots, 2^n$. Now deduce from the above that the inequality $(*)$ holds for any n. Next using the continuity of φ at this point, one gets the usual form $\varphi(\alpha x + \beta y) \leqslant \alpha\varphi(x) + \beta\varphi(y)$ for all $0 \leqslant \alpha, \beta \leqslant 1$, $\alpha + \beta = 1$, and x, y in J, with strict inequality for strictly convex φ, if $0 < \alpha < 1$ and $x \neq y$. [Verify these statements, or consult Hardy, Littlewood, and Pólya (1952, p. 72).] Let (Ω, Σ, μ) be a probability space and $f \in L^1(\mu)$. Prove that, with $J = \mathbb{R}^+$,

$$\varphi\left(\int_\Omega |f|\, d\mu\right) \leqslant \int_\Omega \varphi(|f|)\, d\mu,$$

with equality when φ is strictly convex, iff $|f| =$ constant a.e. [First consider when f is a simple function, and extend. Here the right side may be infinite, and the fact that μ is normalized is essential. This is called *Jensen's inequality*. Since $-\varphi$ is concave iff φ is convex, Jensen's inequality for concave functions is the reverse of the above.]

19. Let $\varphi: \mathbb{R}^+ \to \mathbb{R}^+$ be a continuous nondecreasing function such that (i) $\varphi(x) = 0$ iff $x = 0$ and (ii) $\lim_{x\to\infty}\varphi(x) = +\infty$. It is called a *$\varphi$-function*. If (Ω, Σ, μ) is a measure space and $f: \Omega \to \bar{\mathbb{R}}$ is an a.e. finite measurable function (for Σ), define $\|\cdot\|_\varphi$ by the equation

$$\|f\|_\varphi = \inf\left\{\alpha > 0 : \int_\Omega \varphi\left(\frac{|f|}{\alpha}\right) d\mu \leqslant \alpha\right\}.$$

Let $\mathscr{L}^\varphi(\mu) = \{f : \|f\|_\varphi < \infty\}$. Show that $\|\cdot\|_\varphi$ is a (semi-)metric with $d_\varphi(f, g) = d_\varphi(f - g, 0) = \|f - g\|_\varphi$, and $(\mathscr{L}^\varphi(\mu), \|\cdot\|_\varphi)$ is a complete linear (semi-)metric space, and that in fact $\|\cdot\|_\varphi$ is a metric if a.e. equal functions are identified. Establish, moreover, that if f and g are equimeasurable (see Section 3.3) relative to μ, then $\|f\|_\varphi = \|g\|_\varphi$.

20. In the preceding exercise suppose φ is also convex, and define the function $N_\varphi(\cdot): f \mapsto N_\varphi(f)$ as

$$N_\varphi(f) = \inf\left\{\alpha > 0 : \int_\Omega \varphi\left(\frac{|f|}{\alpha}\right) d\mu \leqslant 1\right\}.$$

Show that if functions equal a.e. are identified, then $N_\varphi(\cdot)$ is a norm and $(\mathscr{L}^\varphi(\mu), N_\varphi(\cdot))$ is a complete (semi-)normed linear space, i.e., $L^\varphi(\mu)$ $[=\mathscr{L}^\varphi(\mu)/\sim]$ is a Banach space with $N_\varphi(\cdot)$ as norm.

21. If $\varphi(x) = x^p$, $x \geqslant 0$, $0 < p < \infty$, $(\mathscr{L}^p(\mu), \|\cdot\|_p)$ is the space introduced in Definition 2, and $\|\cdot\|_\varphi$ is defined as in Exercise 19, show that the two norms are equivalent in the sense that for any measurable $f\colon \Omega \to \overline{\mathbb{R}}$ one has $\|f\|_p < \infty$ iff $\|f\|_\varphi < \infty$, and for any sequence $\{f_n\colon n \geqslant 1\}$ of measurable functions for which these norms are finite one has $\|f_n - f\|_p \to 0$ iff $\|f_n - f\|_\varphi \to 0$. Show also that for convex φ, the functionals $\|\cdot\|_\varphi$ and $N_\varphi(\cdot)$ of Exercises 19 and 20 are equivalent in the sense just defined. [This means that the topologies defined by these functionals are the same, and moreover the spaces $\mathscr{L}^\varphi(\mu)$, called the *(generalized) Orlicz spaces* for (monotone increasing) convex φ-functions, are more extensive than (and include) the Lebesgue $\mathscr{L}^p(\mu)$ spaces for $0 < p < \infty$. This will be very useful for applications. The methods of proof are careful adaptations of the $\mathscr{L}^p(\mu)$ case.]

22. Let $\varphi\colon \mathbb{R}^+ \to \mathbb{R}^+$ be a φ-function as in Exercise 19. It is called *moderated* if $\varphi(2x) \leqslant C\varphi(x)$, $x \geqslant 0$, for some $0 < C < \infty$. Following the method of proof of Theorem 7, show that $(L^\varphi(\mu), \|\cdot\|_\varphi)$ with a moderated φ has simple functions dense in it, and that for any sequence $\{f_n\colon n \geqslant 1\}$ in $L^\varphi(\mu)$, $\int_\Omega \varphi(|f_n|)\, d\mu \to 0$ implies $\|f_n\|_\varphi \to 0$ as $n \to \infty$. If φ is not moderated, give a counterexample for the last statement.

23. A mapping $\varphi\colon \mathbb{R}^+ \to \mathbb{R}^+$ is *s-convex* if $\varphi(x) = \psi(x^s)$, $0 < s \leqslant 1$, and $\psi\colon \mathbb{R}^+ \to \mathbb{R}^+$ is a convex function such that $\psi(x) = 0$ iff $x = 0$. Thus $\varphi(\alpha x + \beta y) \leqslant \alpha^s\varphi(x) + \beta^s\varphi(y)$ for $\alpha \geqslant 0$, $\beta \geqslant 0$, and $\alpha^s + \beta^s = 1$. Verify that for an s-convex φ, $N_\varphi(af) = |a|^s N_\varphi(f)$, where $N_\varphi(\cdot)$ is defined in Exercise 20 for $L^\varphi(\mu)$. Show however that for s-convex φ, $N_\varphi(\cdot)$ and $\|\cdot\|_\varphi$ are equivalent (but not equal), where equivalence is defined in Exercise 21. [Standard, but a *careful* computation is needed.]

24. An extension of absolute continuity of a $\|\cdot\|_p$-norm in the context of φ-functions reads: $f \in L^\varphi(\mu)$ has that property if $N_\varphi(f\chi_{A_n}) \to 0$ for each $A_n \downarrow \varnothing$, $A_n \in \Sigma$. Show that if φ is moderated, then each f in $L^\varphi(\mu)$ has an absolutely continuous norm. Verify by the following example that the assertion is not valid if φ is not moderated. Let (Ω, Σ, μ) be the Lebesgue unit interval, $\varphi(x) = e^x - 1$, and $f(x) = \log\sqrt{1/x}$. Then $N_\varphi(f) = 1$, but $N_\varphi(f\chi_{A_n}) \nrightarrow 0$ if $A_n \downarrow \varnothing$, e.g. $A_n = (0, 1/n)$.

25. Let φ be as above, and $\lim_{x\to 0}\varphi(x)/x = 0$, $\lim_{x\to\infty}\varphi(x)/x = +\infty$. Its *complementary function* ψ is defined on $\mathbb{R}^+$ into $\mathbb{R}^+$ as

$$\psi(y) = \sup\{xy - \varphi(x)\colon x \geqslant 0\}, \qquad y \in \mathbb{R}^+.$$

Verify that ψ is a Borel-measurable convex function such that $\psi(0) = 0$ and $\psi(y) \uparrow \infty$ as $y \uparrow \infty$, even though φ need not be convex. [It is actually continuous on $(0, \infty)$, since one can show that a measurable convex function on an open interval is automatically continuous there.] If $\tilde{\psi}$ is the complementary function to ψ, defined as above, then from the two inequalities

$$xy \leqslant \varphi(x) + \psi(y), \qquad xy \leqslant \psi(y) + \tilde{\psi}(x)$$

show that $\tilde{\psi}(x) \leqslant \varphi(x)$, $x \geqslant 0$, and if $\bar{\psi}(x) \leqslant \varphi(x)$ is any other convex measurable function satisfying the second of the above inequalities with ψ, then $\bar{\psi} \leqslant \tilde{\psi}$, i.e., $\tilde{\psi}$ is the largest convex minorant of φ. If $\varphi(x) = x^p/p$, verify that $\psi(y) = y^q/q$, where $q = p/(p-1)$, $1 < p < \infty$. The inequality (11) prior to Theorem 4 is included here. With this setup, let φ be a φ-function such that its complementary function ψ is moderated. If $L^{\varphi}(\mu)$ and $L^{\psi}(\mu)$ are the corresponding Orlicz spaces and if $\{f_n: n \geqslant 1\} \subset L^{\varphi}(\mu)$ is such that $\lim_n \int_A f_n \, d\mu$ exists as a finite number for each $A \in \Sigma$ of σ-finite measure, then show that $\{f_n: n \geqslant 1\}$ is a weak Cauchy sequence in the sense that $\int_\Omega (f_n - f_m) g \, d\mu \to 0$ as $n, m \to \infty$ for each $g \in L^{\psi}(\mu)$. [The method of proof is the same as in Exercise 13, and uses also the result of Exercise 22. It can be shown that for the φ of Exercise 24, which is not moderated, its complementary function ψ is given by $\psi(y) = (y + 1)\log(y+1) - y$, which is moderated. Thus with their φ-functions the corresponding Orlicz spaces extend the theory of $L^p(\mu)$ spaces.]

26. This exercise combines the decreasing rearrangement f^* of functions f, introduced in Section 3.3, and integration theory to define another class of function spaces generalizing the Lebesgue class L^p. Let (Ω, Σ, μ) be a finite measure space. Taking $g(x) = |x|^p$ and then integrating by parts in Corollary 2.7, we get (verify this)

$$\begin{aligned}
\int_\Omega |f|^p \, d\mu &= p \int_0^\infty y^{p-1} \mu(\{\omega : |f(\omega)| \geqslant y\}) \, dy \\
&= \int_0^\infty [f^*(y)]^p \, dy \\
&= \int_0^\infty [y^{1/p} f^*(y)]^p \frac{dy}{y}.
\end{aligned}$$

This is generalized to the following for $0 < p \leqslant \infty$, $0 < q \leqslant \infty$:

$$\|f\|_{pq} = \begin{cases} \left(\dfrac{q}{p} \displaystyle\int_0^\infty [y^{1/p} f^*(y)]^q \frac{dy}{y} \right)^{1/q}, & 0 < p < \infty, 0 < q < \infty, \\ \sup\{ y^{1/p} f^*(y) : y > 0 \}, & 0 < p \leqslant \infty, \quad q = +\infty, \\ \displaystyle\int_0^\infty [f^*(y)]^q \frac{dy}{y}, & p = \infty, \quad 0 < q < \infty. \end{cases}$$

Let $\mathscr{L}^{p,q}(\mu) = \{f: \Omega \to \overline{\mathbb{R}}$, measurable, and $\|f\|_{p,q} < \infty\}$, and denote by $L^{p,q}(\mu) = \mathscr{L}^{p,q}(\mu)/\sim$, where $f \sim g \Rightarrow f = g$ a.e. The $L^{p,q}(\mu)$ is called a *Lorentz space*. Verify that $L^{p,q}(\mu)$ is linear and $\|\cdot\|_{pq}$ is a metric, so that $(L^{p,q}(\mu), \|\cdot\|_{pq})$ is a linear metric space. Deduce that $\|f\|_p = \|f\|_{pp}$, so that $L^p(\mu) = L^{p,p}(\mu)$. Thus the Lorentz spaces include the Lebesgue class. Develop the properties of $L^{p,q}(\mu)$ using the following steps:

(a) $\|f\|_{\infty q} < \infty \Rightarrow f = 0$ a.e., so $L^{\infty,q}(\mu) = \{0\}$.

(b) $0 < q_1 \leqslant q_2 \leqslant \infty \Rightarrow \|f\|_{pq_2} \leqslant \|f\|_{pq_1}$, so that for $0 < q_1 \leqslant p \leqslant q_2 \leqslant \infty$ we have $L^{p,q_1}(\mu) \subset L^{p,p}(\mu) = L^p(\mu) \subset L^{p,q_2}(\mu) \subset L^{p,\infty}(\mu)$. [If $f = \sum_{i=1}^n a_i \chi_{A_i}$, $0 < a_1 < \cdots < a_n$, then letting $b_1 = a_1$, $b_2 = a_2 - a_1, \ldots, b_n = a_n - a_{n-1}$, $B_i = \bigcup_{j=i}^i A_j$ $(\supset B_{i+1})$, verify that $f^* = \sum_{i=1}^n b_i \chi_{[0,\mu(B_i))}$ and that $(1/t)\int_0^t f^*(s)\,ds$ is decreasing in $0 \leqslant t \leqslant \mu(\Omega)$.]

(c) If $f: \Omega \to \mathbb{R}$ is in $L^1(\mu)$, and f^* is its decreasing rearrangement, let $f^{**}(t)$ be $= (1/t)\int_0^t f^*(s)\,ds$. Now verify that $f^* \leqslant f^{**}$, and for any pair of real μ-integrable functions f, g on Ω, $(f+g)^{**} \leqslant f^{**} + g^{**}$, and hence again $f^* \leqslant f^{**}$. [Verify that $\int_\Omega |fg|\,d\mu \leqslant \int_0^a (f^* g^*)(t)\,dt$, $a = \mu(\Omega)$.]

(d) If $f: \Omega \to \mathbb{R}$ is μ-integrable, show that for $1 < p < \infty$, $1 \leqslant q < \infty$,

$$\|f\|_{pq} \leqslant \|f^{**}\|_{pq} \leqslant \frac{p}{p-1}\|f\|_{pq}.$$

[Use the classical inequality of Hardy: If $q \geqslant 1$, $r > 0$, then for $f \geqslant 0$,

$$\left[\int_0^\infty \left(\int_0^t f(y)\,dy\right)^q \frac{dt}{t^{r+1}}\right]^{1/q} \leqslant \frac{q}{r}\left(\int_0^\infty [yf(y)]^q \frac{dy}{y^{r+1}}\right)^{1/q}.$$

This is obtained with Jensen's inequality (Exercise 18) applied to $(\int_0^t [f(y)/y^\alpha] y^\alpha\,dy)^q$ with $d\mu = (\alpha+1)y^\alpha\,dy/t^{\alpha+1}$.]

(e) $(L^{p,q}, \|\cdot\|_{pq})$ is complete, where $d(f,g) = d(f-g,0) = \|f-g\|_{pq}$, and if $q \neq \infty$, simple functions are dense in it. [For the last statement, by the structure theorem there are simple $f_n \to f$ with $|f_n| \uparrow |f|$, and so $|f - f_n| \to 0$ a.e. and $(f - f_n)^{**} \to 0$ monotonically.]

(f) $L^{1,1}(\mu)$ and $L^{p,q}(\mu)$ for $1 < p \leqslant \infty$, $1 \leqslant q \leqslant \infty$ are Banach spaces, but for other values of $p, q > 0$ they are only complete linear metric (but not Banach) spaces.
[For other properties and applications of these spaces and analysis, see R. A. Hunt (1966) and A. P. Calderón (1966).]

27. Let $\mathscr{B}$ be the Borel σ-algebra of the real line $\mathbb{R}$, and μ be a bounded Lebesgue-Stieltjes measure on $\mathbb{R}$. Show that $\hat{\mu}$ defined by $\hat{\mu}(t) = \int_{\mathbb{R}} e^{itx}\,d\mu(x)$, is uniformly continuous (see Exercise 10 above) and that $\hat{\mu}$

uniquely determines μ. [This is the *uniqueness theorem* for Fourier-Stieltjes transforms. First, by Theorem 2.6.1, $\hat{\mu}(t) = \int_{\mathbb{R}} e^{itx}\, dF(x)$ exists for a bounded nondecreasing left continuous F determined by μ on $\mathbb{R}$. Then verify the formula, by substitution, that

$$F(a+h) - F(a-h) = \lim_{T\to\infty} \frac{1}{2\pi} \int_{-T}^{T} \frac{\sin ht}{t} e^{-ita} \hat{\mu}(t)\, dt,$$

holds for all points $a \pm h$ at which F is continuous. If F_1, F_2 are two such functions determined by μ_1, μ_2, observe that they agree on the semi-ring of left-closed intervals whose end points are continuity points of both F_1, F_2. Theorem 2.3.1 then implies $\mu_1 = \mu_2$ on $\mathscr{B}$. There are a few details to fill in, and this is an extremely useful result.]

5

DIFFERENTIATION AND DUALITY

The preceding work on measures is now extended to scalar set functions in order to analyze their variational properties. Then the Hahn and Lebesgue decompositions are established. After considering concretely the absolutely continuous and completely monotone functions on the line, the Radon-Nikodým theorem and related differentiation results are presented. As applications, a brief account of likelihood ratios of finite measures and the duality of L^p-spaces are given, together with essential properties of conditional expectations.

5.1. VARIATIONS OF SET FUNCTIONS AND THE HAHN DECOMPOSITION

We have seen in Sections 4.4 and 4.5 that there are set functions which arise in the theory of integration (and also abstractly) taking both positive and negative values, called signed measures. For instance, if $f: \Omega \to \mathbb{R}$ is a μ-integrable function on (Ω, Σ, μ), then $\nu: A \mapsto \int_A f\,d\mu$, $A \in \Sigma$, is such a function on Σ. To study further properties of ν than those discussed in Section 4.4 (the Jordan decomposition, boundedness, and unique extension), we introduce their generalized variation concepts as follows:

Definition 1. Let $\mathscr{A}$ be an algebra of subsets of Ω, $\mu: \mathscr{A} \to \overline{\mathbb{R}}^+$ a measure, and $\nu: \mathscr{A} \to \mathbb{R}$ be (σ-)additive. Then

(i) for a given $p \geqslant 1$, ν is said to have *p-bounded variation relative* to μ if $|\nu|_p(\Omega) < \infty$, where, letting $\mathscr{A}_0 = \{A \in \mathscr{A}: 0 < \mu(A) < \infty\}$,

$$|\nu|_p(A) = \sup\left\{ \sum_{i=1}^{n} \left(\left| \frac{\nu(A_i)}{\mu(A_i)} \right| \right)^p \mu(A_i): A \supset A_i \in \mathscr{A}_0, \text{disjoint}, n \geqslant 1 \right\}, \tag{1}$$

and more generally

(ii) if $\varphi: \mathbb{R}^+ \to \overline{\mathbb{R}}^+$ is an increasing function satisfying $\varphi(x) = 0 \Leftrightarrow x = 0$, then ν has *φ-variation finite relative* to μ whenever $|\nu|_\varphi(\Omega) < \infty$, where

$$|\nu|_\varphi(A) = \sup\left\{ \sum_{i=1}^{n} \varphi\left(\left|\frac{\nu(A_i)}{\mu(A_i)}\right|\right)\mu(A_i): \right.$$

$$\left. A \supset A_i \in \mathscr{A}_0, \text{ disjoint}, n \geqslant 1 \right\}. \tag{2}$$

In particular, if $p = 1$ or equivalently $\varphi(x) = x$, then (1) and (2) reduce to

$$|\nu|(A) = \sup\left\{ \sum_{i=1}^{n} |\nu(A_i)|: A \supset A_i \in \mathscr{A}_0, \text{ disjoint}, n \geqslant 1 \right\}, \tag{3}$$

and ν is simply said to have *finite variation* when $|\nu|(\Omega) < \infty$, irrespective of the measure μ.

It is useful to note that the concept in the case $p = 1$ and the more general p-variation case ($p \neq 1$) have a basic difference, since only for $p \neq 1$ does the definition involve an auxiliary measure μ. This distinction results in some distinct properties for the two cases. Let us indicate a few of them.

Proposition 2. *Let ν be an additive set function on an algebra $\mathscr{A}$ of sets of Ω. If ν has finite p-variation on $\mathscr{A}$ relative to an additive μ: $\mathscr{A} \to \mathbb{R}^+$, then for $p \neq 1$, ν is μ-continuous and is σ-additive if μ is. On the other hand, if μ is any measure on $\Sigma = \sigma(\mathscr{A})$, and f is in $L^p(\mu)$, let ν: $A \mapsto \int_A f\,d\mu$, $A \in \Sigma_0$. Then $\|\nu\|_p(\Omega) = \int_\Omega |f|^p\,d\mu$ for all $1 \leqslant p < \infty$ of* (1). [$\Sigma_0 = \{A \in \Sigma: \mu(A) < \infty\}$.]

Proof. If $\epsilon > 0$ is given and $A \in \Sigma$ satisfies $\mu(A) < \epsilon$, then since (1) implies

$$\left|\frac{\nu(A)}{\mu(A)}\right|^p \mu(A) \leqslant |\nu|_p(A) \leqslant |\nu|_p(\Omega) < \infty, \; p > 1,$$

so that $|\nu(A)/\mu(A)|$ is bounded, one has $|\nu(A)|$ of the same order of magnitude as $\mu(A)$, i.e., as $\mu(A) \to 0$, $|\nu(A)| \to 0$. So ν is μ-continuous. If $p = 1$, then since $|\nu|(A)$ does not depend on μ [see (3)], we clearly have to exclude this case from the above.

If μ is a finite measure on $\mathscr{A}$, then the σ-additivity is equivalent to the condition that $\mu(A_n) \to 0$ for any sequence $A_n \downarrow \varnothing$. But then, by the above paragraph, $|\nu(A_n)| \to 0$ also. Since ν is additive, this implies its σ-additivity, because for any disjoint $\{B_n: n \geqslant 1\} \subset \mathscr{A}$,

$$\left|\nu\left(\bigcup_{k=1}^{\infty} B_k\right) - \sum_{k=1}^{n} \nu(B_k)\right| = \left|\nu\left(\bigcup_{k>n} B_k\right)\right| \to 0, \qquad n \to \infty.$$

Next let $\nu_f: A \mapsto \int_A f\,d\mu$, $f \in L^p(\mu)$, $A \in \Sigma$, $0 < \mu(A) < \infty$. Then ν_f is σ-additive and μ-continuous on Σ_0, by Proposition 4.3.8. Also, by (1)

$$|\nu_f|_p(\Omega) = \sup\left\{\sum_{j=1}^{n}\left|\frac{\nu_f(A_j)}{\mu(A_j)}\right|^p \mu(A_j): A_j \in \Sigma_0, \text{ disjoint}, \mu(A_j) > 0\right\}$$

$$= \sup\left\{\sum_{j=1}^{n}\left|\int_{A_j} f\frac{d\mu}{\mu(A_j)}\right|^p \mu(A_j): A_j \text{ as above}\right\}$$

$$\leqslant \sup\left\{\sum_{j=1}^{n}\left(\int_{A_j} |f|^p\frac{d\mu}{\mu(A_j)}\right)\mu(A_j): A_j \text{ as above}\right\}$$

[by Jensen's inequality, since $\mu(\cdot)/\mu(A_j)$ is a probability on A_j (Exercise 4.5.18)]

$$\leqslant \int_\Omega |f|^p\,d\mu. \tag{4}$$

On the other hand, if $f = f_n = \sum_{k=1}^n a_k \chi_{A_k} \in L^p(\mu)$, then taking the disjoint collection in (1) to be $\{A_1, \ldots, A_n\}$ itself, we get

$$|\nu_{f_n}|_p(\Omega) = \int_\Omega |f_n|^p\,d\mu \left[= |\nu_{|f_n|}|_p(\Omega)\right]. \tag{5}$$

But for each $f \in L^p(\mu)$, there exist simple functions $f_n \in L^p(\mu)$ such that $|f_n| \uparrow |f|$ and so $\int_\Omega |f_n|^p\,d\mu \uparrow \int_\Omega |f|^p\,d\mu$, by the monotone convergence theorem. Hence,

$$0 \leqslant |\nu_f|_p(\Omega) - |\nu_{f_n}|_p(\Omega), \qquad \text{since} \quad |\nu_f|_p = |\nu_{|f|}|_p \geqslant |\nu_{|f_n|}|_p$$

$$\leqslant \int_\Omega |f|^p\,d\mu - \int_\Omega |f_n|^p\,d\mu, \qquad \text{by (4) and (5)}$$

$$\to 0 \qquad \text{as} \quad n \to \infty.$$

It follows from this that there is equality in (4), and the proof is complete.

Remark. A similar result holds with the definition (2) if φ there is also convex. We leave this to the reader.

In case $p = 1$, it is possible to present an alternative definition of finite variation which proves to be convenient in computations.

Proposition 3. *Let ν be a signed measure on an algebra $\mathscr{A}$ of subsets of Ω. Then*

$$\sup_{A\in\mathscr{A}} |\nu(A)| \leqslant |\nu|(\Omega) \leqslant 2 \sup_{A\in\mathscr{A}} |\nu(A)|, \tag{6}$$

and 2 is to be replaced by 4 if ν is complex valued.

Proof. The left inequality of (6) is obvious from (3), since a signed measure is bounded (Corollary 4.4.6), $|\nu(A)| \leqslant C < \infty$, $A \in \mathscr{A}$. Consider a disjoint collection $\{A_1, \dots, A_n\} \subset \mathscr{A}$, and for a simplification of (3),

$$\sum_{i=1}^{n} |\nu(A_i)| = \sum_{i=1}^{n_1}{}_{+}\, \nu(A_i) - \sum_{j=1}^{n_2}{}_{-}\, \nu(A_j), \qquad n_1 + n_2 = n, \quad n_i \geqslant 0,$$

where the first sum consists of positive $\nu(A_i)$'s and the second one of negative $\nu(A_j)$'s. Since ν is additive, we get

$$\sum_{i=1}^{n} |\nu(A_i)| = \nu\left(\bigcup_{i=1}^{n_1} A_i\right) - \nu\left(\bigcup_{j=1}^{n_2} A_j\right).$$

Hence (3) becomes

$$|\nu|(\Omega) = \sup\left\{\nu\left(\bigcup_{i=1}^{n_1} A_i\right) - \nu\left(\bigcup_{j=1}^{n_2} A_j\right)\right\} \leqslant 2C = 2\sup_{A\in\mathscr{A}} |\nu(A)|. \tag{7}$$

In the complex case, we consider $\nu = \nu_1 + i\nu_2$, where ν_1, ν_2 are signed measures, and the last statement follows at once from (7).

Since we have already seen (Proposition 4.4.5) that $\nu = \nu^+ - \nu^-$ is the Jordan decomposition of a signed measure, and $\tilde{\nu} = \nu^+ + \nu^-$ is (σ-)additive if ν has that property, it is natural to seek the relations between $|\nu|(\cdot)$ and $\tilde{\nu}$. We next settle this question by proving their identity. This is based on the important Hahn decomposition, for which we present two proofs. The first one is based on the Jordan decomposition proved in Proposition 4.4.5. The second one is independent of that proposition, and from it we deduce the Jordan decomposition for signed measures. The two proofs thus provide an internal equivalence between these two decompositions, and this information illuminates the results.

Definition 4. If (Ω, Σ) is a measurable space and $\nu: \Sigma \to \mathbb{R}$ is a signed measure, then a set $\Omega^+ \in \Sigma$ is called *positive for* ν if $\nu(A \cap \Omega^+) \geqslant 0$ for all $A \in \Sigma$, and $\Omega^- \in \Sigma$ is *negative for* ν if $\nu(A \cap \Omega^-) \leqslant 0$ for all $A \in \Sigma$.

We then have the following

Theorem 5. *Let (Ω, Σ) be a measurable space, and ν a signed measure on the σ-algebra Σ. Then there exists a decomposition of Ω such that $\Omega = \Omega^+ \cup \Omega^-$, where Ω^+ (Ω^-) is a positive (negative) set for ν, $\Omega^+ \cap \Omega^- = \varnothing$, and where $\nu(A) \geqslant 0$ for all $A \subset \Omega^+$, $A \in \Sigma$, and $\nu(B) \leqslant 0$ for all $B \subset \Omega^-$, $B \in \Sigma$. The decomposition in general is not unique, but has the following weaker uniqueness property: if $\Omega = \Omega_1^+ \cup \Omega_1^-$ is another such decomposition into positive and negative sets, then $\nu(A \cap \Omega^+) = \nu(A \cap \Omega_1^+)$ as well as $\nu(B \cap \Omega^-) = \nu(B \cap \Omega_1^-)$, for all A, B in Σ.*

First Proof. By Proposition 4.4.5, $\nu^+ = \sup(\nu, 0)$, $\nu^- = -\inf(\nu, 0)$, and

$$\nu^+(A) = \sup\{\nu(B): B \subset A,\ B \in \Sigma\} \geqslant 0,$$

$$\nu^-(A) = -\inf\{\nu(B): B \subset A,\ B \in \Sigma\} \geqslant 0,$$

since $\nu(\varnothing) = 0$. Also, $\nu^+(\Omega) = \alpha < \infty$ and $\nu^-(\Omega) = \beta < \infty$. Hence there exists a sequence $A_n \in \Sigma$ such that $\nu^+(\Omega) = \lim_n \nu(A_n) = \alpha$, and in fact each A_n may be chosen (by definition of sup) such that $\alpha - 2^{-n} < \nu(A_n) \leqslant \alpha$. Let $\Omega^+ = \liminf_n A_n$, and $\Omega^- = \Omega - \Omega^+$. We claim that $\Omega = \Omega^+ \cup \Omega^-$ is a decomposition of the desired kind.

Since Σ is a σ-algebra, Ω^+ is in Σ. To see that it is a positive set, we need to show that for each $A \subset \Omega^+$, $A \in \Sigma \Rightarrow \nu(A) \geqslant 0$. First note that $\nu^+(\Omega^-) = 0 = \nu^-(\Omega^+)$. In fact, by the choice of A_n,

$$\alpha - 2^{-n} < \nu(A_n) = \nu^+(A_n) - \nu^-(A_n) \leqslant \nu^+(\Omega) - \nu^-(A_n) = \alpha - \nu^-(A_n).$$

Hence $\nu^-(A_n) < 2^{-n}$. Since $\nu^{\pm}$ are σ-additive, we also have

$$\nu^+(A_n^c) = \nu^+(\Omega) - \nu^+(A_n) < \alpha - (\alpha - 2^{-n}) = 2^{-n}. \tag{8}$$

Thus

$$\begin{aligned} \nu^-(\Omega^+) &= \nu^-\Big(\bigcup_{k \geqslant 1} \bigcap_{n \geqslant k} A_n\Big) = \lim_k \nu^-\Big(\bigcap_{n \geqslant k} A_n\Big) \\ &\leqslant \varliminf_k \nu^-(A_k) \leqslant \lim_k 2^{-k} = 0, \end{aligned} \tag{9}$$

and

$$\nu^+(\Omega^-) = \nu^+\left(\bigcap_{k\geqslant 1}\bigcup_{n\geqslant k} A_n^c\right) = \lim_k \nu^+\left(\bigcup_{n\geqslant k} A_n^c\right)$$

$$\leqslant \lim_k \sum_{n\geqslant k} \nu^+(A_n^c) \leqslant \lim_k \sum_{\eta=k}^{\infty} 2^{-n} = 0. \tag{10}$$

Consequently,

$$\nu^+(\Omega) = \nu^+(\Omega^+) + \nu^+(\Omega^-) = \nu^+(\Omega^+) = \alpha$$

and

$$\nu(\Omega^+) = \nu^+(\Omega^+) - \nu^-(\Omega^+) = \nu^+(\Omega^+) = \alpha. \tag{11}$$

If there existed a set $A \subset \Omega^-$, $A \in \Sigma$, $\nu(A) > 0$, then

$$\nu(\Omega^+ \cup A) = \nu(\Omega^+) + \nu(A) > \nu(\Omega^+) = \alpha, \tag{12}$$

and this contradicts the fact that α is the maximum value. Thus $\nu(A) \leqslant 0$, and Ω^+ is a positive set. Similarly Ω^- is shown to be a negative set, since $\nu^+(\Omega^-) = 0$.

The decomposition in general is not unique, since any set $B \subset \Omega^+$ such that $\nu^+(B) = 0 = \nu^-(B)$ can be added to one and subtracted from the other of the sets Ω^+, Ω^-. To see that this is the only exception, let $\Omega = \Omega^+ \cup \Omega^- = \Omega_1^+ \cup \Omega_1^-$ be two such decompositions for ν. If $A \in \Sigma$ is arbitrary, then

$$\nu(\Omega^+ \cap \Omega_1^- \cap A) = \nu(\Omega^- \cap \Omega_2^+ \cap A) = 0, \tag{13}$$

since each of these sets is simultaneously positive and negative for ν. Hence

$$\nu(\Omega^+ \cap A) = \nu((\Omega^+ \cup \Omega_1^+) \cap A) = \nu(\Omega_1^+ \cap A), \tag{14}$$

because $(\Omega_1^+ - \Omega^+) \cap A$ is both a positive $[\subset \Omega_1^+ \cap A]$ and a negative $[\subset (\Omega^+)^c \cap A]$ set for ν. Similarly,

$$\nu(\Omega^- \cap A) = \nu((\Omega^- \cup \Omega_1^-) \cap A) = \nu(\Omega_1^- \cap A). \tag{15}$$

This completes the proof.

The second method does not depend on Proposition 4.4.5 in obtaining (9) and (10). The demonstration explicitly uses Zorn's lemma, but the argument applies to a slightly more general result. (On some equivalents of Zorn's lemma, see the Appendix.)

Second proof. Let $\nu: \Sigma \to \mathbb{R}$ be a given signed measure, and consider $\mathscr{P} = \{A \in \Sigma: \nu(A \cap B) \geqslant 0, \text{ all } B \in \Sigma\}$. Since ν is additive, if $A_i \in \mathscr{P}$, disjoint, then $A_1 \cup A_2 \in \mathscr{P}$. Also, if they are not disjoint, and every measurable subset of $A_1 \cap A_2$ has positive ν-measure, then $A_1 \cap A_2 \in \mathscr{P}$. But $A_1 - A_2 \subset A_1$, and so $\nu((A_1 - A_2) \cap B) = \nu((A_1 - A_2) \cap B \cap A_1) \geqslant 0$, all $B \in \Sigma$. Thus $\mathscr{P}$ is closed under finite unions and differences. It is also closed under countable unions. In fact if $A_n \in \mathscr{P}$, $n \geqslant 1$, $A = \bigcup_{n=1}^{\infty} A_n$, and $B \in \Sigma$ is arbitrary, then $B_n = B \cap (A_n - \bigcup_{k=1}^{n-1} A_k)$, $n > 1$ is a positive set for each n $(B_1 = B \cap A_1)$, and by the σ-additivity of ν,

$$\begin{aligned} \nu(A \cap B) &= \nu\left(A \cap \bigcup_{n=1}^{\infty} B_n\right) \\ &= \sum_{n=1}^{\infty} \nu(A \cap B_n), \qquad \text{since the } B_n \text{ are disjoint} \\ &\geqslant 0, \end{aligned}$$

since $A \cap B_n \subset B_n$ and is a positive set. Thus A is a positive set and $A \in \mathscr{P}$. Consider

$$\alpha_0 = \sup\{\nu(A): A \in \mathscr{P}\}. \tag{16}$$

Then there exists a sequence $A_n \in \mathscr{P}$ such that $\nu(A_n) \to \alpha_0$. Let $\tilde{A} = \bigcup_{n=1}^{\infty} A_n$. Then $\tilde{A}$ is positive, $\tilde{A} \in \mathscr{P}$ by the above, and $\nu(\tilde{A}) \leqslant \alpha_0$. We claim that (i) $\nu(\tilde{A}) = \alpha_0$, and (ii) if $A^+ = \tilde{A}$ and $A^- = \Omega - \tilde{A}$, then A^- is negative, so that $\Omega = A^+ \cup A^-$ is the desired decomposition.

To verify (i), note that $\tilde{A} = \bigcup_{n=1}^{\infty} \tilde{A}_n$, where $\tilde{A}_n = \bigcup_{k=1}^{n} A_k \in \mathscr{P}$. Also $A_n \subset \tilde{A}_n$, and $\tilde{A}_n - A_n$ is a positive set, since by the first part each measurable subset of a positive set is positive. However,

$$\nu(\tilde{A}_n) = \nu(A_n) + \nu(\tilde{A}_n - A_n) \geqslant \nu(A_n).$$

Hence $\nu(\tilde{A}_n) \geqslant \nu(\tilde{A}_{n-1})$, and $\alpha_0 \geqslant \lim_n \nu(\tilde{A}_n)$ (limit exists) $\geqslant \lim_n \nu(A_n) = \alpha_0$. Consequently (i) is true. As for (ii), $A^+ = \tilde{A} \in \mathscr{P}$ being positive, we need to show that A^- is a negative set. This is proved by an indirect argument.

Suppose A^- is not negative. Then there is a $B_0 \subset A^-$, $B_0 \in \Sigma$, such that $\nu(B_0) > 0$. Consider the class

$$\mathscr{C} = \{C \in \Sigma: C \subset B_0, \nu(C) \geqslant \nu(B_0)\}. \tag{17}$$

Then $\mathscr{C}$ is nonempty ($B_0 \in \mathscr{C}$), and we partially order it by writing $C_1 < C_2$ iff $C_2 \subsetneqq C_1$ and $\nu(C_1) < \nu(C_2)$, (and otherwise $C_1 = C_2$). It will be shown that each chain in $(\mathscr{C}, \leqslant)$ has an upper bound, so that $\mathscr{C}$ has a maximal element $C_0 \in \mathscr{C}$, by Zorn's lemma. This implies that $C_0 \in \mathscr{P}$, for in the opposite case there exists a $B \subset C_0$, $B \in \Sigma$, $\nu(B) < 0$, and then $C_0 - B \subset C_0 \subset B_0$, so that

$$\nu(C_0 - B) = \nu(C_0) - \nu(B) > \nu(C_0) \geqslant \nu(B_0).$$

But then by definition of $\mathscr{C}$, $C_0 - B \in \mathscr{C}$ and $C_0 - B \geqslant C_0$. This contradicts the maximality of C_0 in $\mathscr{C}$. Thus $C_0 \in \mathscr{P}$, and then, since $C_0 \subset B_0 \subset A^-$, we have $C_0 \cap A^+ = \varnothing$, $C_0 \cup A^+ \in \mathscr{P}$, so that

$$\nu(C_0 \cup A^+) = \nu(C_0) + \nu(A^+) \geqslant \nu(B_0) + \nu(A^+) > \nu(A^+) = \alpha_0.$$

This again contradicts the definition of α_0 in (16), and so our supposition is not valid. Thus A^- is a negative set.

We now show therefore that if $\mathscr{C}_0 \subset \mathscr{C}$ is a chain, then it is bounded above. Let $\beta = \sup\{\nu(C): C \in \mathscr{C}_0\}$. Then there is a sequence $C_n \in \mathscr{C}_0$ such that $\nu(C_n) \uparrow \beta$, and since $\mathscr{C}_0$ is linearly ordered, $C_n \supset C_{n+1}$ and $\nu(C_n) < \nu(C_{n+1}) < \infty$. Let $C_0 = \cap_n C_n$ and $D_n = C_n - C_{n+1}$. Then $C_m = C_0 \cup \bigcup_{k=m}^{\infty} D_k$ and by the σ-additivity of ν,

$$\nu(C_n) = \nu(C_0) + \sum_{k \geqslant n} \nu(D_k),$$

since the D_k are disjoint. This shows that the right side series is convergent, and then goes to zero as $n \to \infty$. Thus $\beta = \lim_n \nu(C_n) = \nu(C_0)$. But $C_0 \subset C_n \subset B_0$. If $C_0 \in \mathscr{C}_0$, then by definition of ordering, C_0 is the upper bound of C_n. If $C_0 \notin \mathscr{C}_0$, then for each $C \in \mathscr{C}_0$, $\nu(C) < \beta = \nu(C_0)$, so that for some $C_n \in \mathscr{C}_0$, $\nu(C) < \nu(C_n) \leqslant \beta$. Hence by the ordering, $C \supset C_n \supset C_0$, so that C_0 is again an upper bound of $\mathscr{C}_0$. This is just what we had to show to conclude the truth of (ii).

With this, (13)–(15) hold as before, and the second proof of the theorem is complete.

Remark i. The preceding (second) proof allows ν to be extended real valued. By additivity of ν, it can take only one of the two values $-\infty$ or $+\infty$. Indeed, if $\nu(A) = +\infty$ and $\nu(B) = -\infty$, then by additivity

$$\nu(A \cup B) = \nu(A - B) + \nu(B - A) + \nu(A \cap B)$$

must hold. But if $\nu(A \cap B) \in \mathbb{R}$, then the sum of the other two terms will be of the form $\infty - \infty$, so that the right side is undefined. The same difficulty persists between any two of the terms, so that ν cannot be additive unless $\infty \geqslant \nu(A) > -\infty$, $A \in \Sigma$, or $-\infty \leqslant \nu(A) < +\infty$, $A \in \Sigma$. One sees, by reviewing the preceding proof, that the Hahn decomposition holds in this slightly more general case. [But we consider only real values of ν.]

Remark ii. This important result has been given other proofs than the two presented here. Some of them do not make explicit use of Zorn's lemma, but employ a "selection" procedure of a certain type which may be equivalent to this lemma. There are also proofs that, like our first one, do not use Zorn's lemma, and we indicate below one such in Exercise 4.

Since in the second proof we did not use the Jordan decomposition, the latter is deduced from the Hahn decomposition as follows. Let $\nu^+(A) = \nu(A \cap A^+)$, $\nu^-(A) = -\nu(A \cap A^-)$, $A \in \Sigma$, where A^+ (A^-) is the positive (negative) set for ν in Σ. Evidently $\nu^\pm\colon \Sigma(A^\pm) \to \overline{\mathbb{R}}^+$ are measures on the σ-algebra $\Sigma(A) = \{A \cap B\colon B \in \Sigma\}$. Let $|\nu| = \nu^+ + \nu^-$, so $|\nu|$ is a measure on Σ, $\nu = \nu^+ - \nu^-$. Here ν^+, ν^-, and $|\nu|$ are called the *positive*, *negative*, and *total variations* of ν. To show that this is not in conflict with Definition 1, we establish the following:

Proposition 6. *With the notation of* (1) *for* $p = 1$, *we have* $|\nu|_1(\Omega) = \nu^+(\Omega) + \nu^-(\Omega)$. *Thus* $|\nu|_1(\Omega) = |\nu|(\Omega)$.

Proof. Let $\alpha = |\nu|_1(\Omega)$. If $\{A_1, \ldots, A_n\} \subset \Sigma$ is as in (1), then with A^+ (A^-) as a positive (negative) set of ν we have

$$\begin{aligned}\sum_{i=1}^{n} |\nu(A_i)| &= \sum_{i=1}^{n} |\nu(A^+ \cap A_i) + \nu(A^- \cap A_i)| \\ &= \sum_{i=1}^{n} |\nu^+(A_i) - \nu^-(A_i)| \\ &\leqslant \sum_{i=1}^{n} (\nu^+(A_i) + \nu^-(A_i)) = \sum_{i=1}^{n} |\nu|(A_i) \leqslant |\nu|(\Omega).\end{aligned}$$

Taking the supremum on the left of (1) gives $\alpha \leqslant |\nu|(\Omega)$. On the other hand, considering the trivial partition $\{A^+, A^-\}$, $\Omega = A^+ \cup A^-$, we have

$$\alpha \geqslant |\nu(A^+)| + |\nu(A^-)| = \nu^+(\Omega) + \nu^-(\Omega) = |\nu|(\Omega).$$

Consequently $\alpha = |\nu|_1(\Omega) = |\nu|(\Omega)$, as asserted.

Hereafter we drop the suffix, and write $|\nu|$ for the (1-) variation of ν. We note that it is σ-additive on Σ, since $\nu^\pm$ are. Due to this property, if $\nu\colon \Sigma \to \overline{\mathbb{R}}$ is an (extended) signed measure, then we define $\mathscr{L}^p(\nu) = \mathscr{L}^p(|\nu|) = \{f\colon \Omega \to \overline{\mathbb{R}}$, measurable, $\int_\Omega |f|^p \, d|\nu| < \infty\}$. By Section 4.5, $\mathscr{L}^p(|\nu|)$ $[= \mathscr{L}^p(\nu)$ by definition] is a complete linear semimetric space for $0 < p < \infty$, where the norm (or metric) is in terms of the measure $|\nu|(\cdot)$. The Lebesgue spaces $L^p(\nu)$ $[= \mathscr{L}^p(\nu)/\sim]$ of equivalence classes (of measurable functions) are formed in the usual way. The same definition extends to complex functions and complex measures also, and one defines $\mathscr{L}^p(\nu)$ as $\mathscr{L}^p(|\nu|)$. In this connection, the formula (6) will be useful. We also have $\int_\Omega f \, d\nu = \int_\Omega f \, d\nu^+ - \int_\Omega f \, d\nu^-$ for computations.

As an example, let $f \in \mathscr{L}^1(\mu)$, and consider $\nu(A) = \int_A f\,d\mu$. If $\Omega^+ = \{\omega: f(\omega) > 0\}$, $\Omega^- = \{\omega: f(\omega) \leqslant 0\}$, then it is clear that Ω^+ (Ω^-) is a positive (negative) set for ν. If $\Omega_0 = \{\omega: f(\omega) = 0\}$, then $\{\Omega^+ \cup \Omega_0, \Omega^- - \Omega_0\}$ is another Hahn decomposition of Ω for ν. Later in this chapter we shall establish a converse to this representation with $\mu = |\nu|$ [see Exercise 3.8(b)].

EXERCISES

0. Let (Ω, Σ) be a measurable space and $\nu: \Sigma \to \mathbb{R}$ be σ-additive. Show that (a) if $\nu_A = \nu|\Sigma(A)$, $A \in \Sigma$, then $\nu_{A \cup B} = \nu_A + \nu_B - \nu_{A \cap B}$ for A, B in Σ; and (b) if Σ is a σ-ring, but $\sup\{|\nu(A)|: A \in I\} < \infty$, then there is an S in Σ such that $\nu = \nu_S$. (Recall that $\Sigma(A) = \{A \cap B: B \in \Sigma\}$.)

1. Let $\varphi: \mathbb{R}^+ \to \mathbb{R}^+$ be an increasing continuous function such that $\varphi(x) = 0$ iff $x = 0$. If (Ω, Σ, μ) is a measure space, $\Sigma_0 = \{A \in \Sigma: \mu(A) < \infty\}$, $\int_\Omega \varphi(|f|)\,d\mu < \infty$, and $\nu(A) = \int_A f\,d\mu$, show that $\nu: \Sigma_0 \to \mathbb{R}$ is a signed measure and $\|\nu\|_\varphi = \int_\Omega \varphi(|f|)\,d\mu$. [Use (5) appropriately.]

2. Let (Ω, Σ, μ) be a measure space, and $V^p(\mu) = \{\nu: \Sigma \to \mathbb{R}, \sigma\text{-additive}, \|\nu\|_p < \infty\}$, $\|\nu\|_p = |\nu|_p(\Omega)$. If $N_p(\nu) = (\|\nu\|_p)^{1/p}$, show that $N_p(\cdot)$ is a norm on $V_p(\mu)$ and that $(V^p(\mu), N_p(\cdot))$ is a Banach (i.e., a complete normed linear) space, $1 \leqslant p < \infty$. [If $p = 1$, we denoted $V^1(\mu)$, which is independent of μ, as ca(Ω, Σ) in Exercise 4.4.5, and this is an extension. Proposition 2 can also be used.]

3. If $p = 1$, $|\nu|_1(\Omega)$ can be calculated as follows. Show that

$$|\nu|(\Omega) = \sup\left\{\left|\int_\Omega f\,d\nu\right|: \sup_\omega |f(\omega)| \leqslant 1, f \text{ measurable}\right\},$$

where $\int_\Omega f\,d\nu = \int_\Omega f\,d\nu^+ - \int_\Omega f\,d\nu^-$ always exists for $|f| \leqslant 1$ here. [First, $|\int_\Omega f\,d\nu| \leqslant \int_\Omega |f|\,d\nu^+ + \int_\Omega |f|\,d\nu^- \leqslant |\nu|(\Omega)$ for all $|f| \leqslant 1$. If Ω^+ (Ω^-) is a positive (negative) set of ν, then define $f = \chi_{\Omega^+} - \chi_{\Omega^-}$ and obtain the opposite inequality. We thus have a *third* expression for $|\nu|(\Omega)$.]

4. Let (Ω, Σ) be a measurable space, and $\nu: \Sigma \to \overline{\mathbb{R}}$ be σ-additive. Establish a Hahn decomposition with the following steps, which thus gives a somewhat different method than the two already presented.

(a) The ν above attains its maximum and minimum on two sets Ω_1, Ω_2 of Σ, so that $\nu(\Omega_1) = \max\{\nu(A): A \in \Sigma\}$ and $\nu(\Omega_2) = \min\{\nu(A): A \in \Sigma\}$. [If $\nu(A) = +\infty$ for some $A \in \Sigma$, let $\Omega_1 = A$. Otherwise, there exist $A_n \in \Sigma$ with $\alpha = \lim_n \nu(A_n) = \sup\{\nu(A): A \in \Sigma\} \geqslant 0$. If $\tilde{A} = \bigcup_n A_n$, define $B_n = \bigcap_{k=1}^n \tilde{A}_k$, $n \geqslant 1$, where $\tilde{A}_k = A_k$ or $\tilde{A} - A_k$. So $B_n \in \Sigma$, and for each n there are 2^n such B_n's. Let $C_n = \bigcup\{B_{n'}: \nu(B_{n'}) \geqslant 0\}$ with $C_n = \varnothing$ if the union is empty. $C_n \in \Sigma$ and $\nu(C_n) \geqslant \nu(A_n)$, since each $B_m \subset C_n$ or $B_m \cap C_n = \varnothing$ for $m > n$, and A_n is a union of some of the B_n. Thus $\nu(A_n) \leqslant \nu(C_n) \leqslant \nu(\bigcup_{k \geqslant n} C_k)$, so that if $\Omega_1 = \limsup_k C_k$, then

$\alpha = \nu(\Omega_1)$. Similarly Ω_2 is obtained.]

(b) Let $\Omega^+ = \Omega_1$ and $\Omega^- = \Omega_1^c$, where Ω_1 is as in (a). If $\nu^+(A) = \nu(\Omega_1 \cap A)$, $\nu^-(A) = \nu(\Omega^- \cap A)$, $A \in \Sigma$, then $\Omega = \Omega^+ \cup \Omega^-$ is a Hahn decomposition and $\nu = \nu^+ - \nu^-$ is Jordan's decomposition. [First show that $\nu^+(\Omega^-) = 0 = \nu^-(\Omega^+)$, as in the first proof of Theorem 5, and follow the rest of the argument there.]

5. This exercise extends the result of Exercise 3 to p-variation. Let (Ω, Σ, μ) be a measure space, and $\Sigma_0 = \{A \in \Sigma: \mu(A) < \infty\}$ a δ-subring of Σ. If $\nu: \Sigma_0 \to \mathbb{R}$ is additive and has finite p-variation relative to μ, where $p > 1$, show that with $q = p/(p-1)$,

$$\left[|\nu|_p(\Omega)\right]^{1/p} = \sup\left\{\left|\int_\Omega f\,d\nu\right| : f \in L^q(\mu),\ \|f\|_q \leqslant 1\right\}.$$

(Here again, since ν is σ-additive by Proposition 2, the integral is defined.) Verify that conversely, if the right side is finite, then ν has finite p-variation relative to μ as given in Definition 1. Hence this is an alternative expression; it is convenient for some work. [If $\alpha^p = |\nu|_p(\Omega)$ and $f = \sum_{k=1}^n a_k \chi_{A_k} \in L^q(\mu)$, $\|f\|_q \leqslant 1$, then

$$\left|\int_\Omega f\,d\nu\right| = \left|\sum_{k=1}^n a_k \left(\frac{\nu(A_k)}{\mu(A_k)}\right)\mu(A_k)\right| \leqslant \|f\|_q \cdot \alpha,$$

and on the other hand, if $0 < \epsilon < \alpha$, consider

$$g = \sum_{k=1}^n \left(\frac{|a_k|}{\alpha}\right)^{p-1} \operatorname{sgn}(a_k)\,\chi_{A_k}, \text{ where } \quad a_k = \nu(A_k)/\mu(A_k)$$

for $0 < \mu(A_k)$, $A_k \in \Sigma_0$, are chosen such that $\alpha^p - \epsilon < \sum_{k=1}^n |a_k|^p \mu(A_k)$. Then $1 \geqslant \|g\|_q > [1 - (\epsilon/\alpha)^p]^{1/q}$. This gives $|\int_\Omega g\,d\nu| > \alpha - \epsilon'$, where $\epsilon' = \epsilon/\alpha^{p-1}$, and implies the desired result. The converse is similar.]

6. Let $V^p(\mu)$ be the space defined in Exercise 2. A set function of the form $\nu_\pi = \sum_{i=1}^n a_i \mu(A_i \cap \cdot\,)$, $\pi = \{A_1, \ldots, A_n\} \subset \Sigma_0$, A_i disjoint, where $a_i \in \mathbb{R}$, is called *simple*. In particular, if $f_n = \sum_{i=1}^n a_i \chi_{A_i}$, then $\nu_\pi(\cdot) = \int_{(\cdot)} f_n\,d\mu = \sum_{i=1}^n a_i \mu(A_i \cap \cdot\,)$ is of this type. If $1 \leqslant p < \infty$, show that simple functions are dense in $V^p(\mu)$ for the norm $N_p(\cdot)$ defined there. [Use the isomorphism of Proposition 2 and Theorem 4.5.7.]

7. Let $\varphi: \mathbb{R}^+ \to \mathbb{R}^+$ be a monotone increasing function such that $\varphi(x) = 0$ iff $x = 0$ and $\varphi(x)/x \to \infty$ as $x \to \infty$. Let $V^\varphi(\mu) = \{\nu: \Sigma \to \mathbb{R},\ \sigma\text{-additive},\ |\nu|_\varphi(\Omega) < \infty\}$ on (Ω, Σ, μ). Define a functional $\|\cdot\|_\varphi: V^\varphi(\mu) \to \mathbb{R}^+$ as

$$\|\nu\|_\varphi = \inf\{\alpha > 0: |\nu/\alpha|_\varphi(\Omega) \leqslant \alpha\}, \qquad \nu \in V^\varphi(\mu),$$

where $|\nu|_\varphi(\Omega)$ is given by (2) in Definition 1. Show that $d(\nu_1, \nu_2) =$

$\|\nu_1 - \nu_2\|_\varphi$ defines a metric, $(V^\varphi(\mu), \|\cdot\|_\varphi)$ is a complete linear metric space, and if φ is moderated, then simple functions are dense in it. [Recall that φ moderated means $\varphi(2x) \leqslant C\varphi(x)$, $x \geqslant 0$, $0 < C < \infty$. The result can be reduced to Exercise 4.5.22, after using the isomorphism of Proposition 2, but can also be proven directly.]

8. Suppose $\mathscr{A}$ is an algebra of a set Ω, and $\mu: \mathscr{A} \to \mathbb{R}^+$ is merely an additive function. If $\varphi: \mathbb{R}^+ \to \mathbb{R}^+$ is as in the above problem, then the φ-variation of an additive $\nu: \mathscr{A} \to \mathbb{R}$, relative to μ, is again given by (2) in Definition 1. With this, $V^\varphi(\mu) = \{\nu: \mathscr{A} \to \overline{\mathbb{R}}, |\nu|_\varphi(\Omega) < \infty, \nu \text{ additive}\}$ is a vector space, and $\|\cdot\|_\varphi$ of Exercise 7 defines a metric. Show that the conclusions of Exercise 7 hold for $(V^\varphi(\mu), \|\cdot\|_\varphi)$. [The above method of proof does not work here, since $L^\varphi(\mu)$ is not complete when μ is only additive, and so the isomorphism of $L^\varphi(\mu) \to V^\varphi(\mu)$ is *not* onto. Still $V^\varphi(\mu)$, called a *generalized Orlicz space of additive set functions* on $(\Omega, \mathscr{A}, \mu)$, is always complete. The work has to be carried out independently and is difficult but instructive. If φ is also convex, but ν takes values in certain Banach spaces (including the Lebesgue spaces L^p, $1 < p < \infty$, and all Hilbert spaces), the result has been established by J. J. Uhl (1967). In such a study no measurability problems arise. The advantage of working with classes of additive set functions, in lieu of point functions, for some problems is thus exemplified.]

5.2. ABSOLUTE CONTINUITY AND COMPLETE MONOTONICITY OF FUNCTIONS

We first consider the comparison of a pair of additive set functions and then specialize the property to point functions. This gives a feeling for the concept, its deeper implications, and a motivation for the general case, which will be treated in the next two sections.

If (Ω, Σ) is a measurable space and ν, μ are signed measures on Σ, then ν is said to be *absolutely continuous* relative to μ, denoted $\nu \ll \mu$, whenever for any $A \in \Sigma$ one has $|\mu|(A) = 0 \Rightarrow \nu(A) = 0$, where, as usual, $|\mu|(\cdot)$ is the variation of μ. This unsymmetric concept has a symmetric form also, as seen from the following

Proposition 1. *Let μ, ν be signed measures on a measurable space (Ω, Σ). Then the following statements are equivalent:*

(i) $\nu \ll \mu$,

(ii) $\nu^+ \ll |\mu|$ *and* $\nu^- \ll |\mu|$ *(abbreviated as $\nu^\pm \ll \mu$)*,

(iii) $|\nu| \ll |\mu|$,

(iv) $|\nu|$ *is* $|\mu|$*-continuous, i.e., for each* $\epsilon > 0$ *there is a* $\delta_\epsilon > 0$ *such that* $|\mu|(A) < \delta_\epsilon \Rightarrow |\nu|(A) < \epsilon$.

Proof. (i) $\Rightarrow$ (ii): Let $|\mu|(A) = 0$, so that $\nu(A) = 0$ and hence also $|\mu|(B) = 0$, $B \subset A$, $B \in \Sigma \Rightarrow \nu(B) = 0$. So by Proposition 1.3, $|\nu|(A) = 0$. Thus $\nu^+(A) = 0$ and $\nu^-(A) = 0$, which is (ii). That (ii) $\Rightarrow$ (iii) is clear. For (iii) $\Rightarrow$ (iv), by (iii) $|\mu|(A) = 0 \Rightarrow |\nu|(A) = 0$. Since $|\mu|$ and $|\nu|$ are finite measures on Σ, this proves (iv) by Proposition 4.4.8. (Note that the result can be false if ν is an "extended signed measure", since then $|\nu|(\cdot)$ will not be finite, and the above proposition is false for infinite measures.) By the same proposition, (iv) $\Rightarrow$ (i), since $|\mu|(A) = 0 \Rightarrow |\nu|(A) = 0$, so that $\nu(A) = 0$. This completes the proof.

The symmetric definition aluded to is the equivalence of (i) and (iii), though (i) is apparently a weaker concept for a definition. It should however be noted that $\mu(A) = 0 \Rightarrow \nu(A) = 0$ for signed measures is indeed too weak a concept, and this will not be equivalent to any of the conditions of the proposition.

If $\nu \ll \mu$ and $\mu \ll \nu$, then μ and ν have the same class of null sets, and μ, ν are said to be mutually *equivalent*, denoted $\nu \equiv \mu$. For instance, let μ be the Lebesgue measure on $\mathbb{R}$ and $\nu: A \mapsto \int_A f\,d\mu$, where $f(x) = e^{-|x|} > 0$ and A is a Borel set. Then $\mu \equiv \nu$ (but they are *not* equal). If on the other hand, μ, ν are measures on (Ω, Σ) and if there is a set $B \in \Sigma$ such that $\mu(B) = 0$ and $\nu(B^c) = 0$ [or equivalently, $\nu(A) = \nu(A \cap B)$, $A \in \Sigma$], then μ and ν are called mutually *singular* or *orthogonal*, denoted $\mu \perp \nu$. Clearly this is a symmetric relation (i.e., $\mu \perp \nu \Leftrightarrow \nu \perp \mu$), in contrast to absolute continuity. For example, if μ is the Lebesgue measure on $\mathbb{R}$, and if $f(x) = e^{-x}$ for $x \geqslant 0$ and $= 0$ for $x < 0$, then $\nu: A \mapsto \int_A f\,d\mu$ (A Borel) defines a measure such that $\nu \ll \mu$ but $\nu \not\equiv \mu$. On the other hand, let $f: \mathbb{R} \to \mathbb{R}^+$ be given by $f(x) = n$ if $x = n \geqslant 0$, $= 0$ otherwise, so that f is a Borel function, let $\nu(\{x: f(x) = k\}) = \lambda^k$, $0 < \lambda < 1$, $k \geqslant 0$. Then ν is a finite measure on the Borel σ-algebra of $\mathbb{R}$, and if μ is the Lebesgue measure then $\mu \perp \nu$. Thus these are somewhat nonintuitive concepts, and we need to analyze the behavior of measures in relation to those ideas. Note that in the earlier ordering introduced in Section 4.4 for real additive set functions, $\mu \perp \nu$ means that $\inf(\mu, \nu) = 0$.

We now present a first important simplification of the problem, due to H. Lebesgue, and the result is of independent interest.

Theorem 2 (Lebesgue decomposition). *Let (Ω, Σ, μ) be a measure space, and ν be a given σ-finite measure on Σ. Then ν can be uniquely expressed as $\nu = \nu_1 + \nu_2$ where $\nu_1 \ll \mu$ and $\nu_2 \perp \mu$.*

Proof. We first prove the result, assuming $\nu(\Omega) < \infty$, and extend it later to the general case. The method is somewhat similar to that of the second proof of Theorem 1.5, but is shorter. For, let $\mathcal{N} = \{A \in \Sigma: \mu(A) = 0\}$. Then $\mathcal{N}$ is a σ-ring. (It is even an *ideal*, in that it also satisfies $A \in \mathcal{N}$, $B \in \Sigma \Rightarrow A \cap B \in \mathcal{N}$.) Consider $\nu: \mathcal{N} \to \mathbb{R}^+$. Introduce a partial ordering into $\mathcal{N}$ relative to ν, by declaring $A_1 \leqslant A_2$ iff $A_1 \subset A_2$ and $\nu(A_1) \leqslant \nu(A_2)$ for A_1, A_2

in $\mathcal{N}$. We show that each chain $\mathcal{N}_1 \subset \mathcal{N}$ has an upper bound, so that by Zorn's lemma, $\mathcal{N}$ has a maximal element A_0, from which the result will follow.

Thus if $\mathcal{N}_1$ is a chain, let $\alpha = \{\nu(A): A \in \mathcal{N}_1\}$. Then $0 \leqslant \alpha < \infty$, and there exist $A_n \in \mathcal{N}_1$ such that $\nu(A_n) \leqslant \nu(A_{n+1}) \to \alpha$. Since $\mathcal{N}_1$ is linearly ordered, $A_n \subset A_{n+1}$, and if $\tilde{A} = \bigcup_{n=1}^{\infty} A_n$, then $\tilde{A} \in \mathcal{N}$ (being a σ-ring), and if $A \in \mathcal{N}_1$ then $\nu(A) \leqslant \alpha = \nu(\tilde{A})$, so that for some n, $\tilde{A} \supset A \supset A_n$. Thus $\tilde{A}$ is an upper bound for $\mathcal{N}_1$. Hence by Zorn's lemma, $\mathcal{N}$ has a maximal element, say A_0. Define $\nu_i: \Sigma \to \mathbb{R}^+$, $i = 1, 2$, by the equations

$$\nu_2(A) = \nu(A \cap A_0), \quad \nu_1(A) = \nu(A \cap A_0^c), \qquad A \in \Sigma. \tag{1}$$

Then $\nu_1 \ll \mu$, $\nu_2 \perp \mu$, and $\nu = \nu_1 + \nu_2$ is a decomposition. In fact, (1) shows that ν_1, ν_2 are σ-additive. To see that $\nu_1 \ll \mu$, suppose that $\nu_1(A) > 0$ for some $A \in \mathcal{N}$. Then the set $\tilde{A} = A_0 \cup (A_0^c \cap A) \in \mathcal{N}$, and A_0 is a proper subset of $\tilde{A}$. But this contradicts the maximality of A_0, and so $\nu_1(A) = 0$, i.e., $\nu_1 \ll \mu$. Since ν_2 is defined on $\Sigma(A_0) \subset \mathcal{N}$ (ideal property), we have $\nu_2 \perp \mu$.

Now suppose ν is, as given, σ-finite. Then $\Omega = \bigcup_{n=1}^{\infty} \Omega_n$, $\nu(\Omega_n) < \infty$, Ω_n disjoint measurable sets. Let $\nu_n = \nu(\Omega_n \cap \cdot\,)$. Then each ν_n is a finite measure and $\nu = \sum_{n=1}^{\infty} \nu_n$. But by the special case above, $\nu_n = \nu_n^1 + \nu_n^2$, where $\nu_n^1 \ll \mu$ and $\nu_n^2 \perp \mu$, for a pair (ν_n^1, ν_n^2), $n \geqslant 1$. It follows from the definition that a sum of absolutely continuous (singular) measures on Σ, which is clearly a measure, also has the same property. Thus if $\nu^i = \sum_{n=1}^{\infty} \nu_n^i$, then $\nu = \nu^1 + \nu^2$ and $\nu^1 \ll \mu$, $\nu^2 \perp \mu$. The uniqueness of the representation is immediate. Indeed, if $\nu = \nu^1 + \nu^2 = \nu' + \nu''$ are two such representations, then there exist two sets $A_0, \bar{A}_0 \in \mathcal{N}$ such that ν^2 and ν'' vanish outside of these sets, and hence if $B_0 = A_0 \cup \bar{A}_0$, one has $\nu^2(B_0^c) = 0 = \nu''(B_0^c)$. But for any $A \in \Sigma(B_0)$, we have $\nu^1(A) = \nu'(A)$. So $\nu(A) = 0 + \nu^2(A) = 0 + \nu''(A)$; hence $\nu^2 = \nu''$, and then $\nu^1 = \nu'$. This completes the proof of the theorem.

There are other proofs of this result without using Zorn's lemma, but they employ the Radon-Nikodým theorem, to be proved in the next section. For the latter, however, one has to restrict μ to be σ-finite or at least localizable, whereas μ is unrestricted in the above proof. We shall indicate one of these alternative versions later for comparison. The result clearly extends to complex measures.

This theorem is somewhat intriguing; and the structure of these set functions should be analyzed in simpler cases to understand the underlying problem. Thus we consider $\Omega = \mathbb{R}$, with ν a distribution function and μ Lebesgue measure. Then the absolute continuity condition can be expressed as follows. If $f: [a, b] \to \mathbb{R}^+$ is a Lebesgue-measurable function, $F(x) = \int_a^x f(t)\, dt$, then by Proposition 4.4.8, for each $\epsilon > 0$, there is a $\delta_\epsilon > 0$ such that $A = \bigcup_{i=1}^{n} [a_i, b_i]$, $a \leqslant a_i < b_i \leqslant b$, (a_i, b_i), $i = 1, \ldots, n$, disjoint, and $\mu(A) < \delta_\epsilon \Rightarrow$

$$\epsilon > \int_A f\, d\mu = \sum_{i=1}^{n} \int_{a_i}^{b_i} f(t)\, dt = \sum_{i=1}^{n} [F(b_i) - F(a_i)]$$

($d\mu = dt$ here). Thus such an F is absolutely continuous. More generally we have:

Definition 3. A point function $G\colon [a, b] \to \mathbb{R}$ is *absolutely continuous* (relative to the Lebesgue measure, but this relation is usually not mentioned) if for each $\epsilon > 0$ there is a $\delta_\epsilon > 0$ such that for any disjoint intervals $[a_i, b_i)$ from $[a, b]$ satisfying $\sum_{i=1}^n (b_i - a_i) < \delta_\epsilon$ we have $\sum_{i=1}^n |G(b_i) - G(a_i)| < \epsilon$.

The preceding example shows that an indefinite integral of a Lebesgue-integrable function is absolutely continuous. The interesting point here is that the converse is true. This result, due to H. Lebesgue, extends to more general set functions. It will be studied in the next two sections and is of fundamental importance in both theory and applications. We first consider this useful, special but nontrivial, characterization here.

To note the significance, one should remark that not all (even continuous) increasing and bounded functions are absolutely continuous. For instance, the Cantor function defined in Exercise 3.1.7 is of this kind. Recall from calculus that $f\colon [a, b] \to \mathbb{R}$ is of *bounded variation*, namely, if $\pi = \{a = a_0 < a_1 < \cdots < a_n = b\}$ is a partition of $[a, b]$, then

$$v(f; [a, b]) = \sup_{\pi} \sum_{i=1}^{n} |f(b_i) - f(a_i)| < \infty.$$

It follows from Definition 3 that each absolutely continuous real function f on $[a, b]$ is of bounded variation. Indeed, if we take, for instance, $\epsilon = 1$, then there is a $\delta > 0$ such that on every subinterval of length δ, f has variation at most 1. Since $[a, b]$ is compact, we can cover it by a finite number n_0 of intervals of length δ, and hence $v(f; [a, b]) \leqslant n_0 < \infty$. It is also clear that each absolutely continuous function is (uniformly) continuous.

We now state the fundamental result and then discuss its significance:

Theorem 4 (Lebesgue-Vitali). *A function $f\colon [a, b] \to \mathbb{R}$ is absolutely continuous iff it admits a representation as*

$$f(x) = f(a) + \int_a^x f'(t)\, dt, \qquad a \leqslant x \leqslant b, \tag{2}$$

where f' is the derivative of f, which exists at almost all points of $[a, b]$ and is Lebesgue integrable on the interval.

The result is therefore a converse of Proposition 4.4.8 for functions on the line, but gives more information, namely, an absolutely continuous f on the line is the indefinite integral of its derivative, which exists outside of a

Lebesgue null set. Since differentiation is a local operation, the assertion of (2) holds for all intervals on the line, and so on $\mathbb{R}$ itself. After proving (2), we show as application that each continuous convex function has such a representation.

The proof of this result is surprisingly difficult. The basic problem already appears in showing that each indefinite integral on $\mathbb{R}$ is differentiable and that its derivative agrees with the integrand a.e. If the integrand is continuous, except possibly for a finite number of jumps, then the result is the fundamental theorem of elementary calculus. By replacing the continuity of the integrand merely with its integrability, the problem takes on a different character. H. Lebesgue originally solved a special case, and the general result had to wait for the basic covering lemma of G. Vitali. It showed how the classical differential calculus can be extended to the general case and how the geometry of the underlying space plays a crucial role in this work. Thus the geometric measure and integration theories were born, and grew into separate branches of the subject. Because of this circumstance, we present several auxiliary results, some of independent interest, and complete the proof of (2).

Since each absolutely continuous function f on $[a, b]$ is of bounded variation, it is natural to look at the structure of the latter. But we know from advanced calculus [see, e.g., Bartle (1976, p. 226)] that each such f can be written as $f = f_1 - f_2$ with the $f_i: [a, b] \to \mathbb{R}^+$, as bounded nondecreasing functions (the original Jordan decomposition for such f), and our signed measure decomposition stemmed from this result. Hence it suffices to consider a monotone function, and thus we shift the difficult part of proof of (2) to the monotone case. We have

Theorem 5 (Lebesgue). *Let $f: [a, b] \to \mathbb{R}$ be an increasing function. Then f is differentiable a.e. with a Lebesgue-measurable derivative f', and moreover we have*

$$\int_a^b f'(t)\, dt \leqslant f(b) - f(a), \tag{3}$$

with possibly a strict inequality.

The weaker inequality in (3) shows that the absolutely continuous class has a nicer structure than the family of functions of bounded variation and is also a proper subclass (see Exercises 11 and 12). Before proving this we should restate the concept of derivative to avoid ambiguities, since f is not assumed continuous. A general definition due to U. Dini, who worked on the structure of monotone functions before Lebesgue, is as follows. By considering the lim sup and lim inf of the differential quotient from the right and left, one gets four quantities, called the *Dini derivatives* $D^{\pm}f$, $D_{\pm}f$, and if all are equal and finite, then f has a derivative at the point in question. In symbols, if

$f:[a, b] \to \mathbb{R}$ is any function, then we have for $a < x < b$

$$(D^+f)(x) = \overline{\lim_{h \downarrow 0}} \frac{f(x+h) - f(x)}{h},$$

$$(D_+f)(x) = \underline{\lim_{h \downarrow 0}} \frac{f(x+h) - f(x)}{h},$$

$$(D^-f)(x) = \overline{\lim_{h \downarrow 0}} \frac{f(x) - f(x-h)}{h},$$

$$(D_-f)(x) = \underline{\lim_{h \downarrow 0}} \frac{f(x) - f(x-h)}{h}.$$

Thus $f'(x)$ exists at x iff $(D^+f)(x) = (D^-f)(x) = (D_+f)(x) = (D_-f)(x) \in \mathbb{R}$. (Note that $D^{\pm}f$, $D_{\pm}f$ need not be Borel measurable.)

To prove this result we have to establish the previously noted covering lemma. This is technical and unfortunately has no motivation, but it is very plausible. Thus if $A \subset \mathbb{R}$ is any set and $\mathscr{F}$ is a family of closed intervals, then $\mathscr{F}$ is said to be a *Vitali cover* of A whenever it satisfies: (i) $\lambda(I) > 0$ for each $I \in \mathscr{F}$ and (ii) if $x \in A$ and $\epsilon > 0$, there is an $I_\epsilon \in \mathscr{F}$ such that $x \in I_\epsilon$ and $\lambda(I_\epsilon) < \epsilon$, where λ is the Lebesgue length function (or measure). Thus $\mathscr{F}$ has sufficiently many nondegenerate closed intervals of small lengths which cover A. Here we use $d\lambda$ (or dt) to indicate the special nature of the problem.

Theorem 6 (Vitali's covering lemma). *Let $A \subset \mathbb{R}$ be a (not necessarily λ-measurable) subset with $\mathscr{F}$ as a Vitali cover for it. If $\lambda^*(A) < \infty$, then for any $\epsilon > 0$ there exists a finite disjoint collection $\{I_1, \ldots, I_n\}$ of $\mathscr{F}$ such that (λ^* denoting the Lebesgue outer measure)*

$$\lambda^*\left(A - \bigcup_{k=1}^{n} I_k\right) < \epsilon, \tag{4}$$

and if $\lambda^*(A) = +\infty$, there exists a countable disjoint family $\{I_n: n \geqslant 1\}$ of $\mathscr{F}$ such that

$$\lambda^*\left(A - \bigcup_{k=1}^{\infty} I_k\right) = 0. \tag{5}$$

Proof. The argument here is a specialization from the $\mathbb{R}^n$ case given in Saks (1964, p. 109), which goes back to S. Banach. The reader should appreciate the result, since, except for some modifications, it is the only known means of establishing such differentiation theorems, (cf., also Hewitt and Stromberg (1965)).

First let us consider (4), i.e., $\lambda^*(A) < \infty$. Then A is bounded, and so there is an open set $U \supset A$ with $\lambda(U) < \infty$. The fact that $\mathscr{F}$ is a Vitali cover implies that we may assume that each I in $\mathscr{F}$ is contained in U, for otherwise we can replace $\mathscr{F}$ by $\mathscr{F}_0 = \{I \in \mathscr{F}: I \subset U\}$. A disjoint sequence $\{I_n: n \geqslant 1\} \subset \mathscr{F}$ satisfying our conditions will be chosen inductively as follows.

Let $I_1 \in \mathscr{F}$ be arbitrary. If $A \subset I_1$, then we stop. Otherwise, let $I_1, \dots, I_k$ be disjoint intervals of $\mathscr{F}$, chosen inductively, and stop if $A \subset A_k = \bigcup_{j=1}^k I_j$. If not let $\alpha_k = \sup\{\lambda(I): I \subset U - A_k,\ I \in \mathscr{F}\}$. Then $0 < \alpha_k \leqslant \lambda(U) < \infty$. Since $A_k \subset A_{k+1}$, closed, $\subset U$, and $\mathscr{F}$ is a Vitali cover, there exists $I_{k+1} \in \mathscr{F}$ such that $\lambda(I_{k+1}) > \alpha_k/2$ and that is disjoint from $I_1, \dots, I_k$. If the process does not stop, then we get a collection $\{I_n: n \geqslant 1\}$, a disjoint sequence from $\mathscr{F}$, such that

$$\lambda\left(\bigcup_{n=1}^{\infty} I_n\right) = \sum_{n=1}^{\infty} \lambda(I_n) \leqslant \lambda(U) < \infty, \tag{6}$$

since $B = \bigcup_{n=1}^{\infty} I_n \subset U$. Thus $\lim_n \lambda(I_n) = 0$, and $\alpha_k \downarrow 0$. *Claim*: for each ϵ, there exists an n_0 such that $n \geqslant n_0 \Rightarrow$

$$\lambda^*\left(A - \bigcup_{k=1}^{n} I_k\right) < \epsilon. \tag{7}$$

For this choose a larger interval $J_k \supset I_k$, closed, with the same midpoint as I_k, and such that $\lambda(J_k) = 5\lambda(I_k)$, $k \geqslant 1$. Even then, $\bigcup_{k=1}^{\infty} J_k$ is bounded, because (λ is Radon, see Definition 2.3.9 and Theorem 2.2.17)

$$\lambda\left(\bigcup_{k=1}^{\infty} J_k\right) \leqslant \sum_{k=1}^{\infty} \lambda(J_k) = 5 \sum_{k=1}^{\infty} \lambda(I_k) < \infty, \qquad \text{by (6)}.$$

Since λ is a measure, we have $\lim_n \lambda(\bigcup_{k \geqslant n} J_k) = 0$. To prove (7) we first show that $A - B \subset \bigcup_{k \geqslant n} J_k$, $n \geqslant 1$, so that $\lambda(A - B) = 0$. Thus let $x \in A - B \subset A - A_n \subset U - A_n$, $n \geqslant 1$. Then since $U - A_n$ is open, there exists some $I \in \mathscr{F}$ such that $x \in I \subset U - A_n$. Since $\alpha_n \to 0$, there is an n such that $0 < \alpha_n < \lambda(I)$. By definition of α_n, I cannot be $\subset U - A_n$. Let $p = \inf\{n \geqslant 1: \alpha_n < \lambda(I)\}$. Then $A_p \cap I \neq \varnothing$ but $A_{p-1} \cap I = \varnothing$. Since the I_p are disjoint, so that $A_p - A_{p-1} = I_p$, this implies $I_p \cap I \neq \varnothing$. By our choice of p and the selection of the I_n sequence, we have

$$\lambda(I) \leqslant \alpha_{p-1} < 2\lambda(I_p) = \tfrac{2}{5}\lambda(J_p).$$

It therefore follows (draw a picture of these intervals) that

$$x \in I \subset J_p \subset \bigcup_{n \geqslant p} J_n,$$

so that $x \in A - B \subset \bigcup_{n \geqslant p} J_n$. Hence $\lambda^*(A - B) \leqslant \lambda(\bigcup_{n \geqslant p} J_n)$, and the right side goes to zero. From this we now deduce (7).

By (6), for each $\epsilon > 0$ there is an n_0 $[= n_0(\epsilon)]$ such that $n \geqslant n_0$ implies $\sum_{k \geqslant n} \lambda(I_k) < \epsilon$. Hence

$$A - A_n = A - \left(B - \bigcup_{k \geqslant n} I_k\right)$$
$$= (A - B) \cup \left(A \cap \bigcup_{k \geqslant n} I_k\right) \subset (A - B) \cup \bigcup_{k \geqslant n} I_k.$$

Hence $\lambda^*(A - A_n) \leqslant \lambda(\bigcup_{k \geqslant n} I_k) < \epsilon$, since $\lambda^*(A - B) = 0$. This is (7).

For the second part, let $A_n = A \cap (-n, n)$, $0 < n < \infty$, and define $\mathscr{F}_n = \mathscr{F}((-n, n)) = \{I \subset (-n, n): I \in \mathscr{F}\}$. Then $\mathscr{F}_n$ is a Vitali cover of A_n. Since $\lambda^*(A_n) < \infty$, by (4) there is a pairwise disjoint family $\{I_{nk}: k \geqslant 1\} \subset \mathscr{F}_n$, for each n, such that $\lambda^*(A_n - \bigcup_{k=1}^{\infty} I_{nk}) = 0$. If now we consider $\mathscr{I} = \{I_{nk}: k \geqslant 1, 0 < n < \infty\}$, then $\mathscr{I} \subset \mathscr{F}$, countable, and by the above (and σ-subadditivity of λ^*) we get

$$\lambda^*\left(A - \bigcup_n \bigcup_k I_{nk}\right) = 0.$$

This is (5), and hence the proof is complete.

With this technical result at hand, we can now present the

Proof of Theorem 5. Noting that $f: [a, b] \to \mathbb{R}$ is differentiable at x iff all four Dini derivatives are equal and are finite, we show that (i) $(D^+f)(x) = D_+(f)(x) = (D^-f)(x) = (D_-f)(x)$ for a.a. x, and (ii) the common value $(D^+f)(x)$ is finite a.e. Clearly we always have $(D^+f)(x) \geqslant (D_+f)(x)$ and $(D^-f)(x) \geqslant (D_-f)(x)$. So for (i) it suffices to show that $(D^+f)(x) > (D_+f)(x)$ and $(D^-f)(x) > (D_-f)(x)$ hold only on a Lebesgue null net, implying that f'_+ and f'_-, the right and left derivatives, exist a.e. Then one can verify that for *any* real function having right and left derivatives the latter can be different on at most a denumerable set of points of $\mathbb{R}$, which thus is a Lebesgue null set. However, we show directly that, using monotonicity of f (we may assume f increasing), any two of the four Dini derivatives differ only on a Lebesgue null set, and their union is thus a null set, proving (i).

Since the method is the same for all pairs, it suffices to consider one of them, say (D^+f) and (D_-f). Thus let

$$A = \{a \leqslant x \leqslant b: (D^+f)(x) > (D_-f)(x)\}$$
$$= \bigcup_{r, s} \{a \leqslant x \leqslant b: (D^+f)(x) > r > s > (D_-f)(x),\ r, s \text{ rational}\}$$
$$= \bigcup_{r, s} A_{r, s} \qquad \text{(say)}.$$

To show $\lambda^*(A) = 0$, it is enough to verify that $\lambda^*(A_{r, s}) = 0$ for each $r > s$.

Since $\lambda^*(A_{r,s}) \leqslant \lambda([a, b]) = b - a < \infty$ by the fact that λ^* is Carathéodory outer regular, if $\alpha = \lambda^*(A_{r,s})$, then for each $\epsilon > 0$ there is a bounded open set $U \supset A_{r,s}$ such that $\lambda(U) < \alpha + \epsilon$, by Theorem 2.2.17. Defining $f(x) = f(a)$ for $x < a$, and $= f(b)$ for $x > b$, one can find for each $x \in A_{r,s}$ a small interval $\tilde{I}$ such that $x \in \tilde{I} = [x - h, x] \subset U$ for some $h > 0$, and satisfying $0 \leqslant f(x) - f(x - h) < sh$, since f is increasing. But such closed intervals form a Vitali cover, by (4). So there exists a finite family $\{I_1, \ldots, I_m\}$ such that $I_k \subset U$ and

$$\lambda^*\left(A_{r,s} - \bigcup_{k=1}^{m} I_k\right) < \epsilon. \tag{8}$$

If $I_k = [x_k - h_k, x_k]$, $x_k \in A_{r,s}$, and $\mathring{I}_k = (x_k - h_k, x_k)$, then clearly (8) holds if I_k is replaced by $\mathring{I}_k$. Hence $\sum_{k=1}^{m} h_k = \lambda(\bigcup_{k=1}^{m} I_k) \leqslant \lambda(U) < \alpha + \epsilon$, and so

$$0 \leqslant \sum_{k=1}^{m} [f(x_k) - f(x_k - h_k)] < \sum_{k=1}^{m} s h_k < s(\alpha + \epsilon). \tag{9}$$

In a similar manner, for each $y \in A_{r,s}$, we can find a small $h' > 0$ such that $y \in \tilde{J} = [y, y + h'] \subset I_n \subset U_1$ for some n, and $f(y + h') - f(y) > rh'$. Proceeding as above, and using (4) again, we get a finite collection $h'_1, \ldots, h'_{m'}$ such that $\bigcup_{k=1}^{m'} \mathring{J}_k = \bigcup_{k=1}^{m'} (y_k, y_k + h'_k) \subset U$, but $\lambda(\bigcup_{k=1}^{m'} \mathring{J}_k) > \alpha - \epsilon$, and

$$\lambda^*\left(A_{r,s} - \bigcup_{k=1}^{m'} \mathring{J}_k\right) < \epsilon. \tag{10}$$

This gives

$$\sum_{k=1}^{m'} [f(y_k + h'_k) - f(y_k)] > r \sum_{k=1}^{m'} h'_k = r\lambda\left(\bigcup_{k=1}^{m'} J_k\right)$$
$$> r(\alpha - \epsilon). \tag{11}$$

Since $\bigcup_{k=1}^{m} I_k$ and $\bigcup_{k=1}^{m'} \mathring{J}_k$ have measures arbitrarily close to that of $A_{r,s}$ and by construction each $\mathring{J}_k$ is in some I_n, consider all the $\mathring{J}_k \subset I_{k'}$ for a fixed k'. Summing the f increments over those k, we get

$$\sum [f(y_k + h'_k) - f(y_k)] \leqslant f(x_{k'}) - f(x_{k'} - h_{k'}), \tag{12}$$

because these J's are in $(x_{k'} - h_{k'}, x_{k'})$ and f is increasing. Thus summing both sides of (12) for all indices, one gets, with (9) and (11),

$$r(\alpha - \epsilon) < \sum_{k=1}^{m'} [f(y_k + h'_k) - f(y_k)] \leqslant \sum_{k=1}^{m} [f(x_k) - f(x_k - h_k)]$$
$$< s(\alpha + \epsilon).$$

Since $\epsilon > 0$ is arbitrary, we must have $r\alpha \leqslant s\alpha$. But $r > s$ and $0 \leqslant \alpha < \infty$. So this is possible only if $\alpha = 0$. Thus $\lambda^*(A) = 0$, and by our earlier discussion (i) follows.

We show (ii) independently, even though it is a consequence of the last part. Thus let $B = \{a \leqslant x \leqslant b: (D^+f)(x) = +\infty\}$. (Since f is increasing, all the Dini derivatives are nonnegative.) If $x \in B$ and $\delta > 0$ is arbitrary, we can find an $h > 0$ such that $[x, x+h] \subset U$ and

$$f(x+h) - f(x) > \delta h. \tag{13}$$

We again use the same Vitali's lemma, and find $\{I_1, \dots, I_n\}$ such that (4) holds and $\sum_{j=1}^n h_j \leqslant b - a$, $\lambda^*(B - \bigcup_{j=1}^n I_j) < \epsilon$. Thus with (13)

$$\begin{aligned} \delta\lambda(B) &\leqslant \delta\left(\lambda^*\left(\bigcup_{j=1}^n I_j\right) + \epsilon\right) \\ &\leqslant \sum_{j=1}^n \delta h_j + \delta\epsilon \leqslant \sum_{j=1}^n \left[f(x_j + h_j) - f(x_j)\right] + \delta\epsilon \\ &< f(b) - f(a) + \delta\epsilon. \end{aligned} \tag{14}$$

Since $\epsilon > 0$ is arbitrary, (14) implies $\delta\lambda(B) \leqslant f(b) - f(a) < \infty$. But $\delta > 0$ is arbitrary, so we must have $\lambda(B) = 0$. Hence f' exists a.e. and is finite a.e.

For the last part of the theorem, taking $h = 1/n$, we get $f_n(x) = n[f(x + 1/n) - f(x)]$, and by the above, $f_n(x) \to f'(x)$ for a.a. x. Since $f_n(\cdot)$ is a Borel function for each n, we see that f' is a Lebesgue-measurable function, by Proposition 3.2.4. Then by Fatou's lemma, since $f_n \geqslant 0$ a.e.,

$$\begin{aligned} 0 &\leqslant \int_a^b f' \, d\lambda \leqslant \liminf_n \int_a^b f_n \, d\lambda \\ &= \liminf_n \left[n \int_a^b \left[f\left(x + \frac{1}{n}\right) - f(x)\right] dx\right] \\ &= \liminf_n \left[n \int_b^{b+1/n} f(b) \, dx - \int_a^{a+1/n} f(x) \, dx\right] \end{aligned}$$

[since $f(x) = f(b)$ for $x \geqslant b$]

$$\leqslant \liminf_n \left[f(b) - f(a)\right] = f(b) - f(a),$$

because $f(a) \leqslant f(x)$, $x \geqslant a$. This gives (3) and also that f' is finite a.e., which was already shown in (14).

To see there can be strict inequality in (3), we consider the Cantor function (Exercise 3.1.7) on the unit interval. It increases only on the Cantor set and is constant elsewhere. Since this set of increase has measure zero, $f' = 0$ a.e. But $f(1) = 1$, $f(0) = 0$, so that $\int_0^1 f'\,d\lambda = 0$ and $f(1) - f(0) = 1$. Thus (3) cannot be improved in general. This completes the proof.

Before turning to the proof of Theorem 4, we need one more fact, which is also due to Lebesgue and is of independent interest. This is on the differentiation of an indefinite integral. It is a basic result, being an extension of the fundamental theorem of calculus.

Theorem 7. *Let $f\colon [a,b] \to \mathbb{R}$ be a Lebesgue-integrable function. If $F(x) = \int_a^x f\,d\lambda$, then $F'(x) = f(x)$ for a.a. (x). Equivalently,*

$$\lim_{h\downarrow 0} \frac{1}{h}\int_x^{x+h} f\,d\lambda = f(x) \qquad \text{for a.a. } x. \tag{15}$$

Proof. Since for any $h > 0$ we have $(1/h)[F(x+h) - F(x)] = (1/h)\int_x^{x+h} f\,d\lambda$, the equivalence of the two statements is clear. On the other hand, expressing $f = f^+ - f^-$, we get

$$F(x) = \int_a^x f^+\,d\lambda - \int_a^x f^-\,d\lambda = G_1(x) - G_2(x) \qquad \text{(say)}, \tag{16}$$

so that $G_i \uparrow$, and since f (hence $f^{\pm}$) is integrable on $[a,b]$, we have $G_i(b) < \infty$, $i = 1, 2$. Thus F, being the difference of two increasing functions, (is of bounded variation and) has a derivative a.e., i.e., $F'(x) = G_1'(x) - G_2'(x)$ a.e., by Theorem 5. So it suffices to show that $G_1'(x) = f^+(x)$ a.e. and $G_2'(x) = f^-(x)$ a.e. Since the argument is the same in both cases, we only consider G_1.

Set $f_n = \min(f^+, n) \uparrow f^+$. Then $H_n(x) = \int_a^x f_n\,d\lambda \uparrow$ (in x) for each n, and hence $H_n'(x)$ exists a.e. We claim that $H_n'(x) = f_n(x)$ a.e. Indeed, since

$$H_n'(x) = \lim_{h\downarrow 0} \frac{H_n(x+h) - H_n(x)}{h},$$

we have, by letting $h_k \downarrow 0$, for any $a \leqslant x_0 \leqslant b$,

$$\int_a^{x_0} H_n'(t)\,dt = \int_a^{x_0} \lim_{h_k\downarrow 0} \frac{H_n(t+h_k) - H_n(t)}{h_k}\,dt$$

$$= \lim_{h_k\downarrow 0} \int_a^{x_0} \frac{H_n(t+h_k) - H_n(t)}{h_k}\,dt$$

[by the bounded convergence theorem, since $1/h\int_x^{x+h} f_n(t)\,dt \leqslant n$ for all x]

$$= \lim_{h_k \downarrow 0} \frac{1}{h_k}\left(\int_a^{x_0+h_k} H_n(t)\,dt - \int_a^{a+h_k} H_n(t)\,dt - \int_a^{x_0} H_n(t)\,dt\right)$$

$$= \lim_{h_k \downarrow 0} \frac{1}{h_k}\left(\int_{x_0}^{x_0+h_k} H_n(t) - \int_a^{a+h_k} H_n(t)\,dt\right)$$

$$= H_n(x_0) - H_n(a)$$

[since $H_n(\cdot)$ is (absolutely) continuous and we can use the elementary fundamental theorem of calculus]

$$= \int_a^{x_0} f_n(t)\,dt, \qquad \text{by definition.}$$

Thus we have

$$\int_a^{x_0} [H_n'(t) - f_n(t)]\,dt = 0 \qquad \text{for all} \quad a \leqslant x_0 \leqslant b.$$

This implies $H_n'(t) = f_n(t)$ for a.a. t, as shown in Exercise 4.3.12.

Having established the claim, we note that $f^+ - f_n \geqslant 0$, $n \geqslant 1$. So

$$G_1(x) - H_n(x) = \int_a^x (f^+ - f_n)\,d\lambda \geqslant 0, \qquad n \geqslant 1,$$

and both G_1, H_n increase in x. So $G_1 \geqslant H_n$, and they are differentiable a.e. Consequently

$$G_1'(x) \geqslant H_n'(x) = f_n(x) \quad \text{a.e.,} \qquad \text{by the claim.}$$

Letting $n \uparrow \infty$, we get $G_1'(x) \geqslant \lim_n f_n(x) = f^+(x)$ a.e. Thus by Theorem 5 [use the definition of G_1 from (16)]

$$G_1(b) - G_1(a) \geqslant \int_a^b G_1'(x)\,dx \geqslant \int_a^b f^+(x)\,dx = G_1(b) - G_1(a).$$

Hence the nonnegative function $G_1' - f^+$ has zero integral. But this implies $G_1' = f^+$ a.e., and the theorem is proved.

If $f: [a, b] \to \mathbb{R}$ is a function such that f' exists a.e., and $f' = 0$ a.e., then f is called a *singular* function. In case $f' = 0$ everywhere, then from elementary calculus we know that $f=$ constant. However, we have seen above that the Cantor function, which is continuous and increasing (nonconstant) has derivative 0 a.e. It is even known that there are strictly increasing functions with a.e. zero derivative on the line. In this more general context, singular functions exist (continuous, and nonconstant). Thus we have to isolate these, primarily to classify functions. Such a study leads to a digression here, and we are content with the following special but useful characterization.

Proposition 8. *If $f\colon [a,b] \to \mathbb{R}$ is both singular and absolutely continuous, then it is a constant.*

Proof. Even though the assertion seems simple, the proof depends on the Vitali covering lemma, and hence the result is not entirely trivial. We show that $f(a) = f(c)$ for all $a < c \leqslant b$. Consider $A = \{x \in (a,c)\colon f'(x) = 0\}$. Then by hypothesis A has full measure, i.e., $\lambda(A) = \lambda((a,c)) = c - a$. Hence if $x \in A$, and $\epsilon > 0$ is given, we can find $h > 0$ such that $x \in I = [x, x+h] \subset (a,c)$ and $|f(x+h) - f(x)| < \epsilon h$. This shows that such I's form a Vitali covering. Hence if $\delta > 0$ is fixed (to be chosen conveniently later), then by Theorem 6 there is a finite disjoint collection $\{I_1, \ldots, I_n\}$ of intervals $I_k = [x_k, x_k + h_k]$ such that

$$\lambda\left(A - \bigcup_{k=1}^{n} I_k\right) < \delta. \tag{17}$$

On the other hand, since f is absolutely continuous, given $\epsilon > 0$ there exists a $\delta_\epsilon > 0$ (the above δ will be taken $= \delta_\epsilon$) such that, by Definition 3, for any subdivision $a \leqslant y_1 < y_2 < \cdots < y_m \leqslant c$ satisfying $\sum_{i=1}^{m}(y_{i+1} - y_i) < \delta_\epsilon$, one has

$$\sum_{i=1}^{m} |f(y_{i+1}) - f(y_i)| < \epsilon. \tag{18}$$

For a proper choice of the y_i's, using (6), we deduce the desired result.

As shown in the proof of Theorem 6, we have $\lambda(\bigcup_{k=1}^{n} I_k) > (a-c) - \delta$. Since the I_k's are disjoint, we may assume, by relabeling if necessary, that $a < x_1 < x_2 < \cdots < x_n < c$. Hence $A - \bigcup_{k=1}^{n} I_k$ is the following set:

$$A - \bigcup_{k=1}^{n} I_k = (a, x_1) \cup (x_1 + h_1, x_2) \cup \cdots \cup (x_n + h_n, c),$$

and has measure $< \delta = \delta_\epsilon$. So (18) becomes

$$|f(a) - f(x_1)| + \sum_{i=1}^{n-1} |f(x_i + h_i) - f(x_{i+1})| + |f(x_n + h_n) - f(c)| < \epsilon. \tag{19}$$

But from the Vitali condition, we have $\Sigma_{i=1}^{n}|f(x_i + h_i) - f(x_i)| < \epsilon\Sigma_{i=1}^{n}h_i$. So adding and subtracting $f(x_i + h_i)$ in (18), one gets with (19):

$$|f(a) - f(c)| \leqslant [\text{sum in } (19)] + \epsilon \sum_{i=1}^{n} h_i < \epsilon + \epsilon\lambda\left(\bigcup_{k=1}^{n} I_k\right)$$

$$< \epsilon + \epsilon(b - a) = \epsilon(1 + b - a).$$

Since $\epsilon > 0$ is arbitrary, this implies $f(a) = f(c)$, as asserted.

With all this work, we are finally ready to complete the

Proof of Theorem 4. Since an indefinite integral of an integrable function $F: x \mapsto \int_a^x f(t)\,dt$ is absolutely continuous for even more general measures by Proposition 4.3.8, only the converse needs a proof.

If f is absolutely continuous, then it is of bounded variation. So by Theorem 5, f' exists a.e. Let $F(x) = \int_a^x f'(t)\,dt$. Then by Theorem 7, $F'(x) = f'(x)$ for a.a. x, and F is of course absolutely continuous. Thus $F - f$ is absolutely continuous, and since $(F - f)'(x) = F'(x) - f'(x) = 0$ a.e., it is also singular. Hence by Proposition 8, $(F - f)(x) =$ constant. Thus $F(a) - f(a) = F(x) - f(x)$, $a \leqslant x \leqslant b$. By definition $F(a) = 0$. Hence $f(x) = f(a) + F(x)$, $a \leqslant x \leqslant b$, which is (2). Thus the theorem is proved.

Remark. A proof of this result without using the full Theorem 6 has been given by Varberg (1965). He analyzed it much further and gave other characterizations. But one needs Vitali's covering lemma and its conclusion for the abstract case as well as in differentiation. Hewitt and Stromberg (1965) give an extended account.

Two consequences of this important result will now be recorded. The first one plays a key role in probability theory among others.

Proposition 9 (Special Lebesgue decomposition). *Let $F: \mathbb{R} \to \mathbb{R}$ be any bounded monotone (normalized to be left continuous) function. Then it can be uniquely expressed as*

$$F(x) = F_a(x) + F_s(x) + F_d(x), \qquad x \in \mathbb{R}, \tag{20}$$

where F_a is absolutely continuous with $F_a(x) \to 0$ as $x \to -\infty$, F_s is singular, and F_d is discrete, i.e. $F_d(x) = \Sigma_{i<x}a_i$. The sum has at most a countable number of nonzero terms and is absolutely convergent, so that F_d is an elementary function.

Proof. If F is decreasing, then $-F$ is increasing, so we consider only the increasing case. Since F is bounded, by adding a suitable constant if necessary, we may also assume that F is nonnegative, increasing, left continuous, and bounded. If we set $\tilde{F}(x) = \sup\{F(u): u < x\}$, then $\tilde{F}$ has the same character as F and $\tilde{F}(x - 0) = \tilde{F}(x)$. So F is taken as $\tilde{F}$ for convenience. Thus being monotone, F can have at most countably many jumps and no other discontinuities. Let $x_1, x_2, \ldots$ be an enumeration of the discontinuity points of F. Next, letting $a_i = F(x_i + 0) - F(x_i) \geqslant 0$, the jump at x_i, define F_d by

$$F_d(x) = \sum_{i<x} a_i = \sum_{x_i<x} a(x_i), \qquad x \in \mathbb{R}. \tag{21}$$

Then F_d is increasing at the points x_i and $F_d(x) \leqslant F(+\infty) - F(-\infty) < \infty$. So the series converges absolutely. Let $F_c = F - F_d$. Then F_c is increasing, nonnegative, bounded, and continuous. In fact, for any $h > 0$, $x \in \mathbb{R}$,

$$\begin{aligned} F_c(x+h) - F_c(x) &= F(x+h) - F(x) - \sum_{x \leqslant x_i < x+h} a(x_i), \qquad \text{by (21)} \\ &= F(x+h) - F(x+0) - \sum_{x < x_i < x+h} a(x_i) \\ &\downarrow 0 \quad \text{as} \quad h \downarrow 0. \end{aligned}$$

Thus by Theorem 5, $F'(x) = F_c'(x)$ exists a.e. Next, if $F_a(x) = \int_\alpha^x F'(t)\,dt$, where $\alpha \in \mathbb{R}$ is arbitrarily fixed, then F_a is absolutely continuous and $F = F_c - F_a$ is increasing and singular on $[\alpha, \beta] \subset \mathbb{R}$, by Theorem 4. Since the interval $[\alpha, \beta]$ is arbitrary in $\mathbb{R}$, this shows that the same result holds if $\alpha \to -\infty$ and $\beta \to +\infty$. Thus $F = F_c + F_d = F_a + F_s + F_d$ is a decomposition of the required type.

Regarding uniqueness, if $F = F_c + F_d = \tilde{F}_c + \tilde{F}_d$, then $F_c - \tilde{F}_c = \tilde{F}_d - F_d$, and the left side is continuous and the right discrete. So both sides must be zero, whence $F_d = \tilde{F}_d$ and $F_c = \tilde{F}_c$. Similarly, $F_c = F_a + F_s = \tilde{F}_a + \tilde{F}_s$ gives $F_a - \tilde{F}_a = \tilde{F}_s - F_s$. Hence each side is both an absolutely continuous and a singular function. Then by Proposition 8, it must be a constant. But $\lim_{x\to-\infty} F(x) = \lim_{x\to-\infty} F_c(x) = 0 = \lim_{x\to-\infty} F_a(x)$. Hence this constant is zero, and we get $F_a = \tilde{F}_a$ and so $F_s = \tilde{F}_s$. The decomposition is thus unique. Note that if $\lim_{x\to-\infty} F(x) = \alpha > -\infty$, then the same conclusion also obtains, since $(F_a - \tilde{F}_a)(x) \to 0$ as $x \to -\infty$ even then. This time the constant may be added to either F_a or F_s, but adding it to F_s is a matter of convention. With this, uniqueness obtains. This completes the proof.

As another application, we obtain an integral representation for a class of functions which are absolutely continuous. This is the set of convex (hence also the set of concave) functions. They were already used in the study of L^p-spaces (cf. Remark 2 after Proposition 2.4.1). A precise general characterization can now be given as follows.

Theorem 10. *A function $\varphi: (a, b) \to \mathbb{R}$ is convex (concave) iff for each closed subinterval $[c, d] \subset (a, b)$, φ admits a representation as*

$$\varphi(x) = \varphi(c) + \int_c^x f(t)\,dt, \qquad c \leqslant x \leqslant d, \tag{22}$$

where f is a monotone nondecreasing (nonincreasing) left continuous function. Consequently φ has a right and a left derivative at each $a < x < b$, and they are equal at all but at most a countable set of points.

Proof. Since φ convex implies $-\varphi$ concave, we only need to consider one of them. Thus let φ be convex. We show that it is absolutely continuous and its derivative has the stated properties, so that the representation (22) follows as a consequence of (2). [Increasing and nondecreasing are synonymous.]

By definition of convexity, if $0 \leqslant \alpha \leqslant 1$, $\alpha + \beta = 1$, and $c \leqslant x_1, x_2 \leqslant d$, then (cf. the discussion preceding Theorem 4.5.4, and also Exercise 4.5.18)

$$\varphi(\alpha x_1 + \beta x_2) \leqslant \alpha\varphi(x_1) + \beta\varphi(x_2). \tag{23}$$

Let $c \leqslant x < y < z \leqslant d$, and set $\alpha = (y - x)/(z - x)$, $\beta = (z - y)/(z - x)$, and $x_1 = z$, $x_2 = x$ in (23). Then $\alpha x_1 + \beta x_2 = y$, and we get

$$\varphi(y) \leqslant \frac{(y - x)\varphi(z) + (z - y)\varphi(x)}{z - x}.$$

Hence

$$\varphi(y)(z - y + y - x) \leqslant (y - x)\varphi(z) + (z - y)\varphi(x),$$

so that

$$\frac{\varphi(y) - \varphi(x)}{y - x} \leqslant \frac{\varphi(z) - \varphi(y)}{z - y}. \tag{24}$$

Taking $y = x + h$ and $z = y + h'$ here, with $h, h' > 0$ such that x, y, z are in $[c, d]$, one gets

$$\frac{\varphi(x + h) - \varphi(x)}{h} \leqslant \frac{\varphi(y + h') - \varphi(y)}{h'}. \tag{25}$$

Since $x < y$, this shows that the difference quotient is *nondecreasing* in x. Thus (24) and (25) hold in (a, b). Now from (24) and (25) in $[c, d]$, if $c < c_1 \leqslant x < y \leqslant d_1 < d$, then

$$\frac{\varphi(c_1) - \varphi(c)}{c_1 - c} \leqslant \frac{\varphi(y) - \varphi(x)}{y - x} \leqslant \frac{\varphi(d) - \varphi(c)}{d - c}. \tag{26}$$

If we let

$$K_0 = \max\left(\left|\frac{\varphi(c_1) - \varphi(c)}{c_1 - c}\right|, \left|\frac{\varphi(d) - \varphi(c)}{d - c}\right|\right),$$

then one has

$$|\varphi(y) - \varphi(x)| \leqslant K_0|x - y|, \tag{27}$$

so that φ satisfies a *Lipschitz condition* [i.e., (27) holds for $c_1 \leqslant x, y \leqslant d_1$]. But this trivially implies that φ is absolutely continuous, since (27) holds in each closed subinterval of (a, b). Moreover, by (25)

$$(D^+\varphi)(x) = \lim_{h \downarrow 0} \frac{\varphi(x+h) - \varphi(x)}{h} \leqslant \frac{\varphi(d) - \varphi(c)}{d - c} < \infty,$$

and similarly

$$(D^-\varphi)(y) = \lim_{h \downarrow 0} \frac{\varphi(y) - \varphi(y-h)}{h} > -\infty.$$

Thus φ has both a right and a left derivative at *each* point of (a, b). Further, (27) implies for $x < y$

$$(D^+\varphi)(x) \leqslant \frac{\varphi(y) - \varphi(x)}{y - x} \leqslant (D^-\varphi)(y). \tag{28}$$

Hence $(D^+\varphi)(\cdot)$ is increasing, because the first half of (28) gives $(D^-\varphi)(x) \leqslant (D^+\varphi)(x)$; and similarly the second half shows $(D^-\varphi)(\cdot)$ is also increasing. Since a monotone function has at most a countable number of discontinuities, the set of the union of discontinuities of $D^+\varphi$ and $D^-\varphi$ is countable. If x is a continuity point of both of these, then (28) implies that $(D^+\varphi)(x) = (D^-\varphi)(x)$. So Theorem 4 implies (22).

We now consider the converse. Let φ be given by (22). To show that φ is convex, if $[c, d] \subset (a, b)$ and $c < x < d$, we need to verify that any point on the chord joining $(c, \varphi(c))$ and $(d, \varphi(d))$ lies above the arc, i.e., the line $\ell(\cdot)$ satisfies

$$\ell(x) = \varphi(c) + \frac{\varphi(c) - \varphi(d)}{c - d}(x - c) \geqslant \varphi(x). \tag{29}$$

But that is equivalent to showing

$$\frac{\varphi(x) - \varphi(c)}{x - c} \leqslant \frac{\varphi(d) - \varphi(c)}{d - c}.$$

By (22), this follows if we show

$$\frac{1}{x-c}\int_c^x f(t)\,dt \leqslant \frac{1}{d-c}\int_c^d f(t)\,dt. \tag{30}$$

Since f is increasing and $c < x < d$, so that $f(c) \leqslant f(t) \leqslant f(x) \leqslant f(d)$, we have

$$\frac{1}{x-c}\int_c^x f(t)\,dt \leqslant f(x) \leqslant \frac{1}{d-x}\int_x^d f(t)\,dt.$$

Now the right side of (30) can be written as

$$\begin{aligned}\frac{\int_c^x f(t)\,dt + \int_x^d f(t)\,dt}{(d-x)+(x-c)} &\geqslant \min\left(\frac{1}{x-c}\int_c^x f(t)\,dt, \frac{1}{d-x}\int_x^d f(t)\,dt\right)\\ &= \frac{1}{x-c}\int_c^x f(t)\,dt,\end{aligned}$$

because of the following simple numerical inequality for any $\beta_1, \beta_2 > 0$, α_1, α_2 real:

$$\min\left(\frac{\alpha_1}{\beta_1}, \frac{\alpha_2}{\beta_2}\right) \leqslant \frac{\alpha_1+\alpha_2}{\beta_1+\beta_2} \leqslant \max\left(\frac{\alpha_1}{\beta_1}, \frac{\alpha_2}{\beta_2}\right), \tag{31}$$

which is verified at once. Thus (30) is true, and hence φ is convex. This completes the proof.

The preceding analysis shows how Lebesgue's theory on the line enables one to study the structure of classes of measurable functions. In most of these considerations, monotone functions played a key role. From the theory of Stieltjes integrals in advanced calculus, their special importance in integration is perhaps anticipated. To emphasize this point further, we prove another interesting result due to S. N. Bernstein on a subclass of these functions as a final item of this section.

A function $f: J \to \mathbb{R}$ is termed *completely monotone* if it is infinitely differentiable and $(-1)^n f^{(n)}(x) \geqslant 0$, $n \geqslant 0$, for all $x \in J$, where J is an interval in $\mathbb{R}$ and $f^{(n)}$ is the nth derivative of f.

Examples of such functions are $f(x) = \int_0^\infty e^{-xt} h(t)\,dt$, where $h \geqslant 0$ is any bounded measurable function on $J = \mathbb{R}^+$. The interesting point here is that this formula essentially characterizes these functions. More precisely, one has the following

Theorem 11 (Bernstein). *A completely monotone function $f: \mathbb{R}^+ \to \mathbb{R}$ is representable as*

$$f(x) = C + \int_0^\infty e^{-xt}\,dF(t), \qquad x \in \mathbb{R}^+, \tag{32}$$

where $C \geqslant 0$ is a constant and F is a bounded positive, increasing, and left-continuous function.

Even though every function of the form (32) is clearly completely monotone, the converse is nontrivial. For this we need an auxiliary result, called the Helly-Bray theorem, of independent interest. We now establish this with a short proof based on our work on image measures in Section 3.3. An independent proof is also possible, but it is somewhat long and does not show the relation with Lebesgue's dominated convergence theorem. The desired form of the result is as follows.

Theorem 12 (E. Helly, H. E. Bray). *Let $\{G, G_n: n \geqslant 1\}$ be a sequence of left-continuous, nondecreasing, individually bounded functions on $\mathbb{R}$. Suppose that* (i) $\lim_n G_n(x) = G(x)$ *at all continuity points x of G, and* (ii) $\lim_n \lim_{x \to \pm\infty} G_n(x) = \lim_{x \to \pm\infty} G(x)$ [*or briefly* $G_n(\pm\infty) \to G(\pm\infty)$]. *Then for any sequence of bounded continuous functions $h_n: \mathbb{R} \to \mathbb{R}$ such that $h_n \to h$ uniformly as $n \to \infty$, we have*

$$\lim_n \int_{\mathbb{R}} h_n \, dG_n = \int_{\mathbb{R}} h \, dG, \tag{33}$$

where the integrals are in the Lebesgue-Stieltjes sense (cf. Exercise 4.2.5 and Section 2.6 on the latter).

Proof. Since G and G_n are nondecreasing and bounded, for each $\epsilon > 0$ there is an n_ϵ such that $n \geqslant n_\epsilon \Rightarrow$

$$G(-\infty) - \epsilon < G_n(-\infty) \leqslant G_n(x) \leqslant G_n(+\infty) \leqslant G(+\infty) + \epsilon,$$

by (ii). Hence the G_n's are uniformly bounded, say by $M < \infty$, i.e., $|G_n(x)| \leqslant M < \infty$, $x \in \mathbb{R}$. Consider the Lebesgue measure space (Ω, Σ, μ), where $\Omega = (0, M)$, Σ is the Lebesgue σ-algebra of Ω, and μ is Lebesgue measure. Then, as shown in the proof of Theorem 3.3.5, we can define functions X_n, X on Ω as the inverses of G_n, G. The fact that $G_n \to G$ at all continuity points of G becomes: X_n, X are measurable on (Ω, Σ, μ) and $X_n \to X$ a.e.(μ). Thus (33) follows from

$$\lim_n \int_{\mathbb{R}} h_n(x) \, dG_n(x) = \lim_n \int_{\Omega} h_n(X_n) \, d\mu, \quad \text{by Theorem 4.2.6}$$

$$= \int_{\Omega} h(X) \, d\mu, \quad \text{to be established} \tag{34}$$

$$= \int_{\mathbb{R}} h(x) \, dG(x), \quad \text{by Theorem 4.2.6.}$$

Since $h_n \to h$ uniformly (and each h_n is bounded and continuous), h is bounded and continuous, and hence $h_n(X_n) \to h(X)$ a.e. Then (34) holds by the bounded convergence theorem, because μ is a finite measure and $\sup_n |h_n(X_n)| \leqslant K_0 < \infty$ a.e., where K_0 is the common bound of the h_n's. Hence (34) is true and the theorem follows.

The point of Skorokhod's representation and the image measure theorem used above is to reduce the Helly-Bray result to the Lebesgue bounded convergence statement. This is not an obvious but important fact.

With this technical result, we are ready to present the

Proof of Theorem 11. Let $f: \mathbb{R}^+ \to \mathbb{R}$ be completely monotone. Then

$$\begin{aligned} f(x) - f(\infty) &= -\int_x^\infty df = -\int_x^\infty f'(x)\,dx \\ &= -xf'(x)\big|_x^\infty + \int_x^\infty xf'(x)\,dx, \qquad \text{by integrating by parts} \\ &= (-1)^2 \int_x^\infty (u-x) f'(u)\,du, \end{aligned} \tag{35}$$

since $xf'(x) \to 0$ as $x \to \infty$. In fact, we have more generally, on using the complete monotonicity hypothesis, that $(-1)^n f^{(n)}(x)$ is increasing for each $n \geqslant 1$:

$$\begin{aligned} |f^{(n)}(x)| &= (-1)^n f^{(n)}(x) \\ &\leqslant \frac{1}{x} \int_x^{2x} (-1)^n f^{(n)}(x) \\ &= \frac{1}{x} |f^{(n-1)}(2x) - f^{(n-1)}(x)|. \end{aligned}$$

Hence if $n = 1$, then $|xf'(x)| \leqslant |f(2x) - f(x)| \to 0$, since f is monotone, as $x \to \infty$. By induction, we get $|x^n f^{(n)}(x)| \to 0$ as $n \to \infty$, and (35) holds. Indeed, assuming it for n, one has

$$\begin{aligned} f(x) - f(\infty) &= \frac{(-1)^{n+1}}{n!} \int_x^\infty (u-x)^n f^{(n+1)}(u)\,du \\ &= \int_{x/n}^\infty \frac{(-1)^{n+1}}{n!} n \left(1 - \frac{x}{nu}\right)^n (nu)^n f^{(n+1)}(nu)\,du. \end{aligned} \tag{36}$$

Let $g_n(x) = (1 - x/n)^n$, and F_n be defined by

$$F_n(t) = \frac{1}{n!}\int_{1/t}^{\infty}(-1)^{n+1}n(nu)^n f^{(n+1)}(nu)\,du. \tag{37}$$

But g_n is continuous and $g_n(x) \to e^{-x}$ uniformly for $x \geqslant 0$, from advanced calculus, and $F_n \geqslant 0$ for $t > 0$. On the other hand, $F_n(0) = 0$ and

$$\begin{aligned} F_n(\infty) &= \frac{1}{n!}\int_0^{\infty}(-1)^{n+1}(nu)^n f^{(n+1)}(nu)n\,du \\ &= \frac{(-1)^{n+1}}{n!}\int_0^{\infty} t^n f^{(n+1)}(t)\,dt \\ &= -\int_0^{\infty} f'(t)\,dt, \qquad \text{on integrating by parts } n \text{ times} \\ &= f(0) - f(\infty) < \infty, \qquad n \geqslant 1. \end{aligned} \tag{38}$$

Clearly F_n is increasing for each n, and $\{F_n(t): n \geqslant 1\}$ is a bounded set of reals. So by the Bolzano-Weierstrass theorem it has a convergent subsequence with a limit, say $F(t)$ [cf. Theorem 1.2.7(ii)]. However, we have to be more systematic in this selection procedure (also due to E. Helly), since there are infinitely many such subsequences. The argument follows.

Let $\{t_1, t_2, \dots\}$ be an enumeration of the rationals in $\mathbb{R}^+$. Then there is a subsequence $F_{n1}(t_1) \to G_1(t_1)$. Inductively, if $F_{n(k-1)}$ is such a subsequence which converges at $t_1, \dots, t_{k-1}$, choose a subsequence from it, say $F_{nk}(t_i)$, which converges at t_i to $G_k(t_i)$, $1 \leqslant i \leqslant k$. Proceeding in this way, consider the diagonal sequence $\{F_{kk}: k \geqslant 1\}$ which converges at $\{t_1, t_2, \dots\}$, a dense set in $\mathbb{R}^+$. If $t \in \mathbb{R}^+$, define $F(t) = \inf\{G_k(t_k): t_k > t\}$. Then the fact that each F_n is nondecreasing implies that F is nondecreasing, left continuous, and bounded. Also, taking 0 and $+\infty$ in this set, we get $F_{kk}(0) \to F(0)$ and $F_{kk}(\infty) \to F(\infty)$, as well as $F_{kk}(x) \to F(x)$ at all continuity points x of F. (This is a standard computation, and a further detail is left to the reader.)

Now replace F_n by this subsequence in (36), using (37). Because the left side is independent of n, one has

$$\begin{aligned} f(x) - f(\infty) &= \lim_{k\to\infty}\int_0^{\infty} g_k(ux)\,dF_{kk}(u) \\ &= \int_0^{\infty} e^{-ux}\,dF(x), \qquad \text{by Theorem 12.} \end{aligned}$$

Thus

$$f(x) = f(\infty) + \int_0^{\infty} e^{-ux}\,dF(x), \qquad u \geqslant 0.$$

Since $C = f(\infty) \geqslant 0$, by definition of complete monotonicity, this gives (32), and hence the theorem is established.

Many of these results on $\mathbb{R}$ use the structure of the reals, and they are of considerable interest in applications. However, one needs to study the analogs in $\mathbb{R}^n$, $n > 1$. The results are not simple modifications. For instance, there are several other definitions of bounded variation for $f: \mathbb{R}^2 \to \mathbb{R}$, and similarly of absolute continuity. Individual theories exist for differentiation of set functions, if different modifications of Vitali covers (e.g. Besicovitch bases) are employed. For further work, one may consult Saks (1964) or, for a recent survey, Hayes and Pauc (1970).

We now proceed to abstract Radon-Nikodým theory for measures.

EXERCISES

0. Let (Ω, Σ) be a measurable space and μ, ν on Σ be measures. For each $A \in \Sigma$ denote the restrictions μ_A, ν_A as $\mu_A = \mu|\Sigma(A)$, $\nu_A = \nu|\Sigma(A)$. If $A_i \in \Sigma$, $i \in I$ and $A_0 = \cup_i A_i \in \Sigma$, consider $\tilde{\mu} = \Sigma_i \mu_{A_i}$, $\tilde{\nu} = \Sigma_i \nu_{A_i}$. Show that $\tilde{\mu}$ and $\tilde{\nu}$ are measures on Σ and $\tilde{\nu} \ll \tilde{\mu}$ iff $\nu_{A_i} \ll \mu_{A_i}$ for each $i \in I$. Give an example to show that $\tilde{\nu} \ll \tilde{\mu}$ need not imply that $\nu \ll \mu$, but find a sufficient condition for this to hold. How far does the result remain true if μ, ν are signed measures? Is $\tilde{\mu} = \mu_{A_0}$? (Here I indexes Σ.)

1. Let $f: [a, b] \to \mathbb{R}$ be a function of bounded variation. Show that f admits a representation as $f = f_1 + f_2$, where f_1 is absolutely continuous and f_2 is singular, and that the decomposition is unique if we require that the absolutely continuous component at a have a given value, say $f(a)$.

2. If $f: [a, b] \to \mathbb{R}$ is absolutely continuous and $A \subset [a, b]$ is a set of Lebesgue measure zero, show that $f(B)$ is also a Lebesgue null set for each $B \subset A$. [Consider the proof of (i) $\Rightarrow$ (ii) in Proposition 1.]

3. Let $f: [a, b] \to \mathbb{R}$ be a function. If it satisfies a Lipschitz condition [cf. (27) above], then show that f' exists a.e. and is bounded. Give an example to show that the converse is false even when f is monotone.

4. Prove the following lemma, due to Fubini, on the interchange of summation and differentiation: Let $f_n: [a, b] \to \mathbb{R}$ be a sequence of increasing functions such that $S(x) = \Sigma_{n=1}^{\infty} f_n(x)$ exists and $S(x) \in \mathbb{R}$ for a.a. x. Then $S'(x) = \Sigma_{n=1}^{\infty} f_n'(x)$ for a.a. x. [If $S_n = \Sigma_{k=1}^{n} f_k$ and $r_n = S - S_n$, then $r_n \to 0$ a.e., so that $0 \leqslant \Sigma_{n=0}^{\infty} f_n' \leqslant S'$ a.e. Choose a subsequence r_{n_i} such that $\Sigma_{i=1}^{\infty} r_{n_i}$ converges for $x = a$ and $x = b$. So by Corollary 4.3.3, $0 \leqslant \int_a^b \Sigma_{j=1}^{\infty} r_{n_j}' \, dx \leqslant \Sigma_{j=1}^{\infty} [r_{n_j}(b) - r_{n_j}(a)] < \infty$, and $r_{n_j}' \to 0$ a.e., implying $S_{n_j}' \to S'$ a.e.]

5. Let $f: [a, b] \to \mathbb{R}$ be a function of bounded variation. Let $v(f; x)$ be the variation of f on $[a, x]$, $a \leqslant x \leqslant b$. Show that

$$\int_a^b |f'| \, dx \leqslant v(f; b),$$

and there is equality iff f is absolutely continuous. In any case, we always have $v(f; x)' = |f'(x)|$ for a.a. x. [One may use the Jordan decomposition of f here.]

6. A function $f: J \to \mathbb{R}$ is called *absolutely monotone* in an interval $J \subset \mathbb{R}$ if it is infinitely differentiable and its derivatives $f^{(n)}(x) \geqslant 0$, $x \in J$. Show that an absolutely monotone function on $J = \mathbb{R}^- = (-\infty, 0]$ is representable as

$$f(x) = C + \int_0^\infty e^{tx}\, dF(t), \qquad x \in J,$$

where $C \geqslant 0$ is a constant and F is a monotone nondecreasing, left-continuous, bounded function. [Reduce it to Theorem 11.]

7. Let $f: [a, b] \to \mathbb{C}$ be a complex function. With Definition 3, show that f is absolutely continuous iff both $\operatorname{Re} f$ and $\operatorname{Im} f$ have the same property.

8. Let $f: \mathbb{R} \to \mathbb{R}$ be a function. Its symmetric derivative at x in $\mathbb{R}$ is defined as

$$f_s'(x) = \lim_{h \downarrow 0} \frac{f(x+h) - f(x-h)}{2h}$$

Thus if $f(x) = |x|$, then $f_s'(0)$ exists and $= 0$, even though $f'(0)$ does not exist. Show that if f is monotone and bounded on $\mathbb{R}$, then f_s' exists at $x \in \mathbb{R}$ iff $f'(x)$ exists [and then $f'(x) = f_s'(x)$]. [Use Proposition 9.]

9. Let $f: (a, b) \to \mathbb{R}$ be a continuous convex function. Show that its symmetric derivative f_s' exists at *each* point $x \in (a, b)$, and $f_s'(x)\uparrow$. Deduce that its second symmetric derivative exists at a.a. points and is nonnegative. Conversely, the nonnegativity of the second symmetric derivative of f implies the convexity of f in each open subinterval of its domain of definition. [Use Exercise 8 for the converse, and show that f satisfies a Lipschitz condition.]

10. Let $\mathrm{BV}([a, b])$ be the class of all $f: [a, b] \to \mathbb{R}$ of bounded variation, so that $v(f) = v(f: [a, b]) < \infty$. Show that $\mathrm{BV}([a, b])$ is a linear space; that for $a < c < b$ we have $v(f; [a, b]) = v(f; [a, c]) + v(f; [c, b])$; and that if f, g are in $\mathrm{BV}([a, b])$, then $fg \in \mathrm{BV}([a, b])$ also.

11. If $\mathrm{BV}([a, b])$ is as above and $f \in \mathrm{BV}([a, b])$, let $\|f\| = |f(a)| + v(f)$. Show that $(\mathrm{BV}([a, b]), \|\cdot\|)$ is a Banach space. If $\mathscr{A}$ is the algebra generated by the semiring of right-closed subintervals of $[a, b]$, define $\mu_f((c, b]) = f(b) - f(c)$ for $f \in \mathrm{BV}([a, b])$, and extend μ_f by additivity to $\mathscr{A}$. Show that $\tau: f \mapsto \mu_f$ sets up an isometry between $\mathrm{BV}_0([a, b]) = \{f \in \mathrm{BV}([a, b]): f(a) = 0\}$ and $\mathrm{ba}(\Omega, \mathscr{A})$, the space of bounded additive set functions with norm $\|\cdot\|: \mu \mapsto |\mu|(\Omega)$, $\Omega = [a, b]$.

12. If $\Omega = [a, b]$, and μ is the Lebesgue measure, let $F_f(x) = \int_a^x f(t)\, d\mu(t)$, for $f \in L^1(\mu)$ and $a \leqslant x \leqslant b$. Denoting by $\mathrm{AC}_0(\Omega)$ the subspace of absolutely continuous functions of $\mathrm{BV}_0(\Omega)$ of the preceding problem,

show that $(AC_0(\Omega), \|\cdot\|)$ is a closed subspace and that the mapping $\tau: f \mapsto F_f$ defines an isometry between $(AC_0(\Omega), \|\cdot\|)$ and $(L^1(\mu), \|\cdot\|_1)$. Thus $\|F_f\| = v(F_f) = v(\tau f) = \|f\|_1$, and τ maps $AC_0(\Omega)$ onto $L^1(\mu)$. [Use Theorem 4 and Proposition 1.2.]

13. Complete the details of the *Helly selection principle* indicated in the proof of Theorem 11: Let $\{F_n: n \geqslant 1\}$ be a uniformly bounded sequence of nondecreasing left-continuous functions on $\mathbb{R}$. Then there exists at least one subsequence $\{F_{n_i}: i \geqslant 1\}$ such that $F_{n_i} \to F$ at all continuity points of F where F is a bounded left-continuous nondecreasing function.

14. Let $f: [a, b] \to \mathbb{R}$ be a Lebesgue-integrable function. For each $a < x < b$, let $I_h = (x - h, x + h) \subset [a, b]$, with $h > 0$. Show that there is a measurable set $A \subset (a, b)$ of full measure such that for each $x \in A$,

$$\lim_{h \downarrow 0} \frac{1}{2h} \int_{I_h} |f(t) - f(x)|\, dt = 0.$$

Such an A is called a *Lebesgue set* of f. [Apply Theorem 7.]

15. Let $\mathscr{G} = \{B_\alpha: \alpha \in J\}$ be a collection of Lebesgue-measurable sets in $[a, b]$, of positive measure. If $a < x < b$, then $\mathscr{G}$ is said to be *regularly shrinking to* x (x need not be in B_α), denoted $\mathscr{G} \rightsquigarrow x$, provided (i) $d(B_\alpha) = \sup\{|x - y|: x, y \in B_\alpha\} \to 0$ as $\alpha \to \infty$, in the directed set J, and (ii) if $I_h = (x - h, x + h) \supset B_\alpha$ for the smallest $h > 0$, then there is a fixed $0 < k_0 < \infty$ such that $\lambda(I_h) = 2h < k_0 \lambda(B_\alpha)$. Show that for any $f \in L^1(\lambda)$, and for each x in the Lebesgue set of f, we have

$$\lim_{\mathscr{G} \rightsquigarrow x} \frac{1}{\lambda(B_\alpha)} \int_{B_\alpha} |f(t) - f(x)|\, d\lambda = 0.$$

[Apply the preceding result, after using the regularity hypothesis (ii).]

5.3. THE RADON-NIKODÝM THEOREM: SIGMA-FINITE CASE

In the preceding section we just established Lebesgue's theorem on the differentiation of an indefinite integral on the line. That result was extended by J. Radon in 1913 for (what we have called) Radon measures in $\mathbb{R}^n$. This was within a decade after Lebesgue's proof. The final abstract extension took much longer: in 1930 O. M. Nikodým obtained the desired result, which is called the Radon-Nikodým, or sometimes the Lebesgue-Nikodým, theorem. We use all three names to reflect both the establishment and the justification of this crucial result. There are now different methods of proving this theorem, and we present three of them because of its importance as well as their intrinsic interest. All the arguments depend implicitly or explicitly on the Vitali covering lemma. But these are so formulated that the Vitali property is built into the procedure. Thus, the first method uses the Hahn decomposition, which in this context turns out to have the stated property. The second method, due to J. von Neumann, uses a representation of functionals in a Hilbert space. The third is the "probabilist's proof", which is a differentiation using the idea of a martingale.

A few preliminaries are needed for two of the methods and they are proved first. These will be useful later on, and are of independent interest. The results included in Theorems 3 and 5 below are sufficient for most of the usual applications.

Let $L^p(\mu)$ be the Lebesgue space on a measure space (Ω, Σ, μ), $p \geqslant 1$. Then a linear mapping $l: L^p(\mu) \to \mathscr{X}$, a normed linear space [i.e., $l(af + g) = al(f) + l(g)$ for $f, g \in L^p(\mu)$ and scalar a, both spaces being defined on the same scalar field] is *bounded* if there is a $k_0 < \infty$ such that $\|l(f)\|_{\mathscr{X}} \leqslant k_0\|f\|_p$, $f \in L^p(\mu)$. The smallest k_0 for which this holds is denoted by $\|l\|$. Thus $\|l(f)\|_{\mathscr{X}} \leqslant \|l\|\,\|f\|_p$, $f \in L^p(\mu)$. Clearly $\|l\| = \sup\{\|l(f)\|_{\mathscr{X}}: \|f\|_p = 1\} = \sup\{\|l(f)\|_{\mathscr{X}}: \|f\|_p \leqslant 1\}$. However, for linear mappings we find that boundedness, (uniform) continuity, and continuity at a single point are equivalent, thereby showing the restricted (and pleasant) nature of this concept. To see the equivalence, let l be bounded. Then for $f, g \in L^p(\mu)$ we have

$$\|l(f) - l(g)\|_{\mathscr{X}} = \|l(f-g)\|_{\mathscr{X}} \leqslant \|l\|\,\|f-g\|_p. \tag{1}$$

So given $\epsilon > 0$, let $\delta_\epsilon = \epsilon/\|l\|$, excluding the trivial case that $l = 0$. If $\|f - g\|_p \leqslant \delta_\epsilon$, then (1) $\Rightarrow$ $\|l(f) - l(g)\|_{\mathscr{X}} \leqslant \epsilon$ and uniform continuity results, which implies continuity on $L^p(\mu)$ and hence at each given point. Conversely, if l is continuous at f_0, then for $\epsilon > 0$, there is a δ_ϵ $[= \delta(\epsilon, f_0) > 0]$ such that $\|f - f_0\|_p < \delta_\epsilon \Rightarrow \|l(f) - l(f_0)\|_{\mathscr{X}} < \epsilon$. Hence for $0 \neq f \in L^p(\mu)$

$$\left\|\delta_\epsilon \frac{f}{\|f\|_p} + f_0 - f_0\right\|_p \leqslant \delta_\epsilon, \qquad \text{since } \|\cdot\|_p \text{ is a norm,}$$

and

$$\|l(f)\|_{\mathscr{X}} = \frac{\|f\|_p}{\delta_\epsilon}\left\|l\left(\frac{\delta_\epsilon f}{\|f\|_p} + f_0\right) - l(f_0)\right\|_{\mathscr{X}} \leqslant \frac{\epsilon}{\delta_\epsilon}\|f\|_p. \tag{2}$$

Taking $k_0 = \epsilon/\delta_\epsilon$, we get l bounded (by k_0). So we can use these two properties interchangeably. If $\mathscr{X}$ is the scalars (real or complex numbers), then l is called a *linear functional.*

The following result was proved by F. Riesz, but our demonstration in its key part is a specialized one from a more inclusive proof due to E. J. McShane. The general form will also be given later (Theorem 5.5), but the present one is simple and motivates it.

Theorem 1. *For each bounded linear functional l on $L^2(\mu)$, there is an $f_0 \in L^2(\mu)$ such that $l(f_0) = \|l\|$, and moreover, if $\tilde{f}_0 = \|l\| f_0$, then*

$$l(f) = (f, \tilde{f}_0) = \int_\Omega f\bar{\tilde{f}}_0\, d\mu, \qquad f \in L^2(\mu). \tag{3}$$

In both equations (3) *and $l(f_0) = \|l\|$, the f_0 and $\tilde{f}_0$ are unique.*

Proof. The inner product notation is used in the first equation of (3) to indicate that this result is true for any abstract Hilbert space, and we use very little of the measure theoretical properties here. However, this is the most general case, since we will show later (in Chapter 6) that *every (abstract) Hilbert space is isometrically isomorphic to a closed subspace of $L^2(\bar{\mu})$ on some probability space $(\bar{\Omega}, \bar{\Sigma}, \bar{\mu})$* (Theorem 6.5.5).

As noted above, $\|l\| = \sup\{|l(g)|: \|g\|_2 = 1\}$, where the absolute value $|\cdot|$ is the norm of $\mathscr{X}=$ scalars here. Hence there exist $g_n \in L^2(\mu)$, $\|g_n\|_2 = 1$, such that $\|l\| = \lim_n |l(g_n)|$. Let $f_n = [\operatorname{sgn} l(g_n)]g_n$. Then $\|f_n\|_2 = 1$ and $l(f_n) = [\operatorname{sgn} l(g_n)]l(g_n) = |l(g_n)|$, since $\operatorname{sgn} l(g_n)$ is a scalar. [Note that for a complex number z the signum function $\operatorname{sgn}(\cdot)$ is defined as $\operatorname{sgn}(z) = \exp(i \arg z)$ if $z \neq 0$, $= 0$ if $z = 0$, where $\arg z$ is the argument of the complex number z. This formula includes the real case, since $\arg(-x) = \pi$ for $x > 0$.] Thus $\|l\| = \lim_n l(f_n)$. We claim that $\{f_n: n \geqslant 1\}$ is a Cauchy sequence in $L^2(\mu)$. So it converges to an $f_0 \in L^2(\mu)$ as the required element.

Indeed, if $0 < \epsilon < 1$ is given, there is an n_0 $[= n_0(\epsilon)]$ such that $n \geqslant n_0 \Rightarrow l(f_n) \geqslant \|l\|(1-\epsilon)$. Then for $n, m \geqslant n_0$,

$$\|l\| \left\| \frac{f_n + f_m}{2} \right\|_2 \geqslant l\left(\frac{f_n + f_m}{2}\right), \qquad \text{since} \quad l(f_n) \geqslant 0$$

$$= \tfrac{1}{2}[l(f_n) + l(f_m)] \geqslant \|l\|(1-\epsilon). \tag{4}$$

If $\|l\| = 0$, then (3) is trivial; $f_0 = 0$, will satisfy. So let $\|l\| > 0$. Then (4) gives

$$\|f_n + f_m\|_2 \geqslant 2(1-\epsilon).$$

Squaring and using the inner product properties, we get

$$\|f_n\|_2^2 + \|f_m\|_2^2 + 2\operatorname{Re}(f_n, f_m) = \|f_n + f_m\|_2^2 \geqslant 4(1-\epsilon)^2. \tag{5}$$

Hence

$$\|f_n - f_m\|_2^2 = \|f_n\|_2^2 + \|f_m\|_2^2 - 2\operatorname{Re}(f_n, f_m)$$

$$\leqslant 2 - \left[4(1-\epsilon)^2 - 2\right], \qquad \text{by (5)}$$

$$= 8\epsilon - 4\epsilon^2 < 8\epsilon. \tag{6}$$

Thus $\{f_n: n \geqslant 1\}$ is Cauchy, and $f_n \to f_0$ in $L^2(\mu)$, since $L^2(\mu)$ is complete. Also $|\|f_n\|_2 - \|f_0\|_2| \leqslant \|f_n - f_0\|_2 \to 0$, so that $\|f_0\|_2 = 1$, as asserted. But then $|l(f_0) - l(f_n)| = |l(f_0 - f_n)| \leqslant \|l\| \, \|f_0 - f_n\|_2 \to 0$, so that $l(f_0) = \lim_n l(f_n) = \|l\|$. This gives the first part.

For ease of writing we take $\|l\| = 1$ (otherwise $\tilde{l} = l/\|l\|$ will do). The beautiful idea of McShane's is this. For any scalar t, and $f \in L^2(\mu)$, we note that

$$\begin{aligned} 1 = l(f_0) &= l(f_0 + t[f - l(f)f_0]) \\ &\leqslant \| f_0 + t[f - l(f)f_0] \|_2 \qquad \text{for all } t, \end{aligned} \tag{7}$$

since $\|l\| = 1$. If $tl(f) \neq -1$, then

$$f_0 + tf = [1 + tl(f)]\left(f_0 + \frac{t}{1 + tl(f)}[f - l(f)f_0] \right).$$

Since the norm of the right side of (7) is at least 1 and equals 1 at $t = 0$, we have, with the above expression,

$$\begin{aligned} \|f_0 + tf\|_2 - 1 &= |1 + tl(f)| \left\| f_0 + \frac{t}{1 + tl(f)}(f - l(f)f_0) \right\|_2 - 1 \\ &\geqslant |1 + tl(f)| - 1, \qquad \text{by (7)}. \end{aligned}$$

Hence if $t > 0$ and small (use $|z| \geqslant |\text{Re}\, z|$),

$$\frac{\|f_0 + tf\|_2 - 1}{t} \geqslant \frac{|1 + tl(f)| - 1}{t} \geqslant \text{Re}\, l(f), \tag{8}$$

and if $t < 0$ and small, we get the reverse inequality. Thus if $\psi_{f_0, f}(t) = \|f_0 + tf\|_2$, then (8) implies that on letting $t \downarrow 0$ and $t \uparrow 0$,

$$\left(D^- \psi_{f_0, f}\right)(0) \leqslant \text{Re}\, l(f) \leqslant \left(D^+ \psi_{f_0, f}\right)(0). \tag{9}$$

However, $\psi_{f_0, f}^2(t) = 1 + t^2\|f\|_2^2 + 2t\,\text{Re}(f, f_0)$, and so $\psi_{f_0, f}(\cdot)$ is differentiable at $t = 0$, with derivative $2\psi_{f_0, f}(0)\psi'_{f_0, f}(0) = 2\psi'_{f_0, f}(0) = 2\,\text{Re}(f, f_0)$. Thus,

$$\left(D^- \psi_{f_0, f}\right)(0) = \left(D^+ \psi_{f_0, f}\right)(0) = \text{Re}(f, f_0) = \text{Re}\, l(f), \qquad \text{by (9)}.$$

But $\text{Im}\, l(f) = \text{Re}\, l(-if) = \text{Re}(-if, f_0) = \text{Im}(f, f_0)$. It follows that $l(f) = (f, f_0)$.

Finally, to prove uniqueness, let f_0' be another element such that $l(f) = (f, f_0) = (f, f_0')$, or $(f, f_0 - f_0') = 0$, $f \in L^2(\mu)$. Taking $f = f_0 - f_0' \in L^2(\mu)$, we get $\|f_0 - f_0'\|_2 = 0$, or $f_0 = f_0'$ a.e., giving uniqueness. Note that $\|l\| = l(f_0)$ is also for a unique f_0, since if f_0'' is another such element, then $\|l\| = [l(f_0) + l(f_0'')]/2 \leqslant \|l\|\,\|(f_0 + f_0'')/2\|_2 \leqslant \|l\|$, because $\|f_0\|_2 = \|f_0''\|_2 = 1$, and the triangle inequality can be used. So there is equality and $\|(f_0 + f_0'')/2\|_2 = \|f_0\|_2 + \|f_0''\|_2$. This implies $f_0 = \alpha f_0''$ for some $\alpha > 0$, and then $\alpha = 1$, since $\|f_0\|_2 = 1$. This completes the proof.

It should be noted that there are no (σ-)finiteness restrictions on the measure space (Ω, Σ, μ) in the above proof. Further, the argument depends on only two critical facts: (i) for each bounded linear functional l on the space $L^p(\mu)$ there exists an element $f_0 \in L^p(\mu)$ of unit norm such that $l(f_0) = \|l\|$, and (ii) the function $\psi_{f_0, f}: t \mapsto \|f_0 + tf\|_p$ is differentiable at $t = 0$. Our later work shows that these two results hold not only for $L^2(\mu)$, as we have just seen, but for all $L^p(\mu)$, $1 < p < \infty$. (Cf. Theorem 5.5 below.)

In order to present the "probabilistic proof", we need to introduce a concept. Thus if (Ω, Σ, μ) is a measure space, $\Sigma_n \subset \Sigma_{n+1} \subset \Sigma$, $n = 1, 2, \ldots$, is a sequence of σ-subalgebras, let $f_n \in L^1(\mu)$ be such that f_n is measurable relative to Σ_n, $n \geqslant 1$. This means, if we denote by $L^1(\Omega, \Sigma_n, \mu) \subset L^1(\Omega, \Sigma, \mu)$ then $f_n \in L^1(\Omega, \Sigma_n, \mu)$, $n \geqslant 1$. The sequence $\{f_n, \Sigma_n: n \geqslant 1\}$ is called a *martingale* if for each n we have the following system of equations:

$$\int_A f_n \, d\mu = \int_A f_{n+1} \, d\mu, \qquad A \in \Sigma_n. \tag{10}$$

Hence by iteration, on using the fact that $A \in \Sigma_m \subset \Sigma_n$, $n \geqslant m$,

$$\int_A f_m \, d\mu = \int_A f_n \, d\mu, \qquad n > m \text{ and } A \in \Sigma_m. \tag{11}$$

Taking $A = \Omega$, it follows that all the f_n's have the same (total) integrals. However, it is not obvious that there really exist such sequences. We now exhibit a simple but important example of a martingale. To gain some insight, suppose $A \in \Sigma_n$ is an atom for $\mu|\Sigma_n$, $0 < \mu(A) < \infty$. Then f_n is constant on A, and (10) becomes

$$f_n(A)\mu(A) = \int_A f_{n+1} \, d\mu,$$

so that $f_n(A) = [1/\mu(A)] \int_A f_{n+1} \, d\mu$, i.e., on each such atom A, the constant $f_n(\omega)$ is the average of f_{n+1}, and hence also of f_{n+k} for all $k \geqslant 1$, by (11). Similarly if, $A_1, \ldots, A_k$ are k atoms of positive finite measure in Σ_m, then by the same argument, we get

$$f_m(\omega) = \sum_{i=1}^{k} \frac{1}{\mu(A_i)}\left(\int_{A_i} f_{m+n}\, d\mu\right)\chi_{A_i}(\omega), \qquad \omega \in \bigcup_{i=1}^{m} A_i, \quad n \geqslant 1. \tag{12}$$

This suggests a way of generating martingales.

Example A. Let (Ω, Σ, μ) be a measure space with $0 < \mu(\Omega) < \infty$, and $\nu: \Sigma \to \mathbb{R}$ be a signed measure. If Π_1 is any measurable partition of Ω [i.e., $\Pi_1 = \{A_i: i = 1, 2, \dots\}$, $\bigcup_{i=1}^{\infty} A_i = \Omega$, $A_i \in \Sigma$, disjoint, $\mu(A_i) > 0$], then let Π_2 be a refinement of Π_1, and generally let Π_{n+1} be a refinement of Π_n. Thus each member of Π_{n+1} is a subset of some member of Π_n. If $\Pi_n = \{A_{ni}: i = 1, 2, \dots\}$, $n \geqslant 1$, we define (all partitions being measurable)

$$f_n = \sum_{i=1}^{\infty} \frac{\nu(A_{ni})}{\mu(A_{ni})}\chi_{A_{ni}}, \qquad n \geqslant 1. \tag{13}$$

If Σ_n is the σ-algebra generated by the partition Π_n, then the refinement property ensures that $\Sigma_n \subset \Sigma_{n+1}$, and by construction f_n is Σ_n-measurable. We claim that $\{f_n, \Sigma_n: n \geqslant 1\}$ is a martingale. Since each Σ_n is countably generated, it is sufficient to check (10) for the generators A_{ni}. But then these are atoms for μ on Σ_n, and

$$\begin{aligned}\int_{A_{ni}} f_n\, d\mu &= \nu(A_{ni}) = \nu\left(\bigcup_j A_{(n+1)j}\right) \\ &= \sum_j \nu\left(A_{(n+1)j}\right) = \int_{A_{ni}} \sum_j \frac{\nu\left(A_{(n+1)j}\right)}{\mu\left(A_{(n+1)j}\right)}\chi_{A_{(n+1)j}}\, d\mu \\ &= \int_{A_{ni}} f_{n+1}\, d\mu. \end{aligned}\tag{14}$$

Here we used the additivity of ν and the fact that A_{ni} is (at most) a countable union of disjoint sets from Π_{n+1}. Note that in this case the martingale equation holds for any signed measure ν on Σ, and thus we have, for given μ, many martingales (one for each ν) on the space (Ω, Σ, μ). These martingales are sufficient for our work here.

We now establish the following special

Proposition 2. *Let* (Ω, Σ, μ) *be a finite measure space, and* $\nu: \Sigma \to \mathbb{R}^+$ *be a measure. Define* f_n *and* Σ_n *as in* (13) *and* (14) *for this* ν. *Let* $\nu(A) \leqslant \mu(A)$,

$A \in \Sigma$. Then the martingale $\{f_n, \Sigma_n: n \geqslant 1\}$ converges in $L^2(\mu)$ to an f_∞, and f_∞ is measurable relative to $\Sigma_\infty = \sigma(\bigcup_n \Sigma_n) \subset \Sigma$.

Proof. It was already shown in (14) that $\{f_n, \Sigma_n: n \geqslant 1\}$ is a martingale, and each f_n is an elementary function. Also $f_n \in L^2(\mu)$. In fact,

$$\begin{aligned}\int_\Omega f_n^2\, d\mu &= \sum_{k=1}^{\infty} \int_\Omega \left(\frac{\nu(A_{nk})}{\mu(A_{nk})}\right)^2 \chi_{A_{nk}}\, d\mu \\ &= \sum_{k=1}^{\infty} \left(\frac{\nu(A_{nk})}{\mu(A_{nk})}\right)^2 \mu(A_{nk}) \\ &= \sum_{k=1}^{\infty} \frac{\nu(A_{nk})}{\mu(A_{nk})}\nu(A_{nk}) \leqslant \sum_{k=1}^{\infty} \nu(A_{nk}) = \nu(\Omega) < \infty, \end{aligned} \tag{15}$$

since $0 \leqslant \nu(A) \leqslant \mu(A)$ by hypothesis and μ is a finite measure. Thus, the f_n are in the nonnegative part of the ball of radius $\sqrt{\nu(\Omega)}$ in $L^2(\mu)$. To prove that the f_n form a Cauchy sequence, we need to prove two properties: (i) $\int_\Omega f_n^2\, d\mu \leqslant \int_\Omega f_{n+1}^2\, d\mu$, and (ii) $\int_\Omega f_n f_{n+1}\, d\mu = \int_\Omega f_n^2\, d\mu$.

For (i) let $a_n = \int_\Omega f_n^2\, d\mu$ [$\leqslant \nu(\Omega)$ by (15)]. But

$$a_n = \sum_{k=1}^{\infty} \left(\frac{\nu(A_{nk})}{\mu(A_{nk})}\right)^2 \mu(A_{nk}) = \sum_{k=1}^{\infty} \left(\frac{\Sigma_j \nu(A_{(n+1)j}^k)}{\mu(A_{nk})}\right)^2 \mu(A_{nk})$$

(since by refinement each A_{nk} is divided into m_k subsets $A_{(n+1)j}^k$ of positive μ-measure, and $1 \leqslant j \leqslant m_k$)

$$\leqslant \sum_k \sum_j \left(\frac{\nu(A_{(n+1)j}^k)}{\mu(A_{(n+1)j}^k)}\right)^2 \mu(A_{(n+1)j}^k)$$

[since by definition of a convex function φ we have $\varphi(\Sigma_{i=1}^k \alpha_i x_i) \leqslant \Sigma_{i=1}^k \alpha_i \varphi(x_i)$, $\alpha_i > 0$, with $\Sigma_{i=1}^k \alpha_i = 1$, $\varphi(x) = x^2$, $\alpha_i = \mu(A_{(n+1)i}^k)/\mu(A_{nk})$, and $x_i = \nu(A_{(n+1)i}^k)/\mu(A_{(n+1)i}^k)$]

$$= \sum_{j \geqslant 1} \left(\frac{\nu(A_{(n+1)j})}{\mu(A_{(n+1)j})}\right)^2 \mu(A_{(n+1)j})$$

[by rearrangement for the $(n+1)$st partition]

$$= \int_\Omega f_{n+1}^2\, d\mu = a_{n+1}. \tag{16}$$

Thus $0 \leqslant a_n \leqslant a_{n+1} \leqslant \cdots \leqslant \nu(\Omega) < \infty$, so that $a_n \to a \leqslant \nu(\Omega)$ as $n \to \infty$.

Property (ii) is similarly verified. Indeed

$$\int_\Omega f_n f_{n+1}\,d\mu = \sum_{k,\,j\geqslant 1} \frac{\nu(A_{nk})}{\mu(A_{nk})}\frac{\nu(A_{(n+1)j})}{\mu(A_{(n+1)j})}\mu(A_{nk}\cap A_{(n+1)j})$$

$$= \begin{cases} \displaystyle\sum_{k,\,j\geqslant 1} \frac{\nu(A_{nk})}{\mu(A_{nk})}\nu(A_{(n+1)j}) & \text{if } A_{(n+1)j}\subset A_{nk}, \\ 0 & \text{if } A_{(n+1)j}\cap A_{nk}=\varnothing \end{cases}$$

$$= \sum_{k=1}^{\infty} \frac{\nu(A_{nk})}{\mu(A_{nk})}\nu(A_{nk})$$

(since ν is additive, and $A_{nk} = \bigcup_j (A_{(n+1)j})$)

$$= \sum_{k=1}^{\infty} \left(\frac{\nu(A_{nk})}{\mu(A_{nk})}\right)^2 \mu(A_{nk})$$

$$= \int_\Omega f_n^2\,d\mu, \qquad n \geqslant 1. \tag{17}$$

By the same argument (or induction), we get $\int_\Omega f_n f_{n+k}\,d\mu = a_n$. Consider

$$0 \leqslant \int_\Omega (f_n - f_{n+k})^2\,d\mu$$

$$= \int_\Omega f_n^2\,d\mu + \int_\Omega f_{n+k}^2\,d\mu - 2\int_\Omega f_n f_{n+k}\,d\mu$$

$$= a_n + a_{n+k} - 2a_n$$

[by (17) and the induction noted above]

$$= a_{n+k} - a_n \to 0 \qquad \text{as} \quad n, k \to \infty,$$

by (16), since $a_n \to a < \infty$. Thus $\{f_n\colon n \geqslant 1\}$ is a Cauchy sequence in $L^2(\mu)$, and by its completeness, $f_n \to f_\infty$ in $L^2(\mu)$ and hence in measure. Clearly f_∞ is Σ_∞-measurable. This completes the proof.

Remark. The above result for positive martingales in $L^2(\mu)$ is true more generally than for the simple functions f_n. However, the general case uses the properties given by (16) and (17), which for the nonelementary case depend on the Radon-Nikodým theorem. Since our aim is to prove the latter with the preceding result, this is all that we can establish at this time. Fortunately it is enough here. [For the general case, see Proposition 5.18(iii) below.]

We are now ready to prove the fundamental Lebesgue-Radon-Nikodým result for arbitrary *finite* measures. The σ-finite case then follows from it very quickly. The general form is established in the next section. As will be seen here, the finite case is a key step in the whole process. Because of its importance and the hidden Vitali covering system in it (cf. Exercises 5 and 6), we present three different proofs of the theorem, since each one adds something to the understanding of this crucial result.

Theorem 3 (Lebesgue-Radon-Nikodým). *Let* (Ω, Σ) *be a measurable space and* μ, ν *be two finite measures on* Σ. *Then there exists two measures* ν_1, ν_2 *on* Σ *and a set* $A_0 \in \Sigma$ *satisfying*

(i) $\nu_1 \ll \mu$, $\nu_2 \perp \mu$, *and* $\nu = \nu_1 + \nu_2$ *uniquely*; $\mu(A_0) = 0$, *but* $\nu_2(A_0^c) = 0$ (*i.e., the Lebesgue decomposition*),

(ii) *there is a* μ-*unique* $0 \leqslant f_0 \in L^1(\mu)$, *such that* (a) $f_0 = 0$ *on* A_0 *and* (b)

$$\nu(A) = \int_A f_0 \, d\mu + \nu_2(A \cap A_0), \qquad A \in \Sigma. \tag{18}$$

In particular, if $\nu \ll \mu$ *then* $\nu_2 = 0$ *in* (18).

Remark. In the last case f_0 is called the ***Radon-Nikodým derivative*** of ν relative to μ and denoted $f_0 = d\nu/d\mu$. In the general case $f_0 = d\nu_1/d\mu$ is termed the derivative of the absolutely continuous part of ν relative to μ. (ν_1, ν_2 are often denoted: $\nu_1 = \nu^c$, $\nu_2 = \nu^s$.)

First proof (probabilistic). Let $\Pi_1, \Pi_2, \ldots$ be a sequence of refinement partitions (countable at most) of Ω, and $\Sigma_k = \sigma(\Pi_k)$, with $\Sigma_1 \subset \Sigma_2 \subset \cdots \subset \Sigma$. Let $\xi = \nu + \mu: \Sigma \to \mathbb{R}^+$. Then ξ is a finite measure such that $\nu(A) \leqslant \xi(A)$, $A \in \Sigma$. Let f_n be defined by (13) for ν and ξ. Then by Proposition 2, $\{f_n, \Sigma_n: n \geqslant 1\}$ is a positive martingale on (Ω, Σ, ξ) in a ball of $L^2(\xi)$. Hence $f_n \to f_\infty$ in $L^2(\xi)$, and f_∞ is Σ_∞-measurable. Note that Σ_∞ is countably generated, namely, the generators are all the sets in $\{\Pi_k: k \geqslant 1\}$. Since $\xi(\Omega) < \infty$, so that $L^2(\xi) \subset L^1(\xi)$, we have $\{f_n: n \geqslant 1\}$ to be uniformly integrable. Now using (14) one obtains

$$\int_A f_n \, d\xi = \int_A f_{n+k} \, d\xi, \qquad A \in \bigcup_{n=1}^{\infty} \Sigma_n.$$

So letting $k \to \infty$, since $A \in \Sigma_m$ for some m, we get for all large n ($\geqslant m$), that

$$\nu(A) = \int_A f_n \, d\xi = \int_A f_\infty \, d\xi. \tag{19}$$

But if $\tilde{\nu}(A) = \int_A f_\infty \, d\xi$, then $\tilde{\nu}$ is a finite measure $[f_\infty \in L^2(\xi) \subset L^1(\xi)]$, and ν and $\tilde{\nu}$ are finite measures agreeing on the algebra $\bigcup_{n=1}^{\infty} \Sigma_n$. [The fact that this union of monotone σ-algebras is an algebra (but not necessarily a σ-algebra) is clear and was also noted in Chapter 2.] Thus (19) holds for all $A \in \Sigma_\infty$, so that f_∞ is also $\xi|\Sigma_\infty$-unique. Since every countable generated σ-algebra can be so obtained, this holds for all such algebras.

If Σ is not countably generated, let $\mathscr{C}$ be the set of all σ-subalgebras of Σ which are countably generated; partially order $\mathscr{C}$ by inclusion, and index its elements by I. Then by the preceding part, for each $\Sigma_i \in \mathscr{C}$, we have an f_i (Σ_i-measurable) such that

$$\nu(A) = \int_A f_i \, d\xi, \qquad A \in \Sigma_i, \quad i \in I. \tag{20}$$

If $i < i'$ (so $\Sigma_i \subset \Sigma_{i'}$), then $\nu|\Sigma_i = (\nu|\Sigma_{i'})|\Sigma_i$, so that $\{f_i, \Sigma_i : i \in I\}$ is a directed positive martingale in a ball (of radius $[\nu(\Omega)]^{1/2}$), so that it is uniformly integrable, i.e.,

$$\int_\Omega f_i^2 \, d\xi \leqslant \nu(\Omega) < \infty$$

and

$$\lim_{\xi(A) \to 0} \int_A f_i \, d\xi = 0 \qquad \text{uniformly in } i.$$

The latter is a consequence of the CBS inequality, since it is bounded by $[\nu(\Omega)\xi(A)]^{1/2}$. With this we claim that $f_i \to \tilde{f}_\infty$ in $L^2(\xi)$, for some (unique) $\tilde{f}_\infty$ in the latter space.

We prove this using the argument outlined in Exercise 4.3.10. Thus if the assertion is false, there exists an $\epsilon > 0$ and for each $i \in I$ one can find a pair of indices i_1, i_2 such that $\|f_{i_1} - f_{i_2}\|_2 \geqslant \epsilon$. By the directedness, if $j \geqslant i_1, i_2$ ($\Sigma_j \supset \Sigma_{i_1} \cup \Sigma_{i_2}$ will do), then $\|f_{i_1} - f_j\| \geqslant \epsilon/2$ or $\|f_{i_2} - f_j\| \geqslant \epsilon/2$. Suppose it is the former; then denoting this j by j_1, we proceed to obtain $j_2 \geqslant j_1$ such that for all $j' \geqslant j_2$, $\|f_{j_2} - f_{j'}\|_2 \geqslant \epsilon/2$. Then by induction, we get $j_1 \leqslant j_2 \leqslant \cdots$ such that $\|f_{j_k} - f_{j_{k-1}}\|_2 \geqslant \epsilon/2$. So $\{f_{j_k} : k \geqslant 1\}$ does not converge in $L^2(\xi)$. But $\{f_{j_k}, \Sigma_{j_k} : k \geqslant 1\}$ is a uniformly integrable martingale, and each Σ_{j_k} is countably generated, so by the first part $f_{j_k} \to g$ in $L^2(\xi)$. But this is a contradiction to the above inequalities, and hence our supposition cannot hold, so that $f_i \to \tilde{f}$ in $L^2(\xi)$. Then $\tilde{f}$ is Σ-measurable. By the CBS inequality, for each $A \in \bigcup_i \Sigma_i$ (which is an algebra), $f_i \chi_A \to \tilde{f} \chi_A$ in $L^2(\xi)$. Hence

$$\lim_i \int_A f_i \, d\xi = \int_A \tilde{f} \, d\xi.$$

If $i < j$, we have by (14)

$$\int_A f_i \, d\xi = \int_A f_j \, d\xi.$$

Letting "$j \to \infty$", we get $\int_A \tilde{f} d\xi = \int_A f_i \, d\xi = \nu(A)$ if i is large so that $A \in \Sigma_i$. Hence (19) is true, and by the same argument one has

$$\nu(A) = \int_A \tilde{f} d\xi, \qquad A \in \sigma\Big(\bigcup_i \Sigma_i\Big). \tag{21}$$

If $\Sigma \neq \sigma(\bigcup_i \Sigma_i)$, then there will exist a $\Sigma_0 \subset \Sigma - \sigma(\bigcup_i \Sigma_i)$, $\Sigma_0 \in \mathscr{C}$, and an f_{Σ_0} satisfying (19). But by the construction, f_{Σ_0} is one of the f_i, and so $\Sigma_0 \subset \sigma(\bigcup_i \Sigma_i)$, and hence (20) holds for all $A \in \Sigma$.† The uniqueness of $\tilde{f}$ is clear. From (21) we derive (18).

Now (21) can be written as

$$\int_A d\nu = \int_A \tilde{f} d(\nu + \mu) = \int_A \tilde{f} d\nu + \int_A \tilde{f} d\mu. \tag{22}$$

Hence, since $\tilde{f} \geqslant 0$ a.e.,

$$\int_\Omega \chi_A (1 - \tilde{f}) \, d\nu = \int_\Omega \chi_A \tilde{f} d\mu, \qquad A \in \Sigma.$$

By linearity of the integral this implies

$$\int_\Omega f(1 - \tilde{f}) \, d\nu = \int_\Omega f \tilde{f} d\mu \tag{23}$$

for all simple and then, by monotone convergence, for all nonnegative measurable f. Since $\nu(A) \leqslant \xi(A)$, (21) implies that $0 \leqslant \tilde{f} \leqslant 1$ a.e. (ξ). Hence replacing f in (23) by $\chi_A/(1 - \tilde{f})$, where $A \in \Sigma$ and where $\tilde{f}(\omega) < 1$ for $\omega \in A$, then (23) implies

$$\nu(A) = \int_A d\nu = \int_A \frac{\tilde{f}}{1 - \tilde{f}} d\mu. \tag{24}$$

Let $A_0 = \{\omega \colon \tilde{f}(\omega) = 1\}$ and $f_0 = \tilde{f}/(1 - \tilde{f})$. Then f_0 is measurable, $A_0 \in \Sigma$, and by (21) $\nu(A_0) = \xi(A_0) = \mu(A_0) + \nu(A_0)$. So $\mu(A_0) = 0$. Hence, if $\nu_1(A)$

†This also follows from the *general fact* that if $\mathscr{C}$ is a subcollection of 2^Ω, then $\sigma(\mathscr{C}) = \bigcup\{\sigma(\mathscr{C}_i) \colon \mathscr{C}_i \subset \mathscr{C}, \text{countable}\}$. Indeed, if $\mathscr{B}$ is this union, then $\mathscr{B}$ is closed under complements and countable unions $[B_k \in \mathscr{B} \Rightarrow B_k \in \sigma(\bigcup_k \mathscr{C}_k)$ all $k \Rightarrow \bigcup_k B_k \in \sigma(\bigcup \mathscr{C}_k) \subset \mathscr{B} \subset \sigma(\mathscr{C})]$. Since $\mathscr{C} \subset \mathscr{B}$, $\mathscr{B}$ is a σ-algebra. So we get $\sigma(\mathscr{C}) = \mathscr{B}$.

$= \nu(A \cap A_0^c)$ and $\nu_2(A) = \nu(A \cap A_0)$, then $\nu_1 \ll \mu$ and $\nu_2 \perp \mu$. Also $\nu = \nu_1 + \nu_2$, and this, by definition, is a Lebesgue decomposition of ν relative to μ. As in Theorem 2.2, uniqueness is immediate, and (24) gives

$$\nu(B) = \nu_1(B) + \nu_2(B) = \int_B f_0 \, d\mu + \nu_2(B), \qquad B \in \Sigma.$$

This proves (18), and the theorem is established.

Second Proof (J. von Neumann). We need to establish the key relation (22); the rest of the argument, being independent of any special considerations, is the same here also. The early relation (20) already gives us a clue to how one should proceed. It shows that [cf. (21)]

$$l(\chi_A) = \int_\Omega \chi_A \, d\nu = \int_\Omega \chi_A \tilde{f} d\xi, \qquad A \in \Sigma, \quad \xi = \nu + \mu. \tag{25}$$

By linearity, we need to look at $l(f)$, $f \in L^2(\nu)$, and represent it as the right side integral in terms of ξ. Theorem 1 then gives a solution.

With the above observation as motivation, we introduce a functional l: $L^2(\nu) \to \mathbb{C}$ by the defining $l(f) = \int_\Omega f d\nu$, $f \in L^2(\nu)$. Since $L^2(\nu) \subset L^1(\nu)$ [because $\nu(\Omega) < \infty$], $l(\cdot)$ is well defined and is a linear functional on $L^2(\nu)$. But

$$\nu(A) \leqslant \xi(A), \qquad A \in \Sigma,$$

so that $f \in L^2(\xi)$ and we have

$$\infty > \|f\|_{2,\xi}^2 = \int_\Omega |f|^2 \, d\xi = \int_\Omega |f|^2 \, d\nu + \int_\Omega |f|^2 \, d\mu \geqslant \int_\Omega |f|^2 \, d\nu. \tag{26}$$

Thus $f \in L^2(\nu)$, i.e., $L^2(\xi) \subset L^2(\nu)$. Hence $l(\cdot)$ is a linear functional on $L^2(\xi)$ also. Further, it is bounded (or continuous) there, since $f \in L^2(\xi)$ implies

$$|l(f)| = \left| \int_\Omega f d\nu \right| \leqslant \|f\|_{2,\nu} \|1\|_{2,\nu}, \qquad \text{by the CBS inequality}$$

$$= \left[\int_\Omega |f|^2 \, d\nu \cdot \nu(\Omega) \right]^{1/2} \leqslant \|f\|_{2,\xi} \sqrt{\nu(\Omega)}, \qquad \text{by (26)}.$$

Hence $\sup\{|l(f)|: \|f\|_{2,\xi} \leqslant 1\} = \|l\| \leqslant \sqrt{\nu(\Omega)} < \infty$, and $l(\cdot)$ is bounded. Then by Theorem 1, there exists a unique $\tilde{f} \in L^2(\xi)$ such that

$$l(f) = (f, \tilde{f}), \qquad f \in L^2(\xi), \quad \|\tilde{f}\|_{2,\xi} = \|l\|. \tag{27}$$

Thus

$$\int_\Omega f\,d\nu = \int_\Omega f\tilde{f}\,d\xi, \qquad f \in L^2(\xi).$$

Taking $f = \chi_A$, $A \in \Sigma$, this becomes

$$\int_A d\nu = \int_A \tilde{f}\,d\xi = \int_A \tilde{f}\,d\nu + \int_\Omega \tilde{f}\,d\mu. \tag{28}$$

This is (22). Clearly $0 \leqslant \tilde{f} \leqslant 1$ a.e., and setting $f_0 = \tilde{f}/(1 - \tilde{f})$ and $A_0 = \{\omega: \tilde{f}(\omega) = 1\}$ gives the desired pair. The rest is the same as in the preceding (concluding part of) proof, and hence the result follows.

Third proof (classical). This original argument is based on the Hahn decomposition of a signed measure (Theorem 1.5). Let ν and ξ $(= \mu + \nu)$ be the finite measures as before. We again establish the key equation (22) or (28), the rest of the (final) argument being the same as above.

Let $\mathbb{Q}^+$ be the set of positive rationals, and for each $r \in \mathbb{Q}^+$, consider the signed measure $\eta_r = \nu - r\xi$ on Σ. Then by Theorem 1.5, there exist two measurable sets (A_r, B_r) such that $A_r \cap B_r = \varnothing$, $\Omega = A_r \cup B_r$, A_r is a positive set for $\nu - r\xi$, and B_r is its negative set. If $r = 0$, we take $B_0 = \varnothing$ and $A_r = \Omega$. For the negative sets, $B_r - B_s = B_r \cap B_s^c = B_r \cap A_s$. Hence $B_r - B_s$ is both a positive and a negative set for η_r for each $s \in \mathbb{Q}^+$. In particular, if $r < s$, then

$$(\nu - r\xi)(B_r - B_s) \leqslant 0 \quad \text{and} \quad (\nu - s\xi)(B_r - B_s) \geqslant 0.$$

By subtraction of the first from the second inequality we get

$$(r - s)\xi(B_r - B_s) \geqslant 0, \qquad \text{and hence} \quad \xi(B_r - B_s) = 0.$$

Thus B_r is essentially contained in B_s for $s > r$. In fact let $E_s = \bigcup_{r<s} B_r - B_s$ be the "error set" for $r, s \in \mathbb{Q}^+$. So $\xi(E_s) = 0$, and if we let $E = \bigcup_{s \in \mathbb{Q}^+} E_s$, then $E \in \Sigma$, $\xi(E) = 0$. Define $C_s = B_s \cup E$, $s \in \mathbb{Q}^+$. Thus $C_s \subset C_{s'}$ for $s < s'$, since $C_s - C_{s'} = (B_s \cup E) \cap (B_{s'} \cup E)^c = B_s \cap B_{s'}^c \cap E^c \subset E \cap E^c = \varnothing$. Hence by Lemma 3.1.5, there exists an $f: \Omega \to \mathbb{R}^+$ measurable, such that $\omega \in C_s \Rightarrow f(\omega) \leqslant s$ and $\omega \notin C_s \Rightarrow f(\omega) \geqslant s$. Let $\tilde{f}(\omega) = f(\omega)$ if $\omega \in B_s$, $= 0$ if $\omega \in E$. Then $f = \tilde{f}$ a.e., and we claim that $\tilde{f}$ satisfies (22) or (28).

To prove this, let $A \in \Sigma$ be arbitrary, and $n_0 \geqslant 1$ be an integer. We express A as a disjoint union by setting $D_k = A \cap (B_{(k+1)/n_0} - B_{k/n_0})$, $k \geqslant 0$. But $D_k \subset B_{(k+1)/n_0}$ and also $D_k \subset B_{k/n_0}^c = A_{k/n_0}$, so that $\eta_{k/n_0}(D_k) \leqslant 0$ and $\eta_{(k+1)/n_0}(D_k) \geqslant 0$. Thus, if $\omega \in D_k$, we have

$$\frac{k}{n_0} \leqslant \tilde{f}(\omega) \leqslant \frac{k+1}{n_0}$$

by construction, and also

$$\nu(D_k) \geqslant \frac{k}{n_0}\xi(D_k) \quad \text{and} \quad \nu(D_k) \leqslant \frac{k+1}{n_0}\xi(D_k).$$

Combining these inequalities and integrating $\tilde{f}$ on D_k, we get

$$\frac{k}{n_0}\xi(D_k) \leqslant \int_{D_k} \tilde{f}\, d\xi \leqslant \frac{k+1}{n_0}\xi(D_k).$$

This may also be written, using the relations on η_r noted above [i.e., $\eta_{k/n_0}(D_k) \leqslant 0$ and $\eta_{(k+1)/n_0}(D_k) \geqslant 0$] as

$$\nu(D_k) - \frac{1}{n_0}\xi(D_k) \leqslant \int_{D_k} \tilde{f}\, d\xi \leqslant \nu(D_k) + \frac{1}{n_0}\xi(D_k). \tag{29}$$

Note that if $\omega \notin B_s$, $s \in \mathbb{Q}^+$, then $\tilde{f}(\omega) \geqslant s$ for all $s \geqslant 0$ implies $\tilde{f} = +\infty$ a.e. But if $\omega \in (\bigcup_s B_s)^c = \bigcap_{s \in \mathbb{Q}^+} A_s = \bar{A}$ (say), then $\eta_s(\bar{A}) \geqslant 0$ for all $s \in \mathbb{Q}^+$. Hence $\nu(\bar{A}) \geqslant s\xi(\bar{A})$ for all s implies $\xi(\bar{A}) = 0$, since $\nu(\Omega) < \infty$. But $\nu \leqslant \xi \Rightarrow \nu(\bar{A}) = 0$ also. It follows that if $D_\infty = A - \bigcup_k B_{k/n_0}$, then $D_\infty \subset \bar{A}$ and so $\xi(A_\infty) = 0 = \nu(A_\infty)$. Since the D_k's are disjoint, adding on k, (29) becomes

$$\nu(A) - \frac{1}{n_0}\xi(A) \leqslant \int_A \tilde{f}\, d\xi \leqslant \nu(A) + \frac{1}{n_0}\xi(A). \tag{30}$$

But n_0 is arbitrary and ξ is a finite measure, so that (30) is precisely (22) or (28). This completes the proof.

We have given these different demonstrations to show that (3) is really a differentiation result even though for the second proof that is not evident, since it is embedded in Theorem 1. (For another proof, related to the third, see Exercise 14.) The abstract form of the Vitali covering lemma naturally contains the conclusion of Theorem 2.6 as part of the formulation. It is termed a Vitali system and is given as follows.

Definition 4. Let (Ω, Σ, μ) be a Carathéodory regular measure space (so that it has no similar extensions). A π-system $\mathscr{D} \subset \Sigma_0 = \{D \in \Sigma : 0 < \mu(D) < \infty\} \cup \{\varnothing, \Omega\}$ is a *Vitali system* if for any $A \subset \Omega$, and for any covering $\mathscr{E} \subset \mathscr{D}$ of A with the property that $\omega \in A$ and $B \in \mathscr{D}(\omega) = \{C \in \mathscr{D} : \omega \in C\}$ there exists a $D \in \mathscr{E}$ satisfying $\omega \in D \subset B$, then there is a countable disjoint system $\{A_n : n \geqslant 1\} \subset \mathscr{E}$ which almost covers A, i.e., $\mu^*(A - \bigcup_{n \geqslant 1} A_n) = 0$. (Here $\Sigma = \mathscr{M}_{\mu^*}$ and $\mu = \mu^* | \Sigma$.)

In Theorem 2.6, $\Omega = \mathbb{R}$, $\mathscr{D} = \mathscr{F}$, $\mathscr{E}$ is Vitali's cover, and $\mu = \lambda$ is the Lebesgue measure. The requirement of the last part is the conclusion of that result, which was established using the topology of $\mathbb{R}$. In the abstract case without topology this condition is postulated. In $\mathscr{D}$ we need to have sets of arbitrarily small μ^*-measure. In Theorem 3, to get the Radon-Nikodým density f_0 $(= d\nu_1/d\mu)$, we have produced a system of sets $\{B_s: s \in \mathbb{Q}^+\}$ using the Hahn decomposition, and this is essentially a Vitali system. Thus the differentiation procedure is definitely built in, and it is based on (an abstract) Vitali cover. This process is embedded in Theorem 1 for the second proof and in Proposition 2 for the first one (thereby also showing that the martingale method is another form of the same process). Thus Theorem 3 is a vast generalization of Theorem 2.4 when $\nu \ll \mu$, and in fact is regarded as an abstract extension of the fundamental theorem of calculus.

Actual construction of the Radon-Nikodým derivatives is in general a hard problem, and there is no universal recipe. If a certain differentiation system (Vitali's system or others such as Besicovitch's) is used and if the derivative exists, the limit depends usually on both the system and ν. If the Radon-Nikodým derivative exists, and other limits are computed using different Vitali systems, the limits can be different at various points, but they agree a.e.(μ), and if this derivative does not exist, then other methods may apply in some cases, and then the "densities" will be different. The detailed analysis of such methods and the corresponding theory is involved, and has branched out into a separate subject. For a survey of many of these results, the reader may consult, for instance, Hayes and Pauc (1970). We shall not consider this aspect here. Instead, the Radon-Nikodým theory will be extended to more general measures than those admitted in Theorem 3. There is still considerable research to be accomplished in this direction.

Since by Theorem 2.2 every σ-finite measure ν on (Ω, Σ) is uniquely decomposable as $\nu = \nu_1 + \nu_2$ relative to a given measure $\mu: \Sigma \to \overline{\mathbb{R}}^+$, where $\nu_1 \ll \mu$, $\nu_2 \perp \mu$, for extending (18) it is only necessary to consider ν_1. Thus we may assume that $\nu \ll \mu$ itself. For such measures, one has the following:

Theorem 5. *Let (Ω, Σ) be a measurable space, $\mu: \Sigma \to \overline{\mathbb{R}}^+$ be a σ-finite measure, and $\nu: \Sigma \to \overline{\mathbb{R}}$ be either a σ-finite or a signed measure, $\nu \ll \mu$. Then there exists a μ-unique Σ-measurable $f_0: \Omega \to \overline{\mathbb{R}}$, a.e. finite, such that*

$$\nu(A) = \int_A f_0 \, d\mu, \qquad A \in \Sigma. \tag{31}$$

Proof. First let $\nu \geq 0$ and be σ-finite. Then there exist sequences $\Omega_n^i \in \Sigma$ such that $\mu(\Omega_n^1) < \infty$, $\nu(\Omega_n^2) < \infty$, and $\bigcup_n \Omega_n^1 = \Omega = \bigcup_n \Omega_n^2$. Letting $\{\Omega_n: n \geq 1\}$ be the sequence $\{\Omega_n^1 \cap \Omega_m^2: n, m \geq 1\}$ we have $\Omega = \bigcup_n \Omega_n$, $\Omega_n \in \Sigma$, and $\mu(\Omega_n) < \infty$. Hence by Theorem 3, there exists a μ-unique $f_n: \Omega_n \to \overline{\mathbb{R}}^+$ such that

$$\nu(A) = \int_A f_n \, d\mu, \qquad A \in \Sigma(\Omega_n) = \{B \cap \Omega_n: B \in \Sigma\}, \quad n \geq 1, \tag{32}$$

since μ, ν are finite measures on the restriction (or trace) σ-algebra $\Sigma(\Omega_n)$. Now define $f_0(\omega) = f_n(\omega)$, $\omega \in \Omega_n$, $n \geqslant 1$. Then f_0 is well defined, measurable for Σ, and finite a.e., and since $A \in \Sigma \Rightarrow A = \bigcup_{n=1}^{\infty} A \cap \Omega_n$, $A \cap \Omega_n \in \Sigma$, disjoint, we have

$$\nu(A) = \sum_{n=1}^{\infty} \nu(A \cap \Omega_n)$$

$$= \sum_{n=1}^{\infty} \int_{A \cap \Omega_n} f_n \, d\mu = \int_A \left(\sum_{n=1}^{\infty} f_n \chi_{\Omega_n} \right) d\mu$$

[by (32) and Corollary 4.3.3(c)]

$$= \int_A f_0 \, d\mu, \qquad \text{by definition of } f_0.$$

If f_0' is another such measurable function, satisfying (31), let $\tilde{\nu}: A \mapsto \int_A f_0' \, d\mu$. Then $\tilde{\nu}(\cdot)$ is σ-additive and σ-finite. But they agree on each set of finite measure, by Theorem 3. Hence by σ-finiteness, they agree on Σ (cf. Theorem 2.3.1). If $B = \{\omega: f_0(\omega) \neq f_0'(\omega)\}$, then $B \in \Sigma$. If $\mu(B) > 0$, then there exists a measurable $B_1 \subset B$, $0 < \mu(B_1) < \infty$, $\nu(B_1) < \infty$, such that $\nu(B_1) = \int_{B_1} f_0 \, d\mu \neq \int_{B_1} f_0' \, d\mu = \tilde{\nu}(B_1)$, which contradicts the preceding result. Hence $\mu(B) = 0$, f_0 is μ-unique, and (31) is established in this case.

Next let $\nu: \Sigma \to \mathbb{R}$ be a signed measure. Then by the Jordan decomposition, $\nu = \nu^+ - \nu^-$, where $\nu^{\pm}: \Sigma \to \mathbb{R}^+$ are (finite) measures. It follows by Theorem 3 that there exist $f_0^{\pm}: \Omega \to \overline{\mathbb{R}}^+$, μ-integrable, such that if $f_0 = f_0^+ - f_0^-: \Omega \to \overline{\mathbb{R}}$, then

$$\nu(A) = \nu^+(A) - \nu^-(A)$$

$$= \int_A f^+ \, d\mu - \int_A f^- \, d\mu = \int_A f_0 \, d\mu, \qquad A \in \Sigma,$$

and the μ-uniqueness of f_0 follows immediately (cf. Exercise 4.2.3). This completes the proof.

Let us present a pair of useful consequences of these results.

Proposition 6. *Let μ, ν be two σ-finite measures on a measurable space (Ω, Σ). Suppose that $\nu \ll \mu$. Then for any $f: \Omega \to \overline{\mathbb{R}}$, measurable for Σ, we have*

$$\int_\Omega f \, d\nu = \int_\Omega f \left(\frac{d\nu}{d\mu} \right) d\mu, \tag{33}$$

in the sense that if either side exists, so does the other and equality holds.

Proof. Let $\mathscr{K}$ be the class of $f: \Omega \to \overline{\mathbb{R}}$ measurable for Σ and for which (33) holds. By (31) the set $\{\chi_A: A \in \Sigma\} \subset \mathscr{K}$, since $d\nu/d\mu = f_0$ exists by the same result. Then by linearity of the integral, if $f_n = \sum_{i=1}^n a_i \chi_{A_i}$, $a_i \geqslant 0$, $A_i \in \Sigma$, we conclude that $f_n \in \mathscr{K}$. By the monotone convergence theorem, (33) holds for all measurable $f \geqslant 0$. Since a general f can be expressed as $f = f^+ - f^-$, the result follows as stated when f^+ or f^- is integrable. This completes the proof.

Remark. The above argument can obviously be extended to deduce the chain rule: if $\nu_1 \ll \nu_2 \ll \mu$, where ν_i and μ are σ-finite measures on (Ω, Σ), then we have

$$\int_\Omega f\,d\nu_1 = \int_\Omega f\frac{d\nu_1}{d\nu_2}\,d\nu_2 = \int_\Omega \left(f\frac{d\nu_1}{d\nu_2}\frac{d\nu_2}{d\mu}\right) d\mu \tag{34}$$

for all measurable $f \geqslant 0$, and more generally when these quantities are defined. Similarly, if $\nu_i \ll \mu$, $i = 1, 2$, then

$$\frac{d(\nu_1 + \nu_2)}{d\mu} = \frac{d\nu_1}{d\mu} + \frac{d\nu_2}{d\mu} \qquad \text{a.e.}(\mu)$$

can be verified without difficulty.

The next result shows how Theorem 5 is used in some problems of optimization. It is essentially due to U. Grenander, who abstracted it from certain earlier work by J. Neyman and E. Pearson.

Proposition 7. *Let (Ω, Σ, μ) be a σ-finite space, and $\nu: \Sigma \to \mathbb{R}^+$ be a (finite) measure. Then for each $\epsilon > 0$, from the collection $\mathscr{A}_\epsilon = \{A \in \Sigma: \mu(A) \leqslant \epsilon\}$, there exists an $A_\epsilon \in \mathscr{A}_\epsilon$ such that $\nu(A_\epsilon)$ is a maximum in that if $B \in \mathscr{A}_\epsilon$ and $\mu(B) \leqslant \mu(A_\epsilon)$, then $\nu(A_\epsilon) \geqslant \nu(B)$. So A_ϵ is "optimal" in this sense.*

Proof. By Theorem 5 (and 3), $\nu = \nu_1 + \nu_2$ with $\nu_1 \ll \mu$, $\nu_2 \perp \mu$, so that for an $A_0 \in \Sigma$ we have $\mu(A_0) = 0$ and $f = d\nu_1/d\mu \geqslant 0$ a.e.(μ). Then

$$\nu(A) = \int_A f\,d\mu + \nu(A \cap A_0), \qquad A \in \Sigma. \tag{35}$$

Define $A_k = \{\omega: f(\omega) \geqslant k\} \cup A_0$ for $k \geqslant 0$. This unmotivated definition of A_k is now shown to give the desired optimal set. Clearly $A_k \in \Sigma$. Also

$$\mu(A_k) = \int_{A_k} d\mu \leqslant \frac{1}{k}\int_{A_k} f\,d\mu \leqslant \frac{1}{k}\nu(\Omega) < \infty. \tag{36}$$

Hence for an $\epsilon > 0$, choose a $k_\epsilon \geqslant 0$ such that $\mu(A_{k_\epsilon}) \leqslant \epsilon$. So $A_\epsilon = A_{k_\epsilon} \in \mathscr{A}_\epsilon$. We claim that A_ϵ is the desired set.

Indeed, let $B \in \Sigma$ be such that $\mu(B) \leqslant \mu(A_\epsilon) \leqslant \epsilon$. Now if we put $D = A_\epsilon \cap A_0^c$, $E = B \cap A_0^c$, and $F = B \cap A_0$, then (drawing a picture and) elementary calculus shows the following holds:

$$\nu(D) = \nu(D - E \cap D) + \nu(E \cap D)$$

$$= \int_{D - E\cap D} f\,d\mu + \nu(E \cap D), \qquad \text{by (35)}$$

$$\geqslant k_\epsilon \mu(D - E \cap D) + \nu(E \cap D), \qquad \text{since} \quad D \subset A_\epsilon$$

$$= k_\epsilon[\mu(D) - \mu(E \cap D)] + \nu(E \cap D), \qquad \text{since} \quad \mu(D) < \infty,$$

$$= k_\epsilon[\mu(A_\epsilon) - \mu(E \cap D)] + \nu(E \cap D)$$

[since $A_\epsilon = D \cup A_0$ and $\mu(A_0) = 0$]

$$\geqslant k_\epsilon[\mu(B) - \mu(E \cap D)] + \nu(E \cap D), \qquad \text{by choice of } B$$

$$= k_\epsilon[\mu(E \cup F) - \mu(E \cap D)] + \nu(E \cap D)$$

$$\geqslant k_\epsilon[\mu(E) - \mu(E \cap D)] + \nu(E \cap D)$$

$$= k_\epsilon \mu(E - E \cap D) + \nu(E \cap D)$$

$$\geqslant \int_{E - E\cap D} f\,d\mu + \nu(E \cap D), \qquad \text{since } E - E \cap D \subset A_\epsilon^c$$

$$= \nu(E - E \cap D) + \nu(E \cap D) = \nu(E). \tag{37}$$

But $A_\epsilon = D \cup A_0$ and $A_0 \supset F$, so that

$$\nu(A_\epsilon) = \nu(D) + \nu(A_0)$$

$$\geqslant \nu(E) + \nu(A_0), \qquad \text{by (37)}$$

$$\geqslant \nu(E) + \nu(F) = \nu(B),$$

since $B = E \cup F$. This completes the proof.

It is clear that Theorem 5 played a key role here. If $\nu \ll \mu$, then $\nu(A_0) = 0$, so that $f = d\nu/d\mu$. In such cases, f is called a *likelihood ratio*, especially if both μ, ν are probability measures, and the above result is basic in the theory of statistical inference. It also is important in the optimization problems of mathematical economics. However, some other applications demand that we also consider non-σ-finite measures, which are treated in the next section. Finally, there are some complements in the exercises below.

EXERCISES

0. Let $\{\mu_n, n \geqslant 1\}$ be a set of finite measures on a measurable space (Ω, Σ).

(a) Show that there exists a probability measure μ on Σ such that $\mu_n \ll \mu$, $n = 1, 2, \dots$. (This fails if the set is uncountable.)

(b) Let $\mathscr{C} = \{\nu: \nu = \sum_{k=1}^{\infty} a_k \mu_k, \sum_{k=1}^{\infty} |a_k| \leqslant 1\}$, called an "absolutely convex" set generated by the $\{\mu_n, n \geqslant 1\}$. Show that for the μ_n of (a) and the μ there, one can put a one-to-one correspondence between $\mathscr{C}$ and a ball of radius β in $L^1(\Omega, \Sigma, \mu)$, where $\beta = \sup\{\mu_n(\Omega): n \geqslant 1\}$ assumed finite. [This exercise will be extended in the following, and serves as an introduction to the latter.]

1. Let (Ω, Σ) be a measurable space and $\nu: \Sigma \to \mathbb{R}$ be a signed measure. Prove the Hahn decomposition of ν by means of the Radon-Nikodým theorem. (Since the first two proofs of Theorem 3 did not depend on the Hahn decomposition, this is a legitimate third proof of Theorem 1.5. Note that $\nu \ll |\nu|$.)

2. Let ν, μ be two finite measures on (Ω, Σ) and $\Pi_n = \{A_{ni}, i = 1, 2, \dots\}$ be a finite or countable (measurable) partitions of Ω. If Π_{n+1} is a refinement of Π_n, if $f_n(\omega) = \nu(A_{ni})/\mu(A_{ni})$ for $\omega \in A_{ni}$ and $\mu(A_{ni}) > 0$, $= 0$ otherwise, and if $\Sigma_n = \sigma(\Pi_n)$ is the σ-algebra generated by Π_n, show that for each $A \in \Sigma_n$ one has $\int_A f_n \, d\mu \geqslant \int_A f_{n+1} \, d\mu$, $n \geqslant 1$. Thus $\{f_n, \Sigma_n: n \geqslant 1\}$ need not be a martingale. It is called a (positive) *supermartingale*. [Same method of proof as for Proposition 2.]

3. If (Ω, Σ, μ) is a σ-finite measure space, show that there exists a probability measure $\nu: \Sigma \to \mathbb{R}^+$ which is equivalent to μ, i.e., $\nu \ll \mu$ and $\mu \ll \nu$. Indeed, if μ is σ-finite and nonfinite, so that there exist no nonzero constants in $L^1(\mu)$, show (still) that there is an a.e. positive and finite $f_0 \in L^1(\mu)$ such that $\|f_0\|_1 = 1$, called a *weak unit*, and that $\int_\Omega g \, d\nu = \int_\Omega g f_0 \, d\mu$ for all $g \geqslant 0$, measurable for Σ. [By σ-finiteness of μ, there exist $\Omega_n \in \Sigma$ with $0 < \mu(\Omega_n) < \infty$, $\Omega = \bigcup_{n=1}^{\infty} \Omega_n$. Consider $f_0 = \sum_{n=1}^{\infty} \alpha_n \chi_{\Omega_n}/\mu(\Omega_n)$, where $\alpha_n > 0$, $\sum_{n=1}^{\infty} \alpha_n = 1$.]

4. Let (Ω, Σ) be a measurable space and $\mu_n: \Sigma \to \overline{\mathbb{R}}^+$ be σ-finite for each $n \geqslant 1$. Show that there exists a probability measure ν on Σ such that $\mu_n \ll \nu$ for all $n \geqslant 1$. (Thus Exercise 0(a) is extended.) Construct an example to show that there exist $\mu_n \perp \mu_m$, $n \neq m$, but $\mu_n \ll \nu$ can hold in the above. [For the first part, use Exercise 3.]

5. Let (Ω, Σ, μ) be a *diffuse measure space*, $\mu(\Omega) < \infty$, so that $\omega \in \Omega \Rightarrow \{\omega\} \in \Sigma$ and $\mu(\{\omega\}) = 0$. Let Π_1 be a countable partition of Ω from Σ, i.e., $\Pi_1 = \{A_{1i}: i \geqslant 1\} \subset \Sigma$, A_{1i}'s disjoint, $\Omega = \bigcup_{i=1}^{\infty} A_{1i}$. Partition each A_{1i} into a countable disjoint union of measurable sets for each $i \geqslant 1$, and let this be Π_2. Proceeding by induction, let $\mathscr{P} = \bigcup_{n \geqslant 1} \Pi_n$. Suppose that the thus obtained $\mathscr{P}$ is fine enough, in the sense that it is dense in (Σ, d) where $d(A, B) = \mu(A \triangle B)$ and each μ-null set is covered by a set of $\mathscr{P}$ of small measure. [So if $N \in \Sigma$, $\mu(N) = 0$, then for each $\epsilon > 0$ there is an $A \in \mathscr{P}$ such that $N \subset A$ and $\mu(A) < \epsilon$.] Show that such a $\mathscr{P}$ is a Vitali cover in the sense of Definition 4.

6. This exercise shows how the Radon-Nikodým derivative can be computed using a Vitali system of the type described in the preceding. This process, due to R. de Possel, is an involved one, but is worth trying to learn. Let (Ω, Σ, μ) and $\mathscr{P}$ be as in the preceding exercise, and $\nu: \Sigma \to \mathbb{R}$ be a signed measure such that $\nu \ll \mu$. Show that $f = d\nu/d\mu$, the Radon-Nikodým derivative, exists. [Use Jordan decomposition and apply Theorem 3.] Thus $\nu(A) = \int_A f\,d\mu$, $A \in \Sigma$. For each $\epsilon > 0$, $\omega \in \Omega$, let $A_\epsilon(\omega)$ be a set in $\mathscr{P}$ such that $\omega \in A_\epsilon(\omega)$, $0 < \mu(A_\epsilon(\omega)) < \epsilon$. We define

$$(\overline{D}\nu)(\omega) = \varlimsup_{\epsilon \downarrow 0} \frac{\nu(A_\epsilon(\omega))}{\mu(A_\epsilon(\omega))} \quad \text{and} \quad (\underline{D}\nu)(\omega) = \varliminf_{\epsilon \downarrow 0} \frac{\nu(A_\epsilon(\omega))}{\mu(A_\epsilon(\omega))}.$$

If $(\overline{D}\nu)(\omega) = (\underline{D}\nu)(\omega)$ exists and is finite for a.a. $\omega \in \Omega$, this value defines $(D_{\mathscr{P}}\nu)(\cdot)$, the derivative of ν relative to μ along $\mathscr{P}$. Show that for any such ν and $\mathscr{P}$, we have when $\Sigma = \sigma(\mathscr{P})$, a.e.,

$$f(\omega) = \frac{d\nu}{d\mu}(\omega) = (D_{\mathscr{P}}\nu)(\omega) \qquad \text{for a.a. } \omega.$$

Thus we get the same f for any such Vitali system $\mathscr{P}$, if the Radon-Nikodým theorem applies. [Different Vitali systems generally yield different derivatives, but they agree outside of a μ-null set when the above hypothesis is satisfied. The method of proof is an extension of that of Theorem 2.4. Briefly, if $(\overline{D}\nu)(\omega) > \alpha$, or $(\underline{D}\nu)(\omega) < \beta$, on a set A_0, then there is some $A \in \Sigma$, $A \subset A_0$, such that $\nu(A) \geqslant \alpha\mu(A)$, or $\nu(A) \leqslant \beta\mu(A)$, respectively. If $A_{\alpha\beta} = \{\omega: \alpha < f(\omega) < \beta\}$, then one shows that $(\overline{D}\nu)(\omega) \leqslant \beta$ and $(\underline{D}\nu)(\omega) \geqslant \alpha$ for a.a. $\omega \in A_{\alpha\beta}$. This implies $\alpha \leqslant (\underline{D}\nu)(\omega) \leqslant (\overline{D}\nu)(\omega) \leqslant \beta$ for a.a. ω. Letting $\beta - \alpha \to 0$ through rationals, one gets the final result, but the details need care and are not trivial.]

7. Let (Ω, Σ, μ) be a measure space and $f \in L^1(\mu)$. Consider $\nu: A \mapsto \int_A f\,d\mu$. Show that g is (measurable and) integrable for ν iff fg is (measurable and) integrable for μ. When this condition holds, we have the (usual) equation

$$\int_A g\,d\nu = \int_A fg\,d\mu, \qquad A \in \Sigma.$$

[It suffices to establish the result under the condition that $\mu(\Omega) < \infty$.]

8. (a) If μ, ν are two σ-finite measures on a measurable space (Ω, Σ) such that $\nu \ll \mu \ll \nu$ (i.e., they are equivalent), show that the usual differential (symbolic) calculus holds in that $d\nu/d\mu$, $d\mu/d\nu$ exist and are reciprocal to each other a.e.(μ).

(b) If $\nu: \Sigma \to \mathbb{R}$ is a signed measure, $\mu = |\nu|: \Sigma \to \mathbb{R}^+$, show that $\nu(A) = \int_A f\,d\mu$, $f \in L^1(\mu)$, $|f| = 1$, a.e.(μ), that $\nu^\pm(A) = \int_A f^\pm\,d\mu$, $A \in \Sigma$, $\nu = \nu^+ - \nu^-$ is a Jordan decomposition, and that $\Omega^+ = \{\omega: f(\omega) > 0\}$, $\Omega^- = \{\omega: f(\omega) \leqslant 0\}$ is Hahn's.

9. The result of Theorem 5 is not true for arbitrary measures, as seen from the following (Saks's) example: Let $\Omega = [0,1]$; Σ be the Borel σ-algebra of Ω; μ_1, μ_2 be measures on Σ such that $\mu_1(A) = +\infty$ if $A \neq \varnothing$, $= 0$ if $A = \varnothing$; and $\mu_2(A)$ be the number of points in A, $A \in \Sigma$. Thus μ_2 has the finite subset property, but not μ_1. Let ν be the Lebesgue measure on Σ. Show that there do not exist measurable $f_i: \Omega \to \overline{\mathbb{R}}^+$, $i = 1, 2$, such that

$$\nu(A) = \int_A f_i\,d\mu_i, \qquad i = 1,2, \quad A \in \Sigma.$$

Show, however, that there is a (trivial) measurable $f_3: \Omega \to \overline{\mathbb{R}}^+$ such that $\mu_1(A) = \int_A f_3\,d\mu_2$, $A \in \Sigma$. [We shall analyze this phenomenon in detail in the following section and show that there exist "quasifunctions", which are not measurable, but for which the above integral representations, suitably defined, almost always hold.]

10. Let μ, ν be two finite measures on a measurable space (Ω, Σ), and let $\zeta = \mu + \nu$. Let f, g be the Radon-Nikodým derivatives of μ, ν on Σ relative to ζ. If $k \geqslant 0$ is a number and $A_k = \{\omega: (g/f)(\omega) \geqslant k\}$, show that for each $\epsilon > 0$, there is a k_ϵ such that $\mu(A_{k_\epsilon}) \leqslant \epsilon$ and A_{k_ϵ} is the optimal set for ν among all $B \in \Sigma$ with $\mu(B) < \epsilon$, as in Proposition 7. Show moreover that if $\xi: \Sigma \to \overline{\mathbb{R}}^+$ is any σ-finite measure such that $\mu \ll \xi$, $\nu \ll \xi$ and if $\tilde{f} = d\mu/d\xi$, $\tilde{g} = d\nu/d\xi$, then again $A_k = \{\omega: (\tilde{g}/\tilde{f})(\omega) \geqslant k\}$, i.e., A_k does not depend on ξ.

11. Let (Ω, Σ, μ) be a measure space, and $V^p(\mu)$ be the vector space of additive $\nu: \Sigma \to \mathbb{R}$ which have p-variation finite relative to μ, in the sense of Exercise 1.2, with norm $N_p(\cdot)$ defined there. If $1 < p < \infty$, for each $\nu \in V^p(\mu)$ there exists a σ-finite set S_ν for the μ-measure such that $\nu(A) = 0$ for each $A \in \Sigma$, $A \subset \Omega - S_\nu$. We may say that ν has a *support* S_ν which is of σ-finite μ-measure. If $L^p(\mu)$ is the Lebesgue space, show that for each ν there is a unique $f_\nu \in L^p(\mu)$ such that $T: \nu \mapsto f_\nu$ defines an isometric linear isomorphism between these spaces, i.e., $\|T\nu\|_p = \|f_\nu\|_p = N_p(\nu)$, and T is a one to one mapping of $V^p(\mu)$ onto $L^p(\mu)$. (This result is false for $p = 1$, but again holds for $0 < p < 1$ if $N_p(\cdot)$ is replaced by $|\nu|_p(\Omega)$ of Definition 1.1.) [Apply Theorem 5 and Proposition 1.2.]

12. Consider the space $V^1 = \mathrm{ca}(\Omega, \Sigma)$, the vector space of σ-additive signed measures on Σ. Let $\mathscr{V}(\mu) = \{\nu \in V^1: \nu \ll \mu\}$, where μ is a measure on (Ω, Σ). Show that $\tau(L^1(\mu)) \subset \mathscr{V}(\mu)$, where for each $f \in L^1(\mu)$, (τf) in $\mathscr{V}(\mu)$ is defined by $(\tau f)(A) = \int_A f\,d\mu$. Show however, that, there is equality if μ is σ-finite. Thus a restriction on μ is needed for the equality, and one may have strict inclusion in the general case. [Apply Theorem 5 and

Proposition 1.2 appropriately, but use Exercise 9 for a strict inclusion example.]

13. Let (Ω, Σ, μ) be a σ-finite space, and $\mu_1, \mu_2 \in \mathscr{V}(\mu)$ of the above exercise. Using the relation $\mu_1 \leqslant \mu_2$ iff $\mu_1(A) \leqslant \mu_2(A)$, $A \in \Sigma$, introduced for Proposition 4.4.5, show that

$$\max(\mu_1, \mu_2)(A) = \int_A \max\left(\frac{d\mu_1}{d\mu}, \frac{d\mu_2}{d\mu}\right) d\mu, \qquad A \in \Sigma,$$

$$\min(\mu_1, \mu_2)(A) = \int_A \min\left(\frac{d\mu_1}{d\mu}, \frac{d\mu_2}{d\mu}\right) d\mu, \qquad A \in \Sigma.$$

[The mapping τ, and hence $\tau^{-1}: \mathscr{V}(\mu) \to L^1(\mu)$, is order preserving in the above problem, in addition to all the other properties.]

14. Another proof of Theorem 3 is as follows: Let $\mathscr{K} = \{0 \leqslant f: \int_A f\, d\xi \leqslant \nu(A), A \in \Sigma\}$, where $\xi = \mu + \nu$, a finite measure. Then $0 \in \mathscr{K}$, and $f_1, f_2 \in \mathscr{K} \Rightarrow \max(f_1, f_2) \in \mathscr{K}$. $\mathscr{K}$ is ordered by $f_1 \leqslant f_2$ a.e. If $\sup\{\int_\Omega f\, d\xi: f \in \mathscr{K}\} = \alpha \leqslant \nu(\Omega) < \infty$, then show that every chain $\mathscr{C} \subset \mathscr{K}$ has an upper bound in $\mathscr{K}$. [Use monotone convergence.] Then by Zorn's lemma $\mathscr{K}$ has a maximal element f_0, and this gives $\nu(A) = \int_A f_0\, d\xi$. [For, let $\nu_1(A) = \int_A f_0\, d\xi$. Then $\nu_2 = \nu - \nu_1 \geqslant 0$. If $\nu_2 > 0$, then we can find $k > 0$ such that $\xi(\Omega) - k\nu_2(\Omega) < 0$. Next, by the Hahn decomposition, there is $\Omega_1 \in \Sigma$, a positive set, and Ω_2, a negative set, $\Omega_2 = \Omega_1^c$ for $\xi - k\nu_2$. But $\xi(\Omega_2) > 0$, for if $\xi(\Omega_2) = 0$, then $\xi(\Omega_1) - k\nu_2(\Omega_1) = \xi(\Omega) - k\nu_2(\Omega) \geqslant 0$, since $\nu_2(\Omega_2) = 0$ also. This is impossible by the choice of k. Let $f' = (1/k)\chi_{\Omega_2} \geqslant 0$. Verify that $f' \in \mathscr{K}$, and then $f_0 + f' \in \mathscr{K}$ also holds. Since however $f_0 + f' > f_0$, we get a contradiction to the maximality of f_0, and one has $\nu_2 = 0$. In case $\nu \ll \mu$, the above argument—as well as the second proof, which depends on the Hahn decomposition—holds if ν is an arbitrary measure (cf. Remark 1 after the second proof of Theorem 1.5). However, a simple alternative version will be given below as Proposition 4.1.]

15. Let (Ω, Σ) be a measurable space, and μ, ν be two finite measures on Σ. If $\zeta = \mu + \nu$, let $f = d\mu/d\zeta$, $g = d\nu/d\zeta$. Define the functional $H_\alpha(\cdot, \cdot)$ as

$$H_\alpha(\mu, \nu) = \int_\Omega f^\alpha g^{1-\alpha}\, d\zeta, \qquad 0 \leqslant \alpha \leqslant 1.$$

The functional $H_\alpha(\cdot, \cdot)$ is called the *generalized Hellinger distance* between μ and ν. Show that H_α depends only on μ, ν, not on the dominating measure ζ, in the sense that if ζ is replaced by any other finite measure ξ on Σ such that $\mu \ll \xi$, $\nu \ll \xi$, we get the same H_α. Show also that (i) $0 \leqslant H_\alpha(\mu, \nu) \leqslant \max(\mu(\Omega), \nu(\Omega))$, (ii) for $0 < \alpha < 1$, $\mu \perp \nu$ iff $H_\alpha(\mu, \nu) = 0$, (iii) $H_\alpha(\mu, \nu) = \mu(\Omega)$ iff $\mu = \nu$. [Use Hölder's inequality and equality conditions in it for the last part. The case $\alpha = \frac{1}{2}$ is the classical one used by E. Hellinger. Further properties of the function H_α, for $\alpha = \frac{1}{2}$, will be discussed in Chapter 6. See Exercises 6.4.9–6.4.11.]

5.4. THE RADON-NIKODÝM THEOREM: GENERAL CASE

In the preceding section we presented, as Theorem 5, a result for σ-finite measures. Exercises 3.3 and 3.9 indicate two different aspects of this situation. The first one shows that σ-finiteness is not a really general case, and the second one implies that Theorem 5 cannot be expected to hold for arbitrary measures. However, a part of Exercise 3.9 indicates that the Radon-Nikodým derivative does exist for some non-σ-finite cases. Unless one explores this problem here, the mystery surrounding the result will remain. Thus we now consider the general form and obtain optimum conditions for the validity of the Radon-Nikodým theorem. It will be seen in the next section that this generality is actually needed in some important applications.

Briefly stated, we first show that, for two measures μ, ν on a measurable space (Ω, Σ) with $\nu \ll \mu$, the theorem holds if $\mu(\Omega) < \infty$ and $\nu: \Sigma \to \overline{\mathbb{R}}^+$ is arbitrary. Next we establish that the result holds if $\mu(\Omega) = \infty$, but μ is localizable in the sense of Definition 2.3.4. Indeed, that concept was introduced by I. E. Segal (1954) precisely for this purpose. It helps one to understand the counterexample of Exercise 3.9. This phenomenon is also closely related to the inclusion conditions in Exercise 3.12. Because of all these questions, we devote this section to the case where all infinite measures have the finite subset property. Note that the last property is not a real restriction, as shown in Lemma 2.3.3 and the discussion following its proof. *This will be assumed hereafter.*

We start with the first extension noted above. This is obtainable from the third proof of Theorem 3.3, but an independent argument is given for convenience.

Proposition 1. *Let (Ω, Σ, μ) be a finite measure space, and $\nu: \Sigma \to \overline{\mathbb{R}}^+$ be any measure such that $\nu \ll \mu$. Then there exists a measurable (for Σ), not necessarily finite valued, μ-unique function f_0 such that*

$$\nu(A) = \int_A f_0 \, d\mu, \qquad A \in \Sigma. \tag{1}$$

This f_0 may be chosen finite a.e. on each set A which has σ-finite ν-measure.

Proof. Let $\mathscr{C} = \{A \in \Sigma: A \subset \bigcup_{n=1}^{\infty} A_n,\ A_n \in \Sigma,\ \nu(A_n) < \infty\}$. Then it is clear that $\mathscr{C}$ is a σ-ring contained in Σ. Since $\mu(\Omega) < \infty$, if $\alpha = \sup\{\mu(A): A \in \mathscr{C}\}$ $[\leq \mu(\Omega)]$, there exists an increasing sequence $A_n \in \mathscr{C}$ such that $\alpha = \lim_n \mu(A_n)$. If $A_0 = \bigcup_{n=1}^{\infty} A_n$, then $A_n \in \mathscr{C}$ and $\mu(A_0) = \alpha$. We assert that A_0 is a maximal set with the property that ν is σ-finite on it. To see this, consider the trace σ-algebra $\Sigma(A_0^c) = \{A_0^c \cap B: B \in \Sigma\}$. Then we first show that ν does not have the finite subset property on $\Sigma(A_0^c)$.

Suppose the contrary. Then there is a $B \in \Sigma(A_0^c)$, $0 < \nu(B) < \infty$, and by definition of $\mathscr{C}$, $B \in \mathscr{C}$. But then $B \cap A_0 = \varnothing$ and $A_0 \cup B \in \mathscr{C}$. Hence by definition of α, we have $\mu(B) > 0$ (since $\nu \ll \mu$), and

$$\alpha \geqslant \mu(A_0 \cup B) = \mu(A_0) + \mu(B) > \alpha. \tag{2}$$

This contradiction shows that ν does not have the finite subset property on $\Sigma(A_0^c)$. Also note that if $A \in \Sigma(A_0^c)$, then $\mu(A) > 0$ implies $\nu(A) = +\infty$, since otherwise $\nu(A) = 0 \Rightarrow A \in \mathscr{C}$ and we get a contradiction again by (2).

On $\Sigma(A_0)$, ν is σ-finite by construction, and μ is still finite. So by Theorem 3.5 there exists $\tilde{f}_0: A_0 \to \overline{\mathbb{R}}^+$, finite a.e. and μ-unique, such that

$$\int_B \tilde{f}_0 \, d\mu = \nu(B), \qquad B \in \Sigma(A_0).$$

Define $f_0 = \tilde{f}_0$ on A_0, and $= +\infty$ on A_0^c. Then f_0 is measurable for Σ, and for any $A \in \Sigma$ we have

$$\begin{aligned}\int_A f_0 \, d\mu &= \int_{A \cap A_0} f_0 \, d\mu + \int_{A \cap A_0^c} f_0 \, d\mu, \qquad \text{by Proposition 4.1.5}\\ &= \nu(A \cap A_0) + \nu(A \cap A_0^c) = \nu(A),\end{aligned}$$

since ν is a measure. This gives (1).

Finally, if f_0' is another function satisfying (1), then by Theorem 3.5, $f_0 \chi_{A_0} = f_0' \chi_{A_0}$ a.e. To see that $f_0' = \infty$ a.e. on A_0^c must also hold, suppose that on a set $B \in \Sigma(A_0^c)$, $\mu(B) > 0$, we have $f_0'(\omega) < \infty$, $\omega \in B$. Then, as noted before, $\nu(B) = \infty$ must be true. Let $B_n = \{\omega \in B: f_0'(\omega) \leqslant n\}$. We see that $B_n \uparrow B$, $B_n \in \Sigma(A_0^c)$, and hence for some $n = n_0$, one has from $\lim_n \mu(B_n) = \mu(B) > 0$ that $\mu(B_{n_0}) > 0$, so that $\nu(B_{n_0}) = \infty$. Consequently

$$\infty = \nu(B_{n_0}) = \int_{B_{n_0}} f_0' \, d\mu \leqslant n_0 \mu(B_{n_0}) \leqslant n_0 \mu(\Omega) < \infty.$$

This contradiction shows that $f_0 = f_0'$ a.e., and the last statement being clear, the proof is complete.

We next turn to the second, more involved, question. Let us introduce a concept to describe the situation.

Definition 2. If (Ω, Σ) is a measurable space, let $\{\mu_j: j \in J\}$ be a family of measures on Σ such that all the μ_j have a common finite subset property,

$j \in J$. (This is explained further after the definition.) A mapping $f: \Omega \times J \to \overline{\mathbb{R}}$ is called a *quasifunction* relative to $\{\mu_j: j \in J\}$ if for each $J_c \subset J$, J_c countable, there exists a function $f: \Omega \to \overline{\mathbb{R}}$ such that $\mu_j^*(\{\omega: f(\omega, j) \neq f(\omega)\}) = 0$ $(f = f_{J_c})$, $j \in J_c$, μ^*_j being the outer measure generated by (μ_j, Σ). And $f(\cdot, \cdot)$ is called a *measurable quasifunction* if $f(\cdot, j)$ is μ_j-measurable for each $j \in J$.

This definition needs an elucidation. First, we note that the assumption on the μ_j, $j \in J$, is not restrictive. Indeed, if $\{(\Omega, \Sigma, \mu_j): j \in J\}$ is an arbitrary family of measure spaces, let $\Sigma_0 = \{A \in \Sigma: \mu_j(A) < \infty, \ j \in J\}$. If $\tilde{\Sigma} = \{A: A \cap B \in \Sigma_0, \text{ all } B \in \Sigma_0\}$, and $\tilde{\mu}_j: \tilde{\Sigma} \to \mathbb{R}^+$ is given by $\tilde{\mu}_j(A) = \sup\{\mu_j(B): B \subset A, \ B \in \Sigma_0\}$, then it is easily seen that $\tilde{\Sigma}$ is a σ-algebra and $\tilde{\mu}_j$ is σ-additive on $\tilde{\Sigma}$. Thus $\{(\Omega, \tilde{\Sigma}, \tilde{\mu}_j): j \in J\}$ has the property assumed in the definition. The by now familiar computation of σ-additivity of $\tilde{\mu}_j$ is left to the reader. (Cf. Exercise 2.2.12.) Note that now all $\tilde{\mu}_j$'s have the same δ-ring Σ_0 on which they are determined, and each has the finite subset property.

The second point is to exhibit a measurable quasifunction $f: \Omega \times J \to \mathbb{R}$ which is not equivalent to a measurable function. Here is an example. Let $\Omega = (0,1)$, $\Sigma = \{A \subset \Omega: A \text{ countable or } A^c \text{ countable}\}$. Let $J = \{\text{all finite subsets of } \Omega\}$, $\mu(A) = (\text{number of points in } A)$, and $\mu_j(A) = \mu(j \cap A)$, $A \in \Sigma$, $j \in J$. Then $\{(\Omega, \Sigma, \mu_j): j \in J\}$ is a family of measurable spaces. If $f: \Omega \times J \to \mathbb{R}$ is given by $f(\omega, j) = \omega$ for $\omega \in j$, $= 0$ otherwise, then for any countable $A \subset J$, the function $f(\cdot, j)$, $j \in A$, is measurable (for Σ), so that $\{f(\cdot, j): j \in J\}$ is a measurable quasifunction. But the only f_0 for which $f(\omega, j) = f_0(\omega)$ for all $j \in J$ is the identity mapping $f(\omega) = \omega$, $\omega \in \Omega$, and this is not measurable (for Σ). Of course, the trouble here is that Σ is not "rich" enough and that μ is not σ-finite or localizable on Σ even though it has the finite subset property.

We shall present now the desired extension of the Radon-Nikodým theorem, following McShane (1962); an equivalent version was given independently by Zaanen (1961). If (Ω, Σ, μ) is a measure space with the finite subset property, let $\mu_E = \mu(E \cap \cdot\,)$, $E \in \Sigma_0$ $(= J)$. Then for any $E_1 \subset E_2$ $(E_i \in \Sigma_0)$ and any measurable quasifunction $f(\cdot, \cdot): \Omega \times J \to \mathbb{R}$, we have

$$\int_\Omega f(\omega, E_1)\mu_{E_1}(d\omega) = \int_\Omega f(\omega, E_2)\mu_{E_2}(d\omega), \tag{3}$$

where $f(\omega, E_1) = f(\omega, E_2)|_{E_1}$, the restriction to E_1. We say that the net $\{\int_\Omega f(\omega, E)\mu_E(d\omega): E \in \Sigma_0\}$, which is monotone if $f \geq 0$, *converges to* $a_0^f \in \overline{\mathbb{R}}$ if for each neighborhood U of a_0^f in $\overline{\mathbb{R}}$ there exists a set $E_U \in \Sigma_0$ such that for all $F \supset E_U$, $F \in \Sigma_0 \Rightarrow \int_\Omega f(\omega, F)\mu_F(d\omega) \in U$. This limit is denoted as

$$\int_\Omega f(\omega, \cdot)\mu_\cdot(d\omega) = a_0^f = a_0(f). \tag{4}$$

It is clear that if the limit exists, then it is unique and $a_0(\alpha f + g) = \alpha a_0(f) + a_0(g)$, if this sum is defined, $\alpha \in \mathbb{R}$. By monotonicity, if $f \geqslant 0$, the limit always exists. The class $\{f: a_0(f) \in \mathbb{R}\}$ of all measurable quasifunctions is a vector space, and it has the lattice properties of ordinary functions when $f \leqslant g$ is defined as $f(\cdot, j) \leqslant g(\cdot, j)$ a.e.(μ_j), $j \in J$. In case $\mu_j = \mu_E = \mu(E \cap \cdot)$, $E \in \Sigma_0$ (the most important example), the corresponding measurable quasifunction is called a cross-section by Zaanen. Here we use McShane's term, which is somewhat intuitive.

With these preliminaries we can present the following:

Theorem 3 (Radon-Nikodým, general form). *Let* (Ω, Σ) *be a measurable space and* μ, ν *be a pair of measures on* Σ, *having a common determining ring* $\mathscr{A} \subset \Sigma$ *(cf. Definition 2),* $\nu \ll \mu$. *Then there exists a unique measurable quasifunction* $f(\cdot, \cdot): \Omega \times \Sigma_0 \to \mathbb{R}^+$ *relative to* $\{\mu_E: E \in \Sigma_0\}$ *such that*

$$\nu(A) = \int_A f(\omega, \cdot)\,\mu_\cdot(d\omega), \qquad A \in \Sigma, \tag{5}$$

where $J = \Sigma_0 = \{B \subset \Sigma: \mu(B) < \infty\}$. *If* μ *is localizable, then* $f(\cdot, j) = f_0(\cdot)$, $j \in J$, *a.e.*(μ), *where* f_0 *is an a.e.*(μ) *unique measurable (for* Σ*) function, so that the quasifunction* $f(\cdot, \cdot)$ *is determined by a measurable function* f_0.

Proof. Let $E \in \Sigma_0$ be fixed. Then $\mu_E(\cdot)$ is a finite measure on $\Sigma(E)$, and by Proposition 1, there exists a μ_E-unique measurable function $f_E(\cdot)$ such that for all $F \in \Sigma(E)$ we get

$$\nu(F) = \int_F f_E(\omega)\mu_E(d\omega). \tag{6}$$

Here f_E may be taken to be $\geqslant 0$, and $= 0$ on $\Omega - E$. Hence, by uniqueness, if $E_1, E_2 \in \Sigma_0$ are two sets and f_{E_1}, f_{E_2} are the corresponding functions, then $f_{E_1}(\omega) = f_{E_2}(\omega)$ for a.a. ω in $E_1 \cap E_2$, relative to μ. If $E_1 \subset E_2$, then $f_{E_1} = 0$ a.e. on $E_2 - E_1$, so that $f_{E_1} \leqslant f_{E_2}$ a.e.(μ). Thus $\{f_E: E \in \Sigma_0\}$ is a measurable quasifunction. Hence $\{\int_\Omega f_E(\omega)\mu_E(d\omega): E \in \Sigma_0\}$ is a monotone net of nonnegative (extended) real numbers, and so it converges to a unique limit a_0^f as defined by (4). Consequently, for the measurable quasifunction $f(\cdot, E) = f_E(\cdot)$, and since μ, ν are both determined on $\mathscr{A}$ with $\Sigma = \{A: A \cap B \in \mathscr{A}, B \in \mathscr{A}\}$,

$$\nu(A) = \int_A f(\omega, \cdot)\mu_\cdot(d\omega), \qquad A \in \Sigma, \tag{7}$$

and if A is σ-finite for μ, then $f(\omega, A) = f_{\bigcup_n A_n}(\omega) = \sum_{n=1}^\infty f_{A_n}(\omega)$ a.e., $A_n \in \Sigma_0$, disjoint, $A = \bigcup_n A_n$. This proves (5).

On the other hand, if μ is localizable, then the net $\{f_A: A \in \Sigma_0\}$ has a measurable supremum f_0, as shown in Exercise 3.2.9, and then, by the (extended) monotone convergence theorem (cf. Exercise 4.3.13), we get

$$\nu(A) = \int_A f_0 \, d\mu, \qquad A \in \Sigma. \tag{8}$$

The uniqueness part is simple and proved exactly as in Proposition 1. This completes the proof.

We now make an important comment on this theorem.

Remark. By the Saks example given in Exercise 3.9, we had $\Omega = [0,1]$, $\Sigma = \{A \subset \Omega: A$ countable or A^c countable$\}$, $\mu(A) =$ (number of points in A), and $\nu(A) = 0$ if A is countable, $= 1$ if A^c is countable. But there is no f_0 such that (8) holds. Here $\Sigma_0 = \{A: \mu(A) < \infty\}$ is a ring of finite sets on which $\nu(A) = 0$, i.e., $\nu \equiv 0$ on Σ_0. Hence if $A \in \Sigma$, then $\nu(A) = \sup\{\nu(B): B \in \Sigma_0, B \subset A\}$ is not valid for uncountable A, although it is with μ instead of ν. Thus μ, ν do not *simultaneously* satisfy the finite subset property in the form of Lemma 2.3.3. *This property is thus essential for the validity of Theorem* 3 *above*, and it is the best possible condition. (See Theorem 5 below.)

To explain the last remark further we introduce the following:

Definition 4. A measure space (Ω, Σ, μ), or just the measure μ, is said to have the *Radon-Nikodým property* if for any measure $\nu: \Sigma \to \overline{\mathbb{R}}^+$ such that $\nu \ll \mu$, there is a (Σ-) measurable $f_0: \Omega \to \overline{\mathbb{R}}^+$ ($f_0 = f_{0,\nu}$) such that

$$\int_\Omega f \, d\nu = \int_\Omega f f_0 \, d\mu, \qquad f \in L^1(\nu). \tag{9}$$

This concept enables us to check, in some cases, the localizability of a measure. The following result, which is essentially due to I. E. Segal (1954), elaborates this point.

Theorem 5. *Let (Ω, Σ, μ) be a measure space with the finite subset property. Then μ is localizable iff it has the Radon-Nikodým property.*

Proof. Suppose that μ is localizable and $\Sigma_0 = \{A \in \Sigma: \mu(A) < \infty\}$. Let ν be any measure on Σ having the finite subset property with μ, i.e., ν is also determined by Σ_0 (so both μ, ν are determined on Σ_0). If $\nu \ll \mu$, then by Theorem 3 there is a μ-unique measurable (for Σ) $f_0: \Omega \to \overline{\mathbb{R}}^+$ such that

$$\int_\Omega \chi_A \, d\nu = \nu(A) = \int_A f_0 \, d\mu = \int_\Omega \chi_A f_0 \, d\mu. \tag{10}$$

By linearity, (10) shows that (9) holds for all simple $f_n \geqslant 0$, $f_n \in L^1(\nu)$. Then with monotone convergence and structure theorems (9) follows for all $f \in L^1(\nu)$. So μ has the Radon-Nikodým property.

For the converse, we need to show that any subcollection $\mathscr{C} \subset \Sigma$ has a μ-supremum (cf. Definition 2.3.4). Here we may assume that $\mathscr{C} \subset \Sigma_0$ as noted following that definition, and that (9) holds. Thus consider $\{f_E: E \in \mathscr{C}\}$, a measurable quasifunction, where $f_E = \chi_E$, so that $0 \leqslant f_E \leqslant 1$ a.e. If we define

$$\nu: A \mapsto \int_A f(\omega, E)\mu_E(d\omega), \qquad A \in \Sigma, \quad E \in \mathscr{C},$$

then $\nu(\cdot)$ is σ-additive, μ, ν have the finite subset property together, and for each $g = \sum_{i=1}^n a_i \chi_{A_i}$, $a_i \geqslant 0$, $A_i \in \Sigma_0$,

$$\int_\Omega g(\omega)\nu(d\omega) = \int_\Omega g(\omega) f(\omega, \cdot)\mu.(d\omega), \tag{11}$$

for each simple function, and then, by the monotone convergence, for all $g \in L^1(\nu)$. Since μ has the Radon-Nikodým property, there exists an f_0 $(= f_{0,\nu})$, measurable (Σ) and μ-unique, such that

$$\int_\Omega g(\omega)\nu(d\omega) = \int_\Omega g(\omega) f_0(\omega)\mu(d\omega), \qquad g \in L^1(\nu). \tag{12}$$

Thus (11) and (12) imply

$$\int_\Omega g(\omega) f(\omega, \cdot)\mu.(d\omega) = \int_\Omega g(\omega) f_0(\omega)\mu(d\omega) \tag{13}$$

for all simple functions g with support in Σ_0. Since the quasifunction $f(\cdot, \cdot)$ satisfies $0 \leqslant f(\omega, \cdot) \leqslant 1$ and g can be any simple function, (13) implies $0 \leqslant f_0 \leqslant 1$. But now a familiar argument implies that (taking $g = \chi_E$, $E \in \Sigma_0$, etc.) that a.e. $f(\cdot, E) = (f_0 \chi_E)(\cdot)$ for each $E \in \mathscr{C}$. So f is determined by the measurable function f_0. But also (13) gives $f(\cdot, E)\chi_E = \chi_E f_0$ a.e., and since $f(\cdot, E) = \chi_E$, we get $\chi_E f_0 = \chi_E$ a.e. for all $E \in \mathscr{C}$, and $f_0^2 = f_0$ a.e. So $f_0 = \chi_F$, $F \in \Sigma$. This means $E \subset F$ a.e. for all $E \in \mathscr{C}$. Thus F is an upper bound to $\mathscr{C}$. To see that f_0 is the least such, let $\tilde{f}_0$ be any other upper bound $(\tilde{f}_0 = \chi_{\tilde{F}})$. Then $E \subset \tilde{F}$ a.e., and $f(\omega, \cdot)\chi_E \leqslant \tilde{f}_0$ a.e., $E \in \mathscr{C}$. But then we must have $gf(\cdot, E) \leqslant g\tilde{f}_0$ a.e., with $0 \leqslant g$, simple, and by (13) this implies that $f_0 \leqslant \tilde{f}_0$ a.e. So f_0 and hence F is the least upper bound. By Definition 2.3.4, μ is thus localizable. This completes the proof.

The point of this result is that if $\nu \ll \mu$, and ν is to be an indefinite integral of μ, then ν must have the finite subset property *jointly* with μ, and μ must be localizable. The Saks example violates these conditions, and hence the Radon-Nikodým theorem fails for such measures. It should be noted that there are cases, somewhat peripheral, such as those given by the second part of Exercise

3.9, which do not satisfy these conditions. [For the problem there, (9) does not obtain. Exercise 7 below has a similar unpleasant characteristic.]

There are at least two reasons for us to present the above results in Theorems 3 and 5. While the dual space of $L^p(\mu)$ will be shown to be identifiable with $L^q(\mu)$ for any μ if $1 < p < \infty$, (as usual $q = p/(p-1)$), this is no longer true if $p = 1$. In the next section we shall show that the dual of $L^1(\mu)$ is identified with $L^\infty(\mu)$ iff μ is localizable. Thus the preceding work is essential for the important space $L^1(\mu)$. Secondly, the Radon-Nikodým theory itself is not fully comprehended unless its limitations are determined. Further it will be seen in Chapter 9 that a Haar measure on a locally compact group is localizable but not σ-finite, and this is basic for all work in harmonic analysis on such groups.

Since localizability implies strict localizability (cf. Definition 2.3.4 and Exercise 2.3.5), but not conversely, we have the following useful consequence of the above theorem which is stated for reference. (Like σ-finite measures, these automatically have the finite subset property by definition.)

Corollary 6. *Let (Ω, Σ, μ) be a strictly localizable measure space and $\nu: \Sigma \to \overline{\mathbb{R}}^+$ be a measure such that $\nu \ll \mu$, and both ν, μ have the finite subset property. Then there exists a measurable [for Σ] and μ-unique $f: \Omega \to \overline{\mathbb{R}}^+$ such that*

$$\nu(A) = \int_A f\,d\mu, \qquad A \in \Sigma;$$

and $f < \infty$ a.e. on each σ-finite set (for ν) can be arranged.

This result will be used (Chapter 8) in the discussion of a lifting operation, the existence of which turns out to be equivalent to the strict localizability of μ. We now present some complements as exercises before turning to the duality theory of $L^p(\mu)$-spaces.

EXERCISES

0. Let (Ω, Σ) be a measurable space, μ, ν be measures on Σ, and $\Sigma_0, \Sigma_0', \Sigma_0''$ be the rings of sets from Σ of finite measure for μ, ν and simultaneously for μ and ν. Illustrate the differences between Σ_0, Σ_0', and Σ_0'' for μ, ν if they have finite subset properties separately. (Consider Saks' example again.)

1. Let (Ω, Σ) be a measurable space. Let μ, ν be a pair of measures on Σ with the finite subset property, and both be localizable. Show that ν admits a unique decomposition, $\nu = \nu_1 + \nu_2$ where $\nu_1 \ll \mu$ and $\nu_2 \perp \mu$ in the same sense as before. [Note that in Theorem 2.2, μ is somewhat restricted. It is localizable. There does not seem to be a usable Lebesgue decomposition for arbitrary measures. For this problem, we may take $\nu + \mu$ to be localizable.]

2. Let (Ω, Σ, μ) be a measure space, and $\mathscr{A} \subset \Sigma$ be a σ-subalgebra. If $\mu|\mathscr{A}$, the restriction to $\mathscr{A}$, is strictly localizable, then show that μ on Σ is also strictly localizable. (Thus the same result holds as for σ-finite measures.) Does this property hold for all localizable measures?

3. Another form of Theorem 5 is the following. Let (Ω, Σ, μ) be a measure space with the finite subset property. If $\Sigma_0 = \{A \in \Sigma: \mu(A) < \infty\}$, let $f: \Omega \times \Sigma_0 \to \overline{\mathbb{R}}^+$ be a measurable quasifunction relative to $\{\mu_E: E \in \Sigma_0\}$. Show that $f(\omega, E) = g(\omega)\chi_E$ a.e., $E \in \Sigma_0$, for a measurable $g: \Omega \to \overline{\mathbb{R}}^+$ iff μ has the Radon-Nikodým property. [Same argument as in the proof of the theorem, but a little more care is needed.]

4. Let $\{\mu_j: j \in J\}$ be a family of measures on a measurable space (Ω, Σ), all jointly with finite subset properties. The family is said to be *simultaneously localizable* if for any subcollection $\mathscr{C} \subset \Sigma$, there is a set $A \in \Sigma$ which is a simultaneous supremum in the sense that for each $B \in \mathscr{C}$ we have $\mu_j(B - A) = 0$, $j \in J$, and if $\tilde{A} \in \Sigma$ is any set with the same property as A, then $\mu_j(A - \tilde{A}) = 0$, $j \in J$. Thus A is the smallest such set essentially containing all sets of $\mathscr{C}$. Similarly, a class $\mathscr{F}$ of measurable (Σ) real functions on (Ω, Σ) has a *simultaneous supremum* relative to $\{\mu_j: j \in J\}$, if there is a measurable (Σ) function $g: \Omega \to \overline{\mathbb{R}}^+$, such that $g \geqslant f$ a.e.(μ_j), $j \in J$, for all $f \in \mathscr{F}$, and such that if $\tilde{g}$ is another function, measurable (Σ) and dominating $\mathscr{F}$, then $\tilde{g} \geqslant g$ a.e.(μ_j), $j \in J$. Show that if $\{\mu_j: j \in J\}$ is simultaneously localizable, then any such collection $\mathscr{F}$ has a measurable (Σ) simultaneous supremum. [The class $\mathscr{C} = (\{\omega: f(\omega) \geqslant r\}: f \in \mathscr{F}) \subset \Sigma$ has a simultaneous supremum B_r for each rational r. Let $C_r = \bigcup\{B_{r'}: r' \leqslant r, \text{rational}\}$. Then $C_r \in \Sigma$ and increasing. Let $g_r(\omega) = r$ if $\omega \in C_r$, and $= -\infty$ otherwise. If $g = \sup\{g_r, r \text{ rational}\}$, then g is the desired supremum of $\mathscr{F}$.]

5. If $\{(\Omega, \Sigma, \mu_n): n \geqslant 1\}$ is a sequence of finite measure spaces, then $\{\mu_n: n \geqslant 1\}$ is simultaneously localizable. [If $\mathscr{A} \subset \Sigma$ is any subring, and $\alpha_n = \sup\{\mu_n(A): A \in \mathscr{A}\}$, then for each $\epsilon_k > 0$, there is a $B_k \in \mathscr{A}$ such that $\alpha_n - \epsilon_k \leqslant \mu_n(B_k) \leqslant \alpha_n$, $n = 1, 2, \ldots, k$. Let $B = \bigcup_k B_k$. Then B is a simultaneous supremum of $\mathscr{A}$ for all μ_n, $n \geqslant 1$.]

6. Let $\{(\Omega, \Sigma, \mu_j): j \in J\}$ be a family of measure spaces each with the finite subset property. Let $\mathscr{F}_J$ be a collection of measurable quasifunctions $f: \Omega \times J \to \overline{\mathbb{R}}$ such that $g(\cdot, j) \geqslant f(\cdot, j)$ a.e.(μ_j) for $f \in \mathscr{F}_J$, and g is the smallest such relative to $\{\mu_j: j \in J\}$. Show that if each countable subset of $\{\mu_j: j \in J\}$ is simultaneously localizable, then $\mathscr{F}_J$ has a simultaneous supremum, in the sense just defined. [Use the result of Exercise 4 above, and the fact that each μ_j is localizable.]

7. Let $\Omega = \mathbb{R}^2$, the Euclidean plane. If $A \subset \Omega$, let $\mu(A) = +\infty$ for A uncountable, and $= 1$ or 0 according as the origin $0 \in A$ or $\notin A$ when A is countable. Show that μ is an outer measure and $\mathscr{M}_\mu = \mathscr{P}$, the power set of Ω. If $\nu = \mu|_{\mathscr{M}_\mu}$, verify that $(\Omega, \mathscr{M}_\mu, \nu)$ does not have the finite subset property and ν is not localizable. Describe $L^\infty(\mu)$. Show that $L^p(\mu) = L^1(\mu)$ $[\neq L^\infty(\mu)]$ for all $1 < p < \infty$, and $\|f\|_p = |f(0)|$ for $f \in L^p(\mu)$. [This example is from McShane (1950) and will be used for other illustrations later.]

5.5. DUALITY OF L^p-SPACES AND CONDITIONAL EXPECTATIONS

The work of this section contains some important applications of the Radon-Nikodým theorem, exhibiting its influence on some basic parts of analysis. We concentrate on characterizing the dual spaces of the familiar L^p-spaces studied in Section 4.5, and present the concept of conditional expectation, which is fundamental in probability theory and necessary for a general martingale analysis. Also the weak completeness of the L^p-spaces, $1 \leqslant p < \infty$, and other facts of this class are included (some as exercises with copious hints), since they are useful in many mathematical studies.

As shown at the beginning of Section 3, if l is a linear mapping on a normed linear space $\mathscr{X}$ into another normed linear space $\mathscr{Y}$, then it is bounded iff it is continuous at one point and then it is automatically uniformly continuous on $\mathscr{X}$. Also $\|l\| = \sup\{\|l(x)\|_{\mathscr{Y}}: \|x\|_{\mathscr{X}} = 1\}$. It is clear that $\|\cdot\|$ is a norm, and the set of all such bounded linear mappings, often denoted by $B(\mathscr{X}, \mathscr{Y})$, is a normed linear space in this norm. If $\mathscr{Y} = \mathbb{R}$ (or $\mathbb{C}$), then it is denoted by $\mathscr{X}^*$ and called the *dual* (or *adjoint* or *conjugate*) space of $\mathscr{X}$. The elements of $\mathscr{X}^*$ are the continuous linear functionals on $\mathscr{X}$, and not only is $\mathscr{X}^*$ a normed linear space, but it is always complete. In fact let $\{l_n: n \geqslant 1\} \subset \mathscr{X}^*$ be a Cauchy sequence, i.e. $\|l_n - l_m\| \to 0$ as $n, m \to \infty$. Hence for each $x \in \mathscr{X}$, $\|x\|_{\mathscr{X}} = 1$, we have

$$|l_n(x) - l_m(x)| = |(l_n - l_m)(x)| \leqslant \|l_n - l_m\| \cdot \|x\|_{\mathscr{X}} = \|l_n - l_m\| \to 0.$$

Thus $\{l_n(x): n \geqslant 1\}$ is Cauchy in $\mathbb{R}$ (or $\mathbb{C}$), and since the latter space is complete, $l_n(x) \to l(x) \in \mathbb{R}$ ($\in \mathbb{C}$). Clearly $l(x_1 + \alpha x_2) = l(x_1) + \alpha l(x_2)$, so that l is linear. Also $|l(x)| = \lim_n |l_n(x)| \leqslant \lim_n \|l_n\|\,\|x\|$, since $|\|l_n\| - \|l_m\|| \leqslant \|l_n - l_m\| \to 0$. But then $\sup_n \|l_n\| = M < \infty$, because a Cauchy sequence of numbers is bounded. Hence $|l(x)| \leqslant M\|x\|_{\mathscr{X}}$, and l is bounded, so $l \in \mathscr{X}^*$. Moreover

$$|l(x) - l_n(x)| = \lim_m |l_m(x) - l_n(x)| \leqslant \lim_m \|l_m - l_n\|\,\|x\|_{\mathscr{X}}.$$

The right side approaches 0 as $n \to \infty$, uniformly in $\|x\|_{\mathscr{X}} = 1$. So $\|l - l_n\| \to 0$, and l is the limit of $\{l_n: n \geqslant 1\}$. Thus $\mathscr{X}^*$ is always complete, even if $\mathscr{X}$ is not. Note that the same argument shows that $B(\mathscr{X}, \mathscr{Y})$ is always complete if only $\mathscr{Y}$ is.

The above definition of $\mathscr{X}^*$ does not tell us much about the structure of $\mathscr{X}^*$. If $\mathscr{X} = L^p(\mu)$, $p \geqslant 1$, then one can say more about its dual space $[L^p(\mu)]^*$. We show that the latter can be identified with $L^q(\mu)$ if $1 < p < \infty$, $q = p/(p-1)$, and this statement is true for $p = 1$ only under some restrictions on (Ω, Σ, μ)—it must be localizable. The case that $p = +\infty$ is different and must be considered separately. (On the latter, see the last half of Section 10.2 for another perspective.)

In order to characterize $[L^p(\mu)]^*$ for $1 < p < \infty$, without any restrictions on μ, we first establish a geometric property of the Banach space $L^p(\mu)$, namely its "uniform convexity". This is a consequence of a numerical inequality for any pair of numbers, which we now establish; the proof follows McShane (1950). The inequality and its application are not well motivated and are technical, although the tools are elementary.

Lemma 1. *If* $1 < p < \infty$, $\epsilon > 0$, *let* $x, y \in \mathbb{R}$ *be such that* $|x| = 1 \geqslant |y|$ *and* $|x - y| \geqslant \epsilon$. *Then there is a* $\delta_\epsilon > 0$ *such that*

$$\left|\frac{x+y}{2}\right|^p \leqslant \tfrac{1}{2}(1-\delta_\epsilon)(|x|^p + |y|^p). \tag{1}$$

Proof. Since $\varphi(t) = |t|^p$ is strictly convex and $|x - y| \geqslant \epsilon > 0$, so that $x \neq y$, we have

$$\left|\frac{x+y}{2}\right|^p = \varphi\left(\frac{x+y}{2}\right) < \tfrac{1}{2}[\varphi(x) + \varphi(y)] = \tfrac{1}{2}(1 + |y|^p). \tag{2}$$

If (1) is not true (i.e., $\delta_\epsilon > 0$ does not exist), then for some $\{x_n, y_n\} \subset \mathbb{R}$, $|y_n| \leqslant |x_n| = 1$, we get for (1), with the triangle inequality,

$$\begin{aligned} 1 &= \lim_n \frac{\left|\frac{1}{2}(x_n + y_n)\right|^p}{\frac{1}{2}(|x_n|^p + |y_n|^p)} \\ &\leqslant \overline{\lim_n} \frac{\left[\frac{1}{2}(1 + |y_n|)\right]^p}{\frac{1}{2}(1 + |y_n|^p)} \\ &\leqslant 1, \qquad \text{by (2)}. \end{aligned} \tag{3}$$

But $\{|y_n|: n \geqslant 1\}$ must converge. In fact, $|y_n| \leqslant 1$ implies there is a convergent subsequence $|y_{n_i}| \to a$ with $a \leqslant 1$. So for this subsequence the above becomes

$$1 \leqslant \lim_n \left[\frac{\left[\frac{1}{2}(1 + |y_{n_i}|)\right]^p}{\frac{1}{2}(1 + |y_{n_i}|^p)}\right] = \frac{\left[\frac{1}{2}(1 + a)\right]^p}{\frac{1}{2}(1 + a^p)} \leqslant 1, \quad 1 < p < \infty.$$

This and (2) show the equality holds iff $a = 1$. Thus each convergent subsequence of $\{|y_n|: n \geqslant 1\}$ has limit 1, so the whole sequence must converge to 1, i.e., $|y_n| \to 1$. Now let $u_n = y_n/|y_n|$. Then

$$|u_n - y_n| = |y_n|\left(\frac{1}{|y_n|} - 1\right) \to 0 \quad \text{as} \quad n \to \infty, \tag{4}$$

and

$$\epsilon \leqslant |x_n - y_n| \leqslant |x_n - u_n| + |u_n - y_n|,$$

so that

$$\epsilon \leqslant \underline{\lim}_n |x_n - u_n|. \tag{5}$$

We assert that $|x_n + u_n| \to 2$ as $n \to \infty$. If this is not true, there exists a $\delta_0 > 0$ such that $|x_n + u_n|/2 < 1 - \delta_0$ for all large n. Hence

$$1 - \delta_0 > \frac{|y_n + |y_n|x_n|}{2|y_n|} \quad \Rightarrow \quad \overline{\lim}_n \tfrac{1}{2}(|y_n + |y_n|x_n|) \leqslant 1 - \delta_0. \tag{6}$$

But since $|x_n| = 1$, we have

$$|y_n + x_n|y_n|| = |(y_n + x_n) - x_n(1 - |y_n|)| \geqslant |(y_n + x_n) - (1 - |y_n|)|. \tag{7}$$

However, $|y_n| \to 1$, and by (3), $|x_n + y_n| \to 2$, so that (6) and (7) imply

$$2 \leqslant \overline{\lim}_n |y_n + x_n|y_n|| \leqslant 2(1 - \delta_0).$$

This contradiction shows that $|u_n + x_n| \to 2$, so that

$$4 = \lim_n (u_n + x_n)^2 = \lim_n (1 + 1 + 2u_n x_n).$$

Thus $u_n x_n \to 1$. Hence (5) becomes

$$0 < \epsilon^2 \leqslant \underline{\lim}_n (x_n - u_n)^2 = \underline{\lim}_n (1 + 1 - 2u_n x_n) = 0.$$

This contradiction implies that our initial supposition is false, and (1) holds as stated.

As a consequence, if $x, y \in \mathbb{R}$ are arbitrary and not both zero, we may replace x, y by $x/\max(|x|, |y|)$ and y by $y/\max(|x|, |y|)$ to obtain:

Corollary 2. *If* $1 < p < \infty$, $x, y \in \mathbb{R}$ *(not both zero) and* $|x - y| \geqslant \max(|x|, |y|)$, *then we have*

$$\left|\frac{x + y}{2}\right|^p \leqslant \tfrac{1}{2}(1 - \delta)(|x|^p + |y|^p), \tag{8}$$

where $\delta = \delta(|x - y|/\max(|x|, |y|)) \to 0$ *as* $|x - y| \to 0$.

Using this numerical inequality, we can establish a key property for $L^p(\mu)$:

Proposition 3. *If $f, g \in L^p(\mu)$, where (Ω, Σ, μ) is any measure space and $1 < p < \infty$, $\|f\|_p = 1 = \|g\|_p$, then for each $\epsilon > 0$, if $\|f - g\|_p \geqslant \epsilon$, there is a $\delta_\epsilon > 0$ such that*

$$\left\| \frac{f+g}{2} \right\|_p < 1 - \delta_\epsilon, \tag{9}$$

where $\delta_\epsilon \downarrow 0$ as $\epsilon \downarrow 0$.

Proof. We apply (8) with $x = f(\omega)$, $y = g(\omega)$ for ω in the following set A:

$$A = \left\{ \omega : |f(\omega) - g(\omega)|^p \geqslant \frac{\epsilon^p}{4}\left(|f(\omega)|^p + |g(\omega)|^p\right) \right\}$$

$$\subset \left\{ \omega : \frac{|f(\omega) - g(\omega)|^p}{\max\left(|f(\omega)|^p, |g(\omega)|^p\right)} \geqslant \frac{\epsilon^p}{4} \right\}.$$

Hence for $\omega \in A$, by (8) we get

$$\left| \frac{f(\omega) + g(\omega)}{2} \right|^p \leqslant \left[1 - \delta\left(\frac{\epsilon}{4^{1/p}} \right) \right] \frac{|f(\omega)|^p + |g(\omega)|^p}{2}. \tag{10}$$

On integration one has

$$\epsilon^p \leqslant \|f - g\|_p^p = \int_{\Omega - A} |f - g|^p \, d\mu + \int_A |f - g|^p \, d\mu$$

$$\leqslant \frac{\epsilon^p}{4} \int_{\Omega - A} (|f|^p + |g|^p) \, d\mu + \int_A |f - g|^p \, d\mu$$

$$\leqslant \frac{\epsilon^p}{4}(1 + 1) + \|(f - g)\chi_A\|_p^p.$$

This implies

$$\frac{\epsilon}{2^{1/p}} \leqslant \|(f - g)\chi_A\|_p \leqslant \|f\chi_A\|_p + \|g\chi_A\|_p \leqslant 2.$$

Thus one of the terms is at least $(\epsilon/2 \times 2^{1/p})$, and each is at most

$\max(\|f\chi_A\|_p, \|g\chi_A\|_p)$. But $|(f+g)/2|^p \leqslant (|f|^p + |g|^p)/2$ by convexity. Integrating, we have

$$1 - \left\|\frac{f+g}{2}\right\|_p^p = \int_\Omega \left(\frac{|f|^p + |g|^p}{2} - \left|\frac{f+g}{2}\right|^p \right) d\mu$$

$$\geqslant \int_A \left(\frac{|f|^p + |g|^p}{2} - \left|\frac{f+g}{2}\right|^p \right) d\mu$$

(since the integrand $\geqslant 0$)

$$\geqslant \delta\left(\frac{\epsilon}{4^{1/p}}\right) \int_A \frac{|f|^p + |g|^p}{2} \, d\mu, \qquad \text{by (10)}$$

$$\geqslant \delta\left(\frac{\epsilon}{4^{1/p}}\right) \frac{1}{2} \frac{\epsilon^p}{2 \times 2^p} = \delta\left(\frac{\epsilon}{4^{1/p}}\right) \frac{\epsilon^p}{2^{p+2}},$$

by the preceding estimate. Let $\delta_1(\epsilon) = 1 - \delta(\epsilon/4^{1/p})\epsilon^p/2^{p+2}$. Then $\delta_1(\epsilon) > 0$ and $\uparrow 1$ as $\epsilon \downarrow 0$, since $\delta(\epsilon) \downarrow 0$. Thus

$$\left\|\frac{f+g}{2}\right\|_p \leqslant [\delta_1(\epsilon)]^{1/p} = 1 - \bar{\delta}(\epsilon) \quad \text{(say)}.$$

Then $\bar{\delta}(\epsilon) \downarrow 0$ as $\epsilon \downarrow 0$, and this gives (9) with $\bar{\delta}(\epsilon)$ for δ_ϵ there, as asserted.

In order to see this result in its proper perspective, we recall that a Banach space $\mathscr{X}$ is termed *uniformly convex* (or *uniformly rotund*) if for any pair $x, y \in \mathscr{X}$ such that $\|x\| = 1 = \|y\|$, and for any $\epsilon > 0$ such that $\|x - y\| \geqslant \epsilon$, there is $\delta_\epsilon > 0$ (depending only on ϵ) such that $\|(x+y)/2\| \leqslant 1 - \delta_\epsilon$. The preceding proposition shows that $\mathscr{X} = L^p(\mu)$, is uniformly convex for $1 < p < \infty$. This property does not hold for $p = 1$, since we can have $\|(f+g)/2\|_1 = 1$ for positive unit-norm elements f, g. Simple examples show that $L^\infty(\mu)$ is also not uniformly convex.

With the help of the preceding result one can establish a useful fact:

Proposition 4. *Let* (Ω, Σ, μ) *be a measure space,* $0 \neq l \in [L^p(\mu)]^*$, $1 < p < \infty$. *Then there exists a* μ*-unique element* $f_0 \in L^p(\mu)$, $\|f_0\|_p = 1$, *with* $l(f_0) = \|l\|$.

Proof. By definition of $\|l\|$, there exist $f_n \in L^p(\mu)$, $\|f_n\|_p = 1$, such that $|l(f_n)| \to \|l\|$. As noted in the proof of Theorem 3.1, by replacing f_n with $[\operatorname{sgn} l(f_n)]f_n$, if necessary, one may assume that $l(f_n) \geqslant 0$, and so for large enough n, $l(f_n) > 0$. We assert again that $\{f_n : n \geqslant 1\}$ is Cauchy in $L^p(\mu)$.

Indeed, given $\epsilon > 0$, there is a $\delta_\epsilon > 0$ such that for all large m, n

$$\|l\|\left\|\frac{f_n + f_m}{2}\right\|_p \geqslant l\left(\frac{f_n + f_m}{2}\right) = \frac{l(f_n) + l(f_m)}{2} > \|l\|(1 - \delta_\epsilon).$$

This implies $\|(f_n + f_m)/2\|_p \geqslant 1 - \delta_\epsilon$, since $0 < \|l\| < \infty$. By Proposition 3 (and the definition of uniform convexity) we must have $\|f_n - f_m\|_p < \epsilon$ [cf. (9)]. Thus $\{f_n: n \geqslant 1\}$ is Cauchy, and so $f_n \to f_0$ in $L^p(\mu)$, since the latter is a complete space. It is clear that $\|f_0\|_p = \lim_n \|f_n\|_p = 1$ and $\|f_n - f_0\|_p \to 0$. So

$$|l(f_0) - l(f_n)| = |l(f_0 - f_n)| \leqslant \|l\| \, \|f_0 - f_n\|_p \to 0$$

and we have $l(f_0) = \lim_n l(f_n) = \|l\|$.

The uniqueness is easy. If $f_0' \in L^p(\mu)$, $\|f_0'\|_p = 1$, is another element satisfying $l(f_0') = \|l\|$, then $\|f_0 - f_0'\|_p \geqslant \epsilon > 0 \Rightarrow$

$$0 < \|l\| = l\left(\frac{f_0 + f_0'}{2}\right) \leqslant \|l\|\left\|\frac{f_0 + f_0'}{2}\right\|_p \leqslant \|l\|(1 - \delta_\epsilon)$$

for some $\delta_\epsilon > 0$, by (9). This contradiction shows that $f_0 = f_0'$ a.e., completing the proof.

With these auxiliary results, which are also of independent interest, we are ready to present the characterization of $[L^p(\mu)]^*$. It is amusing to note that in 1907, in the same issue of the Paris Academy of Sciences Proceedings, both F. Riesz and M. Fréchet announced the result independently for $[L^2(0,1)]^*$. It was much later, in the 1930s, an abstract measure case for σ-finite measures was obtained by N. Dunford whose proof will be included in the Exercises section (Exercise 7). Here we present McShane's argument, for which we developed the above machinery. For simplicity let us treat the case where all functions are real. Note that *no* restrictions are imposed on μ here.

Theorem 5. *Let* (Ω, Σ, μ) *be a measure space,* $1 < p < \infty$, $q = p/(p-1)$. *Then for each* $l \in [L^p(\mu)]^*$, *there exists a* μ*-unique* $g_l \in L^q(\mu)$ *such that*

$$l(f) = \int_\Omega f g_l \, d\mu, \qquad f \in L^p(\mu), \tag{11}$$

and $\|l\| = \|g_l\|_q$.

Proof. If $l = 0$, then take $g_l = 0$. So let $l \neq 0$. Then by Proposition 4 there is a unique $f_0 \in L^p(\mu)$, $\|f_0\| = 1$, such that $l(f_0) = \|l\|$. Dividing by $\|l\|$ if necessary, we may (and do) assume $\|l\| = 1$. Then proceeding exactly as in the proof of Theorem 3.1, we consider, for $t \in \mathbb{R}$ and $f \in L^p(\mu)$,

$$1 = l(f_0) = l(f_0 + t[f - l(f)f_0]) \leqslant \|f_0 + t[f - l(f)f_0]\|_p. \tag{12}$$

Hence if $tl(f) \neq -1$, we can express $f_0 + tf$ as $[1 + tl(f)]\{f_0 + t[f - l(f)f_0][1 + tl(f)]^{-1}\}$ and get with (12)

$$\|f_0 + tf\|_p - 1 \geqslant |1 + tl(f)| - 1. \tag{13}$$

If $t > 0$ and small we then have from (13)

$$\frac{\|f_0 + tf\|_p - 1}{t} \geqslant l(f),$$

and if $t < 0$ and small, then the opposite inequality holds. Thus if the norm is differentiable, then one gets

$$l(f) = \lim_{t \to 0} \frac{\|f_0 + tf\|_p - 1}{t}. \tag{14}$$

To see that this derivative exists, note that for $p > 1$

$$\frac{d}{dt}|f_0(\omega) + tf(\omega)|^p = p|f_0(\omega) + tf(\omega)|^{p-1}\frac{d}{dt}|f_0(\omega) + tf(\omega)|$$

$$= p|f_0(\omega) + tf(\omega)|^{p-1}\frac{f_0(\omega)f(\omega)}{|f_0(\omega)|}, \tag{15}$$

since $(d/dt)|a + tb| = ab/|a|$ if $a \neq 0$. But now if $|t| < 1$, the right side is dominated by $p(|f_0| + |tf|)^{p-1}|f|(\omega)$ for a.a. ω, and $(|f_0| + |tf|)^{p-1} \in L^q(\mu)$, $f \in L^p(\mu)$, so that this product is integrable (by the Hölder inequality). Hence by the dominated convergence theorem [in (14) we can let $t \to 0$ through a sequence],

$$\frac{d}{dt}\left(\|f_0 + tf\|_p^p\right)\bigg|_{t=0} = \lim_{t \to 0}\int_\Omega \frac{|f_0 + tf|^p - |f_0|^p}{t}\,d\mu$$

$$= \int_\Omega p|f|\,|f_0|^{p-1}\operatorname{sgn} f_0\,d\mu. \tag{16}$$

Hence the norm is differentiable at $t = 0$. So (14) and (16) give

$$p\left(\|f_0 + tf\|_p^{p-1}\frac{d}{dt}\|f_0 + tf\|_p\right)\bigg|_{t=0}$$

$$= pl(f) = p\int_\Omega f|f_0|^{p-1}\operatorname{sgn} f_0\,d\mu.$$

Thus if $g_l = \|l\|\,|f_0|^{p-1}\operatorname{sgn} f_0$, then $g_l \in L^q(\mu)$ and

$$l(f) = \int_\Omega f g_l \, d\mu.$$

But we also have

$$\|l\| = \sup_{\|f\|_p \leqslant 1} |l(f)| = \sup_{\|f\|_p \leqslant 1} \left| \int_\Omega f g_l \, d\mu \right| = \|g_l\|_q,$$

by Proposition 4.5.17. This gives (11).

The uniqueness is again easy. In fact, if $g_l' \in L^p(\mu)$ is another function satisfying (11), one gets

$$\int_\Omega f(g_l - g_l') \, d\mu = 0, \qquad f \in L^p(\mu).$$

Hence taking the supremum on $\|f\|_p \leqslant 1$, we obtain $\|g_l - g_l'\|_q = 0$, so that $g_l = g_l'$ a.e. This establishes the theorem.

The preceding proof contains more information than we have asserted. To state it, let us introduce some concepts. If $\mathscr{X}$ is a Banach space and $S = \{x \in \mathscr{X}: \|x\| = 1\}$, the unit sphere, we say that the norm is *weakly* (*or Gateâux*) *differentiable* if

$$G(x, h) = \lim_{t \to 0} \frac{\|x + th\| - \|x\|}{t}, \qquad x, h \in S, \tag{17}$$

exists at each $x \in S$. The norm is *strongly* (or *Fréchet*) *differentiable* if the limit in (17) holds at each $x \in S$ uniformly in $h \in S$. Then the sphere is also called *smooth* (has a unique tangent plane) at each $x \in S$. The sphere S is *uniformly smooth* if the limit in (17) exists uniformly in $(x, h) \in S \times S$ (has *uniformly strongly differentiable* norm).

Thus what we have shown in (16) is that the sphere of $L^p(\mu)$, $1 < p < \infty$, is uniformly smooth. It is useful to state this for reference.

Corollary 6. *The space $L^p(\mu)$, $1 < p < \infty$, is uniformly convex, and its sphere is uniformly smooth.*

It is actually a well-known result due to V. L. Šmulian that a Banach space $\mathscr{X}$ is uniformly convex iff its dual $\mathscr{X}^*$ is uniformly smooth. Here $\mathscr{X} = L^p(\mu)$, $1 < p < \infty$, gives an important class of Banach spaces which have both these properties simultaneously.

The result of Theorem 5 implies the following statement.

Corollary 7. *If $l \in [L^p(\mu)]^*$, then the mapping $\tau: l \mapsto g_l \in L^q(\mu)$, $1 < p < \infty$, is a one to one norm-preserving transformation of $[L^p(\mu)]^*$ onto $L^q(\mu)$, so that* *the spaces* $[L^p(\mu)]^*$ *and* $L^q(\mu)$, $q = p/(p-1)$, *are isometrically isomorphic, and hence can be identified.*

Proof. By Theorem 5, l and g_l determine each other uniquely, since $l(f) = \int_\Omega fg_l\,d\mu$ clearly implies $l \in [L^p(\mu)]^*$ by the Hölder inequality, and $\|l\| = \|g_l\|_q = \|\tau(l)\|$. Thus τ is linear, one to one, and onto. This proves the statement.

In view of this identification, one says that the dual (or adjoint or conjugate) space of $L^p(\mu)$ is $L^q(\mu)$, $1 < q = p/(p-1) < \infty$. No such general statement can hold for $p = 1$, as illustrated by Exercise 4.7. There $L^1(\mu) = L^p(\mu)$, $1 < p < \infty$. Hence $[L^1(\mu)]^* = [L^p(\mu)]^* = L^q(\mu) = L^1(\mu)$ implies the surprising result that $L^1(\mu)$ is its own dual space, which thus is not $L^\infty(\mu)$ (since $q = \infty$ for $p = 1$). The trouble there is that μ does not even have the finite subset property. [In this example $L^1(\mu) \neq L^\infty(\mu)$.] If μ has the finite subset property, then $L^\infty(\mu)$ can be identified as a closed subspace of $[L^1(\mu)]^*$. We shall see that, under a further restriction on μ, these two spaces *can* be identified. It was first shown by H. Steinhaus in 1918 that $[L^1(0,1)]^*$ may be identified as $L^\infty(0,1)$. This was extended by O. M. Nikodým (also N. Dunford) to (σ-)finite μ, and the general form given below is essentially due to I. E. Segal (1954).

Theorem 8. *Let* (Ω, Σ, μ) *be a measure space with the finite subset property. Then* $l \in [L^1(\mu)]^*$ *can be represented as*

$$l(f) = \int_\Omega fg_l\,d\mu, \qquad f \in L^1(\mu), \tag{18}$$

for a μ*-unique* $g_l \in L^\infty(\mu)$ *iff* μ *is localizable. When this holds, we have* $\|l\| = \|g_l\|_\infty$ *and the mapping* $\tau\colon l \mapsto g_l$ *is an isometric isomorphism between* $[L^1(\mu)]^*$ *and* $L^\infty(\mu)$, *so that they may be identified.*

Proof. In contrast to Theorem 5, this demonstration depends on (and by Theorem 4.5, is equivalent to) the Radon-Nikodým theorem in its general form, so that the work of Section 4 is fully utilized here.

If $l_g\colon L^1(\mu) \to \mathbb{R}$ is defined for each $g \in L^\infty(\mu)$ by the equation

$$l_g(f) = \int_\Omega fg\,d\mu, \qquad f \in L^1(\mu),$$

then by the Hölder inequality, l is linear and bounded, since

$$\sup_{\|f\|_1 \leqslant 1} |l_g(f)| \leqslant \sup_{\|f\|_1 \leqslant 1} \|f\|_1 \|g\|_\infty = \|g\|_\infty < \infty.$$

Thus $l_g \in [L^1(\mu)]^*$, and further, by Proposition 4.5.7, $\|l_g\| = \|g\|_\infty$. Thus

$\tau: g \mapsto l_g$ defines an isometric isomorphism $\|\tau(g)\| = \|g\|_\infty$ of $L^\infty(\mu)$ into $[L^1(\mu)]^*$, so that we have $\tau(L^\infty(\mu)) \subset [L^1(\mu)]^*$ as a closed subspace. The opposite inclusion is involved and is proved in steps, using the localizability of μ.

1. Let $\mu(\Omega) < \infty$. If $l \in [L^1(\mu)]^*$, then let $\nu(A) = l(\chi_A)$, $\chi_A \in L^1(\mu)$. Clearly the additivity of l implies that $\nu(A \cup B) = \nu(A) + \nu(B)$ for disjoint A, B in Σ. So ν is finitely additive. If $A_n \downarrow \varnothing$, $A_n \in \Sigma$, then $|\nu(A_n)| = |l(\chi_{A_n})| \leqslant \|l\|\,\|\chi_{A_n}\|_1 = \|l\|\mu(A_n) \to 0$ as $n \to \infty$, since μ is a finite measure. Thus ν is a signed measure. If $\mu(A) = 0$, then $\chi_A = 0$ a.e., so that $\nu(A) = l(\chi_A) = 0$. By Proposition 5.2.1, $\nu \ll \mu$. So by the Radon-Nikodým theorem, there is a μ-unique g_0 such that $\nu(B) = \int_B g_0\, d\mu$, $B \in \Sigma$. Hence by linearity

$$\sum_{i=1}^{n} a_i l(\chi_{B_i}) = \sum_{i=1}^{n} a_i \int_\Omega \chi_{B_i} g_0\, d\mu = \int_\Omega \left(\sum_{i=1}^{n} a_i \chi_{B_i} \right) g_0\, d\mu. \tag{19}$$

Thus if f is a simple function, then (19) implies

$$l(f) = \int_\Omega f g_0\, d\mu. \tag{20}$$

If $f \in L^1(\mu)$ is arbitrary, then there exist simple $f_n \to f$ a.e. and in $L^1(\mu)$, so that

$$\left| \int_\Omega f_n g_0\, d\mu - \int_\Omega f_m g_0\, d\mu \right| = |l(f_n - f_m)| \leqslant \|l\|\,\|f_n - f_m\|_1 \to 0,$$

and hence $\lim_n \int_\Omega f_n g_0\, d\mu = \lim_n l(f_n) = l(f)$ exists. Also, by (20), and since simple functions are dense in $L^1(\mu)$, we have

$$\left| \int_\Omega f g_0\, d\mu \right| = |l(f)| \leqslant \|l\|\,\|f\|_1.$$

Thus by Proposition 4.5.17, since $\|l\| < \infty$, this implies $g_0 \in L^\infty(\mu)$. Then it follows that

$$\left| \int_\Omega (f_n - f) g_0\, d\mu \right| \leqslant \|g_0\|_\infty \|f_n - f\|_1 \to 0,$$

and hence $\lim_n \int_\Omega f_n g_0\, d\mu = \int_\Omega f g_0\, d\mu = l(f)$. This is (18) in this case. As before, the μ-uniqueness of g_0 and that $\|l\| = \|g_0\|_\infty$ follow immediately.

2. Let μ be localizable. If $\Sigma_0 = \{A \in \Sigma: \mu(A) < \infty\}$, then for each $A \in \Sigma_0$, let $\mu_A = \mu(A \cap \cdot\,)$. By the above case there is a μ_A-unique g_A,

bounded, such that

$$l(f) = \int_\Omega (fg_A)(\omega)\mu_A(d\omega), \qquad f \in L^1(\mu),$$

holds. Since each f in $L^1(\mu)$ vanishes outside a set of σ-finite measure, the integral is well defined, and $\{g_A: A \in \Sigma_0\}$ is a bounded measurable quasifunction relative to $\{\mu_A: A \in \Sigma\}$. Note that by the uniqueness assertion, if $A_1 \subset A_2$, then $g_{A_1} = g_{A_2}$ a.e.[μ], and in general, for any A_1, A_2 in Σ_0, $g_{A_1} = g_{A_2}$ a.e. on $A_1 \cap A_2$. Since μ is localizable, $\{g_A: A \in \Sigma_0\}$ is determined by a μ-measurable function g on Ω, by Theorem 4.3. [Note that we had $g_A \geqslant 0$ a.e. there, but evidently the theorem holds in general, since we could consider that $l(\chi_A) = \nu(A) = \nu^+(A) - \nu^-(A) = l^+(\chi_A) - l^-(\chi_A)$ gives $l = l^+ - l^-$ as a Jordan decomposition, so that $g_A = g_A^+ - g_A^-$ and $\{g_A^\pm: A \in \Sigma_0\}$ is a measurable quasifunction determining $g^\pm$. Thus $g = g^+ - g^-$ is the above one. This simple modification is the assertion.] Hence we have the result

$$l(f) = \int_\Omega fg\, d\mu, \qquad f \in L^1(\mu).$$

Proposition 4.5.17 shows that $\|g\|_\infty = \|l\| < \infty$. Thus $g \in L^\infty(\mu)$, and (18) is proved in this case also.

3. It remains to consider the converse: $[L^1(\mu)]^* = L^\infty(\mu)$ under the identification implies μ is localizable. Now this means every $l \in [L^1(\mu)]^*$ is given by (18) for some $g_l \in L^\infty(\mu)$. We may also assume $l(f) \geqslant 0$ for $f \geqslant 0$, so $g_l \geqslant 0$ a.e. Then if $f = \chi_A \in L^1(\mu)$, we get

$$\nu(A) = l(\chi_A) = \int_A g_l\, d\mu, \qquad A \in \Sigma_0.$$

This implies ν is determined on Σ_0 so that μ, ν have the finite subset property together, and then

$$\int_\Omega f\, d\nu = \int_\Omega fg_l\, d\mu, \qquad f \in L^1(\nu).$$

Consequently μ has the Radon-Nikodým property. By Theorem 4.5, we then conclude that μ must be localizable. This completes the proof of the theorem.

The preceding result shows that the localizability concept is just the right one in the duality theory of these spaces. Theorem 5 can also be proved using the above method in lieu of the differentiation of norms that we employed. However, both these methods are important enough that the reader should indeed know both of them; the above alternative one is given more generally as Exercise 7.

We summarize the preceding results to obtain the following comprehensive statement, due essentially to I. E. Segal (1954, p. 301) in this formulation.

Theorem 9 (Segal). *Let* (Ω, Σ, μ) *be a measure space with the finite subset property, and denote by* $L^1(\mu)$, $L^\infty(\mu)$ *the Lebesgue spaces on it. Then the following are equivalent*:

(i) μ *is localizable.*
(ii) μ *has the Radon-Nikodým property.*
(iii) *The spaces* $[L^1(\mu)]^*$ *and* $L^\infty(\mu)$ *are isometrically isomorphic.*

It only remains to study the dual space of $L^\infty(\mu)$. However, in this case one has to consider the integration relative to *finitely* additive set functions, since the related representation is in terms of such integrals. A brief account is in Exercises 9–11; the corresponding discussion is omitted here.

We turn to some weak convergence results for sequences in the $L^p(\mu)$-spaces. In order to have a satisfactory treatment one needs to know the *uniform boundedness principle*, which is a consequence of the Baire category theorem. It is given here, since it has many other applications.

Theorem 10 (Banach and Steinhaus). *Let* $\mathscr{X}$ *be a Banach space, and* $\{T_\alpha: \alpha \in A\}$ *be a family of continuous linear mappings from* $\mathscr{X}$ *to* $\mathscr{Y}$, *a normed linear space. If* $\{T_\alpha x: \alpha \in A\}$ *is bounded for each* $x \in \mathscr{X}$ *(i.e., pointwise bounded), then* $\{T_\alpha: \alpha \in A\}$ *is uniformly bounded, i.e.,* $\sup_\alpha \|T_\alpha\| < \infty$.

Proof (T. H. Hildebrandt). For each $\epsilon > 0$, $n \geqslant 1$, consider

$$\mathscr{X}_n = \left\{x \in \mathscr{X}: \sup_\alpha \left(\left\|\frac{1}{n}T_\alpha(x)\right\| + \left\|\frac{1}{n}T_\alpha(-x)\right\|\right) \leqslant \epsilon\right\}. \tag{21}$$

Since T is continuous, the right side is an intersection of a family of closed sets, so that $\mathscr{X}_n$ is closed, and it is nonempty ($0 \in \mathscr{X}_n$). Since $\|T_\alpha x\| \leqslant K_x$, $\alpha \in A$, it is clear that $\mathscr{X} = \bigcup_{n=1}^\infty \mathscr{X}_n$. But $\mathscr{X}$ is a complete metric space and hence is of the second category. So by the Baire category theorem (Proposition 4.4.1), $\mathscr{X}_{n_0}$ contains an open ball $B(x_0, r)$ for at least one n_0. Thus for each $x \in B(x_0, r)$ we have (for some x_0 in $\mathscr{X}_{n_0}$)

$$\left\|\frac{1}{n_0}T_\alpha(x_0 + x)\right\| < \frac{\epsilon}{2}, \qquad \alpha \in A, \quad \|x\| < r. \tag{22}$$

By linearity of T_α and the triangle inequality, we get

$$\left\|\frac{1}{n_0}T_\alpha(x)\right\| \leqslant \left\|\frac{1}{n_0}T_\alpha(x_0 + x)\right\| + \left\|\frac{1}{n_0}T_\alpha(-x_0)\right\| \leqslant \epsilon.$$

Hence

$$\|T_\alpha(x/n_0)\| \leqslant \epsilon, \qquad x \in B(x_0, r), \quad \alpha \in A.$$

This implies the result, since $x \mapsto x/n_0$ is a homeomorphism of $\mathscr{X}$ onto itself, i.e., $\sup_\alpha \|T_\alpha(x)\| \leqslant \epsilon n_0$ for all $\|x\| < r$. The proof is complete.

It should be noted that the argument uses only the fact that $\mathscr{X}$ is of second category, and that T_α can be nonlinear. The above proof without change implies the following result, stated for reference [$\mathscr{X}, \mathscr{Y}$ are assumed to have invariant metrics, i.e., $d(x, y) = d(0, x - y)$]:

Proposition 11 (General form of uniform boundedness principle). *Let $\mathscr{X}$ be a linear metric space, [with distance $d(x, y) = d(0, x - y)$], which is of the second category. Suppose $\{V_\alpha: \alpha \in A\}$ is a family of continuous mappings of $\mathscr{X}$ into a similar linear metric space $\mathscr{Y}$ such that ($|\cdot|$ denoting the invariant metric of $\mathscr{Y}$)*

(i) $|V_\alpha(x + y)| \leqslant |V_\alpha(x)| + |V_\alpha(y)|$,
(ii) $|V_\alpha(ax)| = |aV_\alpha(x)|$,

for all $x, y \in \mathscr{X}$, $\alpha \in A$, $a \in \mathbb{R}$ (or $\mathbb{C}$). Then if $\{V_\alpha(x): \alpha \in A\}$ is bounded in $\mathscr{Y}$ for each $x \in \mathscr{X}$, we have $\lim_{x \to 0} V_\alpha(x) = 0$ uniformly in $\alpha \in A$.

Since for any normed linear space $\mathscr{X}$, its dual $\mathscr{X}^*$ is always complete, we have the following consequence.

Corollary 12. *If $x_n \in \mathscr{X}$, $n \geqslant 1$, is weakly Cauchy, i.e., $\{l(x_n): n \geqslant 1\}$ is a Cauchy sequence for each $l \in \mathscr{X}^*$, then there is a constant $M < \infty$ such that $|l(x_n)| \leqslant M\|l\|$.*

Proof. If $T_n(l^*) = l^*(x_n)$, then $T_n: \mathscr{X}^* \to \mathbb{R}$ (or $\mathbb{C}$) is pointwise bounded, and hence the result follows from Theorem 10.

We are now ready to establish a result which was referred to in Exercise 4.5.13 and other places.

Theorem 13. *Let (Ω, Σ, μ) be a measure space, and $L^p(\mu)$, $1 \leqslant p < \infty$, be the Lebesgue space on it. Then $L^p(\mu)$ is weakly sequentially complete in the sense that if $\{f_n: n \geqslant 1\} \subset L^p(\mu)$ is weakly Cauchy, then there is a μ-unique $f \in L^p(\mu)$ such that $f_n \to f$ weakly.*

Proof. The argument is different for $p = 1$ from that for $1 < p < \infty$. It is simpler in the latter case, which we dispose of first. (Note that μ is unrestricted in all cases.)

Since $[L^p(\mu)]^*$ is identifiable with $L^q(\mu)$, $1 < q = p/(p-1) < \infty$, the hypothesis that $\{l(f_n): n \geqslant 1\}$ is a convergent sequence of reals can alternatively be written for $\{\int_\Omega f_n g_l \, d\mu \colon n \geqslant 1\}$. Recall that, by Theorem 5, the l and g_l correspond to each other uniquely. Hence for each $g \in L^q(\mu)$, define $T_n(g) = \int_\Omega f_n g \, d\mu$. Then the Hölder inequality implies that T_n is a continuous linear functional, i.e., $T_n \in [L^q(\mu)]^*$. Also $\{T_n(g) \colon n \geqslant 1\}$ converges for each $g \in L^q(\mu)$, which is a Banach space. So $\{|T_n(g)| \colon n \geqslant 1\}$ is bounded. Hence by Theorem 10, $\sup_n \|T_n\| = K_0 < \infty$. If we set $T(g) = \lim_n T_n(g)$, which exists by hypothesis for each g, it follows that T is uniquely defined and linear on $L^q(\mu)$. Moreover,

$$|T(g)| = \lim_n |T_n(g)| \leqslant M\|g\|_q, \qquad g \in L^q(\mu).$$

So $T \in [L^q(\mu)]^*$. By Theorem 5 again, there exists a μ-unique $f \in L^p(\mu)$, which is identifiable with $[L^q(\mu)]^*$, such that

$$T(g) = \int_\Omega fg \, d\mu, \qquad g \in L^q(\mu).$$

This means for each $l \in [L^p(\mu)]^*$, we have

$$l(f_n) = \int_\Omega f_n g_l \, d\mu \to \int_\Omega f g_l \, d\mu = l(f), \tag{23}$$

and hence $f_n \to f$ weakly. So $L^p(\mu)$, $1 < p < \infty$, is weakly sequentially complete.

Now consider the case $p = 1$. Then we cannot apply Theorem 8 (or 9) directly, since μ is not restricted. However, $\{f_n \colon n \geqslant 1\} \subset L^1(\mu)$. So by the Markov inequality, the set $A_n = \{\omega \colon |f_n(\omega)| > 0\} \in \Sigma$ and is at most σ-finite. Thus $A = \bigcup_{n=1}^\infty A_n$ is σ-finite and all $f_n(\omega) = 0$ a.e., $\omega \in A^c$, so that $\{f_n \colon n \geqslant 1\} \subset L^1(\mu_A)$, where the measure space is $(A, \Sigma(A), \mu_A)$ with $\mu_A(\cdot) = \mu(A \cap \cdot)$, or equivalently $(\Omega, \Sigma(A), \mu_A)$ with $\Sigma(A) = \{A \cap B \in \Sigma \colon B \in \Sigma\}$. Since $L^1(\mu_A)$ is clearly a closed subspace of $L^1(\mu)$ it suffices to show that there exists an $f \in L^1(\mu_A)$ such that $f_n \to f$ weakly. If $l \in [L^1(\mu)]^*$ is any element, let $l_1 = l | L^1(\mu_A)$, the restriction, which is thus continuous. But then μ on $\Sigma(A)$ is σ-finite, so that by Theorem 8,

$$l_1(h) = \int_A h g_{l_1} \, d\mu, \qquad h \in L^1(\mu_A), \quad g_{l_1} \in L^\infty(\mu_A).$$

The hypothesis implies $l_1(f_n) = \int_A f_n g_{l_1} \, d\mu \to$ a limit. Thus, if $g_{l_1} = \chi_B$, $B \in \Sigma(A)$, we get $\int_B f_n \, d\mu \to$ a limit, since $l_1 \leftrightarrow \chi_B \in L^\infty(\mu_A)$. If $\nu_n(B) = \int_B f_n \, d\mu$, then $\nu_n \colon \Sigma(A) \to \mathbb{R}$ converges on the σ-algebra $\Sigma(A)$. Since $\nu_n(\cdot)$ is σ-additive, is bounded for each n, and vanishes on μ-null sets, by Corollary 4.4.7 of Nikodým $\nu(B) = \lim_n \nu_n(B)$ defines ν as a signed measure on $\Sigma(A)$ and $\nu \ll \mu_A$. Hence by the Radon-Nikodým theorem there is a μ-unique f

such that $f(\omega) = 0$ for a.a. ω in A^c and

$$\nu(B) = \int_B f\,d\mu, \qquad B \in \Sigma(A).$$

It follows that, by the linearity of the integral,

$$\int_A f_n h_1\,d\mu \to \int_A f h_1\,d\mu \tag{24}$$

for all step functions h_1 in $L^\infty(\mu_A)$. By Proposition 4.5.8 such functions are dense in $L^\infty(\mu_A)$. Since we have $\|f_n\|_1 \leqslant M < \infty$ for all n and $\|f\|_1 = |\nu|(A) < \infty$, by Corollary 12 for any $\epsilon > 0$, $h \in L^\infty(\mu_A)$, there is a step function h_ϵ such that $\|h - h_\epsilon\|_\infty < \epsilon/(M + \|f\|_1)$. Hence (24) gives

$$\left|\int_A (f_n - f)h\,d\mu\right| \leqslant \left|\int_A (f_n - f)h_\epsilon\,d\mu\right| + \int_A (|f_n| + |f|)|h - h_\epsilon|\,d\mu$$

$$\leqslant \left|\int_A (f_n - f)h_\epsilon\,d\mu\right| + (M + \|f\|_1)\|h - h_\epsilon\|_\infty,$$

by the Hölder inequality. Letting $n \to \infty$, we get by (24)

$$\limsup_n \left|\int_A (f_n - f)h\,d\mu\right| \leqslant \epsilon.$$

Since $\epsilon > 0$ is arbitrary, we deduce that $f_n \to f$ weakly, and hence $L^1(\mu)$ is also weakly sequentially complete. This finishes the proof.

The corresponding result is false for $L^\infty(\mu)$. In fact, the latter is not weakly sequentially complete. (See Exercise 12 below for a weaker result.)

Discussion. It is appropriate to indicate at this point where all these properties lead us. We have seen that if $\mathscr{X}$ is a normed linear space, then its dual $\mathscr{X}^*$ is a Banach space. Hence it makes sense to consider the dual $(\mathscr{X}^*)^*$, simply written $\mathscr{X}^{**}$ and called the *second dual* (space) of $\mathscr{X}$, which is a Banach space. Thus if $T \in \mathscr{X}^{**}$, then $|T(l)| \leqslant \|T\|\,\|l\|$ for each $l \in \mathscr{X}^*$ and $\|T\| = \sup\{|T(l)|: \|l\| \leqslant 1\}$. If $x \in \mathscr{X}$, define $T(l) = l(x)$, $l \in \mathscr{X}^*$. Clearly T is linear and $|T(l)| = |l(x)| \leqslant \|l\|\,\|x\|$, so that $\|T\| \leqslant \|x\| < \infty$ and $T \in \mathscr{X}^{**}$. Thus each x in $\mathscr{X}$ defines an element $T_x \in \mathscr{X}^{**}$, so that each x in $\mathscr{X}$ defines a continuous linear mapping on $\mathscr{X}^*$, and perhaps there are some additional ones. In this way we may identify spaces by means of a natural mapping $\iota: x \mapsto T_x$, $\iota(\mathscr{X}) = \hat{\mathscr{X}} \subset \mathscr{X}^{**}$. (One shows $\|T_x\| = \|x\|$.) Also $\hat{\mathscr{X}} = \mathscr{X}^{**}$ can happen. When it does (in which case $\mathscr{X}$ must necessarily be complete), $\mathscr{X}$ is called *reflexive*. For instance, if $\mathscr{X} = L^p(\mu)$, $1 < p < \infty$, then $\mathscr{X}^* = L^q(\mu)$, $1 < p < \infty$, is reflexive. On the other hand, $L^1(\mu)$ is not reflexive in general. By Corollary 6, $L^p(\mu)$ is further uniformly convex (and uniformly smooth) when $1 < p < \infty$. Moreover, one can show that a uniformly convex Banach space is reflexive (but not conversely). Also, a space $\mathscr{X}$ is uniformly convex iff its dual $\mathscr{X}^*$ is uniformly smooth, but an $\mathscr{X}$ need not possess these two simultaneously. However, each $L^p(\mu)$, $1 < p < \infty$, has all these properties together. Further, a Banach space [e.g. $L^\infty(\mu)$] need not be weakly sequentially complete, although it can be shown that a reflexive space always has this property. But as $L^1(\mu)$ shows, some nonreflexive spaces may have that property. A detailed study of these interrelations is the subject of abstract analysis, and we shall not treat them here. They are indicated to show how the L^p-theory motivates the general study.

We shall now introduce conditional expectations on L^p-spaces as another application of the Radon-Nikodým theorem, and because of their importance in probability theory. Let (Ω, Σ, μ) be a measure space and $\mathscr{B} \subset \Sigma$ be a σ-subalgebra. If $f: \Omega \to \overline{\mathbb{R}}^+$ is a (Σ-)measurable function, $\nu_f: A \mapsto \int_A f\, d\mu$, then $\nu_f \ll \mu$. Suppose that $\mu_{\mathscr{B}} = \mu|\mathscr{B}$, the restriction, (has the finite subset property and) is localizable. Then by Theorem 4.3 applied to $\nu_f|\mathscr{B}$ (since $\nu_f \ll \mu_{\mathscr{B}}$ still), there exists a $\mu_{\mathscr{B}}$-unique $\mathscr{B}$-measurable $\tilde{f}: \Omega \to \overline{\mathbb{R}}^+$ such that, *when ν_f is σ-additive on $\mathscr{B}$*,

$$\nu_f(B) = \int_B \tilde{f}\, d\mu_{\mathscr{B}}, \qquad B \in \mathscr{B}.$$

This may be expressed alternatively as

$$\int_B f\, d\mu = \int_B \tilde{f}\, d\mu_{\mathscr{B}}, \qquad B \in \mathscr{B}. \tag{25}$$

The mapping $T: f \mapsto \tilde{f}$ is well defined. If $f: \Omega \to \overline{\mathbb{R}}$ $(f = f^+ - f^-)$ is any measurable function (for Σ) such that either f^+ or f^- is integrable for μ, then the above procedure applied to f^+, f^- gives $\tilde{f}_1, \tilde{f}_2$, and we can set $Tf = T(f^+ - f^-) = \tilde{f}_1 - \tilde{f}_2 = \tilde{f}$ (say). This is again well defined, leading to the general concept as follows.

Definition 14. Let (Ω, Σ, μ) be a measure space, and $\mathscr{B} \subset \Sigma$ a σ-subalgebra such that $\mu_{\mathscr{B}}$ $(= \mu|\mathscr{B})$ is localizable. If $f: \Omega \to \overline{\mathbb{R}}$ is any measurable function *such that f^+ or f^- is μ-integrable*, then any $\mathscr{B}$-measurable $\tilde{f}: \Omega \to \overline{\mathbb{R}}$ satisfying the system of equations

$$\int_B f\, d\mu = \int_B \tilde{f}\, d\mu_{\mathscr{B}}, \qquad B \in \mathscr{B}, \tag{26}$$

is called (a version of) the *conditional expectation* of f relative to (or given) $\mathscr{B}$, and denoted by $E^{\mathscr{B}}(f) = \tilde{f}$ a.e. (also written $E(f|\mathscr{B}) = \tilde{f}$ a.e.)

By the preceding discussion, $E^{\mathscr{B}}(f)$ always exists under the condition that $\mu_{\mathscr{B}}$ is localizable *and* $f^+(f^-)$ is integrable. Also it is $\mu_{\mathscr{B}}$-unique. Any member of the equivalence class is called a *version* of the conditional expectation. In other words $E^{\mathscr{B}}(f) = d\nu_f/d\mu_{\mathscr{B}}$, the Radon-Nikodým derivative, so that its explicit calculation is in general nontrivial. We indicate some properties of this functional operation $E^{\mathscr{B}}(\cdot)$ in the next result.

Theorem 15. *Let* (Ω, Σ, μ), $\mathscr{B} \subset \Sigma$, *be as above, such that* $\mu_{\mathscr{B}}$ *is localizable. Then* $E^{\mathscr{B}}$: $L^1(\mu) \to L^1(\mu)$ *is a linear order-preserving operator such that*

(i) $\|E^{\mathscr{B}}(f)\|_1 \leqslant \|f\|_1$ (*i.e.*, $E^{\mathscr{B}}$ *is a contraction*);
(ii) $\{f, g, fg\} \subset L^1(\mu) \Rightarrow E^{\mathscr{B}}(fE^{\mathscr{B}}(g)) = E^{\mathscr{B}}(f)E^{\mathscr{B}}(g)$ *a.e.* (*averaging identity*);
(iii) *if* $\mathscr{B}_1 \subset \mathscr{B}_2 \subset \Sigma$ *are* σ-*algebras, and* $\mu_{\mathscr{B}_i}$ *are localizable, then* $E^{\mathscr{B}_1}(E^{\mathscr{B}_2}(f)) = E^{\mathscr{B}_2}(E^{\mathscr{B}_1}(f)) = E^{\mathscr{B}_1}(f)$ *a.e.* (*commutativity*);
(iv) $0 \leqslant f \in L^1(\mu)$, $E^{\mathscr{B}}(f) = 0$ *a.e.* $\Rightarrow f = 0$ *a.e.*, (*faithfulness*);
(v) $E^{\mathscr{B}}(E^{\mathscr{B}}(f)) = E^{\mathscr{B}}(f)$ *a.e.* [*projection property on* $L^1(\mu)$];
(vi) $0 \leqslant f_n \leqslant f_{n+1} \in L^1(\mu) \Rightarrow E^{\mathscr{B}}(f_n) \uparrow E^{\mathscr{B}}(\lim_n f_n)$ *a.e.* (*monotone convergence*).

Proof. The fact that $E^{\mathscr{B}}: f \to \tilde{f}$ given by (26) is linear and positivity preserving is clear. Let us establish the other properties.

(i): Since $-|f| \leqslant f \leqslant |f|$ implies $-E^{\mathscr{B}}(|f|) \leqslant E^{\mathscr{B}}(f) \leqslant E^{\mathscr{B}}(|f|)$ a.e., we have

$$\|E^{\mathscr{B}}(f)\|_1 = \int_\Omega |E^{\mathscr{B}}(f)|\, d\mu_{\mathscr{B}}$$

$$\leqslant \int_\Omega E^{\mathscr{B}}(|f|)\, d\mu_{\mathscr{B}} = \int_\Omega |f|\, d\mu = \|f\|_1, \qquad \text{by (26)}.$$

Thus $E^{\mathscr{B}}$ is a contractive positive linear mapping on $L^1(\Omega, \Sigma, \mu)$ with range $L^1(\Omega, \mathscr{B}, \mu_{\mathscr{B}})$. Also (26) implies $E^{\mathscr{B}}(f) = f$ a.e. if f is in the range of $E^{\mathscr{B}}$, so $E^{\mathscr{B}}$ is the identity on $L^1(\Omega, \mathscr{B}, \mu_{\mathscr{B}})$. This shows (v) is also true.

(ii): Since for each $g \in L^1(\mu)$, $E^{\mathscr{B}}(g)$ is a $\mathscr{B}$-measurable element in $L^1(\mu)$, it suffices to show that for any $g \in L^1(\Omega, \mathscr{B}, \mu_{\mathscr{B}})$ we have $E^{\mathscr{B}}(fg) = gE^{\mathscr{B}}(f)$ a.e. This means we need to establish for all $B \in \mathscr{B}$

$$\int_B E^{\mathscr{B}}(fg)\, d\mu_{\mathscr{B}} = \int_B fg\, d\mu, \qquad \text{by (26)}$$

$$= \int_B gE^{\mathscr{B}}(f)\, d\mu_{\mathscr{B}}, \qquad \text{to be verified.} \tag{27}$$

Let $A \in \mathscr{B}$ and $g = \chi_A$. Since $A \cap B \in \mathscr{B}$, the last two quantities of (27) are equal by (26). Hence if $g = \sum_{i=1}^n a_i \chi_{A_i}$, $A_i \in \mathscr{B}$, then by the linearity of the integral, it holds again for all simple $\mathscr{B}$-measurable functions g. The general case now follows, with the monotone convergence theorem, if we first consider $g \geqslant 0$, $f \geqslant 0$ a.e., and then the result easily obtains by linearity.

(iii): Since $\mathscr{B}_1 \subset \mathscr{B}_2$ so that $E^{\mathscr{B}_1}(f)$ is $\mathscr{B}_2$-measurable and $E^{\mathscr{B}_2}$ is identity on its range (as noted above), it follows that $E^{\mathscr{B}_2}(E^{\mathscr{B}_1}(f)) = E^{\mathscr{B}_1}(f)$ a.e. For the other equality, let $A \in \mathscr{B}_1 \subset \mathscr{B}_2$. Then

$$\begin{aligned}
\int_A E^{\mathscr{B}_1}(E^{\mathscr{B}_2}(f))\, d\mu_{\mathscr{B}_1} &= \int_A E^{\mathscr{B}_2}(f)\, d\mu_{\mathscr{B}_2}, && \text{by (26)} \\
&= \int_A f\, d\mu, && \text{since } A \in \mathscr{B}_2 \\
&= \int_A E^{\mathscr{B}_1}(f)\, d\mu_{\mathscr{B}_1}, && \text{by Definition 14.}
\end{aligned}$$

Since the extreme integrands are $\mathscr{B}_1$-measurable and A in $\mathscr{B}_1$ is arbitrary, we can identify them, so that $E^{\mathscr{B}_1}(E^{\mathscr{B}_2}(f)) = E^{\mathscr{B}_1}(f)$ a.e. Hence (iii) holds. Note that the result need not be true if the stated inclusions between $\mathscr{B}_1$ and $\mathscr{B}_2$ are not maintained.

(iv): This is immediate from (26), since $f \geqslant 0 \Rightarrow 0 = \int_\Omega E^{\mathscr{B}}(f)\, d\mu_{\mathscr{B}} = \int_\Omega f\, d\mu \Rightarrow f = 0$ a.e., by the properties of the Lebesgue integral [see Theorem 4.1.3(ii)].

(vi): Finally let $f_n \uparrow f$ a.e.; since $E^{\mathscr{B}}(f_n) \leqslant E^{\mathscr{B}}(f_{n+1})$ a.e., and $\mathscr{B}$ is a σ-algebra, it follows that $\lim_n E^{\mathscr{B}}(f_n)$ exists and is equivalent to a $\mathscr{B}$-measurable function. Hence for each $A \in \mathscr{B}$ we have

$$\begin{aligned}
\int_A \lim_n E^{\mathscr{B}}(f_n)\, d\mu_{\mathscr{B}} &= \lim_n \int_A E^{\mathscr{B}}(f_n)\, d\mu_{\mathscr{B}}, && \text{by Theorem 4.3.1} \\
&= \lim_n \int_A f_n\, d\mu, && \text{by (26)} \\
&= \int_A f\, d\mu, && \text{by Theorem 4.3.1 again} \\
&= \int_A E^{\mathscr{B}}(f)\, d\mu_{\mathscr{B}}, && \text{by Definition 14.}
\end{aligned}$$

Since the extreme integrands are a.e. $\mathscr{B}$-measurable, and $A \in \mathscr{B}$ is arbitrary, we deduce that $E^{\mathscr{B}}(f) = \lim_n E^{\mathscr{B}}(f_n)$ a.e. This finishes the proof. (See Exercise 22 concerning the assumptions here.)

Regarding the preceding result and Definition 14, the following remarks are in order. If $\mu(\Omega) < \infty$, then $\mu_{\mathscr{B}}$ is a finite measure and no additional assumption on $\mu_{\mathscr{B}}$ is needed. In the general case, $\tilde{f}$ required of (26) need not exist. For example, let $\Omega = \mathbb{R}$, $\Sigma =$ Borel σ-algebra, $\mu =$ Lebesgue measure. If $\mathscr{B} = \{A \in \Sigma: A \text{ countable or } A^c \text{ countable}\}$, then $\mu_{\mathscr{B}}$ takes only 0 or $+\infty$, and hence by Theorem 4.3 there exists no $\mathscr{B}$-measurable $\tilde{f}$. Thus we need to assume that $\mu_{\mathscr{B}}$ is localizable. Note that μ on Σ is even σ-finite here. However, one may ask whether one cannot define a conditional expectation using the result of Theorem 5, which did not depend on the Radon-Nikodým theorem. A positive answer is provided by the following

Proposition 16. *Let (Ω, Σ, μ) be a measure space, $\mathscr{B} \subset \Sigma$ be a σ-algebra, and $L^p(\Omega, \Sigma, \mu)$, $1 < p < \infty$, be the corresponding Lebesgue space. Then for each $f \in L^p(\Omega, \Sigma, \mu)$ there is a unique $\tilde{f} \in L^p(\Omega, \mathscr{B}, \mu_{\mathscr{B}})$ such that*

$$\int_\Omega gf\,d\mu = \int_\Omega g\tilde{f}\,d\mu_{\mathscr{B}}, \qquad g \in L^q(\Omega, \mathscr{B}, \mu_{\mathscr{B}}), \quad q = \frac{p}{p-1}. \tag{28}$$

The mapping $E^{\mathscr{B}}: f \mapsto \tilde{f}$ is a (generalized) conditional expectation, and has similar properties to those given in Theorem 15.

Proof. If $f \in L^p(\mu)$, let $l_f \in [L^q(\mu)]^*$ be defined by

$$l_f(g) = \int_\Omega fg\,d\mu, \qquad g \in L^q(\mu).$$

By the Hölder inequality, it follows that $l_f \in [L^q(\mu)]^*$. Let $\tilde{l}_f = l_f | L^q(\Omega, \mathscr{B}, \mu_{\mathscr{B}})$, the restriction to the subspace shown. Evidently $\|\tilde{l}_f\| \leqslant \|l_f\| = \|f\|_p < \infty$. So $\tilde{l}_f \in [L^q(\Omega, \mathscr{B}, \mu_{\mathscr{B}})]^*$, and hence by Theorem 5, there exists a $\mu_{\mathscr{B}}$-unique $\tilde{f} \in L^p(\Omega, \mathscr{B}, \mu_{\mathscr{B}})$ such that

$$\tilde{l}_f(g) = \int_\Omega fg\,d\mu = \int_\Omega \tilde{f}g\,d\mu_{\mathscr{B}}, \qquad g \in L^q(\Omega, \mathscr{B}, \mu_{\mathscr{B}}).$$

Let $E^{\mathscr{B}}: f \mapsto \tilde{f}$, so that $\|E^{\mathscr{B}}(f)\|_p = \|\tilde{f}\|_p = \|\tilde{l}_f\| \leqslant \|l_f\| = \|f\|_p$. Thus $E^{\mathscr{B}}$ is a contraction. It is evident that it is a positive linear mapping on $L^p(\mu)$ with range $L^p(\Omega, \mathscr{B}, \mu_{\mathscr{B}})$. The other properties in Theorem 15 are verified with essentially the same proof, and the details are left to the reader. This yields the result.

Using the ideas of measurable quasifunctions and the fact that each $f \in L^p(\mu)$ for $1 \leqslant p < \infty$ has σ-finite support, one can extend Theorem 15 to $L^1(\mu)$ also. We shall not pursue the matter. [A detailed account is given by the author in (1975, 1976).]

The concept of a martingale given in (3.10) can be restated as follows. Let (Ω, Σ, μ) be a measure space with the finite subset property, and $\Sigma_n \subset \Sigma_{n+1} \subset \Sigma$ be σ-algebras such that $\mu_n = \mu | \Sigma_n$, $n \geqslant 1$, is localizable. If $\{f_n: n \geqslant 1\} \subset$

$L^1(\mu)$ is such that f_n is measurable for Σ_n, $n \geqslant 1$, then $\{(f_n, \Sigma_n): n \geqslant 1\}$ is called a *martingale* if for each $n \geqslant 1$,

$$E(f_{n+1}|\Sigma_n) = f_n \qquad \text{a.e.} \tag{29}$$

(or $\int_A f_{n+1}\, d\mu_{n+1} = \int_A f_n\, d\mu_n$, $A \in \Sigma_n$). It is called a *submartingale* if $=$ is replaced by $\geqslant$ in (29). It is a *supermartingale* if $=$ is replaced by $\leqslant$ there. Our purpose here is to generalize Proposition 3.2 and prove an a.e. convergence of (sub-)martingales lying in a ball of $L^1(\mu)$. There are many different proofs; the one we present here is measure theoretical, although it is not the shortest. For this we need a pair of auxiliary facts.

Proposition 17. *Let (Ω, Σ, μ) be a measure space, and $L^1(\mu)$ be the Lebesgue space on it. We then have:*

(i) (*Conditional Jensen's inequality*) *If $\{f, \varphi(f)\} \subset L^1(\nu)$, $\mathscr{B} \subset \Sigma$, is a σ-algebra such that $\mu_{\mathscr{B}}$ $(= \mu|\mathscr{B})$ is localizable, where $\varphi: \mathbb{R} \to \mathbb{R}$ is convex, then*

$$E^{\mathscr{B}}(\varphi(f)) \geqslant \varphi(E^{\mathscr{B}}(f)) \quad a.e.(\mu_{\mathscr{B}}). \tag{30}$$

(ii) (*Basic inequalities*) *If $\{f_k, \Sigma_k: k \geqslant 1\} \subset L^1(\mu)$ is a submartingale and $\lambda \in \mathbb{R}$, then*

$$\lambda\mu\left(\left\{\omega: \max_{1 \leqslant k \leqslant n} f_k(\omega) > \lambda\right\}\right) \leqslant \int_{\{\omega:\, \max_{k \leqslant n} f_k(\omega) > \lambda\}} f_n\, d\mu, \tag{31}$$

$$\lambda\mu\left(\left\{\omega: \min_{1 \leqslant k \leqslant n} f_k(\omega) < \lambda\right\}\right) \geqslant \int_\Omega f_1\, d\mu - \int_{\{\omega:\, \min_{k \leqslant n} f_k(\omega) > \lambda\}} f_n\, d\mu. \tag{32}$$

Proof. (i): Let a, x be two points in $\mathbb{R}$. Then by Theorem 2.10,

$$\varphi(x) = \varphi(a) + \int_a^x \varphi'(t)\, dt,$$

(the result holding also if $x < a$), and $\varphi'(\cdot)$ is nondecreasing. Hence

$$\varphi(x) \geqslant \varphi(a) + (x - a)\varphi'(a). \tag{33}$$

Let now $a = E^{\mathscr{B}}(f)(\omega)$, $x = f(\omega)$ in (33), where $\omega \in \Omega$. Then

$$\varphi(f) \geqslant \varphi(E^{\mathscr{B}}(f)) + [f - E^{\mathscr{B}}(f)]\varphi'(E^{\mathscr{B}}(f)) \quad \text{a.e.}(\mu). \tag{34}$$

Next, noting that φ is continuous and φ' increasing, so that the $\varphi(E^{\mathscr{B}}(f))$ and $\varphi'(E^{\mathscr{B}}(f))$ are $\mathscr{B}$-measurable a.e., we get, after applying $E^{\mathscr{B}}$ to both sides of (34),

$$E^{\mathscr{B}}(\varphi(f)) \geqslant \varphi\big(E^{\mathscr{B}}(f)\big) + \varphi'\big(E^{\mathscr{B}}(f)\big)\big(E^{\mathscr{B}}(f) - E^{\mathscr{B}}(f)\big) \qquad \text{a.e.}$$
$$= \varphi\big(E^{\mathscr{B}}(f)\big) \qquad \text{a.e.,}$$

due to Theorem 15(iii). This gives (30).

(ii): We first express the set in braces of (31) conveniently as a disjoint union of n sets that the kth one belongs to the Σ_k. This construction is as follows. Let $H_1^\lambda = \{\omega: f_1(\omega) > \lambda\}$, and for $j > 1$, define $H_j^\lambda = \{\omega: f_j(\omega) > \lambda$ and $f_i(\omega) \leqslant \lambda,\ 1 \leqslant i \leqslant j-1\}$. Thus $H_i^\lambda \in \Sigma_i$, and on H_i^λ the f_k-sequence crosses λ for the first time at i. Clearly H_i^λ are disjoint and $\bigcup_{i=1}^n H_i^\lambda = \{\omega: \max_{1 \leqslant k \leqslant n} f_k(\omega) > \lambda\} = H^\lambda$ (say). Then

$$\int_{H^\lambda} f_n \, d\mu = \sum_{i=1}^n \int_{H_i^\lambda} f_n \, d\mu \geqslant \sum_{i=1}^n \int_{H_i^\lambda} f_i \, d\mu$$

(by definition of a submartingale)

$$\geqslant \lambda \sum_{i=1}^n \mu\big(H_i^\lambda\big), \qquad \text{since} \quad f_i > \lambda \quad \text{on } H_i^\lambda$$
$$= \lambda\mu\big(H^\lambda\big), \qquad \text{since the } H_i^\lambda \text{ are disjoint.}$$

This proves (31).

The second inequality is similar. Here we define $K_1^\lambda = \{\omega: f_1(\omega) < \lambda\}$, and for $j > 1$, $K_j^\lambda = \{\omega: f_j(\omega) < \lambda,\ f_i(\omega) \geqslant \lambda,\ 1 \leqslant i \leqslant j-1\}$. Then $K_j^\lambda \in \Sigma_j$, disjoint, and $\bigcup_{i=1}^n K_i^\lambda = \{\omega: \min_{1 \leqslant k \leqslant n} f_k(\omega) < \lambda\} = K^\lambda$ (say). Also, since $K_j^\lambda \subset (\bigcup_{i=1}^{j-1} K_i^\lambda)^c$, we have

$$\int_\Omega f_1 \, d\mu \leqslant \int_{K_1^\lambda} f_1 \, d\mu + \int_{(K_1^\lambda)^c} f_2 \, d\mu$$

[by the submartingale inequality, since $(K_1^\lambda)^c \in \Sigma_1$]

$$\leqslant \lambda\mu\big(K_1^\lambda\big) + \int_{K_2^\lambda} f_2 \, d\mu$$
$$+ \int_{(K_1^\lambda \cup K_2^\lambda)^c} f_3 \, d\mu, \qquad \text{since} \quad \big(K_1^\lambda \cup K_2^\lambda\big)^c \in \Sigma_2$$
$$\leqslant \big[\mu\big(K_1^\lambda\big) + \mu\big(K_2^\lambda\big)\big] + \int_{(K_1^\lambda \cup K_2^\lambda)^c} f_3 \, d\mu$$
$$\leqslant \cdots \leqslant \lambda \sum_{i=1}^n \mu\big(K_i^\lambda\big) + \int_{(\bigcup_{i=1}^n K_i^\lambda)^c} f_n \, d\mu$$
$$= \lambda\mu\big(K^\lambda\big) + \int_{(K^\lambda)^c} f_n \, d\lambda.$$

This is precisely (32), and the proof is complete.

The second fact that is needed for the a.e. convergence result is given by

Proposition 18. *Let $\{f_n, \Sigma_n: n \geqslant 1\} \subset L^1(\mu)$ be a martingale.*

(i) *If for a continuous convex $\varphi: \mathbb{R} \to \mathbb{R}$, $\varphi(f_n) \in L^1(\mu)$, $n \geqslant 1$, then $\{\varphi(f_n), \Sigma_n: n \geqslant 1\}$ is a submartingale sequence; the same conclusion holds if $\{f_n\}_{n\geqslant 1}$ is a submartingale and φ is also monotone increasing.*

(ii) (*Jordan decomposition*) *If $\sup_n \|f_n\|_1 < \infty$ then there exist unique positive martingales $\{g_n^{(i)}, \Sigma_n: n \geqslant 1\} \subset L^1(\mu)$, $i = 1, 2$, such that $f_n = g_n^{(1)} - g_n^{(2)}$ and $\sup_n \|f_n\|_1 = \|g_n^{(1)}\|_1 + \|g_n^{(2)}\|_1$.*

(iii) *If $\{f_n, \Sigma_n: n \geqslant 1\} \subset L^2(\mu)$, $\mu(\Omega) < \infty$, is a positive submartingale and $\sup_n \|f_n\|_2 < \infty$, then $f_n \to f$ a.e. and in L^2-norm.*

Proof. (i): This is immediate from (i) of the preceding result, since $\varphi(f_n)$ is Σ_n-measurable, and for $n \geqslant 1$,

$$E^{\Sigma_n}(\varphi(f_{n+1})) \geqslant \varphi(E^{\Sigma_n}(f_{n+1})) = \varphi(f_n) \qquad \text{a.e.},$$

the f_n being a martingale sequence. But this is the defining property of a submartingale [cf. (30)]. For f_n a submartingale, the last relation is $\geqslant$ instead of $=$, since $\varphi(\cdot)$ is nondecreasing in addition. So $\varphi(f_n)$ is again of the same kind.

(ii): This is a consequence of Theorem 4.4.3 (or rather Corollary 4.4.7). Indeed, let $\nu_n(A) = \int_A f_n \, d\mu$, $A \in \Sigma$. Then by the martingale property $\nu_n = \nu_{n+1}|\Sigma_n$. Also $|\nu_n|(A) = \int_A |f_n| \, d\mu \leqslant \sup_n \|f_n\|_1 < \infty$. Taking $\varphi(x) = x^+$ in (i), we see that $\{f_n^+, \Sigma_n: n \geqslant 1\}$ is a positive submartingale. Since by the Jordan decomposition of the signed measure $\nu_n: \Sigma_n \to \mathbb{R}$ we have $\nu_n^+(A) = \int_A f_n^+ \, d\mu \leqslant \int_A f_{n+1}^+ \, d\mu = \nu_{n+1}^+(A)$, it follows that for each n, $\xi_n: \Sigma_n \to \mathbb{R}^+$ defined by $\xi_n(A) = \sup_m \nu_m^+(A) = \lim_m \nu_m^+(A)$, $A \in \Sigma_n$, exists, and since each $\nu_m^+ \ll \mu$, it follows from Corollary 4.4.7 that ξ_n is σ-additive on Σ_n and μ-continuous. Thus $g_n^{(1)} = d\xi_n/d\mu$ exists a.e., and it is Σ_n-measurable by the Radon-Nikodým theorem. On the other hand, from the definition it is immediate that $\xi_n = \xi_{n+1}|\Sigma_n$. Hence for each $A \in \Sigma_n$, we get

$$\xi_n(A) = \int_A g_n^{(1)} \, d\mu = \int_A g_{n+1}^{(1)} \, d\mu = \xi_{n+1}(A), \tag{35}$$

so that $\{g_n^{(1)}, \Sigma_n: n \geqslant 1\}$ is a positive martingale. But $\xi_n - \nu_n \geqslant 0$, and if $g_n^{(2)} = g_n^{(1)} - f_n$, $n \geqslant 1$, then $\{g_n^{(2)}, \Sigma_n: n \geqslant 1\}$ is a martingale, since a linear combination of martingales is a martingale. Also, for each $A \in \Sigma_n$,

$$0 \leqslant \xi_n(A) - \nu_n(A) = \int_A (g_n^{(1)} - f_n) \, d\mu = \int_A g_n^{(2)} \, d\mu,$$

and since $g_n^{(2)}$ is measurable for Σ_n, it follows that $g_n^{(2)} \geqslant 0$ a.e (cf. Exercise 4.1.3). Thus a decomposition of the desired kind is obtained.

As for the uniqueness, let $f_n = h_n^{(1)} - h_n^{(2)}$, $n \geq 1$, be another such decomposition, so that $f_n^+ \leq h_n^{(1)}$. By definition of ξ_n and (35) we have

$$\int_A g_n^{(1)}\,d\mu = \xi_n(A) = \lim_m \nu_{n+m}^+(A)$$

$$= \lim_m \int_A f_{n+m}^+\,d\mu$$

$$\leq \lim_m \int_A h_{n+m}^{(1)}\,d\mu = \int_A h_n^{(1)}\,d\mu, \qquad A \in \Sigma_n,$$

since $\{h_n^{(1)}, \Sigma_n: n \geq 1\}$ is a martingale. This implies $g_n^{(1)} \leq h_n^{(1)}$ a.e. Next, replacing f_n by $-f_n$, the same analysis implies $g_n^{(2)} \leq h_n^{(2)}$ a.e. But then we have

$$\|g_n^{(1)}\|_1 + \|g_n^{(2)}\|_1 = \sup_n \|f_n\|_1 = \|h_n^{(1)}\|_1 + \|h_n^{(2)}\|_1, \qquad n \geq 1.$$

This implies $g_n^{(1)} = h_n^{(1)}$ a.e. and $g_n^{(2)} = h_n^{(2)}$ a.e., in view of the preceding inequalities. So the Jordan decomposition is unique under this condition.

(iii): Assume $\mu(\Omega) < \infty$. Since $\sup_n \|f_n\|_2 < \infty$, we have $\{f_n: n \geq 1\}$ uniformly integrable, because

$$\lim_{\mu(A)\to 0} \int_A f_n\,d\mu \leq \lim_{\mu(A)\to 0} \|f_n\|_2 \|\chi_A\|_2, \qquad \text{by the CBS inequality}$$

$$\leq \sup_n \|f_n\|_2 \lim_{\mu(A)\to 0} \sqrt{\mu(A)} = 0,$$

uniformly in n . Also, taking $\varphi(x) = x^2$ in part (i), so that φ is convex and increasing on $\mathbb{R}^+$, we see that $\{f_n^2, \Sigma_n: n \geq 1\}$ is a submartingale. Hence $a_n = \int_\Omega f_n^2\,d\mu \leq \int_\Omega f_{n+1}^2\,d\mu = a_{n+1} \to a \leq \sup_n \|f_n\|_2^2 < \infty$. But for $n > m$

$$0 \leq \int_\Omega (f_n^2 - f_m^2)\,d\mu = \int_\Omega (f_n - f_m)^2\,d\mu + 2\int_\Omega f_m(f_n - f_m)\,d\mu. \qquad (36)$$

Also $E^{\Sigma_m}(f_n) \geq f_m$ a.e., by the submartingale inequality. Since by Theorem 15(ii) we have $f_m^2 \leq f_m E^{\Sigma_n}(f_n) = E^{\Sigma_n}(f_m f_n)$ a.e., we get on integration

$$0 \leq \int_\Omega E^{\Sigma_n}(f_m f_n - f_m^2)\,d\mu = \int_\Omega f_m(f_n - f_m)\,d\mu.$$

Thus both terms on the right in (36) are ≥ 0, and the left side $\to a - a = 0$ as $n \to \infty$ and then $m \to \infty$. So each of the right side terms $\to 0$. In particular $\{f_n: n \geq 1\}$ is Cauchy in $L^2(\mu)$, so that $f_n \to f \in L^2(\mu)$ in norm.

It remains to show that $f_n \to f$ a.e. (Proposition 3.2 is a special case of the present result.) For this let $m > 1$ be fixed. Then $\{f_n - f_m, \Sigma_n: n > m\}$

is a submartingale. So for any $\epsilon > 0$,

$$\mu\left(\left\{\omega: \max_{m<k\leq n} |f_k - f_m|(\omega) > \epsilon\right\}\right)$$

$$\leq \mu\left(\left\{\omega: \max_{m<k\leq n} (f_k - f_m)(\omega) > \epsilon\right\}\right)$$

$$+\mu\left(\left\{\omega: \min_{m<k\leq n} (f_k - f_m)(\omega) < -\epsilon\right\}\right)$$

$$\leq \frac{1}{\epsilon}\left(2\int_\Omega |f_n - f_m|\, d\mu - \int_\Omega |f_{m+1} - f_m|\, d\mu\right),$$

[by (31) and (32)]

$$\leq \frac{2}{\epsilon}(\|f_n - f_m\|_2 + \|f_{m+1} - f_m\|_2)\sqrt{\mu(\Omega)}$$

(by the CBS inequality)

$$\to 0 \quad \text{as} \quad n \to \infty \text{ and then } m \to \infty. \tag{37}$$

Here we have used the fact that $\mu(\Omega) < \infty$ and that $\{f_n: n \geq 1\}$ is Cauchy in $L^2(\mu)$. It follows that $\max_{k>m}|f_k - f_m| < \infty$ a.e., and then

$$\mu\left(\left\{\omega: \overline{\lim_n} f_n - \underline{\lim_n} f_n > \epsilon\right\}\right) \leq 2 \lim_n \mu\left(\left\{\omega: \max_{k>n} |f_k - f_n|(\omega) \geq \frac{\epsilon}{2}\right\}\right)$$

$$= 0, \quad \text{by (37)}.$$

Hence $f_n \to \tilde{f} = \overline{\lim}_n f_n = \underline{\lim}_n f_n$ a.e. Since $f_n \to f$ in norm, and thus in measure, we have $f = \tilde{f}$ a.e. This completes the proof of the proposition.

We are now prepared to establish the general a.e. convergence of martingales as a final major convergence theorem, which was originally obtained by J. L. Doob and, in a different form, by E. S. Andersen and B. Jessen.

Theorem 19. *Let (Ω, Σ, μ) be a measure space and $\{f_n, \Sigma_n: n \geq 1\} \subset L^1(\mu)$ be a martingale such that $\sup_n \|f_n\|_1 < \infty$ (so $\mu_n = \mu|\Sigma_n$ is localizable for each $n \geq 1$). Then $f_n \to f$ a.e. and $\|f\|_1 \leq \underline{\lim}_n \|f_n\|_1$.*

Proof. First suppose $\mu(\Omega) < \infty$. By Proposition 18(ii), since $f_n = g_n^{(1)} - g_n^{(2)}$, where the $\{g_n^{(i)}, \Sigma_n: n \geq 1\}$ are positive martingales, the result will follow if the latter are shown to converge a.e. So let $f_n \geq 0$ a.e. in addition. Then $h_n = \exp(-f_n)$ converges a.e. iff f_n does. Hence the result will be proved if

$\{h_n, \Sigma_n: n \geqslant 1\}$ is shown to converge a.e. But if $\varphi(x) = e^{-x}$, $x \geqslant 0$, then $\varphi: \mathbb{R}^+ \to \mathbb{R}^+$ is convex, so that by Proposition 18(i), $\{h_n, \Sigma_n: n \geqslant 1\}$ is a positive bounded (by 1) submartingale. So $\sup_n \|h_n\|_2 \leqslant \sqrt{\mu(\Omega)} < \infty$, and by Proposition 18(iii), $h_n \to h_\infty$ a.e. Thus $f_n \to f$ a.e., and the result is proved for finite μ.

For the general case let $f_* = \liminf_n f_n$, $f^* = \limsup_n f_n$. If $\Sigma_\infty = \sigma(\bigcup_n \Sigma_n) \subset \Sigma$, then f_* and f^* are measurable for Σ_∞. It is to be shown that $f_* = f^*$ a.e. If this is false, then $f_* < f^*$ on a set of positive μ-measure. Thus $A = \{\omega: f_*(\omega) < f^*(\omega)\} \in \Sigma_\infty$ and $\mu(A) > 0$. By the finite subset property of μ on Σ_n, $n \geqslant 1$, there is a set $A_0 \in \Sigma_\infty$, $A_0 \subset A$, such that $0 < \mu(A_0) < \infty$. Then by Proposition 2.4.1, for each $\epsilon > 0$ there exists a $B_\epsilon \in \bigcup_{n=1}^\infty \Sigma_n$ such that $\mu(A_0 \Delta B_\epsilon) < \epsilon$. Since $|\mu(A_0) - \mu(B_\epsilon)| \leqslant \mu(A_0 \Delta B_\epsilon) < \epsilon$, it follows that there is a set $B_0 \subset A_0 \cap B_\epsilon$ such that $B_0 \in \bigcup_{n=1}^\infty \Sigma_n$ and $\mu(B_0) > 0$. But this means $B_0 \in \Sigma_{n_0}$ for some $n_0 \geqslant 1$. Since $\mu(B_0) < \infty$, and $\{f_n \chi_{B_0}, \Sigma_n: n \geqslant n_0\}$ is a martingale on a finite measure space, we get, by the first part, that $f_n \chi_{B_0} \to f \chi_{B_0}$ a.e. Since $B_0 \subset A$, this is a contradiction. It follows that $\mu(A) = 0$ and $f_* = f^* = f$ a.e. The last inequality is now a consequence of Fatou's lemma. This completes the proof.

As an important consequence of this theorem, we present an interesting complement to the Yosida-Hewitt decomposition (cf. Theorem 4.4.9). First, a simple but useful extension of the above convergence theorem to submartingales will be given. It is also needed for the above complement. Since the negative of a submartingale is a supermartingale, a similar convergence statement holds for the latter sequences.

Theorem 20. *Let* (Ω, Σ, μ) *be a measure space, and* $\{f_n, \Sigma_n: n \geqslant 1\} \subset L^1(\mu)$ *be a submartingale. Then*:

(i) (*Doob decomposition*) $f_n = g_n + \sum_{k=1}^n a_k$ *a.e., where* $\{g_n, \Sigma_n: n \geqslant 1\}$ *is a martingale in* $L^1(\mu)$, *and where* $a_k \geqslant 0$, $k \geqslant 2$, $a_1 = 0$ *a.e., and* a_k *is* Σ_{k-1} *measurable*, $k > 1$.

(ii) $\sup_n \|f_n\|_1 < \infty \Rightarrow f_n \to f$ *a.e., and* $\|f\|_1 \leqslant \underline{\lim}_n \|f_n\|_1$.

Proof. (i): Since $\{f_n: n \geqslant 1\}$ is a submartingale sequence, if $a_1 = 0$ and

$$a_{k+1} = E^{\Sigma_k}(f_{k+1}) - f_k \geqslant 0, \qquad k \geqslant 1, \tag{38}$$

then a_k is Σ_{k-1} measurable. Let $g_n = f_n - \sum_{k=1}^n a_k$. Then $g_n \in L^1(\mu)$ and g_n is Σ_n-measurable. Moreover, for $n \geqslant 1$

$$E^{\Sigma_n}(g_{n+1}) = E^{\Sigma_n}(f_{n+1}) - \sum_{k=1}^{n+1} a_k$$

(since a_{n+1} is Σ_n-measurable)

$$= f_n + a_{n+1} - \sum_{k=1}^{n+1} a_k, \qquad \text{by (38)}$$

$$= f_n - \sum_{k=1}^{n} a_k = g_n \qquad \text{a.e.}$$

Hence $\{g_n, \Sigma_n: n \geqslant 1\}$ is a martingale in $L^1(\mu)$.

(ii): Since the $a_k \geqslant 0$ a.e., we have

$$0 \leqslant \int_\Omega \sum_{k=1}^{n} a_k \, d\mu = \int_\Omega f_n \, d\mu - \int_\Omega g_n \, d\mu \leqslant \|f_n\|_1 - \int_\Omega g_1 \, d\mu$$

(because the integrals are constant for a martingale)

$$\leqslant \sup_n \|f_n\|_1 - \int_\Omega g_1 \, d\mu < \infty, \qquad n \geqslant 1.$$

Thus by the monotone convergence theorem, on letting $n \to \infty$, we have $\sum_{n=1}^{\infty} a_n < \infty$ a.e, since $\|\sum_{n=1}^{\infty} a_n\|_1 \leqslant \sup_n \|f_n\|_1 - \int_\Omega g_1 \, d\mu < \infty$. Also

$$\sup_n \|g_n\|_1 \leqslant \sup_n \|f_n\|_1 + \left\| \sum_{k=1}^{\infty} a_k \right\|_1 < \infty.$$

So $g_n \to g$ a.e. by Theorem 19. Thus $f_n = g_n + \sum_{k=1}^{n} a_k \to g + \sum_{k=1}^{\infty} a_k = f$ a.e. The last statement is a consequence of Fatou's lemma, and the theorem is proved.

The complement noted above can now be presented, and it is a consequence of the preceding work.

Theorem 21. *Let (Ω, Σ, μ) be a measure space with the finite subset property, and $\Sigma_n \subset \Sigma_{n+1} \subset \Sigma$ be σ-algebras such that $\mu_n = \mu|\Sigma_n$ is localizable for each $n \geqslant 1$. Let $\nu: \bigcup_{n=1}^{\infty} \Sigma_n \to \mathbb{R}$ be a bounded additive function which vanishes on μ-null sets. If $\nu = \nu^1 + \nu^2$ is the Yosida-Hewitt decomposition, so that $\nu^1: \bigcup_{n=1}^{\infty} \Sigma_n \to \mathbb{R}$ is σ-additive (and bounded) and ν^2 on the same algebra is purely finitely additive, we let $\nu^1 = \nu_a^1 + \nu_s^1$ be the Lebesgue decomposition of ν^1 for its unique σ-additive extension to Σ_∞ relative to μ_∞. Then the following statements hold*:

(i) (*Andersen and Jessen*) *If $\nu_{an}^1 = \nu_a^1|\Sigma_n$ and $f_n = \dfrac{d\nu_{an}^1}{d\mu_n}$, $f_\infty = \dfrac{d\nu_a^1}{d\mu_\infty}$ a.e., then $f_n \to f_\infty$ a.e.*;

(ii) *If $\nu_n^2 = \nu^2|\Sigma_n$, and ν_{na}^2 is the μ_n-absolutely continuous part of ν_n^2 on Σ_n, then $g_n = d\nu_{na}^2/d\mu_n \to 0$ a.e. Hence if $\nu_n = \nu|\Sigma_n$, and ν_{na}^1 is the μ-*

absolutely continuous part of the σ-additive component of ν_n, $h_n = d\nu_{na}^1/d\mu_n$, *we have* $h_n \to f$ *a.e.*

Sketch of proof. (i): Using the Jordan decomposition (cf. Proposition 4.4.5, $\nu = \nu^+ - \nu^-$), if necessary, we may conveniently assume that $\nu \geqslant 0$ also. Then $\{f_n, \Sigma_n: n \geqslant 1\}$ is a supermartingale in $L^1(\mu)$, because, since $\nu_n^1 = \nu_{n+1}^1|\Sigma_n$, the singular sets A_n, A_{n+1} of ν_n^1, ν_{n+1}^1 for μ_n, μ_{n+1} satisfy $A_n \subset A_{n+1}$ and

$$\int_B f_n \, d\mu_n + \nu_{ns}^1(A_n \cap B) = \nu_n^1(B) = \nu_{n+1}^1(B)$$

$$= \int_B f_{n+1} \, d\mu_{n+1} + \nu_{(n+1)s}^1(A_{n+1} \cap B), \quad B \in \Sigma_n.$$

Since $\nu_{(n+1)s}^1(A_{n+1} \cap B) - \nu_{ns}^1(A_n \cap B) = \nu_{(n+1)s}^1((A_{n+1} - A_n) \cap B) \geqslant 0$, the above equation implies the supermartingale property for the f_n. Also $\sup_n \|f_n\|_1 \leqslant |\nu^1|(\Omega) < \infty$. Hence $f_n \to f$ a.e. by Theorem 19. We now have to show that $f_\infty = d\nu_a^1/d\mu_\infty$ a.e. This is almost clear. One can use the argument of the last part of the proof of Theorem 20 (or else use Scheffé's lemma carefully; cf. Exercise 4.3.14). It will be omitted here.

(ii): Again assuming, as we may, that $\nu \geqslant 0$, we have that $\{g_n, \Sigma_n: n \geqslant 1\}$ is a supermartingale. Also $\sup_n \|h_n\|_1 < \infty$. So $g_n \to g_\infty$ a.e. as before. But we have

$$\int_A g_n \, d\mu_n \leqslant \nu^2(A) \leqslant \nu^2(\Omega), \qquad A \in \Sigma_n.$$

Since $g_n \geqslant 0$, by Fatou's lemma one gets $\int_A g_\infty \, d\mu \leqslant \nu^2(A)$, $A \in \bigcup_{n=1}^{\infty} \Sigma_n$. But $\zeta(\cdot) = \int_{(\cdot)} g_\infty \, d\mu_\infty$ on Σ_∞ is σ-additive, and ν^2 is purely finitely additive. So $\zeta(\cdot)$ vanishes identically. Since $g_n \geqslant 0$ a.e., this implies $g_\infty = 0$ a.e. The last part follows from the preceding, and the proof is completed.

Several other applications of some of the preceding theorems will be included in the exercises below. Related results on this topic are given in Doob (1953), and recent accounts can be found in Meyer (1966), and in the author's books (1981, 1984).

EXERCISES

0. Let μ be a bounded Lebesgue-Stieltjes measure on the Borel σ-algebra $\mathscr{B}$ of the real line $\mathbb{R}$, and let $L^p(\mu)$ be the corresponding Lebesgue space.

(a) If $e_t: \mathbb{R} \to \mathbb{C}$ is defined by $e_t(x) = e^{itx}$, and $\mathscr{S}$ is the linear span of e_t, $t \in \mathbb{R}$, show that $\mathscr{S}$ is dense in $L^2(\mu)$. [Verify that $f \in L^2(\mu)$, $f \perp \mathscr{S}$ $\Rightarrow$ $f = 0$ a.e., on using the uniqueness theorem for Fourier-Stieltjes transforms. See Exercise 4.5.27.]

(b) Extend the above result for $L^p(\mu)$, $1 \leqslant p < 2$. (Use Exercise 4.5.18 so that $\|\cdot\|_p \leqslant k_p\|\cdot\|_2$, for some $0 < k_p < \infty$.)

(c) Prove the same result for $L^p(\mu)$, $p > 2$ also. [Here the above method does not work. Instead verify that $l(\mathscr{S}) = 0 \Rightarrow l = 0$ for any $l \in [L^p(\mu)]^*$, and use Proposition 4 and (b) above.]

1. Prove that $L^p(\mu)$, $1 < p < \infty$, on a measure space (Ω, Σ, μ) is reflexive, using the representation result of Theorem 5.

2. If (Ω, Σ, μ) is an arbitrary measure space (without the finite subset property), then $[L^1(\mu)]^* \neq L^\infty(\mu)$ will happen, and if μ has the finite subset property but is not localizable, then $L^\infty(\mu) \subsetneqq [L^1(\mu)]^*$ still holds. Construct example of this case. [Examine Exercise 4.7 again.]

3. Let $\varphi: \mathbb{R}^+ \to \mathbb{R}^+$ be a continuous convex function such that $\varphi(x) = \theta(x^2)$ where $\theta: \mathbb{R}^+ \to \mathbb{R}^+$ is convex (concave). If $\theta(0) = 0$ and $f, g: \Omega \to \mathbb{R}$ are measurable on (Ω, Σ, μ), show that (with $\geqslant$ for concave θ)

$$\int_\Omega \varphi\left(\left|\frac{f+g}{2}\right|\right) d\mu + \int_\Omega \varphi\left(\left|\frac{f-g}{2}\right|\right) d\mu$$

$$\leqslant \frac{1}{2}\left[\int_\Omega \varphi(|f|)\, d\mu + \int_\Omega \varphi(|g|)\, d\mu\right].$$

If $\varphi(x) = |x|^p$, $p \geqslant 2$, this is one of the classical inequalities due to J. A. Clarkson. [Use Theorem 2.10 and the definition of convexity with $a = (f+g)/2$, $b = (f-g)/2$, so $\theta(a^2) + \theta(b^2) \leqslant \theta(a^2 + b^2) \leqslant \frac{1}{2}[\theta(f^2) + \theta(g^2)]$, with reverse inequalities in the concave case.]

4. The preceding simple inequality is useful in the study of a class of Orlicz spaces introduced in Exercise 4.5.20. Thus let $\varphi: \mathbb{R}^+ \to \mathbb{R}^+$ be a convex φ-function such that $\varphi(x) = \theta(x^2)$, where θ is a moderated [i.e., $\theta(2x) \leqslant c\theta(x)$, $\theta(x) = 0$ iff $x = 0$, $0 < c < \infty$, $x \geqslant 0$) convex φ-function. If $L^\varphi(\mu)$ is the corresponding Orlicz space on a measure space (Ω, Σ, μ), with norm $N_\varphi(\cdot)$, where $N_\varphi(f) = \inf\{\alpha > 0: \int_\Omega \varphi(|f|/\alpha)\, d\mu \leqslant 1\}$, show that $\int_\Omega \varphi(|f|/N_\varphi(f))\, d\mu \leqslant 1$, and using it verify that $N_\varphi(f_n) \to 0$ iff $\int_\Omega \varphi(|f_n|)\, d\mu \to 0$ as $n \to \infty$ for any sequence $\{f_n: n \geqslant 1\} \subset L^\varphi(\mu)$.

5. Let $L^\varphi(\mu)$ be as in the preceding exercise, and $0 \neq l \in [L^\varphi(\mu)]^*$. Show that there exists a μ-unique $f_0 \in L^\varphi(\mu)$ such that $N_\varphi(f_0) = 1$ and $l(f_0) = \|l\|$. [As in Theorem 3.1, there exists $f_n \in L^\varphi(\mu)$, $N_\varphi(f_n) = 1$, and $l(f_n) \to \|l\|$. Hence $N_\varphi(f_n + f_m) \geqslant 2(1 - \epsilon)$, $n, m \geqslant n_0(\epsilon)$, for a given $\epsilon > 0$. If $N_\varphi(f_n - f_m) \geqslant \alpha_0 > 0$, then with Exercise 4, $\int_\Omega \varphi(|f_n - f_m|)\, d\mu \nrightarrow 0$ and so $\geqslant \beta > 0$ for some $0 < \beta < 1$. Use Exercise 3, and deduce that since

$N_\varphi(f_n) = 1$, $\int_\Omega \varphi(|f_n + f_m|/2)\,d\mu \leqslant 1 - \beta$. Hence $N_\varphi((f_n + f_m)/2) < 1 - \beta$, $n, m \geqslant n_0$, and this contradicts the earlier inequality. So $f_n \to f_0$ in $N_\varphi(\cdot)$-norm, $l(f_0) = \|l\|$. Uniqueness follows from the fact that φ is strictly convex.]

6. If $L^\varphi(\mu)$ is as in the preceding exercise, show that $(L^\varphi(\mu), N_\varphi(\cdot))$ is a uniformly convex Orlicz space. [The preceding hints apply here. This is a partial extension of Proposition 4, i.e., only $p \geqslant 2$ is included.]

7. Let $L^\varphi(\mu)$ be as in the preceding exercise, and ψ be a complementary function to φ (cf. Exercise 4.5.25). If $(L^\psi(\mu), N_\psi(\cdot))$ is the corresponding space, show that for each $l \in [L^\varphi(\mu)]^*$ there is a μ-unique $g_l \in L^\psi(\mu)$ such that $l(f) = \int_\Omega f g_l\,d\mu$, $f \in L^\varphi(\mu)$. If $\|g_l\|_\psi = \sup\{|\int_\Omega f g_l\,d\mu| : N_\varphi(f) \leqslant 1\}$, then $\|\cdot\|_\psi$ is a norm and is equivalent to $N_\psi(\cdot)$. Show $\|l\| = \|g_l\|_\psi$ and $[L^\varphi(\mu)]^*$ and $L^\psi(\mu)$ are isometrically isomorphic. [First let $\mu(\Omega) < \infty$, and set $\nu(A) = l(\chi_A)$. Show that $\nu\colon \Sigma \to \mathbb{R}$ is σ-additive and has ψ-variation finite. Now $\nu \ll \mu$, so that $g_l = d\nu/d\mu$ exists a.e., by Theorem 3.3. Also $\|\nu\|_\psi = \int_\Omega \psi(|g_l|) < \infty$, by Exercise 1.1. Deduce that $g_l \in L^\psi(\mu)$, and then $l(f)$ has the integral representation, and $\|l\| = \|g_l\|_\psi$. If μ is not finite, check that the hypothesis on φ implies $\psi(x) = 0$ iff $x = 0$ and $\psi(x)/x \to \infty$ as $x \to \infty$, and that $\|\nu\|_\psi(\Omega) = \|\nu\|_\psi(S)$ for an S which is σ-finite for μ. So there exist $S_n \in \Sigma$ such that $\mu(S_n) < \infty$ and $S_n \uparrow S$. By the special case, there is a unique g_n vanishing outside S_n and satisfying the representation for all $f \in L^\varphi(S_n, \Sigma(S_n), \mu)$. Let $g_l = g_n$ on S_n, $n \geqslant 1$. Then $g_l \in L^\psi(\mu)$, and one gets the general case. This is the alternative method we mentioned after Theorem 8. It should be observed that this and the last three problems include spaces that are not in the list of standard L^p-classes. For instance, $\varphi(x) = a_1|x|^{p_1} + a_2|x|^{p_2} + \cdots + a_n|x|^{p_n}$, $x \in \mathbb{R}$, where $a_k \geqslant 0$, $p_k \geqslant 2$ for $1 \leqslant k \leqslant n$, yields an $L^\varphi(\mu)$-space satisfying our conditions. Other spaces can be constructed.]

8. There is an inequality on norms for $1 < p \leqslant 2$. An elementary version is as follows. Let $q = p/(p-1)$, and $f, g \in L^p(\mu)$ on a measure space (Ω, Σ, μ). Verify that

$$\left\|\frac{f+g}{2}\right\|_p^q + \left\|\frac{f-g}{2}\right\|_p^q \leqslant \tfrac{1}{2}\left(\|f\|_p^p + \|g\|_p^p\right)^{1/(p-1)}.$$

[Let $r = p - 1$, $a = |f+g|/2$, $b = |f-g|/2$. Using the "wrong" Minkowski inequality (Proposition 4.5.6), one gets

$$\|a^q + b^q\|_r^r \geqslant \|a\|_p^q + \|b\|_p^q.$$

Since $q \geqslant 2$, $(a^q + b^q)^r \leqslant [\frac{1}{2}(|f|^q + |g|^q)]^r \leqslant 2^{-r}(|f|^p + |g|^p)$, by the convexity (and concavity) of $|x|^q$ (and $|x|^r$). Integrating this and using the preceding inequality, the result follows. The given inequality can be improved by replacing $\frac{1}{2}$ with 2^{-r}, but this involves a lot more work; that result is the key Clarkson (second) inequality.]

9. In Theorems 5 and 8, the dual space $[L^p(\mu)]^*$, $1 \leqslant p < \infty$, was characterized, and the remaining case of $[L^\infty(\mu)]^*$ will be sketched here. For this we have to define the integral relative to a finitely additive set function. Thus if $\nu \in \mathrm{ba}(\Omega, \Sigma, \mu)$, the space of bounded additive set functions which vanish on μ-null sets, the (Ω, Σ, μ) being a measure space with the finite subset property, then define

$$\int_\Omega f\,d\nu = \sum_{k=1}^{n} a_k \nu(A_k),$$

for each $f = \sum_{k=1}^n a_k \chi_{A_k}$, $A_k \in \Sigma$. If f is a bounded measurable (for Σ) scalar function, then there exists a sequence of step functions f_n (measurable for Σ) such that $f_n \to f$. If $\{\int_\Omega f_n\,d\nu \colon n \geqslant 1\}$ is Cauchy, define

$$\int_A f\,d\nu = \lim_n \int_A f_n\,d\nu, \tag{+}$$

for each A in Σ. Show that the integral is well defined for each step function (i.e., it does not depend on the representation of a step function) and that the limit determines the integral of f on A for ν. So (+) is a unique number and does not depend on the sequence $\{f_n \colon n \geqslant 1\}$. (If $\{f_n' \colon n \geqslant 1\}$ is another such sequence, then combining them gives a new sequence with the same limit.) Finally, show that the integral (+) is a linear operation and $|\int_\Omega f\,d\nu| \leqslant \int_\Omega |f|\,d|\nu| \leqslant \|f\|_\infty |\nu|(\Omega)$, where $\|f\|_\infty$ is the usual L^∞- bound.

10. If $L^\infty(\mu)$ on (Ω, Σ, μ), a measure space as in the above exercise, is given, let $l \in [L^\infty(\mu)]^*$. Show that there is a unique $\nu \in \mathrm{ba}(\Omega, \Sigma, \mu)$ such that $l(f) = \int_\Omega f\,d\nu$, $f \in L^\infty(\mu)$, where the integral is in the sense of the preceding exercise, and that $\|l\| = |\nu|(\Omega)$. [If $f \in L^\infty(\mu)$ and $\epsilon > 0$, there is a step function $f_\epsilon \in L^\infty(\mu)$ such that $\|f - f_\epsilon\|_\infty < \epsilon$, by Theorem 4.5.7. So f is ν-integrable by (+) of Exercise 9, and $\|l\| \leqslant |\nu|(\Omega)$. If $\nu(A) = l(\chi_A)$, then $\nu \in \mathrm{ba}(\Omega, \Sigma, \mu)$. For $\epsilon > 0$, choose $\{A_1, \dots, A_n\}$, a partition of Ω in Σ such that $\sum_{i=1}^n |\nu(A_i)| > |\nu|(\Omega) - \epsilon$. Let $a_i = \operatorname{sgn} \nu(A_i)$ and $f = \sum_{i=1}^n a_i \chi_{A_i}$. Then $f \in L^\infty(\mu)$, $\|f\|_\infty = 1$ and $l(f) = \sum_{i=1}^n |\nu(A_i)| \geqslant |\nu|(\Omega) - \epsilon$, so that $\|l\| \geqslant |\nu|(\Omega)$. Uniqueness follows from this, and the rest is easy.]

11. Let (Ω, Σ, μ) be as in the above exercise, so that $[L^\infty(\mu)]^*$ can be identified with $\mathrm{ba}(\Omega, \Sigma, \mu)$. Let $\mathscr{L} = \{l \in (L^\infty(\mu))^* \colon |l(\chi_{A_n})| \to 0$ for each $A_n \in \Sigma$, $A_n \downarrow \varnothing\}$. Show that $\mathscr{L}$ may be identified with $L^1(\mu)$ iff μ is localizable. Thus, when μ is localizable, $l(f) = \int_\Omega f g_l\,d\mu$ for a unique $g_l \in L^1(\mu)$.

12. A dual Banach space $\mathscr{X}^*$ of $\mathscr{X}$ is *weak*-sequentially complete* if $\{x_n^* \colon n \geqslant 1\}$ is a Cauchy sequence in the weak* topology (i.e., the neighborhood system of 0 in $\mathscr{X}^*$ is given by $N(x_1, \dots, x_n; \epsilon) = \{x^* \colon |x^*(x_i)| < \epsilon,$

$i = 1, \dots, n\}$); then there is an $x^* \in \mathscr{X}$ such that $(x_n^* - x^*)(x) \to 0$ for each $x \in \mathscr{X}$. Show that if μ is localizable, then $L^\infty(\mu)$ is weak*-sequentially complete. [If μ is not localizable, then $L^\infty(\mu)$ need not be a dual space. This is why we need that hypothesis. Note that by Theorem 10, $\sup_n \|l_n\| = M < \infty$, where $\{l_n: n \geq 1\} \subset [L^\infty(\mu)]^*$ is weak* Cauchy. By the latter condition, $l_n(g) = \int_\Omega g f_n \, d\mu \to l_g$ (say) for each $g \in L^1(\mu)$ as $n \to \infty$. Verify that $l_g = l(g)$ gives $l \in [L^1(\mu)]^*$ and so $l(g) = \int_\Omega g f \, d\mu$, for a unique $f \in L^\infty(\mu)$, $\|l\| = \|f\|_\infty \leq M < \infty$. However, $L^\infty(\mu)$ is *not* weakly sequentially complete.]

13. Let Ω be a nonvoid set, $B(\Omega) = \{f: \Omega \to \mathbb{R}, \ f \text{ bounded}\}$. If $\|f\|_u = \sup_\omega |f(\omega)|$, then $\{B(\Omega), \|\cdot\|_u\}$ is a Banach space. (Verify this.) Show that $[B(\Omega)]^* = \mathrm{ba}(\Omega, \Sigma)$, where $\Sigma = \mathscr{P}(\Omega)$, the power set of Ω. Here again the equality is meant in the sense of isometric isomorphism, i.e., $l \in [B(\Omega)]^* \Rightarrow l(f) = \int_\Omega f \, d\nu$ for a unique $\nu \in \mathrm{ba}(\Omega, \Sigma)$, $\|l\| = |\nu|(\Omega)$. [Note that $B(\Omega) = l^\infty(\Omega, \Sigma, \mu)$, where $\mu(A)$ is the number of points in A, and $\Sigma = \mathscr{P}(\Omega)$; then apply Exercise 10.]

14. If (Ω, Σ, μ) is an arbitrary measure space, $L^p(\mu)$, $1 < p < \infty$, is the Lebesgue space, and $\mathscr{B} \subset \Sigma$ is a σ-algebra, then let $E^{\mathscr{B}}$ be the (generalized) conditional expectation, as given by Proposition 16. Show that $E^{\mathscr{B}}$ has all the properties of Theorem 15, i.e., complete the proof of Proposition 16. Show also that, if $f \in L^\varphi(\mu)$, $\mu_{\mathscr{B}}$ localizable, φ a convex φ-function, then $N_\varphi(E^{\mathscr{B}}(f)) \leq N_\varphi(f)$. [On $L^\varphi(\mu)$ and $N_\varphi(\cdot)$, see Exercise 4, and use the inequality of Proposition 17(i).]

15. Let (Ω, Σ, μ) be a measure space with the finite subset property, and $\{f_n, \Sigma_n: n \geq 1\} \subset L^1(\mu)$ be a submartingale. If $\sup_n \|f_n\|_1 < \infty$, show that there exists a positive martingale $\{g_n, \Sigma_n: n \geq 1\} \subset L^1(\mu)$ such that $f_n^+ \leq g_n$ a.e., $n \geq 1$, and $\sup_n \|f_n^+\|_1 = \|g_n\|_1$. [Use the same argument as in the proof of Proposition 18(ii). Note that by definition of a (sub-)martingale $\mu_n = \mu | \Sigma_n$ is localizable, $n \geq 1$.]

16. Show that the Doob decomposition of Theorem 20(i) is a.e. unique.

17. Construct an example each to show that the convergence assertions of Theorems 19 and 20 is not in L^1-norm.

18. Complete the details of the proof of Theorem 21.

19. Let (Ω, Σ, μ) be a measure space with the finite subset property, and $\Sigma_n \subset \Sigma_{n+1} \subset \Sigma$ be σ-algebras such that $\mu_n = \mu | \Sigma_n$ is localizable for each $n \geq 1$. Let $\{f_n, \Sigma_n: n \geq 1\} \subset L^1(\mu)$ be a martingale. If $\nu_n(A) = \int_A f_n \, d\mu$, $A \in \Sigma_n$, then $\nu_n = \nu_{n+1} | \Sigma_n$. Show that there is an additive function $\nu: \bigcup_{n=1}^\infty \Sigma_n \to \mathbb{R}$, vanishing on μ-null sets, such that $\nu | \Sigma_n = \nu_n$, and that ν need not be σ-additive even though each ν_n is. [Analyze Example 4.4.4.]

20. Let ν be as in the preceding exercise. If $\nu = \nu^1 + \nu^2$ is the Yosida-Hewitt decomposition, $\hat{\nu}^1$ is the (unique) extension of ν^1 to $\Sigma_\infty = \sigma(\bigcup_n \Sigma_n)$, and μ_∞ is localizable, let $f_\infty = d\hat{\nu}^1 / d\mu_\infty$ be the Radon-Nikodým derivative. Show that $f_n \to f_\infty$ a.e. [This and the preceding exercise are "partial converses" to Theorem 21, and the same ideas apply here.] Finally establish that if $\mu(\Omega) < \infty$, then $f_n \to f$ in L^1-norm also if $\nu^2 = 0$, i.e., ν is σ-additive iff the martingale is uniformly integrable.

21. Let (Ω, Σ, μ) be a measure space, and $\Sigma \supset \Sigma_n \supset \Sigma_{n+1}$ be a sequence such that $\mu_n = \mu|\Sigma_n$ is localizable. (μ has the finite subset property.) If $\nu: \Sigma \to \mathbb{R}$ is a signed measure, $\nu_n = \nu|\Sigma_n$, and $\nu \ll \mu$, let $f_n = d\nu_n/d\mu_n$. Show that $\{f_n, \Sigma_n: n \geqslant 1\}$ is a decreasing martingale, i.e., $E^{\Sigma_{n+1}}(f_n) = f_{n+1}$ a.e. If $\Sigma_\infty = \bigcap_n \Sigma_n$ is such that $\mu|\Sigma_\infty$ has no finite subset property, verify that $f_n \to 0$ a.e.$[\mu]$. [Show $f_* = \underline{\lim}_n f_n \geqslant 0 \geqslant \overline{\lim}_n f_n = f^*$, by checking that $\mu(\{\omega: f^* > \beta\}) > 0$ is impossible for any $\beta > 0$.]

22. In Definition 14, it is vital that $\nu_f(A) = \int_A f\,d\mu$, $A \in \mathscr{B}$, is σ-additive. Show that $E^{\mathscr{B}}(f) = \tilde{f}$ need not exist in general, by means of the following example. Let $\Omega = \mathbb{R}^2$, $\Sigma =$ (Borel σ-algebra of Ω), and $\mu(A) = \iint_A h(x, y)\,dx\,dy$, where $h(x, y) = (|y|/4)e^{-|y|(1+|x|)}$, $x \in \mathbb{R}$, $y \in \mathbb{R}$. If $\omega = (x, y) \in \Omega$, $X(\omega) = x$, $Y(\omega) = y$, and $\mathscr{B} = \sigma(Y) \subset \Sigma$, then $\int_\Omega X\,d\mu$ does not exist, and $\tilde{X} = E^{\mathscr{B}}(X)$ does not exist, i.e., there is no $\tilde{X}$ such that $\int_B X\,d\mu = \int_B \tilde{X}\,d\mu_{\mathscr{B}}$, $B \in \mathscr{B}$. Deduce that Theorem 15(iii) cannot hold if $\mathscr{B}_2 = \mathscr{B}$ and $\mathscr{B}_1 = \{\varnothing, \Omega\}$.

6

PRODUCT MEASURES AND INTEGRALS

This chapter is devoted to constructions of measures on Cartesian products of measurable spaces, whose marginals agree with the measures given *a priori* on the individual spaces. The work is in two parts. The first is on finite products. Here the major results are the Fubini-Stone and Tonelli theorems together with some extensions. The second half contains the infinite products, and the corresponding results fall into two distinct classes: The Fubini-Jessen theorem is on abstract spaces, while the Kolmogorov-Bochner and Prokhorov theorems are for more general measure systems, but topology intervenes here and certain "regularity" conditions must be satisfied by the measures. As an important application of the latter we present a purely measure theoretical characterization of positive definite functions. Also, a realization of an abstract Hilbert space as an $L^2(\mu)$ is given. In these constructions we use both the measure and integration results developed in the preceding work.

6.1. BASIC DEFINITIONS AND PROPERTIES

To consider finite products, it is enough to treat two spaces. Thus let (Ω_i, Σ_i), $i = 1, 2$, be measurable spaces, and $\Omega = \Omega_1 \times \Omega_2$ their Cartesian product. If $A \in \Sigma_1$, $B \in \Sigma_2$, then $A \times B$ $(\subset \Omega)$ is called a *measurable rectangle* with sides A and B. Let $\mathscr{S}$ denote the collection of all measurable rectangles of Ω. Note that if $E_i = A_i \times B_i \in \mathscr{S}$, $i = 1, 2$, then $E_1 \cap E_2 = (A_1 \cap A_2) \times (B_1 \cap B_2) \in \mathscr{S}$, since $A_1 \cap A_2 \in \Sigma_1$ and $B_1 \cap B_2 \in \Sigma_2$. Thus $\mathscr{S}$ is closed under intersections (so it is a π-class). Clearly $\Omega \in \mathscr{S}$. More is true, and in fact we shall see that $\mathscr{S}$ is a semialgebra, which follows from an easy computation. These properties are given by:

Lemma 1.

(a) *Let $C_i = A_i \times B_i \in \mathscr{S}$, $i = 1, 2$. Then $C_1 = \varnothing$ iff either $A_1 = \varnothing$ or $B_1 = \varnothing$, and when $C_i \neq \varnothing$, $i = 1, 2$, then $C_1 \subseteq C_2$ iff $A_1 \subseteq A_2$ and*

$B_1 \subseteq B_2$, with equality throughout or proper inclusion (for at least one pair.)

(b) *If $C = C_1 \cup C_2$, where C_i's are as above, and $C \in \mathscr{S}$, then $C_1 \cap C_2 = \varnothing$ iff either $A_1 \cap A_2 = \varnothing$ and $B_1 = B_2$, or $A_1 = A_2$ and $B_1 \cap B_2 = \varnothing$.*

(c) *The class $\mathscr{S}$ is a semialgebra of Ω.*

Proof. (a) is obvious (draw a picture). Note that in the second half the necessity will not hold if $C_i = \varnothing$, although the sufficiency will.

(b): First let $C = C_1 \cup C_2 \in \mathscr{S}$ and $C_1 \cap C_2 = \varnothing$. To see that the asserted conditions must hold, let $C = A \times B$. Since $C_i \subset C$, we have, by (a), $A_i \times B_i \subset A \times B \Rightarrow A_i \subset A$, $B_i \subset B$, so that $(A_1 \cup A_2) \times (B_1 \cup B_2) \subset A \times B$. On the other hand, $A_i \subset A_1 \cup A_2$, $B_i \subset B_1 \cup B_2$, $i = 1,2$. So $C_i \subset (A_1 \cup A_2) \times (B_1 \cup B_2)$, so that $C_1 \cup C_2 = C = (A_1 \cup A_2) \times (B_1 \cup B_2)$. Hence $A = A_1 \cup A_2$, $B = B_1 \cup B_2$. Also $\varnothing = C_1 \cap C_2 = (A_1 \cap A_2) \times (B_1 \cap B_2)$, so that by (a), either $A_1 \cap A_2 = \varnothing$ or $B_1 \cap B_2 = \varnothing$. Suppose $A_1 \cap A_2 = \varnothing$. If $B_1 \neq B_2$, then $B_1 - B_2 \neq \varnothing$ or $B_2 - B_1 \neq \varnothing$. In the first case, let $\omega = (\omega_1, \omega_2) \in C = A \times B$ such that $\omega_2 \in B_1 - B_2$ and $\omega_1 \in A_2$ (so that $\omega_1 \notin A_1$). Since $\omega_2 \notin B_2$, we have $(\omega_1, \omega_2) = \omega \notin A_i \times B_i = C_i$, $i = 1,2$. But then $\omega \notin C_1 \cup C_2 = C$, a contradiction. Thus $B_1 = B_2$. Similarly the second possibility is verified.

Conversely, let $A = A_1 \cup A_2$, $A_1 \cap A_2 = \varnothing$, and $B = B_1 = B_2$. Then $A_i \subset A$, so $C_i \subset C$, $i = 1,2$, and $C_1 \cup C_2 \subset C$. For the opposite inclusion, let $\omega = (\omega_1, \omega_2) \in C = A \times B$, so that $\omega_1 \in A = A_1 \cup A_2$ and $\omega_2 \in B = B_1 = B_2$. So $\omega_1 \in A_1$ or $\omega_1 \in A_2$ (but not both). Hence $\omega \in C_1$ or C_2 but not both. Thus $C \subset C_1 \cup C_2$ and $C_1 \cap C_2 = \varnothing$. The other possibility is similar, giving (b).

(c): We have already seen that $\mathscr{S}$ is closed under intersections, $\{\varnothing, \Omega\} \subset \mathscr{S}$. If C_1, C_2 are in $\mathscr{S}$, then (by drawing a picture) we have

$$C_1 - C_2 = (A_1 \times B_1) - (A_2 \times B_2)$$

$$= \left[(A_1 \cap A_2^c) \times B_1\right] \cup \left[(A_1 \cap A_2) \times (B_1 \cap B_2^c)\right]. \tag{1}$$

The right side of (1) is a disjoint union of elements of $\mathscr{S}$. So it is a semialgebra. This proves the lemma.

If $\mathscr{E}$ is the class of all finite disjoint unions of elements of $\mathscr{S}$, then $\mathscr{E}$ is an algebra and the σ-algebras generated by $\mathscr{S}$ and $\mathscr{E}$ are the same Σ, i.e., $\Sigma = \sigma(\mathscr{E}) = \sigma(\mathscr{S})$ (obvious). Since $\mathscr{S}$ is the Cartesian product of the classes Σ_1, Σ_2, so that $\Sigma \supset \mathscr{E} \supset \mathscr{S} = \Sigma_1 \times \Sigma_2$, one would like to know the size of Σ. Indeed, this can be small and there are some new problems to contend with. For instance, if Ω_1 has m points and Ω_2 has n points, then Ω has mn points. The power sets of these spaces respectively contain 2^m, 2^n, and 2^{mn} elements. Thus $\mathscr{P}(\Omega_1) \times \mathscr{P}(\Omega_2)$ has 2^{m+n} elements, while $\mathscr{P}(\Omega_1 \times \Omega_2)$ has 2^{mn} elements. If $m = n = \infty$, is there equality? This is not entirely a trivial problem. To understand the situation, let $\Omega_1 = \Omega_2 = [0,1]$, so that Ω is the unit square. If both the continuum hypothesis and the axiom of choice hold, then B. V.

Rao (1969) has shown that $\mathscr{P}(\Omega_1) \times \mathscr{P}(\Omega_2) = \mathscr{P}(\Omega)$ does hold. [Recall that, at least here, the continuum hypothesis asserts that each infinite subset of $\mathbb{R}$ either is countable or has the cardinality of $\mathbb{R}$ itself. It is also known that this axiom is independent of the other axioms of set theory, including the axiom of choice. See Suppes (1972). We do not use any of this later.] Thus considerations of products of classes of sets involve some such foundational results.

Keeping the preceding points in mind, we proceed to the introduction of a measure on Ω if a pair of measures μ_1, μ_2 are given on Ω_1, Ω_2. One wants, more precisely, to find a probability μ on $\sigma(\mathscr{E})$ such that its Ω_i-marginals coincide with the μ_i in the sense that $\mu(\Omega_1 \times \cdot\,) = \mu_2(\cdot)$ and $\mu(\cdot \times \Omega_2) = \mu_1(\cdot)$. The problem thus stated, however, is too general and does not have a unique solution, because there exist many μ on $\sigma(\mathscr{E})$ satisfying the above condition. So we solve a more restricted version, namely to find an $\alpha: \mathscr{S} \to \overline{\mathbb{R}}^+$ such that $\alpha(A \times B) = \mu_1(A)\mu_2(B)$ for all $A \times B \in \mathscr{S}$. This is an abstraction of the classical (Lebesgue) result that if μ_1, μ_2 are length functions on Ω_1 and Ω_2, where $\Omega_1 = \mathbb{R} = \Omega_2$, then $\alpha(\cdot)$ is the area function on the rectangles of $\Omega = \mathbb{R}^2$. This case is the basic stepping stone of the product measure theory, and so we treat it in detail.

If $\alpha: \mathscr{S} \to \mathbb{R}^+$ is defined as above from the measure spaces $(\Omega_i, \Sigma_i, \mu_i)$, $i = 1, 2$, then we have the elementary but important:

Lemma 2. *If $\alpha(A \times B) = \mu_1(A)\mu_2(B)$, then α is σ-additive on $\mathscr{S}$, i.e., α is a measure on $\mathscr{S}$.*

Proof. It is immediate that $\alpha(\varnothing) = 0$ and $\alpha(C_1) \leqslant \alpha(C_2)$ for $C_1 \subset C_2$, $C_i \in \mathscr{S}$, by Lemma 1(a) and the fact that μ_1, μ_2 are measures. For the σ-additivity, let C_n, C in $\mathscr{S}$ be such that C_n's are disjoint and $C = \bigcup_n C_n$. Let $C_n = A_n \times B_n$ and $C = A \times B$. Then we have

$$\chi_C = \chi_A \chi_B = \sum_{n=1}^{\infty} \chi_{A_n} \chi_{B_n}, \tag{2}$$

and by disjointness for each $\omega \in C$, only one term on the right in (2) is nonvanishing at $\omega = (\omega_1, \omega_2)$. Hence (2) becomes

$$\chi_A(\omega_1)\chi_B(\omega_2) = \sum_{n=1}^{\infty} \chi_{A_n}(\omega_1)\chi_{B_n}(\omega_2), \qquad \omega = (\omega_1, \omega_2) \in \Omega. \tag{3}$$

Fix $\omega_2 \in B$ arbitrarily, and consider (3) as ω_1 varies over A. Integrating this relative to μ_1, we have by Corollary 4.3.3(c)

$$\mu_1(A)\chi_B(\omega_2) = \sum_{n=1}^{\infty} \mu_1(A_n)\chi_{B_n}(\omega_2). \tag{4}$$

Next, as ω_2 varies over B in (4), we may apply the same corollary to this series to obtain again, on integration relative to μ_2 (since μ_2 is σ-additive just as μ_1 is),

$$\alpha(A \times B) = \mu_1(A)\mu_2(B) = \sum_{n=1}^{\infty} \mu_1(A_n)\mu_2(B_n) = \sum_{n=1}^{\infty} \alpha(A_n \times B_n). \quad (5)$$

But this is precisely the σ-additivity of α on $\mathscr{S}$, completing the proof.

The point of this result is that we have been able to generate a σ-additive set function on $\mathscr{S}$ and hence can generate an outer measure μ^* on all the subsets of Ω by the pair $(\alpha, \mathscr{S})$, in accordance with Definition 2.2.7. We are therefore in a position to apply the fundamental theorems of Section 2.2 to obtain the following

Theorem 3. *Let* $(\Omega_i, \Sigma_i, \mu_i)$, $i = 1, 2$, *be a pair of measure spaces,* $\alpha: \mathscr{S}(= \Sigma_1 \times \Sigma_2) \to \overline{\mathbb{R}}^+$ *be the measure determined by Lemma 2, and* μ^* *be the generated outer measure of* $(\alpha, \mathscr{S})$ *with* $\mathscr{M}_{\mu^*}$ *as the class of* μ^**-measurable sets of* Ω $(= \Omega_1 \times \Omega_2)$. *Then writing* $\mu = \mu^* | \mathscr{M}_{\mu^*}$, *one has*

(i) $(\Omega, \mathscr{M}_{\mu^*}, \mu)$ *is a Carathéodory regular measure space, so that for each* $C \subset \Omega$,

$$\mu^*(C) = \inf\{\mu(D): D \in \mathscr{E}_\sigma, D \supset C\}, \quad (6)$$

where $\mathscr{E}$ *is the class of elementary sets obtained from* $\mathscr{S}$, *and* $\mathscr{E}_\sigma$ *is the closure of* $\mathscr{E}$ *under countable unions;*

(ii) *for each* $C \subset \Omega$, *there exists a* $D \in \mathscr{E}_{\sigma\delta}$ *such that* $D \supset C$ *and* $\mu^*(C) = \mu(D)$ *(thus D is a measurable cover of C), and* $\mathscr{E}_{\sigma\delta}$ *is the closure of* $\mathscr{E}_\sigma$ *under countable intersections;*

(iii) $\mathscr{S} \subset \mathscr{E} \subset \mathscr{M}_{\mu^*}$ *and* $\mu | \mathscr{E} = \hat{\alpha}$, $\mu | \mathscr{S} = \alpha$, *so that* $\mu(A \times B) = \mu_1(A)\mu_2(B)$ *for* $A \in \Sigma_1$, $B \in \Sigma_2$, *where* $\hat{\alpha}$ *is the unique* σ*-additive extension of* α *to* $\mathscr{E}$.

Proof. Since μ^* is a generated outer measure, (i) and (ii) are immediate consequences of Theorem 2.2.10. Only one part of (iii) needs a small computation. Note that since α on $\mathscr{S}$ is σ-additive (by Lemma 2), so is $\hat{\alpha}$ on $\mathscr{E}$ and hence $\hat{\alpha} = \mu | \mathscr{E}$, $\alpha = \mu | \mathscr{S}$ by Theorem 2.2.10. Here we use the elementary result that $\hat{\alpha}$ is σ-additive on $\mathscr{E}$, which is seen as follows. Let E, E_n be in $\mathscr{E}$, $E = \bigcup_{n=1}^{\infty} E_n$, a disjoint union. Then by definition of $\mathscr{E}$, there exist disjoint C_i, D_{ni} in $\mathscr{S}$ such that $E = \bigcup_{i=1}^{k} C_i$, $E_n = \bigcup_{i=1}^{k_n} D_{ni}$. Hence for each $1 \leqslant i_0 \leqslant k$,

$$C_{i_0} = E \cap C_{i_0} = \bigcup_{n=1}^{\infty} \bigcup_{i=1}^{k_n} (D_{ni} \cap C_{i_0}), \qquad D_{ni} \cap C_{i_0} \in \mathscr{S}. \quad (7)$$

Since the right side of (7) is a disjoint union in $\mathscr{S}$, one has

$$\alpha(C_{i_0}) = \sum_{n=1}^{\infty}\left(\sum_{i=1}^{k_n}\alpha(D_{ni}\cap C_{i_0})\right), \qquad \text{by Lemma 2}$$

$$= \sum_{n=1}^{\infty}\hat{\alpha}\left(\bigcup_{i=1}^{k_n} D_{ni}\cap C_{i_0}\right), \qquad \text{by definition of } \hat{\alpha},$$

$$= \sum_{n=1}^{\infty}\hat{\alpha}(E_n \cap C_{i_0}).$$

Hence

$$\hat{\alpha}(E) = \sum_{i_0=1}^{k}\alpha(C_{i_0}) = \sum_{n=1}^{\infty}\hat{\alpha}(E_n\cap E) = \sum_{n=1}^{\infty}\hat{\alpha}(E_n), \tag{8}$$

since $\hat{\alpha}$ is additive on $\mathscr{E}$. Thus $\hat{\alpha} = \mu|\mathscr{E}$, by Theorem 2.2.10. Also $\alpha = \mu|\mathscr{S}$ gives, for the rectangle $C = A \times B \in \mathscr{S}$, $\mu(C) = \alpha(A \times B) = \mu_1(A)\mu_2(B)$. This completes the proof of the theorem.

This result gives us a measure μ on Ω such that $\mu(\Omega_1 \times \cdot\,)$ and $\mu(\cdot \times \Omega_2)$ are constant multiples of $\mu_2(\cdot)$ and $\mu_1(\cdot)$. If $\mu_i(\Omega_i) = 1$, then these are precisely μ_2, μ_1, called the *marginals* of μ. Other connections between the measures μ_i and the possible uniqueness (or lack thereof) of μ, [called the *(Cartesian) product measure and denoted* $\mu_1 \otimes \mu_2$] will now be analyzed.

Discussion 4. Following custom, we denote $\Sigma\ [= \sigma(\mathscr{S})]$ as $\Sigma_1 \otimes \Sigma_2$. Thus we have $\Sigma_1 \otimes \Sigma_2$ as the smallest σ-algebra containing all the measurable rectangles $\mathscr{S}$, and $\mathscr{M}_{\mu^*} \supset \Sigma_1 \otimes \Sigma_2$, the class $\mathscr{M}_{\mu^*}$ being the largest σ-algebra on which the generated outer measure μ^*, by $(\alpha, \mathscr{S})$, is σ-additive. Also, $\mu = \mu^*|\mathscr{M}_{\mu^*}$ *is a complete measure*. In general, even if $(\Omega_i, \Sigma_i, \mu_i)$, $i = 1, 2$, are complete, $\mu|\Sigma_1 \otimes \Sigma_2$ need not be. (See Exercise 4.) By the results of Section 2.3, μ is not necessarily the unique measure which coincides with α on $\mathscr{S}$. However, if μ_1, μ_2 are σ-finite, so that there exist disjoint $\Omega_{in} \in \Sigma_i$, $i = 1, 2$, $\Omega_i = \bigcup_{n=1}^{\infty}\Omega_{in}$, $\mu_i(\Omega_{in}) < \infty$, we have

$$\Omega = \bigcup_{n=1}^{\infty}\bigcup_{m=1}^{\infty}(\Omega_{1n}\times\Omega_{2m})\ (\in \mathscr{S}), \tag{9}$$

and α is σ-finite on $\mathscr{S}$. Hence by Theorem 2.3.1, μ on $\Sigma_1 \otimes \Sigma_2$ is a unique σ-finite measure extending α. Moreover, the completion $\overline{\Sigma_1 \otimes \Sigma_2}$ of $\Sigma_1 \otimes \Sigma_2$ is identical with $\mathscr{M}_{\mu^*}$. If no such σ-finiteness of μ is true, then $\overline{\Sigma_1 \otimes \Sigma_2} \subset \mathscr{M}_{\mu^*}$ with a possibly strict inclusion (cf. Proposition 2.3.6). [But see Exercise 6 for a unique extension condition weaker than σ-finiteness.]

We next consider some properties of the new (i.e., nonrectangular) elements of $\Sigma_1 \otimes \Sigma_2$. Let us first define the concept of a section.

Definition 5. If $E \subset \Omega = \Omega_1 \times \Omega_2$, then the set $E(\omega_1) = \{\omega_2 \in \Omega_2: (\omega_1, \omega_2) \in E\}$ is called a *section* of E in Ω_2. The section $E(\omega_2) \subset \Omega_1$ is defined similarly. If $f: \Omega \to \overline{\mathbb{R}}$ is a function, then $f_{\omega_1}: \Omega_2 \to \overline{\mathbb{R}}$ defined by $f_{\omega_1}(\omega_2) = f(\omega_1, \omega_2)$, $\omega_2 \in \Omega_2$, is called a *section of f* on Ω_2. The section $f_{\omega_2}: \Omega_1 \to \mathbb{R}$ is defined similarly.

Regarding this concept one has:

Lemma 6. *Let $E \in \Sigma = \Sigma_1 \otimes E_2$. Then*

(i) $E(\omega_1) \in \Sigma_2$, $E(\omega_2) \in \Sigma_1$ *for each* $\omega_1 \in \Omega_1$, $\omega_2 \in \Omega_2$;
(ii) *if* $f: \Omega \to \overline{\mathbb{R}}$ *is measurable for* Σ, *one has* f_{ω_1} (f_{ω_2}) *measurable for* Σ_2 (Σ_1).

Proof. (i): We establish this by using a monotone class result (Proposition 2.1.3). Thus let $\mathscr{C} = \{E \in \Sigma: E(\omega_1) \in \Sigma_2$ and $E(\omega_2) \in \Sigma_1$ for $\omega = (\omega_1, \omega_2) \in \Omega\}$. Since for $E = A \times B \in \mathscr{S}$ we have $E(\omega_1) = B$ if $\omega_1 \in A$, and $= \varnothing$ if $\omega_1 \in A^c$, and similarly $E(\omega_2) = A$ for $\omega_2 \in B$, and $= \varnothing$ for $\omega_2 \in B^c$, we get $E \in \mathscr{C}$. Hence $\mathscr{S} \subset \mathscr{C} \subset \Sigma$. We now show that $\mathscr{C}$ is a monotone class, so that by Proposition 2.1.3, it is a σ-algebra, and then $\mathscr{C} \supset \sigma(\mathscr{S}) = \Sigma$, implying $\mathscr{C} = \Sigma$, which establishes (i).

Thus let $E_1, E_2, \ldots$ be sets in $\mathscr{C}$. Then for each $\omega_1 \in \Omega_1$,

$$\left\{\omega_2 \in \Omega_2: (\omega_1, \omega_2) \in \bigcup_{i=1}^{\infty} E_i\right\} = \bigcup_{i=1}^{\infty} \{\omega_2 \in \Omega_2: (\omega_1, \omega_2) \in E_i\}$$

$$= \bigcup_{i=1}^{\infty} E_i(\omega_1) \in \Sigma_2, \tag{10}$$

since $E_i(\omega_1) \in \Sigma_2$ and the latter is a σ-algebra. Similarly $E_i(\omega_2) \in \Sigma_1 \Rightarrow \bigcup_{i=1}^{\infty} E_i(\omega_2) \in \Sigma_1$, so that $\bigcup_{i=1}^{\infty} E_i \in \mathscr{C}$. In the same way, replacing unions by intersections, one concludes that $\bigcap_{i=1}^{\infty} E_i \in \mathscr{C}$. In particular $\mathscr{C}$ is closed under monotone limits. Also $E \in \mathscr{C} \Rightarrow (E^c)(\omega_1) = [E(\omega_1)]^c \in \Sigma_2$ and $[E(\omega_2)]^c \in \Sigma_1$. So $\mathscr{C}$ is a λ-class, and in fact a σ-algebra already, as asserted.

(ii): This follows easily. By hypothesis, for each $a \in \overline{\mathbb{R}}$,

$$E^a = \{(\omega_1, \omega_2): f(\omega_1, \omega_2) \geqslant a\} \in \Sigma = \Sigma_1 \otimes \Sigma_2.$$

Hence by (i), $E^a(\omega_1) \in \Sigma_2$, $E^a(\omega_2) \in \Sigma_1$, $\omega_1 \in \Omega_1$, $\omega_2 \in \Omega_2$. Since

$$E^a(\omega_1) = \{\omega_2: f(\omega_1, \omega_2) \geqslant a\} = \{\omega_2: f_{\omega_1}(\omega_2) \geqslant a\},$$

f_{ω_1} is measurable for Σ_2. This completes the proof.

It should be noted that the (joint) measurability of f on Ω relative to $\Sigma_1 \otimes \Sigma_2$ implies the measurability of sections $f_{\omega_1}, f_{\omega_2}$ for Σ_2 and Σ_1, but the converse is generally false, as seen from the remarks following Lemma 1. Thus joint measurability is a stronger condition, and a more involved one to check, than the sectional case. This is an inherent difficulty in the product measure theory, and one has to put up with it in all applications.

We next prove a key inequality to be used in establishing the fundamental theorem of the following section. It is due, essentially, to M. H. Stone.

Lemma 7. *Let $(\Omega_i, \Sigma_i, \mu_i)$, $i = 1, 2$, be measure spaces, and (Ω, Σ, μ) their product as given by Theorem 3. If $f: \Omega \to \overline{\mathbb{R}}^+$ is a (not necessarily measurable) function, then we have*

$$\int_{\Omega_1}^{*}\left(\int_{\Omega_2}^{*} f(\omega_1, \omega_2)\mu_2(d\omega_2)\right)\mu_1(d\omega_1) \le \int_{\Omega}^{*} f(\omega)\mu(d\omega), \tag{11}$$

and

$$\int_{\Omega_2}^{*}\left(\int_{\Omega_1}^{*} f(\omega_1, \omega_2)\mu_1(d\omega_1)\right)\mu_2(d\omega_2) \le \int_{\Omega}^{*} f(\omega)\mu(d\omega). \tag{12}$$

Here the integrals for nonmeasurable functions are as in Section 4.2.

Proof. We prove (11), and (12) is similar. For the proof we may assume that $\int_{\Omega}^{*} f(\omega)\mu(d\omega) < \infty$, (11) being true and trivial otherwise.

Since μ is a generated measure (with (6) and), the definition of our "star integral" (cf. Proposition 4.2.1) implies

$$\int_{\Omega}^{*} f\,d\mu = \inf\left\{\int_{\Omega} g\,d\mu : g \geqslant f,\ g\ \mu\text{-elementary}\right\}. \tag{13}$$

Hence given $\epsilon > 0$, there is a $g_\epsilon \geqslant f$ (g_ϵ is μ-elementary) such that

$$\infty > \int_{\Omega}^{*} f\,d\mu + \frac{\epsilon}{2} > \int_{\Omega} g_\epsilon\,d\mu. \tag{14}$$

But by Theorem 4.1.4 we have, on writing $\tau(E_n \times [0, t_n)) = t_n\mu(E_n)$,

$$\begin{aligned}\int_{\Omega} g_\epsilon\,d\mu = \inf\Bigg\{&\sum_{n=1}^{\infty} \tau(E_n \times [0, t_n)):\\ &\bigcup_{n=1}^{\infty} E_n \times [0, t_n) \supset (\text{region under } g_\epsilon),\ t_n \geqslant 0,\ E_n \in \mathscr{S}\Bigg\},\\ = \inf\Bigg\{&\sum_{n=1}^{\infty} \int_{\Omega} t_n \chi_{E_n}\,d\mu : (\text{same conditions})\Bigg\}\\ = \inf\Bigg\{&\sum_{n=1}^{\infty} \int_{\Omega} f_n\,d\mu : \sum_{n=1}^{\infty} f_n \geqslant g_\epsilon,\ f_n = \sum_{i=1}^{n} a_i^n \chi_{E_i^n},\ E_i^n \in \mathscr{S},\ a_i^n \geqslant 0\Bigg\}.\end{aligned}$$

So there exists a sequence f_n, $n \geqslant 1$, such that $\sum_{n=1}^{\infty} f_n \geqslant g_\epsilon$, f_n are $\mathscr{S}$-step functions, and

$$\int_\Omega g_\epsilon \, d\mu + \frac{\epsilon}{2} > \sum_{n=1}^{\infty} \int_\Omega f_n \, d\mu$$

$$= \sum_{n=1}^{\infty} \sum_{i=1}^{n} a_i^n \mu_1(A_i^n) \mu_2(B_i^n), \qquad \text{with} \quad E_i^n = A_i^n \times B_i^n \in \mathscr{S},$$

$$= \sum_{n=1}^{\infty} \int_{\Omega_1} \left(\int_{\Omega_2} f_n \, d\mu_2 \right) d\mu_1, \qquad \sum_{n=1}^{\infty} f_n \geqslant g_\epsilon \geqslant f. \tag{15}$$

Since $f(\cdot, \omega_2) \leqslant (\sum_{n=1}^{\infty} f_n)(\cdot, \omega_2) = \sum_{n=1}^{\infty} f_n(\cdot, \omega_2)$, and the star integral is order preserving, we get

$$\int_{\Omega_2}^{*} f(\omega_1, \omega_2) \mu_2(d\omega_2) \leqslant \sum_{n=1}^{\infty} \int_{\Omega_2} f_n(\omega_1, \omega_2) \mu_2(d\omega_2), \qquad \omega_1 \in \Omega_1,$$

and hence

$$\int_{\Omega_1}^{*} \left(\int_{\Omega_2}^{*} f(\omega_1, \omega_2) \right) \mu_2(d\omega_2) \mu_1(d\omega_1)$$

$$\leqslant \sum_{n=1}^{\infty} \int_{\Omega_1} \left(\int_{\Omega_2} f_n(\omega_1, \omega_2) \mu_2(d\omega_2) \right) \mu(d\omega_1)$$

[by Lemma 6(ii) and Corollary 4.3.3]

$$< \int_\Omega g_\epsilon \, d\mu + \frac{\epsilon}{2}, \qquad \text{by (15)}$$

$$< \int_\Omega^{*} f \, d\mu + \epsilon, \qquad \text{by (14).}$$

Since $\epsilon > 0$ is arbitrary, this implies (11). The inequality (12) being similar, the proof is finished.

The preceding argument also gives the following result:

Proposition 8. *If $f: \Omega \to \overline{\mathbb{R}}$ is measurable for $\Sigma = \Sigma_1 \otimes \Sigma_2$, and $\int_\Omega |f| \, d\mu < \infty$, then for each $\epsilon > 0$ there exists a μ-simple function $h_\epsilon = \sum_{i=1}^{n} a_i \chi_{E_i}$, $E_i \in \mathscr{S}$, such that $\int_\Omega |f - h_\epsilon| \, d\mu < \epsilon$.*

Proof. Writing $f = f^+ - f^-$, one may consider $f^{\pm}$ separately. Then the result will follow if it is proved for f^+ and f^-. So let us assume that $f \geqslant 0$.

Now by (14) and (15), for each $\epsilon > 0$, there is a $g_\epsilon \geq f$ such that

$$\epsilon + \int_\Omega f\,d\mu > \int_\Omega g_\epsilon\,d\mu + \frac{\epsilon}{2} > \sum_{n=1}^{\infty} \int_\Omega f_n\,d\mu, \qquad \sum_{n=1}^{\infty} f_n \geq g_\epsilon$$

$$\geq \int_\Omega g_\epsilon\,d\mu \qquad \text{(using Corollary 4.3.3)}.$$

So there is an N_0 $[= N_0(\epsilon)]$ such that $\sum_{n=1}^{\infty}\int_\Omega f_n\,d\mu < \sum_{n=1}^{N_0}\int_\Omega f_n\,d\mu + \epsilon/2$. Consequently

$$\int_\Omega g_\epsilon\,d\mu < \int_\Omega \left(\sum_{n=1}^{N_0} f_n\right) d\mu + \frac{\epsilon}{2}. \tag{16}$$

It follows from this that, on letting $h_\epsilon = \sum_{n=1}^{N_0} f_n$,

$$\int_\Omega |g_\epsilon - h_\epsilon|\,d\mu \leq \int_\Omega \left(\sum_{n=N_0+1}^{\infty} f_n\right) d\mu = \sum_{n=N_0+1}^{\infty} \int_\Omega f_n\,d\mu < \frac{\epsilon}{2}.$$

Since h_ϵ is a μ-simple function of the desired kind, and

$$\int_\Omega |f - h_\epsilon|\,d\mu \leq \int_\Omega |f - g_\epsilon|\,d\mu + \int_\Omega |g_\epsilon - h_\epsilon|\,d\mu < \frac{\epsilon}{2} + \frac{\epsilon}{2} = \epsilon,$$

the result follows.

We now present an extension of the formula for sets in $\Sigma_1 \otimes \Sigma_2$ given in Theorem 3(iii), where a set is not necessarily a rectangle. This will indicate a new problem arising in the product measure theory.

Proposition 9. *Let $(\Omega_i, \Sigma_i, \mu_i)$, $i = 1, 2$ be finite measure spaces and (Ω, Σ, μ) be their product, as given by Theorem 3. Then for each $E \in \Sigma$ $(= \Sigma_1 \otimes \Sigma_2)$;*

(i) *the functions $\omega_1 \mapsto \mu_2(E(\omega_1))$, $\omega_2 \mapsto \mu_1(E(\omega_2))$ are Σ_1- and Σ_2-measurable,*

(ii) *we have*

$$\mu(E) = \int_{\Omega_1} \mu_2(E(\omega_1))\mu_1(d\omega_1) = \int_{\Omega_2} \mu_1(E(\omega_2))\mu_2(d\omega_2), \tag{17}$$

(iii) *if $\mu(E) = 0$, then $\mu_1(E(\omega_2)) = 0$ a.e.$[\mu_2]$ and $\mu_2(E(\omega_1)) = 0$ a.e.$[\mu_1]$.*

Proof. Let $\mathscr{C} = \{E \in \Sigma: \text{(i) and (ii) of the statement hold}\}$. Then by Theorem 3, $\mathscr{S} \subset \mathscr{C}$. We now use Proposition 2.1.3(b). Since clearly $\mathscr{S}$ is a

π-class, it suffices to show that $\mathscr{C}$ is a λ-class to use that proposition, according to which $\mathscr{C} \supset \sigma(\mathscr{S}) = \Sigma$, so that $\mathscr{C} = \Sigma$.

If E_1, E_2 are disjoint elements of $\mathscr{C}$, then $(E_1 \cup E_2)(\omega_1) = E_1(\omega_1) \cup E_2(\omega_1)$, and this is a disjoint union in Σ_2. By (i) and (ii)

$$\begin{aligned}\mu(E_1 \cup E_2) &= \int_{\Omega_1} \mu_2(E_1(\omega_1) \cup E_2(\omega_1))\mu_1(d\omega_1)\\ &= \int_{\Omega_1} \mu_2(E_1(\omega_1))\mu_1(d\omega_1) + \int_{\Omega_1} \mu_2(E_2(\omega_1))\mu_1(d\omega_1)\\ &= \int_{\Omega_2} \mu_1(E_1(\omega_2))\mu_2(d\omega_2) + \int_{\Omega_2} \mu_1(E_2(\omega_2))\mu_2(d\omega_2)\\ &= \int_{\Omega_2} \mu_1(E_1(\omega_2) \cup E_2(\omega_2))\mu_2(d\omega_2),\end{aligned}$$

since $E_1(\omega_2) \cap E_2(\omega_2) = \varnothing$. Hence $E_1 \cup E_2 \in \mathscr{C}$. If $E_1 \supset E_2$, $E_i \in \mathscr{C}$, then $E_1 - E_2 \in \Sigma$, so that

$$\begin{aligned}\mu(E_1 - E_2) &= \int_{\Omega_1} \mu_2(E_1(\omega_1) - E_2(\omega_1))\mu_1(d\omega_1)\\ &= \int_{\Omega_1} \mu_2(E_1(\omega_1))\mu_1(d\omega_1) - \int_{\Omega_1} \mu_2(E_2(\omega_1))\mu_1(d\omega_1)\end{aligned}$$

[since μ_2 is finite and $E_1(\omega_1) \supset E_2(\omega_1)$],

$$\begin{aligned}&= \int_{\Omega_2} \mu_1(E(\omega_2))\mu_2(d\omega_2) - \int_{\Omega_2} \mu_1(E_2(\omega_2))\mu_2(d\omega_2), \qquad \text{by (ii)}\\ &= \int_{\Omega_2} \mu_1(E_1(\omega_2) - E_2(\omega_2))\mu_2(d\omega_2),\end{aligned}$$

since μ_1 is a finite measure and $E_1(\omega_2) \supset E_2(\omega_2)$. So $E_1 - E_2 \in \mathscr{C}$. By similar reasoning using the monotone convergence theorem, we conclude that if $E_n \in \mathscr{C}$, $E_1 \subset E_2 \subset \cdots$, then $\bigcup_{n=1}^{\infty} E_n \in \mathscr{C}$. Hence Proposition 2.1.3(b) implies $\mathscr{C} = \Sigma$, so (i) and (ii) are true for all elements of Σ. Then (iii) is an immediate consequence of (ii). This completes the proof.

The preceding result easily extends to σ-finite measures.

Corollary 10. *Let* $(\Omega_i, \Sigma_i, \mu_i)$, $i = 1, 2$, *be σ-finite measure spaces, and* (Ω, Σ, μ) *be their product (which is σ-finite by Discussion* 4). *Then all the statements of Proposition* 9 *hold in this case.*

Let $\Omega_{in} \in \Sigma_i$, $\mu_i(\Omega_{in}) < \infty$, be increasing sequences converging to Ω_i, $i = 1, 2$. Consider $\mathscr{F} = \{E \in \Sigma: E \cap (\Omega_{1n} \times \Omega_{2m}) \in \mathscr{C}$, all $m \geqslant 1$, $n \geqslant 1\}$, where $\mathscr{C}$ is the class of all sets for which (i) and (ii) of Proposition 9 hold. It is easily seen that $\mathscr{F}$ is a λ-class containing $\mathscr{S}$. Hence $\sigma(\mathscr{S}) \subset \mathscr{F} \subset \Sigma$, so that there is equality. The details are left to the reader. (One again uses here the convention that $0 \cdot \infty = 0$, as usual.)

The restriction (of σ-finiteness) on measures cannot be omitted entirely in (17), even though on $\mathscr{S}$, $\mu(A \times B) = \mu_1(A)\mu_2(B)$ holds for any measures. That (17) is false generally for sets which are not necessarily rectangles is seen from the following:

Example 11. Let $\Omega_1 = \Omega_2 = [0, 1]$, $\Sigma_1 = \Sigma_2$ be the Borel σ-algebra, μ_1 be Lebesgue measure, and μ_2 be the counting measure $[\mu_2(A) =$ number of points in A, $A \in \Sigma_2]$. Consider the set $D = \{(\omega_1, \omega_2) \in \Omega: \omega_1 = \omega_2\}$, the diagonal of the unit square Ω. It is not hard to show that D is a Borel set and $D \in \Sigma_1 \otimes \Sigma_2$. In fact, $D \in (\Sigma_1 \times \Sigma_2)_{\sigma\delta}$. However,

$$\int_{\Omega_1} \mu_2(D(\omega_1))\mu_1(d\omega_1) = 1 > 0 = \int_{\Omega_2} \mu_1(D(\omega_2))\mu_2(d\omega_2),$$

since $\mu_2(D(\omega_1)) = 1$ and $\mu_1(D(\omega_2)) = 0$, $D(\omega_i)$ being a singleton. Here μ is not σ-finite (since μ_2 is not) and (17) fails. Note also that our μ on $\Sigma_1 \otimes \Sigma_2$ has the finite subset property, and thus it is not enough to save the result.

In view of this pathology, some authors define μ on $\Sigma_1 \otimes \Sigma_2$ by the equation (17) only for σ-finite measures. However, as seen in Theorem 3, the Carathéodory approach allows us to define the *product measure* μ *directly* without using (17), and we maintain this generality as far as possible.

EXERCISES

0. Let $(\Omega_i, \Sigma_i, \mu_i)$, $i = 1, \ldots, n$ be measure spaces. Show that there is a measure space (Ω, Σ, μ) such that $\Omega = \times_{i=1}^n \Omega_i$, $\Sigma_1 \times \Sigma_2 \times \cdots \times \Sigma_n \subset \Sigma$, and $\mu(A_1 \times \cdots \times A_n) = \mu_i(A_1) \cdots \mu_n(A_n)$ for $A_i \in \Sigma_i$, $i = 1, \ldots, n$. [This is an extension of Theorem 3. Use the fact that $(\Omega_1 \times \Omega_2) \times \Omega_3$ and $\Omega_1 \times (\Omega_2 \times \Omega_3)$ are identifiable.] If $f_i: \Omega_i \to \overline{\mathbb{R}}$ is μ_i-measurable, show that $f = \prod_{i=1}^n f_i$ is μ-measurable. Can we replace "measurable" by "integrable" in both places here? (For a slightly strengthened version, see Exercise 2.1.)

1. Let $(\mathbb{R}, \mathscr{B})$ be the Borelian line. Show that $(\mathbb{R}^2, \mathscr{B} \otimes \mathscr{B})$, the product measurable space, is Borelian, i.e., every Borel set of the plane $\mathbb{R}^2$ is in $\mathscr{B} \otimes \mathscr{B}$. [It suffices to show that each open or closed set is in this class.]

2. Complete the details of proof of Corollary 10. The result is false if μ_i's are only assumed to have the finite subset property. Show this, by the following example: $\Omega_1 = \Omega_2 = [0,1]$, $\Sigma_1 = \Sigma_2 =$ the Borel σ-algebra; let $A \in \mathscr{P}(\Omega_1) - \Sigma_1$ (e.g. a non-Lebesgue-measurable A will do). So A is infinite. Let $\mu_1 = \mu_2: \Sigma_i \to \overline{\mathbb{R}}^+$ be $\mu_i(B) = \Sigma_{x \in B}\mu_i(\{x\})$ with $\mu_i(\{x\}) = 10$, $x \in A$, $= 1$, $x \in \Omega - A$. If $\mu = \mu_1 \otimes \mu_2$, $D = \{(x, x): x \in \Omega_1\}$, then $D \in \Sigma_1 \otimes \Sigma_2$, and if we let $f: x \mapsto \mu_2(D(x)) = \chi_A(x) + 1$, then f is *not* a Borel function.

3. Extend Proposition 9 if there are finitely many σ-finite measure spaces. [Use Corollary 10, and note that $(\Omega_1 \times \Omega_2) \times \Omega_3$ can be identified again with $\Omega_1 \times (\Omega_2 \times \Omega_3)$, etc.]

4. Let $\Omega = [0,1]$, and (Ω, Σ, μ) be the Lebesgue unit interval. Let $\tilde{\Omega} = \Omega \times \Omega$, $\tilde{\Sigma} = \Sigma \otimes \Sigma$, and $\tilde{\mu}$ be the product measure on $\tilde{\Sigma}$, which is $\mu \otimes \mu$, as defined by Theorem 3. Show that $(\tilde{\Omega}, \tilde{\Sigma}, \tilde{\mu})$ is not complete, even though (Ω, Σ, μ) is complete. [The completion $\hat{\Sigma}$ of $\tilde{\Sigma}$ for $\tilde{\mu}$ gives the two dimensional Lebesgue measure $\hat{\mu}$. Thus, if $A \subset \Omega$ is a Lebesgue nonmeasurable set and $C \subset \Omega$ is the Cantor ternary set, then $\mu(C) = 0$, so that $\hat{\mu}(A \times C) = 0$, and $A \times C \in \hat{\Sigma} - \tilde{\Sigma}$. So the product measure space of complete (even probability) measure spaces need not be complete.]

5. Show that the result of Exercise 1 above holds if $\mathbb{R}$ is replaced by an arbitrary separable metric space (with $\mathscr{B}$ as its Borel σ-algebra), and construct an example to show that it does not hold for general (even compact) topological (Borel) measurable spaces $(\Omega_i, \mathscr{B}_i)$, $i = 1, 2$.

6. In Theorem 3, the product measure μ on $\Sigma_1 \otimes \Sigma_2$ is not necessarily unique without some conditions on μ_1, μ_2. Show, however, that if μ has the finite subset property on $\Sigma_1 \otimes \Sigma_2$, or if $\mu_1, \mu_2 \neq 0$ and have finite subset properties, then the extension is unique. [Verify that $\mu(E) = \sup\{\mu((A \times B) \cap E): A \times B \in \Sigma_1 \times \Sigma_2,\ \mu(A) < \infty,\ \mu_2(B) < \infty\}$ in both cases with $E \in \Sigma_1 \otimes \Sigma_2$.]

7. Let (Ω, Σ, μ) be the product measure space of $(\Omega_i, \Sigma_i, \mu_i)$, $i = 1, 2$, so that $\Omega = \Omega_1 \times \Omega_2$, $\Sigma = \Sigma_1 \otimes \Sigma_2$, $\mu = \mu_1 \otimes \mu_2$. Let $\hat{\Sigma}$ and $\hat{\mu}$ denote completions of Σ, μ. If $f: \Omega \to \overline{\mathbb{R}}$ is $\hat{\mu}$-measurable (or equivalently measurable for $\hat{\Sigma}$) and $\hat{\mu}$-integrable, then show that there exists a $g: \Omega \to \overline{\mathbb{R}}$ which is measurable for Σ, satisfies $f = g$ a.e.$(\hat{\mu})$, and satisfies

$$\int_\Omega f\,d\hat{\mu} = \int_\Omega g\,d\mu \left(= \int_\Omega g\,d\hat{\mu} \right).$$

[First note that f vanishes outside a set $\tilde{\Omega} \in \Sigma$ of σ-finite $\hat{\mu}$-measure. Hence we may suppose $\hat{\mu}$ is σ-finite, restricting to $\hat{\Sigma}(\tilde{\Omega})$ if necessary. But then μ is also σ-finite on the class $\Sigma(\tilde{\Omega})$, since each set in $\hat{\Sigma}$ differs from one in Σ by a μ-null set. Now $\Sigma \subset \hat{\Sigma}$ and $\mu = \hat{\mu}|\Sigma$. So by Theorem 5.5.15, the conditional expectation $E(\cdot|\Sigma)$ exists. Let $g = E(f|\Sigma)$. Then by Definition 5.5.14, the desired statement holds. Verify that the statement of this exercise holds for all measure spaces satisfying a σ-finiteness condition if f is $\hat{\mu}$-measurable. It is not difficult to prove the existence of a

$g\,[=f$ a.e.$(\hat{\mu})]$ which is measurable for Σ when f is $\hat{\Sigma}$-measurable without being integrable, because of the special relationship between Σ and $\hat{\Sigma}$. The above equality between the integrals holds if $f \geqslant 0$ (need not be integrable), and μ is σ-finite or even localizable. The argument with $E(\cdot|\Sigma)$ works again. [In light of this result, the measure μ of Exercise 4 can be understood as the (incomplete) *Borel-Lebesgue measure* and its completion $\hat{\mu}$ as the *planar Lebesgue* measure. Similarly, if $(\mathbb{R}^n, \Sigma_1, \mu_1)$ and $(\mathbb{R}^m, \Sigma_2, \mu_2)$ are the Lebesgue spaces, then $(\mathbb{R}^{m+n}, \overline{\Sigma_1 \otimes \Sigma_2, \mu_1 \otimes \mu_2})$, the completed product space, is the $m+n$ dimensional *Lebesgue measure space*.]

8. If $f: \Omega_1 \mapsto \overline{\mathbb{R}}^+$ is a mapping, then one says $U_f = \{(\omega, y): 0 < y \leqslant f(\omega)\}$ and $L_f = \{(\omega, y): 0 < y < f(\omega)\}$ are the *upper* and *lower ordinate sets* of f. Verify the following assertions:

(a) $0 \leqslant f \leqslant g \Rightarrow U_f \subset U_g$ and $L_f \subset L_g$.

(b) $0 \leqslant f_n \uparrow f$ $(f_n \downarrow f \geqslant 0)$ implies $L_f = \bigcup_{n=1}^{\infty} L_{f_n}$ $(U_f = \bigcap_{n=1}^{\infty} U_{f_n})$.

(c) If Σ_1 is a σ-algebra of Ω_1, then f is measurable relative to Σ_1 iff $U_f \in \Sigma_1 \otimes \mathscr{B}$ (or iff $L_f \in \Sigma_1 \otimes \mathscr{B}$), where $\mathscr{B}$ is the Borel σ-algebra of $\overline{\mathbb{R}}^+$. [If f is measurable, then use the structure theorem and the fact that U_f for a simple f is in $\Sigma_1 \otimes \mathscr{B}$. Conversely, if $U_f \in \Sigma_1 \otimes \mathscr{B}$, then each set $\{(\omega, y): (\omega, (y/k) + a) \in U_f,\ y > 0\}$ is in $\Sigma_1 \otimes \mathscr{B}$ for each $a > 0$, $k \geqslant 1$. Hence their union which is $\{(\omega, y): f(\omega) > a,\ y > 0\}$ is also in $\Sigma_1 \times \mathscr{B}$ for each a. So $\{\omega: f(\omega) > a\} \in \Sigma_1$, $a > 0$. Similarly one can deduce that $\{\omega: f(\omega) \geqslant a\} \in \Sigma_1$ if $L_f \in \Sigma_1 \otimes \mathscr{B}$.]

(d) If μ_1 is a measure on Σ_1, and λ is the Lebesgue measure on $\mathscr{B}$, let $\mu = \mu_1 \otimes \lambda$ be the measure on $\Sigma_1 \otimes \mathscr{B}$, and μ^* be the generated outer measure by $(\mu_1 \times \lambda, \Sigma_1 \times \mathscr{B})$. Then establish that

$$\int_{\Omega}^{*} f\, d\mu_1 = \mu^*(U_f),$$

where U_f is the upper ordinate set of f, and if f is measurable, then we can erase the "stars" in the above equation. Thus one has $\mu(U_f) = \mu(L_f)$. [Use our definition of the integral, and Proposition 4.2.1. Here U_f is just the region under f of Section 4.1.]

9. If $f: \Omega \to \overline{\mathbb{R}}$, then the set $G(f) = \{(\omega, y): y = f(\omega)\}$ is called the *graph* of f. If $(\Omega_1, \Sigma_1, \mu_1)$ is a σ-finite complete measure space, $(\mathbb{R}^+, \mathscr{B}, \lambda)$ is the Lebesgue measure space, and $f: \Omega \to \overline{\mathbb{R}}^+$ is measurable for Σ_1, show that $G(f)$ is in $\Sigma_1 \otimes \mathscr{B}$ and that if $\mu = \mu_1 \otimes \lambda$, then $\mu(G(f)) = 0$. [Use Proposition 9 or Corollary 10.] Is the converse implication true?

10. Let $(\Omega_i, \Sigma_i, \mu_i)$, $i = 1, 2, 3$, be measure spaces each with the finite subset property and no μ_i equal to zero. Show that the product measures $(\mu_1 \otimes \mu_2) \otimes \mu_3$ and $\mu_1 \otimes (\mu_2 \otimes \mu_3)$ are the same, so that the parentheses can be removed. (This extends Exercise 2. [Use Exercise 6 in its proof.] The uniqueness questions are discussed by Luther (1967).)

11. Let (Ω, Σ) be a measurable space, $(\Omega \times \Omega, \Sigma \otimes \Sigma)$ its product measurable space, and $D = \{(\omega, \omega): \omega \in \Omega\}$ the diagonal of $\Omega \times \Omega$. Show that $D \in \Sigma \otimes \Sigma$ iff Σ is *countably separated*, which means that there is a countable sequence $\{A_n: n \geqslant 1\}$ of subsets of Ω such that $A_n \in \Sigma$, $n \geqslant 1$, and for any pair ω_1, ω_2 in Ω there is some A_n such that $\omega_1 \in A_n$ but $\omega_2 \notin A_n$. Hence $\{\omega\} \in \Sigma$, $\omega \in \Omega$.

6.2. THE FUBINI-STONE AND TONELLI THEOREMS

The preceding work allows us to establish the fundamental Fubini-Stone theorem. This was proved originally by G. Fubini in 1907 for $\mathbb{R}^{m+n}$; the more refined version, as given here, is due to M. H. Stone in the late 1940s. An essential role in this extension is played by Lemma 1.7.

The basic result is thus the following:

Theorem 1 (Fubini and Stone). *Let* $(\Omega_i, \Sigma_i, \mu_i)$, $i = 1, 2$, *be measure spaces, and* (Ω, Σ, μ) *be their product, as given by Theorem* 1.3. *If* $f: \Omega \to \overline{\mathbb{R}}$ *is measurable for* Σ *and* μ*-integrable, then*:

(i) $g: \omega_1 \mapsto \int_{\Omega_2} f(\omega_1, \omega_2)\mu_2(d\omega_2)$ *is a.e. equal to a* μ_1*-integrable function on* Ω_1,

(ii) $h: \omega_2 \mapsto \int_{\Omega_1} f(\omega_1, \omega_2)\mu_1(d\omega_1)$ *is a.e. equal to a* μ_2*-integrable function on* Ω_2, *and*

(iii) *the following equation holds*:

$$\int_{\Omega} f\, d\mu = \int_{\Omega_1}\left[\int_{\Omega_2} f(\omega_1, \omega_2)\mu_2(d\omega_2)\right]\mu_1(d\omega_1)$$

$$= \int_{\Omega_2}\left[\int_{\Omega_1} f(\omega_1, \omega_2)\mu_1(d\omega_1)\right]\mu_2(d\omega_2).$$

Proof. By hypothesis $\int_{\Omega}|f|\, d\mu < \infty$. So given $\epsilon = 2^{-n}$, there exists a simple function $g_n = \sum_{i=1}^{n} a_i^n \chi_{A_i^n}\chi_{B_i^n}$, $A_i^n \times B_i^n \in \Sigma_1 \times \Sigma_2$, such that

$$\int_{\Omega}|f - g_n|\, d\mu < 2^{-n} \qquad \text{(cf. Proposition 1.8).} \tag{1}$$

The sequence $\{g_m: m \geqslant 1\}$ is Cauchy in $L^1(\mu)$, since

$$\int_{\Omega}|g_m - g_{m'}|\, d\mu \leqslant \int_{\Omega}|g_m - f|\, d\mu + \int_{\Omega}|f - g_{m'}|\, d\mu$$

$$\leqslant 2^{-m} + 2^{-m'} \to 0$$

as $m, m' \to \infty$. We now assert that $\int_{\Omega_2}|f - g_m|(\omega_1, \omega_2)\mu_2(d\omega_2) \to 0$ as $m \to \infty$, for a.a. ω_1, and is μ_1-equal to a Σ_1-measurable function. Since $g - g_m$ is measurable for Σ, and hence $|f - g|(\omega_1, \cdot)$ is measurable for Σ_2 by Lemma 1.6(ii), the above integral relative to μ_2 is defined, and its value for each ω_1 is to be proved μ_1-measurable. For this, consider the number

$$0 \leqslant \tilde{g}(\omega_1) = \sum_{n=1}^{\infty} \int_{\Omega_2} |f - g_n|(\omega_1, \omega_2)\mu_2(d\omega_2), \qquad \omega_1 \in \Omega_1.$$

Then we have

$$\int_{\Omega_1}^{*} \tilde{g}(\omega_1)\mu_1(d\omega_1) \leqslant \sum_{n=1}^{\infty} \int_{\Omega_1}^{*} \left[\int_{\Omega_2} |f - g_n|(\omega_1, \omega_2)\mu_2(d\omega_2) \right] \mu_1(d\omega_1)$$

(since the "star" integral is σ-subadditive)

$$\leqslant \sum_{n=1}^{\infty} \int_{\Omega}^{*} |f - g_n|\, d\mu, \qquad \text{by Lemma 1.7}$$

$$\leqslant \sum_{n=1}^{\infty} 2^{-n} = 1, \qquad \text{by (1)}.$$

Hence $0 \leqslant \tilde{g}(\omega_1) < \infty$ for a.a. ω_1, and $\int_{\Omega}^{*}|f - g_n|\, d\mu \to 0$ as $n \to \infty$. This implies also $\int_{\Omega_2}|f - g_n|(\omega_1, \omega_2)\mu_2(d\omega_2) \to 0$ as $n \to \infty$, a.e.$[\mu_1]$. Moreover, $\{g_n(\omega_1, \cdot): n \geqslant 1\}$ is Cauchy in $L^1(\mu_2)$ for a.a. ω_1, since

$$\int_{\Omega_2} |g_m - g_{m'}|(\omega_1, \omega_2)\mu_2(d\omega_2) \leqslant \int_{\Omega_2} |f - g_m|(\omega_1, \mu_2)\mu_2(d\omega_2)$$

$$+ \int_{\Omega_2} |f - g_{m'}|(\omega_1, \omega_2)\mu_2(d\omega_2)$$

and the right side $\to 0$. Hence $g_m(\omega_1, \cdot) \to u(\omega_1, \cdot)$ in $L^1(\mu_2)$, by completeness of the latter, and $u(\omega_1, \cdot) \in L^1(\mu_2)$. Consequently

$$\int_{\Omega_2} |f(\omega_1, \omega_2) - u(\omega_1, \omega_2)|\mu_2(d\omega_2)$$

$$\leqslant \int_{\Omega_2} |f - g_n|(\omega_1, \omega_2)\mu_2(d\omega_2) + \int_{\Omega_2} |g_n(\omega_1, \omega_2) - u(\omega_1, \omega_2)|\mu_2(d\omega_2) \to 0,$$

so that $f(\omega_1, \cdot) = u(\omega_1, \cdot)$ a.e.$[\mu_2]$, and hence $f(\omega_1, \cdot) \in L^1(\mu_2)$.

Define $g(\omega_1) = \int_{\Omega_2} f(\omega_1, \omega_2)\mu_2(d\omega_2)$, $\omega_1 \in \Omega_1$. We note that g is μ_1-measurable. Indeed, we already have, by Lemma 1.7,

$$\int_{\Omega_1}^{*} |g(\omega_1)|\mu_1(d\omega_1) \leqslant \int_{\Omega_1}^{*}\left[\int_{\Omega_2}|f(\omega_1, \omega_2)|\mu_2(d\omega_2)\right]\mu_1(d\omega_1)$$

$$\leqslant \int_{\Omega}|f|\,d\mu < \infty.$$

Further,

$$\int_{\Omega_1}^{*}\left|g(\omega_1) - \int_{\Omega_2} g_n(\omega_1, \omega_2)\mu_2(d\omega_2)\right|\mu_1(d\omega_1)$$

$$= \int_{\Omega_1}^{*}\left|\int_{\Omega_2}[f(\omega_1, \omega_2) - g_n(\omega_1, \omega_2)]\,\mu_2(d\omega_2)\right|\mu_1(d\omega_1)$$

$$\leqslant \int_{\Omega_1}^{*}\int_{\Omega_2}|f - g_n|(\omega_1, \omega_2)\mu_2(d\omega_2)\mu_1(d\omega_1)$$

$$\leqslant \int_{\Omega}|f - g_n|\,d\mu < 2^{-n} \to 0, \tag{2}$$

by Lemma 1.7 and (1). But by the form of g_n, $\int_{\Omega_2} g_n(\cdot, \omega_2)\mu_2(d\omega_2)$ is μ_1-measurable [cf. Proposition 1.9(i)], and (2) implies $g \in L^1(\mu_1)$, as well as $g(\omega_1) = \lim_n \int_{\Omega_2} g_n(\omega_1, \omega_2)\mu_2(d\omega_2)$ for a.a. ω_1. Thus g is μ_1-measurable and integrable. Since $L^1(\mu_1) = L^1(\Omega_1, \Sigma_1, \mu_1)$, this proves (i).

The limit in (2) also implies

$$\int_{\Omega_1}\left[\int_{\Omega_2} f(\omega_1, \omega_2)\mu_2(d\omega_2)\right]\mu_1(d\omega_1) = \int_{\Omega_1} g(\omega_1)\mu_1(d\omega_1)$$

$$= \int_{\Omega_1}\lim_n\left[\int_{\Omega_2} g_n(\omega_1, \omega_2)\mu_2(d\omega_2)\right]\mu_1(d\omega_1)$$

$$= \lim_n \int_{\Omega_1}\left[\int_{\Omega_2} g_n(\omega_1, \omega_2)\mu_2(d\omega_2\right]\mu_1(d\omega_1)$$

(since the sequence is Cauchy)

$$= \lim_n \int_{\Omega} g_n\,d\mu, \qquad \text{by Proposition 1.9(ii)},$$

$$= \int_{\Omega} f\,d\mu, \qquad \text{by (1)}. \tag{3}$$

This proves the first half of (iii). Since the argument is symmetric in μ_1, μ_2, the same computations imply (ii) and the other half of (iii). Thus (iii) holds as stated, and the theorem is completely proved.

We now explain by (counter-)examples the optimality of this important result in order to understand it better. If μ_1, μ_2 are assumed σ-finite, then the original result of Fubini is obtained. The most important aspect of the hypothesis is not the (joint) measurability of f, but its integrability relative to the product measure μ. There is no easy recipe to verify this hypothesis. Even the joint measurability is not always easy to check, though in a particular problem one usually has enough information (e.g., of a topological nature) to overcome this hurdle. It can happen that there is inequality between the three quantities of Theorem 1(iii) or there is equality between two of the three but not for all. (See Exercise 2 for a general inequality.)

The following examples explain these difficulties:

Example I. Consider the situation of measures given in the example near the end of Section 6.1. If D is the diagonal of the unit square, μ_1 is Lebesgue measure, and μ_2 is counting measure, then $\mu = \mu_1 \otimes \mu_2$ is a measure on the Borel σ-algebra of the unit square, $f = \chi_D$ is jointly measurable, and

$$\int_{\Omega_1}\int_{\Omega_2} f\,d\mu_2\,d\mu_1 = 1 \neq 0 = \int_{\Omega_2}\left[\int_{\Omega_1} f\,d\mu_1\right] d\mu_2.$$

Hence $\int_\Omega f\,d\mu = +\infty$ must hold in this example. Thus all three members of Theorem 1(iii) are different.

Example II. Let $\Omega_1 = \Omega_2 = \mathbb{R}^+$, $\Sigma_1 = \Sigma_2$ be the Borel σ-algebra, and $\mu_1 = \mu_2$ with $\mu_1(\{\omega\}) = 1$ if ω is an integer in $\mathbb{R}^+$, $= 0$ otherwise. Thus μ_1 is σ-finite, and so $\mu = \mu_1 \otimes \mu_2$ on $\Omega = \mathbb{R}^+ \times \mathbb{R}^+$ is σ-finite such that $\mu(\{\omega_1, \omega_2\}) = 1$ if (ω_1, ω_2) is a lattice point with integer coordinates, and $= 0$ otherwise. Let $f\colon \Omega \to \mathbb{R}^+$ be defined as $f(\omega_1, \omega_2) = \alpha - \beta^n$ if $\omega_1 = \omega_2 = n$, $= -\alpha + \beta^n$ if $\omega_1 = n$, $\omega_2 = n + k$, and $= 0$ otherwise, where $k \geqslant 1$ is a fixed integer. Here $\alpha > 1 > \beta > 0$ are given numbers. It is trivial that f is (jointly) measurable for $\Sigma = \Sigma_1 \otimes \Sigma_2$, and that $\int_{\Omega_1} f(\omega_1, \omega_2)\mu_1(d\omega_1) = (\alpha - \beta^n) + (-\alpha + \beta^n)$ $[= 0]$ if $\omega_1 = n$, and $= 0$ otherwise. Similarly $\int_{\Omega_2} f(\omega_1, \omega_2)\mu_2(d\omega_2) = 0$, so that

$$\int_{\Omega_1}\left[\int_{\Omega_2} f(\omega_1, \omega_2)\mu_2(d\omega_2)\right]\mu_1(d\omega_1)$$

$$= 0 = \int_{\Omega_2}\left[\int_{\Omega_1} f(\omega_1, \omega_2)\mu_1(d\omega_1)\right]\mu_2(d\omega_2).$$

But $\int_\Omega f\,d\mu$ does not exist, and $\int_\Omega |f|\,d\mu = +\infty$. Thus there is equality between the last two quantities of Theorem 1(iii) but not for all.

It can also be verified that, if f is *not* jointly measurable but $f_{\omega_1}, f_{\omega_2}$, the sections, are measurable, the functions $\omega_1 \mapsto \int_{\Omega_2} f_{\omega_1}\,d\mu_2$, and $\omega_2 \mapsto \int_{\Omega_1} f_{\omega_2}\,d\mu_1$ are measurable, the iterated integrals exist, the product measure space is finite, and conclusion (iii) is invalid. We give this below as Exercise 4.

If we restrict the measures and functions admitted in Theorem 1 somewhat, the integrability condition may be relaxed. This is the substance of the following complementary result:

Theorem 2 (Tonelli). *Let* $(\Omega_i, \Sigma_i, \mu_i)$, $i = 1, 2$, *be measure spaces, and* (Ω, Σ, μ) *be their product as before. If* $f: \Omega \to \overline{\mathbb{R}}^+$ *is a (jointly) measurable function, suppose that either (i) both* μ_1, μ_2 *are σ-finite, or* (ii) *there exists a sequence of μ-simple (= μ-integrable step) functions* $\{f_n: n \geq 1\}$ *such that* $0 \leq f_n \uparrow f$ *a.e.*$[\mu]$. *Then*

$$\int_\Omega f\,d\mu = \int_{\Omega_1}\left[\int_{\Omega_2} f(\omega_1, \omega_2)\mu_2(d\omega_2)\right]\mu_1(d\omega_1)$$

$$= \int_{\Omega_2}\left[\int_{\Omega_1} f(\omega_1, \omega_2)\mu_1(d\omega_1)\right]\mu_2(d\omega_2),$$

where all quantities are finite and equal or all are infinite. Consequently, if one of the iterated integrals is finite, then f is μ-integrable.

Proof. Suppose (i) holds, so that $\mu = \mu_1 \otimes \mu_2$ is also σ-finite. Since f is measurable, by the structure theorem there exists a sequence $0 \leq f_n \uparrow f$, pointwise, and since μ is σ-finite, we may choose f_n to be μ-integrable. Indeed, by σ-finiteness of μ, there is a sequence $E_n \subset E_{n+1}$, $E_n \in \Sigma$, $\Omega = \bigcup_{n=1}^\infty E_n$, and $\mu(E_n) < \infty$, $n \geq 1$. Now for each $n \geq 1$ and $k \geq 1$, let

$$F_{k,n} = \left\{\omega \in E_n: \frac{k-1}{2^n} \leq f(\omega) < \frac{k}{2^n}\right\}, \qquad k = 1, 2, \ldots, n2^n. \tag{4}$$

Hence $F_{k,n} \in \Sigma$, $\mu(F_{k,n}) < \infty$, disjoint. Let $f_n = \sum_{k=1}^{n2^n}[(k-1)/2^n]\chi_{F_{k,n}}$. Then $0 \leq f_n \uparrow f$ pointwise, and each f_n vanishes outside E_n. Clearly f_n is measurable for Σ. Thus the sequence $\{f_n: n \geq 1\}$ satisfies hypothesis (ii). So we only need to establish the result in this case.

By the Lebesgue monotone convergence theorem one has

$$\int_\Omega f\,d\mu = \lim_n \int_\Omega f_n\,d\mu$$

$$= \lim_n \int_{\Omega_1}\left[\int_{\Omega_2} f_n(\omega_1, \omega_2)\mu_2(d\omega_2)\right]\mu_1(d\omega_1) \tag{5}$$

(by Theorem 1, since f_n is μ-integrable)

$$= \int_{\Omega_1} \lim_n \left[\int_{\Omega_2} f_n(\omega_1, \omega_2) \mu_2(d\omega_2) \right] \mu_1(d\omega_1)$$

$$= \int_{\Omega_1} \left[\int_{\Omega_2} \lim_n f_n(\omega_1, \omega_2) \mu_2(d\omega_2) \right] \mu_1(d\omega_1)$$

(by the monotone convergence theorem)

$$= \int_{\Omega_1} \left[\int_{\Omega_2} f(\omega_1, \omega_2) \mu_2(d\omega_2) \right] \mu_1(d\omega_1). \tag{6}$$

Interchanging the order in (5), which is permissible by Theorem 1, we get the second member. The last statement is clear from (6). This completes the proof.

Remark. In (ii) we could simply ask for any sequence $0 \leqslant f_n \uparrow f$ a.e.[μ], such that f_n is μ-integrable. The sequence need not be μ-simple. In this case (i) is reduced by simply defining $f_n = \chi_{E_n} \min(f, n)$, $n \geqslant 1$. Then $0 \leqslant f_n \uparrow f$ and f_n is μ-integrable. The rest of the argument is identical in arriving at (6).

The following consequence of the above theorem is often conveniently employed in applications when the σ-finiteness hypothesis is available:

Corollary 3. *Let $(\Omega_i, \Sigma_i, \mu_i)$, $i = 1, 2$, be σ-finite measure spaces, and (Ω, Σ, μ) be their product, as in the theorem. If $f: \Omega \to \overline{\mathbb{R}}^+$ is measurable for Σ, and either $\int_{\Omega_1}[\int_{\Omega_2} |f(\omega_1, \omega_2)| \mu_2(d\omega_2)] \mu_1(d\omega_1) < \infty$, or the integral with Ω_1, Ω_2 (and μ_1, μ_2) interchanged is finite, then both are finite and f is integrable, so that one has*

$$\int_\Omega f \, d\mu = \int_{\Omega_1} \left[\int_{\Omega_2} f(\omega_1, \omega_2) \mu_2(d\omega_2) \right] \mu_1(d\omega_1)$$

$$= \int_{\Omega_2} \left[\int_{\Omega_1} f(\omega_1, \omega_2) \mu_1(d\omega_1) \right] \mu_2(d\omega_2).$$

All members are finite.

In fact, by the last part of Theorem 2, $|f|$ being measurable for Σ, it is μ-integrable. But then Theorem 1 can be applied to f, and the conclusion follows.

We show by an explicit evaluation how these results are applied in practice. The problem to be considered is the convolution operation, which is one of the most useful and important techniques in analysis.

Let $(\mathbb{R}, \Sigma, \lambda)$ be the Lebesgue measure space of the line, and consider $L^1(\mathbb{R})$ $[= L^1(\mathbb{R}, \Sigma, \lambda)]$. If f, g are in $L^1(\mathbb{R})$, we consider the new function $h: (x, y) \mapsto f(x-y)g(y)$. One needs to show that (i) h is jointly measurable on $(\mathbb{R}^2, \widetilde{\Sigma \otimes \Sigma}, \widetilde{\lambda \otimes \lambda})$ and (ii) $h_1: x \mapsto \int_{\mathbb{R}} h(x, y)\lambda(dy)$ is well defined and is measurable for Σ. If these two facts are granted, then

$$\begin{aligned}\int_{\mathbb{R}} |h_1(x)|\lambda(dx) &\leqslant \int_{\mathbb{R}}\int_{\mathbb{R}} |h(x, y)|\lambda(dx)\lambda(dy)\\ &= \int_{\mathbb{R}}\int_{\mathbb{R}} |f(x-y)||g(y)|\lambda(dx)\lambda(dy)\\ &= \int_{\mathbb{R}} |g(y)|\left(\int_{\mathbb{R}} |f(x-y)|\lambda(dx)\right)\lambda(dy)\\ &= \|g\|_1\|f\|_1 < \infty. \end{aligned} \tag{7}$$

Hence $h_1 \in L^1(\mathbb{R})$. The function h_1 is called the *convolution* of f and g, and denoted $h_1 = f * g$. Using a change of variables and the fact that $\lambda(\cdot)$ is translation invariant, one sees immediately that $f * g = g * f$. The convolution is also clearly associative, and thus has the properties of a multiplication. Thus we infer the important property that $L^1(\mathbb{R})$ *is closed under convolution, and that this multiplication operation satisfies*

$$\|f * g\|_1 \leqslant \|f\|_1\|g\|_1, \qquad \text{by (7)}. \tag{8}$$

This fact makes the Banach space $L^1(\mathbb{R})$ an algebra, and it is tremendously important in analysis. Also it is a beautiful illustration of Fubini's theorem. We thus establish (i) and (ii).

(i): If $v: (x, y) \mapsto x - y$, then $v: \mathbb{R}^2 \to \mathbb{R}$ is a continuous function, and if $f: \mathbb{R} \to \mathbb{R}$ is Borel measurable, then $f \circ v: \mathbb{R}^2 \to \mathbb{R}$ is a Borel function, as a composition, since $(f \circ v)^{-1}(\mathscr{B}) = v^{-1}(f^{-1}(\mathscr{B})) \subset \mathscr{B}$. But in $L^1(\mathbb{R})$ we admit Lebesgue-measurable functions, so that $f^{-1}(\mathscr{B}) \subset \Sigma$, the Lebesgue σ-algebra, and hence one has to show that $v^{-1}(\Sigma) \subset \widetilde{\Sigma \otimes \Sigma}$, a nontrivial fact. If this is established, the measurability of $f * g$ follows without much difficulty. Let us therefore consider the problem with v, and then treat $f \circ v$.

By the fact that f is Lebesgue measurable, we get $C = f^{-1}(B) \in \Sigma$ for each Borel set B, i.e., $B \in \mathscr{B}$. However, each such element C of Σ can be expressed as $C = B_1 \cup N_1$, where $B_1 \in \mathscr{B}$ and $N_1 \in \Sigma$, $\lambda(N_1) = 0$ (cf. Corollary 2.2.18). Thus $(f \circ v)^{-1}(B) = v^{-1}(f^{-1}(B)) = v^{-1}(B_1 \cup N_1) = v^{-1}(B_1) \cup v^{-1}(N_1)$ and $v^{-1}(B_1) \in \Sigma \otimes \Sigma$ $[= \mathscr{B}(\mathbb{R}^2)]$. So $v^{-1}(B_1) \in \widetilde{\Sigma \otimes \Sigma}$, the Lebesgue class of $\mathbb{R}^2$. It suffices therefore to show that $v^{-1}(N_1)$ is a Lebesgue null set in $\mathbb{R}^2$. By σ-finiteness, it is enough to show that $v^{-1}(N_1) \cap A$ is a null set for each rectangle $A \in \mathscr{B} \times \mathscr{B}$ of finite planar Lebesgue measure. Thus let $A = [a, b] \times [c, d]$, $\lambda \otimes \lambda(A) > 0$. By Definition 2.2.2, given $\epsilon > 0$, there exists a set

$D_\epsilon \in \mathscr{S} \subset \mathscr{B}$ (where $\mathscr{S}$ is the semiring generating $\mathscr{B}$) such that $N_1 \subset D_\epsilon$ and $\lambda(D_\epsilon) < \epsilon/(b-a)$. Now denoting by μ $(= \widetilde{\lambda \otimes \lambda})$ the planar Lebesgue measure, we have, since $v^{-1}(N_1) \cap A \subset v^{-1}(D_\epsilon) \cap A$ (denoting by μ^* the Lebesgue outer measure on $\mathbb{R}^2$),

$$\begin{aligned}
0 &\leqslant \mu^*\big(v^{-1}(N_1) \cap A\big) \\
&\leqslant \mu\big(v^{-1}(D_\epsilon) \cap A\big) \\
&= \int_\Omega \chi_{v^{-1}(D_\epsilon)\cap A}\, d\mu \qquad (\Omega = \mathbb{R}^2 \text{ here}) \\
&= \int_{\mathbb{R}}\int_{\mathbb{R}} \chi_{v^{-1}(D_\epsilon)\cap A}(x, y)\lambda(dx)\lambda(dy), \qquad \text{by Theorem 2} \\
&= \int_{\mathbb{R}}\int_{\mathbb{R}} \chi_A(x, y)\chi_{D_\epsilon} \circ v(x, y)\lambda(dx)\lambda(dy) \\
&= \int_{\mathbb{R}} \chi_{[a,b]}(x)\left[\int_{\mathbb{R}} \chi_{[c,d]}(y)\chi_{D_\epsilon}(x-y)\lambda(dy)\right]\lambda(dx) \\
&\leqslant \int_a^b\left[\int_{-\infty}^{\infty} \chi_{D_\epsilon}(x-y)\lambda(dy)\right]\lambda(dx) \\
&= \int_a^b\left[\int_{-\infty}^{\infty} \chi_{D_\epsilon}(y)\lambda(dy)\right]\lambda(dx),
\end{aligned}$$

[by the translation invariance of λ (cf. Theorem 2.2.17)]

$$= (b-a)\lambda(D_\epsilon) < \epsilon.$$

Since $\epsilon > 0$ is arbitrary, $v^{-1}(N_1) \cap A$ is a μ-null set for all such $A \in \Sigma_1 \otimes \Sigma_2$. By the finite subset property of μ, this implies that $v^{-1}(N_1)$ is μ-null (cf. Lemma 2.3.3), proving (i).

Since g is λ-measurable, $(x, y) \mapsto (f \circ v)(x, y)g(x)$ is μ-measurable also. But the computation in (7) shows that the hypothesis of Corollary 3, namely $\int_{\mathbb{R}}\int_{\mathbb{R}}|f(x-y)g(y)|\lambda(dx)\lambda(dy) < \infty$, is satisfied. Hence (ii) holds, and (8) follows. Thus all the statements on convolution are hereby established. We summarize this result as follows.

Proposition 4. *Let* $(\mathbb{R}, \Sigma, \lambda)$ *be the Lebesgue line, and* $f, g \in L^1(\mathbb{R})$. *Then the mapping* $(x, y) \mapsto f(x-y)g(y)$ *is (jointly)* $\widetilde{\lambda \otimes \lambda}$*-measurable and is in* $L^1(\mathbb{R}^2)$. *Moreover, the convolution of* f, g, *given by*

$$(f * g)(x) = \int_{\mathbb{R}} f(x-y)g(y)\lambda(dy) = \int_{\mathbb{R}} g(x-y)f(y)\, dy = (g * f)(x),$$

exists for a.a. x, *and* $f * g \in L^1(\mathbb{R})$. *Also the inequality* (8) *holds.*

This illustration shows how Theorems 1 and 2 help extend the earlier work to more than one dimension and obtain new results. As a final item of this section we include a (slightly restricted) version of the Lebesgue-Radon-Nikodým theorem for product measures.

Proposition 5. *Let $(\Omega_i, \Sigma_i, \nu_i)$, $i = 1,2$, be measure spaces, $\nu_1 \otimes \nu_2$ with the finite subset property, and $(\Omega_i, \Sigma_i, \mu_i)$, $i = 1,2$, be σ-finite measure spaces. If $\nu_i \ll \mu_i$ then $\nu_1 \otimes \nu_2 \ll \mu_1 \otimes \mu_2$, and if $d\nu_i/d\mu_i = f_i$, $i = 1,2$, are the Radon-Nikodým derivatives (which exist by Theorem 5.4.3), then we have*

$$\frac{d\nu_1 \otimes \nu_2}{d\mu_1 \otimes \mu_2} = f_1 f_2 \qquad \text{a.e.}(\mu_1 \otimes \mu_2). \tag{9}$$

If at least one of the ν_i is orthogonal to μ_i, $i = 1,2$, then (assuming only that $\nu_1 \otimes \nu_2$, $\mu_1 \otimes \mu_2$ have the finite subset property) $\nu_1 \otimes \nu_2 \perp \mu_1 \otimes \mu_2$. Moreover, if the ν_i are σ-finite and $\mu_1 \otimes \mu_2$ has the finite subset property on $\Sigma_1 \otimes \Sigma_2$, then the Lebesgue decomposition of product measures holds:

$$(\nu_1 \otimes \nu_2) = (\nu_1 \otimes \nu_2)_a + (\nu_1 \otimes \nu_2)_s,$$

where $(\nu_1 \otimes \nu_2)_a \ll \mu_1 \otimes \mu_2$, $(\nu_1 \otimes \nu_2)_s \perp \mu_1 \otimes \mu_2$, and

$$(\nu_1 \otimes \nu_2)_a = (\nu_1)_a \otimes (\nu_2)_a \qquad [(\nu_i)_a \ll \mu_i,\ (\nu_i)_s \perp \mu_i,\quad i = 1,2], \tag{10}$$

$$(\nu_1 \otimes \nu_2)_s = (\nu_1)_a \otimes (\nu_2)_s + (\nu_1)_s \otimes (\nu_2)_a + (\nu_1)_s \otimes (\nu_2)_s. \tag{11}$$

Proof. Our hypothesis implies ν_i, μ_i have the finite subset property on Σ_i, $i = 1,2$, because the products $\nu_1 \otimes \nu_2$ and $\mu_1 \otimes \mu_2$ have it and are unique products of the given pairs on $\Sigma_1 \otimes \Sigma_2$ (cf. Exercise 1.6). Let $f: \Omega \to \overline{\mathbb{R}}^+$ be a jointly measurable function which is integrable relative to $\nu_1 \otimes \nu_2$. Then with $\Omega = \Omega_1 \times \Omega_2$, $\nu = \nu_1 \otimes \nu_2$, and $\mu = \mu_1 \otimes \mu_2$, we have

$$\infty > \int_\Omega f\, d\nu$$

$$= \int_{\Omega_1}\left[\int_{\Omega_2} f(\omega_1, \omega_2)\nu_2(d\omega_2)\right]\nu_1(d\omega_1), \qquad \text{by Theorem 1}$$

$$= \int_{\Omega_1}\left[\int_{\Omega_2} f(\omega_1, \omega_2) f_2(\omega_2)\mu_2(d\omega_2)\right] f_1(\omega_1)\mu_1(d\omega_1)$$

(by Theorem 5.4.6, since $\nu_i \ll \mu_i$, $i = 1,2$)

$$= \int_{\Omega_1}\left[\int_{\Omega_2} f(\omega_1, \omega_2) f_1(\omega_1) f_2(\omega_2)\mu_2(d\omega_2)\right]\mu_1(d\omega_1)$$

$$= \int_\Omega (f f_1 f_2)\, d\mu, \tag{12}$$

by Corollary 3, since ff_1f_2 is measurable for $\Sigma_1 \otimes \Sigma_2$ and μ is σ-finite. Now (12) implies that the first and last integrals agree on $L^1(\nu)$. Hence we have $d\nu = f_1f_2\,d\mu$ a.e.$[\mu]$, which is (9) and establishes at the same time that $\nu \ll \mu$.

Next suppose that $\nu_1 \perp \mu_1$. So there is a μ_1-null set A ($\in \Sigma_1$) such that $\nu_1(A^c) = 0$. Since $\mu(A \times \Omega_2) = \mu_1(A)\mu_2(\Omega_2) = 0$, we have

$$(\nu_1 \otimes \nu_2)\big((A \times \Omega_2)^c\big) = (\nu_1 \otimes \nu_2)(A^c \times \Omega_2) = \nu_1(A^c)\nu_2(\Omega_2) = 0.$$

Hence $\nu_1 \otimes \nu_2 \perp \mu_1 \otimes \mu_2$.

If the ν_i are σ-finite, we conclude immediately that $\nu_1 \otimes \nu_2$ is σ-finite, and by hypothesis $\mu_1 \otimes \mu_2$ has the finite subset property, so that by Theorem 5.2.2, $\nu_1 \otimes \nu_2$ admits the Lebesgue decomposition as stated. The corresponding components are obtained by using the bilinearity of the product $\otimes$ as

$$\nu_1 \otimes \nu_2 = \big[(\nu_1)_a + (\nu_1)_s\big] \otimes \big[(\nu_2)_a + (\nu_2)_s\big],$$

which yields (10) and (11) at once. This completes the proof.

Remark. The above result is a restricted version in the sense that it extends Theorem 5.3.3 but not Theorem 5.4.3. The latter result is needed if one wants to use it for products of locally compact groups with Haar measures which are localizable but not necessarily σ-finite. This generalization needs further computations. It is also useful to extend Theorem 2. We present this as a graded exercise (Exercises 7 and 8) with a detailed sketch of the proof. Since the above result suffices for σ-compact spaces, which is the most important case, we shall omit further analysis at this point. Also, integration by parts is given in Exercise 11, as an application of Theorem 1 and Proposition 1.9. An extended discussion on the latter subject may be found in Hewitt (1960).

EXERCISES

0. Let $(\Omega_i, \Sigma_i, \mu_i)$, $i = 1, 2$, be measure spaces and (Ω, Σ, μ) their product. If $f: \Omega \to \bar{\mathbb{R}}$ is measurable (for Σ), and μ-integrable, show that $\nu: A \times B \mapsto \int_A \int_B f\,d\mu_1\,d\mu_2$, $A \in \Sigma_1$, $B \in \Sigma_2$, defines a signed measure on Σ. The converse holds if μ_1, μ_2 are both σ-finite; so f is μ-integrable.

1. Let $(\Omega_i, \Sigma_i, \mu_i)$, $i = 1, 2$, be arbitrary measure spaces with only the finite subset property. Let $\mu = \mu_1 \otimes \mu_2$ be the unique product measure on $\Sigma_1 \otimes \Sigma_2$ of $\Omega_1 \times \Omega_2 = \Omega$. (See Exercise 1.6.) If $g_i: \Omega_i \to \bar{\mathbb{R}}^+$ is measurable for Σ_i, $i = 1, 2$, show that $g = g_1g_2$ is measurable for $\Sigma_1 \otimes \Sigma_2$ and that

$$\int_\Omega g\,d\mu = \int_{\Omega_1}\left[\int_{\Omega_2} g_1(\omega_1)g_2(\omega_2)\mu_2(d\omega_2)\right]\mu_1(d\omega_1)$$

$$= \int_{\Omega_1} g_1\,d\mu_1 \int_{\Omega_2} g_2\,d\mu_2.$$

[Note that $\mu(A \times B) = \mu_1(A)\mu_2(B)$ for $A \in \Sigma_1$, $B \in \Sigma_2$.]

2. Suppose (Ω_i, Σ_i), $i = 1, 2$, are a pair of measurable spaces, and $\mu_1: \Sigma_1 \to \overline{\mathbb{R}}^+$ is a measure. Let $\mu_2(\cdot, \cdot): \Sigma_2 \times \Omega_1 \to \overline{\mathbb{R}}^+$ be a mapping such that (i) $\mu_2(\cdot, \omega_1)$ is a measure on Σ_2, $\omega_1 \in \Omega_1$, and (ii) $\mu_2(A, \cdot)$ is measurable on Ω_1 relative to Σ_1, $A \in \Sigma_2$. If $\mathscr{S} = \Sigma_1 \times \Sigma_2$, the semialgebra of measurable rectangles on $\Omega = \Omega_1 \times \Omega_2$, let $\alpha: \mathscr{S} \to \overline{\mathbb{R}}^+$ be defined as $\alpha(A \times B) = \int_A \mu_2(B, \omega)\mu_1(d\omega_1)$, $A \times B \in \mathscr{S}$. Show that $\alpha(\cdot)$ is σ-additive on $\mathscr{S}$, and let μ be the generated Carathéodory measure, by the pair $(\alpha, \mathscr{S})$, restricted to $\Sigma_1 \otimes \Sigma_2 = \Sigma$. Verify that for any $f: \Omega \to \overline{\mathbb{R}}^+$, one has

$$\int_{\Omega_1}^{*}\left[\int_{\Omega_2}^{*} f(\omega_1, \omega_2)\mu_2(d\omega_2, \omega_1)\right]\mu_1(d\omega_1) \leqslant \int_{\Omega}^{*} f\, d\mu.$$

3. (A Fubini-Stone theorem for non-Cartesian product measures.) Let (Ω_i, Σ_i), $i = 1, 2$, μ_1, μ_2, and (Ω, Σ, μ) be as in the preceding exercise. If $f: \Omega \to \overline{\mathbb{R}}$ is (jointly) measurable for Σ and is μ-integrable, verify that the mapping $\omega_1 \mapsto \int_{\Omega_2} f(\omega_1, \omega_2)\mu_2(d\omega_2, \omega_1)$ is μ_1-measurable and μ_1-integrable. Hence show that (following closely the proof of Theorem 1) we have

$$\int_{\Omega} f\, d\mu = \int_{\Omega_1}\left[\int_{\Omega_2} f(\omega_1, \omega_2)\mu_2(d\omega_2, \omega_1)\right]\mu_1(d\omega_1).$$

If in the above $f \geqslant 0$ is measurable but not necessarily μ-integrable, suppose that μ_1 is σ-finite and $\mu_2(\cdot, \omega_1)$ is σ-finite uniformly in $\omega_1 \in \Omega_1 - N$, $\mu_1(N_1) = 0$. Verify that the corresponding extension of Tonelli's theorem holds.

4. This exercise indicates a standard example showing that the measurability of $f: \Omega \to \overline{\mathbb{R}}^+$ in Theorem 1 (or 2) is also essential for its validity. Recall that if A is a well-ordered set, and B is a linearly ordered set which is order isomorphic to A, then they are said to have the same *order type*, denoted *ord*(A). The isomorphism depends only on the order type and not on the specific sets A or B. When A is well ordered, *ord*(A) is its ordinal number (e.g. $ord(1, 2, \dots, n) = n$, $ord(\varnothing) = 0$, and $ord(\mathbb{N}) = \omega$, the countable nonfinite ordinal number). Let $\tilde{\omega}$ be the first uncountable ordinal number, and let $\Omega = \{x: x \text{ ordinal}, x \leqslant \tilde{\omega}\}$. Here $\leqslant$ is the well-ordering relation between the ordinals. [For more details, see Suppes (1972) Chapter 5.] If $\Sigma = \{A \subset \Omega: A \text{ countable or } A^c \text{ countable}\}$, then Σ is a σ-algebra. Let $\mu: \Sigma \to \mathbb{R}^+$ be defined as $\mu(A) = 0$ if A is countable, and $= 1$ if not. Then (Ω, Σ, μ) is a finite measure space. Let $\overline{\Omega} = \Omega \times \Omega$, $\overline{\Sigma} = \Sigma \otimes \Sigma$, $\bar{\mu} = \mu \otimes \mu$. Consider $C = \{(a, b): a < b\}$. Then verify that for each $a \in \Omega$, $b \in \Omega$, the sets $C(a), C(b)$ are in Σ. Deduce that $f = \chi_C: \overline{\Omega} \to \mathbb{R}^+$ is not measurable and

$$\int_{\Omega}\left[\int_{\Omega} \chi_C(a, b)\mu(da)\right]\mu(db) = 0 < 1 = \int_{\Omega}\left[\int_{\Omega} \chi_C(a, b)\mu(db)\right]\mu(da).$$

5. Let $(\mathbb{R}, \Sigma, \lambda)$ be the Lebesgue line and $(\mathbb{R}^2, \tilde{\Sigma}, \tilde{\lambda})$ be the planar Lebesgue space. If $f: \mathbb{R}^2 \to \mathbb{R}^+$ is (jointly) measurable for $\tilde{\Sigma}$ and integrable relative to $\tilde{\lambda}$, verify that the following integrals exist and that the equality holds:

$$\int_0^4 \left[\int_{(x^2-4)/4}^{\sqrt{25-x^2}} (x^2 - y^2) f(x, y) \lambda(dy) \right] \lambda(dx)$$

$$= \int_{-1}^3 \left[\int_0^{\sqrt{4(1+y)}} (x^2 - y^2) f(x, y) \lambda(dx) \right] \lambda(dy)$$

$$+ \int_3^5 \left[\int_0^{\sqrt{(25-y^2)}} (x^2 - y^2) f(x, y) \lambda(dx) \right] \lambda(dy).$$

$$= \int_\Omega (x^2 - y^2) f(x, y)\, d\tilde{\lambda}.$$

Draw the graph of $\Omega \subset \mathbb{R}^2$, on which the last product integral is defined.

6. In this exercise we present an application of convolutions. Let $1 \leqslant p < \infty$. Let $f \in L^p(\mathbb{R})$ and $g \in L^q(\mathbb{R})$, the Lebesgue spaces on the line $(\mathbb{R}, \Sigma, \lambda)$. Suppose that either (i) $q = 1$ or (ii) $q = p/(p-1)$. Show that in both cases the mapping $(x, y) \mapsto g(x - y) f(y)$ is $\tilde{\lambda}$-measurable ($\tilde{\lambda}$ = planar Lebesgue measure), and that in case (i) the mapping

$$T_g: f \mapsto \int_{\mathbb{R}} g(\cdot - y) f(y) \lambda(dy),$$

is a bounded linear operator on $L^p(\mathbb{R})$ into $L^p(\mathbb{R})$, and in fact $\|T_g\| = \sup\{\|Tf\|_p : \|f\|_p \leqslant 1\} \leqslant \|g\|_1$. In case (ii), the same operator T_g takes $L^p(\mathbb{R})$ into $C_\infty(\mathbb{R})$, the space of continuous functions vanishing at ∞, with uniform (or supremum) norm, and $\|T\| = \sup\{\|Tf\|_u : \|f\|_p \leqslant 1\} \leqslant \|g\|_q$, where $\|\cdot\|_u$ is the norm of $C_\infty(\mathbb{R})$. [(i): First let f be a step function in $L^p(\mathbb{R})$, and $h \in L^r(\mathbb{R})$, $r = p/(p-1)$. Then use Hölder's inequality and Theorem 1 to show that $T_g f$ exists and satisfies the inequality. Since step functions are dense in $L^p(\mathbb{R})$, the same result holds and T_g is defined on all of $L^p(\mathbb{R})$. Here one has to use the translation invariance of λ, as well as Proposition 4.5.17. In case (ii) consider $p = 1$, $p > 1$ separately. In the former case, by Theorem 4.5.9 and Hölder's inequality, one gets $\|T_g\| \leqslant \|g\|_1$, and $(T_g f)(\cdot)$ is uniformly continuous. In the latter, $\|T_g\| \leqslant \|g\|_q$ by Hölder's inequality; verify that $|(T_g f)(x)|$ is small if x is off a suitable compact set C with $\|f\chi_{C^c}\|_p < \epsilon$, $\|g\chi_{C^c}\|_q < \epsilon$.]

7. Let $(\Omega_i, \Sigma_i, \mu_i)$, $i = 1, 2$, be localizable measure spaces, and $(\Omega_i, \Sigma_i, \nu_i)$, $i = 1, 2$, be spaces with the finite subset property where the basic δ-rings of sets of finite μ_i- and ν_i-measures (on which they are determined as explained after Definition 5.4.2) are the *same*. Then (i) if $\mu = \mu_1 \otimes \mu_2$ has

the finite subset property, it is localizable, and (ii) if $\nu = \nu_1 \otimes \nu_2$, $\nu_i \ll \mu_i$, $i = 1, 2$, we have the Radon-Nikodým derivative of ν relative to μ on $\Sigma_1 \otimes \Sigma_2$ given by

$$\frac{d\nu_1 \otimes \nu_2}{d\mu_1 \otimes \mu_2} = \frac{d\nu}{d\mu} = \frac{d\nu_1}{d\mu_1}\frac{d\nu_2}{d\mu_2} \qquad \text{a.e.}[\mu],$$

as in (9). [(i): First, ν and μ are uniquely defined by their finite subset properties. Let $f \in L^1(\nu)$ and assume that $\nu_i \ll \mu_i$. So $\nu \ll \mu$ follows from Proposition 1.9(iii), since $\mu(E) = 0 \Rightarrow \mu_1(E(\omega_2)) = 0 \Rightarrow \nu(E) = 0$. Then by Theorem 1 we have

$$\int_\Omega f\,d\nu = \int_{\Omega_1}\left[\int_{\Omega_2} f(\omega_1, \omega_2) f_2(\omega_2)\mu_2(d\omega_2)\right]\mu_1(d\omega_1)$$

[as in the line preceding Eq. (12)]

$$= \int_\Omega ff^*\,d\mu, \qquad \text{by Theorem 5.4.3,}$$

where f^* is a quasi-function determined by ν and μ. If $E = A \times B \in \Sigma_1 \times \Sigma_2$, $\mu(E) < \infty$, then replacing f by $f\chi_E$, one gets from the above

$$\int_E ff^*\,d\mu = \int_{\Omega_1}\left[\int_{\Omega_2}(f\chi_E)(\omega_1, \omega_2)(f_1\chi_A)(\omega_1)\right.$$

$$\left.\times (f_2\chi_B)(\omega_2)\mu_2(d\omega_2)\right]\mu_1(d\omega_1).$$

If we replace f here by $f\chi_{A_1 \times B_1}$, where $A_1 \subset A$, $B_1 \subset B$, $(\Sigma_i\text{-})$measurable, we get $f_1\chi_A f_2\chi_B = f^*\chi_{A \times B}$, a.e.$[\mu]$, with $f^* = f_1 f_2$. Since $\mu(A \times B) < \infty$, and $f^*\chi_{A\times B}$ is measurable (μ), we get the result. Thus as A, B vary over the finite subsets, using the finite subset property, we deduce that f^* is determined by $f_1 f_2$, and hence is equivalent to this measurable function. Since ν_i are arbitrary, we can apply Theorem 5.4.5, and deduce that μ has the Radon-Nikódym property. Hence it is localizable by that result. For (ii), first one infers that when the μ_i are localizable the mappings $\omega_1 \mapsto \int_{\Omega_2} f(\omega_1, \omega_2)\mu_2(d\omega_2)$ and $\omega_2 \mapsto \int_{\Omega_1} f(\omega_1, \omega_2)\mu_1(d\omega_1)$ are μ_1- and μ_2-measurable for any $f \geqslant 0$ that is μ-measurable. Then use the result of the next problem on an extension of Corollary 3, and complete the proof exactly as in Proposition 5.]

8. Let $(\Omega_i, \Sigma_i, \mu_i)$, $i = 1, 2$, be measure spaces with the finite subset property, and $f\colon \Omega = \Omega_1 \times \Omega_2 \to \overline{\mathbb{R}}^+$ be a measurable function for $\Sigma = \Sigma_1 \otimes \Sigma_2$. If $\mu = \mu_1 \otimes \mu_2$ has the finite subset property and $\omega_1 \mapsto \int_{\Omega_2} f(\omega_1, \omega_2)\mu_2(d\omega_2)$

is μ_1-measurable, then we have $\int_\Omega f\,d\mu = \int_{\Omega_1}[\int_{\Omega_2} f(\omega_1, \omega_2)\mu_2(d\omega_2)]\mu_1(d\omega_1)$ and a similar equality with Ω_1, Ω_2 interchanged if the resulting function $\omega_2 \mapsto \int_{\Omega_1} f(\omega_1, \omega_2)\mu_1(d\omega_1)$ is μ_2-measurable. [When the measures μ_i are not σ-finite or localizable, the additional measurability assumption is not satisfied, in view of the counterexample of Exercise 1.3. Now by Lemma 1.7, we always have $\int_\Omega f\,d\mu \geqslant \int^*_{\Omega_1}[\int_{\Omega_2} f(\omega_1, \omega_2)\mu_2(d\omega_1)]\mu_1(d\omega_1)$, even without the additional measurability hypothesis. (With measurability we can erase the "star".) If there is strict inequality here, then the right side is finite, $= \beta_0$ (say). Consider the set $E_n = \{(\omega_1, \omega_2): f(\omega_1, \omega_2) \geqslant 1/n\} \in \Sigma_1 \otimes \Sigma_2$. Then $E = \bigcup_{n=1}^\infty E_n = \{(\omega_1, \omega_2): f(\omega_1, \omega_2) > 0\}$ and E is σ-finite. Indeed, by the finite subset property, there exist sets $F_{mn} \subset E_n$, $F_{mn} \in \Sigma_1 \otimes \Sigma_2$, $\mu(F_{mn}) < \infty$, and $F_{mn} \subset F_{(m+1)n}$ such that $\mu(E_n) = \lim_m \mu(F_{mn})$ by Lemma 2.3.3. Since $\chi_{F_{mn}} \leqslant nf$ a.e., we have

$$\mu(F_{mn}) = \int_{\Omega_1} \mu_2(F_{mn}(\omega_1))\mu_1(d\omega_1), \qquad \text{by Proposition 1.9(ii)}$$

$$\leqslant n\int_{\Omega_1}\left[\int_{\Omega_2} f(\omega_1, \omega_2)\mu_2(d\omega_2)\right]\mu(d\omega_1) = n\beta_0 < \infty.$$

But since $f = 0$ a.e. on E^c, we get $\int_\Omega f\,d\mu = \int_E f\,d\mu$, and since E is σ-finite, the result is reduced to Corollary 3. Thus Theorem 2 and Proposition 5 hold iff μ_1, μ_2, μ are localizable, and so these are the general forms of the Fubini theorem.]

9. Let $(\mathbb{R}^+, \Sigma, \lambda)$ be the Lebesgue half line, and let $L^2(\mathbb{R}^+)$ be the corresponding Lebesgue space. If $g: \mathbb{R}^+ \to \mathbb{R}^+$ is a measurable function such that $\int_{\mathbb{R}^+}[g(x)/\sqrt{x}]\lambda(dx) < \infty$, show that the mapping T defined on $L^2(\mathbb{R}^+)$ by $(Tf)(x) = \int_{\mathbb{R}^+} g(xy)f(y)\lambda(dy)$, $f \in L^2(\mathbb{R}^+)$, is measurable in x, and then $Tf \in L^2(\mathbb{R}^+)$. Show also that

$$\|T\| = \sup\{\|Tf\|_2: \|f\|_2 \leqslant 1\} = \int_{\mathbb{R}^+} g(x)x^{-1/2}\lambda(dx).$$

[The proof is similar to that of Exercise 6 above. If $h \in L^2(\mathbb{R}^+)$, show that $\int_{\mathbb{R}^+} h(x)(Tf)(x)\lambda(dx)$ exists after using Theorem 1 (or Corollary 3) and the CBS inequality.]

10. Let $(\Omega_i, \Sigma_i, \mu_i)$, $i = 1, 2$, be measure spaces with the finite subset property. If (Ω, Σ, μ) is their product measure space, as in Theorem 1.3, and $h: \Omega \to \overline{\mathbb{R}}$ is measurable for Σ, and μ-integrable, let

$$l(f) = \int_{\Omega_1}\left[\int_{\Omega_2} h(\omega_1, \omega_2)f(\omega_2)\mu_2(d\omega_2)\right]\mu_1(d\omega_1).$$

Show that $l(\cdot)$ is a bounded linear functional on $L^\infty(\Omega_2, \Sigma_2, \mu_2)$, and that if $f_n \to f$ a.e., and boundedly, then $l(f_n) \to l(f)$ as $n \to \infty$. Deduce that $\lambda_0: B \mapsto l(\chi_B)$, $B \in \Sigma_2$, is a signed measure. Also show that the σ-additivity of λ_0 is still true if μ_1, μ_2, and μ are localizable and $h: \Omega \to \overline{\mathbb{R}}^+$ is measurable, but not necessarily integrable. [For the last part use the remark at the end of Exercise 8.]

11. Let $f, g: \mathbb{R} \to \mathbb{R}$ be monotone nondecreasing functions and μ_f, μ_g be Lebesgue-Stieltjes measures determined by them as in Section 2.6. Thus $\mu_f([a, b]) = f(a + 0) - f(a - 0)$. Show that $A = \{(x, y): x \leqslant y\}$ is a Borel set, and that for $-\infty < a < b < \infty$ we have

$$\int_{\mathbb{R}} \mu_g(A(x))\mu_f(dx) = \int_{\mathbb{R}} \mu_f(A(y))\mu_g(dy),$$

where $A(x)$, $A(y)$ are sections of A. [Use Proposition 1.9 and Theorem 1.] Deduce that

$$\int_a^b f(x - 0)\mu_g(dx) = f(b + 0)g(b + 0) - f(a - 0)g(a - 0)$$
$$- \int_a^b g(x + 0)\mu_f(dx).$$

(This is the *integration by parts formula* for Lebesgue-Stieltjes integrals.)

6.3. REMARKS ON NON-CARTESIAN PRODUCTS

In the preceding work we have mostly considered the Cartesian product of measures, and indicated how other (non-Cartesian) products can be obtained (e.g. Exercise 2.3). The latter also find applications in areas such as probability (under the name "transition probabilities"), game theory, and statistical decision functions. The generalized Fubini-Stone theorem noted above plays an important role there. Along these lines, we present an analog of Proposition 2.5, which is of interest in such applications.

The non-Cartesian product Radon-Nikodým statement is not as elegant as the Cartesian one; the result is as follows.

Proposition 1. *Let* (Ω, Σ, μ_i), $i = 1, 2$, *be σ-finite measure spaces, and* $(S, \mathscr{S})$ *be a measurable space. If* $\nu_i: \Omega \times \mathscr{S} \to \overline{\mathbb{R}}^+$, $i = 1, 2$, *are mappings such that* $\nu_i(\omega, \cdot)$ *is a measure for each* $\omega \in \Omega$, *and* $\nu_i(\cdot, B)$ *is a measurable (for* Σ*) function for each* $B \in \mathscr{S}$, $i = 1, 2$, *suppose that* $\nu_i(\omega, \cdot)$ *is uniformly σ-finite for* $\omega \in \Omega$. *Define (non-Cartesian) product measures (which exist, cf. Exercise 2.3) as*

$$\xi_i(A \times B) = \int_A \nu_i(\omega, B)\mu_i(d\omega), \qquad i = 1, 2, \tag{1}$$

for $A \times B \in \Sigma \times \mathcal{S}$, *and denote by the same symbol* ξ_i *its extension to* $\Sigma \otimes \mathcal{S}$. *Then* $\xi_1 \ll \xi_2$ *iff* $\mu_1 \ll \mu_2$ *and* $\nu_1(\omega, \cdot) \ll \nu_2(\omega, \cdot)$ *for almost all* ω *relative to* μ_1. *Moreover, in this case, we have for a.a.* (ω, s) $[\xi_2]$, *on letting* $h = d\mu_1/d\mu_2$,

$$\frac{d\xi_1}{d\xi_2}(\omega, s) = \frac{d\nu_1(\omega, \cdot)}{d\nu_2(\omega, \cdot)}(s)\frac{d\mu_1}{d\mu_2}(\omega) = \rho(\omega, s)h(\omega) \qquad \text{(say)}, \tag{2}$$

and for all $f \in L^1(\xi_1)$,

$$\int_{\Omega \times S} f\,d\xi_1 = \int_\Omega \left[\int_S f(\omega, s)\rho(\omega, s)h(\omega)\nu_2(\omega, ds)\right]\mu_2(d\omega). \tag{3}$$

Proof. In spite of the involved nature of the statement, the proof is straightforward, and we include a sketch. Since ξ_1, ξ_2 exist by Exercise 2.3, it follows that they are σ-finite from the conditions on μ_i's and ν_i's.

First let $\xi_1 \ll \xi_2$ on $\Sigma \otimes \mathcal{S}$. If $\mu_2(A) = 0$ for an $A \in \Sigma$, then (1) implies $\xi_2(A \times B) = 0$ for any $B \in \mathcal{S}$. Hence $\xi_1(A \times B) = 0$ must also hold, since $\xi_1 \ll \xi_2$. Choose a $B_0 \in \mathcal{S}$ such that $\nu_1(\omega, B_0) > 0$ a.e.$[\mu_1]$. Such a B_0 exists by the uniform σ-finiteness of ν_1. Then (1) implies $\mu_1(A) = 0$. Since μ_1, μ_2 are measures, this proves $\mu_1 \ll \mu_2$. Regarding the ν_i, we need to make some computations. Let

$$g(\omega, s) = \frac{d\xi_1}{d\xi_2}(\omega, s), \qquad h(\omega) = \frac{d\mu_1}{d\mu_2}(\omega),$$

the functions being defined a.e.$[\xi_2]$ and $[\mu_2]$ respectively. Then we have

$$\int_{\tilde{\Omega}} f\,d\xi_1 = \int_{\tilde{\Omega}} fg\,d\xi_2, \qquad \text{with } \tilde{\Omega} = \Omega \times S, \quad f \in L^1(\xi_1)$$

$$= \int_\Omega \left[\int_S (fg)(\omega, s)\nu_2(\omega, ds)\right]\mu_2(d\omega) \qquad \text{(cf. Exercise 2.3)}$$

$$= \int_\Omega \left[\int_S f(\omega, s)\frac{\nu_2(\omega, ds)}{h(\omega)}\right]\mu_1(d\omega), \qquad \text{since } h = \frac{d\mu_1}{d\mu_2}. \tag{4}$$

In (4), the set $A = \{\omega: h(\omega) = 0\}(\in \Sigma)$ has μ_1-measure zero, and hence by the defining formulas, $\xi_1(A \times B) = 0$ for $B \in \mathcal{S}$. So $g(\omega, s) = 0$ a.e.$[\xi_1]$, and hence we are allowed to set $(f/h)(\omega) = 0$ without altering (4). On the other hand

$$\int_{\tilde{\Omega}} f\,d\xi_1 = \int_\Omega \left[\int_S f(\omega, s)\nu_1(\omega, ds)\right]\mu_1(d\omega). \tag{5}$$

From (4) and (5), by taking $f(\omega, s) = \chi_A(\omega) f_2(s)$, we get for a.a. ω

$$\int_S f_2(s)\nu_1(\omega, ds) = \int_S f_2(s)\frac{g(\omega, s)}{h(\omega)}\nu_2(\omega, ds). \tag{6}$$

To deduce the desired result from (6), consider the class

$$\mathscr{C} = \left\{ B \in \mathscr{S}: \nu_1(\omega, B) = \int_B \frac{g(\omega, s)}{h(\omega)}\nu_2(\omega, ds) \text{ for a.a. } \omega\ [\mu_1] \right\}.$$

By (6), every $B \in \mathscr{S}$, with $\nu_1(\omega, B) < \infty$, is in $\mathscr{C}$. Since all quantities are $\geqslant 0$, by the monotone convergence theorem $S \in \mathscr{C}$, and $B_1, B_2 \in \mathscr{C}$ disjoint $\Rightarrow$ $B_1 \cup B_2 \in \mathscr{C}$. Since $\nu_1(\omega, \cdot)$ is a measure, we get for B, C in $\mathscr{C}$ that $\nu_1(\omega, C) < \infty \Rightarrow B - C \in \mathscr{C}$. Also $B_1 \cap B_2 \in \mathscr{C}$ with any $B_1, B_2 \in \mathscr{C}$, and $\mathscr{C}$ is closed under monotone increasing sequences. Hence by the π-λ class theorem, $\mathscr{C}$ contains the σ-ring generated by the class of all $\nu_1(\omega, \cdot)$-finite sets for almost all ω $[\mu_1]$. This is clearly sufficient to conclude that $\nu_1(\omega, \cdot) \ll \nu_2(\omega, \cdot)$ for a.a. ω $[\mu_1]$. Since f_2 is arbitrary in (6), this implies also that

$$\rho(\omega, s) = \frac{d\nu_1(\omega, \cdot)}{d\nu_2(\omega, \cdot)}(s) = \frac{g(\omega, s)}{h(\omega)} \qquad \text{for a.a. } (\omega, s)\ [\xi_2]. \tag{7}$$

This is (2), and (3) is immediate from (7) and (4).

We turn to the converse. If $\mu_1 \ll \mu_2$ and $\nu_1(\omega, \cdot) \ll \nu_2(\omega, \cdot)$ for a.a. ω $[\mu_1]$, and if $E \in \Sigma \otimes \mathscr{S}$ with $\xi_2(E) = 0$, then the equation

$$\xi_2(E) = \int_\Omega \int_S \chi_E \nu_2(\omega, ds)\mu_2(d\omega)$$

$$= \int_\Omega \nu_2(\omega, E(\omega))\mu_2(d\omega) = 0$$

implies $\nu_2(\omega, E(\omega)) = 0$ a.e.$[\mu_2]$ whence $\nu_1(\omega, E(\omega)) = 0$ a.e.$[\mu_1]$ also. So $\xi_1(E) = 0$, on using a similar equation for ξ_1. Since these are measures, we deduce that $\xi_1 \ll \xi_2$. But then by the preceding part, $d\xi_1/d\xi_2$ is given by (2), and the result follows.

To specialize the above result, let us state the transition measure precisely in the following

Definition 2. If (Ω_i, Σ_i), $i = 1, 2$, are a pair of measurable spaces, then a transition probability function $P(\cdot, \cdot): \Omega_1 \times \Sigma_2 \to \overline{\mathbb{R}}^+$ is a mapping such that (i) $P(\omega, \cdot)$ is σ-additive, $P(\omega, \Omega_2) = 1$, $\omega \in \Omega_1$, and (ii) $P(\cdot, A)$ is a measurable function (for Σ_1) for each A $(\in \Sigma_2)$.

A physical interpretation of $P(\omega, A)$ is that it is the probability of a particle starting in the "state" ω and landing in the set A in a unit time. Using Exercise 2.3, higher order transitions are defined easily. For instance, for each $\omega_0 \in \Omega_1$, we have

$$P_2(\omega_0, A) = \int_{\Omega_1} P_1(\omega, A) P(\omega_0, d\omega), \tag{8}$$

where $P(\omega_0, \cdot)$ acts as $\mu(\cdot)$. This is interpreted as the probability of starting in ω_0 and visiting some intermediate state ω after a unit time and landing in A at the next instant. Here $P_1(\cdot, \cdot)$ is the transition probability of moving from one state to another in one step. Similarly, a kth order transition can be defined inductively, using the Fubini theorem; see especially Exercise 2.3. Equations such as (8) play a key role in the theory of Markov chains in probability, and all our results of this section are important for such a study.

We have the following consequence of the proposition:

Corollary 3. *Let* (Ω, Σ, μ_i), $i = 1, 2$, *be probability spaces, and let* $\nu_i(\cdot, \cdot)$: $\Omega \times \mathscr{S} \to \mathbb{R}^+$, $i = 1, 2$, *be a pair of transition probability functions, where* $(S, \mathscr{S})$ *is a measurable space. If*

$$P_i(A \times B) = \int_A \nu_i(\omega, B) \mu_i(d\omega), \qquad i = 1, 2, \tag{9}$$

then the P_i*'s extend to probability measures on* $\Sigma \otimes \mathscr{S}$, *and* $P_1 \ll P_2$ *iff* $\mu_1 \ll \mu_2$ *and* $\nu_1(\omega, \cdot) \ll \nu_2(\omega, \cdot)$ *for almost all* ω $[\mu_1]$. *When these conditions hold, the Radon-Nikodým density of* P_1 *relative to* P_2 *is given by*

$$\frac{dP_1}{dP_2}(\omega, s) = \frac{d\nu_1(\omega, \cdot)}{d\nu_2(\omega, \cdot)}(s) \frac{d\mu_1}{d\mu_2}(\omega), \qquad \textit{for a.a. } (\omega, s)\ [P_2]. \tag{10}$$

It should be observed that when the ν_i are independent of ω, then these results reduce to the Cartesian products considered in the preceding section, as expected. However, the symmetry of the formulas of the Cartesian product case is lost in the current extension. The result of this corollary, under the additional condition that $\mathscr{S}$ is countably generated, was first formulated by A. V. Skorokhod (1970), motivated by certain applications. (See Exercise 1.)

The fact that the measures must be finite and normalized for infinite products will become clear in the next section. In that sense, probability measures are inherently related to the study of infinite products, which moreover distinguish between the Cartesian and non-Cartesian cases. We thus turn to a detailed analysis of these product measures in the next section.

EXERCISES

0. Let (Ω, Σ, μ) be a measure space and $\mathscr{B} \subset \Sigma$ a σ-algebra such that $\mu|\mathscr{B}$ $(= \mu_{\mathscr{B}}$ for short) is σ-finite (or localizable). Show that there is a function $\psi: \Omega \times \Sigma \to \mathbb{R}^+$ satisfying the system of equations:

$$\int_B \psi(\omega, A)\mu_{\mathscr{B}}(d\omega) = \mu(A \cap B), \qquad A \in \Sigma,\ B \in \mathscr{B},\ \mu(B) < \infty.$$

Further (i) $0 \leqslant \psi(\omega, A) \leqslant 1$ a.e.$[\mu]$, (ii) $\psi(\cdot, A)$ is $\mathscr{B}$-measurable, and for $A_n \in \Sigma$, disjoint, we have $\psi(\omega, \bigcup_n A_n) = \Sigma_n \psi(\omega, A_n)$, a.e.$[\mu]$. [Such a ψ is called a (generalized) conditional probability, and is a prototype of functions $\nu(\cdot, \cdot)$ considered above.]

1. Let Ω be a separable Hilbert space (i.e., it has a countable basis), and $\mathscr{B}$ be its Borel σ-algebra. So $\mathscr{B}$ is the σ-algebra generated by the open sets. If $P: \mathscr{B} \to \mathbb{R}^+$ is a measure, let $P_a(A) = P(A + a)$, where $A \in \mathscr{B}$ and $A + a = \{x + a: x \in A\}$ is a translate of A, so that it is in $\mathscr{B}$ and $P_a: \mathscr{B} \to \mathbb{R}^+$ is a measure. If $P_a \ll P$, then a is called an *admissible translate* of P. Suppose that P is such that ta is an admissible translate for all $t \geqslant 0$. If $\Omega_1 = \text{sp}\{\alpha a: \alpha \in \mathbb{R}\}$ and $\Omega_2 = \{y \in \Omega: y \perp a\}$, where $y \perp a$ means $(y, a) = 0$, $(\cdot, \cdot)$ being the inner product, then one can express Ω as $\Omega_1 \oplus \Omega_2$, the direct sum, or as a "direct product" $\Omega_1 \times \Omega_2$ as in linear algebra. Let μ be the marginal measure of P, i.e., $\mu(A) = P(\pi^{-1}(A))$, where $\pi: \Omega \to \Omega_1$ is the orthogonal projection and $A \subset \Omega_1$ is a Borel set. We assume the existence of a transition probability $\nu(\cdot, \cdot): \Omega_1 \times \mathscr{B}_2 \to \mathbb{R}^+$ (and this can be shown to hold here) such that

$$P(A \times B) = \int_A \nu(x, B)\mu(dx),$$

for all $A \in \mathscr{B}_1$, $B \in \mathscr{B}_2$, where $\mathscr{B}_i$ is the Borel σ-algebra of Ω_i. Show that for each $t > 0$ [with $\mu_a(A) = \mu(A + a)$],

$$P_{ta}(A \times B) = \int_A \nu(x - ta, B)\mu_{ta}(dx)$$

and that $\nu(x - ta, \cdot) \ll \nu(x, \cdot)$ a.e.$[\mu]$, as well as $\mu_{ta} \ll \mu$.

2. Extend Proposition 1 if ν_1, ν_2 satisfy the same conditions there, but μ_1, μ_2 are localizable and $\mu_1 \otimes \mu_2$ has the finite subset property.

3. Show that the following formula generalizes (1) and (8) to more factors. Let (Ω_i, Σ_i), $i = 1, \ldots, n$, be measurable spaces, $P_1: \Sigma_1 \to \mathbb{R}^+$ be a finite measure, and $\nu_{i,i+1}(\cdot, \cdot): \Omega_i \times \Sigma_{i+1} \to \mathbb{R}^+$, $i = 1, \ldots, n-1$, be transition probability functions. Then $P_n: \Sigma_n \to \mathbb{R}^+$ is a finite measure, where

$$P_n(A) = \int_{\Omega_1} P_1(d\omega_1) \int_{\Omega_2} \nu_{1,2}(\omega_1, d\omega_2) \cdots$$

$$\times \int_{\Omega_{n-1}} \nu_{n-1,n}(\omega_{n-1}, A) \nu_{n-2,n-1}(\omega_{n-2}, d\omega_{n-1}).$$

Also define a transition probability $\nu_{1,n}(\cdot, \cdot): \Omega_1 \times \Sigma_n \to \mathbb{R}^+$, without using P_1. (These formulas admit further extensions, as the ensuing work shows.)

6.4. INFINITE PRODUCT MEASURES

Thus far we considered products of two (and finite numbers of) measure spaces, but in several important cases it is necessary to consider infinite products of measures. They even involve both Cartesian and non-Cartesian products. These measure spaces are, for instance, at the very root of stochastic processes, ergodic theory, and the asymptotic analysis of problems in statistics. In this section we first present the infinite Cartesian product analog of Theorem 2.1 and then consider the corresponding non-Cartesian analogs of Proposition 3.1. The first is called the (Fubini-)Jessen theorem, and its extensions in the second case are the Kolmogorov-Bochner and Prokhorov theorems and their relatives. They have interesting connections with the (differentiation) work of Chapter 5, and are fundamental for the above-noted subjects. We now present a detailed discussion.

The very first problem of the subject is the complexity of notation. To lessen this unavoidable difficulty, we introduce and systematically use a convenient (and hopefully intuitive) symbolism which the reader should always keep in mind. Let $\{(\Omega_i, \Sigma_i, \mu_i): i \in I\}$ be a family of measure spaces, where I is an index set having at least countably many members. Define $\Omega = \times_{i \in I} \Omega_i$, the Cartesian product space. Then the points $\omega = (\omega_i, i \in I) \in \Omega$ are often called *threads*, with ω_i $(\in \Omega_i)$ as the ith component. We let $\pi_i: \Omega \to \Omega_i$ be the coordinate projection mapping, i.e., $\pi_i(\omega) = \omega_i$, the ith component. Clearly $\pi_i(\Omega) = \Omega_i$ [and $\pi_i^{-1}(\Omega_i) = \Omega$]. For the subproducts, let $\beta \subset I$ be a subset and $\Omega_\beta = \times_{i \in \beta} \Omega_i$. We then write $\pi_\beta: \Omega \to \Omega_\beta$ by setting $\pi_\beta(\omega) = \{\omega_i: i \in \beta\} = \omega_\beta \in \Omega_\beta$. Since Ω_β is not a subset of Ω, if $A \subset \Omega_\beta$ we form $B = A \times \times_{i \in I - \beta} \Omega_i$ and, identifying Ω with $\Omega_\beta \times \Omega_{I-\beta}$ (isomorphically, since these products are associative, as explained before), write $B \subset \Omega$. Such a B is termed a *cylinder set* of Ω *with base* A (in Ω_β). These sets play an essential role in our work and correspond to rectangles in finite products. Similarly, if $\beta_1 \subset \beta_2 \subset I$ are nonempty sets, and Ω_{β_2} and Ω_{β_1} are the corresponding products, we define $\pi_{\beta_1\beta_2}: \Omega_{\beta_2} \to \Omega_{\beta_1}$ in the familiar way, i.e., if we write β_2 $(\supset \beta_1)$ as $\beta_2 = \beta_1 \cup (\beta_2 - \beta_1)$ and $\omega_{\beta_2} = (\omega_{\beta_1}, (\omega_i: i \in \beta_2 - \beta_1)) \in$

Ω_{β_2}, $\omega_{\beta_1} \in \Omega_{\beta_1}$, then $\pi_{\beta_1\beta_2}(\omega_{\beta_2}) = \omega_{\beta_1}$. Thus, symbolically, $\pi_\beta = \pi_{\beta\infty}$ if "∞" corresponds to the whole set I. This definition at once implies (i) $\pi_{\beta\beta} =$ identity, (ii) if $\beta_1 \subset \beta_2 \subset \beta_3$ then $\pi_{\beta_1\beta_3} = \pi_{\beta_1\beta_2} \circ \pi_{\beta_2\beta_3}$, and (iii) $\pi_{\beta_1} = \pi_{\beta_1\infty} = \pi_{\beta_1\beta_2} \circ \pi_{\beta_2\infty}$. These properties, which are simple (they are obvious from a picture), are also important. For instance, $B = A \times \Omega_{I-\beta}$, the cylinder set noted above, satisfies $\pi_\beta(B) = A$ and hence $\pi_\beta^{-1}(A) = B$.

To proceed further, let $\mathscr{F}$ be the class of all nonempty *finite* subsets of I. If β_1, β_2 are in $\mathscr{F}$, then we write $\beta_1 < \beta_2$ iff $\beta_1 \subset \beta_2$. Then $<$ is a partial ordering, and if β, γ are any two sets in $\mathscr{F}$, there is a larger set $\tilde{\gamma} \in \mathscr{F}$ such that $\beta < \tilde{\gamma}$, $\gamma < \tilde{\gamma}$ (simply put, $\tilde{\gamma} = \beta \cup \gamma$). Thus $(\mathscr{F}, <)$, together with this maximum property for any finite set of elements, becomes a directed set. This notation will be used for all products in our study.

Using the finite dimensional theory of the preceding sections, if Σ_i is a σ-algebra of Ω_i, then we have the product σ-algebra of Ω_β as $\bigotimes_{i\in\beta}\Sigma_i$, $\beta \in \mathscr{F}$. Since $(\Sigma_1 \otimes \Sigma_2) \otimes \Sigma_3 = \sigma(\Sigma_1 \otimes \Sigma_2 \times \Sigma_3)$, which may be easily seen to be also obtained as $\sigma(\Sigma_1 \times \Sigma_2 \otimes \Sigma_3)$, the products are associative and $\Sigma_\beta = \bigotimes_{i\in\beta}\Sigma_i$ is unambiguously defined in Ω_β. (See Exercise 1.) But $A \times \Omega_2 \in \Sigma_1 \otimes \Sigma_2$ for all $A \in \Sigma_1$, and hence for each pair $\beta_1, \beta_2 \in \mathscr{F}$, $\beta_1 < \beta_2$, we have $\pi_{\beta_1\beta_2}^{-1}(\Sigma_{\beta_1}) \subset \Sigma_{\beta_2}$ as a σ-subalgebra. This makes $\pi_{\beta_1\beta_2}: \Omega_{\beta_2} \to \Omega_{\beta_1}$ also measurable mappings as a bonus. It is this result that motivates the definition of a desirable class in Ω relative to which each π_β should be measurable. Thus let $\Sigma_0 = \bigcup_{\beta\in\mathscr{F}}\pi_\beta^{-1}(\Sigma_\beta)$, which is a class of subsets of Ω.

Regarding the structure of Σ_0, we have

Lemma 1. *For each $\beta \in \mathscr{F}$, $\pi_\beta^{-1}(\Sigma_\beta)$ is a σ-algebra and Σ_0 is an algebra. If I is finite then $\Sigma_0 = \Sigma_I(= \bigotimes_{i\in I}\Sigma_i)$, a σ-algebra.*

Proof. Since $\pi_\beta: \Omega \to \Omega_\beta$ is a mapping, π_β^{-1} preserves all set operations (differences, unions, complements); it follows that $\pi_\beta^{-1}(\Sigma_\beta)$ is a σ-algebra (because Σ_β is). If A, B are in Σ_0, then $A \in \pi_{\beta_1}^{-1}(\Sigma_{\beta_2})$, and since $\mathscr{F}$ is directed, there is a γ $(= \beta_1 \cup \beta_2)$ in $\mathscr{F}$ such that $\pi_{\beta_i}^{-1}(\Sigma_{\beta_i}) \subset \pi_\gamma^{-1}(\Sigma_\gamma)$. Hence A, B are in $\pi_\gamma^{-1}(\Sigma_\gamma)$, so that $A \cup B$, $A - B$, A^c are in (the σ-algebra) $\pi_\gamma^{-1}(\Sigma_\gamma) \subset \Sigma_0$. Hence Σ_0 is an algebra (of cylinder sets). It is however not a σ-algebra, since a countable union of (even increasing) σ-algebras need not be a σ-algebra. The last statement is obvious, since $I \in \mathscr{F}$ and is its largest element. The proof is complete.

To construct a measure on Ω from those given on the Ω_i, we get a trivial case if $\mu_i(\Omega_i) = +\infty$ for more than finitely many i, since then each cylinder with a base of positive "area" will have infinite "volume" (or measure), and those with finite ($\neq 1$) volumes must then be just zeros. So for a nontrivial theory we assume $\mu_i(\Omega_i) = 1$, $i \in I$.

The notation with $\pi_\alpha: \Omega \to \Omega_\alpha$, $\alpha \in \mathscr{F}$, can be more effectively used and streamlined if we restate the definitions. Thus $\Omega_\alpha = \times_{i\in\alpha}\Omega_i$, $\Sigma_\alpha = \bigotimes_{i\in\alpha}\Sigma_i$, so that $\{(\Omega_\alpha, \Sigma_\alpha): \alpha \in \mathscr{F}\}$ is a new family of measurable spaces, and $\pi_{\alpha\beta}: \Omega_\beta$

$\to \Omega_\alpha$, $\alpha < \beta$ (α, β in $\mathscr{F}$) satisfies $\pi_{\alpha\beta}^{-1}(\Sigma_\alpha) \subset \Sigma_\beta$. If $\mu_i: \Sigma_i \to \mathbb{R}^+$, $i \in I$, are probability measures [so $\mu_i(\Omega_i) = 1$, $i \in I$], let $\mu_\alpha = \bigotimes_{i \in \alpha} \mu_i: \Sigma_\alpha \to \mathbb{R}^+$. Then by Theorem 2.1, μ_α is again a probability measure. Since $A_\alpha \in \Sigma_\alpha$ implies that $\pi_{\alpha\beta}^{-1}(A_\alpha) = A_\alpha \times \Omega_{\beta - \alpha}$ is a measurable rectangle (or a cylinder), one has the *compatibility relation*

$$\mu_\beta\left(\pi_{\alpha\beta}^{-1}(A_\alpha)\right) = \mu_\alpha(A_\alpha)\mu_{\beta-\alpha}(\Omega_{\beta-\alpha}) = \mu_\alpha(A_\alpha). \tag{1}$$

Since $\mu_\alpha(A_\alpha) = \mu_{i_1}(A_{i_1}) \cdots \mu_{i_k}(A_{i_k})$ when $\alpha = (i_1, \ldots, i_k) \in \mathscr{F}$, this is precisely the condition imposed for (finite) products of measures in Lemma 1.2. Thus (1) is equivalently expressed as $\mu_\beta \circ \pi_{\alpha\beta}^{-1} = \mu_\alpha$ for $\alpha < \beta$ in $\mathscr{F}$. This alternative formulation will serve as a basis for certain other (non-Cartesian) products.

We have the first fundamental assertion extending Fubini's theorem, as follows:

Theorem 2 (Jessen). *Let $\{(\Omega_i, \Sigma_i, \mu_i): i \in I\}$ be a family of probability spaces; $\mathscr{F}$ the directed (by inclusion) class of all nonempty finite subsets of I; and $\Omega = \times_{i \in I} \Omega_i$, $\Omega_\alpha = \times_{i \in \alpha} \Omega_i$, $\Sigma_\alpha = \bigotimes_{i \in \alpha} \Sigma_i$, $\mu_\alpha = \bigotimes_{i \in \alpha} \mu_i$, and $\pi_\alpha: \Omega \to \Omega_\alpha$ the projection mappings, $\alpha \in \mathscr{F}$. If $\Sigma_0 = \bigcup_{\alpha \in \mathscr{F}} \pi_\alpha^{-1}(\Sigma_\alpha)$, then there exists a unique additive function $\mu: \Sigma_0 \to \mathbb{R}^+$ such that $\mu \circ \pi_\alpha^{-1}(A_\alpha) = \mu_\alpha(A_\alpha)$, $A_\alpha \in \Sigma_\alpha$, $\alpha \in \mathscr{F}$. Moreover, μ is σ-additive on Σ_0 and hence has a unique σ-additive extension $\tilde{\mu}$ to $\Sigma = \sigma(\Sigma_0)$, and if $I = \mathbb{N}$, $A_n \in \Sigma_n$, $n \geq 1$, and $A = \times_{n=1}^\infty A_n$, then $A \in \Sigma$, and*

$$\tilde{\mu}(A) = \lim_{n \to \infty} \prod_{k=1}^n \mu_k(A_k) = \prod_{k=1}^\infty \mu_k(A_k). \tag{2}$$

Proof. First we show that $\mu: \Sigma_0 \to \mathbb{R}^+$ is additive *without using the additional hypothesis that each μ_α is a product measure*. We use only that $\{\mu_\alpha: \alpha \in \mathscr{F}\}$ is a directed family of measures satisfying $\mu_\beta \circ \pi_{\alpha\beta}^{-1} = \mu_\alpha$ for each pair $\alpha < \beta$ in $\mathscr{F}$. This fact will be useful later on.

Let $A \in \Sigma_0$. Then $A \in \pi_\alpha^{-1}(\Sigma_\alpha)$ for some $\alpha \in \mathscr{F}$. Then define μ on Σ_0 by the equation $\mu(A) = \mu_\alpha(A_\alpha)$. We assert that

(i) μ is well defined and does not depend on α, and
(ii) μ is additive on Σ_0.

(i): Since $A \in \Sigma_0 = \bigcup_{\alpha \in \mathscr{F}} \pi_\alpha^{-1}(\Sigma_\alpha)$, there is $\beta \in \mathscr{F}$ with $A = \pi_\beta^{-1}(A_\beta)$, for some $A_\beta \in \Sigma_\beta$. Suppose also that $\pi_\alpha^{-1}(A_\alpha) = \pi_\beta^{-1}(A_\beta)$, for some $A_\alpha \in \Sigma_\alpha$, $\alpha \in \mathscr{F}$. The directedness of $\mathscr{F}$ implies the existence of a γ in $\mathscr{F}$ such that $\alpha, \beta < \gamma$. But we also have $\pi_\alpha = \pi_{\alpha\gamma} \circ \pi_\gamma$ and $\pi_\beta = \pi_{\beta\gamma} \circ \pi_\gamma$, as noted before, and all these mappings are onto. Hence

$$(\pi_{\alpha\gamma} \circ \pi_\gamma)^{-1}(A_\alpha) = \pi_\alpha^{-1}(A_\alpha) = A = \pi_\beta^{-1}(A_\beta) = (\pi_{\beta\gamma} \circ \pi_\gamma)^{-1}(A_\beta).$$

The extreme sets are thus equal, so that

$$\pi_\gamma^{-1}\left(\pi_{\alpha\gamma}^{-1}(A_\alpha) - \pi_{\beta\gamma}^{-1}(A_\beta)\right) = \varnothing,$$

since π_γ^{-1} preserves set operations. Moreover, $\pi_\alpha(\Omega) = \Omega_\alpha$ implies, by Lemma 3.3.1(v), that $\pi_\gamma^{-1}(B) = \varnothing$, so that $B = \varnothing$. Thus $\pi_{\alpha\gamma}^{-1}(A_\alpha) = \pi_{\beta\gamma}^{-1}(A_\beta)$. Consequently

$$\begin{aligned}\mu_\alpha(A_\alpha) &= \mu_\gamma\left(\pi_{\alpha\gamma}^{-1}(A_\alpha)\right), \qquad \text{by (1)} \\ &= \mu_\gamma\left(\pi_{\beta\gamma}^{-1}(A_\beta)\right) = \mu_\beta(A_\beta) = \mu(A). \qquad (3)\end{aligned}$$

Thus $\mu(A)$ does not depend on α (or β), and is well defined.

(ii): Let A, B be disjoint elements of Σ_0. Then there exist α_1, β_1 in $\mathscr{F}$ such that $A \in \pi_{\alpha_1}^{-1}(\Sigma_{\alpha_1})$, $B \in \pi_{\beta_1}^{-1}(\Sigma_{\beta_1})$. Hence there is a $\gamma_1 \in \mathscr{F}$ with $\alpha_1, \beta_1 < \gamma_1$, $\pi_{\beta_1}^{-1}(\Sigma_{\beta_1}) \cup \pi_{\alpha_1}^{-1}(\Sigma_{\alpha_1}) \subset \pi_{\gamma_1}^{-1}(\Sigma_{\gamma_1})$. So A, B are in the last σ-algebra, and there exist disjoint sets C, D ($\in \Sigma_{\gamma_1}$) such that $A = \pi_{\gamma_1}^{-1}(C)$, $B = \pi_{\gamma_1}^{-1}(D)$, and hence $A \cup B = \pi_{\gamma_1}^{-1}(C \cup D)$; furthermore,

$$\mu(A \cup B) = \mu_{\gamma_1}(C \cup D) = \mu_{\gamma_1}(C) + \mu_{\gamma_1}(D) = \mu(C) + \mu(D),$$

since μ_{γ_1} is additive by hypothesis. Thus $\mu: \Sigma_0 \to \mathbb{R}^+$ is additive, and clearly $\mu(\Omega)\ [= \mu_\alpha^{-1}(\Omega_\alpha)] = 1$.

We now assert that, when μ_α is a Cartesian product measure, μ is also σ-additive. This implies, by the Hahn extension Theorem 2.3.1, that μ has a unique σ-additive extension $\tilde{\mu}$ to $\Sigma = \sigma(\Sigma_0)$, since Σ_0 is an algebra by Lemma 1. This will follow if we show that for any $A_n \in \Sigma$, $A_n \downarrow \varnothing \Rightarrow \mu(A_n) \downarrow 0$, or equivalently, $\mu(A_n) \geqslant \delta > 0$ for all $n \geqslant 1$ implies $\bigcap_{n=1}^\infty A_n \neq \varnothing$.

Since $A_n \in \Sigma_0$, there is an $\alpha_n \in \mathscr{F}$ such that $A_n = \pi_{\alpha_n}^{-1}(\Sigma_{\alpha_n})$, $n \geqslant 1$. By the directedness of $\mathscr{F}$, we can find, for each n, a $\beta_n \in \mathscr{F}$ such that $\beta_n \uparrow$ and $\alpha_1 = \beta_1$, $\alpha_2 < \beta_2, \ldots, \alpha_n < \beta_n$. In fact $\beta_n = \bigcup_{i=1}^n \alpha_i \in \mathscr{F}$ will suffice here, and then $A_n \in \pi_{\beta_n}^{-1}(\Sigma_{\beta_n})\uparrow$, so that $A_n = \pi_{\beta_n}^{-1}(B_n)$, $B_n \in \Sigma_{\beta_n}$, and $A_n = B_n \times \Omega_{I-\beta_n}$, $\mu(A_n) = \mu_{\beta_n}(B_n) \geqslant \mu_{\beta_{n+1}}(B_{n+1}) \geqslant \delta > 0$. Suppose that $\beta_n = (i_1, \ldots, i_n)$, where β_k will thus denote simply a finite set of indices i in I. Since μ_{β_n} is a (finite) Cartesian product measure, we may apply Theorem 2.2 to

$$f_n(\omega_{i_1}) = \int_{\Omega_{i_2}} \cdots \int_{\Omega_{i_n}} \chi_{B_n}(\omega_{i_1}, \ldots, \omega_{i_n}) \mu_{i_2}(d\omega_{i_2}) \cdots \mu_{i_n}(d\omega_{i_n}),$$

so that f_n is μ_{i_1}-measurable and integrable, and $\mu_{\beta_n}(B_n) = \int_{\Omega_{i_1}} f_n \, d\mu_{i_1} = \mu(A_n) \geqslant \delta > 0$. Clearly $0 \leqslant f_n \leqslant 1$ and $f_n \downarrow$. Hence there is $\bar{\omega}_{i_1} \in \Omega_{i_1}$ such that $f_n(\bar{\omega}_{i_1}) \nrightarrow 0$. Similarly consider

$$g_n(\omega_{i_2}) = \int_{\Omega_{i_3}} \cdots \int_{\Omega_{i_n}} \chi_{B_n}(\bar{\omega}_{i_1}, \omega_{i_2}, \ldots, \omega_{i_n}) \mu_{i_3}(d\omega_{i_3}) \cdots \mu_{i_n}(d\omega_{i_n}).$$

Then again $0 \leqslant g_n \leqslant 1$, and g_n is measurable and μ_{i_2}-integrable by Theorem 2.2, so that $f_n(\bar{\omega}_{i_1}) = \int_{\Omega_{i_2}} g_n \, d\mu_{i_2} \nrightarrow 0$. Hence there is $\bar{\omega}_{i_2} \in \Omega_{i_2}$ such that $g_n(\bar{\omega}_{i_2}) \nrightarrow 0$. Repeating this process, we can find points $\bar{\omega}_{i_k} \in \Omega_{i_k}$ such that $\chi_{B_n}(\bar{\omega}_{i_1}, \ldots, \bar{\omega}_{i_k}, \omega_{i_{k+1}}, \ldots, \omega_{i_n})$ cannot vanish for all $(\omega_{i_{k+1}}, \ldots, \omega_{i_n}) \in \times_{j=k+1}^{n} \Omega_{i_j}$. Thus for some ω_{i_j}, $k+1 \leqslant j \leqslant n$, we must have $(\bar{\omega}_{i_1}, \ldots, \bar{\omega}_{i_k}, \omega_{i_{k+1}}, \ldots, \omega_{i_n}) \in B_n$. If $\bar{\omega}$ is a point in Ω containing the coordinate set $(\bar{\omega}_{i_1}, \bar{\omega}_{i_2}, \ldots)$, we must be able to find the other components such that for any n, $(\bar{\omega}_{i_1}, \bar{\omega}_{i_2}, \ldots, \bar{\omega}_{i_n}, \omega_{i_{n+1}}, \ldots) \in B_n \times \Omega_{I-\beta_n} = A_n$. Hence in particular $\bar{\omega} \in A_n \subset A_{n-1}$, $n \geqslant 1$. So $\bar{\omega} \in \bigcap_{n=1}^{\infty} A_n$ and $\bigcap_{n=1}^{\infty} A_n \neq \varnothing$. This proves that μ is σ-additive on Σ_0 and it has a unique extension $\tilde{\mu}$ onto Σ.

It remains to establish (2). Since $I = \mathbb{N}$, let $\alpha_n = (1, 2, \ldots, n)$ and $B_n = A_1 \times \cdots \times A_n \in \bigotimes_{i=1}^{n} \Sigma_i$. If $C_n = B_n \times \Omega_{I-\alpha_n}$, then $C_n = \pi_{\alpha_n}^{-1}(B_n) \supset C_{n+1} = \pi_{\alpha_{n+1}}^{-1}(B_{n+1})$. It is clear that $A = \bigcap_{n=1}^{\infty} C_n \in \Sigma = \sigma(\bigcup_{n=1}^{\infty} \pi_{\alpha_n}^{-1}(\Sigma_{\alpha_n}))$, since $C_n \in \pi_{\alpha_n}^{-1}(\Sigma_{\alpha_n})$. Then by Proposition 2.1.1(iii),

$$\tilde{\mu}(A) = \lim_{n \to \infty} \mu(C_n) = \lim_{n \to \infty} \mu_{\alpha_n}(B_n)$$

$$= \lim_{n \to \infty} \prod_{i=1}^{n} \mu_i(A_i).$$

This is (2), and the proof of the theorem is complete.

One should note a conceptual problem in the formation of the products here. We have defined $\Omega = \times_{i \in I} \Omega_i$, $\Omega_\alpha = \times_{i \in \alpha} \Omega_i$, $\alpha \in \mathscr{F}$. Since each $\{i\} \in \mathscr{F}$, $\mathscr{F}$ has "more" elements than I. So if $\tilde{\Omega} = \times_{\alpha \in \mathscr{F}} \Omega_\alpha$, clearly Ω and $\tilde{\Omega}$ are not identical. What is the relation between them? Note that $I = \bigcup_{\alpha \in \mathscr{F}} \alpha$. It is not hard to see that there exists a mapping $u: \Omega \to \tilde{\Omega}$ which is one to one and onto, so that they *can* be identified. We add a proof of this fact, which may not be evident. Since Ω_α is a finite product, with $\pi_{\alpha\beta}(\Omega_\beta) = \Omega_\alpha$ for $\alpha < \beta$ in $\mathscr{F}$, let us temporarily denote the projection from $\tilde{\Omega}$ onto Ω_α by g_α: $g_\alpha(\tilde{\Omega}) = \Omega_\alpha$. Now for $\alpha < \beta$, $\pi_{\alpha\beta} \circ g_\beta = g_\alpha$ as well as $\pi_{\alpha\beta} \circ \pi_\beta = \pi_\alpha$, where $\pi_\alpha: \Omega \to \Omega_\alpha$. Let $\omega = \{\omega_i: i \in I\}$ be an element of Ω, and $\alpha \in \mathscr{F}$. Then $\pi_\alpha(\omega) = \omega_\alpha = \{\omega_i: i \in \alpha\}$. If $\tilde{\omega} = \{\omega_\alpha: \alpha \in \mathscr{F}\}$ is composed of the ω_α coordinates among those of ω, then $\tilde{\omega} \in \tilde{\Omega}$ and we define $u(\omega) = \tilde{\omega}$. This mapping $u: \Omega \to \tilde{\Omega}$ is uniquely defined, since with each such $\omega \in \Omega$ we can associate exactly one $\tilde{\omega}$ in $\tilde{\Omega}$. Thus $g_\alpha \circ u = \pi_\alpha$, and $u(\Omega) \subset \tilde{\Omega}$. To see that u is one to one, let $u(\omega) = \tilde{\omega} = u(\omega')$. Then for each $\alpha \in \mathscr{F}$, we have $\omega_\alpha = g_\alpha(\tilde{\omega}) = g_\alpha \circ u(\omega) = \pi_\alpha(\omega)$, and also $\omega_\alpha = g_\alpha \circ u(\omega') = \pi_\alpha(\omega')$, so that $\pi_\alpha(\omega) = \pi_\alpha(\omega')$ for all $\alpha \in \mathscr{F}$. Thus all coordinates of ω and ω' are the same, implying $\omega = \omega'$. The mapping is also onto, since if $\tilde{\omega} = \{\omega_\alpha: \alpha \in \mathscr{F}\} \in \tilde{\Omega}$ is any element and $i \in I$, then there exists an $\alpha \in \mathscr{F}$ such that $\{i\} < \alpha$. Let $\omega_i = g_{\{i\}\alpha}(\omega_\alpha)$ and let $\omega = \{\omega_i: i \in I\}$. The element ω in Ω is well defined, since if $\beta \in \mathscr{F}$ also satisfies $\{i\} < \beta$ and $\omega_i = g_{\{i\}\beta}(\omega_\beta)$, then by directedness there is a $\gamma \in \mathscr{F}$

larger than both α, β, and we have, on using the composition rules,

$$\omega_i = g_{\{i\}\alpha}(\omega_\alpha) = g_{\{i\}\alpha} \circ g_{\alpha\gamma}(\omega_\gamma) = g_{\{i\}\gamma}(\omega_\gamma)$$
$$= g_{\{i\}\beta} \circ g_{\beta\gamma}(\omega_\gamma) = g_{\{i\}\beta}(\omega_\beta). \tag{4}$$

Thus $\omega \in \Omega$ and $u(\omega) = \tilde{\omega}$. So u is a bijection between Ω and $\tilde{\Omega}$, and we therefore can identify them, as asserted.

With the basic construction accomplished, let us present an interesting consequence on integrating sections of $f: \Omega \to \mathbb{R}$ as they expand or contract. One denotes the product space obtained in Theorem 2 as (Ω, Σ, μ) or $\otimes_{i \in I}(\Omega_i, \Sigma_i, \mu_i)$, where $\Omega = \times_{i \in I}\Omega_i$, $\Sigma = \sigma(\bigcup_{\alpha \in \mathscr{F}}\pi_\alpha^{-1}(\Sigma_\alpha)) = \otimes_{i \in I}\Sigma_i$, and $\mu = \otimes_{i \in I}\mu_i$, so that $\mu(\Omega) = 1$. We have the following, also due to B. Jessen, on the mean and pointwise convergences of these integrated sections:

Theorem 3. *Let (Ω, Σ, μ) be the Cartesian product of $\{(\Omega_i, \Sigma_i, \mu_i): i \in I\}$, a family of probability spaces. Let $f \in L^1(\Omega, \Sigma, \mu)$. For each $\alpha \in \mathscr{F}$, the directed class of all finite subsets of I, let $\alpha^c = I - \alpha$. Define*

$$f_{\alpha^c}(\omega) = \int_{\Omega_\alpha} f(\omega_\alpha, \omega_{\alpha^c})\mu_\alpha(d\omega_\alpha) \tag{5}$$

and

$$f_\alpha(\omega) = \int_{\Omega_{\alpha^c}} f(\omega_\alpha, \omega_{\alpha^c})\mu_{\alpha^c}(d\omega_{\alpha^c}). \tag{6}$$

Then

(i) $f_{\alpha^c} \to \int_\Omega f\,d\mu$, $f_\alpha \to f$ *as* $\alpha \to \infty$ *in* $L^1(\mu)$; *and*
(ii) *if* $I = \mathbb{N}$, $\alpha_n = (1, 2, \ldots, n)$, $n \geqslant 1$, *we have* $f_{\alpha_n^c} \to \int_\Omega f\,d\mu$ a.e. *as well as* $f_{\alpha_n} \to f$ a.e. $[\mu]$.

Remark. It should first be noted that in (5) and (6), $f_\alpha(\omega) = f_\alpha(\omega_\alpha)$, $f_{\alpha^c}(\omega) = f_{\alpha^c}(\omega_{\alpha^c})$, since writing $\omega = (\omega_\alpha, \omega_{\alpha^c})$, the second (first) variable is integrated out and thus $f_\alpha(f_{\alpha^c})$ is constant in $\omega_{\alpha^c}(\omega_\alpha)$. The measure spaces being finite, constants belong to $L^1(\mu)$. Hence f_α, f_{α^c} are in $L^1(\mu)$. Thus (5) and (6) make sense when we identify them as functions defined on Ω which are constants on sections.

An interesting point here is that $\{f_\alpha, \mathscr{B}_\alpha: \alpha \in \mathscr{F}\}$ and $\{f_{\alpha^c}, \mathscr{A}_\alpha: \alpha \in \mathscr{F}\}$ are two types of martingales relative to the directed index set $\mathscr{F}$ and suitable σ-algebras $\mathscr{B}_\alpha \downarrow \subset \Sigma$, $\mathscr{A}_\alpha \uparrow \subset \Sigma$, and then the conclusions can be inferred from the general theory of martingales. However, we have established the latter theory only when indexing is by natural numbers. For this reason, the L^1-mean convergence will be proved using a special argument, of independent interest. But for the case $I = \mathbb{N}$, we obtain the pointwise convergence with the martingale argument, using the results of Section 5.5 (Theorem 5.5.19).

Proof. (i) (mean convergence): We observe that the result follows if it is true for each step function in $L^1(\mu)$. For, given $\epsilon > 0$, by the density of step functions in $L^1(\mu)$, there exists such an f_ϵ in $L^1(\mu)$ that $\|f - f_\epsilon\|_1 < \epsilon/3$. If the result holds for f_ϵ, then there are α_0 and α_1 in $\mathscr{F}$, depending on ϵ, such that

$$\left\|(f_\epsilon)_{\alpha^c} - \int_\Omega f_\epsilon \, d\mu\right\|_1 < \frac{\epsilon}{3}, \qquad \alpha \geqslant \alpha_0, \tag{7a}$$

and

$$\|(f_\epsilon)_\alpha - f\|_1 < \frac{\epsilon}{3}, \qquad \alpha \geqslant \alpha_1, \tag{7b}$$

to be used for (i) and (ii). But then for $\alpha \geqslant \alpha_0$

$$\begin{aligned}\left\|f_\alpha - \int_\Omega f \, d\mu\right\|_1 &\leqslant \left|\int_{\Omega_{\alpha^c}} (f - f_\epsilon) \, d\mu_\alpha\right| \\ &\quad + \left|\int_{\Omega_{\alpha^c}} f_\epsilon \, d\mu_{\alpha^c} - \int_\Omega f_\epsilon \, d\mu\right| + \left|\int_\Omega (f_\epsilon - f) \, d\mu\right| \\ &\leqslant \|f - f_\epsilon\|_1 + \left\|(f_\epsilon)_{\alpha^c} - \int_\Omega f \, d\mu\right\|_1 + \|f - f_\epsilon\|_1, \end{aligned} \tag{8}$$

since for each $g \in L^1(\mu)$, the function $(Tg)(\omega) = \int_{\Omega_{\alpha^c}} g(\omega_\alpha, \omega_{\alpha^c}) \mu_{\alpha^c}(d\omega_{\alpha^c})$ defines an element of $L^1(\mu_\alpha)$, and identifying it as an element of $L^1(\mu)$ whose ω_{α^c} section is a constant, we get $\|Tg\|_1 \leqslant \|g\|_1$ [using $\mu_{\alpha^c}(\Omega_{\alpha^c}) = 1$]. This is what we used for the first term. But then with (7a) and the choice of f_ϵ, the right side of (8) is less than $\epsilon/3 + \epsilon/3 + \epsilon/3 = \epsilon$. Hence the result will hold for f itself. Similarly, if $\alpha > \alpha_1$, with (7b),

$$\begin{aligned}\|f_\alpha - f\|_1 &\leqslant \|f - f_\epsilon\|_1 + \|f_\epsilon - (f_\epsilon)_\alpha\|_1 + \|(f_\epsilon)_\alpha - f_\alpha\|_1 \\ &< \frac{\epsilon}{3} + \frac{\epsilon}{3} + \left|\int_{\Omega_\alpha} (f_\epsilon - f) \, d\mu\right| < \frac{\epsilon}{3} + \frac{\epsilon}{3} + \frac{\epsilon}{3} = \epsilon,\end{aligned}$$

as in the preceding argument. It thus remains to prove (i) for simple f.

If $f = \sum_{i=1}^n a_i \chi_{E_i} \in L^1(\mu)$, by linearity of the integral, the result (i) will follow when it is proved for any χ_E, $E \in \Sigma$. But by Proposition 2.4.1, (Σ, ρ) is a complete (semi-)metric space, where $\rho(E, F) = \mu(E \vartriangle F)$. Since $\Sigma_0 \subset \Sigma$ is an algebra, $\sigma(\Sigma_0) = \Sigma$, and (Σ_0, ρ) is a subspace of (Σ, o), we deduce that the closure of (Σ_0, ρ) is all of the complete space (Σ, ρ). This is clear because for $E \in \Sigma$, there exist a sequence $E_n \in \Sigma_0$ such that $\mu(E_n \vartriangle E) \to 0$. Thus

$E \in \Sigma$ implies that, for each $\delta > 0$, there is an $E_\delta \in \Sigma_0 = \bigcup_\alpha \pi_\alpha^{-1}(\Sigma_\alpha)$, so that $E_\delta \in \pi_{\alpha_2}^{-1}(\Sigma_{\alpha_2})$ for some α_2 satisfying $\rho(E, E_\delta) = \|\chi_E - \chi_{E_\delta}\|_1 < \delta$. By the argument of the preceding paragraph, the result of (i) holds for χ_E if it holds for χ_{E_δ}. But $E_\delta = A \times \Omega_{\alpha_2^c}$ is a cylinder, and let $f = \chi_{E_\delta}$. If $\alpha \in \mathscr{F}$, $\alpha > \alpha_2$, then

$$f_{\alpha^c}(\omega) = \int_{\Omega_\alpha} \chi_{E_\delta}(\omega_\alpha, \omega_{\alpha^c})\mu_\alpha(d\omega_\alpha) = \int_{\Omega_\alpha} \chi_A \cdot 1\, d\mu_\alpha$$

$$= \int_\Omega \chi_{A \times \Omega_{\alpha^c}}\, d\mu = \int_\Omega f\, d\mu.$$

Hence $\|f_{\alpha^c} - \int_\Omega f\, d\mu\|_1 \to 0$ as $\alpha \to \infty$. Similarly, if $\alpha \geqslant \alpha_2$,

$$f_\alpha(\omega) = \int_{\Omega_{\alpha^c}} \chi_{A \times \Omega_{\alpha^c}}(\omega_\alpha, \omega_{\alpha^c})\mu_{\alpha^c}(d\omega_{\alpha^c}) = \chi_A \mu_{\alpha^c}(\Omega_{\alpha^c})$$

$$= \chi_{A \times \Omega_{\alpha^c}} = f(\omega).$$

Thus $\|f_\alpha - f\|_1 \to 0$ as $\alpha \to \infty$. Hence (i) is established.

(ii) (pointwise convergence): For this part we use the martingale argument, as indicated by Doob (1953). Since in (i) the mean convergence of f_α and f_{α^c} is established more generally, we only need to prove the pointwise a.e. convergence with $I = \mathbb{N}$ and $\alpha_n = (1, 2, \ldots, n)$. Set $f_n = f_{\alpha_n^c}$ and $g_n = f_{\alpha_n}$. Let $\mathscr{A}_n = \pi_{\alpha_n}^{-1}(\Sigma_{\alpha_n})$, so that $\mathscr{A}_n \subset \mathscr{A}_{n+1}$ and $\bigcup_{n=1}^\infty \mathscr{A}_n = \Sigma_0$. Similarly, let $\mathscr{B}_n = \pi_{\alpha_n^c}^{-1}(\Sigma_{\alpha_n^c})$, so that $\mathscr{B}_n \supset \mathscr{B}_{n+1}$. We claim that (a) $\{f_n, \mathscr{A}_n : n \geqslant 1\}$ is a martingale, and (b) $\{g_n, \mathscr{B}_n : n \geqslant 1\}$ is a decreasing martingale in that $E^{\mathscr{B}_{n+1}}(g_n) = g_{n+1}$ a.e., where $E^{\mathscr{B}_n}(\cdot)$ is the conditional expectation operator (cf. Definition 5.5.14).

For (a), by the Fubini theorem, $f_n(\omega) = f_n(\omega_\alpha)\chi_{\Omega_{\alpha_n^c}}(\omega_{\alpha_n^c})$ is μ_α-measurable, so that $f_n(\omega_\alpha, \cdot)$ is a constant on the $\omega_{\alpha_n^c}$-section, implying that it is $\pi_{\alpha_n}^{-1}(\Sigma_{\alpha_n}) = \mathscr{A}_n$-measurable. If $\nu(E) = \int_E f\, d\mu$, $E \in \Sigma$, and $\nu_n = \nu | \mathscr{A}_n$, we have to verify that $\nu_n(A) = \int_A f_n\, d\mu$, $A \in \mathscr{A}_n$, so that $E^{\mathscr{A}_n}(f) = f_n$ and the martingale property follows. But this is easy, since $A = A_{\alpha_n} \times \Omega_{\alpha_n^c}$, $A_{\alpha_n} \in \Sigma_{\alpha_n}$, and

$$\nu_n(A) = \int_A f\, d\mu = \int_A f(\omega_{\alpha_n}, \omega_{\alpha_n^c})\, d\mu$$

$$= \int_{A_{\alpha_n}} \int_{\Omega_{\alpha_n^c}} f(\omega_{\alpha_n}, \omega_{\alpha_n^c})\mu_{\alpha_n^c}(d\omega_{\alpha_n^c})\mu_{\alpha_n}(d\omega_{\alpha_n}), \qquad \text{by Theorem 2.1}$$

$$= \int_{A_{\alpha_n}} f_{\alpha_n}(\omega_{\alpha_n})\mu_{\alpha_n}(d\omega_{\alpha_n}), \qquad \text{by (6)}$$

$$= \int_{A_{\alpha_n}} \int_{\Omega_{\alpha_n^c}} f_{\alpha_n}(\omega_{\alpha_n}, \omega_{\alpha_n^c})\mu_{\alpha_n^c}(d\omega_{\alpha_n^c})\mu_{\alpha_n}(d\omega_{\alpha_n})$$

$$= \int_A f_n(\omega)\, d\mu, \qquad \text{by Theorem 2.1.} \tag{9}$$

Since $A \in \mathscr{A}_n$ is arbitrary, this shows $\{f_n, \mathscr{A}_n: n \geqslant 1\}$ is a martingale and $\|f_n\|_1 \leqslant \|f\|_1 < \infty$ for all n. Hence $f_n \to f'$ a.e. by Theorem 5.5.19. But by (i) $f_n \to f$ in $L^1(\mu)$, so $f = f'$ a.e., and (a) follows.

As for (b), if we let $\tilde{\nu}(B) = \int_B f\,d\mu$, $B \in \Sigma$, then using (5) in place of (6), and $\omega = (\omega_{\alpha_n^c}, \omega_{\alpha_n})$ in (9), we find that if $\tilde{\nu}_n = \tilde{\nu}|\mathscr{B}_n$ then $\tilde{\nu}_n(A) = \int_A g_n\,d\mu$, $A \in \mathscr{B}_n$, so that $E^{\mathscr{B}_n}(f) = g_n$ a.e. This is the same as the decreasing martingale statement, since $E^{\beta_{n+1}}(E^{\mathscr{B}_n}(f)) = E^{\mathscr{B}_{n+1}}(f)$ and $E^{\mathscr{B}_{n+1}} \circ E^{\mathscr{B}_n} = E^{\mathscr{B}_{n+1}}$ by Theorem 5.5.15(iii). Thus if it is shown that $g_n \to \tilde{f}$ a.e., then by (i), since $g_n \to \int_\Omega f\,d\mu$ in $L^1(\mu)$, we must have $\tilde{f} = \int_\Omega f\,d\mu$ a.e., and (ii) will follow. The decreasing martingale convergence is proved below, and with it the proof will be complete.

We have the following more general result.

Proposition 4. *Let $\{g_n, \mathscr{B}_n: n \geqslant 1\}$ be a decreasing martingale where $\Sigma \supset \mathscr{B}_n \supset \mathscr{B}_{n+1}$. If $\mathscr{B}_\infty = \bigcap_{n=1}^\infty \mathscr{B}_n$, then $g_n \to g_\infty$ a.e. and in $L^1(\mu)$, and $\{g_n, \mathscr{B}_n: 1 \leqslant n \leqslant \infty\}$ is a martingale, our (Ω, Σ, μ) being a finite measure space on which all these functions are defined.*

Proof. This can be proved using the argument of Theorem 5.5.19, but for variety we present another method following E. S. Andersen and B. Jessen.

Let $g_* = \liminf_n g_n$, $g^* = \limsup_n g_n$. To show that $g_n \to g$ a.e., it suffices to show $\{\omega: g_*(\omega) < g^*(\omega)\}$ is a μ-null set, so that $g = g_* = g^*$ a.e. will result. But since

$$\{\omega: g_*(\omega) < g^*(\omega)\} = \bigcup_{\substack{r_1, r_2 \\ \text{rational}}} \{\omega: g_*(\omega) < r_1 < r_2 < g^*(\omega)\}$$

$$= \bigcup A_{r_1 r_2} \qquad \text{(say)},$$

we only need to show that $\mu(A_{r_1 r_2}) = 0$ for all rationals $r_1 < r_2$. Define sets $C_r = \{\omega: g_*(\omega) < r\}$, $C^s = \{\omega: g^*(\omega) > s\}$. Then $C_r, C^s, A_{r_1 r_2}$ are $\mathscr{B}_\infty$-measurable, and $A_{r_1 r_2} = C_{r_1} \cap C^{r_2}$.

Define $\nu_n(A) = \int_A g_n\,d\mu$, $A \in \mathscr{B}_n$. Then $\nu_n|\mathscr{B}_{n+1} = \nu_{n+1}$ by the martingale property. So $\nu(A) = \nu_n(A)$, $A \in \mathscr{B}_\infty$ is well defined for all n (ν does not involve n), and $\nu: \mathscr{B}_\infty \to \mathbb{R}$ is a signed measure. We now assert that for any a, b in $\mathbb{R}$,

$$\nu(C_a \cap A) \leqslant a\mu(C_a \cap A), \quad \nu(C^b \cap A) \geqslant b\mu(C^b \cap A), \qquad A \in \mathscr{B}_\infty. \quad (10)$$

To prove this consider $\tilde{C}_n = \{\omega: \inf_{k \leqslant n} g_k(\omega) < a\}$. Then $\tilde{C}_n \uparrow C_a$. We disjunctify this set by letting $C_{nn} = \{\omega: g_n(\omega) < a\}$ and $C_{kn} = \{\omega: g_k(\omega) < a, g_j(\omega) \geqslant a, k+1 \leqslant j \leqslant n-1\}$. So C_{kn} is the set where g_j falls below a for the first time at k before n. Then $C_{kn} \in \mathscr{B}_k$ and are disjoint, $\bigcup_{k=1}^n C_{kn} = \tilde{C}_n$. Clearly $C_{kn} \subset \{\omega: g_n(\omega) < a\}$. For each $A \in \mathscr{B}_\infty$ we have, since ν is

σ-additive,

$$\nu(\tilde{C}_n \cap A) = \sum_{k=1}^{n} \nu_k(C_{kn} \cap A) = \sum_{k=1}^{n} \int_{C_{kn} \cap A} g_k \, d\mu$$

$$\leqslant a \sum_{k=1}^{n} \mu(C_{kn} \cap A) = a\mu(\tilde{C}_n \cap A), \qquad \text{since} \quad g_k < a \quad \text{on } C_{kn}.$$

Letting $n \to \infty$, we get the first inequality of (10). Replacing g_n, a by $-g_n, -b$ in this, we get the second inequality of (10).

In (10), let $a = r_1$, $b = r_2$, where $r_1 < r_2$. Then setting $A = A_{r_1 r_2}$, we have

$$r_1 \mu(A_{r_1 r_2}) \geqslant \nu(A_{r_1 r_2}) \geqslant r_2 \mu(A_{r_1 r_2}).$$

Since $\mu \geqslant 0$ and $r_1 < r_2$, this is possible only if $\mu(A_{r_1 r_2}) = 0$ as asserted. Thus $g_n \to g_\infty$ a.e. Since $|g_n| \leqslant E^{\mathscr{B}_n}(|g_1|)$, and μ is finite, this implies that $\{g_n: n \geqslant 1\}$ is uniformly integrable. Hence by Vitali's theorem $g_n \to g_\infty$ in $L^1(\mu)$. Since $\int_A g_n \, d\mu = \int_A g_m \, d\mu$, and since $A \in \mathscr{B}_\infty \subset \mathscr{B}_n$ for $n > m$, letting $n \to \infty$, we get $\int_A g_\infty \, d\mu = \int_A g_m \, d\mu$. Thus $g_\infty = E^{\mathscr{B}_\infty}(g_m)$ a.e. for all m, i.e., $\{g_n, \mathscr{B}_n: 1 \leqslant n \leqslant \infty\}$ is a uniformly integrable martingale. This proves the proposition, and with it Theorem 3 is also demonstrated.

Remark. The most interesting point here to note is that the martingale theory and the sectional integrals of the infinite dimensional Fubini-Jessen theorem are closely related. Since the martingale theory is also at the root of the Radon-Nikodým differentiation, all these three topics are linked at a deeper level.

It is now time to consider the infinite non-Cartesian products and investigate some of the corresponding results. Indeed, a new element enters here, and the measures that can be defined as in the first part of Theorem 2 are finitely additive, but generally fail to be σ-additive. So we need additional structure for the measure spaces. Because of its importance, we present the main result here, due to A. Kolmogorov and (in a more general form) to S. Bochner and Yu. A. Prokhorov. This shows how such infinite product systems are related to a martingale process.

To state the result, it is necessary to recall and define some concepts. Let $\{(\Omega_\alpha, \Sigma_\alpha, \mu_\alpha): \alpha \in \mathscr{F}\}$ be a system of probability spaces where $\mathscr{F}$ is some directed index set. Here we do not demand that $\mathscr{F}$ be the class of all finite subsets of some set I, although that is not excluded. Suppose we are given a family of connecting mappings $g_{\alpha\beta}: \Omega_\beta \to \Omega_\alpha$ for each pair $\alpha < \beta$ in $\mathscr{F}$ satisfying $g_{\alpha\alpha} = \text{identity}$ and $g_{\alpha\beta} \circ g_{\beta\gamma} = g_{\alpha\gamma}$ for $\alpha < \beta < \gamma$. Let $\Omega \subset \times_{\alpha \in \mathscr{F}} \Omega_\alpha$ be a subset with the property that $\omega = \{\omega_\alpha: \alpha \in \mathscr{F}\} \in \Omega$ iff for each $\alpha < \beta$, $g_{\alpha\beta}(\omega_\beta) = \omega_\alpha$. Such a set Ω can be identified with $\times_{i \in I} \Omega_i$ if $g_{\alpha\beta} = \pi_{\alpha\beta}$, $\Omega_\alpha = \times_{i \in \alpha} \Omega_i$, as was done in the preceding work. However, in the more general case under consideration Ω can be a small subset, and even an empty set. Here is an example of the last possibility.

Let $\Omega_n = (0, 1/n)$, $n \geq 1$, and let $g_{nr}: \Omega_r \to \Omega_n$, $r > n$, be the inclusion map. Then it is easy to verify that $\{(\Omega_n, g_{nr}): 1 \leq n < r < \infty\}$, with $I = \mathbb{N}$, satisfies the above condition with $\Omega = \varnothing$. Here the g_{nr} are not onto mappings. The same thing can happen even if $g_{\alpha\beta}$'s are onto.

To avoid the above situation, following Bochner, we stipulate that the $g_{\alpha\beta}$ satisfy a condition: namely, the system $\{(\Omega_\alpha, g_{\alpha\beta}): \alpha < \beta \text{ in } \mathscr{F}\}$ is *sequentially maximal* (s.m.), i.e., for each *sequence* $\alpha_1 < \alpha_2 < \cdots$ from $\mathscr{F}$, and any $\omega = \{\omega_\alpha: \alpha \in \mathscr{F}\} \in \times_{\alpha \in \mathscr{F}} \Omega_\alpha$ for which $g_{\alpha_i \alpha_{i+1}}(\omega_{\alpha_{i+1}}) = \omega_{\alpha_i}$, $i \geq 1$, we have $\omega \in \Omega$. Evidently, the s.m. condition is automatic in the Cartesian product case, and it can be shown to be true if each Ω_α is a nonvoid compact set. (See Exercise 6.) The set Ω is called, in the general case, the *projective limit* of the Ω_α's, denoted $\Omega = \lim_{\leftarrow}(\Omega_\alpha, g_{\alpha\beta})$.

Definition 5. If $\{(\Omega_\alpha, \Sigma_\alpha, \mu_\alpha, g_{\alpha\beta}): \alpha < \beta \text{ in } \mathscr{F}\}$ is a system of probability spaces, and $g_{\alpha\beta}: \Omega_\beta \to \Omega_\alpha$ are the mappings, given above, it is called a *projective system* if $\mu_\beta \circ g_{\alpha\beta}^{-1} = \mu_\alpha$ for each $\alpha < \beta$ in $\mathscr{F}$. If each Ω_α is a topological space, then the $g_{\alpha\beta}$ are required to be continuous also. (The word "inverse" is used for "projective" by some authors.)

In the Cartesian product case the family in Theorem 2 is automatically a projective system. Let $\omega = \{\omega_\alpha: \alpha \in \mathscr{F}\} \in \Omega$, and define $g_\alpha: \omega \mapsto \omega_\alpha$, $\alpha \in \mathscr{F}$. Then $g_\alpha: \Omega \to \Omega_\alpha$ is a mapping which satisfies $g_\alpha = g_{\alpha\beta} \circ g_\beta$ for all $\alpha < \beta$. It is easily seen (cf. Exercise 3) that $g_\alpha(\Omega) = \Omega_\alpha$ if the system is sequentially maximal. Let $\Sigma_0 = \bigcup_{\alpha \in \mathscr{F}} g_\alpha^{-1}(\Sigma_\alpha)$. One verifies as in Lemma 1 that $g_\alpha^{-1}(\Sigma_\alpha) \subset g_\beta^{-1}(\Sigma_\beta)$ for $\alpha < \beta$ and Σ_0 is an algebra of Ω. By the first part of Theorem 2, there exists a unique finitely additive $\mu: \Sigma_0 \to \mathbb{R}^+$ such that $\mu \circ g_\alpha^{-1} = \mu_\alpha$. If μ is σ-additive, then it can be uniquely extended to $\Sigma = \sigma(\Sigma_0)$, just as before, to be a measure again denoted μ. Then the triple (Ω, Σ, μ) is called the *projective limit* of the system, written $(\Omega, \Sigma, \mu) = \lim_{\leftarrow}(\Omega_\alpha, \Sigma_\alpha, \mu_\alpha, g_{\alpha\beta})$.

The substance of Theorem 2 is that for a Cartesian product system of probability spaces, the product measure (its projective limit) always exists. For more general (not necessarily Cartesian product) projective systems this is not true. We now present a basic result for *topological* spaces which includes most of the standard applications and then obtain some characterizations.

Theorem 6 (Bochner). *Let $\{(\Omega_\alpha, g_{\alpha\beta}), \alpha < \beta \text{ in } \mathscr{F}\}$ be a system of topological spaces satisfying the* s.m. *condition, and $\{(\Omega_\alpha, \Sigma_\alpha, \mu_\alpha, g_{\alpha\beta}): \alpha < \beta \text{ in } \mathscr{F}\}$ be a projective system of Radon probability spaces. Then the system admits a unique projective limit (Ω, Σ, μ). (However, the limit μ need not be Radon.)*

Proof. Since the family $\{(\Omega_\alpha, g_{\alpha\beta}): \alpha < \beta \text{ in } \mathscr{F}\}$ satisfies the s.m. condition, $\Omega = \lim_{\leftarrow}(\Omega_\alpha, g_{\alpha\beta})$ exists and there are mappings $g_\alpha: \Omega \to \Omega_\alpha$ such that $g_\alpha = g_{\alpha\beta} \circ g_\beta$ for $\alpha < \beta$. Let $\Sigma_0 = \bigcup_{\alpha \in \mathscr{F}} g_\alpha^{-1}(\Sigma_\alpha)$. Then Σ_0 is an algebra on

Ω, and there exists an additive function $\mu: \Sigma_0 \to \mathbb{R}^+$ such that $\mu_\alpha = \mu \circ g_\alpha^{-1}$ and $\mu(\Omega) = 1$. Because of the Hahn extension theorem, it is sufficient to prove that μ is σ-additive on Σ_0. This in turn holds if it is shown that for any $A_n \in \Sigma_0$, $A_n \supset A_{n+1}$, one has $\bigcap_{n=1}^\infty A_n = \varnothing \Rightarrow \mu(A_n) \downarrow 0$, or equivalently, if $\mu(A_n) \geqslant \epsilon > 0$ for all n implies $\bigcap_{n=1}^\infty A_n \neq \varnothing$. Let us establish the latter.

Since $A_n \in \Sigma_0 = \bigcup_{\alpha \in \mathscr{F}} g_\alpha^{-1}(\Sigma_\alpha)$, there exists an $\alpha_n \in \mathscr{F}$ such that $A_n \in g_{\alpha_n}^{-1}(\Sigma_{\alpha_n})$, $A_n \supset A_{n+1}$, $\mu(A_n) \geqslant \epsilon > 0$, $n \geqslant 1$. By the directedness of $\mathscr{F}$, for each finite set $\{\alpha_1, \alpha_2, \ldots, \alpha_k\}$ there exists a $\beta_k \in \mathscr{F}$ such that $\alpha_i < \beta_k$, $i \leqslant k$, and $\beta_k < \beta_{k+1}$, just as in the proof of Theorem 2. In other words we may assume that $A_n \in g_{\beta_n}^{-1}(\Sigma_{\beta_n})$, $\beta_n < \beta_{n+1}$, $n \geqslant 1$. Let $B_n \in \Sigma_{\beta_n}$ be such that $A_n = g_{\beta_n}^{-1}(B_n)$, i.e., B_n is the "base" of the cylinder A_n. Now we use the hypothesis that each μ_α is a Radon measure. Since $\mu(A_n) = \mu \circ g_{\beta_n}^{-1}(B_n) = \mu_{\beta_n}(B_n)$ by definition, we can use the fact that μ_{β_n} is a Radon probability, implying the existence of a *compact* set $C_n \subset B_n$, $C_n \in \Sigma_{\beta_n}$ giving the approximation

$$\mu_{\beta_n}(C_n) > \mu_{\beta_n}(B_n) - \frac{\epsilon}{2^{n+1}}, \qquad n \geqslant 1. \tag{11}$$

But $D_n = g_{\beta_n}^{-1}(C_n)$ $(\subset A_n)$ need not be monotone. So let $E_n = \bigcap_{k=1}^n D_k \subset A_n$. Clearly $E_n \in \Sigma_0$ and $E_n \supset E_{n+1}$. The continuity of the g_α's and the compactness of C_n imply that D_n is closed, so that E_n is closed. We assert that there exists a compact base $\tilde{C}_n \in \Sigma_{\beta_n}$ such that $E_n = g_{\beta_n}^{-1}(\tilde{C}_n)$. Clearly this is true for $n = 1$; if $n > 1$, consider sets $K_{in} = g_{\beta_i\beta_n}^{-1}(C_i)$. Then each K_{in} is closed, since $g_{\beta_i\beta_n}$ is continuous and C_i is compact; also, $K_{nn} = C_n$, and $K_{in} \subset \Omega_{\beta_n}$, $i = 1, 2, \ldots, n$. If $\tilde{C}_n = \bigcap_{i=1}^n K_{in}$, then $\tilde{C}_n$ is compact and $E_n = g_{\beta_n}^{-1}(\tilde{C}_n) = g_{\beta_n}^{-1}(\bigcap_{i=1}^n K_{in}) = \bigcap_{i=1}^n g_{\beta_n}^{-1}(K_{in}) = \bigcap_{i=1}^n g_{\beta_n}^{-1}(g_{\beta_i\beta_n}^{-1}(C_i)) = \bigcap_{i=1}^n g_{\beta_i}^{-1}(C_i) = \bigcap_{i=1}^n D_i$, as asserted. This implies that $\mu(E_n) = \mu_{\beta_n}(\tilde{C}_n)$. We now show, using (11), that $\bigcap_{n=1}^\infty A_n \supset \bigcap_{n=1}^\infty E_n \neq \varnothing$. Since the inclusion is clear, only the nonempty part is to be verified.

We first note that $\mu(E_n) = \mu_{\beta_n}(\tilde{C}_n) > 0$. In fact, by (11),

$$\mu(D_n) = \mu_{\beta_n}(C_n) > \mu_{\beta_n}(B_n) - \frac{\epsilon}{2^{n+1}} = \mu(A_n) - \frac{\epsilon}{2^{n+1}},$$

so that

$$\mu(A_n) - \mu(D_n) < \epsilon/2^{n+1}. \tag{12}$$

Also $D_n - E_n = D_n - \bigcap_{i=1}^n D_i = \bigcup_{i=1}^{n-1}(D_n - D_i)$, and $D_n \subset A_n \subset A_i$, $i \leqslant n-1$. So

$$\mu(D_n - E_n) \leqslant \sum_{i=1}^{n-1} \mu(D_n - D_i) \leqslant \sum_{i=1}^{n-1} \mu(A_i - D_i) \leqslant \sum_{i=1}^{n-1} \frac{\epsilon}{2^{i+1}}, \qquad \text{by (12).} \tag{13}$$

Hence, using $E_n \subset D_n \subset A_n$, we get

$$\begin{aligned}\mu(A_n) - \mu(E_n) &= \mu(A_n - E_n)\\ &\leqslant \mu(A_n - D_n) + \mu(D_n - E_n)\\ &\leqslant \sum_{i=1}^{n} \frac{\epsilon}{2^{i+1}} < \frac{\epsilon}{2} \qquad \text{by (12) and (13).}\end{aligned}$$

Consequently, $\mu_{\beta_n}(\tilde{C}_n) = \mu(E_n) > \mu(A_n) - \epsilon/2 \geqslant \epsilon/2 > 0$. Thus $\tilde{C}_n$ is non-empty. Let $\omega_n^n \in \tilde{C}_n$ be any point. Since $E_m \supset E_n$ for $m < n$, we get

$$\begin{aligned}g_{\beta_m}^{-1}(\tilde{C}_m) &= \left(g_{\beta_m\beta_n} \circ g_{\beta_n}\right)^{-1}(\tilde{C}_m)\\ &= g_{\beta_n}^{-1}\left(g_{\beta_m\beta_n}^{-1}(\tilde{C}_m)\right) \supset g_{\beta_n}^{-1}(\tilde{C}_n),\end{aligned} \tag{14}$$

and it follows from $g_\beta(\Omega) = \Omega_\beta$ and (14) that $g_{\beta_m\beta_n}^{-1}(\tilde{C}_m) \supset \tilde{C}_n$. Now applying $g_{\beta_m\beta_n}$ to both sides, one has [cf. Lemma 3.3.1(v)]

$$\tilde{C}_m \supset g_{\beta_m\beta_n}\left(g_{\beta_m\beta_n}^{-1}(\tilde{C}_m)\right) \supset g_{\beta_m\beta_n}(\tilde{C}_n).$$

Hence, if $\omega_m^n = g_{\beta_m\beta_n}(\omega_n^n)$, then $\omega_m^n \in \tilde{C}$ for all $n > m$. Fix m, and consider $\{\omega_m^r : r \geqslant m\} \subset \tilde{C}_m$. The compactness of the latter set ensures the existence of a convergent subsequence $\{\omega_m^{r'}\}$ of this sequence with a limit $\omega_m^0 \in \tilde{C}_m$. Since the procedure can be repeated for each $m \geqslant 1$, we get, on using the continuity of the $g_{\alpha\beta}$, for each $s > n \geqslant 1$ (by going to a further subsequence if necessary)

$$\omega_n^0 = \lim_{r'} \omega_n^{r'} = \lim_{r'} g_{\beta_n\beta_s}(\omega_s^{r'}) = g_{\beta_n\beta_s}(\omega_s^0). \tag{15}$$

This implies (the s.m. condition is used again here) that there is an $\omega^0 \in \Omega$ satisfying $g_{\beta_n}(\omega^0) = \omega_n^0$, $n \geqslant 1$. Since $\omega_n^0 \in \tilde{C}_n$ and $g_{\beta_n}^{-1}(C_n) = E_n$, one has that $g_{\beta_k}^{-1}(\{\omega_k^0\}) \subset E_k$, and so the point ω^0 with ω_1^0 as a coordinate is in E_1, that with ω_2^0 as a coordinate is in E_2, etc. Thus for such n, $\omega^0 \in \bigcap_{k=1}^n g_{\beta_k}^{-1}\{\omega_k^0\} \subset \bigcap_{k=1}^n E_k$. Hence $\omega^0 \in \bigcap_{n=1}^\infty E_n \subset \bigcap_{n=1}^\infty A_n$, as desired. Since the uniqueness of μ is obvious, the proof of the theorem is complete.

Even though we started with a projective system of Radon measures, the limit in the above theorem does not have the same topological property. The latter is evidently desirable. A characterization of Radon projective systems that admit Radon limits is a nontrivial task. The appropriate condition has been found by Yu. A. Prokhorov, and it sharpens the above result of Bochner's. We present it in two forms. The following is only a special case of the subsequent general theorem, but its proof is much simpler, and it is sufficient for many applications. Unless both these results are established, the significance of the projective limit theory—indeed, the infinite product measure treatment itself—will not be fully appreciated.

Theorem 7 (Prokhorov). *Let* $\{(\Omega_\alpha, \Sigma_\alpha, \mu_\alpha, g_{\alpha\beta}): \alpha < \beta \text{ in } \mathscr{F}\}$ *be a projective system of Radon probability spaces where* Ω_α *is a completely regular space,* $\alpha \in \mathscr{F}$. *Suppose that* $\Omega = \lim_{\leftarrow}(\Omega_\alpha, g_{\alpha\beta})$ *is rich in the sense that there are continuous onto mappings* $g_\alpha: \Omega \to \Omega_\alpha$ *satisfying* $g_\alpha = g_{\alpha\beta} \circ g_\beta$ *for each* $\alpha < \beta$ *in the directed index set* $\mathscr{F}$, *(so that the continuous mappings* $g_{\alpha\beta}: \Omega_\beta \to \Omega_\alpha$ *are also onto) and for each* $\omega \neq \omega'$ *in* Ω, *there is an* $\alpha \in \mathscr{F}$ *with* $g_\alpha(\omega) \neq g_\alpha(\omega')$ *(i.e., the* g_α*'s separate the points of* Ω*). Then the projective limit* (Ω, Σ, μ) *of the given system exists and is Radon iff the following condition holds: Given* $\epsilon > 0$, *there is a compact set* $K_\epsilon \subset \Omega$ *such that*

$$\mu_\alpha(\Omega_\alpha - g_\alpha(K_\epsilon)) < \epsilon, \qquad \alpha \in \mathscr{F}. \tag{16}$$

Proof. The necessity of (16) is very easy, and we establish it for any Radon system without assuming the complete regularity of the Ω_α's. Thus suppose that the limit space (Ω, Σ, μ) is Radon. Then by the inner regularity of μ, for each $\epsilon > 0$ there is a compact $K_\epsilon \subset \Omega$ (the latter space with the projective limit topology, i.e. $\{g_\alpha^{-1}(V): V \subset \Omega_\alpha, \text{ open}, \alpha \in \mathscr{F}\}$, as a base for its topology), such that $\mu(\Omega - K_\epsilon) < \epsilon$. Also $g_\alpha(K_\epsilon) \subset \Omega_\alpha$, compact, in Σ_α, and g_α is onto. So $g_\alpha(g_\alpha^{-1}(g_\alpha(A))) = g_\alpha(A)$ by Lemma 3.3.1, $g_\alpha^{-1}(\Omega_\alpha - g_\alpha(K_\epsilon)) \subset \Omega - K_\epsilon$, and

$$\mu(\Omega_\alpha - g_\alpha(K_\epsilon)) = \mu \circ g_\alpha^{-1}(\Omega_\alpha - g_\alpha(K_\epsilon)) \leqslant \mu(\Omega - K_\epsilon) < \epsilon, \qquad \alpha \in \mathscr{F}.$$

Hence (16) holds. Thus it is the sufficiency part that is nontrivial.

For the converse, we first suppose that each Ω_α is compact. Then the fact that $\mu_\alpha(\Omega_\alpha) = 1$ implies each Ω_α is nonempty, and $\Omega = \lim_{\leftarrow}(\Omega_\alpha, g_{\alpha\beta})$ is also nonempty and compact. (This is not difficult to establish. For a sketch, see Exercise 6.) Next, $g_\alpha = g_{\alpha\beta} \circ g_\beta \Rightarrow g_\alpha^{-1}(\Sigma_\alpha) = g_\beta^{-1}(g_{\alpha\beta}^{-1}(\Sigma_\alpha)) \subset g_\beta^{-1}(\Sigma_\beta)$. Since continuous functions on $\Omega_\beta \to \Omega_\alpha$ are $(\Sigma_\alpha, \Sigma_\beta)$-measurable and so the $g_{\alpha\beta}$ are, we have that $\mathscr{C} = \bigcup\{g_\alpha^{-1}(\Sigma_\alpha): \alpha \in \mathscr{F}\}$ is an algebra. But $\mu_0: \mathscr{C} \to [0, 1]$, defined by $\mu_0 \circ g_\alpha^{-1} = \mu_\alpha$, $\alpha \in \mathscr{F}$, is unambiguous and additive, as seen in the preceding proof. We now assert that μ_0 is inner regular relative to the compact sets of $\mathscr{C}$. Indeed, let $A \in \mathscr{C}$, so that $A = g_\alpha^{-1}(B)$ for some $B \in \Sigma_\alpha$. By (16), for each $\epsilon > 0$ there is a compact set $K_\epsilon \subset \Omega$ satisfying $\mu_\alpha(g_\alpha(K_\epsilon)) > 1 - (\epsilon/2)$. Let $K_1 = g_\alpha(K_\epsilon)$. Then $\tilde{K}_1 = g_\alpha^{-1}(K_1) \in \mathscr{C}$ is closed, so that it is compact, because Ω itself is compact (as noted above). Note that K_ϵ need not be in $\mathscr{C}$. Since μ_α is Radon, there exists a compact set $C_\epsilon \subset B$ such that $\mu_\alpha(B - C_\epsilon) < \epsilon/2$. Thus

$$\begin{aligned}\mu_\alpha(C_\epsilon \cap K_1) &= \mu_\alpha(C_\epsilon) + \mu_\alpha(K_1) - \mu_\alpha(C_\epsilon \cup K_1)\\ &> \mu_\alpha(B) - \frac{\epsilon}{2} + 1 - \frac{\epsilon}{2} - 1 = \mu_\alpha(B) - \epsilon.\end{aligned} \tag{17}$$

Now $g_\alpha^{-1}(C_\epsilon \cap K_1) = g_\alpha^{-1}(C_\epsilon) \cap \tilde{K}_1 \in \mathscr{C}$, compact, and this is a subset of $g_\alpha^{-1}(B) \cap \tilde{K}_1 \subset A$. Hence by (17)

$$\mu_0\big(g_\alpha^{-1}(C_\epsilon) \cap \tilde{K}_1\big) = \mu_\alpha(C_\epsilon \cap K_1) > \mu_\alpha(B) - \epsilon = \mu_0(A) - \epsilon. \quad (18)$$

Since $\epsilon > 0$ is arbitrary, and $g_\alpha^{-1}(C_\epsilon) \cap \tilde{K}_1 \subset A$ is compact, we conclude that μ_0 is inner regular on the algebra $\mathscr{C}$ relative to its compact sets. Hence the Henry extension theorem is applicable for $(\mu_0, \mathscr{C})$, so that there is a Radon measure μ on the Borel σ-algebra Σ of Ω extending μ_0. Since $\mathscr{C}$ contains a base for the topology of Ω, this extension is also unique. (See Exercise 2.3.11, but uniqueness here is simple.) Hence the theorem holds if each Ω_α is compact.

Suppose now that each Ω_α is completely regular. Recall that a topological space is completely regular if points are closed and each closed set and a point not in it can be separated by a continuous function. What is of interest here with these Ω_α's is that each such Ω_α can be embedded in a compact Hausdorff space $\tilde{\Omega}_\alpha$ as a dense subset. (This is the Stone-Čech compactification. See the Appendix for a description of the result.) Moreover, each continuous function on Ω_α into any compact space has a unique continuous extension to $\tilde{\Omega}_\alpha$. If $\tilde{g}_{\alpha\beta}: \tilde{\Omega}_\beta \to \tilde{\Omega}_\alpha$ are the extensions of $g_{\alpha\beta}: \Omega_\beta \to \tilde{\Omega}_\alpha$, then by the above properties $\{(\tilde{\Omega}_\alpha, \tilde{g}_{\alpha\beta}): \alpha < \beta \text{ in } \mathscr{F}\}$ becomes a (compact) projective system. If $\tilde{\Omega} = \lim_{\leftarrow} (\tilde{\Omega}_\alpha, \tilde{g}_{\alpha\beta})$, then $\tilde{\Omega}$ is compact (and nonempty). Since each compact projective system automatically satisfies the s.m. condition (cf. the same Exercise 6), by the discussion following Definition 5 there exist $\tilde{g}_\alpha: \tilde{\Omega} \to \tilde{\Omega}_\alpha$, $\alpha \in \mathscr{F}$, such that $\tilde{g}_\alpha = \tilde{g}_{\alpha\beta} \circ \tilde{g}_\beta$. It follows that $\tilde{g}_\alpha$ is an extension of g_α. Similarly, if $i_\alpha: \Omega_\alpha \to \tilde{\Omega}_\alpha$ is the injection map, then $\tilde{\mu}_\alpha$ defined by the equation $\mu_\alpha = \tilde{\mu}_\alpha \circ i_\alpha$ is additive and inner regular on the algebra $i_\alpha(\Sigma_\alpha)$ of $\tilde{\Omega}_\alpha$, and since Ω_α is dense in $\tilde{\Omega}_\alpha$, it follows that $\tilde{\mu}_\alpha$ has a unique σ-additive extension to be a Radon measure on $\tilde{\Omega}$. (This is also a simple consequence of Theorem 2.3.10.) Then the fact that $\tilde{g}_{\alpha\beta} \circ \tilde{g}_{\beta\gamma} = \tilde{g}_{\alpha\gamma}$ for $\alpha < \beta < \gamma$ implies $\tilde{\mu}_\alpha \circ \tilde{g}_{\alpha\beta}^{-1} = \tilde{\mu}_\beta$, so that $\{(\tilde{\Omega}_\alpha, \tilde{\Sigma}_\alpha, \tilde{\mu}_\alpha, \tilde{g}_{\alpha\beta}): \alpha < \beta \text{ in } \mathscr{F}\}$ is a compact projective system. By the preceding paragraph, there exists a unique Radon measure $\tilde{\mu}$ on $\tilde{\Omega}$ such that $\tilde{\mu} \circ \tilde{g}_\alpha^{-1} = \tilde{\mu}_\alpha$, $\alpha \in \mathscr{F}$.

Thus far we have not used the hypothesis (16) specifically. With this we now assert that $\tilde{\mu}$ actually has its support in Ω. So (Ω, Σ, μ) with Σ as the Borel σ-algebra will be the desired projective limit.

By (16), $\mu_\alpha(g_\alpha(K_\epsilon)) > 1 - \epsilon$, $\alpha \in \mathscr{F}$, where K_ϵ is the compact subset of Ω there. But $\tilde{\mu}_\alpha = \tilde{\mu} \circ \tilde{g}_\alpha^{-1}$. Hence restricting to Ω_α, one has

$$1 - \epsilon < \mu_\alpha\big(g_\alpha(K_\epsilon)\big) = \tilde{\mu}_\alpha\big(\tilde{g}_\alpha(K_\epsilon)\big) = \tilde{\mu}\big(\tilde{g}_\alpha^{-1}(\tilde{g}_\alpha(K_\epsilon))\big). \quad (19)$$

Since $\tilde{g}_\alpha = \tilde{g}_{\alpha\beta} \circ \tilde{g}_\beta$, $\alpha < \beta$, we get for any $A \subset \Omega_\alpha$

$$\tilde{g}_\alpha^{-1}(\tilde{g}_\alpha(A)) = \tilde{g}_\beta^{-1} \circ \tilde{g}_{\alpha\beta}^{-1} \circ \tilde{g}_{\alpha\beta} \circ \tilde{g}_\beta(A) \supset \tilde{g}_\beta^{-1}\big(g_\beta(A)\big) \supset A,$$

by Lemma 3.3.1 applied to $\tilde{g}_{\alpha\beta}$ and to g_β. Hence $\{\tilde{g}_\alpha^{-1}\tilde{g}_\alpha(A): \alpha \in \mathscr{F}\}$ is a decreasing family for each $A \subset \Omega$. Consequently, with $A = K_\epsilon$ and letting $\tilde{K} = \bigcap\{\tilde{g}_\alpha^{-1} \circ \tilde{g}_\alpha(K_\epsilon): \alpha \in \mathscr{F}\}$, we get $\tilde{K} \supset K_\epsilon$, and if $\omega \in \tilde{K}$, then $\tilde{g}_\alpha(\omega) \in \tilde{g}_\alpha(K_\epsilon)$, $\alpha \in \mathscr{F}$. So applying this result for the closed set $A = \{\omega'\} \subset K_\epsilon$, we get $\tilde{g}_\alpha(\omega) = \tilde{g}_\alpha(\omega')$, $\alpha \in \mathscr{F}$. Since $\tilde{g}_\alpha$'s separate points of Ω (and $\tilde{\Omega}$), it follows that $\omega = \omega'$, and this implies $\tilde{K} = K_\epsilon$. Hence (19) becomes, on taking limits for α in $\mathscr{F}$,

$$1 - \epsilon \leqslant \lim_\alpha \tilde{\mu}\left(\tilde{g}_\alpha^{-1} \circ \tilde{g}_\alpha(K_\epsilon)\right)$$

$$= \lim_\alpha \int_\Omega \chi_{\tilde{g}_\alpha^{-1} \circ \tilde{g}_\alpha(K_\epsilon)}\, d\tilde{\mu} = \int_\Omega \chi_{K_\epsilon}\, d\tilde{\mu} = \tilde{\mu}\left(\tilde{K}_\epsilon\right),$$

since $\tilde{\mu}$ is a (finite) Radon measure and hence is localizable. (See Exercise 2.3.13.) Finally, let $\epsilon = 1/n$, so that $K_{\epsilon_n} \subset K_{\epsilon_{n+1}}$ can be assumed and hence $\tilde{\mu}$ is concentrated on a σ-compact subset of Ω, whence $\tilde{\mu}(\Omega) = 1$. Thus (Ω, Σ, μ), the limit measure of the given system, is Radon ($\mu = \tilde{\mu}|\Omega$), and the proof is complete.

The above demonstration already shows the intricate nature of the arguments needed in establishing the finer structure of the limit measure. The problem is that the cylinder algebra $\mathscr{C}$, determined by the system, may not have enough (compact or closed) sets, so that $\sigma(\mathscr{C})$ may be small (or a proper) σ-subalgebra of the Borel σ-algebra of Ω. This is why the conditions of Theorem 6 have to be strengthened further so that the limit σ-algebra $\sigma(\mathscr{C})$ can be enlarged to contain *all* the compact subsets of Ω, and then μ will be inner regular on *all* Borel sets. The general case to be treated now clarifies this point vividly, and rounds out our treatment of the (not necessarily Cartesian) product measure theory. Since no such embedding is now possible, the method of proof below is different. However, the preceding and the following demonstrations imitate Schwartz's (1973), with more detail and with some important differences. The general case is now given by:

Theorem 8 (Extended Prokhorov). *Let* $\{(\Omega_\alpha, \Sigma_\alpha, \mu_\alpha, g_{\alpha\beta}): \alpha < \beta \text{ in } \mathscr{F}\}$ *be a projective system of Radon probability spaces satisfying the hypothesis of Theorem 7 except that each* Ω_α *is merely Hausdorff. Then the system admits a Radon probability limit* (Ω, Σ, μ) *iff the* (ϵ, K_ϵ) *condition* (16) *holds.*

Proof. Since in the necessity proof of the preceding result the additional complete regularity is not used, it applies here verbatim. Hence only the sufficiency has to be demonstrated, under the sole condition (16). We present the many details of the argument in steps for convenience.

1. As before, let $\mathscr{C} = \bigcup_\alpha g_\alpha^{-1}(\Sigma_\alpha)$ be the cylinder algebra and $\mu_0: \mathscr{C} \to [0, 1]$ be the set function given by $\mu_0 \circ g_\alpha^{-1} = \mu_\alpha$, $\alpha \in \mathscr{F}$. Then we assert that μ_0 is σ-additive on $\mathscr{C}$, and hence has a unique σ-additive extension $\tilde{\mu}$ to $\sigma(\mathscr{C})$. [This

is just the conclusion of Theorem 6, under the s.m. condition, and we claim the same under the condition (16).] For, using the notation of the proof of Theorem 6, we only need to verify for $A_n \downarrow \varnothing$, $A_n \in \mathscr{C}$, that $\mu_0(A_n) \downarrow 0$, since its additivity is clear. Equivalently, if $\epsilon > 0$ and $\mu_0(A_n) \geqslant \epsilon > 0$, then it suffices to show that $\bigcap_{n=1}^{\infty} A_n \neq \varnothing$. We can follow the same argument of that proof until (13) and conclude

$$\mu_0(E_n) > \epsilon/2, \qquad n \geqslant 1, \tag{20}$$

where $E_n = \bigcap_{i=1}^{n} g_{\beta_i}^{-1}(C_i) \subset A_n$. Here one has to use (16) to obtain the desired conclusion. Hence for this $\epsilon > 0$, there is a compact set $K_\epsilon \subset \Omega$ such that $\mu_\alpha(g_\alpha(K_\epsilon)) > 1 - (\epsilon/3)$, $\alpha \in \mathscr{F}$, by (16), since $g_\alpha(K_\epsilon) \in \Sigma_\alpha$. Now each E_n is closed and $E_n \supset E_{n+1}$. So it is enough to show $\bigcap_n E_n \cap K_\epsilon \neq \varnothing$. However, this follows from the compactness of $E_n \cap K_\epsilon$, if it is shown that $E_n \cap K_\epsilon \neq \varnothing$ for each n. But $g_{\beta_n}(K_n) \cap g_{\beta_n}(E_n) \neq \varnothing$, $n \geqslant 1$, since, using the notation preceding (12), we have $E_n = g_{\beta_n}^{-1}(\tilde{C}_n)$, or $g_{\beta_n}(E_n) = g_{\beta_n}(g_{\beta_n}^{-1}(\tilde{C}_n)) = \tilde{C}_n$, by the fact that g_α is onto (see Lemma 3.3.1). Hence by (20),

$$\begin{aligned}\mu_{\beta_n}\big(g_{\beta_n}(E_n) \cap g_{\beta_n}(K_\epsilon)\big) &= \mu_{\beta_n}\big(g_{\beta_n}(E_n)\big) + \mu_{\beta_n}\big(g_{\beta_n}(K_\epsilon)\big) \\ &\quad - \mu_{\beta_n}\big(g_{\beta_n}(E_n) \cup g_{\beta_n}(K_\epsilon)\big) \\ &> \frac{\epsilon}{2} + 1 - \frac{\epsilon}{3} - 1 = \frac{\epsilon}{6} > 0. \end{aligned} \tag{21}$$

So $g_{\beta_n}(E_n) \cap g_{\beta_n}(K_\epsilon) \neq \varnothing$, $n \geqslant 1$. If $E_n \cap K_\epsilon = \varnothing$ for some $n = n_0$, then $K_\epsilon \subset E_{n_0}^c = (g_{\beta_{n_0}}^{-1}(\tilde{C}_{n_0}))^c = g_{\beta_{n_0}}^{-1}(\tilde{C}_{n_0}^c)$. Thus $g_{\beta_{n_0}}(K_\epsilon) \subset \tilde{C}_{n_0}^c$, $g_{\beta_{n_0}}$ being onto. Hence on noting that $\mu_0(E_n) = \mu_{\beta_n}(\tilde{C}_n)$, $n \geqslant 1$, one gets

$$1 - \frac{\epsilon}{3} < \mu_{\beta_{n_0}}\big(g_{\beta_{n_0}}(K_\epsilon)\big) \leqslant \mu_{\beta_{n_0}}\big(\tilde{C}_{n_0}^c\big) < 1 - \frac{\epsilon}{2}.$$

This contradiction shows that $E_n \cap K_\epsilon \neq \varnothing$, $n \geqslant 1$, and so $(\bigcap_n E_n) \cap K_\epsilon \neq \varnothing$. Consequently $\varnothing \neq \bigcap_n E_n \subset \bigcap_n A_n$, and μ_0 is σ-additive, as asserted. We denote by $\tilde{\mu}$ the unique extension of μ_0 to $\sigma(\mathscr{C})$.

2. The measure $\tilde{\mu}$ on $\mathscr{C}$ is inner regular relative to the closed sets of $\mathscr{C}$. For this one can repeat the argument of the compact case in the converse part of the proof of Theorem 7. Here we only have to note that, by the continuity of g_α (and $g_{\alpha\beta}$), $g_\alpha^{-1}(C)$ [and $g_{\alpha\beta}^{-1}(C)$] is closed for each closed or compact C. Hence $g_\alpha^{-1}(C_\epsilon) \cap \tilde{K}_1$ ($\subset A$) is closed, and (18) holds for it. This means every set $A \in \mathscr{C}$ can be approximated from inside by a closed set from $\mathscr{C}$ in computing its measure, i.e., $\tilde{\mu}$ is inner regular on $\mathscr{C}$ for its closed sets, as asserted. We need to extend this to conclude that the result holds with compact sets in order to invoke Theorem 2.3.10, and that is nontrivial, amounting to the rest of the demonstration.

3. The idea of the proof is as follows. If μ^* is the outer measure generated by $(\tilde{\mu}, \sigma(\mathscr{C}))$, then $\tilde{\mu} = \mu^*|\sigma(\mathscr{C})$ by Theorem 2.2.10. If $\mathscr{A}$ is the lattice of all compact sets of Ω, then it is not necessarily true that $\mathscr{A}$ is contained in $\sigma(\mathscr{C})$, and not even in $\mathscr{M}_{\mu^*}$. Hence we show that (step 4) μ^* has the inner regular property on $\sigma(\mathscr{C})$ relative to its closed sets using the similar property of $\tilde{\mu}$ on $\mathscr{C}$ (established above), (step 5) μ^* is outer regular on $\mathscr{A}$ for open sets of Ω, (step 6) μ^* satisfies an (ϵ, K_ϵ) condition, and then (step 7) it is "strongly additive" on $\mathscr{A}$. This will show, with a key theorem of Pettis, that μ^* is additive on the algebra generated by $\mathscr{A}$ and is inner regular there for compact sets. These facts will be sufficient to imply that μ^* has an extension to the Borel σ-algebra Σ of Ω which contains $\mathscr{A}$, $\sigma(\mathscr{C})$, and thus it is the desired Radon measure μ. This will be the Radon limit space (Ω, Σ, μ). We thus establish all these properties in the next six steps.

4. μ^* is inner regular on $\sigma(\mathscr{C})$ relative to the closed sets of $\mathscr{C}$. Indeed, let $\mathscr{D}$ be the class of sets from $\sigma(\mathscr{C})$ with the stated property. Then by step 2, $\mathscr{C} \subset \mathscr{D} \subset \sigma(\mathscr{C})$. We note that $\mathscr{D}$ is a monotone class. Since $\tilde{\mu}$ $[= \mu^*|\sigma(\mathscr{C})]$ is a measure, if $A_n \in \mathscr{D}$, $A_n \uparrow A$, then for each $\epsilon > 0$ there is n_0 such that $\mu(A) - (\epsilon/2) < \tilde{\mu}(A_{n_0})$; and since, by definition of $\mathscr{D}$, there is a closed set $C_\epsilon \subset A_{n_0}$ such that $\tilde{\mu}(A_{n_0} - C_\epsilon) < \epsilon/2$, we get $\tilde{\mu}(A - C_\epsilon) < \epsilon$. Hence $A \in \mathscr{D}$. If $A_n \downarrow A$, then for each n choose a closed set $C_n \subset A_n$ such that $\tilde{\mu}(A_n - C_n) < \epsilon/2^n$. Now $C = \bigcap_n C_n$ $(\subset A = \lim_n A_n)$ is closed and

$$\tilde{\mu}(A - C) \leqslant \tilde{\mu}\left(\bigcup_n (A_n - C_n)\right) \leqslant \sum_{n=1}^{\infty} \tilde{\mu}(A_n - C_n) < \epsilon.$$

Thus $A \in \mathscr{D}$, and the latter is a monotone class. Hence by Proposition 2.1.3(a), $\mathscr{D} = \sigma(\mathscr{C})$, and the assertion follows.

5. μ^* is outer regular on $\mathscr{A}$, i.e., for each compact set $K \subset \Omega$,

$$\mu^*(K) = \inf\{\tilde{\mu}(V): V \in \sigma(\mathscr{C}), \text{open}, V \supset K\}. \tag{22}$$

For, by Theorem 2.2.10, there exists a set $B \in (\sigma(\mathscr{C}))_{\sigma\delta} = \sigma(\mathscr{C})$ which is a measurable cover of K, i.e., $K \subset B$ and $\mu^*(K) = \tilde{\mu}(B)$. Since $\mathscr{C}$ is also a base of topology for Ω, there is an open set $U \in \sigma(\mathscr{C})$ such that $B \subset U$. But then $U - B \in \sigma(\mathscr{C})$, and by the preceding step, for each $\epsilon > 0$, there is a closed set $C_\epsilon \in \sigma(\mathscr{C})$ such that $C_\epsilon \subset U - B$ and $\tilde{\mu}(C_\epsilon) + \epsilon > \tilde{\mu}(U - B)$. Consider the open set $V = U - C_\epsilon \in \sigma(\mathscr{C})$. Then $V \supset B$ (draw a picture), and since $U = (U - B) \cup B = V \cup C_\epsilon$ (disjoint unions), one has

$$\tilde{\mu}(U) = \tilde{\mu}(B) + \tilde{\mu}(U - B) = \tilde{\mu}(V) + \tilde{\mu}(C_\epsilon) > \tilde{\mu}(V) + \tilde{\mu}(U - B) - \epsilon.$$

Thus $\tilde{\mu}(V) \leqslant \tilde{\mu}(B) + \epsilon = \mu^*(K) + \epsilon$, and since $V \supset K$, this gives (22).

6. For each $\epsilon > 0$, there exists a compact set $K_\epsilon \subset \Omega$ such that μ^* satisfies the (ϵ, K_ϵ) condition, i.e., (16) holds for μ^* also. For, by (16), there is a

compact set $K_\epsilon \subset \Omega$ such that $\mu_\alpha(g_\alpha(K_\epsilon)) \geqslant 1 - \epsilon$ for all $\alpha \in \mathscr{F}$. Using the fact that $\mathscr{C}$ is a base for topology, we will show that there is an open set $V \in \mathscr{C}$, $V \supset K$. So $V = g_{\alpha_0}^{-1}(U_0)$ for some $\alpha_0 \in \mathscr{F}$, $U_0 \in \Sigma_{\alpha_0}$, and $\tilde{\mu}(V) = \mu_{\alpha_0}(U_0) \geqslant \mu_{\alpha_0}(g_{\alpha_0}(K_\epsilon)) \geqslant 1 - \epsilon$. Hence by (22) we get $\mu^*(K_\epsilon) \geqslant 1 - \epsilon$. Note that, since $\mathscr{C}$ is an algebra, there is always such a V, e.g., $V = \Omega$. Even if we find an open $V \notin \mathscr{C}$, $V \supset K$, we can find a "better" open $\tilde{V} \in \mathscr{C}$ such that $K \subset \tilde{V} \subset V$ as follows. For each $\omega \in V$, choose an (open) neighborhood of ω, say $O_\omega \subset V$, $O_\omega \in \mathscr{C}$, so that $\{O_\omega : \omega \in V\}$ is an open cover of K_ϵ. Hence there is an $n \geqslant 1$ such that $K_\epsilon \subset \bigcup_{i=1}^n O_{\omega_i} = \tilde{V} \subset V$, and so the preceding argument applies to $\tilde{V}$. Thus $\tilde{\mu}(K_\epsilon) \geqslant 1 - \epsilon$ holds.

7. μ^* on $\sigma(\mathscr{C})$ is inner regular for the compact sets of Ω, i.e., for $\mathscr{A}$. For, let $\epsilon = 1/n$, and K_n be the corresponding compact set in step 6, so that $\mu^*(K_n) \geqslant 1 - 1/n$. It may be assumed that $K_n \subset K_{n+1}$, and we let $S = \bigcup_n K_n$. Then by Theorem 2.2.9(iv), $1 = \lim_n \mu^*(K_n) = \mu^*(S) \leqslant 1$, so that the σ-compact set S has full measure. We assert that

(a) both $\tilde{\mu}$ and μ^* have their supports in S, in the sense that for each $A \in \sigma(\mathscr{C})$ (or $A \subset \Omega$) such that $S \cap A = \varnothing$ we have $\tilde{\mu}(A) = 0 = \mu^*(A)$, and

(b) for any $B \in \sigma(\mathscr{C})$, $\tilde{\mu}(B) = \sup\{\mu^*(C): C \subset B, \text{compact}\}$.

We now establish (a) and (b) together. For this let $\epsilon > 0$ and $B \in \sigma(\mathscr{C})$. Then by step 4, there is a closed set $D_0 \in \sigma(\mathscr{C})$ such that $D_0 \subset B$ and $\tilde{\mu}(B - D_0) < \epsilon/2$. Next observe that $\mu^*(D \cap S) = \mu^*(D)$ for any $D \in \sigma(\mathscr{C})$ from which (a) follows. Indeed, since $\mu^*(D \cap S) \leqslant \mu^*(D)$ and $D \cap S \subset D$, let E be a measurable cover of $D \cap S$. We may take E to satisfy $D \cap S \subset E \subset D$, replacing it with $D \cap E$ if necessary. Thus $\mu^*(D \cap S) = \tilde{\mu}(E)$, and since $S = (S \cap E) \cup (S \cap E^c) \subset E \cup D^c$, we get $1 \leqslant \mu^*(S) \leqslant \tilde{\mu}(E \cup D^c) = \tilde{\mu}(E) + \tilde{\mu}(D^c) \leqslant 1$, by the additivity of $\tilde{\mu}$ on $\sigma(\mathscr{C})$ and the disjointness of E and D^c. Thus there is equality, and then $\tilde{\mu}(E) = 1 - \tilde{\mu}(D^c) = \tilde{\mu}(D)$. Consequently, one has $\tilde{\mu}(D) = \tilde{\mu}(E) = \mu^*(D \cap S)$. Hence if $D \cap S = \varnothing$, then $\tilde{\mu}(D) = 0$ so that $\tilde{\mu}$ has support in S. Since μ^* is generated by $(\tilde{\mu}, \sigma(\mathscr{C}))$, the same is true of μ^*. But $K_n \uparrow S$, so $K_n \cap D_0 \uparrow S \cap D_0$ with $D = D_0$ chosen above. By Theorems 2.2.9(v) and 2.2.10 we get $\mu^*(K_n \cap D_0) \uparrow \mu^*(S \cap D_0) = \tilde{\mu}(D_0)$. Hence for our $\epsilon > 0$, there is n_0 such that $n \geqslant n_0 \Rightarrow \mu^*(K_n \cap D_0) + \epsilon/2 > \tilde{\mu}(D_0) > \tilde{\mu}(B) - \epsilon/2$. Since $K_n \cap D_0$ is compact (D_0 being closed) and $K_n \cap D_0 \subset B$, we get (b) from the inequality $\mu^*(K_n \cap D_0) > \tilde{\mu}(B) - \epsilon$. Thus μ^* is inner regular on $\sigma(\mathscr{C})$ for $\mathscr{A}$.

8. μ^* on $\mathscr{A}$ is strongly additive, i.e., for K_1, K_2 compact we have

$$\mu^*(K_1 \cup K_2) + \mu^*(K_1 \cap K_2) = \mu^*(K_1) + \mu^*(K_2). \tag{23}$$

For, by (22) there are open sets $V_i \in \sigma(\mathscr{C})$, $V_i \supset K_i$, $i = 1, 2$. Then

$$\tilde{\mu}(V_1) + \tilde{\mu}(V_2) = \tilde{\mu}(V_1 \cup V_2) + \tilde{\mu}(V_1 \cap V_2)$$

[since $\tilde{\mu}$ is a measure on $\sigma(\mathscr{C})$]

$$\geqslant \mu^*(K_1 \cup K_2) + \mu^*(K_1 \cap K_2),$$

since $K_1 \cup K_2 \subset V_1 \cup V_2 \in \sigma(\mathscr{C})$ and $K_1 \cap K_2 \subset V_1 \cap V_2 \in \sigma(\mathscr{C})$. Taking infimum as V_i vary, one has

$$\mu^*(K_1) + \mu^*(K_2) \geqslant \mu^*(K_1 \cup K_2) + \mu^*(K_1 \cap K_2). \tag{24}$$

To get the opposite inequality, since $K_1 \cup K_2$ and $K_1 \cap K_2$ are in $\mathscr{A}$, there exist open sets $U_i \in \sigma(\mathscr{C})$, $U_1 \supset K_1 \cup K_2$, and $U_2 \supset K_1 \cap K_2$ by (22). One may assume $U_1 \supset U_2$ (replacing U_2 by the intersection of these sets if necessary). If $D = \overline{U_1 - U_2}$ (closed), define $D_i = D \cap K_i$, $i = 1, 2$. Clearly $D_i \subset U_1$, $i = 1, 2$. But since Ω is Hausdorff and D_1, D_2 are disjoint compact sets, there exist disjoint open sets $\tilde{V}_i \supset D_i$, $i = 1, 2$, which (by intersection with U_1 if necessary) can be taken as subsets of U_1. Using the procedure of step 6 above, we may further select open $V_i \in \sigma(\mathscr{C})$ such that $D_i \subset V_i \subset \tilde{V}_i \subset U_1$. Finally consider $G_i = V_i \cup U_2$, $i = 1, 2$. Then $G_i \in \sigma(\mathscr{C})$, and $G_1 \cup G_2 \subset U_1$, $G_1 \cap G_2 = U_2$, [since $V_1 \cap V_2 = \varnothing$], open. Hence

$$\tilde{\mu}(G_1) + \tilde{\mu}(G_2) = \tilde{\mu}(G_1 \cup G_2) + \tilde{\mu}(G_1 \cap G_2)$$

[since $\tilde{\mu}$ is a measure on $\sigma(\mathscr{C})$]

$$\leqslant \tilde{\mu}(U_1) + \tilde{\mu}(U_2).$$

Hence first varying $G_i \supset K_i$, and then $U_1 \supset K_1 \cup K_2$, $U_2 \supset K_1 \cap K_2$, we get with (22)

$$\mu^*(K_1) + \mu^*(K_2) \leqslant \mu^*(K_1 \cup K_2) + \mu^*(K_1 \cap K_2). \tag{25}$$

Thus (24) and (25) imply (23). Note that, in particular, if K_1, K_2 are disjoint, then μ^* is additive on $\mathscr{A}$.

9. μ^* is inner regular (for compact sets) on the algebra generated by $\mathscr{A}$. For, the property (23) of μ^* on the lattice $\mathscr{A}$ of compact sets turns out to be necessary and sufficient for μ^* to be extendable to a measure on the ring generated by $\mathscr{A}$. This fact is a consequence of an important result of Pettis, to be proved in a more general context (Theorem 9.2.8). To see the regularity of μ^* on the algebra $\tilde{\mathscr{A}}$ generated by $\mathscr{A}$, we use the argument of step 4. Thus let $\mathscr{E}$ be the class of subsets of Ω which are inner regular for μ^* and $\mathscr{A}$. Then $\mathscr{A} \subset \mathscr{E}$, and $\mathscr{A}$ is a π-class. We intend to use Proposition 2.1.3(b). If $A, B \in \mathscr{E}$, $A \cap B = \varnothing$, then clearly $A \cup B \in \mathscr{E}$ by the additivity of μ^*. Also $\varnothing \in \mathscr{E}$ and $\mu^*(\Omega - S) = 0$, where $S = \bigcup_n K_n$ of step 7. So given $\epsilon > 0$, there is n_0 such that $\mu^*(K_{n_0}) > 1 - \epsilon$. Since, by step 7 again, μ^* is supported by S, and $\mu^*(D \cap S) = 0$ for $D \subset S^c$, we get $\mu^*(\Omega - K_{n_0}) \leqslant \mu^*(\Omega - S) + \mu^*(S -$

$K_{n_0}) < \epsilon$, and $\Omega \in \mathscr{E}$. The same step also shows that $\mathscr{E}$ is closed under monotone increasing limits. Finally, if $A, B \in \mathscr{E}$, $A \supset B$, then we claim that $A - B \in \mathscr{E}$ again. In fact, there exist measurable covers $M_1, M_2 \in \sigma(\mathscr{C})$ such that $M_1 \supset A$, $M_2 \supset B$, and (as we can choose) $M_1 \supset M_2$, and such that $\mu^*(A) = \tilde{\mu}(M_1)$, $\mu^*(B) = \tilde{\mu}(M_2)$. Consequently, by step 7, for each $\epsilon > 0$ there is a $C_\epsilon \in \mathscr{A}$, $C_\epsilon \subset M_1 - M_2$, such that

$$\mu^*(C_\epsilon) + \epsilon > \tilde{\mu}(M_1 - M_2) = \tilde{\mu}(M_1) - \tilde{\mu}(M_2),$$

[since $\tilde{\mu}$ is a measure on $\sigma(C)$]

$$= \mu^*(A) - \mu^*(B). \tag{26}$$

However, it may not be that $C_\epsilon \subset A - B$. So consider $C = C_\epsilon \cap A$. [If $\mu^*(A) = \mu^*(B)$, then take $C = \varnothing$, and define C as above in the general case.] Then $C \subset A - B$ and $C_\epsilon \cup A \subset M_1$, so that $\mu(A) \leqslant \mu^*(C_\epsilon \cup A) \leqslant \tilde{\mu}(M_1) = \mu^*(A)$. Also, by (23),

$$\mu^*(C_\epsilon) + \mu^*(A) = \mu^*(C_\epsilon \cup A) + \mu^*(C), \qquad \text{since } C_\epsilon \cap A = C$$

$$= \mu^*(A) + \mu^*(C)$$

by the preceding computation. Hence $\mu^*(C_\epsilon) = \mu^*(C)$. This and (26) show that

$$\mu^*(C) + \epsilon \geqslant \mu^*(A) - \mu^*(B) = \mu^*(A - B), \tag{27}$$

since μ^* is a measure on the ring $\tilde{\mathscr{A}}$. Hence $A - B \in \mathscr{E}$. So $\mathscr{E}$ is a λ-class. By Proposition 2.1.3(b), $\mathscr{E}$ then contains the σ-algebra generated by the π-class $\mathscr{A}$. If this is $\mathscr{A}_1$, then we have shown that $\mu^*\colon \mathscr{A}_1 \to \mathbb{R}^+$ is a measure which is inner regular for its compact sets. Hence by (Henry's) Theorem 2.3.10, μ^* extends to a Radon measure μ on the Borel σ-algebra Σ of Ω, which thus contains $\mathscr{A}$, $\sigma(\mathscr{C})$, and $\mu|\sigma(\mathscr{C}) = \tilde{\mu}$. Then (Ω, Σ, μ) is the desired limit Radon space. Since $\mathscr{C}$ is a base for the topology, the measure μ is also unique. Thus the theorem is completely demonstrated.

Discussion. It is now clear that the topological properties of the limit measure obtained in Theorem 6 need considerable additional work. Once (23) and (27) are established, the conclusion also is a consequence of Theorem 9.2.9. There are other proofs—one based on Bourbaki's method given in the author's book (1981), and another given in Schwartz (1973). Still a different proof is obtained by using Exercise 9.2.14, which is based on Theorem 9.2.11. However, none is simpler or noticeably shorter if all details are included. Yu. A. Prokhorov's original result was given for complete separable metric spaces in his fundamental paper in *Theory of Probability and its Applications*, **1** (1956), 157–214, where its power and usefulness in stochastic applications was demonstrated. Thus his work corresponds to Theorem 7. The general form of Theorem 8 is really due to Bourbaki. The basic theorem has justly given fame and position to its originator. Part of the result was also independently obtained by J. R. Choksi (especially the compact case), but the key (ϵ, K_ϵ) condition (16) and the ensuing applications are due to Prokhorov.

The projective limit theory sprang up in the context of stochastic processes, and the first basic result was given by A. Kolmogorov in the early 1930s. (See Corollary 10 below.) That was adequate for most of the early work, but its abstraction as a projective system, necessitated by later applications, was formulated by S. Bochner in the early 1950s, when he established Theorem 6. The general study was stimulated by his work.

To understand the significance of the preceding three results, we make two specializations and present a related theorem. Let $\{\Omega_i: i \in I\}$ be a class of Hausdorff spaces, and let $\Omega = \times_{i \in I} \Omega_i$, $\Omega_\alpha = \times_{i \in \alpha} \Omega_i$, $\alpha \in \mathscr{F}$, be Cartesian products with $\mathscr{F}$ as the directed (by inclusion) set of all finite subsets of I. Let $g_{\alpha\beta} = \pi_{\alpha\beta}$, $g_\alpha = \pi_\alpha$ be the coordinate projections on these spaces, as defined at the beginning of this section. If $\{\mu_\alpha: \alpha \in \mathscr{F}\}$ is a family of Radon probabilities such that $\mu_\alpha = \mu_\beta \circ \pi_{\alpha\beta}^{-1}$ for $\alpha < \beta$ in $\mathscr{F}$ (so it is a *consistent* class), then we have the following result, which is intermediate between Theorem 6 and Kolmogorov's original one. For this reason it will be referred to as a Kolmogorov-Bochner theorem. One should note, however, that no regularity of the limit measure is claimed in this or in the next (Kolmogorov's original) result, and they are really consequences of Theorem 6 itself. The easy analogs of the condition (16) will be left to the reader.

Corollary 9 (Kolmogorov-Bochner). *Let $\{(\Omega_i, \Sigma_i): i \in I\}$ be Borel measurable spaces where each Ω_i is a nonempty Hausdorff space with a countable base. Let $\Sigma_\alpha = \bigotimes_{i \in \alpha} \Sigma_i$, and let μ_α be any Radon probability on Σ_α for each α in $\mathscr{F}$ such that $\mu_\alpha = \mu_\beta \circ \pi_{\alpha\beta}^{-1}$, $\alpha < \beta$. Then there exists a unique probability μ on $\Sigma = \bigotimes_{i \in I} \Sigma_i$ such that $\mu_\alpha = \mu \circ \pi_\alpha^{-1}$, $\alpha \in \mathscr{F}$, the class of finite (nonempty) subsets of I.*

Here the separability of each Ω_i implies that each Σ_α is also a Borel σ-algebra on Ω_α (verify this) and each closed set of Ω_α is in Σ_α. Without this hypothesis, Σ_α can be strictly smaller than the Borel algebra of Ω_α, and then the class of probabilities admitted to satisfy the Radon (regularity) hypothesis will be correspondingly restricted. What is essential here is that $\mu_\alpha(A)$ should be approximable by μ_α applied to compact sets from Σ_α contained in A, as (11) demands in the proof of Theorem 6. The following result shows that the hypothesis of Corollary 9 is automatic in many familiar cases.

Let $I \subset \mathbb{R}$, and if $t_1 < t_2 < \cdots < t_n$, $t_i \in I$, are any n (ordered) points, suppose we have a family $\{F_{t_1,\ldots,t_n}: n \geqslant 1, t_i \in I\}$ of probability distribution functions, i.e., $0 \leqslant F_{t_1,\ldots,t_n}(x_1, x_2, \ldots, x_n) \leqslant 1$, $n \geqslant 1$; each is left (or right) continuous and nondecreasing [i.e., the increments $\Delta F \geqslant 0$—e.g., if $n = 2$ and if $x_i \leqslant y_i$, $i = 1, 2$, then $(\Delta F)(x, y) = F(y_1, y_2) - F(x_1, y_1) - F(x_2, y_1) +$

$F(x_1, x_2)]$; and

$$\lim_{x_i \to -\infty} F_{t_1, \ldots, t_n}(x_1, \ldots, x_i, \ldots, x_n) = 0,$$

$$\lim_{x_1 \to \infty, \ldots, x_n \to \infty} F_{t_1, \ldots, t_n}(x_1, \ldots, x_n) = 1.$$

If $\alpha = (t_1, \ldots, t_n)$, define for each Borel set $A \subset \mathbb{R}^n$

$$\mu_\alpha(A) = \int \cdots \int_A dF_{t_1, \ldots, t_n} \tag{28}$$

as a Lebesgue-Stieltjes integral. Here $\Omega_t = \mathbb{R}$ and $\Omega = \mathbb{R}^I = \times_{t \in I} \mathbb{R}_t$, where $\mathbb{R}_t = \mathbb{R}$; $\Sigma_t = \mathscr{B}_t$, the Borel σ-algebra of $\mathbb{R}_t$; and $\mathscr{B} = \bigotimes_{t \in I} \mathscr{B}_t$. Then $(\mathbb{R}^\alpha, \mathscr{B}_\alpha, \mu_\alpha)$ is a Radon probability for each $\alpha \in \mathscr{F}$, and $\mu_\alpha = \mu_\beta \circ \pi_{\alpha\beta}^{-1}$ becomes, for $(t_1, \ldots, t_m) = \alpha \subset \beta = (t_1, \ldots, t_m, \ldots, t_n)$,

$$\begin{aligned} \mu_\alpha(A) &= \mu_\beta\left(\pi_{\alpha\beta}^{-1}(A)\right) = \mu_\beta(A \times \mathbb{R}^{\beta - \alpha}) \\ &= \int \cdots \int_A \left[\int \cdots \int_{\mathbb{R}^{\beta-\alpha}} \right] dF_{t_1, \ldots, t_n} \\ &= \int \cdots \int_A dF_{t_1, \ldots, t_m}. \end{aligned} \tag{29}$$

Thus we can translate the *compatibility condition* in terms of F's, for each $t_1 < t_2 < \cdots < t_n$ of I, $n \geqslant 1$, as $[(i_1, \ldots, i_n)$ denotes a permutation of $(1, \ldots, n)]$

$$\lim_{x_n \to +\infty} F_{t_1, \ldots, t_n}(x_1, \ldots, x_n) = F_{t_1, \ldots, t_{n-1}}(x_1, \ldots, x_{n-1}), \tag{30a}$$

$$F_{t_{i_1}, \ldots, t_{i_n}}(x_{i_1}, \ldots, x_{i_n}) = F_{t_1, \ldots, t_n}(x_1, \ldots, x_n). \tag{30b}$$

The second condition, which is not explicit in (29), comes from the fact that $\mu_\alpha = \mu_\beta \circ \pi_{\alpha\beta}^{-1}$ should be true for all $\beta \supset \alpha$ and this should not depend on the order structure of the t_i's selected from I. This precise form was originally discovered by Kolmogorov in 1933, and all the later work on this problem is essentially based on his result. With this reduction, Corollary 9 takes the following form: (here $\mathscr{B} = \bigotimes_{i \in I} \mathscr{B}_i$, $\mathscr{B}_i$ the Borel σ-algebra of $\mathbb{R}$)

Corollary 10 (Kolmogorov). *Let* $\{F_{t_1, \ldots, t_n}: t_i \in I,\ n \geqslant 1\}$ *be a family of distribution functions on* $\mathbb{R}^n$, $n \geqslant 1$, *satisfying the compatibility relations* (30). *Then on the measurable space* $(\mathbb{R}^I, \mathscr{B})$ *there exists a unique probability measure* μ *such that for each Borel set* $A \subset \mathbb{R}^n$, $n \geqslant 1$, $\alpha = (t_1, \ldots, t_n) \in \mathscr{F}$, *we have*

$$\mu\left(\pi_\alpha^{-1}(A)\right) = \int \cdots \int_A dF_{t_1, \ldots, t_n}. \tag{31}$$

Note that if $F_{t_1,\ldots,t_n}(x_1,\ldots,x_n) = \prod_{i=1}^n F_{t_i}(x_i)$, $x_i \in \mathbb{R}$, is a product of one dimensional distributions, then this result reduces to Theorem 2. But the latter is *not* a special case of this result, since there was no topological hypothesis in that theorem. Using the ideas of proof of Theorem 2 and of Exercise 2.3, it is possible to formulate a non-Cartesian product result without topology, but supplemented with a suitable hypothesis. This was done by C. Ionescu Tulcea, and it can be given here as:

Corollary 11 (Tulcea). *Let $\{(\Omega_n, \Sigma_n): n \geq 1\}$ be a collection of abstract measurable spaces, $\alpha_n = (1, 2, \ldots, n)$, and $\Omega_{\alpha_n} = \times_{i \in \alpha_n} \Omega_i$, $\Sigma_{\alpha_n} = \otimes_{i \in \alpha_n} \Sigma_i$. Suppose that $\nu_{\alpha_n}: \Omega_{\alpha_{n-1}} \times \Sigma_{\alpha_n} \to \mathbb{R}^+$, with $\nu_{\alpha_1} = \nu_1$ as the initial measure on Σ_1, are given, where ν_{α_n} is a product-like (or transition type) probability, i.e., there are $\nu_n(\cdot, \cdot)$ such that $\nu_n(\cdot, A_n)$ is measurable for $\Sigma_{\alpha_{n-1}}$ for each $A_n \in \Sigma_n$, and $\nu_n(\omega_1, \ldots, \omega_{n-1}; \cdot)$ is a probability measure on Σ_n for each $(\omega_1, \ldots, \omega_{n-1}) \in \Omega_{\alpha_{n-1}}$ in terms of which we have*

$$\nu_{\alpha_n}(A_1 \times \cdots \times A_n) = \int_{A_1} \nu_1(d\omega_1) \int_{A_2} \nu_2(\omega_1; d\omega_2) \cdots$$

$$\times \int_{A_n} \nu_n(\omega_1, \ldots, \omega_{n-1}; d\omega_n), \tag{32}$$

for each rectangle $A_1 \times \cdots \times A_n \in \Sigma_{\alpha_n}$. Then there exists a unique probability ν on $(\times_{k=1}^{\infty} \Omega_k, \otimes_{k=1}^{\infty} \Sigma_k)$ such that $\nu_{\alpha_n} = \nu \circ \pi_{\alpha_n}^{-1}$, $n \geq 1$, where as usual $\pi_{\alpha_n}: \times_{k=1}^{\infty} \Omega_k \to \Omega_{\alpha_n}$ is the coordinate projection.

The key point of this result is the formulation of the measures ν_{α_n} by the exact condition (32), which reduces to Theorem 2 if each $\nu_n(\cdot, A_n)$ is a constant, i.e., $\nu_n(\omega_1, \ldots, \omega_{n-1}; A) = \tilde{\nu}_n(A)$ does not depend on the ω_k's. When once this is recognized, the proof of this result is the same as that of Jessen's theorem, and the details are omitted (see Exercise 4). The transition probabilities of Section 3 and the $\nu_n(\cdot, \cdot)$ here exist and are extensively used in the theory of Markov processes. Thus the non-Cartesian product measures given in the preceding six results play an important role in such applications, and form a basis of other extensions. [See, e.g., the author's book (1984), Chapter 3.]

Finally, it should be noted that all infinite product measures are closely related to martingale processes. To see this, let $\{f_n, \Sigma_n: n \geq 1\}$ be a martingale on a probability space (Ω, Σ, μ). Let $\nu_n(A) = \int_A f_n \, d\mu$, $A \in \Sigma_n$. Then $\nu_n = \nu_{n+1}|\Sigma_n$, by the martingale property. Hence letting $\Omega_\alpha = \Omega$, $\Sigma_\alpha = \Sigma_n$ for $\alpha = (1, 2, \ldots, n)$, $g_{\alpha\beta}$ = identity for $\alpha < \beta$, we get that $\{(\Omega_\alpha, \Sigma_\alpha, \nu_\alpha, g_{\alpha\beta}): \alpha < \beta \text{ in } \mathscr{F}\}$ is a (signed) projective system with $\mathscr{F}$ as a directed set of α's and $\nu_\alpha = \nu_n$. Here $\{\Sigma_\alpha: \alpha \in \mathscr{F}\}$ is called a *base* of the martingale, and $\{\nu_\alpha, \Sigma_\alpha: \alpha \in \mathscr{F}\}$ a *set martingale*. Conversely, each projective system defines a set martingale. Indeed, let $\{(\Omega_\alpha, \Sigma_\alpha, \mu_\alpha, g_{\alpha\beta}): \alpha < \beta \text{ in } \mathscr{F}\}$ be a projective system of measure spaces with the s.m. property, so that $\Omega = \lim_{\leftarrow}(\Omega_\alpha, g_{\alpha\Omega})$ exists. Let $g_\alpha: \Omega \to \Omega_\alpha$ be the mappings determined by the $g_{\alpha\beta}$, and define $\Sigma_\alpha^* = g_\alpha^{-1}(\Sigma_\alpha)$, $\Sigma_0 = \bigcup_{\alpha \in \mathscr{F}} \Sigma_\alpha^*$. Then we know from Theorem 2 that there exists a unique finitely additive $\mu: \Sigma_0 \to \mathbb{R}^+$ such that $\mu_\alpha = \mu \circ g_\alpha^{-1}$. If we let $\mu_\alpha^* = \mu | \Sigma_\alpha^*$, then $\mu_\alpha^* \circ g_\alpha^{-1} = \mu_\alpha$ and μ_α^* is σ-additive, $\mu_\alpha^* = \mu_\beta^* | \Sigma_\alpha^*$, $\alpha < \beta$, so that $\{\mu_\alpha^*, \Sigma_\alpha^*: \alpha \in \mathscr{F}\}$ is a set martingale. When there is a fixed measure P on Σ_0 such that $P | \Sigma_\alpha^*$ is localizable, $\mu_\alpha^* \ll P$, let $f_\alpha = d\mu_\alpha^*/dP$, the Radon-Nikodým derivative; then $\{f_\alpha, \Sigma_\alpha^*: \alpha \in \mathscr{F}\}$ is a martingale. The identification of these two theories makes Theorem 3 especially illuminating and takes the product measure theory into probability: these areas are deeply interrelated. We have to leave the subject at this point, as it leads to specializations which can not be pursued here.

EXERCISES

0. Let $\{(\Omega_\alpha, \Sigma_\alpha, \mu_\alpha, g_{\alpha\beta}): \alpha < \beta \text{ in } I\}$ be a projective system of probability spaces satisfying the sequential maximality condition. Show that the system admits a projective limit iff each (directed) countable subsystem of the given system admits such a limit. Thus it is the countable systems that exhibit all the difficulties. [Since every subdirected system of a projective system is again projective, the sufficiency is easy. For the converse, if every countable system admits a limit, then observe that the limit is uniquely defined and does not depend on the choice of the suffixes of the countable set. If all countable (directed) systems of I are again directed by inclusion, then these limit families form a projective system. Observing that the σ-algebra $\sigma(\bigcup_\alpha g_\alpha^{-1}(\Sigma_\alpha))$ is also the same as $\bigcup_N \sigma(g_N^{-1}(\Sigma_N))$, where Σ_N is the σ-algebra of a countable index $N = \{\alpha_1, \alpha_2 \dots\} \subset I$, (cf. footnote at the end of the first proof of Theorem 5.3.3) the result follows since the countable additivity of the measure uses only a countable directed system.]

1. Suppose that $\{(\Omega_i, \Sigma_i, \mu_i): i \in I\}$ is a family of measure spaces such that $\mu_i(\Omega_i) = 1$ for all but a finite number of i in the index set I, whose cardinality is at least that of the integers. Suppose that the nonfinite μ_i have the finite subset property. Show that there exists a unique (Cartesian) product measure μ (with the finite subset property) on the product measurable space (Ω, Σ) of Theorem 2, such that $\mu \circ \pi_\alpha^{-1} = \mu_\alpha$ for each finite subset α of I. (This extends Theorems 2 and 2.1, in a simple way.)

2. After proving Theorem 2, we showed that there exists a one to one and onto mapping $u: \Omega = \times_{\iota \in I} \Omega_\iota \to \tilde{\Omega} = \times_{\alpha \in \mathscr{F}} \Omega_\alpha$, where $\mathscr{F}$ is the directed set of all finite (nonempty) subsets of I. Show that the same u takes the cylinder σ-algebra $\Sigma = \bigotimes_{\iota \in I} \Sigma_\iota$ onto the cylinder σ-algebra $\tilde{\Sigma} = \bigotimes_{\alpha \in \mathscr{F}} \Sigma_\alpha$ in the sense that each element of Σ is mapped onto a unique element of $\tilde{\Sigma}$ and conversely.

3. Let $\{(\Omega_\alpha, g_{\alpha\beta}): \alpha < \beta \text{ in } \mathscr{F}\}$ be a directed system of spaces satisfying the sequential maximality condition. Let $\Omega = \lim_{\leftarrow}(\Omega_\alpha, g_{\alpha\beta})$, and $g_\alpha: \Omega \to \Omega_\alpha$ be defined for each thread $\omega = \{\omega_\alpha: \alpha \in \mathscr{F}\} \in \Omega$ by $g_\alpha: \omega \mapsto \omega_\alpha$. Show g_α is an onto mapping that verifies $g_\alpha = g_{\alpha\beta} \circ g_\beta$ for each pair $\alpha < \beta$ in the directed index set $\mathscr{F}$.

4. Complete the details of proof of Corollary 11, following the pattern of Theorem 2.

5. Complete the proof that each projective system with the s.m. property defines a set martingale, and that if the projective system admits a projective limit (Ω, Σ, μ) and if there is a probability measure P on this space dominating the measure μ, then there is associated a point martingale which converges in the $L^1(P)$-mean. (The convergence part is not easy but is worthy of an attempt.)

6. Let $\{(\Omega_\alpha, g_{\alpha\beta}): \alpha < \beta \text{ in } \mathscr{F}\}$ be a projective system of nonempty compact Hausdorff spaces. Show that $\Omega = \lim_{\leftarrow}(\Omega_\alpha, g_{\alpha\beta})$ is a nonempty compact space and the s.m. condition is automatically satisfied by the system. [If $C_{\alpha\beta} = \{\omega = (\omega_\alpha, \alpha \in \mathscr{F}) \in \times_{\alpha \in \mathscr{F}} \Omega_\alpha = \Omega_{\mathscr{F}}: \omega_\alpha = g_{\alpha\beta}(\omega_\beta)\}$, then $C_{\alpha\beta} = G_{\alpha\beta} \times \Omega_{\mathscr{F} - \{\alpha,\beta\}}$, where $G_{\alpha\beta} = \{(\omega_\alpha, \omega_\beta): \omega_\alpha = g_{\alpha\beta}(\omega_\beta)\}$ is the graph of $g_{\alpha\beta}$. Then $\Omega = \bigcap\{C_{\alpha\beta}: \alpha < \beta \text{ in } \mathscr{F}\}$ is compact (Tychonov's theorem), and is nonempty because each $C_{\alpha\beta}$ is. Next verify that $D_\alpha = g_\alpha^{-1}(\{\omega_\alpha\}) \subset \Omega$ is closed and nonempty, and $D_{\alpha_i} \supset D_{\alpha_{i+1}}$, $\alpha_n < \alpha_{n+1}$, $n \geqslant 1$. So any point $\omega \in \bigcap_{n=1}^\infty D_{\alpha_n}$ satisfies $g_{\alpha_n}(\omega) = \omega_{\alpha_n}$, and the s.m. condition holds.]

7. This exercise contains a different property of the space constructed in Exercise 1.2.3 to describe an infinite sequence of tosses of a coin. Let Ω_i be the two point space $\{0,1\}$ for $i = 1, 2, \ldots$, let $\Sigma_i = \{\varnothing, \{0\}, \{1\}, \{0,1\}\}$, and let μ_i on Σ_i be a probability given by $\mu(\{0\}) = p$, $0 < p < 1$. If $\Omega = \times_{i=1}^\infty \Omega_i$, $\Sigma = \otimes_{i=1}^\infty \Sigma_i$, and $\mu = \otimes_{i=1}^\infty \mu_i$, show that

$$\mu \circ \pi_{\alpha_n}^{-1}(A_n) = \mu_{\alpha_n}(I_n) = \binom{n}{k} p^k q^{n-k},$$

where $q = 1 - p$, $\alpha_n = (1, 2, \ldots, n)$, $A_n = I_n \times \Omega_{\alpha_n^c}$, and $I_n \subset \Omega_{\alpha_n} = \times_{i \in \alpha_n} \Omega_i$ is the set consisting of all n-tuples of k zeros and $n - k$ ones. If $g: \Omega \to \{0, 1, 2, \ldots\}$ is defined as $g(\omega) =$ (the number of 0's in ω), show that with the power set of the range space as its σ-algebra, g is measurable for Σ. If $f: \mathbb{R}^+ \to \mathbb{R}^+$ is a Borel function, show $f \circ g: \Omega \to \mathbb{R}^+$ is measurable for Σ and find $\int_\Omega f(g)\, d\mu$.

8. Let (Ω, Σ, μ) be the product space as in the above exercise. If $h: \Omega \to \{0, 1, 2, \ldots\}$ is a mapping defined by $h(\omega) =$ (the number of 0's in ω before the first 1), verify that h is measurable for Σ, and show that $\int_\Omega h^2\, d\mu = (1 + p)/q^2$. [The series $\sum_{n=1}^\infty n^k t^n$ is uniformly and absolutely convergent for $|t| < 1$, and differentiation and summation commute.]

9. If (Ω, Σ) is a measurable space, and μ_1, μ_2 are two probability measures, then the *Hellinger distance* between μ_1, μ_2, denoted $H(\mu_1, \mu_2)$ (cf. Exercise 5.3.15) is $H(\mu_1, \mu_2) = \int_\Omega (f_1 f_2)^{1/2}\, d\mu$, where μ is any σ-finite measure on Σ such that $\mu_i \ll \mu$ (e.g., $\mu = \mu_1 + \mu_2$ will do), and $f_i = d\mu_i/d\mu$, $i = 1, 2$. Verify that $0 \leqslant H(\mu_1, \mu_2) = H(\mu_2, \mu_1) \leqslant 1$, and H depends on μ_1, μ_2 but not on μ; that $H(\mu_1, \mu_2) = 0$ iff $\mu_1 \perp \mu_2$, and $H(\mu_1, \mu_2) = 1$ iff $\mu_1 = \mu_2$. Show moreover that if $\Sigma_1 \subset \Sigma$ is a σ-subalgebra and $\mu_i' = \mu_i|\Sigma_1$, then $H(\mu_1, \mu_2) \leqslant H(\mu_1', \mu_2')$. [First verify that $H(\mu_1, \mu_2) = \inf\{\Sigma_n [\mu_1(A_n)\mu_2(A_n)]^{1/2} : A_n \in \Sigma, \text{ disjoint}, \Omega = \bigcup_n A_n\}$, and use the fact that Σ_2 has fewer partitions of Ω than Σ.]

10. (Continued) Let

$$\left\{\left(\Omega_\alpha, \Sigma_\alpha, \begin{matrix}\nu_\alpha \\ \mu_\alpha\end{matrix}, g_{\alpha\beta}\right) : \alpha < \beta \text{ in } \mathscr{F}\right\}$$

be two projective systems of probability spaces. Verify that for each $\alpha < \beta$, $H(\mu_\alpha, \nu_\alpha) \geqslant H(\mu_\beta, \nu_\beta)$. Suppose that

$$\mu = \underset{\leftarrow}{\lim}\, \mu_\alpha \qquad \text{and} \qquad \nu = \underset{\leftarrow}{\lim}\, \nu_\alpha$$

exist. Then show that $\lim_\alpha H(\mu_\alpha, \nu_\alpha) = H(\mu, \nu)$, and deduce that $\mu \perp \nu$ iff this limit is zero. In particular, $\mu \perp \nu$ if $\mu_\alpha \perp \nu_\alpha$ for some $\alpha \in \mathscr{F}$. [Use the hint in the preceding exercise on a computation of $H(\mu_\alpha, \nu_\alpha)$; since $H(\mu_\alpha, \nu_\alpha)$ is monotone and bounded, it suffices to prove the limit for finite partitions. By the directedness of $\mathscr{F}$, this reduces to showing the result for some sufficiently large α_0 in $\mathscr{F}$. A careful ϵ-δ argument is needed to finish the result. This is slightly involved but quite instructive.]

11. Suppose in the above exercise both the systems are Cartesian products of classes and consider

$$\left\{\left(\Omega_n, \Sigma_n, \begin{matrix}\mu_n \\ \nu_n\end{matrix}\right) : n \geqslant 1\right\}.$$

Also assume that for each n, $\nu_n \equiv \mu_n$ (i.e., they are equivalent), so that by Theorem 2 (or rather Proposition 2.5) we have

$$u_n = \frac{d\nu_{\alpha_n}}{d\mu_{\alpha_n}} = \prod_{i=1}^{n} \frac{d\nu_i}{d\mu_i} > 0, \qquad \text{where} \quad \alpha_n = (1, 2, \ldots, n).$$

Deduce that $H(\nu, \mu) = \prod_{i=1}^{\infty} H(\nu_i, \mu_i)$. With this prove that $\nu \equiv \mu$ iff $H(\mu, \nu) > 0$. [If $H(\nu, \mu) = 0$, then $\nu \perp \mu$ by the preceding result, even if the (ν_α, μ_α) are not Cartesian products. In the latter specialized case, the dichotomy occurs, i.e., if $H(\nu, \mu) \neq 0$, then ν, μ are equivalent. For this verify that if $u_n^* = u_n \circ \pi_{\alpha_n} : \Omega \to \mathbb{R}^+$, then for $m < n$ (π_{α_n} here is the usual coordinate projection)

$$\int_{\Omega}\left(\sqrt{u_n^*} - \sqrt{u_m^*}\right)^2 d\mu = 2\left(1 - \prod_{i=m+1}^{n} H(\nu_i, \mu_i)\right).$$

Thus $\{\sqrt{u_n^*} : n \geqslant 1\} \subset L^2(u)$ is a Cauchy sequence, since $H(\mu, \nu) > 0 \Rightarrow \prod_{i=m+1}^{n} H(\nu_i, \mu_i) \to 1$ as $n, m \to \infty$. This will imply, after some computation, that $\int_A u\, d\mu = \nu(A)$, $A \in \Sigma$, where $u_n^* \to u$ in $L^2(\mu)$. The precise product formulation (the dicohotomy) was obtained by S. Kakutani in 1948; its generalization of the singular case of the preceding exercise was given by E. Brody in 1971.]

6.5. TWO APPLICATIONS OF INFINITE PRODUCTS

We present here two novel applications of the preceding section to (i) a characterization of positive definite functions on the integers and the real line, and (ii) realizing an abstract Hilbert space concretely.

Let G be either $\mathbb{R}$ or $\mathbb{Z}$ (the integers). Then a function $\varphi: G \to \mathbb{C}$ is *positive definite* if for $a_i \in \mathbb{C}$, $t_i \in G$, $i = 1, \ldots, n$, $n \geqslant 1$,

$$\sum_{i=1}^{n} \sum_{j=1}^{n} \varphi(t_i - t_j) a_i \bar{a}_j \geqslant 0. \tag{1}$$

Since (1) is an infinite set of conditions, one would like to have a characterization of φ which is better suited for further work. The inequalities (1) contain a large amount of information on φ. For instance, (1) implies the boundedness of φ on G, and its continuity at the identity point 0 of G implies the uniform continuity on G (Exercise 1). Such functions φ appear frequently in Fourier analysis and convolution operations. The main characterization, proved differently by Bochner, shows that the positive definite functions on $\mathbb{R}$ are precisely those which are Fourier transforms of nonnegative integrable functions (or of finite measures). Now we intend to obtain this important result using Corollary 4.9. There are several other proofs, each of which is at least as difficult as this one. But here we illustrate the connection between the projective limit theory and an apparently unrelated result. What is more, this constitutes a purely measure theoretical proof of Bochner's fundamental theorem.

We now start with the simple case of $G = \mathbb{Z}$, due to G. Herglotz:

Proposition 1. *Let $\varphi: \mathbb{Z} \to \mathbb{C}$ be a positive definite function and $\varphi(0) = 1$. Then there exists a unique finite Lebesgue-Stieltjes measure μ on $(0, 2\pi]$ such that*

$$\varphi(k) = \int_0^{2\pi} e^{ikx}\mu(dx), \quad k \in \mathbb{Z}, \qquad \mu((0, 2\pi]) = 1, \tag{2}$$

and conversely, every φ given by (2) is positive definite and satisfies $\varphi(0) = 1$.

Proof. Let φ be positive definite. Then by (1), taking $a_k = e^{ikx}$, $0 < x \leqslant 2\pi$ (x fixed), we have

$$f_n(x) = \frac{1}{2\pi n} \sum_{j,k=1}^{n} \varphi(k-j)e^{i(j-k)x} \geqslant 0,$$

and the set function μ_n given by

$$\mu_n(A) = \int_A f_n(x)\,dx, \tag{3}$$

for each Borel set $A \subset (0, 2\pi]$, defines a probability measure, since $\varphi(0) = 1$. On the other hand, using again the orthogonality relations of the exponentials e^{ijx} for $0 < x \leqslant 2\pi$, one gets

$$\int_0^{2\pi} e^{ijx}\mu_n(dx) = \begin{cases} \varphi(j)(1 - |j|/n) & \text{if } \quad -n < j < n, \\ 0 & \text{otherwise.} \end{cases} \tag{4}$$

But $\{\mu_n : n \geqslant 1\}$ are bounded measures, and hence by the selection principle (Exercise 5.2.13) and the Helly-Bray theorem (Theorem 5.2.12), there is a subsequence $\mu_{n_j}(A) \to \mu(A)$ for each Borel set A. Hence, taking limits in (4) as $n_j \to \infty$, one gets (2).

For the uniqueness, let $\tilde{\mu}$ be another such measure, so that

$$\int_0^{2\pi} e^{ikx}\mu(dx) = \int_0^{2\pi} e^{ikx}\tilde{\mu}(dx), \qquad k \geqslant 1. \tag{5}$$

Clearly the same integrals hold for all trigonometric polynomials by linearity, and then, by the classical Weierstrass theorem asserting that every continuous function on $[0, 2\pi]$ is uniformly approximable by such polynomials, (5) holds for all continuous functions. If $f = \chi_A$, where $A = (a, b) \subset (0, 2\pi]$ is an open interval, then χ_A is lower semicontinuous, so that it can be approximated by continuous functions $0 \leqslant g_n \uparrow f$, by classical results (in advanced calculus). Hence, with the Lebesgue monotone convergence, (5) holds if the integrand is χ_A. Thus $\mu(A) = \tilde{\mu}(A)$ on the σ-algebra generated by such intervals. But this is precisely the Borel σ-algebra (see Proposition 2.1.5). Thus μ in the representation (2) is unique, and every such μ is a Lebesgue-Stieltjes measure. Hence the sufficiency follows.

Conversely, if φ is given by (2), then $\varphi(0) = 1$, since μ is a probability. Let $a_1, \ldots, a_n$ be arbitrary complex numbers. Then (1) becomes

$$\sum_{j=1}^{n}\sum_{k=1}^{n} \varphi(j-k)a_j\bar{a}_k = \int_0^{2\pi}\left(\sum_{j=1}^{n} a_j e^{ijx}\right)\left(\sum_{k=1}^{n} a_j e^{ikx}\right)^{-}\mu(dx) \geqslant 0,$$

since the integrand is nonnegative. Hence φ is positive definite. This completes the proof.

The result admits an easy extension to more than one dimension. We present the precise statement, omitting its proof. (See Exercise 2.)

Proposition 2. *Let $\mathbb{Z}^k$ denote the integer lattice of $\mathbb{R}^k$, $k > 1$, so that each point of $\mathbb{Z}^k$ is a k-tuple of integers and forms a group under componentwise addition. Let $\varphi: \mathbb{Z}^k \to \mathbb{C}$ be a positive definite function with $\varphi(\mathbf{0}) = 1$. Then there is a unique probability μ on the k-torus $\mathbb{T}^k = (0, 2\pi]^k$ such that*

$$\varphi(\mathbf{j}) = \varphi(j_1, \ldots, j_k) = \underset{\mathbb{T}^k}{\int \cdots \int} e^{i\sum_{r=1}^k j_r x_r} \mu(dx_1 \cdots dx_k). \tag{6}$$

Conversely, every φ defined by (6) *is positive definite and* $\varphi(\mathbf{0}) = 1$.

In order to extend the above result, replacing $\mathbb{Z}$ by $\mathbb{R}$, we need to generalize Proposition 2 slightly and obtain its analog using Corollary 4.9. As the following argument (as well as every other known proof) shows, the transition from $\mathbb{Z}$ to $\mathbb{R}$ is delicate. [The 1-torus is denoted simply as $\mathbb{T}$.]

For any set S let $\mathbb{T}^S = \times_{i \in S} \mathbb{T}_i$, with $\mathbb{T}_i = \mathbb{T}$ for all i. Then $\mathbb{T}^S$ may be regarded as the space of all functions on S with values in $\mathbb{T}$. Thus $g \in \mathbb{T}^S$ means $g = \{g(s): s \in S\}$ with $g(s) \in \mathbb{T}$, $s \in S$. If $\pi_s: \mathbb{T}^S \to \mathbb{T}_s$ is the coordinate projection, then we consider $\{\pi_s^{-1}(U): U \subset \mathbb{T}_s, \text{ open}, s \in S\} \subset \mathbb{T}^S$ as a class of sets declared open in $\mathbb{T}^S$, which thus define a base for the (product) topology in $\mathbb{T}^S$ and make each $\pi_s(\cdot)$ continuous. Since $\mathbb{T}$ is compact in this topology, $\mathbb{T}^S$ becomes a compact Hausdorff space. This is A. Tychonov's theorem and, as shown by J. L. Kelley in 1953 is equivalent to a form of the axiom of choice, already used. (Cf. Appendix.) This particular space $\mathbb{T}^S$ becomes a commutative group if for $g_1, g_2 \in \mathbb{T}^S$ we define $g_1 + g_2 = \{g_1(s) + g_2(s) \pmod{2\pi}: s \in S\} \in \mathbb{T}^S$. Thus when S is a one point (finite) set, we have in Proposition 1 (or 2) that φ, defined on $\mathbb{Z}$ (or $\mathbb{Z}^k$), has as its representation space $\mathbb{T}$ (or $\mathbb{T}^k$). Here $\mathbb{T}$ (or $\mathbb{T}^k$) is called the *dual* group of $\mathbb{Z}$ (or $\mathbb{Z}^k$) when addition (componentwise) is the group operation. Notice that $\mathbb{T}^k$ is the same as $\mathbb{T}^S$ with S a finite set. It is also useful to note that if we let $\mathscr{F}$ be the class of all (nonempty) finite subsets of S, then $\mathbb{T}^S$ is identifiable with $\lim_{\leftarrow} \{(\mathbb{T}^F, \pi_F): F \in \mathscr{F}\}$, where $\pi_F: \mathbb{T}^S \to \mathbb{T}^F$, is the coordinate projection in the notation introduced for Theorem 4.6. If $\mathscr{B}_s$ is the Borel σ-algebra of $\mathbb{T}_s$, $\mathscr{B} = \bigotimes_{s \in S} \mathscr{B}_s$, then $\mathscr{B}$ is the cylinder σ-algebra $[= \sigma(\bigcup_{F \in \mathscr{F}} \pi_F^{-1}(\mathscr{B}_F))$ in the same notation, $\mathscr{B}_F = \bigotimes_{s \in F} \mathscr{B}_s]$. Thus $\pi_s: \mathbb{T}^S \to \mathbb{T}_s = \mathbb{T}$ is a measurable function, and in fact each continuous function on $\mathbb{T}^S$ into $\mathbb{T}^F$ is measurable. (If S is countable, then $\mathscr{B}$ is the Borel σ-algebra of $\mathbb{T}^S$.) Let $\mathscr{I}$ be the set of all integer valued mappings on S with finite supports. Then $\mathscr{I}$, under componentwise addition, is a commutative group. (It turns out that $\mathscr{I}$ is the dual of the group $\mathbb{T}^S$. This is not simple, and we do not need it.) For each $F \in \mathscr{F}$, let $\mathscr{I}_F = \{g \in \mathscr{I}: g(S - F) = 0\}$. Then $\mathscr{I}_F$ is a group isomorphic to $\mathbb{Z}^k$, where $k = \text{card}(F)$. By Proposition 2, each positive definite $\varphi: \mathbb{Z}^k (\cong \mathscr{I}_F) \to \mathbb{C}$, $\varphi(0) = 1$, determines uniquely a measure μ_F on $\mathbb{T}^k$ $(\cong \mathbb{T}^F)$, $\mu_F(\mathbb{T}^F) = 1$. From this uniqueness, we deduce with (6) that if $F_1 \subset F_2$ in $\mathscr{F}$, and $\varphi_{F_1} = \varphi_{F_2}|\mathbb{Z}^{\text{card}(F_1)}$, then by integration (and Fubini's theorem) $\mu_{F_1} = \mu_{F_2} \circ \pi_{F_1}^{-1}$. Thus $\{(\mathbb{T}^F, \mathscr{B}_F, \mu_F, \pi_{F_1F_2}): F_1 \subset F_2 \text{ in } \mathscr{F}\}$ is a projective system of Lebesgue-Stieltjes (or Radon) probability spaces which (being compact spaces and also Cartesian product spaces) satisfy sequential maximality (Exercise 4.6). Hence by Theorem 4.7 there exists a unique Radon probability μ on $\mathscr{B}$ such that $\mu \circ \pi_F^{-1} = \mu_F$, $F \in \mathscr{F}$. We summarize the above result with an additional fact in the following:

Proposition 3. *Let $\xi: \mathscr{I} \to \mathbb{C}$ be a positive definite function, with $\xi(\mathbf{0}) = 1$ where $\mathbf{0}$ is the neutral element of $\mathscr{I}$ which maps S into 0. Let $\xi_F = \xi|\mathscr{I}_F$, $F \in \mathscr{F}$, the directed set of all finite (nonempty) subsets of S. Then ξ_F is positive definite, $\xi_F(\mathbf{0}) = 1$, and if μ_F is the corresponding (Radon) probability on $\mathbb{T}^F$, given by (6), then $\{(\mathbb{T}^F, \mathscr{B}_F, \mu_F, \pi_{FG}): F \subset G \text{ in } \mathscr{F}\}$ is a projective system of (Radon) probability spaces with the s.m. property, and so* $\lim_{\leftarrow} (\mathbb{T}^F, \mathscr{B}_F, \mu_F, \pi_{FG}, \mathscr{F}) = (\mathbb{T}^S, \mathscr{B}, \mu)$ *is Radon. Moreover, if $S = \mathbb{R}$, we have*

$$\int_{\mathbb{T}^{\mathbb{R}}} e^{i\sum_{s \in \mathbb{R}} n(s)\theta(s)} \mu(d\theta) = \xi(n), \qquad n \in \mathscr{I}. \tag{7}$$

Proof. It only remains to establish (7). Since for each $n \in \mathscr{I}$, the sum in the exponent of the integrand is finite [$n(\mathbb{R} - F) = 0$ for some finite $F \subset \mathbb{R}$, *depending on n*], the function $= 1$ on $\mathbb{T}^{R-F}$. So considering $\mathbb{T}^R$ as $\mathbb{T}^{\mathbb{R}-F} \times \mathbb{T}^F$, we need to express μ as a (non-Cartesian) product of measures on $\mathbb{T}^{\mathbb{R}-F}$ and $\mathbb{T}^F$, and apply Proposition 3.1 to deduce (7). Let μ_F $(= \mu \circ \pi_F^{-1})$ be the "marginal" measure of μ on $\mathbb{T}^F$. If $A \in \mathscr{B}_F$, $B \in \bigotimes_{s \in F^c} \mathscr{B}_s$, then let $A_1 = A \times \mathbb{T}^F$, $B_1 = B \times \mathbb{T}^{F^c}$, $F^c = \mathbb{R} - F$. Thus $\mu(\pi_F^{-1}(\cdot) \cap B_1)$ is a measure on $\mathscr{B}_F$ which is dominated by μ_F. Since these are finite measures, by the Radon-Nikodým theorem there is $\tilde{\mu}$ with

$$\mu\big(\pi_F^{-1}(A) \cap B_1\big) = \int_A \tilde{\mu}(\omega, B_1)\mu_F(d\omega), \tag{8}$$

for a $\mathscr{B}_F$-measurable $\tilde{\mu}(\cdot, B_1)$. However, it is not evident that $\tilde{\mu}: \Omega_F \times \mathscr{B}_{F^c} \to \mathbb{R}^+$ is always a transition measure as required by Proposition 3.1. In the present case, since $0 \leqslant \mu(\omega_2, B_1) \leqslant 1$ a.e.$[\mu_F]$, it is indeed possible to find a $\bar{\mu}$, in place of $\tilde{\mu}$, as the desired transition function [this follows from the lifting theorem to be proved later, in Chapter 8, with $\bar{\mu}(\cdot, B_1) = \rho(\tilde{\mu}(\cdot, B_1))$, $B_1 \in \mathscr{B}_F$, ρ being the lifting map], and thus we may apply the proposition.

Consequently, writing $\theta = (\theta_1, \theta_2) \in \mathbb{T}^F \times \mathbb{T}^{F^c} = \mathbb{T}^{\mathbb{R}}$, we have

$$\int_{\mathbb{T}^{\mathbb{R}}} e^{i\Sigma_{s\in\mathbb{R}} n(s)\theta(s)} \mu(d\theta) = \int_{\mathbb{T}^{F^c}} \int_{\mathbb{T}^F} e^{i\Sigma_{s\in F} n(s)\theta(s)} \bar{\mu}(\theta_1, d\theta_2)\mu_F(d\theta_1)$$

$$= \int_{\mathbb{T}^F} e^{i\Sigma_{r=1}^n j_r\theta_r} \left(\int_{\mathbb{T}^{F^c}} 1 \cdot \bar{\mu}(\omega_1, d\omega_2) \right) \mu_F(d\omega_1),$$

$$\omega_1 = (\theta_1, \ldots, \theta_r), \quad n = (j_1, \ldots, j_n)$$

$$= \int_{\mathbb{T}^F} e^{i\Sigma_{r=1}^n j_r\theta_r} \mu_F(d\theta_1, \ldots, d\theta_n),$$

$$\text{since} \quad \bar{\mu}\left(\omega_1, \mathbb{T}^{F^c}\right) = 1$$

$$= \xi(n), \qquad \text{by (6)}.$$

This establishes (7), and the proof is completed.

Remark. We should note that to deduce (7) from (6) it is necessary to show the existence of $\bar{\mu}(\cdot, \cdot)$, the transition function. This is essentially a "disintegration" of μ relative to F, $F \in \mathscr{F}$, and the relevant result comes from the key lifting theory to be developed later. In the present context, it is also known as the "regular conditional probability", and it is again a nontrivial result. In such a case, the underlying measure space [$(\mathbb{T}^{\mathbb{R}}, \mathscr{B}, \mu)$ here] should have regularity properties, either topological or other, guaranteeing the absence of pathological sets in the measurable class considered.

Utilizing the preceding work, we now establish the basic result:

Theorem 4 (Bochner). *Let $\varphi: \mathbb{R} \to \mathbb{C}$ be a positive definite and continuous function with $\varphi(0) = 1$. Then there exists a unique probability measure P on the Borel σ-algebra $\mathscr{R}$ of $\mathbb{R}$ such that*

$$\varphi(x) = \int_{\mathbb{R}} e^{ixy} P(dy), \qquad x \in \mathbb{R}. \tag{9}$$

Conversely, if φ is defined by (9), *then it is positive definite and continuous on $\mathbb{R}$, and $\varphi(0) = 1$.*

[It will follow from Theorem 9.2.5 that P of (9) is automatically a Radon measure.]

Proof. As in the last proposition, let $\mathscr{I}$ be the class of all integer valued functions on $\mathbb{R}$ with finite supports. If φ is a given positive definite continuous function with $\varphi(0) = 1$, define an auxiliary $\xi: \mathscr{I} \to \mathbb{C}$ by the equation

$$\xi(n) = \varphi\left(\sum_{s \in \mathbb{R}} n(s)s\right), \qquad n \in \mathscr{I}. \tag{10}$$

Then on the group $\mathscr{I}$, ξ is positive definite, $\xi(\mathbf{0}) = 1$, and by (7)

$$\xi(n) = \int_{\mathbb{T}^{\mathbb{R}}} \exp\left(i \sum_{s \in \mathbb{R}} n(s)\theta(s)\right)\mu(d\theta), \qquad n \in \mathscr{I}. \tag{11}$$

This is a key step in finding the measure for the representation (9), and we need the product measure theory here. Since $\mathbb{T}^{\mathbb{R}}$ can also be thought of as all complex functions on $\mathbb{R}$ of absolute value one, called the space of "characters of $\mathbb{R}$", and since for each point of $\mathbb{R}$ there is a distinct such function in $\mathbb{T}^{\mathbb{R}}$, one can identify the spaces $\mathbb{T}^{\mathbb{R}}$ and $\mathbb{R}$, and the identification is an isomorphism preserving the group operations. [Recall that if $f_1, f_2 \in \mathbb{T}^{\mathbb{R}}$ then $(f_1 + f_2)(s) = f_1(s) + f_2(s) \pmod{2\pi}$. So $f_1 + f_2 \in \mathbb{T}^{\mathbb{R}}$, and also $\mathbb{R}$ is a group under addition.] Thus the further work from (12) to (17) is needed, as it involves

(i) expressing $e^{i\theta(s)}$ as $e^{ix_\theta s}$, where $h: \theta \mapsto x_\theta \in \mathbb{R}$ sets up a one-to-one correspondence between θ and x_θ, and

(ii) using the above stated isomorphism $h: \mathbb{T}^{\mathbb{R}} \to \mathbb{R}$ to transform the measure μ into P.

Here are the computations:

(i): Let $(\mathbb{R}, \mathscr{R}, \lambda)$ be the Lebesgue line and $(\Omega, \hat{\mathscr{B}}, \mu)$ be the completed projective limit measure, $\Omega = \mathbb{T}^{\mathbb{R}}$, obtained in Proposition 3. Consider now $Z(\cdot, \cdot): \mathbb{R} \times \Omega \to \mathbb{C}$ defined by $Z(t, \theta) = e^{i\theta(t)}$. We assert that $Z(\cdot, \cdot)$ is continuous in μ-measure, i.e., for each $\epsilon > 0$,

$$\lim_{t \to t_0} \mu(\{\theta: |Z(t, \theta) - Z(t_0, \theta)| > \epsilon\}) = 0, \qquad t_0 \in \mathbb{R}, \tag{12}$$

from which the desired fact on $Z(\cdot, \cdot)$ is deduced.

Since $n(\cdot)$ in (11) vanishes outside finite sets, let s_1, s_2, and $s_3 = s_1 + s_2$ be points from $\mathbb{R}$, and choose $n(\cdot)$ such that $n(s_1) = n(s_2) = 1$ and $n(s_1 + s_2) = -1$. Then with (10), we have for (11) $\xi(n) = \varphi(n(s_1)s_1 + n(s_2)s_2 + n(s_1 + s_2)(s_1 + s_2)) = \varphi(0) = 1$, so that $\sum_{s \in \mathbb{R}} n(s)\theta(s) = \theta(s_1) + \theta(s_2) - \theta(s_1 + s_2)$,

$$\int_\Omega e^{i(\theta(s_1) + \theta(s_2) - \theta(s_1 + s_2))}\mu(d\theta) = \xi(n) = 1 = \int_\Omega 1 \cdot \mu(d\theta). \tag{13}$$

Since the integrand in the left integral of (13) is a complex exponential, (13) implies

$$\int_\Omega \sin^2 \tfrac{1}{2}[\theta(s_1) + \theta(s_2) - \theta(s_1 + s_2)]\,\mu(d\theta) = 0.$$

But using the group operation in Ω, we deduce that this holds iff

$$\theta(s_1 + s_2) = \theta(s_1) + \theta(s_2) \pmod{2\pi} \qquad \text{a.e.}[\mu]. \tag{$*$}$$

The exceptional set here depends on s_1, s_2 also. Otherwise, since s_1, s_2 are arbitrary in $\mathbb{R}$, and θ is bounded, and since this is the classical Cauchy functional equation, its only nontrivial solution is $\theta(s) = x_\theta s$ for a unique $x_\theta \in \mathbb{R}$. We have to show therefore that there is a *fixed* μ-null set outside of which the above Cauchy equation holds and hence its solution is the desired result in our argument. This is the needed detail noted above, and we give this here. It requires (unfortunately) an additional argument.†

Taking $n(t) = 1$, we get from (10) and (11) that $\int_\Omega Z(t, \theta)\mu(d\theta) = \varphi(t)$, $t \in \mathbb{R}$. Using this, one has

$$\begin{aligned}\int_\Omega |Z(t,\theta) - Z(t_0,\theta)|^2 \mu(d\theta) &= \int_\Omega 2(1 - \operatorname{Re} e^{i[\theta(t)-\theta(t_0)]})\mu(d\theta)\\ &= 2 - 2\operatorname{Re}\varphi(t - t_0)\\ &\to 0 \qquad \text{as} \quad t \to t_0.\end{aligned} \tag{14}$$

This clearly implies (12). To show that $Z(\cdot,\cdot)$ is $\lambda \otimes \mu$-measurable on $\mathbb{R} \times \Omega$, it suffices to consider $I \times \Omega$, where I is a compact interval of $\mathbb{R}$, since $\mathbb{R}$ is a countable union of such intervals. Let D be the set of rationals of I (or any countable dense set will do). By the compactness of I, we can cover it by a finite collection of open sets $\{J_i^n\}_1^{k_n}$, each of diameter at most $1/n$, for each given $n \geqslant 1$. Disjunctify $\{J_i^n\}$ to $\{\tilde{J}_i^n\}$. Choose $t_i^n \in D \cap \tilde{J}_i^n$, and let

$$V_n(t,\theta) = \sum_{i=1}^{k_n} Z(t_i^n, \theta)\chi_{\tilde{J}_i^n}(t), \qquad (t,\theta) \in I \times \Omega. \tag{15}$$

Then V_n is $\lambda \otimes \mu$-measurable; only one term is nonzero on the right. So

$$\mu(\{\theta: |V_n(t,\theta) - Z(t,\theta)| \geqslant \epsilon\}) \to 0 \qquad \text{as} \quad n \to \infty, \quad (\epsilon > 0)$$

†Although the detail is lengthy, the result is clear: Let $\mathbb{Q}$ be the rationals of $\mathbb{R}$. Since $(*)$ holds for θ in $A^c(s_1, s_2)$, and since $\mu(A(s_1, s_2)) = 0$, let $A = \bigcup\{A(s_1, s_2): s_1, s_2 \in \mathbb{Q}\}$. Assume (12), and define $\tilde{\theta}(t) = \theta(t)$ on A^c if $t \in \mathbb{Q}$; $= \lim_{t_n \downarrow t}\theta(t_n)$ on A^c, $t_n \in \mathbb{Q}$, $t \notin \mathbb{Q}$; $= 0$ on A. Then $(*)$ holds for all $\tilde{\theta}$ and t, so $\tilde{\theta}(t) = x_\theta t$. Here the limit exists by (12).

by (12). Hence $\{V_n(t, n): n \geq 1\}$ is Cauchy in measure and is uniformly bounded [by (1)]. So we can apply Corollary 2.3 to conclude

$$\int_{I\times\Omega} |V_n - V_m| \, d(\lambda \otimes \mu) = \int_I \left(\int_\Omega |V_n(t,\theta) - V_m(t,\theta)| \mu(d\theta) \right) \lambda(dt)$$

$$\to 0 \quad \text{as} \quad n, m \to \infty, \tag{16}$$

by (15) and the bounded convergence theorem, since $\lambda \otimes \mu(I \times \Omega) < \infty$. Hence $V_n \to V$ in $\lambda \otimes \mu$-measure, so that $V_{n_i} \to V$ pointwise on N^c, with $\lambda \otimes \mu(N) = 0$. We may adjust N so that, by (14), this gives $Z(t, \cdot) = V(t, \cdot)$ on $N(t)^c$. Since $\mu(N(t)) = 0$ (by the completeness of μ), and by (16) there is a λ-null set M such that on M^c the integral goes to zero, it follows that $Z = V$ on $N^c \cap (M^c \times \Omega) = A$, $\lambda \otimes \mu(A) = \lambda(I)$. It results immediately that Z is equivalent to the measurable function $V(\cdot, \cdot)$, and that for all $\theta \in A(t)$ $[\mu(A(t)) = 1]$ we have

$$V(t, \theta) = Z(t, \theta) = e^{i\theta(t)} \qquad \text{for a.a. } t. \tag{17}$$

Let $\theta(t) = 0$ for $\theta \in A(t)^c$, so that (17) holds for all $\theta \in \Omega$ and all t. Hence our earlier argument on the functional equation applies, and we deduce that $\theta(t) = x_\theta t$ for a unique $x_\theta \in \mathbb{R}$. The mapping $h: \theta \mapsto x_\theta$ being an isomorphism between Ω and $\mathbb{R}$, $h(\Omega) = \mathbb{R}$, $h^{-1}(\mathscr{R}) = \mathscr{B}$, where $\mathscr{R}$ is the σ-algebra of $\mathbb{R}$, and $\mathscr{B}$ of Ω.

(ii): Now define $P = \mu \circ h^{-1}: \mathscr{R} \to \mathbb{R}^+$. Then P is a probability on the Borel σ-algebra $\mathscr{R}$ of $\mathbb{R}$, and

$$\varphi(t) = \int_\Omega Z(t, \theta) \mu(d\theta) = \int_\Omega e^{i\theta(t)} \mu(d\theta)$$

$$= \int_{\mathbb{R}} e^{ix_\theta t} (\mu \circ h^{-1})(dx) = \int_{\mathbb{R}} e^{ixt} P(dx),$$

by the image measure theorem (Theorem 4.2.6). This gives (9).

The uniqueness argument will be omitted. (See e.g., Exercise 4.5.27.) The converse is the same as in Proposition 1, and in fact is trivial. This completes the proof of the theorem.

Remark. The original proof of this result by Bochner involves defining and using several properties of Fourier transforms of functions in $L^2(\mathbb{R})$ and then obtaining (9) by a careful limit argument from Proposition 1. Another method is to use results from Banach algebras, and yet another one is to use a theorem of P. Levy in probability theory. The above argument using the projective limit theory is essentially that of M. S. Bingham and K.R. Parthasarathy (1968). As with others, this also admits an extension if $\mathbb{R}$ is replaced by a locally compact second countable group (λ being replaced by a Haar measure). It is interesting that the Fubini-Jessen product version (Theorem 4.3) is not sufficient for this proof, and the general Theorem 4.7, or at least Corollary 4.9, is essential.

We terminate our study of product measures after giving another useful application to realizing an abstract Hilbert space as a closed subspace of an $L^2(\mu)$ on a probability space (Ω, Σ, μ).

Recall that if $\mathscr{H}$ is a Hilbert space having at least one nonzero element, then a collection $\{e_i: i \in I\} \subset \mathscr{H}$, $(e_i, e_j) = \delta_{ij}$, is *orthonormal*; if for all $i \in I$ one has $(x, e_i) = 0$, $x \in \mathscr{H} \Rightarrow x = 0$, then the family is *complete*. [δ_{ij} is Kronecker's delta, and $(\cdot, \cdot)$ is the inner product.]

An arbitrary Hilbert space can be realized concretely.

Theorem 5. *Let $\mathscr{H}$ ($\neq \{0\}$) be a Hilbert space. Then there exists a probability space (Ω, Σ, P) such that $\mathscr{H}$ is isometrically isomorphic to a closed subspace of $L^2(P)$.*

Proof. First observe that $\mathscr{H}$ has a complete orthonormal system of elements $\{e_i: i \in I\}$. In fact, let $\mathscr{K}$ be the set of all orthonormal systems of $\mathscr{H}$. Since there is $0 \neq x \in \mathscr{H}$, we have $\{e = x/\|x\|\} \in \mathscr{K}$, so that $\mathscr{K}$ is nonempty. Introduce a partial order into $\mathscr{K}$ by declaring $\{e_i: i \in J\} \prec \{e'_j: j \in J'\}$ if the left side is contained in the right side collection, for any pair of elements of $\mathscr{K}$. That $\{\mathscr{K}, \prec\}$ is a directed set is evident. Every chain $\mathscr{C} \subset \mathscr{K}$ has an upper bound, namely their union. Hence by Zorn's lemma, $\mathscr{K}$ contains a maximal set, which is denoted as $\{e_i: i \in I\} = \mathscr{C}_0$. To see that $\mathscr{C}_0$ is complete, let $x \in \mathscr{H}$ and $(x, e_i) = 0$ for all $i \in I$. If $x \neq 0$, then $\tilde{e} = x/\|x\|$ may be adjoined to $\mathscr{C}_0$ to get $\mathscr{C}_1 = \mathscr{C}_0 \cup \{\tilde{e}\}$. Then $\mathscr{C}_1 \in \mathscr{K}$ and $\mathscr{C}_1 \supsetneqq \mathscr{C}_0$, so that the maximality of $\mathscr{C}_0$ is contradicted. Thus $\mathscr{C}_0$ must be complete.

Let $(\Omega_i, \Sigma_i, P_i)$ be a probability space to be specified further below, for each $i \in I$. Then by Theorem 4.2 (Fubini-Jessen), $(\Omega, \Sigma, P) = \bigotimes_{i \in I}(\Omega_i, \Sigma_i, P_i)$ is a product probability space, where $\Omega = \times_{i \in I}\Omega_i$, $\Sigma = \bigotimes_{i \in I}\Sigma_i$ (the σ-algebra generated by the cylinder sets of Ω), and $P = \bigotimes_{i \in I} P_i$ (the product probability). If $\omega = \{\omega(i): i \in I\} \in \Omega$, then $\omega(i) \in \Omega_i$, $i \in I$. We take $\Omega_i = \mathbb{R}$ and $\Sigma_i = \mathscr{B}$, the Borel σ-algebra of $\mathbb{R}$, for all i, so that $X_i: \Omega \to \mathbb{R}$ defined by $X_i(\omega) = \omega(i)$ is a family of functions (the coordinate functions) which are measurable (for Σ). Choose P_i such that each $X_i \in L^2(P)$. In fact we can do better, namely $\|X_i\|_2 = 1$ and X_i orthonormal, if we take

$$P_i: A \mapsto \frac{1}{\sqrt{2\pi}} \int_A e^{-t^2/2}\, dt, \qquad A \in \mathscr{B}. \tag{18}$$

This choice of P_i is also not obvious or well motivated, but it is a favorite probability measure.

Then by the image measure theorem (Corollary 4.2.7) and the fact that $P \circ \Pi_i^{-1} = P_i$, where $\Pi_i: \Omega \to \Omega_i$ is the coordinate projection, we have

$$\int_\Omega X_i^2 \, dP = \int_{\Omega_i} X^2 \, dP_i(\omega_i) = \int_{\mathbb{R}} x^2 e^{-x^2/2} \frac{dx}{\sqrt{2\pi}} = 1, \tag{19}$$

and similarly, for $i \neq j$,

$$\int_\Omega X_i X_j \, dP = \int_{\Omega_i \times \Omega_j} X_i X_j \, d(P_i \otimes P_j)(\omega_i, \omega_j),$$

$$= \int_{\mathbb{R}} \int_{\mathbb{R}} x_i x_j e^{-(x_i^2 + x_j^2)/2} \frac{dx_i \, dx_j}{2\pi} = 0. \tag{20}$$

Let $\{\varphi_i: i \in I\} \subset \mathscr{H}$ be a complete orthonormal set, and consider $\{X_i: i \in I\} \subset L^2(P)$, which is also orthonormal by (19) and (20). If $\tau: \varphi_i \to X_i$, $i \in I$, and τ is extended linearly to all of $\mathscr{H}$, then $(\tau\varphi_i, \tau\varphi_j)_{\mathscr{H}} = (X_i, X_j)_{L^2(P)} = 1 = (\varphi_i, \varphi_j)_{\mathscr{H}}$. [This comes from the polarization identity.] Thus τ is well defined and is an isometric isomorphism of $\mathscr{H}$ onto $\overline{\text{sp}}\{X_i: i \in I\}$ of $L^2(P)$. This completes the proof.

Remark. The space $\mathscr{G} = \overline{\text{sp}}\{X_i: i \in I\} \subset L^2(P)$ constructed above has the property that each linear combination $Y_n = \sum_{k=1}^n a_k X_{i_k}$ $(\in \mathscr{G})$ also has the same image measure:

$$P[\{\omega: Y_n(\omega) < x\}] = \frac{1}{\sqrt{2\pi\sigma_n^2}} \int_{-\infty}^{x} e^{-t^2/2\sigma_n^2} \, dt, \tag{21}$$

where $\sigma_n^2 = \sum_{k=1}^n a_k^2$. This may be verified with an application of Corollary 4.2.7 and is left to the reader. The probability measure given by (18) or (21) on $\mathbb{R}$ is called *Gaussian*, and the subspace $\mathscr{G}$ $[\subset L^2(P)]$, each of whose elements has a Gaussian distribution, is termed a *Gaussian subspace*. Thus we have shown that any abstract Hilbert space is linearly isomorphic (and isometric) to a Gaussian space.

It is easily shown that any two complete orthonormal sets of a Hilbert space have the same cardinality. This common number is called the *dimension* of the Hilbert space. It is also easy to verify that any two Hilbert spaces of the same dimension are isomorphic. Thus the above theorem shows how different the actual spaces can be while being isomorphic. If we did not demand that $\mathscr{G}$ be a subspace of $L^2(P)$ on a *probability space* (Ω, Σ, P), the realization of $\mathscr{H}$ as a sequence space $l^2(I)$ would be simpler than the above (cf. Exercise 9). A few other applications are given below as exercises.

We also remark that if such a nice probability measure P is not desired in the above, one can then obtain an isomorphism (and a topological equivalence) of an abstract Hilbert space with some $L^2(P)$ on a probability space (Ω, Σ, P). Here P will not be Gaussian in general. This construction is obtainable with results from abstract analysis dealing with characterizations of ranges of contractive projections in L^p-type spaces. They fall outside our scheme and are not considered in this book.

EXERCISES

0. Show that there exist positive definite functions $\varphi: \mathbb{R} \to \mathbb{C}$, which are not measurable (much less continuous), so that the continuity hypothesis in Bochner's representation theorem is not redundant. [Try $\varphi(t) = e^{if(t)}$, where $f: \mathbb{R} \to \mathbb{R}$ is a function satisfying the Cauchy functional equation, $f(x+y) = f(x) + f(y)$ with f not bounded on any nondegenerate interval.] If a positive definite $\varphi: \mathbb{R} \to \mathbb{C}$ ($\varphi(0) = 1$) is (Lebesgue) measurable, then it is actually continuous on $\mathbb{R} - N$ where N is a Lebesgue null set. [This last assertion is not easy, but is worth a serious attempt.]

1. If $\varphi: \mathbb{R} \to \mathbb{C}$ is a positive definite function, show that the following properties are consequences of the definition: (i) $\varphi(0) \geqslant 0$; (ii) $|\varphi(x)| \leqslant \varphi(0)$, $\varphi(-x) = \bar{\varphi}(x)$, $x \in \mathbb{R}$, and hence $\bar{\varphi}$ is also positive definite; (iii) if φ is continuous at $t = 0$, then it is uniformly continuous on $\mathbb{R}$. [Consider the $n \times n$ matrix $B_n = (b_{ij})$, $b_{ij} = \varphi(x_i - x_j)$, adjoint B_n^*, and a vector $\mathbf{a} \in \mathbb{R}^n$. Note that $(B_n\mathbf{a}, \mathbf{a}) = (B_n^*\mathbf{a}, \mathbf{a})$ and then deduce by the polarization identity that $(B_n\mathbf{a}, \mathbf{b}) = (B_n^*\mathbf{a}, \mathbf{b})$, where $(\mathbf{x}, \mathbf{y}) = \sum_{i=1}^n x_i \bar{y}_i$ is the inner product. Taking $\mathbf{a} = \mathbf{e}_1$, the first basis vector, we get $\varphi(0) \geqslant 0$, and considering the 2×2 principal minors, (ii) follows. Assuming $\varphi(0) = 1$ (by normalization), consider a 3×3 determinant $|B_3|$ which is nonnegative. Choose the t_1, t_2, t_3 as $t_1 = 0$, $t_2 = t$, $t_3 = t'$, and simplify the determinant to get $|\varphi(t) - \varphi(t')|^2 \leqslant 1 - |\varphi(t - t')|^2 + 2|\varphi(t)\bar{\varphi}(t')|\,|\varphi(t - t') - 1|$, which gives (iii).]

2. Complete the proof of Proposition 2. [Define

$$f_n(x_1, \ldots, x_k) = \frac{1}{(2\pi n)^k} \sum_{\substack{1 \leqslant j_r, j_r' \leqslant n \\ r = 1, \ldots, k}} \varphi(j_r' - j_r) \exp\left\{ i \sum_{r=1}^{k} (j_r - j_r') x_r \right\},$$

and proceed as in Proposition 1.]

3. Supply the details of the uniqueness of the representation in Theorem 4.

4. Our proof of Theorem 4 extends if $\varphi: \mathbb{R}^n \to \mathbb{C}$ is a positive definite function which is continuous at $\mathbf{0} \in \mathbb{R}^n$, and $\varphi(\mathbf{0}) = 1$. Complete the details of the proof, and show that

$$\varphi(t_1,\ldots,t_n) = \int_{R^n}\cdots\int e^{i(t_1x_1+\cdots+t_nx_n)}\mu(dx_1,\ldots,dx_n)$$

for a unique probability measure μ on the Borel sets of $\mathbb{R}^n$.

5. With the preceding exercise, let $\{\varphi_{\alpha_1,\ldots,\alpha_n}(\cdot,\ldots,\cdot): n \geqslant 1\}$ be an indexed family of positive definite functions on $\mathbb{R}^n$, $n \geqslant 1$, satisfying the conditions there for each index $(\alpha_1,\ldots,\alpha_n) \in \mathbb{R}^n$, $\alpha_1 < \alpha_2 < \cdots < \alpha_n$. Let $\mu_{\alpha_1,\ldots,\alpha_n}$ be the corresponding probability measures. Show that $\{\mu_{\alpha_1,\ldots,\alpha_n}: n \geqslant 1\}$ is a compatible family (i.e., $\mu_\alpha = \mu_\beta \circ \pi_{\alpha\beta}^{-1}$ for $\alpha < \beta$, $\pi_{\alpha\beta}: \mathbb{R}^\beta \to \mathbb{R}^\alpha$ are the usual coordinate projections) iff (i) $\lim_{u_i \to 0}\varphi_{\alpha_1,\ldots,\alpha_n}(u_1,\ldots,u_n) = \varphi_{\alpha_1,\ldots,\alpha_{n-1}}(u_1,\ldots,u_{n-1})$ and (ii) $\varphi_{\alpha_{i_1},\ldots,\alpha_{i_n}}(u_{i_1},\ldots,u_{i_n}) = \varphi_{\alpha_1,\ldots,\alpha_n}(u_1,\ldots,u_n)$, where $(i_1,\ldots,i_n)$ is a permutation of $(1,2,\ldots,n)$. [Compare with (17) and (18), and use the uniqueness of Theorem 4 and Exercise 4 above.]

6. Let $L^1(\mathbb{R})$ be the space of all Lebesgue-integrable functions $f: \mathbb{R} \to \mathbb{R}$, and define $\hat{f}: t \mapsto \int_{\mathbb{R}} e^{itx} f(x)\lambda(dx)$, where λ is the Lebesgue measure on $\mathbb{R}$. If $\varphi: \mathbb{R} \to \mathbb{C}$ is a positive definite function, continuous at $t = 0$, and $\varphi(0) = 1$, show that $\sup\{|\int_{\mathbb{R}} f(t)\varphi(t)\lambda(dt)|: \sup_{u\in\mathbb{R}}|\hat{f}(u)| \leqslant 1\} = 1$. [Verify and use the fact that $f \in L^1(\mathbb{R}) \Rightarrow \widehat{f * f} = |\hat{f}|^2$, where $f * f$ is the convolution product.]

7. Let ν be a signed measure on $\mathbb{R}$, and define $\psi: t \mapsto \int_{\mathbb{R}} e^{itx}\nu(dx)$. Show that $\psi: \mathbb{R} \mapsto \mathbb{C}$ is well defined and is the difference of two positive definite continuous functions on $\mathbb{R}$. Verify that if this ψ is used in place of φ in Exercise 1, then $\sup\{|\int_{\mathbb{R}}\psi(t)f(t)\lambda(dt)|: \sup_u|\hat{f}(u)| \leqslant 1,\ f \in L^1(\mathbb{R})\}$ is finite and is at most $|\nu|(\mathbb{R})$, the variation of ν.

8. The converse of the preceding statement is true. Let $\psi: \mathbb{R} \to \mathbb{C}$ be a bounded continuous function such that $\sup\{|\int_{\mathbb{R}}\psi(t)f(t)\lambda(dt)|: \sup_u|\hat{f}|(u) \leqslant 1,\ f \in L^1(\mathbb{R})\} \leqslant C_0 < \infty$. Then there is a unique signed measure ν on the Borel sets of $\mathbb{R}$ such that $\psi(t) = \int_{\mathbb{R}} e^{itx}\nu(dx)$ and $|\nu|(\mathbb{R}) \leqslant C_0$. [This assertion is significantly more difficult than the preceding two, and to use the method of proof of Theorem 4, an extension of Theorem 4.6 for signed measures should be obtained. It is worth a serious effort, and the rewards are high. Of course, the problem can also be solved using some facts from Fourier analysis. A direct application of Theorem 4 is not possible, but each of Propositions 1–3 has an analog here, and thus a parallel argument should be developed.]

9. Let $\mathscr{H}$ be an abstract Hilbert space of dimension α. Let I be a point set of cardinality α. If $l^2(I) = \{x: I \to \mathbb{C},\ \Sigma_{i\in I}|x(i)|^2 < \infty\}$, with inner product $(x, y) = \Sigma_{i\in I}x(i)\bar{y}(i)$, show that $\mathscr{H}$ and $l^2(I)$ are isomorphic.

10. Extend Theorem 5 for complex Hilbert spaces. [In (18) take $dP(z) = (1/2\pi)\exp[-\frac{1}{2}(x^2 + y^2)]$, where $z = x + iy$.]

7

CAPACITY THEORY AND INTEGRATION

This chapter is devoted to a comparative study of a measure and an analogous capacity function and then to a proof of Choquet's fundamental theorem on the capacitability of analytic sets. As an important application of this work, we obtain the Daniell method of integration rather quickly; then the Daniell-Stone theorem connecting the latter with the Lebesgue integral of Chapter 4 is established. The work enables a unification of various points of view and helps in further developments of the theory. A few complements are included as exercises. The subject is placed here only because we use a few elementary properties of Cartesian products, and the experience of the last chapter (and not its results) will give facility. Except for that, one may study it following Chapter 4.

7.1. PRELIMINARIES ON ANALYTIC SETS

Since the main result of the subject is about the capacitability of analytic sets, it is appropriate that we start with a discussion of the latter. This will also indicate that there are analytic sets which are not Borel but which are Lebesgue measurable, amplifying Theorem 2.2.19.

It was seen before (Proposition 2.1.3 and 2.1.5) that countable unions, intersections, and complements of the familiar classes such as the closed, open, or compact intervals of $\mathbb{R}$ do not lead us beyond the Borel class. So one has to perform a somewhat more involved operation to generate (the larger class of) analytic sets. In 1917, this was achieved by R. Souslin, and it is as follows: A *determining system* $\mathscr{D} = \{A_h: h = (n_1, \dots, n_k): n_i \geqslant 1\}$ is a class of sets of a given space Ω, where n is a finite vector index of integers $\geqslant 1$, for each $k \geqslant 1$. If $A_s = \bigcap_{k \geqslant 2} A_{n_1, \dots, n_k}$, where $s = (n_1, n_2, \dots)$, then the set A, denoted

$N(\mathscr{D})$, defined by

$$A = \bigcup_s A_s = \bigcup_{n_1, n_2, \ldots} \left(A_{n_1} \cap A_{n_1, n_2} \cap \cdots\right), \tag{1}$$

is called the *nucleus* of $\mathscr{D}$. The operation from a determining system to nucleus is called *Souslin's operation* (A). If the system $\mathscr{D}$ is Borel, then the resulting class given by (1) is termed (by N. Luzin) the *analytic sets* of Ω, for the case that $\Omega = \mathbb{R}$.

This concept, as introduced above, is not convenient for further analysis. (This work involves generally uncountable operations.) An alternative definition is better suited for applications, and here we follow G. Choquet (1955). As a motivation, note that if S denotes the space of all ordered sequences $s = (n_1, n_2, \ldots)$ in (1), and if S is endowed with the order topology (the order being lexicographic), then one can verify that (1) is simply $A = \pi(B)$, where $B \subset \Omega \times S$, and $\pi: \Omega \times S \to \Omega$ is the coordinate projection, and $B = \bigcap_{n=1}^{\infty} B_n$, with $B_n = \bigcup_h (A_h \times I_h)$, $I_h \subset S$ being an interval. (This will be proved below.) To make this observation operational, we develop some of its properties more systematically. In this approach, we will primarily follow Choquet, incorporating some simplifications by Sion (1963) and by Dellacherie and Meyer (1978).

We need additional terminology for this discussion.

Definition 1. Let Ω be a set, and $\mathscr{C}$ be a collection of subsets of Ω containing the empty set. Then $\mathscr{C}$ is termed a *paving*, and the pair $(\Omega, \mathscr{C})$ a *paved space*. If $\{\Omega_i, \mathscr{C}_i): i \in I\}$ is an indexed family of paved spaces, then $(\times_{i \in I} \Omega_i, \times_{i \in I} \mathscr{C}_i)$ is a *product paving* where $\times_{i \in I} \mathscr{C}_i$ is the collection of all "rectangles", $\times_{i \in I} A_i$, $A_i \in \mathscr{C}_i$. (Here and hereafter we always assume, without further mention, that the spaces Ω_i are all nonempty.)

Abstracting the concept of compactness from Theorem 1.2.8, we say that a paved space $(\Omega, \mathscr{C})$, or just the paving $\mathscr{C}$ of Ω, is *compact* if it has the property that each family $\mathscr{F} \subset \mathscr{C}$ having the finite intersection property (i.e., if $\mathscr{F} = \{K_i: i \in I\}$ is some indexing, and $I_0 \subset I$ is any finite set, then $\bigcap_{i \in I_0} K_i \neq \varnothing$) has a nonempty intersection. This is utilized in the alternative:

Definition 2. Let $(\Omega, \mathscr{C})$ be a paved space. Then a set $A \subset \Omega$ is called *$\mathscr{C}$-analytic* (or analytic relative to $\mathscr{C}$) if there exists a separable compact Hausdorff space E and a set $B \subset E \times \Omega$, $B \in (\mathscr{K} \times \mathscr{C})_{\sigma\delta}$, such that $A = \pi(B)$, where $\pi: E \times \Omega \to \Omega$ is the coordinate projection and $\mathscr{K}$ is the class of closed subsets of E. We denote the class of all $\mathscr{C}$-analytic sets of Ω by $\mathscr{A}(\mathscr{C})$.

Our first task is to show that the two concepts given by (1) and the above definition agree, so that we can use either definition at will.

Proposition 3. *Let* $(\Omega, \mathscr{C})$ *be a paved space. Then* $A \in \mathscr{A}(\mathscr{C})$ *iff* A *is given by the Souslin operation* (1) *on* $\mathscr{C}$.

Proof. Suppose that A is the nucleus of $\mathscr{C}$ relative to a determining system from $\mathscr{C}$. To show that $A \in \mathscr{A}(\mathscr{C})$, we express (1) as follows. Let $\mathbb{N}$ be the set of natural numbers, which with the discrete topology becomes a locally compact space. So $\mathbb{N} \cup \{\infty\} = \bar{\mathbb{N}}$ is compact, and the product space $E = \bar{\mathbb{N}}^{\mathbb{N}}$ is thus also compact (by Tychonov's theorem). Moreover it has clearly a countable neighborhood base, since $\mathbb{N}$ has one, and the point ∞ has a countable system of neighborhoods according to the definition of $\bar{\mathbb{N}}$. This E qualifies for that of Definition 2. To use the terminology introduced after (1) more effectively, denote $I_r = \{s \in S: s \succ r\}$, where $r \in R$, the collection of all *finite* sets of integers from $\mathbb{N}$, and where $s \succ r$ (or $r \prec s$) means that r is the initial segment of $s = (n_1, n_2, \ldots)$. If r_1, r_2 are not comparable, then $I_{r_1} \cap I_{r_2} = \varnothing$, and if $r_1 \prec r_2$ then $I_{r_1} \cap I_{r_2} = I_{r_2}$. Let us define B as follows (here and below, $r \in R$, $r_i \in R$):

$$B = \bigcap_{k=1}^{\infty} \bigcup_{|r|=k} I_r \times A_r, \tag{2}$$

where $|r| = k$ is the number of terms in r, and A_r in $\mathscr{C}$ is an arrangement giving a determining system. (Different arrangements of $\mathscr{C}$ give different systems, and the totality produces all $\mathscr{C}$-analytic sets.) We assert that $\pi(B) = A$. This follows immediately if π commutes with intersections (it always commutes with unions), since then we get

$$\pi(B) = \bigcap_{k \geqslant 1} \bigcup_{|r|=k} \pi(I_r \times A_r) = \bigcap_{k \geqslant 1} \bigcup_{|r|=k} A_r = A.$$

To justify this commutation we use the structures of E and $\mathscr{K}$. Since $\pi(\cap \cdot) \subset \cap \pi(\cdot)$ always, let $v \in \bigcap_{k \geqslant 1} \pi(C_k)$, $C_k = \bigcup_{|r|=k} I_r \times A_r$. Then there exists a $y_k \in C_k$ such that $y_k \in \pi^{-1}(\{v\}) \cap C_k$ for each k, and also $\pi^{-1}(\{v\}) \cap \bigcap_{k=1}^{n} C_k \neq \varnothing$, $n \geqslant 1$. We may assume that $\mathscr{C} \not\equiv \{\varnothing\}$, since otherwise everything is trivial. Let $C = \bigcap_{k=1}^{\infty} C_k$. Note that $\pi^{-1}(\{v\})$ is the cylinder $S \times \{v\}$. Thus one has, from the special form of the C_k's,

$$\begin{aligned} \varnothing \neq \pi^{-1}(\{v\}) \cap C_k &= (S \times \{v\}) \cap \bigcup_{|r|=k} I_r \times A_r \\ &= \bigcup_{|r|=k} I_r \times \{v\} \qquad \text{(omitting the empty sets)} \\ &= \Big(\bigcup_{|r|=k} I_r\Big) \times \{v\}. \end{aligned} \tag{3}$$

Moreover, as noted above, they have the finite intersection property. Taking the intersection of sets of (3) as $k \geqslant 1$ varies, we get

$$\pi^{-1}(\{v\}) \cap C = \bigcap_{k=1}^{\infty}\left(\bigcup_{|r|=k} I_r\right) \times \{v\}$$

$$= \bigcup_{s}\bigcap_{k} I_{s(k)} \times \{v\}$$

[$s(k)$ being the initial part of s with $|s(k)| = k$]

$$\supseteq (\{s\} \times \{v\}).$$

Hence there exists at least one $y \in \pi^{-1}(\{v\}) \cap C$ such that $\pi(y) = v$, and so $\pi(C) \supset \bigcap_{k=1}^{\infty}\pi(C_k)$, as desired. Thus (2) holds. Since $I_r \times A_r \in \mathscr{K} \times \mathscr{C}$ (note also that I_r is compact in the topology of E), (2) implies that $B \in (\mathscr{K} \times \mathscr{C})_{\sigma\delta}$. Thus A is $\mathscr{C}$-analytic by Definition 2.

Conversely, if A is $\mathscr{C}$-analytic, they by Definition 2, it can be expressed as $A = \pi(B)$ for some $\mathscr{K}$ and a $B \in (\mathscr{K} \times \mathscr{C})_{\sigma\delta}$. If $(E, \mathscr{K})$ is $(\overline{\mathbb{N}}^{\mathbb{N}}, \mathscr{K})$, one can reverse the above steps. In general, let $(E, \mathscr{K})$ be any compact paved space, and let $B \in (\mathscr{K} \times \mathscr{C})_{\sigma\delta}$. We show that $\pi(B)$ satisfies (1). Consider

$$B = \bigcap_{n=1}^{\infty}\bigcup_{m=1}^{\infty} K_{nm} \times C_{nm}, \qquad K_{nm} \in \mathscr{K}, \quad C_{nm} \in \mathscr{C}$$

$$= \bigcup_{s}\bigcap_{n=1}^{\infty} K_{ns(n)} \times C_{ns(n)}, \qquad \text{with} \quad s \in S, \quad \text{as before.} \tag{4}$$

Let $A_r = C_{ns(n)}$, $|r| = n$, if $\bigcap_{j=1}^{n} K_{js(j)} \neq \varnothing$, and set $A_r = \varnothing$ otherwise. Hence

$$\pi(B) = \bigcup_{s}\pi\left(\bigcap_{n=1}^{\infty}\left(K_{ns(n)} \times C_{ns(n)}\right)\right)$$

$$= \bigcup_{s}\bigcap_{n=1}^{\infty} A_{s(n)}, \tag{5}$$

since $\pi(\bigcap_{j=1}^{n} K_{js(j)} \times C_{js(j)}) = \bigcap_{j=1}^{n} A_{s(j)}$ by definition. The case with $n = \infty$ follows from the compactness of $\mathscr{K}$, as seen from the proposition below. But (5) is just (1), and so $\pi(B)$ is a nucleus of $\mathscr{C}$. This finishes the proof when we establish the next result.

Remark. The above proposition also shows that in our Definition 2, the auxiliary compact paving $(E, \mathscr{K})$ can be taken as $E = \overline{\mathbb{N}}^{\mathbb{N}}$, and $\mathscr{K}$ as its compact sets, once and for all. The class $\mathscr{A}(\mathscr{C})$ remains the same and thus does not contain an ambiguity because of using an unknown space E.

The above needed assertion is easily proved as in the case of (3). But we state it separately, since it has some independent interest.

Proposition 4. *Let $(E, \mathscr{K})$ be a paved space, and $f\colon E \to F$ be a function with the property that for each $x \in F$ the paving $\mathscr{E} = \{f^{-1}(\{x\}) \cap A\colon A \in \mathscr{K}\}$ is compact. Then for any sequence $A_n \in \mathscr{K}$, $A_n \supset A_{n+1}$, we have*

$$f\left(\bigcap_{n=1}^{\infty} A_n\right) = \bigcap_{n=1}^{\infty} f(A_n). \tag{6}$$

Proof. Since the left side is always contained in the right side for any f, we need an additional hypothesis only for the opposite inclusion. Thus let $x \in \bigcap_{n=1}^{\infty} f(A_n)$. Then $f^{-1}(\{x\}) \cap A_n \neq \varnothing$, $n \geqslant 1$, and since the sequence is decreasing, $\{f^{-1}(\{x\}) \cap A_n\colon n \geqslant 1\}$ has the finite intersection property. Since $\mathscr{E}$ is compact, so is this class, and hence it has a nonempty intersection. Then we pick a $y \in f^{-1}(\{x\}) \cap A_n$, $n \geqslant 1$, implying $x = f(y)$ and $y \in \bigcap_{n=1}^{\infty} A_n$. Thus the right side is contained in the left side, and (6) holds as stated. It also implies (5) if $f = \pi$, $A_n = \bigcap_{j=1}^{n} A_{s(j)}$. This completes the proof.

Using the preceding propositions and the equivalent definitions, one can present some properties of analytic sets. Here we indicate a few of them, since that will give some feeling for the structure of $\mathscr{A}(\mathscr{C})$.

Proposition 5. *Let $(\Omega, \mathscr{C})$ be a paved space, and $\mathscr{A}(\mathscr{C})$ be the class of $\mathscr{C}$-analytic subsets of Ω. Then*

(i) $\mathscr{C} \subset \mathscr{A}(\mathscr{C})$,
(ii) $\sigma(\mathscr{C}) \subset \mathscr{A}(\mathscr{C})$ *iff* $A \in \mathscr{C} \Rightarrow A^c \in \mathscr{A}(\mathscr{C})$, *and*
(iii) *if* $\Omega = \mathbb{R}^n$ *and* $\mathscr{C}$ *denotes the compact subsets of* Ω, *then the Borel* σ*-algebra* $\mathscr{B}$ *of* $\mathbb{R}^n$ *satisfies* $\mathscr{B} \subset \mathscr{A}(\mathscr{C})$ *as well as* $\mathscr{A}(\mathscr{B}) = \mathscr{A}(\mathscr{C})$.

Proof. By (1), it is evident that (i) holds. Regarding (ii), if $\mathscr{A}(\mathscr{C})$ is closed under complements, then it becomes a σ-algebra, since one may verify that any $\mathscr{A}(\mathscr{C})$ is always closed under countable unions and intersections. This fact is obtained easily with the remark following Proposition 3. Thus let $A_n \in \mathscr{A}(\mathscr{C})$. Then there exist $B_n \in (\mathscr{K} \times \mathscr{C})_{\sigma\delta}$ such that $A_n = \pi(B_n)$, where $\pi\colon E \times \Omega \to \Omega$ is the coordinate projection and $E = \overline{\mathbb{N}}^{\mathbb{N}}$ is the *fixed* compact space used there. Then $\bigcup_n A_n = \pi(\bigcup_n B_n)$. But $B_n = \bigcap_k B_{nk}$, $B_{nk} \in (\mathscr{K} \times \mathscr{C})_\sigma$. Hence $\bigcup_n B_n = \bigcap_n \bigcup_k (\bigcap_n B_{nk})$ and $\bigcup_n B_{nk} \in (\mathscr{K} \times \mathscr{C})_\sigma$. Thus $\bigcup_n B_n \in (\mathscr{K} \times \mathscr{C})_{\sigma\delta}$, so that $\bigcup_n A_n \in \mathscr{A}(\mathscr{C})$. The argument with intersections is similar, and in fact follows from Proposition 4. Hence $\mathscr{A}(\mathscr{C})$ is a σ-algebra. Since $\mathscr{C} \subset \mathscr{A}(\mathscr{C})$, it follows that $\sigma(\mathscr{C})$ is contained in $\mathscr{A}(\mathscr{C})$. The converse is evident, since $\sigma(\mathscr{C})$ is closed under complements of $\mathscr{C}$ and contains it.

(iii): Since $\mathscr{B}$ is the Borel σ-algebra of $\mathbb{R}^n$, we have $\mathscr{C} \subset \mathscr{B}$; by (i), $\mathscr{A}(\mathscr{C}) \subset \mathscr{A}(\mathscr{B})$. But $\mathscr{B}$ is also generated by $\mathscr{C}$ [i.e., $\mathscr{B} = \sigma(\mathscr{C})$], and the complement of each compact set in $\mathbb{R}^n$ is σ-compact, so that by (ii) it is in $\mathscr{A}(\mathscr{C})$ and hence $\mathscr{B} \subset \mathscr{A}(\mathscr{C})$. Finally $\mathscr{A}(\mathscr{B}) \subset \mathscr{A}(\mathscr{A}(\mathscr{C}))$. However, using (1) it is not hard to verify that one again has $\mathscr{A}(\mathscr{A}(\mathscr{C})) \subset \mathscr{A}(\mathscr{C})$ and hence equality holds. In other words, $\mathscr{A}: \mathscr{C} \mapsto \mathscr{A}(\mathscr{C})$ is an idempotent operation. So we get

$$\mathscr{A}(\mathscr{C}) \subset \mathscr{A}(\mathscr{B}) \subset \mathscr{A}^2(\mathscr{C}) = \mathscr{A}(\mathscr{C}), \tag{7}$$

completing the proof. [This last equality in (7) can also be proved directly from (1) after some complicated manipulations, identification of $\mathbb{N}$ with $\mathbb{N} \times \mathbb{N}$ bijectively, etc.]

We still have not shown that in $\mathbb{R}^n$, $\mathscr{A}(\mathscr{C}) - \mathscr{B}$ is nonempty. It is precisely to solve this problem that Souslin's operation (A) was invented. But for the proof one needs further detailed analysis using the topology of $\mathbb{R}^n$, and we omit it here. It should be mentioned that one way of showing this is to establish the following two facts:

(i) Let $\Omega \subset \mathbb{R}^n$ be a compact set. Then for each analytic set $E \subset \Omega$ there exists a continuous onto mapping $f: \mathbb{N}^{\mathbb{N}} \to E$.

(ii) If $B \subset \Omega$ is Borel, then there exists a closed subset $F \subset \mathbb{N}^{\mathbb{N}}$ and a one to one continuous mapping g on F such that $g(F) = B$.

Thus a function g which is, or is not, one to one distinguishes the two types. Forgoing further discussion, we proceed to capacity functions in the next section.

EXERCISES

1. Complete the details of the proof of the last part of Proposition 5.

2. Let $(\Omega, \mathscr{C})$ be a paved space, and $(\mathbb{R}^n, \mathscr{K})$ be the usual Euclidean space with $\mathscr{K}$ as the compact sets of $\mathbb{R}^n$. If $\mathscr{B}$ is the Borel σ-algebra of $\mathbb{R}^n$, verify that $\sigma(\mathscr{C}) \otimes \mathscr{B} \subset \mathscr{A}(\sigma(\mathscr{C}) \times \mathscr{K})$. [Use the same argument as for Proposition 5.] Show also that $\mathscr{A}(\Sigma) = \mathscr{A}(\sigma(\mathscr{C}) \times \mathscr{K})$, where $\Sigma = \sigma(\mathscr{C}) \otimes \mathscr{B}$.

3. If $(\Omega_i, \mathscr{C}_i)$, $i = 1, 2$, are any pair of paved spaces, show that $\mathscr{A}(\mathscr{C}_1) \times \mathscr{A}(\mathscr{C}_2) \subset \mathscr{A}(\mathscr{C}_1 \times \mathscr{C}_2)$, by using the fact that $\mathscr{A}(\mathscr{C})$ is closed under countable unions and intersections.

4. Let $(\Omega_i, \mathscr{C}_i)$, $i = 1, 2$, be a pair of paved spaces such that $\Omega_2 \subset \Omega_1$ and $\mathscr{C}_2 = \mathscr{C}_1 \cap \Omega_2$. Show that $\mathscr{A}(\mathscr{C}_2) = \mathscr{A}(\mathscr{C}_1) \cap \Omega_2$, which improves Proposition 5(i) in this case. As usual, $\mathscr{C}_1 \cap \Omega_2 = \{C \cap \Omega_2 : C \in \mathscr{C}_1\}$, and $\mathscr{A}(\mathscr{C}_2)$ is defined similarly.

5. Let $(\Omega_i, \mathscr{B}_i)$, $i = 1, 2$, be a pair of Borelian spaces where each Ω_i is a compact metric space. If $f: \Omega_1 \to \Omega_2$ is $(\mathscr{B}_1, \mathscr{B}_2)$ measurable (i.e. $f^{-1}(\mathscr{B}_2)$

$\subset \mathscr{B}_1$), then show that for each $\mathscr{B}_1$-analytic set B, the image $f(B) \subset \Omega_2$ is $\mathscr{B}_2$-analytic. (Use Proposition 5 carefully.) [The result also holds if Ω_1 here is a complete separable metric space and Ω_2 is simply a separable metric space. But then the argument is more involved.]

6. Let $(I, \mathscr{B})$ be the Borelian unit interval, ($I = [0,1]$). If $\mathscr{C}$ is the class of all closed sets of I, and $\mathscr{A}(\mathscr{C})$ is the collection of $\mathscr{C}$-analytic sets, show that $\mathscr{B} \subset \mathscr{A}(\mathscr{C})$, and since the latter is not a σ-algebra, deduce that there must exist analytic subsets of I which are not Borel. Show however that $\mathscr{A}(\mathscr{B}) = \mathscr{A}(\mathscr{C})$.

7.2. CAPACITY: A CONSTRUCTION AND CHOQUET'S THEOREM

The work of the preceding section on analytic sets will be used in a calculation of certain "sizes", analogous to (but distinct from) the outer measures studies in Chapter 2. The function is called a capacity, and there are important nonmeasurable sets (for a Radon measure) in a topological space which are capacitable. To discuss the situation further, let us first introduce the general concept, following Choquet (1955).

Definition 1. Let Ω be a nonempty set and $\mathscr{C}(\ni \phi)$ be subsets of Ω:

(a) Let $(\Omega, \mathscr{C})$ be a paved space in which $\mathscr{C}$ is also closed under finite unions and countable intersections. A (Choquet) *$\mathscr{C}$-capacity* on Ω [i.e. on the power set $\mathscr{P}(\Omega)$] is a mapping I: $\mathscr{P}(\Omega) \to \overline{\mathbb{R}}$ such that

(i) $A \subset B \;\Rightarrow\; I(A) \leqslant I(B)$,

(ii) $A_n \subset A_{n+1} \subset \Omega \;\Rightarrow\; I(\bigcup_n A_n) = \lim_n I(A_n)$, and

(iii) $A_n \in \mathscr{C},\ n \geqslant 1,\ A_n \supset A_{n+1} \;\Rightarrow\; I(\bigcap_n A_n) = \lim_n I(A_n)$.

(b) A set $C \subset \Omega$ is *capacitable* relative to the $\mathscr{C}$-capacity I if

$$I(C) = \sup\{I(B): B \subset C,\ B \in \mathscr{C}_\delta\}, \tag{1}$$

so that $I(C)$ can be approximated from below with sets in $\mathscr{C}_\delta$.

It is clear that for outer measures condition (a)(iii) is seldom satisfied. But if I is a finite measure on a σ-algebra containing $\mathscr{C}$, then Proposition 2.1.1 implies that we can derive a capacity from such functions. On the other hand, Theorem 2.6.1 shows that finite Lebesgue-Stieltjes measures on $\mathbb{R}$, with $\mathscr{C}$ as its compact sets, satisfy (1) for all Borel sets, and hence they are capacities. Similarly certain metric measures of Section 2.5 are candidates. However, there are other capacity functions—such as the classical Newtonian potential, from which the capacity theory initially emerged—which are not measures. Thus a separate treatment of capacities is necessary, and as we later show, many of the developments in integration can be obtained from this work.

As a motivation, let us present a simple method for abstract capacities derivable from finite measures of Chapter 2.

Proposition 2. *Let (Ω, Σ, μ) be a finite measure space, and $\mathscr{C} \subset \Sigma$ be a paving which is closed under finite unions and countable intersections. Let μ^* be the generated outer measure by (μ, Σ). Then $I = \mu^*$ is a $\mathscr{C}$-capacity, and every I-capacitable set $A \subset \Omega$ is μ-measurable. If the given measure space is complete, then $\mathscr{A}(\mathscr{C}) \subset \Sigma$.*

Proof. By Theorem 2.2.9, μ^* is Carathéodory outer regular, and since $\mu(\Omega) < \infty$, μ^* satisfies the conditions of Definition 2, in view of Proposition 2.1.1. Thus $I = \mu^*$ is a $\mathscr{C}$-capacity. Further, by Theorem 2.2.10, for each $A \subset \Omega$ there exists a set $B_0 \in (\Sigma)_{\sigma\delta} = \Sigma$ such that $A \subset B$ and $\mu^*(A) = \mu(B_0)$. If A is also I-capacitable, then by (1) and the finiteness of $I(\Omega) = \mu(\Omega)$, for each $\epsilon > 0$ there exists a $B_\epsilon \in (\Sigma)_\delta = \Sigma$ such that $B_\epsilon \subset A$ and

$$\mu^*(A) = I(A) < \mu(B_\epsilon) + \epsilon. \tag{2}$$

Hence $B_\epsilon \subset A \subset B_0$, and we get

$$\mu(B_0) = I(A) < \mu(B_\epsilon) + \epsilon, \tag{3}$$

so that $\mu(B_0 - B_\epsilon) < \epsilon$. This with Proposition 2.2.16 implies that A is μ-measurable. If Σ is complete, then Theorem 2.3.1 shows that every μ-measurable set is in Σ, and the result follows.

The above proof used the positivity, additivity, and boundedness of μ on Σ. Let us now extend the result by relaxing the last two conditions. Although some ideas of Section 2.2 may be used, the results themselves can not be invoked. Yet it is important to generate different classes of capacity functions for applications.

If $\mathscr{C}$ is a paving which is closed under finite unions and intersections, then an increasing $\overline{\mathbb{R}}^+$-valued function I on $\mathscr{C}$ is termed *strongly subadditive* whenever it satisfies

$$I(A \cup B) + I(A \cap B) \leqslant I(A) + I(B), \qquad A, B \in \mathscr{C}. \tag{4}$$

It is *strongly additive* if there is equality in (4). Note that $I(\varnothing) = +\infty$ is possible. In case $I(\varnothing) = 0$, then (since $I \geqslant 0$) this is related to the usual (sub-)additivity. (Cf. Section 2.2 and the proof of Theorem 6.4.8.) This concept plays a key role in Section 9.2 also.

Alternative forms of (4) will now be given.

Lemma 3. *A set function I on a paving $\mathscr{C}$ of Ω is strongly subadditive iff for any sets A, B, C, D from $\mathscr{C}$ with $A \subset C$, $B \subset D$ we have*

$$I(A) + I(B) + I(C \cup D) \leqslant I(A \cup B) + I(C) + I(D). \tag{5}$$

Proof. That (5) $\Rightarrow$ (4) is simple. Indeed, if $I: \mathscr{C} \to \{\infty\}$, then both (4) and (5) are true and trivial. Otherwise, suppose $I(E) < \infty$ for some $E \in \mathscr{C}$. Let $C = E$, $D = F$, $A = E \cap F$, $B = E$ in (5). Then

$$I(E \cap F) + I(E \cup F) + I(E) \leqslant I(E) + I(E) + I(F). \tag{6}$$

Canceling the finite term $I(E)$ from both sides in (6), it reduces to (4).

To see that (4) $\Rightarrow$ (5), for any P, Q, R in $\mathscr{C}$, put $A = P \cup R$ and $B = Q \cup R$ in (4) to get

$$I(P \cup Q \cup R) + I((P \cap Q) \cup R) \leqslant I(P \cup R) + I(Q \cup R).$$

Since I is increasing, $I((P \cap Q) \cup R) \geqslant I(R)$, so that this gives

$$I(P \cup Q \cup R) + I(R) \leqslant I(P \cup R) + I(Q \cup R). \tag{7}$$

Now let $P = C$, $R = B$, and $Q = D$ in (7), add $I(A)$ to both sides, and take $B \subset D$, $A \subset C$. Then (7) becomes

$$\begin{aligned} I(C \cup D) &+ I(B) + I(A) \\ &\leqslant I(C \cup B) + I(D) + I(A) \\ &= [I(C \cup B \cup A) + I(A)] + I(D), \qquad \text{since} \quad A \subset C \\ &\leqslant I(A \cup B) + I(A \cup C) + I(D), \qquad \text{by} \quad (7) \\ &= I(A \cup B) + I(C) + I(D), \qquad \text{since} \quad A \subset C. \end{aligned}$$

This is (5), and the lemma follows.

Using induction, we may extend (5) for any finite number of terms. We state this for later use as:

Corollary 4. *Let $\{A_i, B_i: i = 1, \ldots, n\} \subset \mathscr{C}$ be sets with $A_i \subset B_i$, $i = 1, \ldots, n$, the paving $\mathscr{C}$ being as in the lemma. Then*

$$I\left(\bigcup_{i=1}^{n} B_i\right) + \sum_{i=1}^{n} I(A_i) \leqslant I\left(\bigcup_{i=1}^{n} A_i\right) + \sum_{i=1}^{n} I(B_i), \tag{8}$$

and if $I: \mathscr{C} \to \mathbb{R}^+$ (i.e., I is finite), then (8) may be expressed symmetrically as

$$I\left(\bigcup_{i=1}^{n} B_i\right) - I\left(\bigcup_{i=1}^{n} A_i\right) \leqslant \sum_{i=1}^{n} [I(B_i) - I(A_i)]. \tag{9}$$

We are now prepared to present a construction of a positive capacity function, analogous to that of a measure from an outer measure. This will also be of special interest in applications considered in the next section.

Theorem 5. *Let* $(\Omega, \mathscr{C})$ *be a paved space where* $\mathscr{C}$ *is closed under finite unions and intersections. Suppose that* $I: \mathscr{C} \to \overline{\mathbb{R}}^+$ *is an increasing strongly subadditive mapping such that for each* $A_n \subset A_{n+1}$, $A_n \in \mathscr{C}$, *with* $\bigcup_{n=1}^{\infty} A_n \in \mathscr{C}$, *we have*

$$I\left(\bigcup_n A_n\right) = \lim_n I(A_n). \tag{10}$$

Define for each $A \in \mathscr{C}_\sigma$ *a set function* I_* *by*

$$I_*(A) = \sup\{I(B): B \subset A, B \in \mathscr{C}\},$$

and define the outer capacity I^* *by*

$$I^*(A) = \inf\{I_*(B): B \supset A, B \in \mathscr{C}_\sigma\}, \qquad A \subset \Omega. \tag{11}$$

Then:

(i) $I^*: \mathscr{P}(\Omega) \to \overline{\mathbb{R}}^+$ *is increasing and regular, i.e., for* $A_n \subset A_{n+1} \subset \Omega$ *we have* $I^*(\bigcup_n A_n) = \lim_n I^*(A_n)$. *It is strongly countably subadditive, i.e.,* (8) *holds with* $n = +\infty$ *and* $A_i \subset B_i \subset \Omega$, $i \geq 1$.

(ii) *The function* I^* *need not be a capacity in general, and it will be so iff for each* $\{B_n: n \geq 1\} \subset \mathscr{C}$, $B_n \supset B_{n+1}$, *one has* $I^*(\bigcap_n A_n) = \lim_n I(A_n)$.

Proof. We present the various properties in steps for clarity. It is evident that both I_* and I^* are increasing. Other parts need some detailed computations. We first show that I_* satisfies (10), I^* is subadditive, and then it is regular as well as strongly countably subadditive, in that order. Note that we also have by definition $I_* = I^*|\mathscr{C}_\sigma$ and $I = I^*|\mathscr{C}$.

1. I_* is normal on $\mathscr{C}_0$, i.e., $A_n \in \mathscr{C}_\sigma$, $A_n \uparrow A \Rightarrow I_*(A) = \lim_n I_*(A_n)$, so that I_* on $\mathscr{C}_\sigma$ satisfies (10). For, since $A = \bigcup_n A_n \in \mathscr{C}_\sigma$, $I_*(A)$ is defined, and since $I_*(A) \geq \lim_n I_*(A_n)$ follows from monotonicity, we only have to establish the opposite inequality. By definition of $\mathscr{C}_\sigma$, there exist $A_{nm} \in \mathscr{C}$ such that $A_n = \bigcup_m A_{nm}$. The fact that $\mathscr{C}$ is closed under finite unions can be used (by replacing each A_{nm} with $\bigcup_{i=1}^m A_{ni}$, $m \geq 1$, if necessary) to take $A_{nm} \subset A_{n(m+1)}$ for each $n \geq 1$. By the same argument $A_n \subset A_{n+1}$ can be assumed in the representation of A. Thus $A = \bigcup_{n=1}^{\infty}\bigcup_{m=1}^{\infty} A_{nm} = \bigcup_{m=1}^{\infty}\bigcup_{n=1}^{\infty} A_{nm}$. Using similar partial unions in n also, it can be assumed that A_{nm} from $\mathscr{C}$ is increasing in both m and n, and $A_n = \bigcup_{m=1}^{\infty} A_{nm}$. But in the definition of I_*, we may find an increasing sequence $B_n \subset \tilde{A} \in \mathscr{C}_\sigma$, so that $I_*(\tilde{A}) = \lim_m I(B_m)$, $B_m \in \mathscr{C}$. This and the preceding argument applied to each A_n give (with A_{nm}

in place of the B_m)

$$I_*(A_n) = \lim_m I(A_{nm}). \tag{12}$$

Next note that the increasing property of A_{nm} implies $A = \bigcup_n \bigcup_m A_{nm} = \bigcup_n A_{nn}$, where A_{nn} is also increasing. It follows from this and (12) that

$$I_*(A) = \lim_n I(A_{nn}) \leqslant \lim_n \lim_m I(A_{mn}) = \lim_n I_*(A_n). \tag{13}$$

With the earlier inequality, this shows (10) holds for I_* on $\mathscr{C}_\sigma$.

2. The outer capacity I^* is strongly subadditive on $\mathscr{P}(\Omega)$. For, with the preceding step we first deduce that I_* is strongly subadditive on $\mathscr{C}_\sigma$. Indeed, let $A, B \in \mathscr{C}_\sigma$, and represent them as $A = \bigcup_n A_n$, $B = \bigcup_n B_n$, $A_n, B_n \in \mathscr{C}$, and increasing. Then evidently $A_n \cup B_n \uparrow A \cup B$, $A_n \cap B_n \uparrow A \cap B$, and for each $n \geqslant 1$, $A_n \cup B_n \in \mathscr{C}$, $A_n \cap B_n \in \mathscr{C}$. Using (4) for I and taking the limits as $n \to \infty$, we get by step 1

$$\begin{aligned} I_*(A \cup B) + I_*(A \cap B) &= \lim_n \left[I(A_n \cup B_n) + I(A_n \cap B_n) \right] \\ &\leqslant \lim_n \left[I(A_n) + I(B_n) \right] = I_*(A) + I_*(B). \end{aligned} \tag{14}$$

If $\tilde{A}, \tilde{B}$ are arbitrary sets of $\mathscr{P}(\Omega)$, and $\tilde{A} \subset A$, $\tilde{B} \subset B$ are sets from $\mathscr{C}_\sigma$, then (14) holds for A, B, and hence, taking infimum over all such A, B in (14), we get by (11)

$$I^*(\tilde{A} \cup \tilde{B}) + I^*(\tilde{A} \cap \tilde{B}) \leqslant I^*(\tilde{A}) + I^*(\tilde{B}).$$

So I^* has the asserted property.

3. I^* is regular on $\mathscr{P}(\Omega)$, so that (10) holds for I^*. For, let $A_n \subset A_{n+1} \subset \Omega$ be a sequence. Since I^* is monotone, we already have $I^*(\bigcup_n A_n) \geqslant \sup_n I^*(A_n)$, and the result holds if the right side $= +\infty$. So let $\sup_n I^*(A_n) < \infty$. By (11), given $\epsilon > 0$, there exists a $B_n \in \mathscr{C}_\sigma$ such that $A_n \subset B_n$ and

$$I_*(B_n) = I^*(B_n) < I^*(A_n) + \frac{\epsilon}{2^n}. \tag{15}$$

But $\{B_n: n \geqslant 1\} \subset \mathscr{C}_\sigma$ need not be increasing to apply the result of step 1 to I_*, which would give the desired inequality. Consequently, we consider the partial unions $C_n = \bigcup_{k=1}^n B_k$ ($\in \mathscr{C}_\sigma$) and show that this is also a good approximation to the sequence A_n which satisfies an analog of (15). Indeed, if $n = 1$, $C_1 = B_1 \supset A_1$, then (15) is available. To use induction, we write it as

$$I_*(C_n) = I^*(C_n) < I^*(A_n) + \epsilon\left(1 - \frac{1}{2^n}\right). \tag{16}$$

Suppose it is true for n. So for $n+1$, $C_{n+1} = C_n \cup B_{n+1}$ and

$$I_*(C_{n+1}) \leqslant I_*(C_n) + I_*(B_{n+1}) - I_*(C_n \cap B_{n+1}), \qquad \text{by step 2,}$$

$$\leqslant I_*(B_{n+1}) + [I_*(C_n) - I^*(A_n)]$$

(since $I_* = I^*$ on $\mathscr{C}_\sigma$ and $A_n = A_n \cap A_{n+1} \subset C_n \cap B_{n+1}$)

$$< I_*(B_{n+1}) + \epsilon\left(1 - \frac{1}{2^n}\right)$$

[by (16) and the inductive hypothesis]

$$< I^*(A_{n+1}) + \frac{\epsilon}{2^{n+1}} + \epsilon\left(1 - \frac{1}{2^n}\right), \qquad \text{by (15)}$$

$$= I^*(A_{n+1}) + \epsilon\left(1 - \frac{1}{2^{n+1}}\right).$$

Hence the induction is complete, so that step 1 implies

$$I^*\left(\bigcup_n A_n\right) \leqslant I^*\left(\bigcup_n C_n\right) = I_*\left(\bigcup_n C_n\right) = \lim_n I_*(C_n)$$

$$\leqslant \lim_n I^*(A_n), \qquad \text{by (16).}$$

This shows that I^* is regular on $\mathscr{P}(\Omega)$.

4. I^* is strongly countably subadditive on $\mathscr{P}(\Omega)$. For, it suffices to show that (8) holds with $n = \infty$. So let $\{(A_i, B_i): i \geqslant 1\} \subset \mathscr{P}(\Omega)$ be such that $A_i \subset B_i$, $i \geqslant 1$. By step 2, for each n we have (8), i.e.,

$$I^*\left(\bigcup_{i=1}^n B_i\right) + \sum_{i=1}^n I^*(A_i) \leqslant I^*\left(\bigcup_{i=1}^n A_i\right) + \sum_{i=1}^n I^*(B_i). \tag{17}$$

Now let $n \to \infty$ and by step 3 we can interchange I^* and lim. That shows the validity of (17) with $n = \infty$ there. Hence I^* is strongly σ-subadditive, by definition. This establishes (i).

The preceding work shows that the generated function I^* is an outer capacity, and (i)–(ii) of Definition 1(a) hold for it. However, (iii) is not a consequence of the work done thus far, and so I^* is a capacity iff the limit and I^* can be interchanged for decreasing sequences from $\mathscr{C}$. This finishes the proof.

There are other methods of constructing outer capacities. We outline one such, namely the one based on the classical Newtonian capacity, following

Sion (1963), since it shows how different a capacity function can be from a standard measure. Indeed, one finds that even a positive capacity function need *not* be strongly subadditive. [See Exercise 4(b).] To explain this difference further consider the following:

Definition 6. Let Ω be a locally compact Hausdorff space, and $\mathbf{M}(\Omega) = \mathbf{M}$ be the set of all Radon measures on Ω (Definition 2.3.9). If $\mathscr{M}_\mu$ is the class of all μ-measurable subsets of Ω, and $\mathscr{B}_0 = \bigcap\{\mathscr{M}_\mu: \mu \in \mathbf{M}\}$, then $\mathscr{B}_0$ is called a *universally* (or *absolutely*) measurable σ-algebra of Ω. Let $\Omega = \mathbb{R}^3$, $\tilde{I}: A \mapsto \sup\{\mu(A): \mu \in \mathbf{M}_0\}$, and $\tilde{I}^*: A \mapsto \inf\{\tilde{I}(B): B \supset A, B \text{ open}\}$, where $\mathbf{M}_0 = \{\mu \in \mathbf{M}(\mathbb{R}^3): \int_{\mathbb{R}^3}|x-y|^{-1}\mu(dy) \leqslant 1,\ x \in \mathbb{R}^3\}$ with $|x-y|^2 = \sum_{i=1}^3 |x_i - y_i|^2$ as the distance in $\mathbb{R}^3$. Let $\mathscr{C}$ be the compact subsets of Ω. Then a set $A \subset \Omega$ is capacitable, relative to $\mathscr{C}$ (and $\tilde{I}$), if $\tilde{I}^*(A) = \sup\{\tilde{I}(C): C \subset A,\ C \in \mathscr{C}\}$. [$\tilde{I}^*$ turns out to be an *outer capacity*; see Proposition 7 below.]

In the classical theory, $\Omega = \mathbb{R}^3$, one defines the capacity of each closed set $E \subset \mathbb{R}^3$, which is an equilibrium set, to be the total charge of a charge distribution producing an equilibrium potential. [We shall not define the various terms here, since that is not needed for our work. We mention the matter for motivation. The function $U_u: x \mapsto \int_{\mathbb{R}^3}|x-y|^{-1}\mu(dy)$ in the above definition is called the *potential* of μ at $x \in \mathbb{R}^3$.]

We then have the following assertion about this $\tilde{I}$.

Proposition 7. *With the above notation, $\tilde{I}^*$ is an outer capacity, and a set $A \subset \mathbb{R}^3$ is $\tilde{I}$-measurable (in the sense of Carathéodory) iff $\tilde{I}(A) = 0$ or $\tilde{I}(A^c) = 0$.*

The last part of this result shows that $\tilde{I}$ (or $\tilde{I}^*$) can be quite distinct from a standard measure (in the sense of Chapter 2). Nevertheless, we have a positive capacity function out of it. Thus we need to study the properties of these functions from a different point of view. The proof of this result is omitted here, but a sketch will be given as Exercise 7. Also, Exercises 10 and 11 show how for topological paved spaces, outer capacities with a continuity condition sometimes become capacities.

Before going to the main result on the capacitability of concrete classes of sets, we need a technical result of a topological nature, to use with compact pavings. Recall that a nonempty set $\mathscr{F}$ of subsets of Ω is called a *filter* if (i) $\varnothing \notin \mathscr{F}$, (ii) $A \in \mathscr{F}$, $A \subset B \subset \Omega \Rightarrow B \in \mathscr{F}$, and (iii) $A, B \in \mathscr{F} \Rightarrow A \cap B \in \mathscr{F}$. If $\mathscr{F}_1, \mathscr{F}_2$ are two filters in Ω, than $\mathscr{F}_2$ is *finer* than $\mathscr{F}_1$ (or $\mathscr{F}_2$ *refines* $\mathscr{F}_1$) if $\mathscr{F}_1 \subset \mathscr{F}_2$, and $\mathscr{F}_1$ is an *ultrafilter* if for any filter $\mathscr{F}_2$ we have $\mathscr{F}_1 \subset \mathscr{F}_2 \Rightarrow \mathscr{F}_1 = \mathscr{F}_2$. A collection $\mathscr{F}_0$ of subsets of Ω is a *filter base* if (i) $\varnothing \notin \mathscr{F}_0$, (ii) for all $A_1, A_2 \in \mathscr{F}_0$, there is $A_3 \in \mathscr{F}_0$ such that $A_3 \subset A_1 \cap A_2$. Every filter base $\mathscr{F}_0$ generates a filter $\mathscr{F}$, namely, $\mathscr{F} = \{B \subset \Omega: B \supset A$ for at least one $A \in \mathscr{F}_0\}$.

The facts we are after can be stated in the following:

Proposition 8.

(a) (*H. Cartan*) *Every filter of subsets of* Ω *can be refined by an ultrafilter of* Ω.

(b) *If* $\mathscr{F}$ *is an ultrafilter of* Ω *and* $A \cup B \in \mathscr{F}$, *then either* $A \in \mathscr{F}$ *or* $B \in \mathscr{F}$. *In particular, if* $A \cap B = \varnothing$ *then only one of the sets belongs to* $\mathscr{F}$.

(c) (*J. W. Alexander*) *Let* $\mathscr{E}$ *be a compact paving of* Ω (*see discussion after Definition* 1.1). *If* $\tilde{\mathscr{E}}$ *is a paving obtained under finite unions and arbitrary intersections of* $\mathscr{E}$, *then* $\tilde{\mathscr{E}}$ *is also compact.*

Proof. (a): Let $\mathscr{F}_0$ be the given filter, and $\mathscr{G}$ be the collection of all filters of Ω that include (= refine) $\mathscr{F}_0$. The refinement relation is a partial order on $\mathscr{G}$. To use Zorn's lemma, consider $\mathscr{G}_1 \subset \mathscr{G}$, a chain. We assert that $\mathscr{G}_1$ has an upper bound in $\mathscr{G}$. Indeed, let $\mathscr{F}_1 = \{A \subset \Omega: A \in \mathscr{F} \in \mathscr{G}_1\}$. This is an upper bound, since it is a filter. Indeed, (i) $\varnothing \notin \mathscr{F}_1$; (ii) $A \in \mathscr{F}_1$ and $B \supset A \Rightarrow A \in \mathscr{F}$ for some $\mathscr{F} \in \mathscr{G}_1$ and $B \in \mathscr{F}$, so that $B \in \mathscr{F}_1$; and (iii) $A_1, A_2 \in \mathscr{F}_1 \Rightarrow A_1 \in \mathscr{F}$, $A_2 \in \mathscr{F}'$ for some pair in $\mathscr{G}_1$. Now since $\mathscr{G}_1$ is a chain, either $\mathscr{F} \subset \mathscr{F}'$ or $\mathscr{F}' \subset \mathscr{F}$, so that in either case $A_1 \cap A_2 \in \mathscr{F}'$ (or $\in \mathscr{F}$) and hence $A_1 \cap A_2 \in \mathscr{F}_1$. It is clear that every $\mathscr{F}$ in $\mathscr{G}_1$ is $\subset \mathscr{F}_1$. In particular, $\mathscr{F}_0 \subset \mathscr{F}_1$. So $\mathscr{F}_1 \in \mathscr{G}$, and $\mathscr{F}_1$ is an upper bound of $\mathscr{G}_1$. Thus by Zorn's lemma $\mathscr{G}$ has a maximal element $\tilde{\mathscr{F}}$, which is clearly an ultrafilter, and it refines $\mathscr{F}_0$.[†]

(b): Let $A \cup B \in \mathscr{F}$, an ultrafilter on Ω. If $A \notin \mathscr{F}$ and $B \notin \mathscr{F}$, let $\tilde{\mathscr{F}} = \{C \subset \Omega: A \cup C \in \mathscr{F}\}$. It is clear that $\tilde{\mathscr{F}}$ is a filter and $B \in \tilde{\mathscr{F}}$. Also $\mathscr{F} \subset \tilde{\mathscr{F}}$, the inclusion being proper. But then $\tilde{\mathscr{F}}$ refines $\mathscr{F}$, which is impossible, since $\mathscr{F}$ is an ultrafilter. Hence either $A \in \mathscr{F}$ or $B \in \mathscr{F}$ must be true. The last statement is clear, since if both $A, B \in \mathscr{F}$ then $\varnothing = A \cap B \in \mathscr{F}$, which is impossible.

(c) [this proof is due to G. Mokobodzki; see Dellacherie and Meyer (1978)]: Let $\mathscr{E}_0 = \{A: A = \bigcup_{i=1}^n A_i,\ A_i \in \mathscr{E}\}$, and $\tilde{\mathscr{E}} = \{\tilde{A}: \tilde{A} = \bigcap_{i \in I} A_i,\ A_i \in \mathscr{E}_0\}$. Thus $\tilde{\mathscr{E}}$ is obtained by closing $\mathscr{E}_0$ under arbitrary intersections. We first show that the compactness of $\mathscr{E}$ implies that of $\mathscr{E}_0$, from which we deduce immediately the same property of $\tilde{\mathscr{E}}$.

Thus let $\mathscr{E}$ be compact, and consider an arbitrary family $\{K_i: i \in I\}$ from $\mathscr{E}_0$ having the finite intersection property. Then the collection of all finite intersections of this family forms a filter base, and hence by (a) there is an ultrafilter $\mathscr{F}$ containing each K_i, $i \in I$. By definition of $\mathscr{E}_0$, for each $i \in I$

[†] The existence of an ultrafilter is established through Zorn's lemma. It is known, however, that without the use of the latter, this existence cannot be proved with other axioms of set theory. Moreover, the existence result is weaker than Zorn's lemma in that the existence of an ultrafilter does not imply Zorn's lemma.

there is a finite set J_i such that $K_i = \bigcup_{j \in J_i} K_{ij}$, $K_{ij} \in \mathscr{E}$. By (b) above, there exists at least one set $K_{ij_i} \in \mathscr{F}$, and by the fact that $\mathscr{F}$ is a filter we must have $\bigcap_{i \in I_0} K_{ij_i} \neq \varnothing$, $I_0 \subset I$ finite (since $\varnothing \notin \mathscr{F}$). Thus $\{K_{ij_i}: i \in I\}$ has the finite intersection property in $\mathscr{E}$. So $\bigcap_{i \in I} K_{ij_i} \neq \varnothing$ by the compactness of $\mathscr{E}$. But $K_{ij_i} \subset K_i$, so that $\bigcap_{i \in I} K_i \supset \bigcap_{i \in I} K_{ij_i} \neq \varnothing$, and $\mathscr{E}_0$ is a compact paving.

This implies the result for $\tilde{\mathscr{E}}$, since if $\{F_i: i \in I\} \subset \tilde{\mathscr{E}}$ with the finite intersection property, then $F_i = \bigcap_{j \in J_i} F_{ij}$, $F_{ij} \in \mathscr{E}_0$, as above, $i \in I$. Now if $I_0 \subset I$ is any finite subset, then $\varnothing \neq \bigcap_{i \in I_0} F_i = \bigcap_{i \in I_0} \bigcap_{j \in J_i} F_{ij}$, so that $\{F_{ij}: j \in J_i,\ i \in I\}$ has the finite intersection property. The preceding paragraph then implies that $\varnothing \neq \bigcap_{i \in I} \bigcap_{j \in J_i} F_{ij} = \bigcap_{i \in I} F_i$ and $\tilde{\mathscr{E}}$ is compact, which completes the proof of the proposition.

With this result, we are ready to establish the main

Theorem 9 (Choquet). *Let $(\Omega, \mathscr{C})$ be a paved space with $\mathscr{C}$ closed under finite unions and intersections. If $I: \mathscr{P}(\Omega) \to \mathbb{R}$ is any $\mathscr{C}$-capacity, then each $\mathscr{C}$-analytic set is capacitable for I.*

Proof. We present the demonstration in steps for clarity.

1. Every element of $\mathscr{C}_{\sigma\delta}$ is capacitable for I. For, let $A \in \mathscr{C}_{\sigma\delta}$. If $I(A) = -\infty$, then by Definition 1, $I(B) = -\infty$ for each $B \subset A$, and $I(A) = \sup\{I(B): B \subset A,\ B \in \mathscr{C}_\delta\} = -\infty$, so that A is capacitable for I. Thus we only need to consider $I(A) > -\infty$.

Since $A \in \mathscr{C}_{\sigma\delta}$, there exist $A_n \in \mathscr{C}_\sigma$, $A_n \supset A_{n+1}$, and $A = \bigcap_n A_n$; and for each n, there are $A_{nm} \in \mathscr{C}$ such that $A_n = \bigcup_m A_{nm}$, $A_{nm} \subset A_{n(m+1)}$. To obtain the approximation from below, we use the same ideas as in the proof of Theorem 2.2.17(ii). Thus let $-\infty < t < I(A)$. One has to find a set B in $\mathscr{C}_\delta$ such that $B \subset A$ and $I(B) \geqslant t$. Let us construct a sequence $B_n \in \mathscr{C}$ such that $B_n \subset A$, $B_n \downarrow B$, and $I(A \cap B_n) > t$, $n \geqslant 1$. Now, since $A_n \supset A$, consider

$$t < I(A) = I(A \cap A_1) = I\Big(A \cap \bigcup_m A_{1m}\Big)$$

$$= I\Big(\bigcup_m (A \cap A_{1m})\Big) = \lim_m I(A \cap A_{1m}), \qquad \text{by Definition 1(a)(ii).}$$

Thus there exists $m_0 \geqslant 1$ such that $m \geqslant m_0 \Rightarrow I(A \cap A_{1m}) > t$. Let $B_1 = A_{1m}$. Using induction, if $B_1, \ldots, B_n$ are constructed with $I(A \cap B_n) > t$, $B_n \subset A$, consider A_{n+1} ($\supset A$) and repeat the above procedure with $A \cap B_n$ in place of A above. Hence

$$t < I(A \cap B_n) = I(A \cap B_n \cap A_{n+1})$$

$$= I\Big(A \cap B_n \cap \bigcup_m A_{(n+1)m}\Big)$$

$$= \lim_m I\big(A \cap B_n \cap A_{(n+1)m}\big). \tag{18}$$

Choose m_1 such that $m \geqslant m_1 \Rightarrow I(A \cap B_n \cap A_{(n+1)m}) > t$, and set $B_{n+1} = B_n \cap A_{(n+1)m}$. Thus $B_{n+1} \in \mathscr{C}$, $\{B_n: n \geqslant 1\} \subset \mathscr{C}$, and $B_n \supset B_{n+1}$. If $B = \bigcap_n B_n$, then $B \in \mathscr{C}_\delta$, $B \subset A$, and $I(A) \geqslant I(B) = \lim_n I(B_n) \geqslant \lim_n I(A \cap B_n) \geqslant t$. It follows that (since $t < I(A)$ is arbitrary)

$$I(A) = \sup\{I(B): B \subset A, B \in \mathscr{C}_\delta\}. \tag{19}$$

2. This is an intermediate construction to extend the above step. Let A be $\mathscr{C}$-analytic. Then by Definition 1.2 and Proposition 1.3, there is an auxiliary compact separable space E with compact sets $\mathscr{K}$ of E as the paving such that for a $B \in (\mathscr{K} \times \mathscr{C})_{\sigma\delta}$ we have $A = \pi(B)$, where $\pi: E \times \Omega \to \Omega$ is the coordinate projection. Let $\mathscr{G}_0 = \mathscr{K} \times \mathscr{C}$. Then $\mathscr{G}_0$ is a compact paving of $E \times \Omega$, and if $\mathscr{G}$ is the paving obtained from $\mathscr{G}_0$ by closing it under finite unions and intersections, then by Proposition 8(c), $\mathscr{G}$ is still compact. Define $J: \mathscr{P}(E \times \Omega) \to \overline{\mathbb{R}}$ by the equation $J(H) = I(\pi(H))$. We then have:

3. J is a $\mathscr{G}$-capacity on $\mathscr{P}(E \times \Omega)$. For, J is clearly monotone, since I is. Also if $H_n \subset H_{n+1}$ is any sequence, then $\pi(\bigcup_n H_n) = \bigcup_n \pi(H_n)$ implies $J(\bigcup_n H_n) = \lim_n J(H_n)$ by the corresponding property of I. We now have to use the compactness property of $\mathscr{G}$ for the decreasing case. Thus if $B_n \supset B_{n+1}$, $B_n \in \mathscr{G}$, then for each $\omega \in \Omega$, $\{\pi^{-1}(\{\omega\}) \cap B_n: n \geqslant 1\} \subset \mathscr{G}$ has the finite intersection property, so that by the compactness of $\mathscr{G}$, it has a nonempty intersection. Consequently by Proposition 1.4, $\pi(\bigcap_n B_n) = \bigcap_n \pi(B_n)$, and hence $J(\bigcap_n B_n) = \lim_n J(B_n)$, since I has that property. Thus J is a $\mathscr{G}$-capacity.

4. Every $\mathscr{C}$-analytic A is I-capacitable. For, by definition $A = \pi(B)$, $B \in (\mathscr{K} \times \mathscr{C})_{\sigma\delta}$. Since J is a $\mathscr{G}$-capacity, by step 1, B is J-capacitable. Thus given $t < J(B)$, there is a $C \in \mathscr{G}_\delta$ such that $C \subset B$ and $J(C) > t$. Here we are excluding the case $J(B) = -\infty$, which is trivial (and true), as noted at the beginning [since then $J(B) = I(\pi(B)) = I(A) = -\infty$]. Hence setting $A_0 = \pi(C) \subset \pi(B) = A$, one has

$$I(A) \geqslant I(A_0) = J(C) > t, \qquad A_0 \in \mathscr{C}_\delta = \pi(\mathscr{G}_\delta). \tag{20}$$

Letting $t \uparrow I(A)$, we get $I(A) = \sup\{I(B): B \subset A, B \in \mathscr{C}_\delta\}$, and hence A is I-capacitable. The theorem is completely proved.

This is a considerable generalization of Proposition 2, from outer measures to arbitrary capacities. Some important applications of Choquet's theorem will be given in the next section.

EXERCISES

1. Let $\mathscr{S}$ be a paving of a set Ω. Then $\mathscr{S}$ is called *semicompact* if every countable family of elements of $\mathscr{S}$ with the finite intersection property has a nonempty intersection. This concept is weaker than the compact paving. If $(\Omega, \mathscr{S})$ is (a compact or) a semicompact paved space, show that $\tilde{\mathscr{S}} = \mathscr{F}_\delta$ is once again a semicompact paving, where $\mathscr{F} = \{A \cup B: A, B \in \mathscr{S}\}$.

2. Verify that Proposition 8(c) is valid if "compactness" is replaced by "semicompactness" in its hypothesis and conclusion.

3. Let $(\Omega, \mathscr{C})$ be a paved space. Let us say $A \subset \Omega$ is $\mathscr{C}$-analytic in the "generalized sense" if there is an auxiliary set E, some semicompact paving $\mathscr{E}$ on E, and a set $B \in (\mathscr{E} \times \mathscr{C})_{\sigma\delta}$ such that $A = \pi(B)$, where π is the coordinate projection on $E \times \Omega \to \Omega$. Show that, following Proposition 1.3, A is $\mathscr{C}$-analytic iff it is $\mathscr{C}$-analytic in the generalized sense. (Thus this additional qualification is redundant and will be dropped.)

4. (a) Without using Exercise 3, complete the details of proof of Theorem 9 with semicompactness in the definition of analyticity and using Exercise 2.

   (b) If $\Omega = \{a, b, c, d\}$, Σ is the power set, and P_1, P_2 are two measures given by $P_1(\{b\}) = P_1(\{c\}) = P_2(\{d\}) = \frac{1}{5}$, $P_1(\{d\}) = P_2(\{b\}) = P_2(\{c\}) = \frac{1}{10}$, $P_1(\{a\}) = \frac{1}{2}$, $P_2(\{a\}) = \frac{3}{5}$, then let $I = \max(P_1, P_2)$. Show that I is a capacity on Σ, but it is not strongly subadditive. [Consider the subadditivity for $A = \{a, b\}$, $B = \{a, c\}$.]

5. In the proof of Theorem 9 [see (20) there] we used a special property of projected analytic sets. A more general statement can be made. Let $(\Omega, \mathscr{C})$ be a paved space, and $(E, \mathscr{E})$ be a compact or semicompact paved space. Consider $\Omega_1 = E \times \Omega$, $\mathscr{C}_1 = \mathscr{E} \times \mathscr{C}$, and $\pi: \Omega_1 \to \Omega$ as the coordinate projection. Show that for each $\mathscr{C}_1$-analytic set $A_1 \subset \Omega_1$, $\pi(A_1)$ $(\subset \Omega)$ is a $\mathscr{C}$-analytic set, i.e., $\pi(\mathscr{A}(\mathscr{C}_1)) \subset \mathscr{A}(\mathscr{C})$. [Use Propositions 1.3 and 1.4 carefully.]

6. Let (Ω, Σ) be a measurable space, where Σ is a σ-algebra, and let $\mathscr{C}$ be a paving, $\mathscr{C} \subset \Sigma$, closed under finite unions and intersections. Let $I: \mathscr{P}(\Omega) \to \overline{\mathbb{R}}^+$ be a $\mathscr{C}$-capacity which is strongly subadditive on Σ. Show that for each Σ-analytic set A, there exist A_0 and A_1 such that $A_1 \subset A \subset A_0$ and $A_1 \in \Sigma$, $A_0 \in (\mathscr{C})_{\sigma\delta}$, $I(A_0) = I(A) = I(A_1)$. [This is a slight extension of Proposition 2, and the same argument, together with Theorem 5, can be used. Verify that each $\mathscr{C}$-capacitable set is also Σ-capacitable with the same value.]

7. We outline the proof of Proposition 7 here. Since $\tilde{I}$ is an outer measure, verify that $\tilde{I}(A) = \sup\{\mu(A): \mu \in \mathbf{M}_0(A)\}$, where $\mathbf{M}_0(A)$ is the subset of measures from $\mathbf{M}$ which vanish outside A [i.e., $\mu(A^c) = 0$ for $\mu \in \mathbf{M}_0(A)$, with the notation of Definition 6] for each universally measurable set A. [Since $\mu_A = \mu|A$ is also Radon, and since $\tilde{I}(A) \geqslant \mu(A)$, one has $\mu_A \leqslant \mu$, $\mu_A \in \mathbf{M}_0(A)$, and $\mu_A(A) = \mu(A)$.] If $\tilde{I}_*(A) = \sup\{\tilde{I}(B): B \subset A,\ B \in \mathscr{C}\}$, then $\tilde{I}_*(A) = \tilde{I}(A)$ for each universally measurable A, so that each element of $\mathscr{B}_0$ is capacitable for $\tilde{I}$. [Note that μ is Radon, so that $\mu(A) = \sup\{\mu(B): B \subset A,\ B \in \mathscr{C}\}$, and then $\tilde{I}(A) = \sup\{\mu(A): \mu \in$

$\mathbf{M}_0\} = \tilde{I}_*(A)$.] Next note that $\tilde{I}(A_n) = 0 \Rightarrow \tilde{I}(\bigcup_n A_n) = 0 \Rightarrow \tilde{I}_*(B_n) = 0$ for $B_n \in \mathscr{B}_0$, $n \geq 1 \Rightarrow \tilde{I}_*(\bigcup_n B_n) = 0$. Deduce that $\tilde{I}$ is an outer capacity. Next use a classical result, from potential theory (this is not obvious), that for each $C \in \mathscr{C}$, there exists a measure $\mu \in \mathbf{M}$ such that $\mu(C) = \tilde{I}(C)$, and that if $C_1, C_2 \in \mathscr{C}$ disjoint, then $\tilde{I}(C_i) > 0 \Rightarrow \tilde{I}(C_1 \cup C_2) < \tilde{I}(C_1) + \tilde{I}(C_2)$. Taking $B = C_1 \cup C_2$, deduce that if $A \subset \Omega$ is measurable relative to $\tilde{I}$, [i.e., $\tilde{I}(B) = \tilde{I}(A \cap B) + \tilde{I}(A^c \cap B)$, all $B \in \mathscr{C}$], then both $\tilde{I}_*(A) > 0$ and $\tilde{I}_*(A^c) > 0$ is impossible. Thus if $\tilde{I}_*(A) = 0$, for each $C \in \mathscr{C}$ choose a $\mu \in \mathbf{M}(C)$ with $\mu(C) = \tilde{I}(C)$. From this conclude that A is μ-measurable and $\mu(A) > 0$ is impossible [that $\mu(A) > 0 \Rightarrow \tilde{I}_*(A) > 0$]. Thus $\tilde{I}(C) = 0$ for each compact set C. Since $\Omega = \mathbb{R}^3$ is σ-compact, this implies $\tilde{I}(A \cap C) = 0$ and $\tilde{I}(A) = 0$.

8. Let $(\Omega, \mathscr{C})$ be a paved space with $\mathscr{C}$ closed under finite unions and intersections. If $I: \mathscr{C} \to \overline{\mathbb{R}}^+$ is additive, increasing, and $I(\bigcup_n A_n) = \lim_n I(A_n)$ for $A_n \subset A_{n+1}$, with $\bigcup_n A_n \in \mathscr{C}$, then show, with the procedure of Theorem 5, that the outer capacity I^* derived from I is actually σ-additive on $\mathscr{C}$-analytic sets $\mathscr{A}(\mathscr{C})$, and has a σ-additive extension to $\sigma(\mathscr{A}(\mathscr{C}))$. [This is a sort of converse to Proposition 2. It is enough to show that I^* is additive on $\mathscr{A}(\mathscr{C})$.]

9. Let $(\Omega, \mathscr{C})$ be a paved space with $\mathscr{C}$ as a ring. Let $\varphi: \mathbb{R} \to \mathbb{R}^+$ be an increasing, nonzero function, $\varphi(0) = 0$, and I be a strongly subadditive, nonnegative, $\mathscr{C}$-capacity on $\mathscr{P}(\Omega)$. If $f: \Omega \to \mathbb{R}$ is any mapping, say it is (I, φ)-summable whenever $I_\varphi(f) = I(\{\omega: \varphi(|f|)(\omega) > 0\}) < \infty$. Let $L_\varphi(I) = \{f: \Omega \to \mathbb{R}: I_\varphi(f/k) < \infty \text{ for some } k > 0\}$. Let I^* be the outer capacity generated by $(\mathscr{C}, I)$.

(a) Show that $L_\varphi(I^*)$ is a vector space over $\mathbb{R}$.

(b) If $d_\varphi(f) = \inf\{k > 0: I_\varphi^*(f/k) \leq 1\}$, verify that $d_\varphi(\cdot)$ is a metric on $L_\varphi(I^*)$, where I_φ^* is I_φ with I^* in place of I. [$d_\varphi(f) = d_\varphi(0, f)$ really.]

(c) Establish that $(L_\varphi(I^*), d_\varphi(\cdot))$ is a complete metric space. [The fact that I^* is increasing, strongly countably subadditive, and regular is needed in this result. The procedure of Theorem 4.5.7 works here. A further extension of these ideas is possible.]

10. In view of Theorem 9, it is useful to have a more refined form of Theorem 5 indicating that an outer capacity is easily recognizable as a capacity function. This can be done if there is a topology on the underlying space, and so the original set function I can be related to that topology. We indicate how this type of result can be obtained in this and the next exercise. Let Ω be a Hausdorff topological space, and $\mathscr{G}, \mathscr{C}$ be pavings of Ω containing the open and compact sets respectively. A function $I: \mathscr{G} \to \overline{\mathbb{R}}^+$ [$I: \mathscr{C} \to \overline{\mathbb{R}}^+$] which is increasing is termed *left continuous* on open sets [*right continuous* on compact sets] if for each open set G [compact set K] such that $I(G) > \alpha$ [$I(K) < \beta$], for any α [β] we can find a compact set $K_0 \subset G$ [open set $G_0 \supset K$] such that $I(\tilde{G}) > \alpha$ [$I(\tilde{K}) < \beta$] for each open set $\tilde{G} \supset K_0$ [compact set $\tilde{K} \subset \tilde{G}$]. This corresponds to the left and right continuities of an increasing function on $\mathbb{R}$. Show that if $I: \mathscr{G} \to \overline{\mathbb{R}}^+$ is an increasing, strongly subadditive, left-continuous function on open sets, then for each $A_n \subset A_{n+1}$ from $\mathscr{G}$, $\lim_n I(A_n) = I(\bigcup_n A_n)$, and if I^* is the outer capacity defined as in Theorem 5, then I^* is a capacity for sets of $\mathscr{C}$, i.e., if $K_n \supset K_{n+1}$ are sets of $\mathscr{C}$, then $I^*(\bigcap_n K_n) = \lim_n I^*(K_n)$.

11. With the notation and terminology of the preceding exercise, let $I: \mathscr{C} \to \overline{\mathbb{R}}^+$ be an increasing, strongly subadditive, right-continuous function on compact sets. Then again for each $A_n \subset A_{n+1}$ from $\mathscr{C}$, with $A = \bigcup_n A_n \in \mathscr{C}$, verify that $I(A) = \lim_n I(A_n)$. If I^* is the outer capacity defined as in Theorem 5 and $J = I^*|\mathscr{G}$, show that J is a left-continuous, strongly subadditive increasing function. Then I^* is actually a $\mathscr{C}$-capacity, and J^* is a $\mathscr{G}$-capacity satisfying the capacity condition again on $\mathscr{C}$. In general $J^*(A) \geqslant I^*(A)$, but the $\mathscr{C}$-analytic sets are the same for both. [If U, V are open and $K \subset U \cup V$, then $K_1 = K - V \subset U$, $K_2 = K - U \subset V$ are compact disjoint sets. So there exist disjoint open sets G_i containing K_i, $i = 1, 2$. We may assume $G_1 \subset U$, $G_2 \subset V$ (by considering the relative topology on K, and so $K_i \subset K$ also). Let $C = K - G_2$, $D = K - G_1$, so that C, D are compact, $C \subset U$, $D \subset V$, and $C \cup D = K - (G_1 \cap G_2) = K$. Use this fact to show that J is strongly subadditive on $\mathscr{G}$. Indeed, let U, V be open sets, and consider $\alpha < J(U \cap V)$, $\beta < J(U \cup V)$. Choose compact sets $L \subset U \cap V$ and $K \subset U \cup V$ such that $I(L) > \alpha$, $I(K) > \beta$, which is possible because $J(G) = \sup\{I(C): C \subset G, C \in \mathscr{C}\}$, $G \in \mathscr{G}$. We can replace K by $K \cup L = \tilde{K}$. Then there exist $C \subset U$, $D \subset V$ compact, $\tilde{K} = C \cup D$, and we may assume that $L \subset C$, $L \subset D$ also. Hence $\alpha + \beta \leqslant I(L) + I(K) \leqslant I(C \cap D) + I(C \cup D) \leqslant I(C) + I(D) \leqslant J(U) + J(V)$, since I is increasing and strongly subadditive. Taking the supremum over C and D in U, V on the left and letting $\alpha \to J(U \cap V)$, $\beta \to J(U \cup V)$, we get the strong subadditivity of J. But J is left continuous on $\mathscr{G}$, and the result follows therefore from the preceding exercise. In the context of measures on topological spaces, these ideas are explored in depth later. In particular, see the proofs of Theorems 9.2.6, 9.2.8, and 9.2.9.]

12. Choquet's theorem can be used to prove measurability of certain sets. Let (Ω, Σ, μ) be a complete finite measure space and $(\mathbb{R}, \mathscr{B})$ be the Borelian line. If $\pi: \mathbb{R} \times \Omega \to \Omega$ is the coordinate projection and $A \in \mathscr{B} \otimes \Sigma$, show that $\pi(A) \in \Sigma$. [If $\mathscr{C}$ is the class of compact subsets of $\mathbb{R}$, and $I(B) = \mu^*(\pi(B))$, then I is a $\mathscr{C} \times \Sigma$-capacity by Proposition 2, and by the same result A is I-capacitable, since Theorem 9 actually implies $\mathscr{B} \otimes \Sigma \subset \{I$-capacitable sets$\}$. As in (2) and (3), for each $\epsilon \geqslant 0$, there is $A_\epsilon \in (\mathscr{C} \times \Sigma)_\delta$ such that $I(A) \leqslant I(A_\epsilon) + \epsilon$ $(A_\epsilon \subset A)$. Thus $\pi(\bigcup_n A_n)$ approximates $\pi(A)$ if $\epsilon = 1/n$, and $A_\epsilon = A_n$, so that $\pi(\bigcup_n A_n) \in \Sigma$.]

7.3. APPLICATION TO THE DANIELL INTEGRAL

This section is devoted to an application of the capacity theorem in deducing the central results and some consequences of Daniell's integration. The capacity approach gives a short cut to the main representation theorem connecting Daniell's method with the Lebesgue-Carathéodory point of view by associating a unique measure with each linear functional satisfying a fundamental (continuity) requirement. Let us fill in the details.

The Daniell method starts with a positive linear functional on a vector lattice of (enough) real functions on a set Ω. Thus if $\mathcal{L}$ is a vector space of functions $f: \Omega \to \mathbb{R}$, then (i) $f_1, f_2 \in \mathcal{L} \Rightarrow af + bg \in \mathcal{L}$, $a, b \in \mathbb{R}$, and (ii) $f \in \mathcal{L} \Rightarrow |f| \in \mathcal{L}$. Hence $f^+ = (|f| + f)/2 \in \mathcal{L}$, $f^- = f - f^+ \in \mathcal{L}$. Equivalently, $\mathcal{L}$ satisfies (i) and (ii$'$) $f_1, f_2 \in \mathcal{L} \Rightarrow f_1 \vee f_2 = \max(f_1, f_2) \in \mathcal{L}$, since then $f^+ = f \vee 0 \in \mathcal{L}$, $f^- = f - f^+ \in \mathcal{L}$. On such a space $\mathcal{L}$, it is assumed (in lieu of a measure in the Lebesgue approach) that a nonnegative linear functional, obeying the Daniell conditions, is given. Thus there is an $I: \mathcal{L} \to \mathbb{R}$ such that

(1) $I(af + bg) = aI(f) + bI(g)$, $a, b \in \mathbb{R}$, $f, g \in \mathcal{L}$;
(2) $I(f) \geqslant 0$, if $0 \leqslant f \in \mathcal{L}$;
(3) if $0 \leqslant f_n \in \mathcal{L}$, $f_n(\omega) \downarrow 0$ as $n \to \infty$, then $\lim_n I(f_n) = 0$.

Such an I is called a *Daniell integral*; it is an extension of the idea of the Riemann integral, in which $\mathcal{L} = C[a, b]$, the space of real continuous functions on the compact interval $[a, b]$, and $I(f) = \int_a^b f(x)\,dx$, $f \in C[a, b]$. Of course the Lebesgue integral is also a candidate if we restrict to finite valued integrable functions. The point of this method is that we can extend I to a larger class and in fact represent it as a Lebesgue integral for a certain measure, thereby deducing the latter from the former.

The traditional extension procedure is to consider the upper class $\mathcal{L}_u$ which is obtained from $\mathcal{L}$ by closing it with monotone increasing limits of sequences, and to define I on it, just as a measure is extended from a semiring to its elementary class; then one defines an upper and lower Daniell integral on $\mathcal{L}_u$, call the class $\mathcal{L}'$ on which these agree as integrable. That class $\mathcal{L}'$ is proved to be a vector lattice, and one proceeds to show that the extended I includes results analogous to dominated convergence, Fatou's lemma, and the like. Thus the procedure is long and is similar to, but not deducible from, the work of Chapter 2. (This method will be sketched in exercises.) Since we have established some results in the preceding section for a more general (not necessarily additive) capacity function, they will be used here to deduce the Daniell integral directly in a short and quick manner. Our aim is a representation of I as an integral relative to a measure μ, defined on $\mathcal{L}'$ above. We start with an auxiliary result on measurability of elements of $\mathcal{L}$ to use in the main representation theorem later on.

We first establish the following auxiliary [as usual, $\vee$ ($\wedge$) is max (min)]:

Proposition 1. *Let $\mathcal{L}$ be a vector lattice of real functions on a set Ω such that* (i) $1 \in \mathcal{L}$ *(i.e. constants are in $\mathcal{L}$), and* (ii) $f_n \in \mathcal{L}$, $0 \leqslant f_n \uparrow f \leqslant K_0 < \infty$

pointwise $\Rightarrow f \in \mathscr{L}$ *(i.e., pointwise limits of nonnegative bounded sequences are in* $\mathscr{L}$*). If* $\mathscr{K} \subset \mathscr{L}$ *is a vector subspace that is closed under the operation* $\wedge$*, and* $f \in \mathscr{K} \Rightarrow f \wedge 1 \in \mathscr{K}$*, then all real bounded functions on* Ω *which are measurable relative to* $\sigma(\mathscr{K})$ *are contained in* $\mathscr{L}$*, where* $\sigma(\mathscr{K})$ *is the* σ*-algebra generated by the elements of* $\mathscr{K}$.

Proof. Let $\mathscr{A}$ be the collection of all vector spaces $\mathscr{C} \subset \mathscr{L}$ of the following type: $f, g \in \mathscr{C} \Rightarrow f \vee g \in \mathscr{C}$, $1 \in \mathscr{C}$, and $\mathscr{C}$ satisfies (ii) of $\mathscr{L}$. Thus $(-f) \vee (-g) = -(f \wedge g) \in \mathscr{C}$, so that $f \wedge g \in \mathscr{C}$ also. Since $\mathscr{A}$ contains the space generated by $\mathscr{K}$ and 1, and the bounded increasing limits, $\mathscr{A}$ is nonempty. It is partially ordered by inclusion, and each chain has an upper bound (its union). So by Zorn's lemma there exists a maximal element $\mathscr{C}_0$ in $\mathscr{A}$.

Let $\mathscr{S} = \{A \subset \Omega: \chi_A \in \mathscr{C}_0\}$. We observe that $\mathscr{S}$ is a σ-algebra and assert that $\sigma(\mathscr{K}) \subset \mathscr{S}$ and every element of $\mathscr{C}_0$ is $\mathscr{S}$-measurable. Since $\mathscr{C}_0 \subset \mathscr{L}$, this will imply the proposition. Indeed, $1 \in \mathscr{C}_0 \Rightarrow \Omega \in \mathscr{S}$; and $A_n \in \mathscr{S}$, $A_n \uparrow A \Rightarrow \chi_{A_n} \in \mathscr{C}_0$ and $\chi_{A_n} \uparrow \chi_A$. So $A \in \mathscr{S}$. Also $1 - \chi_A \in \mathscr{C}_0$ if $\chi_A \in \mathscr{C}_0$. Hence $A^c \in \mathscr{S}$. This shows that $\mathscr{S}$ is a σ-algebra. Since $\mathscr{C}_0$ is a vector lattice, for each $f \in \mathscr{C}_0$ and $a \in \mathbb{R}$ we have $(f - a)^+ \in \mathscr{C}_0$, and also $g_n = 1 \wedge [n \times (f - a)^+] \in \mathscr{C}_0$. Hence $0 \leqslant g_n \uparrow g \leqslant 1$ and $g \in \mathscr{C}_0$. But $g = \chi_{A_a}$ where $A_a = \{\omega: f(\omega) > a\}$. Hence $A_a \in \mathscr{S}$ for each a, so that f is $\mathscr{S}$-measurable. The maximality of $\mathscr{C}_0$ implies $\mathscr{K} \subset \mathscr{C}_0$, so that $\sigma(\mathscr{K}) \subset \mathscr{S}$, and the result follows.

Remark. Instead of $1 \in \mathscr{L}$, we may replace 1 by any function $f > 0$ on Ω which is in $\mathscr{L}$. Even a weaker requirement—that there exists a class $\{\chi_{A_\alpha}: \alpha \in J\} \subset \mathscr{L}$ such that $\bigcup_\alpha A_\alpha = \Omega$, called a *weak unit*—can be substituted in the above argument. Here $\chi_{A_\alpha} \neq 0$, $\alpha \in J$, and J is countable. (Cf. Exercise 1.)

Another technical property of the class of analytic sets is needed. Although this could have been established earlier, its use is in the main result here, and so we prove it now.

Proposition 2. *If* $(\Omega, \mathscr{C})$ *is a paved space and* $\mathscr{A}(\mathscr{C})$ *is the class of* $\mathscr{C}$*-analytic sets, then* $\mathscr{A}(\mathscr{C})$ *is closed under countable unions and intersections.*

Proof. By Proposition 1.3, we can use the Souslin operation (A) or Definition 1.2 as needed. If $A_1, A_2, \ldots$ are $\mathscr{C}$-analytic sets, let $A = \bigcup_{k=1}^\infty A_k$, $B = \bigcap_{k=1}^\infty A_k$. By the Souslin scheme then, for each A_k there exist sets $A_k(n_1, n_2, \ldots, n_i) \in \mathscr{C}$ such that

$$A_k = \bigcup_s \bigcap_{i=1}^\infty A_k(n_1, \ldots, n_i), \qquad s = (n_1, n_2, \ldots). \tag{1}$$

Define $\tilde{A}(k, n_1, \ldots, n_i) = A_k(n_1, \ldots, n_i)$. Then we have

$$A = \bigcup_{k=1}^\infty A_k = \bigcup_{\tilde{s}} \bigcap_{i=1}^\infty \tilde{A}(k, n_1, \ldots, n_i), \qquad \tilde{s} = (k, n_1, n_2, \ldots). \tag{2}$$

Since each $\tilde{A} \in \mathscr{C}$, this is again a Souslin operation, so that $A \in \mathscr{A}(\mathscr{C})$, by Proposition 1.3.

For the intersections, we use the other equivalence. Thus for each n, there exists a compact space E_n, a compact paving $\mathscr{K}_n$ of E_n, and a set $C_n \in (\mathscr{K}_n \times \mathscr{C})_{\sigma\delta}$ such that $A_n = \pi_n(C_n)$, where π_n: $E_n \times \Omega \to \Omega$ is the coordinate projection. Let $E = \times_{n=1}^{\infty} E_n$, $\mathscr{K} = \times_{n=1}^{\infty} \mathscr{K}_n$. Then with the (Cartesian) product topology E is compact, and $\mathscr{K}$ is a compact paving of E. Let $B_n = p_n^{-1}(C_n) \subset E \times \Omega$, be a cylinder with base C_n where p_n: $E \times \Omega \to E_n \times \Omega$ is a projection (of the E_n's leaving Ω fixed), so that we have

$$A_n = \pi_n \circ p_n(B_n) = \Pi(B_n). \qquad \begin{array}{ccc} E \times \Omega & \xrightarrow{p_k} & E_k \times \Omega \\ & \searrow^{\Pi = \pi_k \circ p_k} & \downarrow \pi_k \\ & & \Omega \end{array}$$

Even though Π is factored through E_k, $k \geqslant 1$, it remains the same after composition. Thus

$$B = \bigcap_{n=1}^{\infty} A_n = \bigcap_{n=1}^{\infty} \Pi(B_n) = \Pi\left(\bigcap_{n=1}^{\infty} B_n\right), \qquad \text{by Proposition 1.4.} \qquad (3)$$

Since $B_n \in \mathscr{K} \times \mathscr{C} \subset (\mathscr{K} \times \mathscr{C})_{\sigma}$, we have $\bigcap_{n=1}^{\infty} B_n \in (\mathscr{K} \times \mathscr{C})_{\sigma\delta}$, and so, by Definition 1.2, $B \in \mathscr{A}(\mathscr{C})$. This completes the proof.

We are ready for the main result, originally due to P. J. Daniell (1918 and 1920) and modified and refined by M. H. Stone (1948 and 1949). Taking also some later simplifications into account, we present the proof as a consequence of the capacity theory developed in the last section, as it gives a short cut.

Theorem (Daniell and Stone). *Let $\mathscr{L}$ be a vector lattice of real functions, on a set Ω, containing constants. If I: $\mathscr{L} \to \mathbb{R}^+$ is a linear functional satisfying conditions* (1)–(3) *at the beginning of this section, then I can be extended to $\bar{I}$ on a larger class $\mathscr{L}'$ satisfying analogous conditions, where $\mathscr{L}'$ is the class of real functions on Ω containing the bounded measurable functions relative to $\sigma(\mathscr{L})$, the σ-algebra generated by the elements of $\mathscr{L}$. Further, there exists a unique measure μ: $\sigma(\mathscr{L}) \to \mathbb{R}^+$ such that*

$$\bar{I}(f) = \int_{\Omega} f\,d\mu, \qquad f \in L^1(\mu)\,[= L^1(\Omega, \sigma(\mathscr{L}), \mu)], \qquad (4)$$

and $f \in L^1(\mu)$ iff $f \in \tilde{\mathscr{L}} = \{f \in \mathscr{L}': |\bar{I}(f)| < \infty\}$.

Proof. The idea here is to associate with I a capacity function J on $E = \Omega \times \mathbb{R}^+$, and show that the desired class $\mathscr{L}'$ has elements, the graphs of whose members are capacitable relative to J. That class will be seen to have the correct properties and contain $\mathscr{L}$. The integral representation will then be obtained. This is how the original extension procedure is replaced by the work on capacity theory of the preceding section. We proceed in steps for convenience. (The original method is outlined in Exercises 6–8 for comparison.)

1. Let $E = \Omega \times \mathbb{R}^+$, and for each $h: \Omega \to \mathbb{R}^+$ define the lower ordinate set $L_h = \{(\omega, x): h(\omega) > x\}$, as in Exercise 6.1.8. Thus L_h just excludes the graph of h [for all $h(\omega) < \infty$]. Since $L_{h_1} = L_{h_2}$ implies that h_1, h_2 have the same graph, we conclude that the mapping $h \mapsto L_h$ is one to one. Let $\mathscr{C} = \{L_h: h \in \mathscr{L}^+\}$. Since $L_{h_1} \cup L_{h_2} = L_{h_1 \vee h_2}$, $L_{h_1} \cap L_{h_2} = L_{h_1 \wedge h_2}$, and $L_h = \varnothing$ if $h = 0$, it follows that $\mathscr{C}$ is a paving which is closed under finite unions and intersections. Define $J: \mathscr{C} \to \overline{\mathbb{R}}^+$ by the equation $J(L_h) = I(h)$, $h \in \mathscr{L}^+$. Then J is nonnegative and increasing, and from $h_1 \vee h_2 + h_1 \wedge h_2 = h_1 + h_2$ we have

$$J(L_{h_1} \cup L_{h_2}) + J(L_{h_1} \cap L_{h_2}) = I(h_1 \vee h_2) + I(h_1 \wedge h_2)$$

$$= I(h_1 + h_2) = J(L_{h_1}) + J(L_{h_2}), \quad (5)$$

since I is additive. We conclude that J on $\mathscr{C}$ is strongly additive. To generate an outer capacity from J, with an application of Theorem 2.5, it is necessary to verify (10) of that result. If $A_n \in \mathscr{C}$, $A_n \subset A_{n+1}$, and $A = \bigcup_{n=1}^{\infty} A_n \in \mathscr{C}$, then we need to show that $J(A) = \lim_n J(A_n)$. We deduce this from the third condition of the Daniell functional I. By definition of $\mathscr{C}$, functions h_n and h exist in $\mathscr{L}^+$ with $A_n = L_{h_n}$, $A = L_h$. Then $h_n \uparrow h$ and so (since $h - h_n \downarrow 0$) we have $I(h - h_n) = I(h) - I(h_n) \downarrow 0$. Hence $J(A) = I(h) = \lim_n I(h_n) = \lim_n J(A_n)$, as desired. Thus let $J^*: \mathscr{P}(E) \to \overline{\mathbb{R}}^+$ be the outer capacity generated by J and $\mathscr{C}$, which is strongly countably additive, as guaranteed by Theorem 2.5(i). (This σ-additivity is simple and was already noted in Exercise 2.8.)

2. The generated outer capacity J^* is actually a capacity. For this we need to verify that for each $B_n \in \mathscr{C}$, $B_n \supset B_{n+1} \Rightarrow J^*(\bigcap_n B_n) = \lim_n J^*(B_n)$.

Note that even though $B_n = L_{g_n}$ for some $g_n \in \mathscr{L}^+$, and $B_n \downarrow B = \bigcap_n B_n \Rightarrow g_n \downarrow g \geqslant 0$, still g is not necessarily in $\mathscr{L}^+$, so that $I(g_n) \downarrow \alpha = I(g)$ is *not* a consequence of the Daniell conditions. So one needs to take an alternative approach to verify the hypothesis of Theorem 2.5(ii). Since $L_g \in \mathscr{P}(E)$ for each $g \geqslant 0$ on Ω, let $I^*(g) = J^*(L_g)$, which is well defined and satisfies $I^*|\mathscr{L} = I$. Thus the desired condition is equivalent to showing $I^*(\lim_n g_n) = \lim_n I(g_n)$ for each $g_n \downarrow g$, $g_n \in \mathscr{L}^+$. Now by the definition of J^* of Theorem 2.5(i), this is obtained first by approximation from below using elements of $\mathscr{C}$, and then by approximation from above using elements of $\mathscr{C}_\sigma$. That translates to I^* as follows: let $f_n \in \mathscr{L}^+$ be an increasing sequence such that $\lim_n f_n = f$ (which is in $\mathscr{L}_u^+$) with $f \geqslant g$; then $I^*(g) = \inf\{I^*(f)$: all such sequences $\{f_n\}_1^\infty\}$. In other words, the capacity condition follows if we show that for any increasing sequence $\{f_n\}_1^\infty$ from $\mathscr{L}^+$ such that $\lim_n f_n \geqslant \lim_n g_n$, where $\{g_n\}_1^\infty$ is a decreasing sequence also from $\mathscr{L}^+$, then $I^*(\lim_n f_n) \geqslant \lim_n I(g_n)$. Since $\lim_n I(g_n) \geqslant I^*(\lim_n g_n)$ is always true, by varying the sequences f_n we get $\inf_{\{f_n\}} I^*(\lim_n f_n) = I^*(\lim_n g_n)$, and by the above-noted result the assertion will follow. Hence we need to show

$$I^*\Big(\lim_n f_n\Big) \geqslant \lim_n I(g_n). \qquad (+)$$

Consider $h_n = g_1 - g_n \geqslant 0$. Then $h_n \in \mathscr{L}^+$ and $h_n \uparrow g_1 - g$. Since $\lim_n f_n \geqslant g$, we have $\lim_n f_n \geqslant g_1 - \lim_n h_n$, or $\lim_n (f_n + h_n) \geqslant g_1$. Consequently

$$I(g_1) \leqslant I^*\Big(\lim_n (f_n + h_n)\Big) = \lim_n I^*(f_n + h_n)$$

(by definition of J^* and hence I^*),

$$= \lim_n I(f_n + h_n),$$

(since $f_n, h_n \in \mathscr{L}^+$, $I^*|\mathscr{L} = I$)

$$= \lim_n \big[I(f_n) + I(h_n)\big],$$

(since I is additive on $\mathscr{L}^+$)

$$= \lim_n I^*(f_n) + \lim_n I(h_n),$$

(since $I = I^*|\mathscr{L}^+$)

$$= \lim_n I(h_n) + I^*\Big(\lim_n f_n\Big),$$

since $f_n \uparrow$.

However, $I(h_n) \leqslant I(g_1) < \infty$. It results therefore that

$$\lim_n \big[I(g_1) - I(h_n)\big] \leqslant I^*\Big(\lim_n f_n\Big). \qquad (6)$$

But $g_1 = h_n + g_n$, so that $I(g_1) = I(h_n) + I(g_n)$. Substituting this in the left side of (6), the desired inequality $(+)$ follows. This proves that J^* is a $\mathscr{C}$-capacity function on $\mathscr{P}(E)$.

3. Let $\mathscr{A}(\mathscr{C})$ be the class of all $\mathscr{C}$-analytic sets of E. We assert that for each A in $\sigma(\mathscr{L})$,

$$L_h \in \mathscr{A}(\mathscr{C}), \qquad h = a\chi_A, \quad a \geqslant 0.$$

For, observe first that the extended I^*, for all (positive) functions on Ω, is positively linear and homogeneous. This is obvious, since J^* is additive and I is positively homogeneous on $\mathscr{L}^+$, implying $I^*(ah) = aI^*(h)$. Let $\Sigma = \{A \subset \Omega: L_h \in \mathscr{A}(\mathscr{C})$ for $h = a\chi_A$ or $h = a\chi_{A^c}\}$. Then $\mathscr{L}$ is linear, $1 \in \mathscr{L} \Rightarrow \Omega \in \Sigma$, and if $A_1, A_2 \in \Sigma$, then since $\mathscr{L}$ is also closed under $\wedge$ and $\vee$, we have $A_1 \cup A_2 \in \Sigma$, $A_1 \cap A_2 \in \Sigma$, as well as $A_i^c \in \Sigma$, $i = 1, 2$. Thus Σ is an algebra. We now note that Σ is a σ-algebra. This will follow if it is shown to be closed under countable unions or intersections. This is obtained at once from Proposition 2, because $A_n \in \Sigma$ implies $L_{h_n} \in \mathscr{A}(\mathscr{C})$ where $h_n = \chi_{A_n}$; and since $\bigcup_{n=1}^{\infty} A_n \in \Sigma$ iff $\bigcup_{n=1}^{\infty} L_{h_n} = L_{\vee_{n=1}^{\infty} h_n} \in \mathscr{A}(\mathscr{C})$. Thus Σ is a σ-algebra.

Finally we claim that $\sigma(\mathscr{L}) \subset \Sigma$. Indeed, since $\mathscr{L}$ is a lattice, and $\sigma(\mathscr{L})$ is generated by sets of the form $\{\omega: f(\omega) > a\}$, $a \in \mathbb{R}$, $f \in \mathscr{L}$, and since this set is the same as $\{\omega: (f-a)^+(\omega) > 0\}$, we assert that for each $g \in \mathscr{L}^+$, $\{\omega: g(\omega) > 0\} \in \Sigma$. That will prove the desired inclusion. However, this follows from the definition of L_g, with $h = a\chi_A$, $A = \{\omega: g(\omega) > 0\}$, since

$$L_h = \bigcup_n L_{a(1 \wedge ng)} \in \mathscr{A}(\mathscr{C}). \tag{7}$$

4. Let $\mathscr{L}^*$ be the space of all real measurable functions for $\sigma(\mathscr{L})$, and $\mathscr{L}' \subset \mathscr{L}^*$ consist of those f such that L_{f^+}, L_{f^-} are J^*-capacitable, where $f = f^+ - f^-$. Then by the preceding work $\mathscr{L} \subset \mathscr{L}'$. Taking $\mathscr{L}$ of Proposition 1 as this $\mathscr{L}$, we get $\mathscr{L}'$ to contain all bounded $\sigma(\mathscr{L})$-measurable functions. Set $\bar{I}(f) = I^*(f^+) - I^*(f^-)$, $f \in \mathscr{L}'$, if the right side makes sense. It is clear that $\bar{I}|\mathscr{L} = I$, since I^* has that property. The additivity of I^* on $(\mathscr{L}^*)^+$ shows that $\bar{I}$ is additive. Let $\tilde{\mathscr{L}} = \{f \in \mathscr{L}': |\bar{I}(f)| < \infty\}$. Then $\tilde{\mathscr{L}}$ is at once seen to be a vector lattice and $1 \in \tilde{\mathscr{L}}$, $\mathscr{L} \subset \tilde{\mathscr{L}}$, and monotone convergence holds for $\bar{I}$.

5. Let $\mu(A) = \bar{I}(\chi_A)$, $A \in \sigma(\mathscr{L})$. Then μ is additive and $\mu(A) \geqslant 0$. It is increasing, since I^* has that property. Also μ is σ-additive. Indeed, let $\{A_n: n \geqslant 1\} \subset \sigma(\mathscr{L})$ be a disjoint sequence. Then

$$\mu\left(\bigcup_{n=1}^{\infty} A_n\right) = \sum_{i=1}^{n} \mu(A_i) + \mu\left(\bigcup_{i>n} A_i\right). \tag{8}$$

But $\mu(\bigcup_{i>n} A_i) = \bar{I}(\chi_{\bigcup_{i>n} A_i}) \to 0$ as $n \to \infty$, by the Daniell property of I (hence of I^* and $\bar{I}$). Thus μ is a measure. Note that $I(1) = \mu(\Omega) < \infty$. If $0 \leqslant f \in \tilde{\mathscr{L}}$, then there exist step functions $0 \leqslant \varphi_n \uparrow f$, $\varphi_n = \sum_{i=1}^{n} a_i \chi_{A_i}$, $A_i \in \sigma(\mathscr{L})$, $a_i \geqslant 0$. Thus by the linearity of $\bar{I}$,

$$\bar{I}(\varphi_n) = \sum_{i=1}^{n} a_n \bar{I}(\chi_{A_i}) = \sum_{i=1}^{n} a_n \mu(A_i) = \int_{\Omega} \varphi_n \, d\mu.$$

But a monotone convergence result for the μ-integrals follows from that of $\bar{I}$ since the μ-integral is uniquely given by the following:

$$\bar{I}(f) = \lim_n \bar{I}(\varphi_n) = \lim_n \int_{\Omega} \varphi_n \, d\mu = \int_{\Omega} f \, d\mu. \tag{9}$$

Thus $f \in L^1(\mu)$. This holds for all $f \in \tilde{\mathscr{L}}$, so that $\tilde{\mathscr{L}} \subset L^1(\mu)$. Conversely, if $0 \leqslant f \in L^1(\mu)$, there are step functions $0 \leqslant \varphi_n \uparrow f$ and $\varphi_n \in \tilde{\mathscr{L}}$ such that

$$\infty > \int_\Omega f \, d\mu = \lim_n \int_\Omega \varphi_n \, d\mu = \lim_n \bar{I}(\varphi_n) = \bar{I}(f).$$

Hence, taking the finite version, $f \in \tilde{\mathscr{L}}$, and since both $\tilde{\mathscr{L}}$ and $L^1(\mu)$ are lattices, this shows that $L^1(\mu) \subset \tilde{\mathscr{L}}$ and so $\tilde{\mathscr{L}} = L^1(\mu)$ when finite versions of elements of $L^1(\mu)$ are used. This proves (4).

Regarding uniqueness, let $\bar{I}(f) = \int_\Omega f \, d\nu$ for some measure ν on $\sigma(\mathscr{L})$. Then $\nu(\Omega) = \bar{I}(1) < \infty$, and taking $f = \chi_A$, $A \in \sigma(\mathscr{L})$, we see that $\nu(A) = \bar{I}(\chi_A) = \mu(A)$, so that $\mu = \nu$, and the representation (4) holds for a unique measure μ. This proves the theorem completely.

Remark. In the above proof the fact that $1 \in \mathscr{L}$ did not play a decisive part. In fact, any weak unit, as noted above, will do, but this time μ need not be a finite measure. It will be σ-finite (or, in the general case of weak unit such as $\{\chi_{A_\alpha}: \alpha \in J\}$, we have $\sum_{A_\alpha \subset A} \chi_{A_\alpha} < \infty$ with $0 < I(\chi_A) < \infty$, so that the set of α's is at most countable; then μ can be non-σ-finite but strictly localizable). However, something like the weak unit cannot be omitted completely. The uniqueness will not hold in general. (For a counterexample, see Exercise 4.)

We present a few consequences of this theorem, indicating a deeper impact of our considerations. Let Ω be a locally compact Hausdorff space, and $C_c(\Omega)$ be the space of real continuous functions with compact supports, i.e., $f \in C_c(\Omega)$ iff the closure of $\{\omega: f(\omega) \neq 0\}$ in Ω, called the *support* of f (= supp(f)), is compact and f is continuous. If $\mathscr{B}$ is the σ-algebra generated by the elements of $C_c(\Omega)$ [i.e., $\mathscr{B} = \sigma(C_c(\Omega))$], then $\mathscr{B}$ is called the *Baire* σ-algebra and each member of $\mathscr{B}$ a *Baire set*. A measure $\mu: \mathscr{B} \to \bar{\mathbb{R}}^+$ is called a *Baire measure* if for each $C \in \mathscr{B}$, C compact, we have $\mu(C) < \infty$. A detailed study of measures on topological spaces is a separate subject in itself, and some aspects of the problem will be considered in Chapter 9.

The following result contains a crucial part of the Riesz representation of a continuous linear functional on $C_c(\Omega)$.

Theorem 4. *Let Ω be a σ-compact space, and $I: C_c(\Omega) \to \mathbb{R}$ be a positive linear functional. Then there is a Baire measure μ on Ω such that*

$$I(f) = \int_\Omega f \, d\mu, \qquad f \in C_c(\Omega), \tag{10}$$

and if Ω is compact then μ is also unique.

Proof. We verify that I satisfies Daniell's conditions (1)–(3), given at the outset of this section. Now $\Omega = \bigcup_{n=1}^\infty \Omega_n$, where each Ω_n is compact, by hypothesis, and then by Urysohn's lemma there exist functions $f_n \in C_c(\Omega)$ such that $f_n | \Omega_n = 1$, to serve as a weak unit. Hence Theorem 3 can be applied.

Since I is a positive linear functional by hypothesis, we need to verify only the key third condition: $f_n \in C_c(\Omega)$, $f_n \downarrow 0 \Rightarrow I(f_n) \to 0$. For this let $C_1 = \operatorname{supp}(f_1)$. Then the support C_1 is compact and $\operatorname{supp}(f_n) \subset C_1$. Again by Urysohn's lemma, there exists a $g_0 \in C_c(\Omega)$ such that $g_0|C_1 = 1$. Let $\epsilon > 0$ be given, and consider $F_n = \{\omega: f_n(\omega) \geqslant \epsilon g_0(\omega)\} \subset C_1$. Clearly each F_n is closed and hence compact (because C_1 is), and $F_n \supset F_{n+1}$. Since $f_n \downarrow 0$, it follows that $\bigcap_n F_n = \varnothing$. By compactness (see Theorem 1.2.7(v)) there is an n_0 such that $F_n = \varnothing$ for $n \geqslant n_0$. Hence $f_n(\omega) < \epsilon g_0(\omega)$, $\omega \in \Omega$, $n \geqslant n_0$. So $I(f_n) \leqslant \epsilon I(g_0)$, $n \geqslant n_0$. Since $\epsilon > 0$ is arbitrary and $I(g_0) < \infty$, this gives $\lim_n I(f_n) = 0$. Thus I is a Daniell integral satisfying the hypothesis of Theorem 3 (and the remark following its proof), so that there is a measure μ on $\sigma(C_c(\Omega)) = \mathscr{B}$ such that the representation (10) holds. But for each compact G_δ-set A in $\mathscr{B}$, we find an $f_0 \in C_c(\Omega)$ with $f_0|A = 1$, and hence

$$\infty > I(f_0) = \int_\Omega f_0 \, d\mu \geqslant \int_A 1 \, d\mu = \mu(A).$$

Thus A is a Baire measure. If Ω is compact, then $1 \in C_c(\Omega) = C(\Omega)$, and the uniqueness follows from the theorem itself. This completes the proof.

Remark. Further work shows that each continuous linear functional on $C_c(\Omega)$ admits a Jordan decomposition into a difference of positive linear functionals for which the above result can be applied. The σ-compactness can also be dropped, but then the uniqueness need not hold. The uniqueness can be proved in the σ-compact case by considering the restrictions $I_n = I|C(\Omega_n)$, but we postpone that study to Chapter 9. (See Theorems 9.3.1 and 9.3.4.)

Another interesting consequence of Theorem 3 is given by:

Proposition 5. *Let Ω be a compact metric space, and $\mathscr{B}$ be its Borel σ-algebra. If μ: $\mathscr{B} \to \mathbb{R}^+$ is a measure, then it is "inner regular", i.e.,*

$$\mu(A) = \sup\{\mu(C): C \subset A, C \text{ compact}\}, \qquad A \in \mathscr{B}. \tag{11}$$

Proof. Since $\mu(\Omega) < \infty$, Proposition 2.2 shows that the generated outer measure μ^* by $(\mu, \mathscr{B})$ is a $\mathscr{C}$-capacity, where $\mathscr{C}$ is the class of all compact sets of Ω. Since Ω is a compact metric space, it is separable, and every set $A \in \mathscr{A}(\mathscr{C})$ is capacitable for μ^*. Since $\mu^*|\mathscr{B} = \mu$, we get (11) for each such A with μ^* on the left side. However, Proposition 1.5(iii), which clearly applies here also, shows that $\sigma(\mathscr{C}) \subset \mathscr{A}(\mathscr{C})$. But $\sigma(\mathscr{C}) = \mathscr{B}$. Hence for each $A \in \mathscr{B}$, (11) holds, as stated.

It may be noted that if $\mathscr{B}$ is replaced by its completion $\hat{\mathscr{B}}$ relative to μ, then also (11) holds because of Proposition 2.3.6. [A more general result will be proved in Proposition 9.2.7.] It is thus interesting to see that each bounded measure on the Borel σ-algebra of a compact metric space automatically has the regularity property (11), and this fact is not at all obvious. (Borel measures will be studied in Chapter 9.)

EXERCISES

1. Prove Proposition 1 if $\mathscr{L}$ has a weak unit, instead of having 1 in it.
2. Establish the following analog of Proposition 1. Let Ω be a set, and $B(\Omega)$ the space of real bounded functions on Ω. Let $\mathscr{L}$ $[\subset B(\Omega)]$ be a lattice containing constants, and $0 \leqslant f_n \uparrow f$, $f_n \in \mathscr{L}$, $f \in B(\Omega)$ $\Rightarrow$ $f \in \mathscr{L}$. If $\mathscr{C} \subset \mathscr{L}$ is a class closed under multiplication, then every $f \in B(\Omega)$ which is measurable relative to $\sigma(\mathscr{C})$ is in $\mathscr{L}$.
3. For the uniqueness part of Theorem 3, show that we can replace the existence of 1 in $\mathscr{L}$ there by the existence of a weak unit.
4. Let $\Omega = \mathbb{R}$ and $K \subset \mathbb{R}$ be a countable dense set (e.g. rationals). Consider $\mathscr{L} = \{f \in L^1(\mathbb{R}): f|K = 0\}$, where $L^1(\mathbb{R})$ is the Lebesgue space, on the usual Lebesgue line $\mathbb{R}$. Let $\mathscr{B} = \sigma(\mathscr{L})$, and define $I: \mathscr{L} \to \mathbb{R}$ by the equation $I(f) = \int_{\mathbb{R}} f\,d\mu$, where μ is the Lebesgue measure on $\mathbb{R}$. Verify that $I(\cdot)$ satisfies the Daniell conditions, but $I(f) = \int_{\mathbb{R}} f\,d\nu$ also holds if $\nu = \mu + P$, where P is the counting measure on K and vanishes outside K. (Thus $\mu \perp P$, and there is no uniqueness, since any $\mu + Q$ with $Q \perp \mu$ will also represent I.)
5. Let (Ω, Σ, P) be a finite measure space, and $(\mathbb{R}^+, \mathscr{B})$ be the Borelian half line. Let $\mathscr{C} = \Sigma \otimes \mathscr{B}$ on $\Omega \times \mathbb{R}^+$. If $A \subset \Omega \times \mathbb{R}^+$ is $\mathscr{C}$-analytic, show that the function $D_A: \Omega \to \overline{\mathbb{R}}^+$, called the *debut* of A, defined as $D_A(\omega) = \inf\{x \geqslant 0: (\omega, x) \in A\}$, with $\inf\{\varnothing\} = +\infty$, is measurable relative to the completion $\hat{\Sigma}$ of Σ for P, where P is any finite measure. [If $\pi: \Omega \times \mathbb{R}^+ \to \Omega$ is the coordinate projection, then $\{\omega: D_A(\omega) < x\} = \pi(\{(\omega, x) \in A: x < \alpha\})$ and is Σ-analytic. Apply Proposition 2.2 or Exercise 2.12.]
6. This and the next two exercises give an alternative (i.e., the original) derivation of the extension procedure of Daniell's integration method, which may thus be compared with Theorem 3. Let $\mathscr{L}$ be a vector lattice of real functions on a set Ω, and $I: \mathscr{L} \to \mathbb{R}$ be a positive linear functional such that $I(f_n) \to 0$ for $f_n \in \mathscr{L}$, $f_n \downarrow 0$. If $\mathscr{L}_u$ is the class of functions $f: \Omega \to \mathbb{R}$ such that $f = \lim_n \varphi_n$, $\varphi_n \in \mathscr{L}$, $\varphi_n \leqslant \varphi_{n+1}$, then set $\tilde{I}(f) = \lim_n I(\varphi_n)$, $I^*(g) = \inf\{\tilde{I}(h): g \leqslant h,\ h \in \mathscr{L}_u\}$ for any $g: \Omega \to \mathbb{R}$. Show that $\tilde{I}$ is additive on $\mathscr{L}_u$, $I = \tilde{I}|\mathscr{L}$, and I^* is positively homogeneous as well as countably subadditive on $\mathbb{R}^\Omega$.
7. With the notation of the above exercise, if $\mathscr{L}' = \{f: I^*(f) = -I^*(-f)$, finite$\}$, verify that $\mathscr{L}'$ is a vector lattice and $\mathscr{L} \subset \mathscr{L}'$. Moreover, $I^*|\mathscr{L} = I$, and I^* is a positive linear functional on $\mathscr{L}'$. [This needs a careful computation, and one uses also $I^*(f) \geqslant I_*(f) = -I^*(-f)$, which holds.] Verify that for I^* on $\mathscr{L}'$ one has the Fatou lemma and the dominated convergence theorem for sequences satisfying the same hypotheses as in the case of measures.

8. Writing I for I^* on $\mathscr{L}'$ also in the above, we say that $f \in \mathscr{L}'$ is a *null* function if $I(|f|) = 0$, and $f = g$ in $\mathscr{L}'$ *nearly everywhere* (n.e.) if $f - g$ is a null function. Verify that the set of all $g: \Omega \to \overline{\mathbb{R}}$ such that $f = g$ n.e. (denoted $f \sim g$), defines an equivalence class; it is written as $[f]$. If $\|[f]\|_1 = I(|f|)$, show that $\|\cdot\|_1$ is a norm on the equivalence class $\mathscr{L}^1 = \mathscr{L}'/\sim$, and $(\mathscr{L}^1, \|\cdot\|_1)$ is a complete normed linear space.

9. Let $\varphi: \mathbb{R}^+ \to \mathbb{R}^+$ be a continuous increasing function such that $\varphi(0) = 0$. If $f: \Omega \to \overline{\mathbb{R}}$ is a function, and I is an extended Daniell integral as in the preceding exercise, let $\mathscr{L}^\varphi = \{f: \varphi(k|f|) \text{ is in } (\mathscr{L}^1, \|\cdot\|_1) \text{ for some } k > 0\}$. If $\|f\|_\varphi = \inf\{k > 0: I(\varphi(|f|/k)) \leqslant k\}$ show that $(\mathscr{L}^\varphi, \|\cdot\|_\varphi)$ is a complete linear metric space. (Note that in all these four problems, we did not define, and do not need, a measure.)

10. Let $(\mathscr{L}^1, I)$ be as in Exercise 8. Define $\mu^*: A \mapsto \mu^*(A) = \inf\{I(f): \chi_A \leqslant f \in \mathscr{L}^1\}$ with $\inf\{\varnothing\} = +\infty$. Show $\mu^*: \mathscr{P}(\Omega) \to \overline{\mathbb{R}}^+$ is an outer measure. If $\mu = \mu^*|\mathscr{M}_{\mu^*}$, where $\mathscr{M}_{\mu^*}$ is the usual class of all μ^*-measurable sets, show that $L^1(\mu) \supset \mathscr{L}^1$; and if $1 \in \mathscr{L}^1$, the original lattice on Ω, verify that this μ is the same as that given by Theorem 3 and that there is equality between the above spaces. (Equality can also hold if the given $\mathscr{L}^1$ satisfies the *Stone axiom*: $f \in \mathscr{L}^1 \Rightarrow f \wedge 1 \in \mathscr{L}^1$. It is possible to give other conditions for this equality. The point here is that one is able to develop a considerable amount of integration theory without explicitly mentioning the measure, at least in most of the earlier parts of the work. However, much effort is needed to show the equivalence of this method with the measure approach, and the natural conditions such as σ-finiteness and localizability take different forms which are not entirely intuitive. So it appears that the measure approach emphasized in this book is a natural (and elementary) one compared with the Daniell method.)

11. This exercise presents a *complex extension* of the Daniell functional. Let $C_c(\Omega)$ be the space of real continuous functions on a locally compact Hausdorff space Ω with compact supports. Suppose $I: C_c(\Omega) \to \mathbb{C}$ is a complex linear functional such that for each compact set $K \subset \Omega$ there is a constant $M_K > 0$ such that $|I(f)| \leqslant M_K\|f\|_u$ for all $f \in C_c(\Omega)$ which vanish outside K, called the continuity condition. Here $\|f\|_u = \sup_\omega |f(\omega)|$. Show that (a) I can be written as $I = I_1^+ - I_1^- + i(I_2^+ - I_2^-)$, where $I_j^\pm: C_c(\Omega) \to \mathbb{R}^+$, $j = 1, 2$, are positive linear functionals satisfying the Daniell conditions (1)–(3) at the beginning of Section 3, and (b) conversely, each pair of real linear functionals $I_j: C_c(\Omega) \to \mathbb{R}$, satisfying the continuity condition above defines, by the equation $I = I_1 + iI_2$, a functional which is complex linear and continuous on $C_c(K)$ for each compact set $K \subset \Omega$. [(a) needs more work.]

12. Let $I: C_c(\Omega) \to \mathbb{C}$ be a complex linear functional, in the sense of the preceding exercise, and let Ω be σ-compact. Show that there exists a complex Baire measure μ on the Baire σ-algebra of Ω such that

$$I(f) = \int_\Omega f\, d\mu, \qquad f \in C_c(\Omega). \tag{$*$}$$

Conversely, verify that each complex Baire measure μ defines a complex linear functional I on $C_c(\Omega)$, where Ω is locally compact, satisfying the continuity condition. Here I is defined by $(*)$, i.e., by the integral. [Note that the complex measure $\mu = \mu_1 - \mu_2 + i(\mu_3 - \mu_4)$ is Baire if each μ_j is a Baire measure for $j = 1, 2, 3, 4$. Letting $\tilde{\mu} = \operatorname{Re} \mu$, $\tilde{\tilde{\mu}} = \operatorname{Im} \mu$, one has that $\tilde{\mu}$, $\tilde{\tilde{\mu}}$ are signed measures in the sense of Section 5.1. The conclusion holds if Ω is locally compact, since one may show that $C_c(\Omega)$ satisfies the Stone axiom for real functions, and the local compactness of Ω and the continuity condition of the linear functional I (see Exercise 11) imply the Daniell conditions. For this reason N. Bourbaki identifies I and μ in this representation and *defines* a complex measure as a complex linear functional on $C_c(\Omega)$ which is continuous on $C_c(K)$ for each compact set $K \subset \Omega$. Thus, our presentation shows not only the equivalence of these two approaches in this case, but the motivation behind the definition. For these reasons—and this becomes more evident in the specialized analysis of the following two chapters—we continue to focus on the measure approach.]

8

THE LIFTING THEOREM

After introducing the problem, an elementary proof of the main lifting theorem, first for finite and then for decomposable measures, is presented here. Also a topology introduced by the lifting operation in an abstract (decomposable) measure space is considered for a better understanding of the problem. An extension of the lifting map for vector functions is included.

8.1. THE PROBLEM, MOTIVATION, AND PRELIMINARIES

We already noted in Chapter 2 that the facility afforded by the Lebesgue measure theory in creating null sets, and hence equivalence classes of objects (sets or functions), carried with it a nontrivial problem of selecting a member from each equivalence class obeying the order-preserving or multiplicative (linear) operations. Let us explain a possible difficulty while introducing the problem at the same time.

To motivate the subject, let G be a topological group and $H \subset G$ be a closed subgroup. If $F = G/H$, the factor space of all left cosets $\{xH: x \in G\}$, then the quotient or canonical map $\pi: G \to F$ defined by $\pi: x \mapsto xH$ ($\in F$), is onto. One introduces a topology into F by declaring that $A \subset F$ is open iff $\pi^{-1}(A) \subset G$ is open, making π continuous at the same time. For many studies, both in topology and elsewhere, it is important to know whether there exists a mapping $h: F \to G$ such that $(\pi \circ h)(y) = y$ for all $y \in F$, called the *cross-section of* π. Since π is onto, we know that $\pi(\pi^{-1}(\{y\})) = y$ for each $y \in F$, by Lemma 3.3.1(v). However, $\pi^{-1}(\{y\})$ is generally a set. So π^{-1} cannot be taken as the desired h above except in the rare case that π is one to one. In our context, $G = \mathscr{L}^p(\Omega, \Sigma, \mu)$, or $\mathscr{L}^p(\mu)$, and $H = \{f: \|f\|_p = 0\}$, so that H is a closed subgroup of the Abelian topological group G, and $F = L^p(\Omega, \Sigma, \mu)$, the Lebesgue space of equivalence classes of measurable functions $\tilde{f}$ such that $\|\tilde{f}\|_p < \infty$. The desired mapping $h: F \to G$ is thus the one which selects a function f from its equivalence class $\tilde{f}$, i.e., $h(\tilde{f}) = f$. Can we not take h as the selector function of the axiom of choice? In fact, the latter gives a mapping for each equivalence class, and the difficulty is to have a *single h* that works for *all* such classes preserving the lattice properties. In the general group situation, this mapping should preserve the group operations. To understand the difficulties involved, we first show that such a mapping need not exist for the $\mathscr{L}^p(\mu)$-spaces if $0 \leqslant p < \infty$, and then prove the existence for the case that $p = +\infty$, which is nontrivial and is also the most important one in applications. The desired mapping is the lifting, which we now define for our case; then we establish its properties. In the following, as usual, we say that $f = g$ a.e., and denote it $f \equiv g$ if $\{\omega: f(\omega) \neq g(\omega)\}$ is a μ-null set.

Definition 1. Let (Ω, Σ, μ) be a complete measure space, and $\mathscr{L}^0(\Omega, \Sigma, \mu)$, or $\mathscr{L}^0(\mu)$ for short, be the space of real measurable functions on (Ω, Σ, μ). Then a mapping ρ: $\mathscr{L}^0(\mu) \to \mathscr{L}^0(\mu)$ is called

(a) a *linear lifting* if (i) $\rho(f) \equiv f$, (ii) $\rho(f) = \rho(g)$ whenever $f \equiv g$, (iii) $\rho(af + bg) = a\rho(f) + b\rho(g)$ for all real a, b, (iv) if $f \geqslant 0$ a.e. then $\rho(f) \geqslant 0$ everywhere, and (v) $\rho(1) = 1$; and

(b) a *lifting* if in addition ρ satisfies (vi) $\rho(fg) = \rho(f)\rho(g)$.

Note that if ρ is a lifting, then taking $f = g = \sqrt{u} \geqslant 0$, (vi) $\Rightarrow$ (iv); if μ is finite, then taking $g = 1$ in (vi), we also get (v). But for a linear lifting satisfying (i)–(v), we may allow extended real valued functions, and (iii) should then hold whenever the terms are defined.

Observing that if μ is a discrete measure or more generally an atomic measure, so that $\mu(A) > 0$ and $B \subset A$, $B \in \Sigma \Rightarrow \mu(B) = \mu(A)$ or $\mu(B) = 0$, then $\mathscr{L}^p(\mu) \subset \mathscr{L}^\infty(\mu)$ for $0 < p < \infty$, and we only need to consider measures which are not purely atomic in showing the nonexistence of ρ. This is the content of the following:

Proposition 2. *Let (Ω, Σ, μ) be a not purely atomic, complete measure space. Then a linear lifting ρ on $\mathscr{L}^p(\mu)$, $0 \leqslant p < \infty$, does not exist.*

Proof. Suppose $A \in \Sigma$, $\mu(A) > 0$, is a set on which μ is nonatomic. But then for each $0 < \alpha < \mu(A)$, there exists a $B \subset A$ such that $B \in \Sigma$, $\mu(B) = \alpha$, by the nonatomicity. Also, if there is a linear lifting ρ on $\mathscr{L}^p(\mu)$, then it is easily seen that $\rho|\mathscr{L}^p(B, \Sigma(B), \mu)$ is also a linear lifting on this subspace, where $\Sigma(B) = \{B \cap C: C \in \Sigma\}$ is the trace σ-algebra. Hence for this proof one can conveniently assume that (Ω, Σ, μ) is a *finite* nonatomic complete measure space.

Another simplification may be noted. It is clear that for any $0 \leqslant p_1 \leqslant p_2 \leqslant \infty$ one has $\mathscr{L}^{p_1}(\mu) \supset \mathscr{L}^{p_2}(\mu)$, since μ is now finite. By the preceding remark, if there exists a linear lifting ρ on $\mathscr{L}^p(\mu)$ for some $p \geqslant 0$, then for any $p_1 > p$, $\rho_1 = \rho|\mathscr{L}^{p_1}(\mu)$ will be such a mapping on $\mathscr{L}^{p_1}(\mu)$. In other words, if there did *not* exist a linear lifting on some $\mathscr{L}^{p_1}(\mu)$ for some $p_1 > 0$, then there cannot exist such mappings on $\mathscr{L}^p(\mu)$ for all $0 \leqslant p < p_1$. Hence it suffices to show the nonexistence of such a lifting on $\mathscr{L}^p(\mu)$ for some $p > 0$. We claim that for any $1 \leqslant p < \infty$ such a ρ on $\mathscr{L}^p(\mu)$ does not exist. We use an indirect argument involving the representation result of Theorem 5.5.5.

Suppose then there exists a linear lifting ρ on $\mathscr{L}^p(\mu)$ for some $1 \leqslant p < \infty$. Consider $L^p(\mu) = \mathscr{L}^p(\mu)/\mathscr{N}_p$, where $\mathscr{N}_p = \{f \in \mathscr{L}^p(\mu): \int_\Omega |f|^p \, d\mu = 0\}$. Then $L^p(\mu)$ is the Lebesgue space defined before, and $[L^p(\mu)]^* = L^q(\mu)$ with $q = p/(p-1)$. Also, if $f \in \mathscr{L}^p(\mu)$ and $\tilde{f}$ is its equivalence class, so that $f \in \tilde{f} \in L^p(\mu)$, then $\tilde{\rho}(\tilde{f}) = \rho(f)$ unambiguously defines $\tilde{\rho}$ on $L^p(\mu)$ into $\mathscr{L}^p(\mu)$ and is a linear lifting. In particular $\tilde{f} \geqslant 0 \;\Rightarrow\; \tilde{\rho}(\tilde{f}) \geqslant 0$. For each $\omega \in \Omega$, $l_\omega(\tilde{f}) = \tilde{\rho}(\tilde{f})(\omega)$ is thus a positive linear functional on $L^p(\mu)$. But each positive linear mapping l on $L^p(\mu)$ is automatically in $[L^p(\mu)]^*$, i.e., is continuous. This is easy: otherwise there exist $0 \leqslant f_n \in L^p(\mu)$ such that $\|f_n\|_p = 1$ and $l(f_n) \geqslant 2^n$, $n \geqslant 1$. If $f = \sum_{k=1}^{\infty} f_k/2^k$, then the completeness of $L^p(\mu)$ implies $f \in L^p(\mu)$, since

$$\|f\|_p^p = \lim_n \int_\Omega \left(\sum_{k=1}^{n} \frac{f_k}{2^k} \right)^p d\mu \leqslant \liminf_n \left(\sum_{k=1}^{n} \frac{\|f_k\|_p}{2^k} \right)^p = 1,$$

by the monotone convergence. However, $f \geqslant \sum_{k=1}^{n} (f_k/2^k)$ and $\infty > l(f) \geqslant \sum_{k=1}^{n} l(f_k)/2^k \geqslant n$. Since this is impossible for arbitrary n, we must have $l(f) \leqslant \|l\| < \infty$. So l is continuous. Hence by Theorem 5.5.5, for our l_ω, there exists a unique $g_\omega \in \mathscr{L}^q(\mu)$ such that

$$l_{\omega_0}(\tilde{f}) = \int_\Omega f g_{\omega_0} \, d\mu, \qquad \omega_0 \in \Omega. \tag{1}$$

But by the nonatomicity of μ, for each n we can express Ω as a disjoint union: $\Omega = \bigcup_{k=1}^{n} \Omega_k^{(n)}$, $\Omega_k^{(n)} \in \Sigma$, and $\mu(\Omega_k^{(n)}) = \mu(\Omega)/n$. Consider the set $A^{(n)} = \{\omega: \rho(\chi_{\Omega_k^{(n)}})(\omega) = 1$ for at least one $k = 1, 2, \ldots, n\}$. Since by assumption ρ is a linear lifting, $\rho(\chi_{\Omega_k^{(n)}}) \equiv \chi_{\Omega_k^{(n)}})$, so that

$$\mu(A^{(n)}) = \sum_{k=1}^{n} \int_\Omega \rho(\chi_{\Omega_k^{(n)}}) \, d\mu = \sum_{k=1}^{n} \mu(\Omega_k^{(n)}) = \mu(\Omega), \qquad n \geqslant 1. \tag{2}$$

But $A^{(n)} \subset \Omega$, and it follows that $\mu(\bigcap_{n=1}^{m} A^{(n)}) = \mu(\Omega)$ for each m and all n. Hence letting $m \to \infty$ and using Proposition 1.2.2, we get $\mu(\bigcap_{n=1}^{\infty} A^{(n)}) = \mu(\Omega)$. So for each $\omega_0 \in \bigcap_{n=1}^{\infty} A^{(n)}$, we have $\rho(\chi_{\Omega_k^{(n)}})(\omega_0) = 1$ for at least one $k = 1, 2, \ldots, n$. This and (1) imply, for such k,

$$\begin{aligned} 1 &= \rho(\chi_{\Omega_k^{(n)}})(\omega_0) \\ &= l_{\omega_0}(\chi_{\Omega_k^{(n)}}) \leqslant \|\chi_{\Omega_k^{(n)}}\|_p \|g_{\omega_0}\|_q \\ &= \left[\mu(\Omega_k^{(n)})\right]^{1/p} \|g_{\omega_0}\|_q = \left(\frac{\mu(\Omega)}{n} \right)^{1/p} \|g_{\omega_0}\|_q \to 0 \end{aligned} \tag{3}$$

as $n \to \infty$, since $p < \infty$. This contradiction proves that our supposition is wrong and the result holds as stated.

Note that the contradiction in (3) will not be obtained if $p = +\infty$, since $\|\chi_{\Omega_k^{(n)}}\|_\infty = 1$. In view of this and the proposition itself, we now analyze in detail the situation for $p = +\infty$. Indeed the picture changes radically, and this chapter is devoted to proving the *existence* of a lifting on $L^\infty(\mu)$ in a series of simplified steps as propositions in order to present a relatively elementary demonstration of the desired result.

Let $\mathscr{L}^\infty(\mu)$ be the space of all essentially bounded real measurable functions on a complete measure space (Ω, Σ, μ) with the norm $\|\cdot\|_\infty$, i.e.,

$$\|f\|_\infty = \inf\{\alpha > 0: \mu(\{\omega: |f(\omega)| > \alpha\}) = 0\}, \qquad f \in \mathscr{L}^\infty(\mu). \tag{4}$$

Then $(\mathscr{L}^\infty(\mu), \|\cdot\|_\infty)$ is an algebra, and if $\mathscr{N}^\infty = \{f \in \mathscr{L}^\infty: \|f\|_\infty = 0\}$, it is a subalgebra, so that $L^\infty(\mu) = \mathscr{L}^\infty(\mu)/\mathscr{N}^\infty$ is the Lebesgue space of equivalence classes of essentially bounded measurable functions. This quotient space may be regarded as the space of cosets $\tilde{f} = f + \mathscr{N}^\infty$, $f \in \mathscr{L}^\infty(\mu)$. Hence $f, g \in \tilde{f} \Rightarrow f - g \in \mathscr{N}^\infty$. If $\rho: \mathscr{L}^\infty(\mu) \to \mathscr{L}^\infty(\mu)$ is a lifting, then $\tilde{\rho}(\tilde{f}) \equiv f$ $[\in \mathscr{L}^\infty(\mu)]$, i.e., if $\tilde{f} = f + \mathscr{N}^\infty$, then $\tilde{\rho}(\tilde{f}) = f$. Since $(fg)\tilde{} = \tilde{f}\tilde{g}$, and $\tilde{f} \geqslant 0$ if $f \geqslant 0$ a.e. and, $a\tilde{f} + b\tilde{g} = (af)\tilde{} + (bg)\tilde{}$, it is clear that $\tilde{\rho}$ is a lifting map. So we may simply write $\tilde{\rho}$ as ρ, since $\tilde{\rho}(\tilde{f}) = \rho(f)$ everywhere. Thus the existence of ρ on $\mathscr{L}^\infty(\mu)$ and that of $\tilde{\rho}$ on $L^\infty(\mu)$ [into $\mathscr{L}^\infty(\mu)$] are equivalent assertions. We simplify this further by considering induced set functions as follows.

Let $\mathscr{N} = \{A \in \Sigma: \mu(A) = 0\}$, so that $\mathscr{N}$ is a σ-subring of Σ with the property that $A \in \mathscr{N}$, $B \in \Sigma \Rightarrow A \cap B \in \mathscr{N}$, i.e., $\mathscr{N}$ is a *σ-subideal* of Σ. If $\tilde{\Sigma} = \Sigma/\mathscr{N}$, so that $\tilde{A} \in \tilde{\Sigma}$ and A, B in $\tilde{A}$ imply their symmetric difference $A \triangle B$ is in $\mathscr{N}$, then $\tilde{\Sigma}$ is an algebra, called the *measure algebra*. We now define a mapping $\lambda: \tilde{\Sigma} \to \Sigma$ induced by (and inducing) ρ, a lifting of $\mathscr{L}^\infty(\mu)$. If $A \in \Sigma$, so that $\chi_A \in \mathscr{L}^\infty(\mu)$, consider the function $\rho(\chi_A) \in \mathscr{L}^\infty(\mu)$, where ρ is a lifting. Then $\rho(\chi_A) = \rho(\chi_A)\rho(\chi_A)$, so that $\rho(\chi_A)$ is also a two valued (or indicator) function, $\chi_{A'}$ (say). If $\lambda: A \mapsto A'$, then λ is well defined, $\rho(\chi_A) = \chi_{\lambda(A)}$, and we have the following assertion, where $A \equiv B$ is written for $A \triangle B \in \mathscr{N}$.

Proposition 3. *If* $\rho: \mathscr{L}^\infty(\mu) \to \mathscr{L}^\infty(\mu)$ *is a lifting with* (Ω, Σ, μ) *as a complete measure space, then* $\lambda: \Sigma \to \Sigma$ *defined by* $\rho(\chi_A) = \chi_{\lambda(A)}$, $A \in \Sigma$, *is a mapping with the properties*

(i) $A \equiv \lambda(A)$,
(ii) $A \equiv B \Rightarrow \lambda(A) = \lambda(B)$,
(iii) $\lambda(A \cap B) = \lambda(A) \cap \lambda(B)$, *and*
(iv) $\lambda(A^c) = [\lambda(A)]^c$.

Conversely, if $\lambda: \Sigma \to \Sigma$ *is a mapping with the above four properties, then a mapping* ρ *defined by* $\rho(\chi_A) = \chi_{\lambda(A)}$, $A \in \Sigma$, *admits a unique extension to* $\mathscr{L}^\infty(\mu)$ *to be a lifting, i.e., satisfies Definition* 1(b).

Proof. Let ρ be a lifting on $\mathscr{L}^\infty(\mu)$, and λ be given by $\rho(\chi_A) = \chi_{\lambda(A)}$ as in the statement. Then by Definition 1(b), $\chi_A \equiv \chi_{\lambda(A)}$ so that $A \triangle \lambda(A) \in \mathscr{N}$, which is (i). Also, if $A \equiv B$, then $\chi_A \equiv \chi_B$ and hence $\rho(\chi_A) = \rho(\chi_B)$. Thus $\chi_{\lambda(A)} = \chi_{\lambda(B)}$, so that $\lambda(A) = \lambda(B)$ as sets, and (ii) is true. Since $\chi_{\lambda(A\cap B)} = \rho(\chi_{A\cap B}) = \rho(\chi_A)\rho(\chi_B) = \chi_{\lambda(A)}\chi_{\lambda(B)}$, we get $\lambda(A\cap B) = \lambda(A)\cap\lambda(B)$ for all A, B in Σ. This gives (iii). Finally $\chi_{\lambda(A^c)} = \rho(\chi_{A^c}) = \rho(1-\chi_A) = \rho(1) - \rho(\chi_A) = 1 - \chi_{\lambda(A)} = \chi_{[\lambda(A)]^c}$ by Definition 1(b) itself. Hence $\lambda(A^c) = [\lambda(A)]^c$ and (iv) holds. So only the converse is less trivial.

Let $\lambda: \Sigma \to \Sigma$ be given, satisfying relations (i)–(iv) of the statement. First note that, on taking $B = A^c$ in (iii), we have $\lambda(\varnothing) = \varnothing$ by (iv), and then (with $A = \varnothing$), $\lambda(\Omega) = \Omega$. Hence

$$\lambda(A\cup B) = \lambda((A^c\cap B^c)^c) = [\lambda(A^c)\cap\lambda(B^c)]^c = \lambda(A)\cup\lambda(B).$$

Define ρ_1 on step functions $f = \sum_{i=1}^n a_i\chi_{A_i}$, $A_i \in \Sigma$, disjoint, by the equation

$$\rho_1(f) = \sum_{i=1}^n a_i\chi_{\lambda(A_i)} \in \mathscr{L}^\infty(\mu).$$

It is clear that ρ_1 is well defined and does not depend on the representation of the step function f. Also, for each such step function,

$$\|\rho_1(f)\|_u = \sup_\omega|\rho_1(f)|(\omega) = \max_{1\leqslant i\leqslant n}|a_i| = \|f\|_\infty, \qquad \text{by (4)}. \tag{5}$$

Hence the mapping ρ_1 is linear and (i)–(iv) imply the conditions of Definition 1(a). By using the fact that $\lambda(\Omega) = \Omega$, we get $\rho_1(1) = 1$ and $\rho_1(fg) = \rho_1(f)\rho_1(g)$ for step functions f and g. Thus ρ_1 is a lifting on $\mathscr{S}(\mu)$, the class of μ-step functions on (Ω, Σ, μ). But by Proposition 4.5.8 the set $\mathscr{S}(\mu)$ is dense in $\mathscr{L}^\infty(\mu)$ in the (semi-)metric $\|\cdot\|_\infty$ of (4). Since by (5), ρ_1 is a continuous linear mapping on $\mathscr{L}^\infty(\mu)$ which is in fact a lifting, it can be uniquely extended to the closure of $\mathscr{S}(\mu)$, by the principle of extension by uniform continuity. But this closure is $\mathscr{L}^\infty(\mu)$, so that the resulting function, denoted ρ, is the desired lifting map. This completes the proof.

In general, the relation (5) for a lifting ρ on $\mathscr{L}^\infty(\mu)$ satisfies an inequality. Thus if ρ exists on $\mathscr{L}^\infty(\mu)$, then since $f \in \mathscr{L}^\infty(\mu)$ implies $|f(\omega)| \leqslant \|f\|_\infty$ a.e. by (4), we have with $\rho(f) \equiv f$ by Definition 1(b), so that

$$\|\rho(f)\|_\infty = \|f\|_\infty \leqslant \sup_\omega|f(\omega)| = \|f\|_u.$$

Hence ρ on $\mathscr{L}^\infty(\mu) \to \mathscr{L}^\infty(\mu)$ is automatically a continuous linear mapping.

The point of the above result is that the existence of a lifting can be proved either for ρ on $\mathscr{L}^{\infty}(\mu)$ or equivalently for λ on Σ. Let us call $\lambda: \Sigma \to \Sigma$ satisfying (i)–(iv) a *set lifting*. Occasionally, one calls λ also a lifting, by a slight abuse of terminology, and this is not serious because of the above proposition. A related concept is that a mapping $\tilde{\lambda}: \mathscr{A} \to \mathscr{A}$ ($\mathscr{A} \subset \Sigma$ is an algebra) is called a *lower* (*upper*) *density* if for any pair A, B in $\mathscr{A}$, one has

(1) $\tilde{\lambda}(A) \equiv A$,
(2) $A \equiv B \Rightarrow \tilde{\lambda}(A) = \tilde{\lambda}(B)$,
(3) $\tilde{\lambda}(\varnothing) = \varnothing$, $\tilde{\lambda}(\Omega) = \Omega$, and
(4) $\tilde{\lambda}(A \cap B) = \tilde{\lambda}(A) \cap \tilde{\lambda}(B)$
[(4′) $\tilde{\lambda}(A \cup B) = \tilde{\lambda}(A) \cup \tilde{\lambda}(B)$].

It is necessary to study only one of these density mappings, since if $\tilde{\lambda}$ and λ' denote the lower and upper densities, then $\lambda'(A) = [\tilde{\lambda}(A^c)]^c$ transforms one into the other. Thus a *set lifting is simultaneously both a lower and an upper density*.

The reason for introducing such a density mapping is that it can always be extended to a set lifting, as seen from the following:

Proposition 4. *Let (Ω, Σ, μ) be a complete measure space and $\mathscr{A} \subset \Sigma$ be any algebra containing all the μ-null sets $\mathscr{N}$. If $\tilde{\lambda}: \mathscr{A} \to \mathscr{A}$ is a lower density mapping, then there exists a set lifting $\lambda: \mathscr{A} \to \mathscr{A}$ satisfying the relation*

$$\tilde{\lambda}(A) \subset \lambda(A) \subset [\tilde{\lambda}(A^c)]^c, \qquad A \in \mathscr{A}. \tag{6}$$

Proof. The mapping λ will be constructed using Proposition 7.2.8. Let $\mathscr{F}_0 \subset \mathscr{A}$ be a *filter base* (i.e., $\varnothing \notin \mathscr{F}_0$, and $A_1, A_2 \in \mathscr{F} \Rightarrow$ there is $A_3 \in \mathscr{F}_0$ such that $A_3 \subset A_1 \cap A_2$), and let $\mathscr{F} = \{A \subset \Omega: A \supset B$ for some $B \in \mathscr{F}_0\}$, so that $\mathscr{F}$ is the filter generated by $\mathscr{F}_0$. Then there is an ultrafilter refining $\mathscr{F}$ by the above mentioned proposition. We now choose $\mathscr{F}_0$ and get an ultrafilter $\mathscr{F}$ as just noted. Consider for each $\omega \in \Omega$

$$\mathscr{F}_0(\omega) = \{A \in \mathscr{A}: \omega \in \tilde{\lambda}(A)\}.$$

The fact that $\tilde{\lambda}$ is a lower density implies that $\mathscr{F}_0(\omega)$ is a filter base. Let $\mathscr{F}(\omega)$ be an ultrafilter containing $\mathscr{F}_0(\omega)$. Define $\lambda: \mathscr{A} \to \mathscr{A}$ as

$$\lambda(A) = \{\omega \in \Omega: A \in \mathscr{F}(\omega)\}, \qquad A \in \mathscr{A}. \tag{7}$$

We now verify that λ is a set lifting of $\mathscr{A} \to \mathscr{A}$ satisfying (6).

Let $A, B \in \mathscr{A}$. Since $\mathscr{F}(\omega)$ is an ultrafilter, either A or A^c is in $\mathscr{F}(\omega)$, but not both. Hence $\omega \in \lambda(A) \Rightarrow A \in \mathscr{F}(\omega)$, so $A^c \notin \mathscr{F}(\omega)$, and $\omega \in [\lambda(A^c)]^c$. Thus $\lambda(A) \subset [\lambda(A^c)]^c$. By reversing the argument we get $[\lambda(A^c)]^c \subset \lambda(A)$, so that $[\lambda(A)]^c = \lambda(A^c)$. Also, since $A \cap B \in \mathscr{F}(\omega)$ iff both $A, B \in \mathscr{F}(\omega)$ for an ultrafilter, we get by a similar argument that $\lambda(A \cap B) = \lambda(A) \cap \lambda(B)$. These two properties together imply $\lambda(A \cup B) = \lambda(A) \cup \lambda(B)$. Next, to see (6) holds, let $\omega \in \tilde{\lambda}(A)$. Hence $A \in \mathscr{F}_0(\omega) \subset \mathscr{F}(\omega)$ so that $\omega \in \lambda(A)$. Thus $\tilde{\lambda}(A) \subset \lambda(A)$. Replacing A by A^c and using the above result, one gets $\tilde{\lambda}(A^c) \subset \lambda(A^c) = [\lambda(A)]^c$, so that $\lambda(A) \subset [\tilde{\lambda}(A^c)]^c$, and (6) follows. Further, $\tilde{\lambda}(A) \equiv A = (A^c)^c \equiv [\tilde{\lambda}(A^c)]^c$. Substituting this in (6) gives $\lambda(A) \equiv A$. On the other hand, if $A_1 \equiv \varnothing$, then $\tilde{\lambda}(A_1) = \varnothing$ $[= \tilde{\lambda}(\varnothing)]$ and $\tilde{\lambda}(A_1^c) = \tilde{\lambda}(\Omega) = \Omega$. Using this in (6), we get $\lambda(A_1) = \varnothing$, so that $\lambda(\varnothing) = \varnothing$ and $\lambda(\Omega) = \Omega$. Finally, if $A \equiv B$, so that $A \triangle B \equiv \varnothing$, we get

$$\begin{aligned}
\varnothing &= \lambda(A \triangle B) = \lambda((A \cap B^c) \cup (A^c \cap B)) \\
&= \lambda(A \cap B^c) \cup \lambda(A^c \cap B) \\
&= [\lambda(A) \cap \lambda(B^c)] \cup [\lambda(A^c) \cap \lambda(B)] \\
&= [\lambda(A) \cap [\lambda(B)]^c] \cup [[\lambda(A)]^c \cap \lambda(B)] = \lambda(A) \triangle \lambda(B).
\end{aligned}$$

Hence $\lambda(A) = \lambda(B)$. Thus λ satisfies (i)–(iv) of Proposition 3, so that λ is a set lifting, completing the proof.

The preceding two propositions imply that, for proving the existence of a lifting on $\mathscr{L}^\infty(\mu)$, it suffices to establish that a lower density $\tilde{\lambda}: \Sigma \to \Sigma$ exists. This is carried out inductively from simple subalgebras step by step to all of Σ, the maximal extension being obtained with an application of Zorn's lemma. This is done in the next section. Here we show that the existence of a set lifting on Σ implies that a corresponding map ρ' for vector valued function spaces exists as well.

The intended extension holds for functions which take values in any Banach space or its adjoint space with any of its topologies (i.e., norm, weak or weak*, the last being the weakest) or any metric space. In order to cover all these cases we consider $f: \Omega \to S$, where S is a completely regular topological space for range, and this has important applications in analysis. Recall that S is *completely regular* if points are closed and for each closed set $A \subset S$ and any point $x_0 \in S - A$, there is a continuous function $f: S \to [0,1]$ such that $f(x_0) = 0$ and $f(x) = 1$, $x \in A$. We need to introduce a suitable concept of measurability to employ in this context.

Thus if (Ω, Σ, μ) is a measure space and S is completely regular, then a mapping $g: \Omega \to S$ is said to be *weakly measurable* if $g^{-1}(A) \in \Sigma$ for each $A \in \mathscr{S}$, the smallest σ-algebra of S relative to which each continuous real function on S is measurable. Hence if $C(S)$ is the space of real continuous functions and $\mathscr{B}$ is the Borel σ-algebra of the line $\mathbb{R}$, then $\mathscr{S} = \sigma\{\bigcup f^{-1}(\mathscr{B}): f \in C(S)\}$. This $\mathscr{S}$ is the *Baire σ-algebra* of S. It should be noted that not all open sets of S need be in $\mathscr{S}$. But the generally larger σ-algebra generated by the open sets of S is the *Borel σ-algebra*, and the two coincide only under some additional conditions. If, for instance, S is a (separable) metric space, they do coincide. However, $\mathscr{S}$ has sufficiently many elements for our purposes.

The space $\mathscr{L}^\infty(\mu; S)$ corresponding to $\mathscr{L}^\infty(\mu)$ $[=\mathscr{L}^\infty(\mu; \mathbb{R})]$ is defined as the set of all mappings $f: \Omega \to S$ such that (i) $A \in \mathscr{S} \Rightarrow f^{-1}(A) \in \Sigma$, and (ii) $f(\Omega)$ is a precompact set in S, i.e., the closure of $f(\Omega)$ is compact in S. Let $f, g \in \mathscr{L}^\infty(\mu; S)$. We say that $f \overset{w}{\equiv} g$ if $h(f) \equiv h(g)$, $h \in C(S)$ [i.e., $h(f) = h(g)$ a.e.$[\mu]$]. The significance of w here is that $f \overset{w}{\equiv} g$ does not generally imply that $\{\omega: f(\omega) \neq g(\omega)\}$ is a μ-null set (it may not even be in Σ). This will be a μ-null set under a supplementary condition such as the existence of a countable set of functions in $C(S)$ which distinguish points of S.

The vector lifting can be introduced as follows:

Definition 5. Let (Ω, Σ, μ) be a complete measure space admitting a set lifting λ. If S is completely regular and $\mathscr{L}^\infty(\mu; S)$ is the space of S-valued weakly measurable functions on Ω, introduced above, then a mapping $\rho': \mathscr{L}^\infty(\mu; S) \to \mathscr{L}^\infty(\mu; S)$ associated with λ, or with its induced lifting ρ on $\mathscr{L}^\infty(\mu)$, given by

$$\rho(h(f)) = h(\rho'(f)), \qquad f \in \mathscr{L}^\infty(\mu, S), \quad h \in C(S), \tag{8}$$

is termed a *lifting* on $\mathscr{L}^\infty(\mu; S)$.

To justify this definition, we need to show (i) that ρ' is uniquely determined by (8), (ii) that it has the other properties (linearity etc.) of a lifting map suitable for the $\mathscr{L}^\infty(\mu; S)$ context, and above all, (iii) the existence of such a mapping. Regarding (i), if ρ'' is another mapping satisfying (8), then $h(\rho'(f)) = h(\rho''(f))$ for all $h \in C(S)$. Since $C(S)$ distinguishes points of S, $\rho'(f) = \rho''(f)$. Thus if there is a ρ' satisfying (8), then there is only one, and Definition 5 is meaningful. As for (ii), if ρ is a lifting, we have

$$h(\rho'(f)) = \rho(h(f)) = h(f) \quad \text{a.e.} \quad \Rightarrow \quad \rho'(f) \overset{w}{\equiv} f. \tag{9}$$

Similarly, if $f \overset{w}{\equiv} g$, so that $h(f) \equiv h(g)$ and $\rho(h(f)) = \rho(h(g))$, we get, by the definition of ρ' in (8), that $h(\rho'(f)) = h(\rho'(g))$, and so

$$f \overset{w}{\equiv} g \Rightarrow \rho'(f) = \rho'(g) \qquad \text{everywhere.} \tag{10}$$

Also, for all $h \in C(S)$, $[\rho'(f)]^{-1}[h^{-1}(\mathscr{B})] = [h \circ \rho'(f)]^{-1}(\mathscr{B}) = [\rho(h \circ f)]^{-1}(\mathscr{B})$, which is a sub-σ-algebra of Σ. Thus $\rho'(f)$ is again weakly measurable. Since h is continuous, so that $h(\overline{f(\Omega)})$ is compact, we deduce that $h(f)$ is bounded and $\rho(h(f)) \in \mathscr{L}^\infty(\mu)$. Hence $h(\rho'(f))$ is bounded and

$\rho'(f) \in \mathscr{L}^\infty(\mu; S)$. Thus we have now shown that if ρ' of Definition 5 exists, then it is a mapping of $\mathscr{L}^\infty(\mu; S)$ into itself. Let us therefore establish its existence whenever (Ω, Σ, μ) has a density mapping, i.e., point (iii) noted above.

Theorem 6. *Let* (Ω, Σ, μ) *be a complete measure space admitting a (set) lifting map* $\lambda: \Sigma \to \Sigma$. *If* S *is any completely regular space and* $\mathscr{L}^\infty(\mu; S)$ *is the corresponding space of bounded weakly measurable* S*-valued functions, then there exists (uniquely) a lifting* $\rho': \mathscr{L}^\infty(\mu; S) \to \mathscr{L}^\infty(\mu; S)$ *induced by* λ, *as required by Definition* 5.

Proof. Since it is already noted above that there can be at most one ρ' associated with λ, it suffices to exhibit one such mapping. Let ρ be the lifting of $\mathscr{L}^\infty(\mu)$ determined by λ, and $f \in \mathscr{L}^\infty(\mu; S)$ be any element. If $K \supset$ [closure $f(\Omega)$] is a compact set, let $C(K)$ denote the space of real continuous functions on K. By hypothesis $\rho(h(f)) \in \mathscr{L}^\infty(\mu)$ for each $h \in C(K)$. If ${}_fl_\omega(h) = \rho(h(f))(\omega)$, $\omega \in \Omega$, then the fact that ρ is a lifting on $\mathscr{L}^\infty(\mu)$ implies ${}_fl_\omega$ is multiplicative. Indeed,

$$\begin{aligned} {}_fl_\omega(h_1h_2) &= \rho(h_1(f)h_2(f))(\omega) \\ &= \rho(h_1(f))(\omega)\rho(h_2(f))(\omega) \\ &= {}_fl_\omega(h_1)\,{}_fl_\omega(h_2). \end{aligned} \tag{11}$$

Also let $h \in C(K)$ be an element such that $h|\text{closure}(f(\Omega)) = 1$. (Such elements exist by Urysohn's lemma.) Then ${}_fl_\omega(h) = \rho(h(f))(\omega) = \rho(1)(\omega) = 1$. In particular ${}_fl_\omega(1) = 1$. Thus ${}_fl_\omega$ is a continuous linear functional on $C(K)$ which is multiplicative and preserves constants [hence it has norm unity on the Banach space $C(K)$, with supremum norm]. We now assert that each such functional is an evaluation on $C(K)$, i.e., ${}_fl_\omega(h) = h(b_{f,\omega})$ for a unique $b_{f,\omega} \in K$. A direct proof of this known fact is included for completeness.

Consider the set M of all multiplicative linear functionals x^* in $[C(K)]^*$, the adjoint space of $C(K)$ [the space of all continuous linear functionals on $C(K)$] with $x^*(1) = 1$. If $E = \{e_a: e_a(h) = h(a),\ h \in C(K), \text{ and } a \in K\}$, then the set E of evaluation functionals is contained in M. Moreover, the mapping $a \mapsto e_a$ is one to one and embeds K into E. To see this, if a_1, a_2 are distinct points of K, let $A \subset K$ be an open set such that $a_1 \in A$ and $a_2 \in K - A$. Then by definition of complete regularity, there is an $h \in C(K)$ such that $h(a_2) = 1$ and $h|A = 0$. Hence $e_{a_1}(h) = 0 \neq e_{a_2}(h) = 1$, and so $\tau: a \mapsto e_a$ is one to one. We now assert that $\tau(K) = E = M$. It suffices to verify that $M \subset \tau(K)$. We endow the following topology (termed the weak* topology) for M. Since by hypothesis $M \subset [C(K)]^*$, let the neighborhood of 0 be defined by

$$\{x^*: |x^*(h_i)| < \epsilon,\ i = 1, 2, \ldots, n,\ h_i \in C(K)\}.$$

Now if $M - \tau(K) \neq \varnothing$, then there exists an $0 \neq x_0^* \in M - \tau(K)$ such that for some $\epsilon > 0$ and $h_0 \in C(K)$, $\{x^*: |x^*(h_0) - x_0^*(h_0)| < \epsilon\} \cap \tau(K) = \varnothing$. Hence $h' = h_0 - x_0^*(h_0) \cdot 1 \in C(K)$ and $|x^*(h')| < \epsilon$ for all x^* in the above set, contained in $[\tau(K)]^c$. But then for all $a \in K$, $|\tau(a)(h')| = |e_a(h')| = |h'(a)| \geq \epsilon > 0$ because $x^* = e_a \in \tau(K)$. Since K is compact, $h' \in C(K)$, and h' never vanishes, we must have $1/h' \in C(K)$. However, since $x_0^*(h') = 0$, we get

$$1 = x_0^*(1) = x_0^*(h'/h') = x_0^*(h')x_0^*(1/h') = 0,$$

a contradiction. Thus $\tau(K) = M\ (= E)$ must hold.

Define $\rho'(f): \Omega \to S$ by the expression $\rho'(f)(\omega) = b_{f,\omega} = \tau^{-1}({}_f l_\omega)$, where ${}_f l_\omega$ is given above and shown to be in M [see (11)], and by the preceding paragraph ${}_f l_\omega \in E$, so that for a unique $b_{f,\omega} \in K$ we have $\tau(b_{f,\omega}) = {}_f l_\omega$. Since τ is one to one, $\rho'(f)$ is well defined and $\rho'(f)(\Omega) \subset K$. Also

$$\begin{aligned} h(\rho'(f))(\omega) &= h(b_{f,\omega}) = l_{b_{f,\omega}}(h) \\ &= {}_f l_\omega(h) = \rho(h(f))(\omega), \qquad h \in C(K). \end{aligned} \tag{12}$$

But $f \in \mathscr{L}^\infty(\mu, S)$ is arbitrary, and hence the desired ρ' satisfying Definition 5 is obtained, as asserted.

The importance of the preceding result is that a lifting on $\mathscr{L}^\infty(\mu; S)$ is determined by the basic measure space (Ω, Σ, μ), as long as S is completely regular, even if it is not necessarily a linear space. The crucial point now is to show the existence in $\mathscr{L}^\infty(\mu)$ of ρ (or a lower density $\tilde{\lambda}$), which we have not yet done. This is the subject of the next section.

EXERCISES

1. Let (Ω, Σ, μ) be a complete measure space and, $\mathscr{A} \subset \Sigma$ be a (σ-)subalgebra on which a set lifting map λ exists. Show that $\lambda(\mathscr{A})$ is a subalgebra contained in the class $\{A \in \mathscr{A}: \mu(A) > 0\} \cup \{\varnothing, \Omega\}$.

2. Let (Ω, Σ, μ) be as in Exercise 1. $\lambda: \Sigma \to \Sigma$ be a lifting, and $\mathscr{F} = \lambda(\Sigma)$. If $A_n \in \mathscr{F}$, and $A = \bigcup_{n \geq 1} A_n$ (so $A \in \Sigma$, but not necessarily in $\mathscr{F}$), consider $\lambda(A)$. Show that $\lambda(A)$ contains all A_n in the sense that $A_n - \lambda(A) = \varnothing$, $n \geq 1$.

3. Establish the analog of Proposition 4 for an upper density λ', and hence deduce that the existence of upper density is enough to find a lifting on $\mathscr{L}^\infty(\mu)$.

4. Let $\rho: \mathscr{L}^\infty(\mu) \to \mathscr{L}^\infty(\mu)$ be a lifting. If h is any bounded continuous function, show that $\rho(h(f)) = h(\rho(f))$, $f \in \mathscr{L}^\infty(\mu)$. Verify that the word "bounded" can be dropped in the preceding sentence, and deduce that $|\rho(f)| = \rho(|f|)$, $f \in \mathscr{L}^{\infty}(\mu)$. [Use the classical Weierstrass theorem to approximate h by suitable polynomials.] Extend this result to n variables: $\mathscr{L}^{\infty}(\mu; S) \times \cdots \times \mathscr{L}^{\infty}(\mu; S)$, i.e., $\rho(h(f_1, \ldots, f_n)) = h(\rho'(f_1), \ldots, \rho'(f_n))$ for all $h: S \times \cdots \times S \to \mathbb{R}$, continuous, S being completely regular and ρ' as in Theorem 6.

5. Let $\mathscr{X}, \mathscr{Y}$ be Banach spaces and $L(\mathscr{X}, \mathscr{Y})$ be the vector space of linear (not necessarily continuous) mappings from $\mathscr{X}$ to $\mathscr{Y}$. If (Ω, Σ, μ) is a measure space such that $\mathscr{L}^{\infty}(\mu)$ admits a lifting ρ, and if $T_i: \Omega \to L(\mathscr{X}, \mathscr{Y})$ we say that $\rho'(T_i) = T_i$ provided $\langle T_i x, y^* \rangle(\cdot) \in \mathscr{L}^{\infty}(\mu)$ for each $x \in \mathscr{X}$, $y^* \in \mathscr{Y}^*$, and $\rho(\langle T_i x, y^* \rangle) = \langle T_i x, y^* \rangle$, $i = 1, 2$. We write $T_1 \equiv T_2$ if $\langle T_1 x, y^* \rangle = \langle T_2 x, y^* \rangle$ a.e. Show that (i) $T_1 \equiv T_2$, $\rho'(T_i) = T_i$, $i = 1, 2 \Rightarrow T_1 = T_2$; and (ii) $\rho'(T_1) = T_1 \Rightarrow \|T_1\|(\omega) \leqslant \|T_1\|_{\infty}$, $\omega \in \Omega$, where $\|T_1\|(\omega) = \sup\{|\langle T_1(\omega)x, y^* \rangle|: \|x\| \leqslant 1, \|y^*\| \leqslant 1\}$ and $\|T_1\|_{\infty} = \sup\{\|\langle T_1(\cdot)x, y^* \rangle\|_{\infty}: \|x\| \leqslant 1, \|y^*\| \leqslant 1\}$.

8.2. EXISTENCE PROOF FOR THE LIFTING MAP

In this section the lifting theorem will be established in two stages. First a relatively simple demonstration of the existence result is given in the key finite measure case, following essentially Traynor's (1974) method. Next this is extended to the general case, which actually establishes the equivalence of the existence of a lifting with the strict localizability of the underlying measure space.

In what follows, unless the contrary is stated, the triple (Ω, Σ, μ) is assumed to be a complete finite measure space. To employ the induction method, we start with the following special but a key case.

Proposition 1. *Let $\mathscr{N} \subset \mathscr{A} \subset \Sigma$, where $\mathscr{A}$ is a σ-algebra and $\mathscr{N}$ is the σ-ideal of μ-null sets of Σ. If $A \in \Sigma - \mathscr{A}$, and $\lambda: \mathscr{A} \to \mathscr{A}$ is a set lifting, then there exists a lifting $\bar{\lambda}: \bar{\mathscr{A}} \to \bar{\mathscr{A}}$ which is an extension of λ, $\bar{\mathscr{A}}$ being the σ-algebra generated by A and $\mathscr{A}$.*

Proof. Let $\mathscr{F} = \lambda(\mathscr{A})$, so that $\mathscr{F}$ is an algebra. Define two collections $\mathscr{F}_1 = \{B \in \mathscr{F}: A \cap B \in \mathscr{N}\}$ and $\mathscr{F}_2 = \{C \in \mathscr{F}: C \cap A^c \in \mathscr{N}\}$. We assert that there exists a largest A_i in $\mathscr{F}_i$, $i = 1, 2$. Indeed, let $\alpha_i = \sup\{\mu(B): B \in \mathscr{F}_i\}$, $i = 1, 2$. Since $\mu(\Omega) < \infty$, we get $\alpha_i < \infty$, and hence there exist $C_i \in (\mathscr{F}_i)_{\sigma}$ such that $\mu(C_i) = \alpha_i$, $i = 1, 2$. Thus $C_i \in \mathscr{A}$. Let $A_i = \lambda(C_i)$ $(\in \mathscr{F})$. To see that these A_i are the largest elements, let $B \in \mathscr{F}_1$. Now $A_i = \lambda(C_i) \equiv C_i$, and $\{B, C_1\} \subset (\mathscr{F}_1)_{\sigma} \subset \mathscr{A}$. One gets $\mu(B - A_1) = \mu(B - C_1) = 0$, since C_1 is the largest set in $(\mathscr{F}_1)_{\sigma}$. But $\{B, A_1\} \subset \mathscr{F}$, so $B - A_1 = \varnothing$. Because of the arbitrariness of B in $\mathscr{F}_1$ and the fact that $\mu(A_1 \cap A) = \mu(C_1 \cap A) = \lim_n \mu(C_{1n} \cap A) = 0$ [where, by definition of C_1, there exist $B_n \in \mathscr{F}_1$, such that $C_{1n} = \bigcup_{k=1}^{n} B_k \uparrow C_1$ and $\mu(C_{1n} \cap A) \leqslant \sum_{k=1}^{n} \mu(B_k \cap A) = 0$], we get

$A_1 \in \mathscr{F}_1$. A similar argument shows that A_2 is the largest in $\mathscr{F}_2$. Thus $A_1 \cap A \equiv A_2 \cap A^c \equiv \varnothing$. This last relation implies that $\chi_{A_1}\chi_A \equiv \chi_{A_2}\chi_{A^c} \equiv 0$. Expanding the second relation $\chi_{A_2} \equiv \chi_{A_2}\chi_A$, we get $\chi_{A_1}\chi_{A_2} \equiv \chi_{A_1}\chi_{A_2}\chi_A \equiv 0$. Hence $A_1 \cap A_2 \equiv \varnothing$. Since A_1, A_2 are in $\mathscr{F}$, this implies $A_1 \cap A_2 = \varnothing$.

Define $\tilde{A} = (A \cup A_2) - A_1$. Then $\tilde{A} = (A \cap A_1^c) \cup A_2 \equiv A \cup A_2$, since $A \cap A_1 \in \mathscr{N}$. Hence $\tilde{A} \supset A_2$ and $\tilde{A} \cap A^c \equiv A_2 \cap A^c$, so that

$$\tilde{A} \triangle A = (\tilde{A} \cap A^c) \cup (A \cap \tilde{A}^c) \equiv (A_2 \cap A^c) \cup (A \cap \tilde{A}^c) \in \mathscr{N},$$

since $A_2 \in \mathscr{F}_2$ and $\tilde{A}^c \cap A \equiv \varnothing$. Thus $\tilde{A} \equiv A$. But the σ-algebra $\bar{\mathscr{A}} = \sigma(\mathscr{A} \cup \{A\})$ can be characterized as (see p. 73 for a similar representation):

$$\bar{\mathscr{A}} = \{(B \cap \tilde{A}) \cup (C \cap \tilde{A}^c) : B, C \in \mathscr{A}\}. \tag{1}$$

Define $\bar{\lambda}$ on $\bar{\mathscr{A}}$ by the equation

$$\bar{\lambda}((B \cap \tilde{A}) \cup (C \cap \tilde{A}^c)) = [\lambda(B) \cap \tilde{A}] \cup [\lambda(C) \cap \tilde{A}^c], \qquad B, C \in \mathscr{A}. \tag{2}$$

To see that $\bar{\lambda}$ is well defined by the equation (2), let a set $F \in \bar{\mathscr{A}}$ have two representations as

$$F = (B \cap \tilde{A}) \cup (C \cap \tilde{A}^c) \equiv (B_1 \cap \tilde{A}) \cup (C_1 \cap \tilde{A}^c) \tag{3}$$

for some B, B_1, C, C_1 in $\mathscr{A}$, so that either (i) $B \cap \tilde{A} \equiv B_1 \cap \tilde{A}$, or (ii) $C \cap \tilde{A}^c \equiv C_1 \cap \tilde{A}^c$, or both. If (i) is the case, then by definition of a (set) lifting, $\lambda(B) \cap \tilde{A} \equiv \lambda(B_1) \cap \tilde{A}$ also holds. For a unique definition in (2) it is enough to show that the latter are actually equal sets. Indeed

$$0 = \mu([\lambda(B) \triangle \lambda(B_1)] \cap \tilde{A}) = \mu([\lambda(B) \triangle \lambda(B_1)] \cap A),$$

since $\tilde{A} \equiv A$, and, then by definition of the largest set A_1 earlier,

$$\lambda(B) \triangle \lambda(B_1) \subset A_1 \subset (A_1 \cap A_2^c) \cup (A_2^c \cap A^c) = A_2^c \cap (A_1 \cup A^c)$$

$$= [(A \cap A_1^c) \cup A_2]^c = \tilde{A}^c.$$

Hence $[\lambda(B) \triangle \lambda(B_1)] \cap \tilde{A} = \varnothing$, so that $\chi_{\tilde{A}}|\chi_{\lambda(B)} - \chi_{\lambda(B_1)}| = 0$. This shows $\lambda(B) \cap \tilde{A} = \lambda(B_1) \cap \tilde{A}$. If (ii) holds, one verifies similarly that $\lambda(C) \cap \tilde{A}^c = \lambda(C_1) \cap \tilde{A}^c$. Thus the right side of (2) is uniquely defined. This computation further implies that for $F_1, F_2 \in \bar{\mathscr{A}}$ with $F_1 \equiv F_2$, we must have $\bar{\lambda}(F_1) = \bar{\lambda}(F_2)$. Taking $B = C = \varnothing$ and $B = C = \Omega$ in (2), we get $\bar{\lambda}(\varnothing) = \varnothing$, $\bar{\lambda}(\Omega) = \Omega$. Since $\lambda(B) \equiv B$ and $\lambda(C) \equiv C$, one has $\bar{\lambda}(F) \equiv F$.

Finally, let $F_1, F_2 \in \bar{\mathscr{A}}$, so that

$$F_i = \left(B_i \cap \tilde{A}\right) \cup \left(C_i \cap \tilde{A}^c\right), \qquad i = 1, 2, \quad B_i, C_i \in \mathscr{A}. \tag{4}$$

Then $F_1 \cap F_2 = (B_1 \cap B_2 \cap \tilde{A}) \cup (C_1 \cap C_2 \cap \tilde{A}^c)$, and $\mathscr{A}$ being an algebra,

$$\bar{\lambda}(F_1 \cap F_2) = \left(\lambda(B_1 \cap B_2) \cap \tilde{A}\right) \cup \left(\lambda(C_1) \cap \lambda(C_2) \cap \tilde{A}^c\right)$$

(since λ is a lifting of $\mathscr{A}$)

$$= \bar{\lambda}(F_1) \cap \bar{\lambda}(F_2), \qquad \text{on simplification.} \tag{5}$$

Hence $\bar{\lambda}$ is a lifting of $\bar{\mathscr{A}}$. Further, if $B \in \mathscr{A}$, so that $B = (B \cap \tilde{A}) \cup (B \cap \tilde{A}^c) \in \bar{\mathscr{A}}$, we have

$$\bar{\lambda}(B) = \left[\lambda(B) \cap \tilde{A}\right] \cup \left[\lambda(B) \cap \tilde{A}^c\right] = \lambda(B).$$

Thus $\bar{\lambda}|\mathscr{A} = \lambda$. This completes the proof.

The next extension is crucial for the induction step, and is, in a sense, the main part of the proof. It is related to the differentiation of set functions, which can also be obtained using the martingale convergence theorem after some work. The proof presented here is the simplest of the known ones, although it is still technical.

Theorem 2. *Let (Ω, Σ, μ) be a finite complete measure space, and $\mathscr{N} \subset \mathscr{A}_n \subset \mathscr{A}_{n+1} \subset \Sigma$ be a sequence of σ-algebras containing all the null sets. Suppose that $\lambda_n\colon \mathscr{A}_n \to \mathscr{A}_n$ is a (set) lifting such that $\lambda_n = \lambda_{n+1}|\mathscr{A}_n$. If $\mathscr{A} = \sigma(\bigcup_{n=1}^{\infty} \mathscr{A}_n)$ is the generated σ-algebra, then there exists a lifting $\lambda\colon \mathscr{A} \to \mathscr{A}$ such that $\lambda_n = \lambda|\mathscr{A}_n$, $n \geqslant 1$.*

Proof (After Traynor). We construct a lower density $\tilde{\lambda}$ on $\mathscr{A}$ such that $\tilde{\lambda}|\mathscr{A}_n = \lambda_n$. Since by Proposition 1.4 such a $\tilde{\lambda}$ extends to a lifting on $\mathscr{A}$, the result will follow.

There is no immediate motivation for the construction to be given, but it is suggested by the Vitali covering and differentiation systems (cf., e.g. Theorem 5.2.6 and Exercise 5.3.6). However, no previous knowledge of these results is utilized in this proof. Thus let $\mathscr{F}_n = \lambda_n(\mathscr{A}_n)$ and $0 \leqslant r < 1$. If $A \in \mathscr{A}$, define a class for each $n \geqslant 1$ as

$$\mathscr{E}_n(A; r) = \left\{B \in \mathscr{F}_n\colon B \supset C \text{ and } \mu(A \cap C) \geqslant r\mu(C), \text{all } C \in \mathscr{F}_n\right\},$$

and consider the set

$$\mathscr{D}_n(A; r) = \bigcup\{B\colon B \in \mathscr{E}_n(A; r)\}$$

and the mapping

$$\tilde{\lambda}: A \mapsto \tilde{\lambda}(A) = \left\{\liminf_n \mathscr{D}_n(A; r) \text{ for all } r < 1\right\}$$

$$= \bigcap_{0<r<1} \bigcup_{k \geqslant 1} \bigcap_{n \geqslant k} \mathscr{D}_n(A; r).$$

We claim that (i) $\mathscr{D}_n(A; r) \in \mathscr{F}_n$ and (ii) $\tilde{\lambda}$ is a lower density of $\mathscr{A}$. This will establish the desired result.

(i): Consider the collection $\mathscr{E}_n(A; r)$. If $B_1, B_2 \in \mathscr{E}_n(A; r)$, then from the definition it follows that $B_1 \cap B_2$, $B_1 \cap B_2^c$, and $B_1 \cup B_2$ are in $\mathscr{E}_n(A; r)$, because $\mathscr{F}_n$ is an algebra. Since $\varnothing \in \mathscr{E}_n(A; r)$, it is nonempty and is a ring. Now if $\{B_k: k \geqslant 1\} \subset \mathscr{E}_n(A; r)$ is a disjoint sequence, and $\mathscr{P}_n$ is the collection of all such sequences, it can be ordered by inclusion. Then each chain in $\mathscr{P}_n$ has an upper bound, namely their union. Note that this upper bound also has at most a countable number of elements, since $\mu(\Omega) < \infty$. Thus by Zorn's lemma there is a maximal element, say $\mathscr{C}_n$ $(\in \mathscr{P}_n)$; whence $\mathscr{C}_n \subset \mathscr{E}_n(A; r)$ and $\mathscr{C}_n$ is countable. Let $B_0 = \bigcup_k \{B_k: B_k \in \mathscr{C}_n\}$ and $B = \lambda_n(B_0)$. Then $B \in \mathscr{F}_n$, $\mu(A \cap B) \geqslant r\mu(B)$, and so $B \in \mathscr{E}_n(A; r)$. We observe that B is the largest element of $\mathscr{E}_n(A; r)$, so that $B = \mathscr{D}_n(A; r)$. Indeed, if $E \in \mathscr{E}_n(A; r)$ is arbitrary, then $E \in \mathscr{F}_n$, and by the maximality of $\mathscr{C}_n$, $E - B_0 \in \mathscr{N}$. Since $B_0 \equiv B$, $E - B \in \mathscr{N}$. But $E, B \in \mathscr{F} \Rightarrow E - B = \varnothing$. Hence B is the largest set, which is $\mathscr{D}_n(A; r)$ in $\mathscr{F}_n$, as asserted. But for each n, $\mathscr{D}_n(A; r) \supset \mathscr{D}_n(A; s)$ if $0 \leqslant r < s < 1$. So we can restrict r to the rationals, and hence $\tilde{\lambda}(A) \in \mathscr{A}$.

(ii): From the definition, it is immediate that $\tilde{\lambda}(\varnothing) = \varnothing$ and $\tilde{\lambda}(\Omega) = \Omega$. Also, if $A_1 \equiv A_2$ in $\mathscr{A}$, then [by definition again] $\mathscr{E}_n(A_1; r) = \mathscr{E}_n(A_2; r)$ so that $\tilde{\lambda}(A_1) = \tilde{\lambda}(A_2)$. Thus we need to verify only that (a) $\tilde{\lambda}(A) \equiv A$ and (b) $\tilde{\lambda}(A_1 \cap A_2) = \tilde{\lambda}(A_1) \cap \tilde{\lambda}(A_2)$ for A, A_1, A_2 in $\mathscr{A}$, and this needs a detailed computation. First consider (b), and let $A_1, A_2 \in \mathscr{A}$. Then for $F \in \mathscr{F}_n$ such that $F \subset \mathscr{D}_n(A_1; (r+1)/2) \cap \mathscr{D}_n(A_2; (r+1)/2)$ (since $(r + 1) < 2$), we get

$$\mu(A_1 \cap A_2 \cap F) = \mu(A_1 \cap F) + \mu(A_2 \cap F) - \mu((A_1 \cup A_2) \cap F),$$

[since $\mu(A \cup B) = \mu(A) + \mu(B) - \mu(A \cap B)$]

$$\geqslant \frac{r+1}{2}[\mu(F) + \mu(F)] - \mu(F)$$

$$= r\mu(F) \tag{6}$$

Thus $F \subset \mathscr{D}_n(A_1 \cap A_2; r)$. Hence

$$\liminf_n \left(\mathscr{D}_n\left(A_1; \frac{r+1}{2}\right) \cap \mathscr{D}_n\left(A_2; \frac{r+1}{2}\right)\right) \subset \liminf_n \mathscr{D}_n(A_1 \cap A_2; r).$$

Letting $r \uparrow 1$ through rationals, since these are monotone sequences, we get $\tilde{\lambda}(A_1) \cap \tilde{\lambda}(A_2) \subset \tilde{\lambda}(A_1 \cap A_2)$. But for each n and r, the sets $\mathscr{D}_n(A; r)$ are increasing in A, so that $\mathscr{D}_n(A_1 \cap A_2; r) \subset \mathscr{D}_n(A_i; r)$, $i = 1, 2$, and so $\subset \mathscr{D}_n(A_1; r) \cap \mathscr{D}_n(A_2; r)$. Taking lim inf on n and then letting $r \uparrow 1$, we get $\tilde{\lambda}(A_1) \cap \tilde{\lambda}(A_2) \supset \tilde{\lambda}(A_1 \cap A_2)$, so that (b) holds. It remains to verify (a).

Consider $\lambda'(A) = \limsup_n \mathscr{D}_n(A; r)$ for some $0 < r < 1$. Thus $\lambda'(A) = \bigcup_{0<r<1} \bigcap_{k \geq 1} \bigcup_{n \geq k} \mathscr{D}_n(A; r)$ $(\in \mathscr{A})$. We claim: (α) $\lambda'(A) \cap A^c \equiv \varnothing$, ($\beta$) $\lambda'(A^c) \cap A \equiv \varnothing$, and ($\gamma$) $\lambda'(A^c) \supset [\tilde{\lambda}(A)]^c$, $\lambda'(A) \supset \tilde{\lambda}(A)$. This will imply

$$\tilde{\lambda}(A) \,\Delta\, A = \left[\tilde{\lambda}(A) \cap A^c\right] \cup \left[[\tilde{\lambda}(A)]^c \cap A\right]$$
$$\subset \left[\lambda'(A) \cap A^c\right] \cup \left[\lambda'(A^c) \cap A\right] \in \mathscr{N},$$

and hence (a) will follow.

Let $0 < r < 1$, and $B \in \mathscr{A}_k$, $k \geq 1$. Writing $D_k = \mathscr{D}_k(A; r) \in \mathscr{E}_k(A; r)$, we have, with $F = \lambda_k(B) \in \mathscr{F}_k$ so that $D_k \cap \lambda_k(B) \in \mathscr{F}$ [cf. (6)],

$$\mu(A \cap B \cap D_k) = \mu(A \cap \lambda_k(B) \cap D_k)$$
$$\geq r\mu(\lambda_k(B) \cap D_k) = r\mu(B \cap D_k), \tag{7}$$

since $\lambda_k(B) \equiv B$. The arbitrariness of B in $\mathscr{A}_k$ shows that this inequality is valid for all $B \in \bigcup_{k \geq 1} \mathscr{A}_k$. Hence for each such B in the algebra $\bigcup_{k \geq 1} \mathscr{A}_k$ such that $B \in \mathscr{A}_m$ for some $m \geq 1$, consider the sequence $\{B \cap D_k : k \geq m\}$. Let $\{C_k : k \geq m\}$ be a disjunctification, i.e., $C_m = B \cap D_m$ and $C_n = B \cap D_n - \bigcup_{k=m}^{n-1} B \cap D_k$ for $n > m$. Then

$$\mu\left(A \cap B \cap \bigcup_{k \geq m} D_k\right) = \mu\left(A \cap \bigcup_{k \geq m} C_k\right) = \sum_{k \geq m} \mu(A \cap C_k)$$
$$\geq r \sum_{k \geq m} \mu(C_k), \qquad \text{by (7)}$$
$$= r\mu\left(B \cap \bigcup_{k \geq m} D_k\right).$$

Letting $m \to \infty$, and using the finiteness of μ, we get

$$\mu\left(A \cap B \cap \bigcap_{m \geq 1} \bigcup_{k \geq m} D_k\right) \geq r\mu\left(B \cap \bigcap_{m \geq 1} \bigcup_{k \geq m} D_k\right). \tag{8}$$

Consider the class $\mathscr{C}$ of all B which satisfy (8). Then $\bigcup_{k=1}^{\infty} \mathscr{A}_k \subset \mathscr{C}$, and it is closed under monotone limits. Hence by the monotone class theorem, $\mathscr{A} \subset \mathscr{C}$. Thus (8) holds for all $B \in \mathscr{A}$. If $B = A^c$, then since $0 < r < 1$, we get $\mu(A^c \cap \bigcap_{n \geq 1} \bigcup_{k \geq n} D_k) = 0$. Since $D_k = \mathscr{D}_k(A; r)$, we have

$\mu(A^c \cap \bigcup_{0<r<1} \limsup_n \mathscr{D}_n(A; r)) = 0$, so that $\lambda'(A) \cap A^c \equiv \varnothing$, which is ($\alpha$). Replacing A by A^c in this computation, we get $\lambda'(A^c) \cap A \equiv \varnothing$, which is ($\beta$).

Regarding (γ), since by definition $\tilde{\lambda}(A) \subset \lambda'(A)$, we only need to verify $\lambda'(A^c) \supset [\tilde{\lambda}(A)]^c$. Since

$$\begin{aligned} \lambda'(A^c) &= \bigcup_{0<r<1} \bigcap_{k\geqslant 1} \bigcup_{n\geqslant k} \mathscr{D}_n(A^c; r) = \bigcup_{0<r<1} \bigcap_{k\geqslant 1} \bigcup_{n\geqslant k} \mathscr{D}_n(A^c; 1-r), \\ (\tilde{\lambda}(A))^c &= \bigcup_{0<r<1} \bigcap_{k\geqslant 1} \bigcup_{n\geqslant k} [\mathscr{D}_n(A; r)]^c, \end{aligned} \tag{9}$$

(γ) clearly will follow if we show that $[\mathscr{D}_n(A; r)]^c \subset \mathscr{D}_n(A^c; 1-r)$. For this, let $E \subset [\mathscr{D}_n(A; r)]^c$ be any nonempty set such that $E \in \mathscr{F}_n$. Since $\mathscr{D}_n(A; r)$ is the largest element of $\mathscr{E}_n(A; r)$, as shown in (i), we have $E \notin \mathscr{E}_n(A; r)$. Hence we can find an $F \in \mathscr{E}_n(A; r)$ such that $\mu(A \cap F) < r\mu(F)$, and $F \subset E$. If $\tilde{\mathscr{E}}_n$ is the collection of all such F, we may find a maximal (countable) disjoint collection $\tilde{\mathscr{C}}_n$ of all such elements of $\tilde{\mathscr{E}}_n$, as in (i). By the same argument, using the maximality of $\tilde{\mathscr{C}}$, we get

$$G = \bigcup \{F\colon F \in \tilde{\mathscr{C}}_n\} \in \mathscr{A}_n \text{ and } E - \lambda_k(G) = \varnothing, \text{or } E = \lambda_k(G).$$

Further

$$\begin{aligned} \mu(A \cap E) = \mu(A \cap G) &= \sum_{k\geqslant 1} \mu(A \cap F_k), \qquad F_k \in \tilde{\mathscr{C}}, \\ &\leqslant r \sum_{k\geqslant 1} \mu(F_k) = r\mu(G) = r\mu(E). \end{aligned}$$

Hence

$$\mu(E) - \mu(A \cap E) = \mu(E \cap A^c) \geqslant (1-r)\mu(E). \tag{10}$$

Since $E \subset [\mathscr{D}_n(A; r)]^c$ is arbitrary, we get $[\mathscr{D}_n(A; r)]^c \in \mathscr{E}_n(A^c; 1-r)$. Hence $[\mathscr{D}(A; r)]^c \subset \mathscr{D}_n(A^c; 1-r)$. This and (9) show that (γ) is true. Thus (α), (β), (γ) hold, so that $\tilde{\lambda}(A) \equiv A$ and (a) follows. This shows that $\tilde{\lambda}$ is a lower density, and hence the proof is finished.

Note that the above argument uses only standard measure theoretical ideas and results, although the details need much care. In this sense it is "elementary", since it does not need functional analysis results, as originally used by Tulcea (1970), or the differentiation technique of Sion (1973) [see e.g., the author (1979), Section III.2)]. However, the above proof is really a simplification of the differentiation method, and is a refinement of the classical ideas on the subject (cf. Chapter 5).

With the preceding results we can quickly establish the main result for finite measures.

Theorem 3. *Let* (Ω, Σ, μ) *be a finite complete measure space. Then there exists a (set) lifting* λ *on* Σ.

Proof. Let $\mathscr{P}$ be the set of pairs $(\mathscr{A}, \lambda)$ where $\mathscr{N} \subset \mathscr{A} \subset \Sigma$ is a σ-algebra on which there is a (set) lifting λ. If $\mathscr{A} = \mathscr{A}_0$ is the trivial algebra, $[A \in \mathscr{A}_0$ iff $A \equiv \Omega$ or $\equiv \varnothing]$, define $\lambda_0(\Omega) = \Omega$ and $\lambda_0(\varnothing) = \varnothing$. Then $(\mathscr{A}_0, \lambda_0) \in \mathscr{P}$, so that the class $\mathscr{P}$ is nonempty. Indeed, by Proposition 1 there are many more such pairs unless Σ is $\mathscr{A}_0$, the trivial case. We order $\mathscr{P}$ by the following relation: if $(\mathscr{A}_i, \lambda_i) \in \mathscr{P}$, then $(\mathscr{A}_1, \lambda_1) \prec (\mathscr{A}_2, \lambda_2)$ whenever $\mathscr{A}_1 \subset \mathscr{A}_2$ and $\lambda_1 = \lambda_2 | \mathscr{A}_1$. Thus $(\mathscr{P}, \prec)$ is a partially ordered set. Let $\mathscr{C} = \{(\mathscr{A}_i, \lambda_i): i \in I\} \subset \mathscr{P}$ be a chain. Then either $\mathscr{C}$ has a countable cofinal subfamily or it does not. In the latter case, let $\mathscr{A} = \bigcup_{i \in I} \mathscr{A}_i$, and $\lambda(A) = \lambda_i(A)$ for $A \in \mathscr{A}_i$, $i \in I$. By Theorem 2, this defines λ on $\mathscr{A} \to \mathscr{A}$ uniquely. Since I has no cofinal subset, $\mathscr{A}$ is a σ-algebra and $\mathscr{N} \subset \mathscr{A} \subset \Sigma$. It is also immediate that λ is a (lower) density on $\mathscr{A}$, so that $(\mathscr{A}, \lambda) \in \mathscr{P}$ and is an upper bound for $\mathscr{C}$. In the first case let $J = \{i_n: n \geqslant 1\} \subset I$ be a cofinal subset of I. Then let $\mathscr{A} = \sigma(\bigcup_{n \geqslant 1} \mathscr{A}_{i_n})$. By Theorem 2, there exists a density mapping $\lambda: \mathscr{A} \to \mathscr{A}$ such that $\lambda | \mathscr{A}_{i_n} = \lambda_{i_n}$, and hence $\lambda | \mathscr{A}_i = \lambda_i$, $i \in I$. So $(\mathscr{A}, \lambda)$ is again an upper bound of $\mathscr{C}$. Hence by Zorn's lemma there is a maximal element $(\bar{\mathscr{A}}, \bar{\lambda})$ in $\mathscr{P}$. If $\bar{\mathscr{A}} \subsetneqq \Sigma$, let $A \in \Sigma - \bar{\mathscr{A}}$. Again by Proposition 1, we can extend $\bar{\lambda}$ to $\bar{\bar{\mathscr{A}}} = \sigma(\bar{\mathscr{A}}, \{A\}) \subset \Sigma$, denoted $\bar{\bar{\lambda}}$, and hence $(\bar{\bar{\mathscr{A}}}, \bar{\bar{\lambda}}) \in \mathscr{P}$ and majorizes $(\bar{\mathscr{A}}, \bar{\lambda})$, contradicting the maximality of the latter. Thus $\bar{\mathscr{A}} = \Sigma$ must hold, and $\bar{\lambda}$ gives a lifting of Σ. This completes the proof of the theorem.

Remark. Since Zorn's lemma enters crucially in the proof, and the maximal element produced by this lemma is seldom unique, it follows that a lifting map is not generally unique, i.e., each (Ω, Σ, μ) has many liftings if it has one (and Σ is infinite). [By Proposition 1.4, we are justified in calling a lower density a lifting.]

We now extend the preceding theorem for nonfinite measures. Recall that a complete measure space (Ω, Σ, μ), or simply μ, with the finite subset property is *strictly localizable* (or has the *direct sum property*) if there is a disjoint family $\{A_i: i \in I\} \subset \Sigma$ such that $\mu(A_i) < \infty$, $i \in I$, $\Omega - \bigcup_{i \in I} A_i \in \mathscr{N}$, and for each set $A \in \Sigma$, $\mu(A) < \infty$, we have $\{i \in I: \mu(A \cap A_i) > 0\}$ at most countable. This, of course, includes σ-finiteness. The following is the desired extension.

Theorem 4. *Let* (Ω, Σ, μ) *be a complete measure space. Then there exists a (set) lifting map* λ *on* Σ *iff* μ *is strictly localizable.*

Proof. Let μ be strictly localizable. Then we extend the preceding theorem in deducing the existence of λ on Σ in this general case. Consider $\{A_i: i \in I\}$ as in the definition of strict localizability, where we may take $0 < \mu(A_i) < \infty$, $i \in I$, and $\Omega - \bigcup_{i \in I} A_i \in \mathscr{N}$. Let $\mu_i = \mu(A_i \cap \cdot)$. Then $(\Omega, \Sigma(A_i), \mu_i)$ is a finite complete measure space, where $\Sigma(A_i) = \{A_i \cap B: B \in \Sigma\}$, the trace σ-algebra. Then by Theorem 3 there exists $\lambda_i: \Sigma(A_i) \to \Sigma(A_i)$, a lifting map.

For each $A \in \Sigma$, $0 < \mu(A) < \infty$, one has an at most a countable set $J_A \subset I$ such that $\mu_i(A) > 0$. Define $\bar{\lambda}: \Sigma \to \Sigma$ by the expression

$$\bar{\lambda}(A) = \bigcup_{i \in J_A} \lambda_i(A \cap A_i), \qquad A \in \Sigma. \tag{11}$$

Since this is a disjoint union, $\bar{\lambda}$ is well defined: $\bar{\lambda}(A) \in \Sigma^+ = \{A \in \Sigma: 0 < \mu(A) < \infty\} \cup \{\varnothing, \Omega\}$. If $\lambda(B) = \bigcup_{i \in I} \bar{\lambda}(B \cap A_i)$, then $\lambda: \Sigma \to \Sigma$, since $\lambda(B) \cap A \in \Sigma$ for each $0 < \mu(A) < \infty$; and this implies measurability of $\lambda(A)$, by Lemma 2.2.13, since μ is a complete measure. Using the properties of $\bar{\lambda}$ [e.g., $\bar{\lambda}(A) \equiv \bigcup_{i \in J_A} A \cap A_i \equiv A$ etc.], it is easily shown that λ is a lifting on Σ. This simple computation is left to the reader. Thus there is a lifting on a strictly localizable (Ω, Σ, μ).

For the converse, suppose there exists a lifting on a complete measure space (Ω, Σ, μ). To exclude a triviality, we assume that $\mu \neq 0$. Consider the class $\mathscr{P}$ of all disjoint collections $\{\tilde{A}_i: i \in I\} \subset \Sigma$ such that $0 < \mu(\tilde{A}_i) < \infty$. By the finite subset property of μ, $\mathscr{P}$ is nonempty. We order $\mathscr{P}$ by inclusion, i.e. $\{A'_i: i \in I'\} \prec \{A''_j: j \in I''\}$ if each A'_i is a union of some elements (at most countable) of the A''_j. Then with Zorn's lemma we can find a maximal element of $\mathscr{P}$, denoted $\{A_i: i \in I\}$. It is asserted that this family satisfies the definition of strict localizability.

Since the A_i are disjoint and $0 < \mu(A_i) < \infty$, $i \in I$, $\lambda(A_i) \equiv A_i$, we set $\Omega_0 = \bigcup_{i \in I} \lambda(A_i)$. It is to be shown that (a) if $B \in \Sigma$, $\mu(B) < \infty$, then $\mu(B \cap A_i) > 0$ for at most countably many indices, (b) $\Omega_0 \in \Sigma$, and (c) $\Omega_0 \equiv \Omega$, to complete the argument. Note that the maximality of the collection $\{A_i: i \in I\}$ and the finite subset property of μ imply (c) when (a) and (b) are established. We thus prove these two properties.

Let $0 < \mu(B) < \infty$, and consider $J_B = \{i \in I: \lambda(B) \cap \lambda(A_i) \neq \varnothing\}$. Then $\mu(\lambda(B) \cap \lambda(A_i)) = \mu(B \cap A_i)$ is a positive finite number for each $j \in J_B$. Considering all finite subsets $\mathscr{J}$ of J_B, ordered by inclusion, which is thus a directed set, we get

$$\sum_{i \in \alpha} \mu(A_i \cap B) = \mu\left(\bigcup_{i \in \alpha} A_i \cap B\right) \leqslant \mu(B), \qquad \alpha \in \mathscr{J}. \tag{12}$$

Taking the supremum on the left (cf. Exercise 2.2.12), we get

$$\sum_{i \in J_B} \mu(A_i \cap B) \leqslant \mu(B) < \infty.$$

Since each term in the sum is positive, this implies that J_B is at most countable, which is (a). And then

$$\Omega_0 \cap \lambda(B) = \bigcup_{i \in J_B} \lambda(A_i) \cap \lambda(B) \in \Sigma.$$

Since B, $0 < \mu(B) < \infty$, is arbitrary and the measure space is complete, we may invoke Lemma 2.2.13 again, and conclude that $\Omega_0 \in \Sigma$. This proves (b).

Letting $\Omega_i = \lambda(A_i)$ for $i \neq i_0$ ($i_0 \in I$ is arbitrarily fixed), and $\Omega_{i_0} = \lambda(A_{i_0}) \cup (\Omega - \Omega_0)$, we get $\{\Omega_i: i \in I\} \subset \Sigma$, to satisfy the definition of strict localizability, and the result follows.

Remark. Since strict localizability is a stronger concept than localizability, as noted in Exercise 2.3.5(c), and since localizability is equivalent to the Radon-Nikodým property of a measure space (Theorem 5.4.5) one can conclude that the class of measure spaces having the Radon-Nikodým property contains properly the class for which (set) liftings exist. This observation will be particularly useful in certain integral representations of linear operators on function spaces with measures, for which both the Radon-Nikodým theorem and the lifting theorem are often used.

To understand this remark further, we present a sufficient condition for both the localizability concepts to coincide. It is due to McShane (1962). As a consequence, we see that these two concepts always agree on separable metric spaces, and may also coincide on some (but not all) other general spaces.

Theorem 5. *Let* (Ω, Σ, μ) *be a complete measure space with the finite subset property. Let* $\mathscr{C} = \{A_i: i \in I\} \subset \Sigma$ *be a maximal family of a.e. disjoint collection with* $0 < \mu(A_i) < \infty$, $i \in I$. *(As seen in the above proof, such a collection exists.) If the cardinarity of* I *is at most that of the continuum, then* μ *is localizable iff it is strictly localizable [or equivalently* (Ω, Σ, μ) *has the Radon-Nikodým property iff it admits a lifting map on* Σ*].*

Proof. Since it has already been seen (Exercise 2.3.5) that strict localizability implies localizability without additional conditions on I, we only need to prove the converse, assuming that the cardinality of I is now at most that of the continuum. Suppose then this holds. So there exists an $h: \mathscr{C} \to C \subset (0,1)$ where h is one to one and onto C. [If card($\mathscr{C}$) is that of the continuum, then $C = (0,1)$.] Consider the class $\mathscr{H} = \{h(A)\chi_A: A \in \mathscr{C}\} \subset \mathscr{L}^\infty(\mu)$. Then $\mathscr{H}$ is a bounded subset, and since μ is localizable, this set has a supremum f_0, $0 \leqslant f_0(\omega) \leqslant 1$, and $g \leqslant f_0$ a.e. for all $g \in \mathscr{H}$ (Exercise 3.2.9). We may assume that $1 \geqslant f_0(\omega) \geqslant 0$, $\omega \in \Omega$, by a redefinition of f_0 on a μ-null set, if necessary. It is now claimed that (i) $h(A) = f_0(\omega)$ for almost all $\omega \in A \in \mathscr{C}$, and (ii) $\mathscr{C}_0 = \{A_r = f_0^{-1}(\{r\}): r \in C\} \subset \Sigma$ qualifies for the collection in the definition of strict localizability of μ. This will imply the result.

(i): Define a function $u: \omega \mapsto h(A)\chi_A(\omega) + f_0(\omega)\chi_{A^c}(\omega)$, $A \in \mathscr{C}$, fixed. By the fact that f_0 is a supremum of $\mathscr{H}$, it follows that $u \leqslant f_0$ a.e., and that u is measurable (for Σ). To get the opposite inequality, consider, for each $B \in \mathscr{C}$, the set $D_B = \{\omega: h(B)\chi_B(\omega) > u(\omega)\}$. Thus $D_B \in \Sigma$, and if $B = A$ then $D_B = \varnothing$. Since $u(\omega) = f_0(\omega)$ for $\omega \in A^c$, and f_0 is the supremum of $\mathscr{H}$, it follows that $D_B \cap A^c$ is a μ-null set. Also, when $B \neq A$, since both A, B are in $\mathscr{C}$, they are a.e. disjoint, i.e., $\mu(A \cap B) = 0$, and so $D_B \cap A \subset A \cap B$, and thus $D_B = (D_B \cap A) \cup (D_B \cap A^c)$ is μ-null. Hence $u \geqslant h(B)\chi_B$, $B \in \mathscr{C}$. By the definition of supremum, then $u \geqslant f_0$ a.e. so that $u(\omega) = f_0(\omega)$ a.e. In particular, $f_0(\omega) = h(A)$ for $\omega \in A$.

(ii): Consider the family $\mathscr{C}_0$. This is a disjoint collection of measurable sets of Σ. Since h is one to one, the mapping $A \mapsto A_r$, where $r = h(A)$, is also one to one. Further, by (i), $f_0(\omega) = h(A)$ on A, so $\mu(A_r) = \mu(f_0^{-1}(h(A))) = \mu(A)$, since $(h^{-1} \circ f_0)(A)$ is A. Consequently all elements of $\mathscr{C}_0$ have the same measure as those of the corresponding ones of $\mathscr{C}$, and the two classes are in a one to one correspondence. Thus $\mathscr{C}_0$ is also a maximal family, so that $\tilde{\Omega}_0 = \bigcup_{r \in C} A_r$ differs from Ω on a μ-null set. Adding this to any one member A_r, we get $\mathscr{C}_0$ to satisfy the hypothesis of strict localizability, proving the theorem.

EXERCISES

1. Let $\mathscr{M}(\Sigma)$ be the class of extended real valued measurable functions on a strictly localizable space (Ω, Σ, μ). If $\mathscr{G} \subset [\mathscr{M}(\Sigma)]^+$ is a subset which is directed upwards (or increasing), show that $\mathscr{G}$ has a supremum $f_0 \in [\mathscr{M}(\Sigma)]^+$, and if ρ is a lifting on the space $\mathscr{M}(\Sigma)$ such that $\rho(f) \geqslant f$ for each $f \in \mathscr{G}$, then $\rho(f_0) \geqslant f_0$ also holds. (Compare with Exercise 4.3.13.)

2. Let (Ω, Σ, μ) be a complete measure space, and ρ be a lifting of $\mathscr{L}^\infty(\mu)$. If S is a completely regular space and $\mathscr{L}^\infty(\mu; S)$ is the space introduced in Section 1, let ρ' be the lifting induced by ρ on this space. If $\rho'(f) = f$ (everywhere), show that for each closed set $C_0 \subset S$, $B = f^{-1}(C_0) \in \Sigma$ (even though C_0 need not be a Baire set) and $\rho(\chi_B) \geqslant \chi_B$ holds. [Apply Exercise 1 to $\mathscr{C} = \{h \circ f : h \in C(S), 0 \leqslant h \leqslant 1, h|C_0 = 1\}$, after noting that $\inf(\mathscr{C}) = \chi_B$, and $\mathscr{C}$ is ordered downwards.]

3. Construct a proof of Theorem 2 using an upper density mapping $\lambda' : \Sigma \to \Sigma$ instead of the lower density $\tilde{\lambda}$ employed there, i.e., modify the demonstration there (with suitable interchanges of inclusions and complements) with the mapping λ' used everywhere for $\tilde{\lambda}$.

4. Theorem 2 is also valid if we weaken the hypothesis slightly. Thus show that if $\mathscr{N} \subset \mathscr{A}_n \subset \mathscr{A}_{n+1} \subset \Sigma$ is a sequence of σ-algebras in a complete space (Ω, Σ, μ) and if $\tilde{\lambda}_n : \mathscr{A}_n \to \mathscr{A}_n$ is a (lower) density such that $\tilde{\lambda}_{n+1}|\mathscr{A}_n = \tilde{\lambda}_n$, then there exists a (lower) density $\tilde{\lambda}$ on $\mathscr{A} = \sigma(\bigcup_{n=1}^\infty \mathscr{A}_n)$ such that $\tilde{\lambda}_n = \tilde{\lambda}|\mathscr{A}_n$, $n \geqslant 1$. A similar result obtains with λ'_n in place of $\tilde{\lambda}_n$ everywhere, λ'_n being the upper density.

5. Show that $\tilde{\lambda}$, defined by the expression (11), is indeed a (set) lifting of Σ.

8.3. TOPOLOGIES INDUCED BY LIFTING AND RELATED CONCEPTS

In the above sections a (set) lifting λ was defined and its existence established on abstract strictly localizable spaces (Ω, Σ, μ). Here we show that with such a λ on $\Sigma \to \Sigma$, it is possible to introduce topologies on Ω in terms of which continuous functions are measurable, and that a measurable function is equal a.e. to a continuous function. This presents a basic relation between measure and topology. Further, a discussion of liftings on Radon spaces will be included to obtain a better appreciation of different measures that we studied in the earlier chapters.

Recall that a complete measure space (Ω, Σ, μ) admitting a (set) lifting λ, or a lifting operator ρ on $\mathscr{L}^\infty(\mu)$, has a family of sets $\{\lambda(A): A \in \Sigma\}$ for which conditions (i)–(iv) of Proposition 1.3 hold. But these form a *base* for a topology, as customarily defined in the latter subject. Let this topology be denoted by $\mathscr{T}_1$. It is also possible to consider a class $\{\lambda(A) - N: A \in \Sigma, N \in \mathscr{N}\}$ which, when $\mathscr{N}$ is replaced by $\{\varnothing\}$, reduces to the preceding one. This new class, having more members, also defines a base for a (stronger) topology, denoted $\mathscr{T}_2$. These definitions clearly imply $\mathscr{T}_1 \subset \mathscr{T}_2$. Since by the notion of a base, each element of $\mathscr{T}_1, \mathscr{T}_2$ is obtained by forming arbitrary unions and finite intersections of the elements of their bases, we now verify that $\mathscr{T}_i \subset \Sigma$, $i = 1, 2$, and present some relevant properties. As usual, $\mathscr{N}$ is the ring of μ-null sets of Σ.

Theorem 1. *Let $\mathscr{T}_i$, $i = 1, 2$, be as defined above. Then they are topologies in fact (they are closed under finite intersections and arbitrary unions), and $\mathscr{T}_i \subset \Sigma$. Moreover, a set $U \in \mathscr{T}_i$, $i = 1, 2$, is nonvoid iff U is not μ-null, and the closure of U in $\mathscr{T}_i$ is $\lambda(U)$, so that $\lambda(U)$ is closed and open (thus $\mathscr{T}_i$ is an "extremally disconnected" topology[†]). Further, $f: \Omega \to \overline{\mathbb{R}}$ is measurable for Σ iff it coincides a.e. with a continuous (for $\mathscr{T}_i$) function on Ω, and the correspondence preserves algebraic operations where they are defined.*

Proof. We present the argument in steps for convenience.

1. Consider $\mathscr{T}_1$. If $U \in \mathscr{T}_1$, then $U = \bigcup_{i \in I} \lambda(A_i)$ for some $\{A_i: i \in I\} \subset \Sigma$, and $\lambda(A_i)$ is a basis set for $\mathscr{T}_1$. Since the existence of λ implies the strict localizability of μ, the collection $\{A_i: i \in I\}$ has a supremum A in Σ, i.e., $A \supset A_i$ a.e., and A is the smallest such set in Σ. Hence for each $B \in \Sigma$, $\mu(B) < \infty$,

$$\|\chi_{A \cap B}\|_1 = \sup_{i \in I} \|\chi_{A_i \cap B}\|_1 = \lim_n \|\chi_{A_{i_n} \cap B}\|_1 \tag{1}$$

[†] This is the classical terminology. In the current treatments this topology is called 0-*dimensional*. Except for the name, we do not use any special properties until Chapter 10. There we use it for compact spaces in which case both concepts coincide. See, e.g., the book by Gillman and Jerison (1960, p. 247).

for some subsequence $i_1, i_2, \ldots$ such that $\chi_{A_{i_n} \cap B} \to \chi_{A \cap B}$ a.e. It follows that $A \cap B$ is the supremum of $\{A_{i_n} \cap B\}$, and that $\lambda(A) \cap B \supset \lambda(A_{i_n}) \cap B$ for all n. Hence $\lambda(A) \cap B \supset \bigcup_{n=1}^{\infty} \lambda(A_{i_n}) \cap B \equiv \bigcup_{n=1}^{\infty} A_{i_n} \cap B \equiv A \cap B$. Since $\lambda(A) \cap B \equiv A \cap B$, and also $\lambda(A) \supset \lambda(A_i)$ for all i, we get $\lambda(A) \cap B \supset \bigcup_{i \in I} \lambda(A_i) \cap B \equiv \bigcup_{n=1}^{\infty} A_{i_n} \cap B$. Consequently $\lambda(A) \cap B \equiv U \cap B$, so that $U \cap B \in \Sigma$. But B is an arbitrary set of finite positive measure, and so $U \equiv \lambda(A)$, by Lemma 2.2.13. Since $A - \bigcup_{n=1}^{\infty} A_{i_n} \in \mathscr{N}$, we deduce that $\bigcup_{i \in I} A_i \in \Sigma$ and $U \in \Sigma$. This also implies that $\mathscr{T}_1$ is closed under arbitrary unions and $\mathscr{T}_1 \subset \Sigma$. That it is closed under finite intersections is obvious. Thus $\mathscr{T}_1$ is a topology relative to which each open set is measurable for Σ.

2. The same argument with a few modifications can be applied for $\mathscr{T}_2$ also. We include the details. To see that $\mathscr{T}_2$ is closed under finite intersections, let $\lambda(A_i) - N_i \in \mathscr{T}_2$, $i = 1, 2$. Then

$$[\lambda(A_1) - N_1] \cap [\lambda(A_2) - N_2] = \lambda(A_1 \cap A_2) - (N_1 \cup N_2) \in \mathscr{T}_2.$$

($\mathscr{N}$ is also a ring!) Thus by induction, this holds for a finite number of sets. Regarding unions, note that for $A \subset B$ in Σ, $\lambda(A) = \lambda(A \cap B) = \lambda(A) \cap \lambda(B) \subset \lambda(B)$, so that λ is monotone (even when it is only a lower density). But then (also for lower densities), on using that $\lambda(A_i) \subset \lambda(A_1 \cup A_2)$ we have

$$\begin{aligned}\lambda(A_1) \cup \lambda(A_2) - (N_1 \cup N_2) &\subset [\lambda(A_1) - N_1] \cup [\lambda(A_2) - N_2] \\ &\subset \lambda(A_1) \cup \lambda(A_2) \subset \lambda(A_1 \cup A_2).\end{aligned}$$

Since the extreme terms differ by an element of $\mathscr{N}$, it follows that $\mathscr{T}_2$ is closed under finite unions as well as finite intersections. If $\lambda(A_i) - N_i \in \mathscr{T}_2$, then to show that $\mathscr{T}_2$ is closed under arbitrary unions, we can use the argument of the preceding paragraph and find a supremum of $\{A_i : i \in I\}$, say A, in Σ. If $B \in \Sigma$, $\mu(B) < \infty$, then $A \cap B$ is the supremum of $\{A_i \cap B: i \in I\}$, and as before, there exists a sequence $\{A_{i_n} \cap B: n \geqslant 1\}$ having $A \cap B$ as its supremum. Note that adding or subtracting N_{i_n} to these sets does not change the supremum. So

$$\lambda(A) \cap B \equiv \bigcup_{n=1}^{\infty} [\lambda(A_{i_n}) - N_{i_n}] \cap B.$$

Hence

$$\begin{aligned}\lambda(A) \cap B &\supset \bigcup_{i \in I} \lambda(A_i) \cap B \supset \bigcup_{i \in I} [\lambda(A_i) - N_i] \cap B \\ &\supset \bigcup_{n=1}^{\infty} [\lambda(A_{i_n}) - N_{i_n}] \cap B \equiv \lambda(A) \cap B. \qquad (2)\end{aligned}$$

This and Lemma 2.2.13 imply that $\bigcup_{i \in I}[\lambda(A_i) - N_i] \in \mathscr{T}_2$ and $\mathscr{T}_2 \subset \Sigma$. Also, in both cases, if $U \in \mathscr{T}_i$ then $U \neq \varnothing \Rightarrow \mu(U) > 0$, since otherwise, $U \in \mathscr{N}$, and then $U \neq \varnothing$ and open for $\mathscr{T}_i$, so that $\lambda(U) \neq \varnothing$, which is impossible.

3. Regarding the second part, let $U \in \mathscr{T}_1$. So $\lambda(U) = [\lambda(U^c)]^c$. But $\lambda(U^c)$ is open, being an element of the base for $\mathscr{T}_1$. Hence as its complement, $\lambda(U)$ is closed in $\mathscr{T}_1$. But $U = \bigcup_{i \in I}\lambda(A_i')$ for some $\{A_i': i \in I\} \subset \Sigma$, and so $\lambda(U) \supset \lambda(A_i')$, $i \in I \Rightarrow \lambda(U) \supset U$ always, and hence its closure $\text{cl}(U) \subset \lambda(U)$. We now claim that $\text{cl}(U)$ is also open. This will follow if it is shown to be $\lambda(U)$, which is open by definition of $\mathscr{T}_1$. Since $\lambda(U)$ is also $\mathscr{T}_1$-closed, it suffices to show that each neighborhood of $\omega \in \lambda(U)$ intersects $\text{cl}(U)$, so that $\omega \in \text{cl}(U)$ and $\lambda(U) \subset \text{cl}(U)$ follows. Thus let V be a neighborhood of ω. Then $V = \lambda(B)$ for some $B \in \Sigma$. Hence $\omega \in \lambda(U) \cap \lambda(B) = \lambda(U \cap B)$, and this is nonvoid. So by the last part of the preceding paragraph $\lambda(U \cap B)$ is not μ-null, and since $U \cap V \in \mathscr{T}_1 \subset \Sigma$, we have $\mu(U \cap V) = \mu(U \cap \lambda(B)) = \mu(\lambda(U) \cap \lambda(B)) = \mu(\lambda(U \cap B)) > 0$ and $U \cap V \neq \varnothing$, so that $\text{cl}(U) \supset \lambda(U)$, and hence $\lambda(U) = \text{cl}(U)$. Thus every element of $\mathscr{T}_1$ is closed and open. This is the property of a topology that is termed *extremally disconnected* (or rather, 0-*dimensional*).

4. The same argument applies to $\mathscr{T}_2$ also. In fact, $N \in \mathscr{N}$ implies $N^c = \Omega - N = \lambda(\Omega) - N \in \mathscr{T}_2$, so N^c is open and hence every element of $\mathscr{N}$ is $\mathscr{T}_2$-closed. Since $\mu(N) = 0$, the $\mathscr{T}_2$-closed set N must be nowhere dense, by the first part. Hence if $U \in \mathscr{T}_2$, then $U = \lambda(A) - N$ for some $A \in \Sigma$ and $N \in \mathscr{N}$, and $\lambda(U) = \lambda(A) \cap [\lambda(N)]^c = \lambda(A) - \lambda(N) = \lambda(A)$, since $\lambda(N) = \varnothing$. So $\lambda(U)$ is $\mathscr{T}_2$-open as well as $\mathscr{T}_1$-open. But $\lambda(A) = [\lambda(A^c)]^c$, and since $\mathscr{T}_1 \subset \mathscr{T}_2$, $\lambda(A^c)$ is also $\mathscr{T}_2$-open, so that $\lambda(A)$ and hence $\lambda(U)$ is $\mathscr{T}_2$-closed. By the same argument $U = \bigcup_{i \in I}[\lambda(A_i) - N_i] \supset \lambda(A_i) - N_i \Rightarrow \lambda(U) \supset \lambda(A_i) \supset \lambda(A_i) - N_i \Rightarrow \lambda(U) \supset U$ again. If $\omega \in \lambda(U)$, and V is a neighborhood of ω in $\mathscr{T}_2$, so $V = \lambda(B) - N$, we get $\lambda(U) \cap V = \lambda(U) \cap \lambda(B) \cap N^c$, and $\mu(U \cap V) = \mu(\lambda(U) \cap V) = \mu(\lambda(U \cap B) \cap N^c) = \mu(\lambda(U \cap B)) > 0$, since $\lambda(V) \cap V$ is nonempty. Thus $U \cap V$ is nonempty, and $\text{cl}(U) = \lambda(U)$, as before.

5. For the last part, let $f: \Omega \to \overline{\mathbb{R}}$ be measurable for Σ. To show that it is equal a.e. to a continuous function, consider $\tilde{f}: \Omega \to \overline{\mathbb{R}}$ given by

$$\tilde{f}(\omega) = \sup\left\{r: \omega \notin \lambda\left(f^{-1}([-\infty, r))\right)\right\}, \qquad \omega \in \Omega. \tag{3}$$

Then we have

$$\tilde{f}^{-1}([-\infty, r)) = \bigcup_{t<r} \lambda\left(f^{-1}([-\infty, t))\right) \tag{4}$$

and similarly

$$\tilde{f}^{-1}((r, +\infty]) = \bigcup_{t<r}\left[\Omega - \lambda\left(f^{-1}((-\infty, t])\right)\right]. \tag{5}$$

Since $A = f^{-1}([-\infty, t)) \in \Sigma$ and $\lambda(A)$ is open in $\mathscr{T}_1$, it follows that the sets on the right sides of (4) and (5) are open, and hence $\bar{f}^{-1}$ maps open sets of $\overline{\mathbb{R}}$ into $\mathscr{T}_1$-open sets. [In (5) we use the fact that $\lambda(A)$ is a closed and open set, established above.] Thus $\bar{f}$ is continuous. Since t may be restricted to vary over the rationals, it follows that the symmetric difference of $\bar{f}^{-1}([-\infty, r))$ and $f^{-1}([-\infty, r))$ is in $\mathscr{N}$, so that $\bar{f} = f$ a.e. Note that if $\tilde{f}$ is another continuous function agreeing a.e. with f, then $\bar{f} - \tilde{f}$ is continuous and $(\bar{f} - \tilde{f})^{-1}(a, b)$ is open. If it is nonempty in $\mathscr{T}_1$, then by the definition of this topology, there exists a non-μ-null set $\lambda(A)$ of the base contained in this open set. But then $\mu(\lambda(A)) > 0$, and $\bar{f} = f = \tilde{f}$ a.e. is violated. Hence $\bar{f} = \tilde{f}$, and $f \to \bar{f}$ is a unique representation, so that the algebraic operations are preserved whenever they are defined on these functions.

Conversely, if f is $\mathscr{T}_1$-continuous, then $\{\omega: f(\omega) > r\} \in \mathscr{T}_1 \subset \Sigma$ for all $r \in \mathbb{R}$, so that f is measurable for Σ.

6. The argument for $\mathscr{T}_2$ is essentially the same. If $f: \Omega \to \overline{\mathbb{R}}$ is $\mathscr{T}_2$-continuous, then for any $N \in \mathscr{N}$,

$$\{\omega: f(\omega) > r\} = [\{\omega: f(\omega) > r\} \cap N] \cup [\{\omega: f(\omega) > r\} \cap N^c]. \quad (6)$$

Since N is $\mathscr{T}_2$-closed, the second set on the right of (6) is open and so is in $\mathscr{T}_2$. The first set is in $\mathscr{N}$ and hence is again in $\mathscr{T}_2$. Thus the left side is in $\mathscr{T}_2 \subset \Sigma$, so that f is measurable for Σ.

Conversely, let f be measurable for Σ. Then by the preceding step $f = \bar{f}$ a.e., where $\bar{f}$ is $\mathscr{T}_1$-continuous, and $\bar{f}$ is given by (3). Since $\mathscr{T}_1 \subset \mathscr{T}_2$ (as already noted before), $\bar{f}$ is also $\mathscr{T}_2$-continuous, i.e., $\{\omega: \bar{f}(\omega) > r\} \in \mathscr{T}_1 \subset \mathscr{T}_2$, so that the result follows. Alternatively this can be proved independently of the preceding, first assuming f to be bounded. Then given $\epsilon > 0$, there exists a step function $f_\epsilon = \sum_{i=1}^n a_i \chi_{E_i}$ such that $\|f - f_\epsilon\|_\infty < \epsilon$, χ_{E_i} is a constant on the open sets $\lambda(E_i) \cap E_i$, and $\lambda(E_i^c) \cap E_i^c$. (In fact, since $E_i - \lambda(E_i) \in \mathscr{N}$ and $\lambda(E_i^c) - E_i^c \in \mathscr{N}$, each of these is closed. Also $\lambda(E_i)$, $\lambda(E_i^c)$ are open, and we get the result by noting $\lambda(E_i) \cap E_i = \lambda(E_i) \cap [E_i - \lambda(E_i)]^c$ etc.) Thus χ_{E_i} and hence f_ϵ is $\mathscr{T}_2$-continuous. So $f_\epsilon \to f$ uniformly outside of a null set, and f is continuous there. This result applied to $\arctan(f)$ gives the general case. The theorem is thus completely proved.

The last part of the above theorem shows that a function $f: \Omega \to \overline{\mathbb{R}}$ is measurable for Σ iff it is a.e. equal to a (unique) continuous function for the topologies $\mathscr{T}_1$ and $\mathscr{T}_2$, even though $\mathscr{T}_2$ is stronger than $\mathscr{T}_1$. One should compare this theorem with Luzin's (Theorem 3.2.9). Thus it is natural here to compare the classes of continuous functions for both these topologies. The next result provides a simple answer to this question even when the functions are vector (or completely regular space) valued.

Theorem 2. *Let S be a completely regular space, (Ω, Σ, μ) be a strictly localizable space, and $\mathscr{L}^\infty(\mu; S)$ be as in Section 1. If $\mathscr{C}_i$ are the classes of all $f: \Omega \to S$ which are $\mathscr{T}_i$-continuous and bounded [i.e., $f(\Omega)$ is contained in a compact subset of S], $i = 1, 2$, then $\mathscr{C}_1 = \mathscr{C}_2$, and this common class is characterized as $\mathscr{M}(S) = \{f \in \mathscr{L}^\infty(\mu; S): \rho'(f) = f\}$, where ρ' is the lifting associated with a lifting of (Ω, Σ, μ) (which exists by Theorem 2.4 and Proposition 1.6). ($S = \mathbb{R}$ or $\overline{\mathbb{R}}$ admits a slight simplification.)*

Proof. Since by definition of the topologies, $\mathscr{C}_1 \subset \mathscr{C}_2$ is always true, we shall show that $\mathscr{M}(S) \subset \mathscr{C}_1$ and $\mathscr{C}_2 \subset \mathscr{M}(S)$ to complete the argument.

Thus let $f \in \mathscr{M}(S)$ and $h \in \mathscr{C}(S)$, the space of real continuous functions on S. If $\rho: \mathscr{L}^\infty(\mu) \to \mathscr{L}^\infty(\mu)$ is a lifting, which exists by Theorem 2.4, let ρ' be its associated lifting of $\mathscr{L}^\infty(\mu; S)$. Then

$$\rho(h(f)) = h(\rho'(f)) = h(f), \qquad h \in C(S), \quad f \in \mathscr{M}(S). \tag{7}$$

But then $h(f) \in \mathscr{L}^\infty(\mu)$, and by Theorem 1, it is $\mathscr{T}_1$-continuous for each $h \in C(S)$. However, for any topology $\mathscr{T}$ on a space T, $f: T \to S$ is $\mathscr{T}$-continuous iff $h(f): T \to \mathbb{R}$ is $\mathscr{T}$-continuous for each $h \in C(S)$, which follows from the definition of the topology of S. Thus f is $\mathscr{T}_1$-continuous, so that $\mathscr{M}(S) \subset \mathscr{C}_1$.

Next let $f \in \mathscr{C}_2$. Then f is bounded and $\mathscr{T}_2$-continuous, so that $h(f)$ is $\mathscr{T}_2$-continuous. Then by Theorem 1, $h(f)$ is measurable for Σ, $h \in C(S)$, so that $f \in \mathscr{L}^\infty(\mu: S)$. Also $h(f)$ is bounded and $\mathscr{T}_2$-continuous for each $h \in C(S)$, and hence $h(f) \in \mathscr{L}^\infty(\mu)$. But $\rho(h(f)) \equiv h(f)$, so that $\rho(\rho(h(f))) = \rho(h(f))$. Consequently $\rho(h(f)) \in \mathscr{M}(\mathbb{R})$. Again by Theorem 1, $\rho(h(f))$ is equal a.e. to a $\mathscr{T}_2$-continuous function, and $h(f)$ is one such. By the preceding paragraph, since $\rho(h(f))$ is in $\mathscr{M}(\mathbb{R})$, which is contained in the class of $\mathscr{T}_1$-continuous (hence $\mathscr{T}_2$-continuous) functions, $\rho(h(f))$ is $\mathscr{T}_2$-continuous. Thus the set $\{\omega: \rho(h(f))(\omega) \neq h(f)(\omega)\}$ is $\mathscr{T}_2$-open, and since $\rho(h(f)) \equiv h(f)$ also, it is μ-negligible. By the preceding theorem, this must be an empty set. Thus $\rho(h(f)) = h(f)$, and so

$$h(f) = \rho(h(f)) = h(\rho'(f)), \qquad h \in C(S). \tag{8}$$

By the earlier comment on the topologies for S, (8) implies $\rho'(f) = f$ and hence $f \in \mathscr{M}(S)$. Thus $\mathscr{C}_2 \subset \mathscr{M}(S)$, completing the proof of the theorem.

Since $\rho'(f) = f$ plays an important part in identifying the classes of $\mathscr{T}_1$ and $\mathscr{T}_2$ continuous functions, it is of interest to look into those lifting maps for which such an equality holds when Ω itself has a topology to begin with. We shall add some remarks on this problem, after introducing an appropriate concept. (See also Section 10.3.)

Specializing Definition 2.3.9 to locally compact spaces, we let (Ω, Σ, μ) be a Radon measure space with Ω locally compact, and Σ a complete σ-algebra containing all the open sets of Ω. Assume here that μ has support Ω (cf. Exercise 2.3.10; thus $\Omega = [\bigcup\{G: \mu(G) = 0, G \text{ open}\}]^c$). In this case it can be verified that (Ω, Σ, μ) is strictly localizable. (For a sketch, see Exercise 3 below.)

Definition 3. Let Ω be locally compact, and (Ω, Σ, μ) be a Radon measure space with $\text{supp}(\mu) = \Omega$. A lifting ρ of $\mathscr{L}^\infty(\mu)$, which exists by Theorem 2.4, is termed (*almost*) *strong* if $\rho(f) = f$ (outside a μ-null set) for all bounded continuous functions f [which always belong to $\mathscr{L}^\infty(\mu)$]. If $B(\Omega)$ denotes the subspace of $\mathscr{L}^\infty(\mu)$ consisting of Borel-measurable functions, then ρ is termed a *Borel lifting* if $\rho(\mathscr{L}^\infty(\mu)) \subset B(\Omega)$, i.e., $\rho(f) \in B(\Omega)$ for each $f \in \mathscr{L}^\infty(\mu)$.

The existence of almost strong lifting maps on $\mathscr{L}^\infty(\mu)$ will be useful in proving results on Borel liftings. Even though $\mathscr{L}^\infty(\mu)$ on a Radon space (Ω, Σ, μ) admits a lifting by the work of the last section, it is not known whether strong (or Borel) liftings exist on all Radon spaces where Ω is locally compact. On the other hand, if Ω is a locally compact group with μ as a (left) Haar measure which is Radon (cf. the next chapter), then not only do strong liftings exist, but they have additional properties: they commute with translations, i.e. $\rho(\tau_s f) = \tau_s \rho(f)$ where $(\tau_s f)(t) = f(s^{-1}t)$, $s, t \in \Omega$, and ρ is a strong lifting. There is also a question of integrating and disintegrating measures, i.e., $\lambda: \omega \mapsto \lambda_\omega$. For measures on (Ω, Σ, μ) one should define $\int_\Omega \lambda_\omega(\cdot)\mu(d\omega)$ as a "weak" integral. Conversely a measure λ on (Ω, Σ, μ) should be so represented (or "disintegrated") relative to a family $\{\lambda_\omega(\cdot): \omega \in \Omega\}$ of measures with some desirable properties regarding their supports. It can be shown that this "disintegration problem" is equivalent to that of the existence of a strong lifting, which in turn is related to the existence of "regular conditional probabilities." Many of these questions lead to specialized studies of the subject. An extended account of these and related results as well as an explanation of the preceding statements are given in the monograph by Tulcea (1969).

These remarks should be supplemented with the following. The introduction of null sets into integration theory by Lebesgue contributed to an enormous growth of both the theory and applications in analysis, since treating equivalence classes in lieu of individual functions sufficed, in contrast with Riemann's method. When the individual functions have to be selected, preserving the algebraic operations, one needs the lifting, and this chapter shows that such a mapping, its existence, and various properties are intricate problems, which are however quite important for many applications. This should show both the ease and difficulty introduced by the μ-null sets in this theory. We do not undertake a study of specializations of the subject here, and turn in the next chapter to an interplay of measure and topology, complementing the above results.

EXERCISES

1. Let (Ω, Σ, μ) be a strictly localizable measure space, and ρ be a lifting of $\mathscr{L}^{\infty}(\mu)$. Verify that $l_{\omega}: L^{\infty}(\mu) \to \mathbb{R}$ defined by $l_{\omega}(\tilde{f}) = \rho(\tilde{f})(\omega)$, is a multiplicative linear functional with $l_{\omega}(1) = 1$. If $\mathscr{T}_1$ is the topology on Ω induced by ρ, show that it is Hausdorff iff the mapping $\omega \mapsto l_{\omega}$ is one to one from Ω into $[L^{\infty}(\mu)]^*$, the adjoint space.

2. Show that when the above condition for Hausdorffness holds, then $(\Omega, \mathscr{T}_1)$ is actually completely regular. [If $A \subset \Omega$ is $\mathscr{T}_1$-closed, $\omega_0 \in \Omega - A$, then, by definition, the open set $\Omega - A = \bigcup_{i \in I} \lambda(A_i)$ for some $\{A_i: i \in I\} \subset \Sigma$. Hence $\omega_0 \in \lambda(A_{i_0})$ for some i_0. Then $f = \chi_{\lambda(A_{i_0})}$ is $\mathscr{T}_1$-continuous, since $\lambda(A_{i_0})$ is open and closed, $f(\omega_0) = 1$, and $f|A = 0$.]

3. Let Ω be a locally compact space and (Ω, Σ, μ) be a Radon measure space on it. Show that μ is strictly localizable, using the following steps. (A more general case is in Section 9.3.)

(a) Since μ is Radon, for compact sets $C \subset \Omega$, $\mu(C) < \infty$. Verify that a set $A \subset \Omega$ is locally μ-null (i.e., $A \cap V_{\omega}$ is μ-null for each neighborhood V_{ω} of $\omega \in A$) iff $\mu(A \cap C) = 0$ for each compact set C.

(b) Consider the collections $\mathscr{P}$ of pairwise disjoint compact sets of Ω. If these are ordered by inclusion, show that each $A \in \Sigma$, $\mu(A) < \infty$, is the union of members of $\mathscr{P}$ having at most countably many non-μ-null sets.

(c) Show that, by Zorn's lemma, $\mathscr{P}$ has a maximal element $\mathscr{C}_0$.

(d) If $B = \bigcup\{A: A \in \mathscr{C}_0\}$, show that $B \cap C \in \Sigma$ for each compact C and hence $B \in \Sigma$.

(e) Using the inner regularity of μ and the maximality of $\mathscr{C}_0$, show that $\Omega - B$ must be μ-null. Deduce that μ is strictly localizable, so that $\mathscr{L}^{\infty}(\mu)$ admits a lifting on it.

4. If $\tilde{\mathscr{T}}_2$ is defined as $\mathscr{T}_2$ but with a lower density, show, by a modification of the argument of step 2 in the proof of Theorem 1 that $\tilde{\mathscr{T}}_2$ is a topology and $f: \Omega \to \overline{\mathbb{R}}$ is measurable for Σ iff it is $\tilde{\mathscr{T}}_2$-continuous outside a μ-null set.

5. Establish the parenthetical remark at the end of Theorem 2 in the following form: Let (Ω, Σ, μ) be a strictly localizable space, and $\mathscr{T}_i$, $i = 1, 2$, be the two associated topologies (as in Theorem 1). Then each extended real measurable function on (Ω, Σ, μ) is almost everywhere equal to a $\mathscr{T}_i$-continuous (extended real) function, $i = 1, 2$ (with a simplified proof).

9

TOPOLOGICAL MEASURES

This chapter is devoted to an introduction to measures on Hausdorff spaces. After discussing the regularity properties of a measure on such a general space, it is first shown that a Radon measure is strictly localizable, and the remaining work is largely for Borel measures on locally compact spaces. One of the results here is an integral representation of local functionals, a concept introduced by Gel'fand and Vilenkin. Two proofs of the classical Riesz-Markov theorem for linear functionals are given. A treatment of the existence of a (left) Haar measure on a locally compact group and extensions of measures on lattices based on general topological spaces are also included.

9.1. INTRODUCTION AND PRELIMINARIES

In the preceding chapters we have already encountered Radon measures and discussed some of their properties. However, a systematic analysis of measure functions on σ-rings of topological spaces generated by some special classes of sets, such as the closed (or open), compact, G_δ, or F_σ, has not been made. Regularity properties of measures on these particular classes will be useful in many parts of analysis, and we include some aspects of the subject in this chapter.

To avoid ambiguity, let us recall that a set A contained in a topological space Ω is a G_δ if there exists a sequence of open sets U_n of Ω such that $A = \bigcap_{n \geqslant 1} U_n$. Considering partial products, we may as well say that A is a *G_δ-set* if there is a decreasing sequence of open sets U_n such that $U_n \downarrow A$. Similarly, B is termed an *F_σ-set* if there exists a sequence of closed sets F_n of Ω such that $B = \bigcup_{n \geqslant 1} F_n$, or alternatively an increasing sequence of closed sets F_n such that $F_n \uparrow B$. Thus the class of all G_δ-sets (or F_σ-sets) of Ω forms a lattice. The smallest σ-algebra of Ω containing all the open (equivalently, closed) sets is the *Borel σ-algebra*, as noted in the preceding chapter. Similarly, if $C(\Omega)$ $[C_c(\Omega)]$ is the space of real continuous functions [with also compact supports] on Ω, then the σ-algebra generated by the functions of $C(\Omega)$ is the

Baire σ-algebra. If Ω is completely regular, then this concept of Baire σ-algebra is the same as that defined earlier. Letting $\mathscr{B}$ and $\tilde{\mathscr{B}}$ be the Baire and Borel σ-algebras of Ω, it is clear that $\mathscr{B} \subset \tilde{\mathscr{B}}$. The inclusion generally is proper. The elements of $\mathscr{B}$ and $\tilde{\mathscr{B}}$ are termed *Baire* and *Borel* sets respectively. The definition of Baire class is often limited to locally compact spaces, for reasons detailed below. Since every open (closed) set is a G_δ (F_σ), $\tilde{\mathscr{B}}$ is also generated by all G_δ (or F_δ) sets of Ω. For instance, let $\mathscr{B}_0$ and $\tilde{\mathscr{B}}_0$ be the σ-algebras generated by the class of all compact G_δ and compact subsets of Ω. They are termed *restrictedly Baire and Borel* classes respectively. If Ω is a σ-compact space itself, then the word "restricted" is redundant, and $\mathscr{B}_0 = \mathscr{B}$, $\tilde{\mathscr{B}}_0 = \tilde{\mathscr{B}}$. These are the types which are often used in applications. We present some elementary properties for motivation and later use. Observe that $\mathscr{B}_0 = \sigma(C_c(\Omega))$, as can also be seen below from Proposition 1 (cf. Exercise 12). It should be mentioned that some authors omit the word "restricted" for $\mathscr{B}_0$ and $\tilde{\mathscr{B}}_0$, and do not consider $\mathscr{B}$ and $\tilde{\mathscr{B}}$ at all when considering locally compact Ω.

Recalling that a set $A \subset \Omega$ is (σ-)*bounded* if it is contained in a (σ-)compact subset of Ω, we have the following result, which justifies the introduction of the "restricted" classes above.

Proposition 1. *Let Ω be a locally compact Hausdorff space, and $f: \Omega \to \mathbb{R}$ be a continuous function. Then $A = \{\omega: f(\omega) \geqslant \alpha\}$, $B = \{\omega: f(\omega) \leqslant \alpha\}$, and $C = \{\omega: f(\omega) = \alpha\}$ are closed G_δ-sets, $\alpha \in \mathbb{R}$. If moreover f has compact support, and $\alpha > 0$, then A and C are also compact. Given a compact G_δ-set K, there exist continuous functions $f_n: \Omega \to \mathbb{R}^+$, with compact supports such that $f_n \downarrow \chi_K$ pointwise. Also, every restrictedly Baire set B has the property that either B or its complement B^c is σ-bounded, and the restricted Baire algebra $\mathscr{B}_0$ is generated as well by the real continuous functions on Ω with compact supports.*

Proof. For each $\alpha \in \mathbb{R}$ we have

$$A = \{\omega: f(\omega) \geqslant \alpha\} = \bigcap_{n \geqslant 1} \left\{\omega: f(\omega) > \alpha - \frac{1}{n}\right\},$$

$$B = \{\omega: f(\omega) \leqslant \alpha\} = \bigcap_{n \geqslant 1} \left\{\omega: -f(\omega) > -\alpha - \frac{1}{n}\right\},$$

$$C = A \cap B.$$

Since f is continuous, the above relations show that A, B, C are closed G_δ-sets. If $\alpha > 0$, then A and C are closed subsets of the support $S(f)$ of f [$S(f) = \text{cl}\{\omega: f(\omega) \neq 0\}$]. If $S(f)$ is compact, then A, C are also compact.

Let now $K \subset \Omega$ be a compact G_δ-set. Then there exist open sets $U_n \downarrow K$. To produce the desired sequence $f_n \downarrow \chi_K$, we first prove a *claim*: For each n, there exists a continuous $g_n: \Omega \to [0,1]$, of compact support, such that $g_n|_K = 1$ and $g_n|_{\Omega - U_n} = 0$, using the local compactness hypothesis of Ω.

In fact, for each ω ($\in K \subset \Omega$) there exists a relatively compact open set O_ω containing ω, by definition of the neighborhood system of a locally compact space, so that $\{O_\omega: \omega \in K\}$ covers K. By compactness of K, (a finite subcover) $K \subset V = \bigcup_{i=1}^{n} O_{\omega_i}$ exists, and V is relatively compact. By Uryshon's lemma, for each $\omega \in K \subset V$, there is a continuous $h_\omega: \Omega \to [0,1]$ such that $h_\omega(\omega) = 1$, $h_\omega|_{\Omega - V} = 0$, so that $S(h_\omega) \subset \bar{V}$, i.e., h_ω has compact support, since $h|_{\Omega - \bar{V}} = 0$ also. Since $\tilde{O}_\omega = \{x: h_\omega(x) > 1/n\} \subset V$ and open, $\{\tilde{O}_\omega: \omega \in K\}$ is an open cover of K, there exists an $m \geqslant 1$ with $V \supset \bigcup_{j=1}^{m} O_{\omega_j} \supset K$. If we set $g = h_{\omega_1} + \cdots + h_{\omega_m}$, then $g > 1$ on K, $= 0$ off $\Omega - V$. Hence $g_0 = \min(g, 1)$ is continuous and of compact support, and $g_0: \Omega \to [0,1]$, $g_0|_K = 1$, $g_0|_{\Omega - V} = 0$. Now in the sequence $\{U_n: n \geqslant 1\}$, replacing it by $\{U_n \cap V: n \geqslant 1\}$ if necessary, we may assume that each U_n is a relatively compact open set. Then by what has been established, for each n we take U_n to be the V and g_n as the corresponding g. Thus $g_n|_K = 1$, $g_n|_{\Omega - U_n} = 0$, and $g_n: \Omega \to [0,1]$ satisfies the requirements of the claim.

To prove the proposition, let $f_n = \min(g_i, 1 \leqslant i \leqslant n)$, so that $f_n \geqslant f_{n+1}$, $f_n|_K = 1$, $f_n|_{\Omega - U_n} = 0$, $f_n: \Omega \to [0,1]$ and continuous. Clearly $f_n \downarrow \chi_K$ on K. If $\omega \notin K = \bigcap_{n \geqslant 1} U_n$, then $\omega \notin U_{n_0}$ for some $n_0 \geqslant 1$ and hence $f_{n_0}(\omega) = 0$. Thus $f_n \downarrow \chi_K$ pointwise.

If $\mathscr{B}_1$ is the smallest σ-algebra relative to which each real continuous function on Ω, of compact support, is measurable, then the above result shows that $\mathscr{B}_0 \subset \mathscr{B}_1 = \sigma(C_c(\Omega))$. On the other hand, if $f \in C_c(\Omega)$, then $\{\omega: f^+(\omega) \geqslant \alpha\}$ is a compact G_δ for each $\alpha > 0$, and so is in $\mathscr{B}_0$. Hence f^+ (and similarly f^-) is $\mathscr{B}_0$-measurable. Thus f is also $\mathscr{B}_0$-measurable and $\mathscr{B}_1 \subset \mathscr{B}_0$, i.e., $\mathscr{B}_0 = \mathscr{B}_1$.

Finally, let $\mathscr{C} = \{B: B \text{ or } B^c \text{ is a } \sigma\text{-bounded Baire set}\}$. Then clearly $\mathscr{C}$ is a σ-algebra. If B is compact, then $B \in \mathscr{C}$, and hence each compact G_δ-set is in $\mathscr{C}$. Thus $\mathscr{B}_0 \subset \mathscr{C}$, completing the proof.

The preceding proof has additional information, and yields the following result which is important for the ensuing work.

Corollary 2. *Let $K \subset U \subset \Omega$ be such that K is compact, U is open, and Ω is locally compact. Then there exist $\{V, D\} \subset \mathscr{B}_0$ such that $K \subset V \subset D \subset U$ with V an open F_σ and D a compact G_δ-set.*

Proof. By the claim in the above proof, there exists an $f \in C_c(\Omega)$ such that $f|_K = 1$, $f|_{\Omega - U} = 0$, $0 \leqslant f \leqslant 1$. Let $V = \{\omega: f(\omega) > \frac{1}{2}\}$ and $D = \{\omega: f(\omega) \geqslant \frac{1}{2}\}$. Then $V = \bigcup_{n \geqslant 1}\{\omega: f(\omega) \geqslant \frac{1}{2} + 1/n\}$ is an open F_σ, and D is a compact G_δ (by the proposition). The result follows.

Remark. One of the reasons for the popularity of Baire sets in a locally compact space is that they form a base for the topology of Ω. Indeed, if $O \subset \Omega$ is open, since $\{\omega\} \subset \Omega$ is compact, by the corollary there is an open F_σ-Baire set V such that $\{\omega\} \subset V \subset O$, so that open Baire sets form a subbase for the topology. Each such V is even a countable union of compact G_δ-sets. (See Exercise 6.)

In the general Hausdorff topological case of Ω, if $\mathscr{B}$ is its Borel σ-algebra and $\mu: \mathscr{B} \to \overline{\mathbb{R}}^+$ is a measure, then by Definition 2.3.9, it is a Radon measure provided (i) μ is finite on compact sets, (ii) for any $A \in \mathscr{B}$, $\mu(A) = \inf\{\mu(B): B \supset A,\ B \text{ open}\}$ (or μ is *outer regular*), and (iii) μ is *inner regular* for compact sets, i.e., $\mu(B) = \sup\{\mu(C): C \subset B,\ C \text{ compact}\}$ for each $B \in \mathscr{B}$. The condition that μ is to be Radon is considerably restrictive and makes μ well behaved even though no other hypothesis is imposed on Ω. This is further illuminated by the following result, due to R. Godement, which extends the result of Exercise 8.3.3 to more general spaces.

Theorem 3. *If μ is a Radon measure on the Borel σ-algebra of Ω, a Hausdorff space, then μ is strictly localizable. In fact there exists a family $\{K_i: i \in I\}$ of compact sets such that $\mu(\Omega - \bigcup_{i\in I} K_i) = 0$, and for each Borel set B of finite measure, $\{i: \mu(B \cap K_i) > 0\}$ is at most countable.*

Proof. The argument is somewhat similiar to that for (the converse part of) Theorem 8.2.4. By definition $\mu \neq 0$, since $\mu(A) > 0$ for open sets $A \neq \varnothing$. Consider the class $\mathscr{P}$ of compact disjoint collections $\{A_i: i \in J\}$ of subsets of Ω. We can assume that $\mu(A_i) > 0$, and indeed even that each neighborhood of ω $(\in A_i)$ has a relatively compact open subset of positive measure, by the regularity hypotheses of μ. Clearly $\mathscr{P}$ is nonempty (there is a compact subset of Ω, of positive μ-measure—compare with the finite subset property in the earlier case). One orders $\mathscr{P}$ by inclusion, as before, and by Zorn's lemma we can find a maximal element of $\mathscr{P}$, denoted $\{K_i: i \in I\}$. It is claimed that (a) for each open set V with $\mu(V) < \infty$, $J_V = \{i: \mu(K_i \cap V) > 0\}$ is at most countable, (b) $\Omega_0 = \bigcup_{i\in I} K_i$ is μ-measurable, and (c) $N = \Omega - \Omega_0$ is μ-null. This will imply all the assertions of the theorem.

Let $\omega \in \Omega$ be arbitrary, and $V = V_\omega$ be a neighborhood of ω, of finite μ-measure. As in the proof of Theorem 8.2.4, if $\alpha \subset J_V$ is a finite subset, then $\sum_{i\in\alpha} \mu(K_i \cap V) \leqslant \mu(V)$, and this being true for all finite subsets $\alpha \subset J_V$, one has

$$\sum_{i\in J_V} \mu(K_i \cap V) \leqslant \mu(V) < \infty.$$

It follows that $\mu(K_i \cap V) > 0$ for at most a countable number of i. Also $\mu(K_i \cap V) = 0 \Rightarrow K_i \cap V = \varnothing$, since by choice each K_i is contained in the support of μ. Thus $\Omega_0 \cap V = \bigcup_{i\in I}(K_i \cap V)$ is μ-measurable, and V being an arbitrary neighborhood of finite measure of Ω, we conclude that Ω_0 is μ-measurable (by Lemma 2.2.13). This gives both (a) and (b). If $N = \Omega - \Omega_0$ is not μ-null, then it contains an open neighborhood of some $\omega \in N$, and hence by the inner regularity has a compact subset of positive measure. But then we can add this set to the collection $\{K_i: i \in I\}$ and enlarge it. Since this contradicts the maximality of that collection, $\mu(N) = 0$ must hold. Hence μ is strictly localizable, completing the proof.

This result shows how the regularity properties of a measure automatically imply several other hypotheses of the abstract case. Moreover, the compact subsets of Ω play distinct and key roles in the general (topological) study of measures. This aspect motivates the primary study of measures on (locally) compact spaces, since then one can stipulate reasonable and natural conditions on the behavior of measure functions for their study on arbitrary topological spaces. Consequently, we devote the next sections for an analysis of measures on locally compact spaces leading up to a comprehensive representation theorem of certain functionals, and obtain the Riesz-Markov theorem from it. On the way, we also include some important extension theorems for set functions on lattices in a general setting in which the local compactness hypothesis does not play a primary part.

EXERCISES

1. Let Ω and S be topological spaces, and $f: \Omega \to S$ be a continuous function. If $A \subset \Omega$ is a set and $B = f^{-1}(A)$, show that B is an F_σ (respectively a G_δ) set whenever A has the same property. Deduce that $\mathscr{B} = f^{-1}(\mathscr{A})$ is a (restrictedly) Baire or Borel σ-algebra of Ω if $\mathscr{A}$ has the corresponding property in S and f is a continuous open onto mapping. [Here f open means that $f(U) \subset S$ is open for each open $U \subset \Omega$.]
2. Show that a countable union (intersection) and a finite intersection (union) of F_σ (G_δ) sets is an F_σ (G_δ). Verify that a restricted Baire σ-algebra of a σ-compact space Ω is the smallest σ-algebra containing the monotone class formed by all compact G_δ-sets of Ω. (Note that $A \subset \Omega$ is a G_δ-set iff A^c is an F_σ.)
3. Let $(\Omega, \bar{d})$ be a metric space with $\bar{d}: \Omega \times \Omega \to \mathbb{R}^+$ as its metric. If $f: \Omega \to \mathbb{R}$ is a function, verify that the set of continuity points of f (i.e., $\{\omega: f \text{ is continuous at } \omega\}$) is a G_δ-set. Deduce that each complete subset of Ω is a G_δ and hence that the *restricted* Baire and Borel σ-algebras of a complete metric space coincide. [If C is a complete subset of Ω, then $C = \bigcap_{n \geqslant 1} \bigcup_{x \in C} B(x, 1/n)$, where $B(x, \alpha)$ is the open ball at x of radius α.]
4. Let Ω be a topological space and (S, d) a metric space. Then according to H. Hankel, a point $\omega_0 \in \Omega$ is a discontinuity of $f: \Omega \to S$ iff $F(\omega_0) > 0$, where

$$F(\omega) = \inf\{\sup\{d(f(u), f(v)): u, v \in U\}: U \in \mathscr{U}(\omega)\},$$

the class $\mathscr{U}(\omega)$ being a neighborhood base at ω of Ω. If $\Delta(f)$ is the set of discontinuity points of f, show that it is a Borel set, in fact an F_σ-set. [If $\Delta_n(f) = \{\omega: f(\omega) \geqslant 1/n\}$, then $\Delta_n(f)^c$ is open and $\Delta(f) = \bigcup_{n \geqslant 1}\Delta_n(f)$. In general $\Delta(f)$ need not be a Baire set, and the Baire σ-algebra can be strictly smaller than the Borel σ-algebra. However, the two coincide if $S = \mathbb{R}^n$ and Ω is a separable metric space, for instance.]

5. Show that the open F_σ-set V in Corollary 2 is a Baire set.

6. Generalize the preceding as follows. Let Ω be locally compact, and $V \subset \Omega$ be an open F_σ such that $V \subset \bigcup_{n \geqslant 1} K_n$, where K_n is compact. Show that $V = \bigcup_{n \geqslant 1} D_n$, where D_n is a compact G_δ-set, and deduce that V is a Baire set.

7. Let $\mathscr{C}$ be the class of all compact (or compact G_δ) sets of Ω, so that it is a *lattice* (i.e. closed under finite unions and intersections and with $\varnothing \in \mathscr{C}$). If $\mathscr{R}$ is the ring generated by $\mathscr{C}$, show that it is characterized as the class of all finite disjoint unions of sets of the form $\bigcup_{i=1}^n (A_i - B_i)$, $A_i \supset B_i$, $\{A_i, B_i\} \subset \mathscr{C}$, and $A_i - B_i$ disjoint. [If $\mathscr{R}_0$ is the latter ring, and $A, B \in \mathscr{C}$, then $A - B \in \mathscr{R}_0$. Now verify that $E, F \in \mathscr{R} \Rightarrow E \cap F$ and $E - F$ are in $\mathscr{R}_0$, so that $\mathscr{R} \subset \mathscr{R}_0$.] Deduce that for locally compact Ω, $\sigma(\mathscr{R})$ is the restricted Borel (or Baire) σ-ring.

8. Recall that a function $f: \Omega \to \mathbb{R}$ is lower [upper] semicontinuous, or l.s.c. [u.s.c.], if $\{\omega: f(\omega) > a\}$ $[\{\omega: f(\omega) < a\}]$ is open for each $a \in \mathbb{R}$, where Ω is a Hausdorff space. For any function f, one defines lower and upper envelopes for the finite l.s.c. and u.s.c. functions in the usual manner:

$$f^* = \inf\{h: h \geqslant f,\ h \text{ is u.s.c.}\},$$

$$f_* = \sup\{g: g \leqslant f,\ g \text{ is l.s.c.}\}.$$

Verify that f_* (f^*) is l.s.c. (u.s.c.) and that $\Delta(f) = \{\omega: f^*(\omega) > f_*(\omega)\}$, the set of discontinuity points of f, is a Borel set and in fact it is an F_σ.

9. Let Ω be a locally compact space, and $A \subset \Omega$ be a compact Baire set. Show that A must be a G_δ-set as well. [This is not easy. By a consequence of Urysohn's lemma, A is G_δ iff $A = f^{-1}(\{0\})$, a zero set of a continuous $f: \Omega \to [0,1]$. Since $A \in \mathscr{B}_0$, there exists a sequence $\{A_n: n \geqslant 1\}$ of compact G_δ-sets such that $A \in \sigma(A_n,\ n \geqslant 1)$, where $A_n = f_n^{-1}(\{0\})$. If $d(\omega, \omega') = \sum_{n \geqslant 1} |f_n(\omega) - f_n(\omega')| 2^{-n}$, then $d(\cdot, \cdot)$ is a semimetric on Ω. If $\tilde{\Omega} = \Omega/\sim$, where $\omega \sim \omega'$ when $d(\omega, \omega') = 0$, and $\pi: \Omega \to \tilde{\Omega}$, then $\tilde{d}(\tilde{\omega}, \tilde{\omega}') = d(\omega, \omega')$, for $\omega \in \tilde{\omega} \in \tilde{\Omega}$, $\omega' \in \tilde{\omega}' \in \tilde{\Omega}$, gives a metric on $\tilde{\Omega}$. Thus $A = \pi^{-1}(\tilde{A})$ for a closed (hence G_δ) set $\tilde{A} \subset \tilde{\Omega}$. Since π is continuous, A is a G_δ.]

10. Let Ω_1 and Ω_2 be a pair of locally compact spaces, and $\Omega = \Omega_1 \times \Omega_2$ be their Cartesian product with the product topology (so Ω is also locally compact). If $\mathscr{B}_{i0}$ is the restricted Baire σ-ring of Ω_i, and $\mathscr{B}_0$ that of Ω, show that $\mathscr{B}_0 = \mathscr{B}_{10} \otimes \mathscr{B}_{20}$. [Note that the neighborhood system of Ω is of the form $U_1 \times U_2$, where U_i is a neighborhood of $\omega_i \in \Omega_i$, $i = 1, 2$.

Thus if $A_i \subset \Omega_i$ is a compact G_δ-set, then $A_1 \times A_2 \subset \Omega$ is a compact G_δ by Proposition 1. On the other hand Corollary 2 implies that open Baire sets of the form $V_1 \times V_2$ constitute a subbase for the topology of Ω, where $V_i \subset \Omega_i$ is an open Baire set.]

11. Let $(\Omega_i, \mathscr{B}_{i0})$ be a restricted Baire-measurable space, as in the above problem, and $\mu_i: \mathscr{B}_{i0} \to \overline{\mathbb{R}}^+$, $i = 1, 2$, be a measure. If $\mu = \mu_1 \otimes \mu_2$ is the product measure, show that the finiteness of μ_i on compact G_δ-sets of $\mathscr{B}_{i0}$ implies the same property of μ on $\mathscr{B}_0$, the product σ-ring of $\mathscr{B}_{10}$ and $\mathscr{B}_{20}$, so that μ is a Baire measure on $\mathscr{B}_0$. [Use the result of the preceding problem.]

12. Let the space Ω be locally compact metric, and $C_c(\Omega)$ the space of real continuous functions with compact supports. Show that the restricted Borel σ-algebra $\tilde{\mathscr{B}}_0$ is the one generated by $C_c(\Omega)$.

13. Let $\mathscr{B}_0$ be the restricted Baire *σ-ring* of a locally compact Ω, and $\mu: \mathscr{B}_0 \to \overline{\mathbb{R}}^+$ be a Baire measure. Verify that each $A \in \mathscr{B}_0$ is σ-finite for μ.

14. Let K be a compact subset of a locally compact space Ω. Show that there is an open set $U \supset K$ and $0 \leqslant f \in C_c(\Omega)$, such that $f|_K = 1$ and $f|_{U^c} = 0$. If further K is a G_δ, then $K = \{\omega: f(\omega) = 1\}$ [so that $f(\omega) < 1$ for $\omega \in K^c$] can be demanded. [For the second part, use the first for $U_n \downarrow K$, so that there are f_n such that $f_n|_K = 1$, and $f_n|_{U_n^c} = 0$. Set $f = \sum_{n=1}^{\infty} f_n / 2^n$.]

9.2. REGULARITY OF MEASURES

It is already evident from the introduction and brief use of Radon measures that certain approximations with (open and) compact sets play important roles in our study. These are the so-called regularity properties, and there are several types of regularities, based on the topologies of underlying spaces. Here we make a fairly detailed study of measures on Baire and Borel σ-rings, often restricted to *locally compact (Hausdorff) spaces*. It will be shown in the next chapter that some general abstract measures can be studied using the present work on locally compact spaces.

If $(\Omega, \mathscr{A}, \mu)$ is a measure space with Ω as a (Hausdorff) topological space, let $\mathscr{A}$ be the *Baire (Borel)* σ-algebra. Then μ is called a *Baire* (respectively *Borel*) measure when $\mu(C) < \infty$ for each compact $C \in \mathscr{A}$. Thus for a Baire measure μ in a locally compact space Ω, μ is finite on the class of all compact G_δ-sets (Exercise 1.9). To gain some facility with operations on these measures and to motivate the introduction of regularities of interest here, we start with the following:

Proposition 1. *Let Ω be a σ-compact space. If μ_1, μ_2 are a pair of Baire measures on $(\Omega, \mathscr{B})$ ($\mathscr{B}$ a Baire σ-algebra) and $\mu_1(C) = \mu_2(C)$ for each compact C of $\mathscr{B}$, then $\mu_1 = \mu_2$ on $\mathscr{B}$.*

Proof. Since Ω is σ-compact, it is clear that $\mathscr{B}_0 = \mathscr{B}$, and hence it is generated by the class of all compact G_δ-sets. Let $\mathscr{S} = \{A: \mu_1(A) = \mu_2(A)\}$. By hypothesis each compact G_δ-set is in $\mathscr{S}$. Also, if C_1, C_2 are two compact G_δ-sets, then clearly $C_1 \cap C_2 \in \mathscr{S}$ and $C_1 \cup C_2 \in \mathscr{S}$. Further

$$\mu_1(C_1 - C_2) = \mu_1(C_1) - \mu_1(C_1 \cap C_2) = \mu_2(C_1) - \mu_2(C_1 \cap C_2)$$
$$= \mu_2(C_1 - C_2).$$

So $C_1 - C_2 \in \mathscr{S}$. Moreover, $\Omega \in \mathscr{S}$, since $\Omega = \bigcup_{k=1}^{\infty}\Omega_k$, Ω_k compact, and Ω_k may be assumed G_δ by Corollary 1.2. By the preceding computation, we have $\Omega = \bigcup_{k=1}^{\infty}\tilde{\Omega}_k$, $\tilde{\Omega}_1 = \Omega_1$, and $\tilde{\Omega}_k = \Omega_k - \bigcup_{i=1}^{k-1}\Omega_i$ for $k > 1$, so that

$$\mu_1(\Omega) = \sum_{k=1}^{\infty} \mu_1(\tilde{\Omega}_k) = \sum_{k=1}^{\infty} \mu_2(\tilde{\Omega}_k) = \mu_2(\Omega).$$

It follows that the complement of each G_δ set is also in $\mathscr{S}$. Hence the algebra (in fact the δ-ring) $\mathscr{A}_0$ generated by the compact G_δ-sets is in $\mathscr{S}$, on which both μ_1, μ_2 are σ-finite.

Let $\mathscr{M}_{\mu_i}$ be the class of all μ_i-measurable sets, $i = 1, 2$. Then $\sigma(\mathscr{A}_0) \subset \mathscr{M}_{\mu_1} \cap \mathscr{M}_{\mu_2}$, by Theorem 2.2.10. But evidently $\mathscr{B} = \mathscr{B}_0 = \sigma(\mathscr{A}_0)$, and by the Hahn extension (Theorem 2.3.1) $\mu_1 = \mu_2$ on $\sigma(\mathscr{A}_0)$, since they agree on $\mathscr{A}_0$ $(\subset \mathscr{S})$. This completes the proof.

Remark *i*. Since Ω is σ-compact, every set $A \subset \Omega$ is σ-bounded. But the above proof implies that if Ω is merely a locally compact space and $\mathscr{B}_0$ is the restricted Baire σ-algebra, then the hypothesis of μ_1, μ_2 agreeing on compact G_δ-sets implies that they agree on all σ-bounded Baire sets of Ω and hence on $\mathscr{B}_0$ also. The details are left to the reader. (See Exercise 1.) However, μ_1 and μ_2 need not agree on the general Baire σ-algebra $\mathscr{B}$ itself. This is one of the motivating factors for restricting to $\mathscr{B}_0$, in many cases.

Remark *ii*. Now $\mu: \mathscr{B} \to \mathbb{R}$ is a *signed Baire measure* if each compact $C \in \mathscr{B}$ has $|\mu|(C) < \infty$, where $|\mu|(\cdot)$ is the variation measure of μ. A similar definition is possible for *signed Borel measures*. The preceding result thus implies that if μ_1, μ_2 are a pair of signed Baire measures agreeing on all compact G_δ sets, then they agree on $\mathscr{B}_0$.

To eliminate some of the difficulties, we now introduce the regularity concepts appropriate to the present spaces.

Definition 2. Let (Ω, Σ, μ) be a measure space, where Ω is a Hausdorff space and Σ is a Baire (Borel) σ-algebra of Ω. Then a (*Baire*) *Borel* μ is:

(i) *outer regular at* $A \in \Sigma$, or A is outer regular relative to μ, if $\mu(A) = \inf\{\mu(O): O \supset A, O \text{ an open } (\sigma\text{-compact}) \text{ set in } \Sigma\}$, (1)

(ii) *inner regular* at $A \in \Sigma$, or A is inner regular relative to μ, if $\mu(A) = \sup\{\mu(C)\ C \subset A, C \text{ a compact set in } \Sigma\}$, and (2)

(iii) *regular* at A if it is both inner and outer regular on A. Thus μ is *regular* on Σ if it is regular on each member of Σ.

This definition reminds one of right and left continuity of a monotone function on a line (or a directed set), and the properties of such functions (and their extensions to those of bounded variation) motivate us to seek analogous results in the present context. To make this precise, let us state the corresponding concepts for signed measures as well.

Definition 3. Let μ be a signed (*Baire*) *Borel* measure on the (Baire) Borel σ-algebra Σ of a Hausdorff space Ω. It is termed *outer regular* on A for μ if for each $\epsilon > 0$, there exists an open (σ-compact) set O in Σ such that for every $B \in \Sigma$ satisfying $A \subset B \subset O$, we have

$$|\mu(A) - \mu(B)| < \epsilon. \tag{3}$$

It is *inner regular* on A for μ if for each $\epsilon > 0$, there is a compact set C in Σ such that for all $B \in \Sigma$ satisfying $C \subset B \subset A$, one has (3). Thus μ is *regular* on A if it is both inner and outer regular on A, and regular on Σ if it is so on each A of Σ.

It is possible to introduce a topology on Σ, preserving its directedness, in terms of which μ is continuous from above and below (or right and left), corresponding to the regularity concepts introduced above. In fact, on the power set $\mathscr{P}$ of Ω, let $\mathscr{I}(K, O) = \{A \in \mathscr{P}: K \subset A \subset O\}$, an "interval" associated with each compact set K and open (σ-compact) $O \supset K$. Set $\mathscr{I}(K, O) = \varnothing$ if $K \not\subset O$. Let $\mathscr{I}(K_1, O_1) \cap \mathscr{I}(K_2, O_2) = \mathscr{I}(K_1 \cup K_2, O_1 \cap O_2)$. Then the class of all such intervals forms a base for the topology of $\mathscr{P}$. If the restriction of this topology to Σ is considered, and denoted by $\mathscr{T}$, then $\mu: \Sigma \to \mathbb{R}$ is regular on Σ when it is continuous for $\mathscr{T}$ at each element of Σ. A proof of this and a few other properties are omitted here (see Exercise 3).

An equivalent form of (3) is as follows. Let $(\Omega, \mathscr{B}, \mu)$ be a signed (Baire) Borel measure space. Then μ is regular iff for each $A \in \mathscr{B}$ and $\epsilon > 0$, there exist compact K and open O (σ-compact open set O) such that $K \subset A \subset O$ and

$$|\mu(B)| < \epsilon, \qquad B \subset O - K, \qquad B \in \mathscr{B}. \tag{4}$$

Indeed, let (4) be true for each $\epsilon > 0$, suitable K, O, and all $B \subset O - K$. If $A_1 \in \mathscr{I}(K, O)$, then $A - A_1$ and $A_1 - A$ are contained in $O - K$, and they are $\mathscr{B}$-measurable if A_1 and A are. Assuming the latter, we have

$$\begin{aligned} |\mu(A) - \mu(A_1)| &= |\mu(A) - \mu((A_1 - A) \cup (A_1 \cap A))| \\ &= |\mu(A) - \mu(A_1 - A) - \mu(A_1 \cap A)| \end{aligned}$$

(since μ is additive and $A_1 \cap A$, $A_1 - A$ are disjoint)

$$=|\mu(A - A_1) - \mu(A_1 - A)|$$

[since $A = (A - A_1) \cup (A_1 \cap A)$ is a disjoint union]

$$\leqslant |\mu(A - A_1)| + |\mu(A_1 - A)| < 2\epsilon,$$

by (4), since $A - A_1$ and $A_1 - A$ qualify for a B there. This shows (4) $\Rightarrow$ (3).

On the other hand, suppose (3) holds for each $\epsilon > 0$ and $A, A_1 \in \mathscr{I}(K, O)$ that are in $\mathscr{B}$. Then for any $B \subset O - K$, $B \in \mathscr{B}$, note that $B = (A \cup B) - (A - B)$, and $A_1 = A \cup B$, $\tilde{B} = A - B \in \mathscr{I}(K, O)$, so that

$$|\mu(B)| = |\mu(A \cup B) - \mu(A - B)|$$

[by (3) with (A, B) there as the pairs (A_1, A) and $(A, \tilde{B})$ here]

$$\leqslant |\mu(A \cup B) - \mu(A)| + |\mu(A) - \mu(A - B)|$$

$$< \epsilon + \epsilon = 2\epsilon.$$

This is (4), and hence the equivalence of (3) and (4) follows.

The same argument applied to the one-sided conditions shows that μ is inner regular iff for each $A \in \mathscr{B}$, $\epsilon > 0$, there is a compact $K \subset A$ (depending on A and ϵ) such that $K \in \mathscr{B}$, and for each $B \subset A - K$, $B \in \mathscr{B}$, we have $|\mu(B)| < \epsilon$; and μ is outer regular iff there is an $O \supset A$ (depending on A and ϵ) such that $O \in \mathscr{B}$ and for each $B \subset O - A$, $B \in \mathscr{B}$, we have $|\mu(B)| < \epsilon$. The necessary computations are left to the reader.

The next result shows that we may restrict ourselves to (positive) measures in the regularity search, on a given Hausdorff space, and obtain the signed case from it.

Proposition 4. *Let $(\Omega, \mathscr{B}, \mu)$ be a Borel (or Baire) signed measure space. Then μ is regular on A $(\in \mathscr{B})$ iff its variation $|\mu|(\cdot)$ has the same property on A $(\in \mathscr{B})$. Again Ω is merely a Hausdorff space.*

Proof. If $|\mu|$ is regular, then the fact that $|\mu(A)| \leqslant |\mu|(A)$, $A \in \mathscr{B}$, implies, by (4), that μ is also regular. (We consider both measures simultaneously.)

For the converse, let μ be regular. Since a signed measure is bounded, we have, by definition of the variation $|\mu|(\cdot)$ of μ (Definition 5.1.1), that for each $\epsilon > 0$ there exist disjoint $\{A_i: 1 \leqslant i \leqslant n\} \subset \mathscr{B}(A)$ [$\mathscr{B}(A)$ is the trace of $\mathscr{B}$ on A] such that

$$|\mu|(A) < \sum_{i=1}^{n} |\mu(A_i)| + \epsilon. \tag{5}$$

But μ is regular on each A_i of $\mathscr{B}$. So there exist compact K_i and open O_i in $\mathscr{B}$ such that $A_i \in \mathscr{I}(K_i, O_i)$ and (3) holds with ϵ replaced by ϵ/n for each $A_i, \tilde{A}_i \in \mathscr{I}(K_i, O_i)$ which are in $\mathscr{B}$, $i = 1, \ldots, n$. We first show that $|\mu|$ is inner regular on $\mathscr{B}$ and then verify that it is also outer regular.

Let $K = \bigcup_{i=1}^n K_i$. Then $K \subset A$, $K \in \mathscr{B}$, compact. Let $\tilde{A} \in \mathscr{B}$ be such that $K \subset \tilde{A} \subset A$. If $\tilde{A}_i = A_i \cap \tilde{A}$, then $K_i \subset \tilde{A}_i \subset A_i$ and

$$0 \leqslant |\mu|(A) - |\mu|(\tilde{A}) \leqslant \sum_{i=1}^{n} |\mu(A_i)| + \epsilon - |\mu|\left(\bigcup_{i=1}^{n} \tilde{A}_i\right)$$

$$\leqslant \sum_{i=1}^{n} |\mu(A_i)| + \epsilon - \sum_{i=1}^{n} |\mu|(\tilde{A}_i)$$

(since the $\tilde{A}_i$ are disjoint and $\tilde{A}_i \in \mathscr{B}$)

$$\leqslant \sum_{i=1}^{n} \left(|\mu(A_i)| - |\mu(\tilde{A}_i)|\right) + \epsilon$$

$$\leqslant \sum_{i=1}^{n} |\mu(A_i) - \mu(\tilde{A}_i)| + \epsilon$$

$$< \epsilon + \epsilon = 2\epsilon.$$

By (3) this implies the inner regularity of $|\mu|$ on $\mathscr{B}$.

To demonstrate the outer regularity, let $A \in \mathscr{B}$. Since μ is also outer regular (on $\mathscr{B}$ and hence on A), there exists an open set $O \in \mathscr{B}$ such that $A \subset O$. By the inner regularity of $|\mu|$ on $\mathscr{B}$ (hence on $O - A$), shown above, for each $\epsilon > 0$ there is a compact $C \in \mathscr{B}(O - A)$ such that for each $B \in \mathscr{B}$ we have $B \subset (O - A) - C$, $|\mu|(B) < \epsilon$.

Consider $O_1 = O - C$, an open set containing A. Then

$$A = O - (O - A) \subset O - C = O_1 \quad \text{and} \quad O_1 - A = (O - A) - C.$$

Hence for each $B \in \mathscr{B}(O_1 - A)$, one has $B \subset (O - A) - C$, so that $|\mu|(B) < \epsilon$. Since $A \subset O_1 \in \mathscr{B}$, this shows that A is outer regular by (4). Thus $|\mu|$ is both inner and outer regular, and hence regular on $\mathscr{B}$, completing the proof.

The above result is improvable for Baire and Borel measures on a locally compact space Ω. An important property of Baire measures is given by the following result, which also presents an additional reason for studying these set functions.

Theorem 5. *Every Baire measure* $\mu: \mathscr{B}_0 \to \overline{\mathbb{R}}^+$ *is regular, where* $\mathscr{B}_0$ *is the restricted Baire σ-ring of a locally compact space* Ω. *In more precise terms, for* *every σ-bounded Baire set $A \subset \Omega$, we have*

$$\mu(A) = \inf\{\mu(O): O \supset A, O \text{ an open } \sigma\text{-compact Baire set}\}, \qquad (6)$$

$$\mu(A) = \sup\{\mu(C): C \subset A, C \text{ a compact } G_\delta \text{ set}\}. \qquad (7)$$

Proof. To demonstrate (6) which denotes outer regularity, note that the result is trivial if $\mu(A) = \infty$, since $A \subset \bigcup_{i=C_i}^{\infty}$, $C_i \in \mathscr{B}_0$, $C_i \subset \Omega$ compact, because A is σ-bounded. Hence, by Corollary 1.2 there are $C_i \subset V_i \subset D_i \subset U_i$ where V_i is an F_σ Baire set, D_i a compact G_δ, and U_i an open set. Consequently $A \subset \bigcup_{i=1}^{\infty} C_i \subset \bigcup_{i=1}^{\infty} V_i = V$ is a σ-compact open Baire set and $\infty = \mu(A) \leqslant \mu(V)$, so that (6) holds. Thus we consider the nontrivial case that $\mu(A) < \infty$, and prove the result in two steps: (a) if A is a compact G_δ and (b) if A is an element of the ring generated by compact G_δ-sets. The result will then follow from these two results.

(a): Let A be a compact G_δ-set. Thus there are open $U_n \downarrow A$, and by Corollary 1.2, for each n one can find an open σ-compact set V_n and a compact G_δ-set D_n such that $A \subset V_n \subset D_n \subset U_n$. Hence $A \subset \bigcap_{n \geqslant 1} V_n \subset \bigcap_{n \geqslant 1} U_n = A$, so that one may assume that $V_n \downarrow A$. Thus $A \subset V_n \subset V_1 \subset D_1 \subset U_1$, and μ is a Baire measure, implying $\mu(V_n) \leqslant \mu(D_1) < \infty$, $n \geqslant 1$. But then $\mu(A) = \lim_n \mu(V_n)$, so that for any $\epsilon > 0$, there is an n_0 such that $\mu(A) > \mu(V_{n_0}) - \epsilon$. Thus (6) is true in this case.

(b): Consider the ring $\mathscr{R}$ generated by the lattice of compact G_δ-sets $\mathscr{G}$ of Ω. Then $\mathscr{R}$ is characterized as the class of all finite disjoint unions of the form $\bigcup_{i=1}^{n}(C_i - D_i)$, $C_i \supset D_i$, with C_i, D_i in $\mathscr{G}$, $C_i - D_i$ disjoint (Exercise 1.7). Thus let $A = C_1 - D_1$ from $\mathscr{G}$ with $C_1 \supset D_1$. Then, by (a), given $\epsilon > 0$, there exists an open σ-compact Baire set $V_1 \supset C_1$ with

$$\mu(V_1) < \mu(C_1) + \epsilon. \qquad (*)$$

Let $O = V_1 - D_1$, which is an open F_σ set; then O is σ-compact ($\subset V_1$). But $O \in \mathscr{B}_0$ and

$$O = V_1 - D_1 \supset C_1 - D_1 = A, \qquad \mu(A) \leqslant \mu(O) \leqslant \mu(V_1) < \infty.$$

Also $O - A = (V_1 \cap D_1^c) \cap (C_1 \cap D_1^c)^c = V_1 \cap C_1^c$, since $D_1 \subset C_1$. So

$$0 \leqslant \mu(O) - \mu(A) = \mu(O - A) = \mu(V_1 - C_1) = \mu(V_1) - \mu(C_1) < \epsilon,$$

by $(*)$. This implies that (6) is true for the A. It is evident that if A_1, A_2 are two such disjoint sets for which (6) is true, then it is also true for $A_1 \cup A_2$. It follows that (6) holds for all elements of $\mathscr{R}$. Note that $\mathscr{R}$ cannot be an algebra unless Ω is σ-compact.

Consider the outer measure μ^* generated by $(\mu, \mathscr{R})$. Then by Theorem 2.2.10(iii), $\mathscr{R} \subset \mathscr{M}_{\mu^*}$, and μ^* is σ-additive on $\mathscr{M}_{\mu^*}$. Also we have $\mu^*|\mathscr{R} = \mu$.

But $A \in \mathscr{B}_0 \subset \mathscr{M}_{\mu^*}$, and since $\mu(A) < \infty$ in our case, $\mu^*(A) < \infty$ and hence there is $(A \subset)$ $B \in \mathscr{R}_{\sigma\delta} \subset \mathscr{B}_0 \subset \mathscr{M}_{\mu^*}$ such that $\mu^*(A) = \mu^*(B)$. Let $\epsilon > 0$. Now, μ^* being Carathéodory regular, we have $\mu^*(A) = \inf\{\mu(B): B \in \mathscr{R}_\sigma, B \supset A\}$. To get an open σ-compact $V \in \mathscr{B}_0$, $V \supset A$, choose $B_\epsilon \in \mathscr{R}_\sigma$ with $\mu^*(A) > \mu(B_\epsilon) + (\epsilon/2)$. Now $B_\epsilon = \bigcup_{i=1}^\infty E_i$ for some $E_i \in \mathscr{R}$, disjoint. So by the preceding paragraph there are σ-compact open Baire $V_i \supset E_i$ such that $\mu(V_i) < \mu(E_i) + (\epsilon/2^{i+1})$. Letting $V = \bigcup_{i=1}^\infty V_i$ $(\in \mathscr{B}_0)$, σ-compact, we have $V \supset B_\epsilon \supset A$ and $\mu(V - B_\epsilon) \leqslant \sum_{i=1}^\infty \mu(V_i - E_i) < \epsilon/2$. Hence (6) holds for $A \in \mathscr{B}_0$, σ-bounded, in all cases.

Next we show that (7) holds. It suffices to consider bounded Baire sets. For, if A is σ-bounded, then $A \subset \bigcup_{i=1}^\infty C_i$, C_i compact, so that $A = \bigcup_{i=1}^\infty (A \cap C_i)$ and each $A_n = \bigcup_{i=1}^n (A \cap C_i)$ is bounded. By Corollary 1.2 again, we may assume that each C_i is a compact G_δ, so that $A_n \in \mathscr{B}_0$, bounded, and $A_n \uparrow A$. Hence $\mu(A) = \lim_n \mu(A_n) = \sup\{\mu(B): B \in \mathscr{B}_0, \text{ bounded}\}$. It follows that (7) holds for all σ-bounded Baire sets if it holds for bounded Baire sets. We now establish the last result, which is slightly easier than (6).

Thus let A be a bounded Baire set. Then as noted above, there is a compact G_δ-set $C \supset A$ because of Corollary 1.2. Since $C - A$ is a bounded Baire set, (by the first part) for each $\epsilon > 0$ there is an open σ-compact Baire $V \supset C - A$ satisfying

$$\mu(C - A) > \mu(V) - \epsilon. \tag{+}$$

Let $D = C - V \subset C - (C - A) = C \cap A = B$, since $A \subset C$. But V is an F_σ, so that V^c is a G_δ and D is a compact G_δ-set. Since $C \subset V \cup D$, we have (note that $D \subset A$)

$$\mu(C) \leqslant \mu(V) + \mu(D) < \mu(C - A) + \epsilon + \mu(D)$$

$$= \mu(C) - \mu(A) + \mu(D) + \epsilon.$$

Hence $\mu(A) < \mu(D) + \epsilon$. Since D is a compact G_δ-set, this implies (7) for all bounded Baire sets. In the general case let $t > -\infty$ be a number such that $t + \epsilon < \mu(A) = \lim_n \mu(A_n)$. This is evidently possible. So there is n_0 such that $t + \epsilon < \mu(A_{n_0})$. By what was proved for (7), there is a compact G_δ-set $D \subset A_{n_0}$ satisfying $\mu(D) \geqslant \mu(A_{n_0}) - \epsilon > t$. Hence $\sup\{\mu(D): D \subset A, \text{ compact } G_\delta\} > t$. Since $t < \mu(A)$ is arbitrary, we get (7) by letting $t \uparrow \mu(A)$. This completes the proof.

Remark. There exist nonregular Baire measures on the Baire σ-algebra of certain locally compact spaces (Exercise 5), and thus the above result is just about the most general statement that can be made on regularity. In order to avoid the repetitions and certain complications, some authors work with Baire measures on σ-rings generated by compact G_δ-sets, calling them the Baire rings. However, the general case is also of interest and is the subject of intense research activity. Recent results on Baire measures on completely regular spaces, as well as extensions, have been surveyed by Wheeler (1983), and those on Borel measures by Gardner and Pfeffer (1984).

It is natural now to investigate the corresponding result for regular Borel measures μ. Fortunately the behavior of such regular measures is determined by the Baire σ-ring contained in $\tilde{\mathscr{B}}_0$ when Ω is locally compact. Then each Borel set of finite μ-measure differs from a Baire set by a Borel set of zero μ-measure. More precisely, we have the following basic result.

Theorem 6. *Let μ be a regular Baire measure on a locally compact space Ω, and define a set function ν as follows: if $O \subset \Omega$ is open,*

$$\mu_*(O) = \sup\{\mu(C): C \subset O,\, C \text{ compact } G_\delta\}, \tag{8}$$

and for each Borel set $A \subset \Omega$ (i.e., $A \in \tilde{\mathscr{B}}_0$)

$$\nu(A) = \inf\{\mu_*(O): O \supset A,\, O \text{ an open set}\}. \tag{9}$$

Then $\nu: \tilde{\mathscr{B}}_0 \to \overline{\mathbb{R}}$ is a measure which is inner regular on sets A in $\tilde{\mathscr{B}}_0$ of finite ν-measure and is an extension of μ. More explicitly,

$$\nu(A) = \sup\{\nu(C): C \subset A,\, C \text{ compact}\}, \qquad \nu(A) < \infty, \tag{10}$$

and to each $A \in \tilde{\mathscr{B}}_0$, $\nu(A) < \infty$, there corresponds a Baire set B such that $A \triangle B$ is a ν-null Borel set. Hence one has $\nu(A) = \nu(B) = \mu(B)$. The Borel extension ν verifying (8)–(10) *is unique, so that ν, μ_*, and μ all agree on the matching parts of their domains. Thus $\mu = \nu|\mathscr{B}_0$.*

Remark. The last property shows that ν agrees with the completion $\hat{\mu}$ of μ on the δ-ring of bounded Borel sets, and each compact set is $\hat{\mu}$-measurable.

***Proof* (*After Royden*).** The idea of proof is to show that μ_* of (8) is σ-subadditive on open sets, to construct a class of "μ_*-thin" sets $\mathscr{N}$, and to extend μ to $\tilde{\mu}$ on the σ-algebra generated by the Baire sets and $\mathscr{N}$. This is shown to determine our Borel σ-ring and $\tilde{\mu} = \nu$ on this class, in which $\mathscr{N}$ will be the ideal of ν-null sets. Even though the procedure parallels the proofs of Theorem 2.2.9 and Proposition 2.3.6, one has to rework the details using the regularity of Baire measures. We present the argument in steps, for convenience, starting with an auxiliary fact about locally compact spaces.

1. If $O_i \subset \Omega$ are open, and $K \subset \Omega$ is a compact G_δ such that $K \subset \bigcup_{i=1}^{\infty} O_i$, then there are compact G_δ-sets $K_i \subset O_i$, $i = 1, 2, \ldots, n$, satisfying $K = \bigcup_{i=1}^{n} K_i$. A similar statement holds if "G_δ" is omitted throughout the statement. For, let $\omega \in K$, and U_ω be a neighborhood of ω with compact closure, $\overline{U}_\omega \subset O_i$ for some i. By the compactness of K, a finite number $U_{\omega_1}, \ldots, U_{\omega_n}$ of such sets cover it. Let C_i be the union of those $\overline{U}_\omega$'s that are contained in O_i. Then C_i is

compact, and by Urysohn's lemma there is a continuous $g_i: \Omega \to [0,1]$ such that $g_i = 1$ on C_i and $= 0$ on O_i^c. If $g = \sum_{i=1}^n g_i$ and $f_i = g_i/g$, then f_i is continuous with support in O_i, and $\sum_{i=1}^n f_i = 1$ on K, $f_i \geqslant 0$. Since the C_i's need not be all G_δ's, set $\tilde{K}_i = \{\omega: f_i(\omega) \geqslant 1/n\}$, $K_i = K \cap \tilde{K}_i$. By Proposition 1.1., each K_i is a compact G_δ whenever K is, and $K = \bigcup_{i=1}^n K_i$. This implies both assertions.

2. The μ_* of (8) is σ-subadditive on the lattice of open sets of Ω. Indeed, if O, O_i are open subsets of Ω such that $O \subset \bigcup_{i=1}^\infty O_i$, let V_ω be a neighborhood of $\omega \in O$, with compact closure, $\overline{V}_\omega \subset O$. Then by Corollary 1.2, there is a compact G_δ-set D such that $\overline{V}_\omega \subset D \subset O$. Thus $D \subset \bigcup_{i=1}^n O_i$ for a finite number n, and by step 1 there are compact G_δ-sets D_i such that $D = \bigcup_{i=1}^n D_i$, $D_i \subset O_i$. Since, by definition, $\mu_*(A) = \mu(A)$ for any compact G_δ or an open Baire set A, the σ-additivity of μ on Baire sets implies its subadditivity, so that

$$\mu(D) \leqslant \sum_{i=1}^{n} \mu(D_i) \leqslant \sum_{i=1}^{\infty} \mu_*(O_i).$$

Varying D over the compact G_δ-subsets of O, and taking the sup, we obtain $\mu_*(O) \leqslant \sum_{i=1}^\infty \mu_*(O_i)$.

3. Let $\mathcal{N} = \{A \subset \Omega$: for each $\epsilon > 0$, there is an open $O_\epsilon \subset \Omega$ such that $A \subset O_\epsilon$ and $\mu_*(O_\epsilon) < \epsilon\}$. If A is in the domain of μ_*, then the monotonicity of μ_* implies that A must be μ_*-null. In the general case, it is clear that $A \in \mathcal{N}$, $B \subset A \Rightarrow B \in \mathcal{N}$ and that $A_i \in \mathcal{N} \Rightarrow \bigcup_{i=1}^\infty A_i \in \mathcal{N}$, since, taking $\epsilon_i = \epsilon/2^i$ and $\mu_*(O_i) < \epsilon_i$, $O_i \supset A_i$, one gets $\bigcup_i A_i \subset \bigcup_i O_i$ and $\mu_*(\bigcup_{i=1}^\infty O_i) < \epsilon$ by step 2. Thus $\mathcal{N}$ is an *hereditary* σ-ring, i.e., it is a σ-ring and $A \in \mathcal{N}$, $B \subset \Omega \Rightarrow A \cap B \in \mathcal{N}$. Also, if $A \in \mathcal{N}$ and A is a Baire set, then $\mu(A) = 0$ by the regularity of μ. In fact, by (7),

$$\mu(A) = \sup\{\mu(C): C \subset A, C \text{ compact } G_\delta\} \leqslant \mu_*(O_\epsilon) < \epsilon,$$

where $O_\epsilon \supset A$ by definition of $\mathcal{N}$. This means $\mu(C) < \epsilon$ for each compact G_δ-set $C \subset A$. Since $\epsilon > 0$ is arbitrary, $\mu(C) = 0$, so that $\mu(A) = 0$. Let $\tilde{\mathcal{B}}$ be the σ-algebra generated by the Baire σ-algebra $\mathcal{B}_0$ and $\mathcal{N}$. Then $\tilde{\mathcal{B}}$ can be characterized as the class

$$\tilde{\mathcal{B}} = \{A \subset \Omega: A = B \triangle N,\ B \in \mathcal{B}_0,\ N \in \mathcal{N}\}. \tag{11}$$

This follows from the obvious fact that $\tilde{\mathcal{B}}$ is a σ-algebra containing $\mathcal{B}_0$ and $\mathcal{N}$, and is contained in the σ-algebra generated by $\mathcal{B}_0$ and $\mathcal{N}$.

4. Let $\hat{\mu}: \tilde{\mathcal{B}} \to \overline{\mathbb{R}}^+$ be defined as $\hat{\mu}(A) = \mu(B)$, where $A = B \triangle N$. We claim that $\hat{\mu}$ is well defined and σ-additive on $\tilde{\mathcal{B}}$ (cf. Proposition 2.3.6). For, if $A = B_1 \triangle N_1 = B_2 \triangle N_2$ are two representations, then by the associativity of the operation $\triangle$, $(B_1 \triangle B_2) = N_1 \triangle N_2$, and since $N_1 \triangle N_2 \in \mathcal{N}$, it is a Baire set, and $\mu(B_1 \triangle B_2) = 0$, so that $\mu(B_1) = \mu(B_2)$. Thus $\hat{\mu}$ is well defined and

$\hat{\mu}|\mathscr{B}_0 = \mu$. To prove its σ-additivity, let $A_n \in \tilde{\mathscr{B}}$, disjoint, and $A = \bigcup_{i=1}^{\infty} A_i$. Let $A_i = B_i \triangle N_i$ be a representation. Then $B_i = A_i \triangle N_i$, so that $B_i \cap B_j = (A_i \cap A_j) \triangle (A_i \cap N_j) \triangle M$, where $M = N_i \cap (A_j \triangle N_j) \in \mathscr{N}$. Since $A_i \cap A_j = \varnothing$ $(i \neq j)$, this shows that $B_i \cap B_j \in \mathscr{N}$. But $B_i \cap B_j$ is a Baire set, and hence $\mu(B_i \cap B_j) = 0$ by the preceding step. Thus the B_i-sequence has a pairwise μ-null intersection. Consider a disjunctification $\{B_i' : i \geqslant 1\}$ of $\{B_i : i \geqslant 1\}$, i.e., $B_1' = B_1$ and $B_n' = B_n - \bigcup_{i=1}^{n-1} B_i$, $n > 1$. Then $B_1 \triangle B_1' = \varnothing$, and for $i > 1$,

$$M_i = B_i \triangle B_i' = B_i - B_i' = \bigcup_{j=1}^{i-1} (B_i \cap B_j) \in \mathscr{N},$$

since $B_i \cap B_j \in \mathscr{N}$ for $i \neq j$ and $\mathscr{N}$ is a ring. But then $\mu(B_i \triangle B_i') = 0$ and $\mu(B_i) = \mu(B_i')$. Consequently $A_i = B_i \triangle N_i = B_i' \triangle M_i'$, where $M_i' = M_i \triangle N_i \in \mathscr{N}$. Hence $A = \bigcup_{i=1}^{\infty} A_i = \bigcup_{i=1}^{\infty} (B_i' \triangle M_i') = \bigcup_{i=1}^{\infty} B_i' \triangle M$ for an $M \subset \bigcup_{i=1}^{\infty} M_i' \in \mathscr{N}$, the latter being closed under countable unions. Let $B = \bigcup_i B_i'$, a Baire set. Then $A = B \triangle M$, $M \in \mathscr{N}$, and

$$\hat{\mu}(A) = \mu(B) = \mu\left(\bigcup_{n=1}^{\infty} B_n'\right)$$

$$= \sum_{n=1}^{\infty} \mu(B_n') = \sum_{n=1}^{\infty} \mu(B_n) = \sum_{n=1}^{\infty} \hat{\mu}(A_n). \tag{12}$$

Here we have used the preceding simplifications and the fact that μ is σ-additive on the Baire σ-algebra $\mathscr{B}_0$ and vanishes on $\mathscr{B}_0 \cap \mathscr{N}$. Thus $\hat{\mu}$ is a measure on $\tilde{\mathscr{B}}$.

5. If $O \subset \Omega$ is an open set with $\mu_*(O) < \infty$, then $O \in \tilde{\mathscr{B}}$. For, by definition of μ_* in (8), there exists an increasing sequence of compact G_δ-sets $K_n \subset O$ such that $\mu_*(O) = \lim_n \mu(K_n)$. Also, for arbitrarily fixed K_n of the sequence, we have from (8)

$$\mu_*(O) = \sup\{\mu(K) : K \subset (O - K_n) \cup K_n,\ K \text{ compact } G_\delta\}$$

$$\geqslant \sup\{\mu(K_n \cup K) : K \subset O - K_n,\ K \text{ compact } G_\delta\}$$

(since the supremum is on a smaller collection)

$$= \sup\{\mu(K_n) + \mu(K) : K \subset O - K_n,\ K \text{ compact } G_\delta\},$$

(since μ is additive on $\mathscr{B}_0$)

$$= \mu(K_n) + \sup\{\mu(K) : K \subset O - K_n,\ K \text{ compact } G_\delta\}$$

$$= \mu(K_n) + \mu_*(O - K_n), \qquad \text{by (8)}.$$

Hence $0 \leqslant \mu_*(O - K_n) \leqslant \mu_*(O) - \mu(K_n) \to 0$ as $n \to \infty$. If now $B = \bigcup_{n=1}^{\infty} K_n$, then $B \subset O$; but $B \in \mathscr{B}_0$, and $O - B \subset O - K_n$ for each n, and by the monotonicity of μ_* one has for the open set $O - K_n$

$$\mu_*(O - B) \leqslant \mu_*(O - K_n) \leqslant \mu_*(O) - \mu(K_n). \tag{13}$$

By choosing n large, the right side can be made arbitrarily small, so that $O - B \in \mathscr{N}$. Consequently $O = B \cup (O - B) \in \tilde{\mathscr{B}}$, and $\hat{\mu}(O) = \mu(B)$ $(= \lim_n \mu(K_n) = \mu_*(O))$.

6. Let $\overline{\mathscr{B}} = \{A \subset \Omega: A \cap B \in \tilde{\mathscr{B}}, \text{ all } B \in \tilde{\mathscr{B}}, \hat{\mu}(B) < \infty\}$. Then $\overline{\mathscr{B}}$ is obviously a σ-algebra. We claim that $\overline{\mathscr{B}}$ contains all Borel sets of Ω. For, if $B \in \tilde{\mathscr{B}}$, then $B = B_0 \Delta N_0$ for some $B_0 \in \tilde{\mathscr{B}}$ and $N_0 \in \mathscr{N}$ with $\hat{\mu}(B) = \mu(B_0)$. Hence $\hat{\mu}(B) < \infty \Rightarrow \mu(B_0) < \infty$, and by regularity of μ, there exists an open Baire set $O_1 \supset B_0$ such that $\mu(O_1) < \mu(B_0) + \epsilon$. But by definition of $\mathscr{N}$, there is an open set $O_2 \supset N_0$ such that $\mu_*(O_2) < \epsilon$. If $O = O_1 \cup O_2$, then $\mu_*(O) < \infty$ and $B \subset O$, open, and by the preceding step $O \in \tilde{\mathscr{B}}$. So $A \in \overline{\mathscr{B}}$ iff $A \cap O \in \tilde{\mathscr{B}}$ for each O, open, $\mu_*(O) < \infty$. In particular, if $A \subset \Omega$ is any open set, then $A \in \overline{\mathscr{B}}$, so that every Borel set of Ω is in $\overline{\mathscr{B}}$. We extend $\hat{\mu}$ to $\tilde{\mu}: \overline{\mathscr{B}} \to \overline{\mathbb{R}}^+$ by defining $\tilde{\mu}(A) = \hat{\mu}(A)$ for $A \in \tilde{\mathscr{B}}$, $= +\infty$ for $A \in \overline{\mathscr{B}} - \tilde{\mathscr{B}}$. Then $\tilde{\mu}$ is a measure on $\overline{\mathscr{B}}$ and is an extension of $\hat{\mu}$. Note that $\tilde{\mu}(N) = 0$ for each $N \in \mathscr{N}$.

7. The restriction of $\tilde{\mu}$ to the Borel σ-algebra of Ω is a Borel measure and agrees with ν of (9). For, if $K \subset \Omega$ is a compact set, it is in $\overline{\mathscr{B}}$, since it is Borel, and by Corollary 1.2, there is a compact G_δ-set D such that $K \subset D$ and so $\tilde{\mu}(K) \leqslant \tilde{\mu}(D) = \hat{\mu}(D) = \mu(D) < \infty$, since μ is a Borel measure. So $\tilde{\mu}$ is finite on compact sets and hence is a Borel measure. Also, if $O \subset \Omega$ is any open set, then $\tilde{\mu}(O) = \mu_*(O)$ by definition, and if A is a Borel set in $\overline{\mathscr{B}} - \tilde{\mathscr{B}}$ and $O \supset A$, then $\tilde{\mu}(O) \geqslant \tilde{\mu}(A) = \hat{\mu}(A) = +\infty$. Hence (9) holds on $\overline{\mathscr{B}} - \tilde{\mathscr{B}}$. If $A \in \tilde{\mathscr{B}}$, then $A = B \Delta N$ for some $B \in \tilde{\mathscr{B}}_0$, $N \in \mathscr{N}$. By the (outer) regularity of μ, there is an open Baire set $O_1 \supset B$, and by definition of $\mathscr{N}$ another open set $O_2 \supset N$, such that $\mu(O_1) < \mu(B) + \epsilon/2$ and $\mu_*(O_2) < \epsilon/2$. Hence $O = O_1 \cup O_2 \supset A$ and $\hat{\mu}(O) \leqslant \tilde{\mu}(A) + \epsilon$, so that $\tilde{\mu}$ is outer regular, i.e., (9) holds. Consequently $\tilde{\mu}|(\text{Borel sets}) = \nu$ in all cases.

8. $\tilde{\mu}$ is also inner regular on the Borel sets, i.e., (10) holds. For, let $A \subset \Omega$ be a Borel set with $\tilde{\mu}(A) < \infty$. Then $A \in \tilde{\mathscr{B}}$, and by the last part of the preceding paragraph (using the outer regularity of μ) one can find an open set $O \supset A$ such that $\mu_*(O) < \infty$ and $\tilde{\mu}(O) \leqslant \tilde{\mu}(A) + \epsilon$, so that $\tilde{\mu}(O - A) < \epsilon$. But by (8) there is a compact G_δ-set $C \subset O$ such that $\tilde{\mu}(O - C) < \epsilon/2$, since $\tilde{\mu}$ is a measure on $\tilde{\mathscr{B}}$. By (9) there exists an open set $U \supset O - A$ such that $\tilde{\mu}(U) < \epsilon/2$. If $K = C - U$, then C is compact (not necessarily a G_δ), and

$$K \subset C - (O - A) = C \cap (O^c \cup A) = C \cap A = A,$$

since $C \subset O$. Finally

$$\tilde{\mu}(K) = \tilde{\mu}(C - U) = \tilde{\mu}(C) - \tilde{\mu}(C \cap U)$$

$$> \tilde{\mu}(C) - \frac{\epsilon}{2}, \quad \text{since} \quad \tilde{\mu}(U) < \frac{\epsilon}{2},$$

$$= \tilde{\mu}(O) - \tilde{\mu}(O - C) - \frac{\epsilon}{2} > \tilde{\mu}(O) - \epsilon$$

$$= \tilde{\mu}(A) + \tilde{\mu}(O - A) - \epsilon > \tilde{\mu}(A) - \epsilon.$$

This proves (10), and hence $\tilde{\mu}|_{\mathscr{B}} = \hat{\mu} = \nu$, and ν is inner regular. Since the last statement has already been verified at the end of step 5, and since by (8) and (9) the values of a Borel measure are completely specified on all open (hence Borel) sets when once they are known on compact G_δ-sets, the uniqueness follows. Thus the theorem is established.

Remark. This result shows that a regular Borel measure on a locally compact space is the same as the Radon measure, as defined in Chapter 2 (Definition 2.3.9). For this reason some authors use these terms interchangeably. An extension of Theorem 6 is given in Theorem 11 below.

A Baire measure μ on a locally compact space Ω is said to be *completion regular* if for each bounded Borel set $B \subset \Omega$, there exist Baire sets A_1 and A_2 such that $A_1 \subset B \subset A_2$ and $\mu(A_2 - A_1) = 0$. A Borel measure is *completion regular* on Ω if its restriction to the Baire class has the same property, as just defined.

The preceding theorem implies the following (cf. Proposition 2.2.16).

Proposition 7. *Let μ be a Baire (or a regular Borel) measure on a locally compact space Ω. Then it is completion regular iff every Borel set A is $\bar{\mu}$-measurable and $\mu(A) = 0 \Rightarrow \mu(B) = 0$ for some Baire set $B \supset A$, where $\bar{\mu}$ is the completion of μ.*

Proof

1. First let μ be a Baire measure. If it is completion regular, then for each bounded Borel set $B \subset \Omega$, there are Baire sets $A_1 \subset B \subset A_2$ with $\mu(A_2 - A_1) = 0$. Expressing B as $B = A_1 \cup (B - A_1)$, we get $A_1 \in \mathscr{B}_0$ and $B - A_1 \subset A_2 - A_1$, so that $B - A_1 \in \mathscr{N}$ in the notation of step 3 of the last proof, since $A_2 - A_1$ is a Baire set of measure zero. So by definition of the completion $\bar{\mu}$ of μ, B is $\bar{\mu}$-measurable.

Conversely, let B be a bounded Borel set. Then it is $\bar{\mu}$ measurable, so that $B = A \Delta N$, where $A \in \mathscr{B}_0$ and $N \in \mathscr{N}$ by the last theorem. This may be expressed as $B = A \cup (N - A) = A \cup N_1$, where $N_1 \subset N_0$, with N_0 as a Baire null set. Letting $A_1 = A$, $A_2 = A \cup N_0$, we get $A_1 \subset B \subset A_2$ and $\mu(A_2 - A_1) = 0$, so that μ is completion regular.

2. Next let μ be a regular Borel measure on Ω and μ_1 be its Baire restriction. If μ is completion regular and $\mu(A) = 0$ for a Borel set A, then by step 1 above, $A_1 \subset A \subset A_2$ for a pair of Baire sets A_1, A_2, with $\mu_1(A_2 - A_1) = 0$. Hence also $\mu(A_2 - A_1) = 0$ and $\mu(A_1) = \mu(A_2) = \mu_1(A) = 0$.

In the opposite direction, consider an arbitrary compact set $K \subset \Omega$. Then by regularity of μ, there are open sets $O_n \supset K$ such that $\mu(K) = \inf_n \mu(O_n)$, by (9). But by Corollary 1.2 there exists $K \subset D_n \subset O_n$ with D_n compact G_δ such that $K \subset \bigcap_{n=1}^{\infty} D_n = D \in \mathscr{B}_0$, and $\mu(K) \leqslant \mu(D) \leqslant \inf_n \mu(O_n) = \mu(K)$. Thus $\mu(D - K) = 0$, and D is compact G_δ. By the current hypothesis we can find a Baire set $B \supset D - K$ satisfying $\mu_1(B) = 0$. It follows that $K = (K \cap B) \cup (D \cap B^c)$ (draw a picture), and $D \cap B^c \in \mathscr{B}_0$ and $K \cap B \in \mathscr{N}$. Hence K is $\hat{\mu}_1$-regular, so that μ is completion regular, since compact sets determine, by (10), all (bounded) Borel sets. Thus both the assertions are demonstrated.

Theorems 5 and 6, as well as the preceding result, show how topology (especially compact sets of Ω) plays a key role in obtaining Baire (and their extensions to Borel) measures. But we have not presented a simple method to generate these regular measures which possess desirable properties, because there is no such method. However, the proof of Theorem 6 is constructive, and an analysis shows that we only used the following properties of the Baire measure μ: (i) μ is finite on compact G_δ-sets, (ii) μ is monotone and finitely additive on the lattice of such sets, and (iii) it is also subadditive there. Any nonnegative set function μ on the lattice of compact subsets of Ω with these properties can be taken to define μ_* and with it to generate a regular Borel measure. Such an auxiliary function is often called a *content*; thus given a content, one can produce a regular Borel measure on Ω. If such a content has the additional property that $\mu(K) = \inf\{\mu(\overline{O}): K \subset O, O \text{ open and } \overline{O} \text{ compact}\}$, then μ is termed a *regular content*. For instance, a regular Borel measure, restricted to the lattice of compact sets, produces a regular content. Conversely, a regular content can be used, as in Theorem 6, to produce a unique regular Borel measure. We shall not repeat the details here. (See Exercise 9, and also Corollary 10.)

As another key item of this section we present an extension theorem and its consequences for additive set functions on lattices of sets. This will complement the extension results of Section 2.3, and the preceding comments on contents, with a link to important applications. Such a theorem was first formulated by B. J. Pettis (1951), and the following result is essentially a version of his. (We have already used it in step 9 of the proof of Theorem 6.4.8.)

In Section 7.2 strong additivity and subadditivity of a nonnegative set function on a paving were defined. Thus one says that $\mu: \mathscr{A} \to \mathbb{R}$ is *strongly additive* (sometimes termed *modular*) if for any pair A, B of $\mathscr{A}$ we have $\mu(A \cup B) + \mu(A \cap B) = \mu(A) + \mu(B)$, where $\mathscr{A}$ is a lattice of sets, i.e., $\varnothing \in \mathscr{A}$ and $\mathscr{A}$ is closed under finite unions and intersections. The use of this concept was seen in the construction of capacity functions in Chapter 7. [It was called a paving in an abstract context.] The next result shows its importance in extending (abstract) set functions from lattices to rings generated by them.

Theorem 8 (Pettis). *Let $\mathscr{A}$ be a lattice of sets of Ω, and $\mu: \mathscr{A} \to \mathbb{R}$ be a mapping satisfying $\mu(\varnothing) = 0$. Let $\mathscr{R}$ be the ring generated by $\mathscr{A}$. Then μ can be uniquely extended to be an additive function $\hat{\mu}: \mathscr{R} \to \mathbb{R}$ iff μ is strongly additive (or modular) on $\mathscr{A}$, or equivalently iff for each A, B in $\mathscr{A}$ with $A \supset B$, $\mu(A) - \mu(B)$ depends only on the difference $A - B$.*

Remark. Actually Pettis proved the first part, and the second equivalence, suggested by the work of J. Kisyński (1968), was isolated and used by J. L. Kelley and T. P. Srinivasan (1971), with further streamlining by A. S. Sastry and K. P. R. Sastry (1977) among others. Thus the proof of this theorem and the next one are essentially based on this collective effort.

Proof. To begin with, let us consider the first equivalence.

If $\mu: \mathscr{A} \to \mathbb{R}$ has an additive extension $\hat{\mu}$ to $\mathscr{R} \to \mathbb{R}$, then it is seen that μ must be strongly additive. Indeed, if A, B in $\mathscr{A}$ are any two elements, then, writing $A = (A - A \cap B) \cup (A \cap B)$ and $A \cup B = (A - A \cap B) \cup B$ as disjoint unions of elements of $\mathscr{R}$, one has

$$\mu(A) = \hat{\mu}(A) = \hat{\mu}(A - A \cap B) + \mu(A \cap B),$$

$$\mu(A \cup B) = \hat{\mu}(A \cup B) = \hat{\mu}(A - A \cap B) + \mu(B), \qquad \text{since } \hat{\mu}|\mathscr{A} = \mu,$$

and subtraction gives $\mu(A \cap B) + \mu(A \cap B) = \mu(A) + \mu(B)$.

As for uniqueness, let $\hat{\nu}: \mathscr{R} \to \mathbb{R}$ be another additive function satisfying $\hat{\nu}|\mathscr{A} = \mu = \hat{\mu}|\mathscr{A}$. First observe that $\mathscr{D} = \{A - B: A, B \in \mathscr{A}\}$ is a semiring, so that $\mathscr{R}$ is the class of finite disjoint unions of $E_i \in \mathscr{D}$. In fact, if E_1, E_2 are in $\mathscr{D}$, then $E_i = A_i - B_i$, $i = 1, 2$, for some A_i, B_i in $\mathscr{A}$, and since the latter is a lattice, $E_1 \cap E_2 = A_1 \cap A_2 - (B_1 \cup B_2) \in \mathscr{D}$ and $E_1 - E_2 = [A_1 - (A_2 \cup B_1)] \cup (A_1 \cap B_2 - B_1)$, so that it is a union of two elements of $\mathscr{D}$. If $B_i \subset A_i$ (which may be assumed), this is a *disjoint* union. Since $\varnothing \in \mathscr{A} \subset \mathscr{D}$, Definition 1.3.1 implies $\mathscr{D}$ is a semiring, and $\mathscr{R}$ is the smallest ring containing $\mathscr{D}$ by Proposition 1.3.2. Moreover, $C \in \mathscr{R}$ can be written as $C = \bigcup_{i=1}^{n}(A_i - B_i)$, a disjoint union with $A_i \supset B_i$, $(A_i, B_i) \subset \mathscr{A}$. Hence for $\hat{\mu}$ and $\hat{\nu}$ we have

$$\hat{\mu}(C) = \sum_{i=1}^{n} \hat{\mu}(A_i - B_i) = \sum_{i=1}^{n} [\mu(A_i) - \mu(B_i)]$$

$$= \sum_{i=1}^{n} \hat{\nu}(A_i - B_i) = \hat{\nu}(C), \tag{14}$$

so that $\hat{\mu} = \hat{\nu}$ on $\mathscr{R}$ and uniqueness follows. It is the converse part that is not simple.

Thus let $\mu: \mathscr{A} \to \mathbb{R}$ be strongly additive, and $\mathscr{D}$ be the semiring of differences of elements of $\mathscr{A}$. We first define $\tilde{\mu}: \mathscr{D} \to \mathbb{R}$ and show it to be additive. Then Proposition 1.2.6, which did not use the positivity of μ or other properties of $\mathscr{D}$ than that it is a semiring, will complete the proof. Hence, for any $C \in \mathscr{D}$ where $C = A - B$, $A \supset B$, A, B in $\mathscr{A}$, define $\tilde{\mu}(C) = \mu(A) - \mu(B)$. We claim that $\tilde{\mu}$ is well defined and is additive, i.e., is the desired function.

Indeed, taking $B = \varnothing \in \mathscr{A}$, we get $\tilde{\mu}(A) = \mu(A)$, so $\tilde{\mu}|\mathscr{A} = \mu$. To see that $\tilde{\mu}$ is finitely additive on $\mathscr{R}$, let C_1, C_2 be disjoint in $\mathscr{D}$, and $C = C_1 \cup C_2 \in \mathscr{D}$. Let $C = A - B$ and $C_i = A_i - B_i$, where $A \supset B$, $A_i \supset B_i$, all the A's and B's being from $\mathscr{A}$. Then $A = B \cup C_1 \cup C_2$ is a disjoint union, and since B is disjoint from C_2 (and $A_1 - B_1$), we let $\tilde{A} = A \cup B_1 = B \cup A_1 \cup C_2$. Thus $A_0 = B \cup A_1 \in \mathscr{A}$ and $\tilde{A} = A_0 \cup C_2 \in \mathscr{A}$. By the strong additivity of μ on $\mathscr{A}$, and the fact that $\tilde{A} \cup B_2 = A_0 \cup A_2$, we have

$$\begin{aligned}\mu(\tilde{A} \cup B_2) &= \mu(\tilde{A}) + \mu(B_2) - \mu(\tilde{A} \cap B_2)\\ &= \mu(\tilde{A}) + \mu(B_2) - \mu(A_0 \cap B_2),\end{aligned}$$

since $\tilde{A} \cap B_2 = A_0 \cap B_2$. The left side

$$\begin{aligned}&= \mu(A_0 \cup A_2)\\ &= \mu(A_0) + \mu(A_2) - \mu(A_0 \cap A_2). \qquad (15)\end{aligned}$$

The equal right sides of (15) give, on transposition,

$$\begin{aligned}\mu(\tilde{A}) - \mu(A_0) &= \tilde{\mu}(C_2) - \tilde{\mu}(A_0 \cap (A_2 - B_2)), \qquad \text{by definition of } \tilde{\mu}\\ &= \tilde{\mu}(C_2) - \tilde{\mu}(A_0 \cap C_2). \qquad (16)\end{aligned}$$

Returning to the original relations, $A = B \cup C_1 \cup C_2$, we have, on noting $C_1 = A_1 - B_1$, that $A \cup B_1 = B \cup A_1 \cup C_2$, and hence, by (16) and the strong additivity of μ,

$$\begin{aligned}&\mu(A) + \mu(B_1) - \mu(A \cap B_1)\\ &\quad = \mu(A \cup B_1)\\ &\quad = \mu(A_1 \cup B \cup C_2)\\ &\quad = \mu(A_1 \cup B) + \tilde{\mu}(C_2) - \tilde{\mu}((A_1 \cup B) \cap C_2),\end{aligned}$$

$[$by (16), in which $A_0 = A_1 \cup B$, $\tilde{A} = A_0 \cup C_2]$,

$$= \mu(A_1) + \mu(B) - \mu(A_1 \cap B) + \tilde{\mu}(C_2) - \tilde{\mu}(A_1 \cap C_2),$$

by the strong additivity of μ and the fact that $B \cap C_2 = \varnothing$. This simplifies to give [use $\tilde{\mu}(C_1) = \mu(A_1) - \mu(B_1)$]

$$\begin{aligned}\tilde{\mu}(C) &= \mu(A) - \mu(B)\\ &= \tilde{\mu}(C_1) + \tilde{\mu}(C_2) + \left[\mu(A \cap B_1) - \mu(A_1 \cap B) - \tilde{\mu}(A_1 \cap C_2)\right]. \quad (17)\end{aligned}$$

However,

$$\begin{aligned}A \cap B_1 &= (B \cap B_1) \cup (C_1 \cap B_1) \cup (C_2 \cap B_1)\\ &= \left[B \cap (C_1 \cup B_1)\right] \cup \varnothing \cup \left[C_2 \cap (B_1 \cup C_1)\right]\end{aligned}$$

(since $C_1 \cap C_2 = C_1 \cap B = \varnothing$)

$$= (B \cap A_1) \cup (C_2 \cap A_1), \qquad \text{since} \quad A_1 = B_1 \cup C_1.$$

Identifying $\tilde{A}$ with $A \cap B_1$, A_0 with $B \cap A_1$, and "C_2" there with $C_2 \cap A_1$ in (16), and using the latter in (17), we get, on eliminating $\mu(A \cap B_1)$,

$$\begin{aligned}\tilde{\mu}(C) &= \tilde{\mu}(C_1) + \tilde{\mu}(C_2)\\ &\quad + \left[\mu(B \cap A_1) + \tilde{\mu}(C_2 \cap A_1) - 0 - \mu(A_1 \cap B) - \tilde{\mu}(A_1 \cap C_2)\right],\\ &\qquad\qquad \text{since} \quad B \cap C_2 = \varnothing\\ &= \tilde{\mu}(C_1) + \tilde{\mu}(C_2). \quad (18)\end{aligned}$$

It remains to show that the "two-additive" $\tilde{\mu}: \mathscr{D} \to \mathbb{R}$ [i.e., $\tilde{\mu}$ satisfying (18)] is finitely additive. For this we use induction with (18). Thus let $C_1, \ldots, C_n$ be disjoint elements of $\mathscr{D}$ such that $C = \bigcup_{i=1}^{n} C_i \in \mathscr{D}$. Assume that the result holds for $n = m$, and consider the case $n = m + 1$. Then $C - C_{m+1} = \bigcup_{i=1}^{m} C_i \in \mathscr{D}$, so that $C - C_{m+1} = D_1 \cup D_2$, $D_i \in \mathscr{D}$, and disjoint (as seen at the beginning about such $\mathscr{D}$). Hence $D_1 = (C - C_{m+1}) \cap D_1 = \bigcup_{i=1}^{m}(C_i \cap D_1)$, and by the induction hypothesis

$$\tilde{\mu}(D_1) = \sum_{i=1}^{m} \tilde{\mu}(C_i \cap D_1), \qquad \text{since} \quad C_i \cap D_1 \in \mathscr{D}, \quad \text{disjoint.}$$

Interchanging D_1 and D_2 here, one gets

$$\tilde{\mu}(D_2) = \sum_{i=1}^{m} \tilde{\mu}(C_i \cap D_2).$$

Thus one finds

$$\begin{aligned}\tilde{\mu}(D_1 \cup D_2) &= \tilde{\mu}(D_1) + \tilde{\mu}(D_2)\\ &= \sum_{i=1}^{m} [\tilde{\mu}(C_i \cap D_1) + \tilde{\mu}(C_i \cap D_2]\\ &= \sum_{i=1}^{m} \tilde{\mu}(C_i),\end{aligned} \tag{19}$$

since $\tilde{\mu}$ satisfies (18). Adding $\tilde{\mu}(C_{m+1})$ to both sides and noting that C_{m+1} and $D_1 \cup D_2$ are disjoint elements of $\mathscr{D}$ on which $\tilde{\mu}$ is two-additive, we get

$$\begin{aligned}\tilde{\mu}(C) &= \tilde{\mu}(C_{m+1} \cup (D_1 \cup D_2))\\ &= \tilde{\mu}(C_{m+1}) + \tilde{\mu}(D_1 \cup D_2)\\ &= \sum_{i=1}^{m+1} \tilde{\mu}(C_i), \qquad \text{by (19)}.\end{aligned}$$

Hence the result holds for $n = m + 1$, so $\tilde{\mu}$ is finitely additive. Now for each $A = \bigcup_{i=1}^{k} C_i \in \mathscr{R}$, with disjoint $C_i \in \mathscr{D}$, define $\hat{\mu}(A) = \sum_{i=1}^{k} \tilde{\mu}(C_i)$. Then, as noted before $\hat{\mu}\colon \mathscr{R} \to \mathbb{R}$ is a unique additive function, as asserted.

For the second equivalence, it is to be shown that A_i, B_i in $\mathscr{A}$, $A_i \supset B_i$, and $A_1 - B_1 = A_2 - B_2$ imply $\mu(A_1) - \mu(B_1) = \mu(A_2) - \mu(B_2)$ iff μ is strongly additive. Thus if the latter holds and A_i, B_i are as given, first note that $A_1 \cup A_2 = A_1 \cup B_2 = A_2 \cup B_1$ and $A_1 \cap B_2 = B_1 \cap A_2$ hold. For instance, using the indicator function calculus, one has

$$\begin{aligned}\chi_{A_1 \cup A_2} &= \chi_{A_1} + \chi_{A_2} - \chi_{A_1 \cap A_2}\\ &= \chi_{B_1} + \chi_{A_2} + \chi_{A_2 - B_2} - \chi_{A_1 \cap A_2}, \qquad \text{since } A_1 - B_1 = A_2 - B_2\\ &= \chi_{B_1 \cup A_2} + \chi_{B_1 \cap A_2} - \chi_{A_1 \cap A_2} + \chi_{A_2 - B_2}\\ &= \chi_{B_1 \cup A_2} - \chi_{(A_1 - B_1) \cap A_2} + \chi_{A_2 - B_2}, \qquad \text{since } A_1 \supset B_1\\ &= \chi_{B_1 \cup A_2} - \chi_{(A_2 - B_2) \cap A_2} + \chi_{A_2 - B_2} = \chi_{B_1 \cup A_2}.\end{aligned} \tag{20}$$

The other relations are similarly verified. Now if μ is strongly additive, then

$$\mu(A_1 \cup B_2) = \mu(A_1) + \mu(B_2) - \mu(A_1 \cap B_2),$$

$$\mu(A_2 \cup B_1) = \mu(A_2) + \mu(B_1) - \mu(A_2 \cap B_1).$$

Since the left sides are equal and the last terms on the right sides are also equal, we get $\mu(A_1) - \mu(B_1) = \mu(A_2) - \mu(B_2)$.

Conversely, suppose this last condition holds for pairs (A_i, B_i) of $\mathscr{A}$, with $A_i \supset B_i$, $i = 1, 2$. Let $A, B \in \mathscr{A}$ be arbitrary, and set $A_1 = A \cup B$, $A_2 = A$, $B_1 = B$, $B_2 = A \cap B$, all elements of $\mathscr{A}$. Then $A_i \supset B_i$, and

$$\mu(A_1) - \mu(B_1) = \mu(A \cup B) - \mu(B) = \mu(A_2) - \mu(B_2)$$
$$= \mu(A) - \mu(A \cap B).$$

Hence $\mu(A \cup B) + \mu(A \cap B) = \mu(A) + \mu(B)$, and the strong additivity follows. Consequently, the second part is reduced to the first one, for which the result was already established. This completes the proof of the theorem.

The strong additivity or its equivalent condition given in the above theorem may not be easy to verify in practice. Here we present a relatively simple sufficient condition, called *tightness*, which can be substituted in some applications. Thus if $\mathscr{A}$ is a lattice of sets of Ω, and $\mu: \mathscr{A} \to \mathbb{R}$ is a mapping, it is termed *tight* if $\mu(\varnothing) = 0$ and

$$\mu(A) - \mu(B) = \lim\{\mu(C): C \subset A - B,\ A \supset B,\ A, B, C \in \mathscr{A}\}, \quad (21)$$

where the net $\{\mu(C): C \in \mathscr{A}]$ is directed by inclusion, the existence of the limit being part of the definition. In case μ is nonnegative real valued, then lim can be weakened to sup in (21). Note that Ω need not be locally compact in the following topological results.

Theorem 9. *Let $\mathscr{A}$ be the set of compact subsets of a topological space Ω, and $\mu: \mathscr{A} \to \mathbb{R}$ be additive* [*hence* $\mu(\varnothing) = 0$]. *If μ is tight, then there exists a unique σ-additive $\hat{\mu}^{\pm}: \sigma(\mathscr{A}) \to \bar{\mathbb{R}}^+$ such that $\mu^{\pm} = \hat{\mu}^{\pm}|\mathscr{A}$, $\hat{\mu}^{\pm}$ is Radon, and $\hat{\mu} = \hat{\mu}^+ - \hat{\mu}^-$ extends μ to the σ-ring of $\mathscr{A}$.*

Proof. By hypothesis, μ is tight. So by (21), for any $A \supset B$, $A, B \in \mathscr{A}$, the difference $\mu(A) - \mu(B)$ depends only on the difference $A - B$. Then by the preceding theorem, μ is strongly additive, so that it has a unique additive extension $\tilde{\mu}$ to the ring $\mathscr{R}$ generated by $\mathscr{A}$. Hence the variation $|\tilde{\mu}|(\cdot)$ is additive on $\mathscr{R}$. Since $\tilde{\mu}(A - B) = \mu(A) - \mu(B)$ in the definition of $\tilde{\mu}$ on the semiring $\mathscr{D}$ of differences of sets of $\mathscr{A}$, we assert that $\tilde{\mu}$ is also tight. Indeed, we have to show [cf. also Definition 3, since tightness is the same as inner regularity for additive set functions on (semi-)rings] that for each $\epsilon > 0$, $A \in \mathscr{R}$, there is a compact set $C \subset A$ (i.e., $C \in \mathscr{A}$), such that for all $B \in \mathscr{R}$, $C \subset B \subset A$ we have $|\tilde{\mu}(A) - \tilde{\mu}(B)| < \epsilon$.

For this, observe that $A = \bigcup_{i=1}^{n} E_i$, $E_i \in \mathscr{D}$, disjoint, and by the tightness of μ, there is $C_i \in \mathscr{A}$, $C_i \subset E_i$, such that

$$|\tilde{\mu}(E_i) - \mu(H_i)| < \frac{\epsilon}{4n}, \quad C_i \subset H_i \subset E_i, \quad H_i \in \mathscr{A}, \qquad 1 \leqslant i \leqslant n. \quad (22)$$

Take $C = \bigcup_{i=1}^n C_i \in \mathscr{A}$; this C is asserted to be the desired compact set. For a proof, let $C \subset \tilde{C} \subset A$, $\tilde{C} \in \mathscr{A}$. Then $\mathscr{A} \subset \mathscr{D} \Rightarrow \tilde{C} \cap E_i \in \mathscr{D}$, and hence by (21) there is a set $\tilde{H}_i \in \mathscr{A}$, $\tilde{H}_i \subset \tilde{C} \cap E_i$, such that for all $H_i' \in \mathscr{A}$, $\tilde{H}_i \subset H_i' \subset \tilde{C} \cap E_i$,

$$|\tilde{\mu}(\tilde{C} \cap E_i) - \mu(H_i')| < \frac{\epsilon}{4n}, \qquad 1 \leqslant i \leqslant n. \tag{23}$$

Consequently,

$$\begin{aligned} &|\tilde{\mu}(A) - \mu(\tilde{C})| \\ &\quad \leqslant \sum_{i=1}^{n} |(\tilde{\mu}(E_i) - \tilde{\mu}(\tilde{C} \cap E_i)| \\ &\quad \leqslant \sum_{i=1}^{n} |\tilde{\mu}(E_i) - \mu(C_i \cap E_i)| + \sum_{i=1}^{n} |\tilde{\mu}(\tilde{C}_i \cap E_i) - \mu(C \cap E_i)| \\ &\quad < \frac{\epsilon}{4n} \cdot n + \frac{\epsilon}{4n} \cdot n = \frac{\epsilon}{2}, \end{aligned} \tag{24}$$

since $C_i \cap E_i \subset E_i$, and also $\subset \tilde{C} \cap E_i$, so it is taken for H and H' in (22) and (23). Now let $B \in \mathscr{R}$ be as given, i.e., $C \subset B \subset A$. Then using this B for A of (24), we find a $C_\epsilon \in \mathscr{A}$ such that $C \subset C_\epsilon \subset B \Rightarrow |\tilde{\mu}(B) - \mu(C_\epsilon)| < \epsilon/2$. Since $C_\epsilon \subset C \cup C_\epsilon \subset B \subset A$, we get

$$\begin{aligned} |\tilde{\mu}(A) - \tilde{\mu}(B)| &\leqslant |\tilde{\mu}(A) - \mu(C \cup C_\epsilon)| + |\mu(C \cup C_\epsilon) - \tilde{\mu}(B)| \\ &< \frac{\epsilon}{2} + \frac{\epsilon}{2} = \epsilon. \end{aligned}$$

This implies the tightness of $\tilde{\mu}$ of $\mathscr{R}$. But then by Proposition 4 above, $|\tilde{\mu}|$, $\tilde{\mu}^{\pm}$ also have the same property, and these are additive on $\mathscr{R} \to \mathbb{R}^+$.

We could apply the Henry extension theorem *if $\mathscr{R}$ were an algebra*, e.g., Ω were compact. But this is not the case, so one must proceed differently, involving an alternative argument.

By treating $\tilde{\mu}^+$ (and $\tilde{\mu}^-$ similarly), we extend it as an inner regular function. Writing $\tilde{\nu}$ for $\tilde{\mu}^+$, define $\nu_*(A) = \sup\{\tilde{\nu}(C): C \subset A, C \in \mathscr{A}\}$, $A \subset \Omega$. We claim that (i) ν_* is strongly superadditive in that $\nu_*(A \cup B) + \nu_*(A \cap B) \geqslant \nu_*(A) + \nu_*(B)$, (ii) the class $\mathscr{M}_{\nu_*} = \{A: \tilde{\nu}(C) = \nu_*(C \cap A) + \nu_*(C \cap A^c), C \in \mathscr{A}\}$ is a σ-algebra containing $\mathscr{A}$, and $\hat{\nu} = \nu_*|\mathscr{M}_{\nu_*}$ is an inner regular measure relative to the compact sets, and it extends $\tilde{\nu}$ or $\tilde{\mu}^+$. This proves the result.

Clearly $\nu_*|\mathscr{A} = \tilde{\nu}$, and ν_* is monotone. Regarding strong superadditivity, we may assume that $\nu_*(A_i) < \infty$, $i = 1, 2$. Then for each $\epsilon > 0$, there are

$C_\epsilon^i \subset A^i$, $C_\epsilon^i \in \mathscr{A}$, such that $\nu_*(A^i) - (\epsilon/2) < \tilde{\nu}(C_\epsilon^i)$, $i = 1, 2$. So $\nu_*(A_1) + \nu_*(A_2) - \epsilon < \tilde{\nu}(C_\epsilon^1 \cup C_\epsilon^2) + \tilde{\nu}(C_\epsilon^1 \cap C_\epsilon^2) \leqslant \nu_*(A_1 \cup A_2) + \nu_*(A_1 \cap A_2)$, since $\tilde{\nu}$ is strongly additive (by tightness). So (i) is true. Since $\mathscr{A} = \mathscr{A}_\delta$ now, we get $C_n \downarrow \varnothing$, $C_n \in \mathscr{A} \Rightarrow \tilde{\nu}(C_n) \downarrow 0$. With this we claim that ν_* has the same property, i.e., for all $A_n \downarrow A$, $A_n \subset \Omega$, $\nu_*(A_n) < \infty$ we have $\nu_*(A) = \lim_n \nu_*(A_n)$. Indeed, by monotonicity, it is enough to verify that $\nu_*(A) \geqslant \lim_n \nu_*(A_n)$. By definition, given $\epsilon > 0$, there is $C_n \in \mathscr{A}$ such that $C_n \subset A_n$ and $\nu_*(A_n) - (\epsilon/2^n) < \tilde{\nu}(C_n)$. Let $D_n = \bigcap_{k=1}^n C_k \in \mathscr{A}$, so that by the strong (super-)additivity of $(\nu_*)\tilde{\nu}$, since $D_{n-1} \cup C_n \subset A_{n-1}$, we get

$$
\begin{aligned}
\nu_*(A_n) - \tilde{\nu}(D_n) &= \nu_*(A_n) - \tilde{\nu}(D_{n-1} \cap C_n) \\
&= \nu_*(A_n) - \left[\tilde{\nu}(D_{n-1}) + \tilde{\nu}(C_n) - \tilde{\nu}(D_{n-1} \cup C_n)\right] \\
&\leqslant \left[\nu_*(A_n) - \tilde{\nu}(C_n)\right] + \left[\nu_*(A_{n-1}) - \tilde{\nu}(C_{n-1})\right], \\
&< \frac{\epsilon}{2^n} + \frac{\epsilon}{2^{n-1}} < \cdots = \sum_{k=1}^{n} \frac{\epsilon}{2^k} < \epsilon.
\end{aligned}
$$

Hence $\lim_n \nu_*(A_n) \leqslant \lim_n \tilde{\nu}(D_n) + \epsilon = \tilde{\nu}(\bigcap_n D_n) + \epsilon \leqslant \nu_*(A) + \epsilon$, as asserted. Then for each $A \in \mathscr{M}_{\nu_*}$, and any $B \subset \Omega$, we have

$$
\begin{aligned}
\nu_*(B) &= \sup\{\tilde{\nu}(C) \colon C \in \mathscr{A}\} = \sup\{\tilde{\nu}(C \cap A) + \tilde{\nu}(C \cap A^c) \colon C \in \mathscr{A}\} \\
&\leqslant \sup\{\nu_*(B \cap A) + \nu_*(B \cap A^c)\} = \nu_*(B \cap A) + \nu_*(B \cap A^c). \quad (24')
\end{aligned}
$$

By the strong superadditivity of ν_*, $\nu_*(B \cap A) + \nu_*(B \cap A^c) \leqslant \nu_*(B)$, since $\nu_*(\varnothing) = 0$. Hence ν_* is additive on $\mathscr{M}_{\nu_*}$ and is inner regular for $\mathscr{A}$. However, the Carathéodory process yields the class of sets on which a set function (ν_* here) is additive is an algebra (cf. Exercise 2.2.14). Since $A_n \downarrow A$, $\nu_*(A_n) \downarrow \nu_*(A)$, $\nu_*(A_n) < \infty$, we have $A \in \mathscr{M}_{\nu_*}$ by (24′). If $\tilde{\mathscr{A}}$ is the hereditary ring determined by $\mathscr{A}$ (i.e., $\tilde{\mathscr{A}}$ is closed under finite unions, and $A \in \tilde{\mathscr{A}}$, $B \subset \Omega \Rightarrow A \cap B \in \tilde{\mathscr{A}}$, so that $\tilde{\mathscr{A}} \supset \mathscr{A}$), then $\bar{\nu}_* = \nu_* | \tilde{\mathscr{A}}$ is additive and inner regular relative to $\tilde{\mathscr{A}}$. Since $\bar{\nu}_*$ on $\tilde{\mathscr{A}}$ is finite, it follows that the class of $\bar{\nu}_*$-measurable sets is $\mathscr{M}_{\bar{\nu}_*} = \{A \colon \tilde{\nu}(C) = \bar{\nu}_*(C \cap A) + \bar{\nu}_*(C \cap A^c), C \in \mathscr{A}\} \supset \mathscr{M}_{\nu_*}$. But $\bar{\nu}_*$ being the restriction of ν_* to $\tilde{\mathscr{A}}$, there is equality: $\mathscr{M}_{\bar{\nu}_*} = \mathscr{M}_{\nu_*}$. Now $\mathscr{M}_{\bar{\nu}_*}$ is a σ-algebra, and hence so is the class of ν_*-measurable sets.

To see that $\mathscr{A} \subset \mathscr{M}_{\nu_*}$, let $A \in \mathscr{A}$; then for any $B \in \mathscr{A}$, by the tightness of $\tilde{\nu}$ ($= \tilde{\mu}^+$ on $\mathscr{A}$), we get for $A \supset A \cap B \in \mathscr{A}$

$$
\begin{aligned}
\tilde{\nu}(A) - \tilde{\nu}(A \cap B) &= \sup\{\tilde{\nu}(D) \colon D \in \mathscr{A} = \mathscr{A}_\delta,\ D \subset A - B\} \\
&= \nu_*(A - B) \leqslant \nu_*(A) - \nu_*(A \cap B)
\end{aligned}
$$

(by the strong superadditivity of ν_*)

$$
= \tilde{\nu}(A) - \tilde{\nu}(A \cap B), \qquad \text{since} \quad \tilde{\nu} = \nu_* | \mathscr{A}.
$$

Hence there is equality throughout, and $A \in \mathcal{M}_{\nu_*}$. This proves our assertions, and $\hat{\nu} = \nu_* | \mathcal{M}_{\nu_*}$ is the inner regular extension of $\tilde{\nu}$.

Applying the same procedure to $\tilde{\mu}^-$, we get a pair of Radon (i.e. inner regular) measures on the Borel σ-algebra of $\mathscr{A}$, which are complete, and if we let $\hat{\tilde{\mu}}^{\pm}$ be these extensions, then $\hat{\tilde{\mu}}^{\pm} | \mathscr{A} = \mu^{\pm}$. Clearly $\hat{\tilde{\mu}} = \hat{\tilde{\mu}}^+ - \hat{\tilde{\mu}}^-$ is unambiguous on the σ-ring generated by $\mathscr{A}$. It is a signed Radon measure. This completes the proof of the theorem.

Observe that if $\mathscr{A}$ is as above, then $\lambda: \mathscr{A} \to \mathbb{R}^+$ is a *tight* (or *regular*) *content* on Ω if it is finitely additive, subadditive, and monotone, and (21) holds. Thus the above theorem implies:

Corollary 10. *A tight content in a Hausdorff space Ω has a unique tight extension onto the σ-ring generated by its compact sets (i.e., the Borel σ-ring).*

If Ω were locally compact, we could start with $\mathscr{A}$ as the lattice of all compact G_δ-sets, and then the same conclusion would obtain as a consequence of Theorem 6 above. However, in the non-locally-compact case, that result does not imply this one. Another useful extension of Theorem 6 for contents on certain smaller lattices than the above one (to correspond to the G_δ-class) will now be given, because it also has interesting applications. Our proof is somewhat different from the original one.

Theorem 11 (Kisyński). *Let $\mathscr{A}$ be a lattice of subsets of a Hausdorff space Ω with the property that* (i) *for each compact set C and open set U of Ω such that $C \subset U$, there is an $A \in \mathscr{A}$ satisfying $C \subset A \subset U$, and* (ii) *for any disjoint compact sets C_1, C_2 there are disjoint elements A_1, A_2 in $\mathscr{A}$ such that $C_i \subset A_i$, $i = 1, 2$. Let $\lambda: \mathscr{A} \to \mathbb{R}^+$ be a monotone finitely subadditive and additive function (i.e., a content) such that for each $\epsilon > 0$ and $A \in \mathscr{A}$ there exist compact C and open U, $C \subset A \subset U$, with the approximation property that $|\lambda(A) - \lambda(B)| < \epsilon$ for all $B \in \mathscr{A}$, $C \subset B \subset U$.*

Then there exists a unique tight measure μ on the Borel σ-ring of Ω such that for each $A \in \mathscr{A}$, there are Borel subsets F, G of Ω, satisfying $F \subset A \subset G$ and $\mu(F) = \lambda(A) = \mu(G)$, or equivalently, there is a Borel set H such that $A \Delta H$ is a μ-null set and the completion $\hat{\mu}$ of μ extends λ.

Proof. In view of Corollary 10, it suffices to produce a tight content $\tilde{\lambda}$ out of λ, using the present hypothesis. Since the crucial property is the inner approximation (or inner regularity), Theorem 6 essentially indicates a way to proceed. If $\mathscr{C}$ is the class of compact subsets of Ω (so $\mathscr{C}$ is a lattice), define $\lambda': \mathscr{C} \to \mathbb{R}^+$ as

$$\lambda'(C) = \inf\{\lambda(A): A \supset C,\ A \in \mathscr{A}\}, \qquad C \in \mathscr{C}. \tag{25}$$

Then the conditions on $\mathscr{A}$, especially the fact that disjoint compact sets have disjoint covers from $\mathscr{A}$, imply that λ' is a content on $\mathscr{C}$. Next define $\lambda'_*: \mathscr{U} \to \mathbb{R}^+$, where $\mathscr{U}$ is the lattice of open sets of Ω, as

$$\lambda'_*(U) = \sup\{\lambda'(C): C \subset U, C \in \mathscr{C}\}. \tag{26}$$

Then the additional condition on $\mathscr{A}$ about approximation shows that

$$\lambda'(C) = \inf\{\lambda'_*(U): U \in \mathscr{U}, U \supset C\}, \qquad C \in \mathscr{C}. \tag{27}$$

Indeed, by (25), for each $\epsilon > 0$ and $C \in \mathscr{C}$, there is a set $A \in \mathscr{A}$ such that $C \subset A$ and $\lambda'(C) > \lambda(A) - (\epsilon/3)$. For this A, again by the approximation condition, there are $C_1 \in \mathscr{C}$, $U_1 \in \mathscr{U}$, $C_1 \subset A \subset U_1$ such that for *all* $B \in \mathscr{A}$, if $C_1 \subset B \subset U_1$ then $|\lambda(A) - \lambda(B)| < (\epsilon/3)$, or $\lambda(A) > \lambda(B) - (\epsilon/3)$. But by (26), for this U_1 there is a $C_2 \in \mathscr{C}$, $C_2 \subset U_1$, such that $\lambda'_*(U_1) < \lambda'(C_2) + (\epsilon/3)$. Hence $C_2 \subset C \cup C_1 \cup C_2 \subset U_1$, and for all $C \cup C_1 \cup C_2 \subset B \subset U_1$, $B \in \mathscr{A}$, the preceding approximation between A and B holds, and since $\lambda'(C_2) \leqslant \lambda(B)$, we get $\lambda_*(U_1) < \lambda'(C_2) + (\epsilon/3) \leqslant \lambda(B) + (\epsilon/3) \leqslant \lambda(A) + (2\epsilon/3) \leqslant \lambda'(C) + \epsilon \leqslant \lambda_*(U_1) + \epsilon$. This implies (27).

We now assert that λ' is tight. For this it suffices to show that for each $C_1 \subset C_2$, $C_i \in \mathscr{C}$, we have $\lambda'(C_2) - \lambda'(C_1) = \sup\{\lambda'(C): C \subset C_2 - C_1, C \in \mathscr{C}\}$. But by the monotonicity of λ', $\lambda'(C_2) - \lambda'(C_1) \geqslant \lambda'(C)$ for all $C \subset C_2 - C_1$, $C \in \mathscr{C}$. On the other hand, by the preceding paragraph, for each $\epsilon > 0$ there exist $U_i \in \mathscr{U}$ such that $U_i \supset C_i$ and $\lambda'(C_i) + \epsilon > \lambda_*(U_i) \geqslant \lambda'(C)$ for all $C \in \mathscr{C}$ satisfying $C \subset U_i$ [cf. (27)]. Now $C_2 \subset U_1 \cup V_1$, where U_1, $V_1 = U_2 \cap C_1^c$ are open sets. As shown in the detailed hints of Exercise 7.2.11, C_2 can be expressed as $C_2 = \tilde{C} \cup \tilde{C}'$, $\tilde{C} \subset U_1$, $\tilde{C}' \subset V_1$, a union of compact sets. Hence

$$\lambda'(C_2) \leqslant \lambda'(\tilde{C}) + \lambda'(\tilde{C}')$$

(by the subadditivity of λ' on $\mathscr{C}$)

$$\leqslant \lambda_*(U_1) + \lambda'(\tilde{C}')$$

$$\leqslant \lambda'(C_1) + \lambda'(\tilde{C}') + \epsilon.$$

Hence $\lambda'(C_2) - \lambda'(C_1) \leqslant \lambda'(\tilde{C}') + \epsilon$, where $\tilde{C}' \subset C_2 \cap V_1 \subset C_2 \cap C_1^c = C_2 - C_1$. Since $\tilde{C}'$ is compact, this shows that λ' is tight. Hence by Corollary 10, there is a tight measure μ on the Borel σ-ring generated by $\mathscr{C}$ such that μ extends λ'. We next connect μ and λ as in the statement.

Let $A \in \mathscr{A}$ and $\epsilon > 0$ be given. Then there exist sets $C \in \mathscr{C}$, $U \in \mathscr{U}$, $C \subset A \subset U$ such that $|\lambda(A) - \lambda(B)| < \epsilon$ for all $C \subset B \subset U$, $B \in \mathscr{A}$, by the approximation condition. Also, in establishing (27) we have obtained $\lambda_*(U) = \sup\{\lambda(B): B \in \mathscr{A}, B \subset U\}$. Hence there is a $\tilde{B} \in \mathscr{A}$ such that $C \subset \tilde{B} \subset U$, $\lambda_*(U) < \lambda(\tilde{B}) + \epsilon$, and $|\lambda(A) - \lambda(\tilde{B})| < \epsilon$. Since $\mu|\mathscr{C} = \lambda'$ and also $\lambda'(C) \geqslant \lambda(A) - \epsilon$, we get

$$\lambda(A) - \epsilon < \lambda'(C) = \mu(C) \leqslant \lambda(A)$$
$$\leqslant \lambda_*(U) \leqslant \lambda(\tilde{B}) + \epsilon \leqslant \lambda(A) + 2\epsilon. \tag{28}$$

Since $C \subset A \subset U$, $\epsilon > 0$ is arbitrary, and μ is a measure, this gives the asserted result that the completion $\hat{\mu}$ of μ extends λ.

Finally, to show the uniqueness, since μ is a Radon measure, if ν is any other measure satisfying the conditions of the theorem, it is asserted that $\nu|\mathscr{C} = \mu|\mathscr{C}$, so that $\mu = \nu$. The constructed $\mu|\mathscr{C}$ is λ'. We show that $\nu|\mathscr{C} = \lambda'$ also. Since ν extends λ, for each $C \in \mathscr{C}$ and each $U \in \mathscr{U}$, $C \subset U$ there exists an $A \in \mathscr{A}$, $C \subset A \subset U$, satisfying $\nu(C) \leqslant \nu(A) = \lambda(A)$; and $\nu(A) \leqslant \nu(U)$, because ν is monotone. Taking the supremum over all $A \in \mathscr{A}$, $A \subset U$, and all $C \in \mathscr{C}$, $C \subset U$, gives $\lambda'_*(U) = \nu(U)$, since ν is also tight, so that it is inner regular. Hence by (27) one has, with $\lambda'_* = \nu$ on $\mathscr{U}$,

$$\lambda'(C) = \inf\{\nu(U): C \subset U \in \mathscr{U}\} \geqslant \nu(C), \qquad C \in \mathscr{C}. \tag{29}$$

Thus $\nu(C) < \infty$, and for each $C \in \mathscr{C}$, there is $U_1 \in \mathscr{U}$, $C \subset U_1$, with $\nu(U_1) < \infty$, and since ν is a measure,

$$\nu(C) = \nu(U_1) - \nu(U_1 - C)$$
$$= \nu(U_1) - \sup\{\nu(B): B \subset U_1 - C,\ B \in \mathscr{C}\}$$

(by tightness of ν),

$$= \inf\{\nu(U_1 - B): B \subset U_1 - C, C \in \mathscr{C}\}$$
$$\geqslant \inf\{\nu(U): C \subset U \in \mathscr{U}\} = \lambda'(C),$$

by (27) and the fact that $\nu = \lambda'_*$ on $\mathscr{U}$. Hence $\nu|\mathscr{C} = \lambda'$, so that μ and ν agree on $\mathscr{C}$. Since a tight measure is uniquely determined by its values on compact sets, the proof is finished.

In the above work, we have been asserting that the extension is given on the σ-ring generated by the compact sets. As seen in the proof of Theorem 6, the statements hold on the σ-algebras of Borel sets if we define μ_1 on $\mathscr{P}(\Omega)$, called the *canonical extension*, as

$$\mu_1(A) = \sup\{\mu(B): B \in \mathscr{A},\ B \subset A\}, \tag{30}$$

$\mathscr{A}$ being the lattice of compact sets.

EXERCISES

1. If μ_1 and μ_2 are two Baire measures on the restricted Baire σ-algebra $\mathscr{B}_0$ of a locally compact space agreeing on the lattice of compact G_δ-sets, verify that they also agree on $\mathscr{B}_0$ itself. Deduce that the same conclusion holds on the Baire σ-algebra $\mathscr{B}$ if Ω is σ-compact.

2. Let Ω be locally compact, $\mathscr{B}$ its Baire (or Borel) σ-ring, and $\mu: \mathscr{B} \to \mathbb{R}$ be a Baire (or Borel) signed measure. Show that μ is regular iff μ^+ and μ^- are regular, where μ^+ (μ^-) is the positive (negative) variation measure of μ.

3. Verify that the class of "intervals" $\{\mathscr{I}(K, O): K \subset O \subset \Omega,\ K$ compact and O open$\}$ of a topological space Ω forms a base for a topology in $\mathscr{P}(\Omega)$. If $\mathscr{T}$ denotes this topology, show that the lattices of all compact and of all open sets are "dense" in this topology, when Ω is locally compact. Here "dense" means this: a class $\mathscr{A}$ is dense in $\mathscr{P}(\Omega)$ for $\mathscr{T}$ if each interval $\mathscr{I}(K, O)$ contains an element of $\mathscr{A}$.

4. Let $(\Omega, \mathscr{B}_0)$ be a Baire-measurable space, and $\mu: \mathscr{B}_0 \to \overline{\mathbb{R}}^+$ be a Baire measure. Show that μ is inner (or outer) regular iff it is left (or right) continuous for the topology $\mathscr{T}$ on $\mathscr{B}_0$ (of the above exercise).

5. There exist nonregular Baire measures. We present a standard example consisting of the *Dieudonné measure space* built up by using elementary properties of ordinal numbers. Let S be the collection of all ordinal numbers less than or equal to the first uncountable ordinal α_0. The class of all intervals of S of the form $\{s: a < s \leqslant b\}$, together with $\{0\}$, form a base for the topology of S in terms of which S becomes a compact Hausdorff space. If for each Baire (Borel) set $A \subset S$, define $\mu(A) = 0$ or 1 according as A does or does not have an unbounded closed subset of $S - \{\alpha_0\}$. Show that μ is a Baire (Borel) measure which is not regular. Verify also that the Baire σ-algebra of S is strictly smaller than its Borel σ-algebra. (Note that $S - \{\alpha_0\}$ is locally compact but not σ-compact.)

6. Let μ be a regular Baire or Borel measure on a locally compact space Ω, i.e., on the corresponding σ-algebra. Show that μ is strictly localizable, i.e., has the direct sum decomposition property. (Compare with Theorem 1.3. and Exercise 1.13.)

7. Let μ be a Borel measure on $(\Omega, \tilde{\mathscr{B}})$ as in the above exercise, and let μ_1 be the restriction of μ to the Baire σ-algebra $\mathscr{B}_0 \subset \tilde{\mathscr{B}}$. If μ_1^* is the outer measure generated by the pair $(\mu_1, \mathscr{B}_0)$, show that μ_1^* is a regular Borel measure on $\tilde{\mathscr{B}}$, and it is identical with μ on $\tilde{\mathscr{B}}$ if μ is itself regular.

8. Let $\mu: \hat{\mathscr{B}} \to \mathbb{R}$ be a signed measure where $\hat{\mathscr{B}}$ is a Borel (or Baire) σ-algebra. Verify that the following are equivalent statements: (i) μ is regular, (ii) μ is inner regular, (iii) μ is outer regular. (See Definition 3 about these concepts. Here $\hat{\mathscr{B}}$ refers to the class on a locally compact space Ω.)

9. (a) Let μ be a content on the lattice $\mathscr{C}$ of compact sets of a locally compact space Ω, so that $\mu: \mathscr{C} \to \mathbb{R}^+$ is monotone increasing, additive, and subadditive. Define μ_* as in (8) of Theorem 6. With a procedure analogous to the proof of the theorem, show that $\mu_*: \mathscr{U} \to \overline{\mathbb{R}}^+$ is monotone, countably additive, and countably subadditive, $\mu_*(\varnothing) = 0$, and $\mu_*(U) < \infty$ for bounded $U \in \mathscr{U}$, where $\mathscr{U}$ is the lattice of open sets of Ω. [U bounded means $\bar{U}$ is compact as usual.]

(b) If μ^* is defined by (9) of Theorem 6 with this μ_*, show μ^* is an outer measure on $\mathscr{P}(\Omega)$ and is an extension of μ_*, and that the class $\mathscr{M}_{\mu^*}$ of μ^*-measurable sets contains the Borel σ-ring $\tilde{\mathscr{B}}_0$. Verify that $\hat{\mu} = \mu^* | \tilde{\mathscr{B}}_0$ is a regular Borel measure. (Proceed to establish various properties, as in Theorem 2.2.9, involving many details.)

10. Show that the following version of Theorem 6 is deducible from Henry's theorem (Theorem 2.3.10): Let $(\Omega, \mathscr{B}_0, \mu)$ be a Baire measure space with Ω a σ-compact space. Then there exists a unique regular Borel measure ν on the Borel σ-algebra $\tilde{\mathscr{B}}$ of Ω, extending μ. [Use the results of Exercise 8 above and Exercise 2.3.11. However, the proof of Theorem 6 does not rely on Zorn's lemma, and its statement thus has more information on the structure of $\tilde{\mathscr{B}}$.]

11. Let $\mathscr{U}$ be the lattice of open sets of a Hausdorff space Ω, and $\lambda: \mathscr{U} \to \mathbb{R}^+$ be a mapping such that (i) $U_n \subset U_{n+1}$ in $\mathscr{U}$, $\lambda(\bigcup_n U_n) = \lim_n \lambda(U_n)$, and (ii) $\lambda(U) - \lambda(\tilde{U}) = \inf\{\lambda(V): V \supset U - \tilde{U}, V \in \mathscr{U}\}$ for $U \supset \tilde{U}$ ($U, \tilde{U}$ in $\mathscr{U}$). Show that λ has a unique extension μ to be a measure on the ring $\mathscr{R}$ generated by $\mathscr{U}$, and that μ is outer regular for $\mathscr{U}$. [Verify that μ is σ-subadditive on $\mathscr{U}$.]

12. Extensions on lattices do not hold in general, as seen from the following simple example. Let $\mathscr{A}$ be the class of all compact subsets of the unit square $S = [0,1] \times [0,1]$. Let $A \subset \operatorname{int}(S)$ be a closed set with nonempty interior. If $\lambda: \mathscr{A} \to \mathbb{R}^+$ is defined as $\lambda(B) = 1$ for all $B \in \mathscr{A}$, $B \supset A$, and $= 0$ if $B \not\supset A$, then show, by checking that λ is not strongly additive, that λ is σ-additive on $\mathscr{A}$, but that there is no measure $\mu: \sigma(\mathscr{A}) \to \mathbb{R}^+$ such that $\mu | \mathscr{A} = \lambda$. [As usual, int = interior here.]

13. Let $\mathscr{A}$ be a lattice of subsets of a set Ω, and $\lambda: \mathscr{A} \to \mathbb{R}$ be a strongly additive function. If $\mathscr{R}$ is the ring generated by $\mathscr{A}$, and $B_n \supset B_{n+1}$, $B_n \in \mathscr{R}$, then show that for each $\epsilon > 0$ there exist $A_n \in \mathscr{A}$, $A_n \supset A_{n+1}$, such that $A_n \subset B_n$ and $|\tilde{\mu}(B_n) - \lambda(A_n)| < \epsilon$, where $\tilde{\mu}$ is the additive extension of μ to $\mathscr{R}$. In case λ is continuous at $\varnothing$ from above [i.e., $C_n \downarrow \varnothing \Rightarrow \lambda(C_n) \to 0$, $C_n \in \mathscr{A}$], then $\tilde{\mu}$ is σ-additive. [Use the method and result of proof of Theorem 9 until (24), and then complete the argument for the assertion here.]

14. Let Ω be a Hausdorff space and $\mathscr{A}$ be a lattice of subsets with the following "density" property: if ω_1, ω_2 are any two distinct points of Ω, then there exist disjoint sets $A_i \in \mathscr{A}$ with $\omega_i \in \operatorname{int}(A_i)$, $i = 1, 2$. Thus $\mathscr{A}$ separates points of Ω. Let $\lambda: \mathscr{A} \to \mathbb{R}^+$ be a content such that (i)

$\sup\{\lambda(A): A \in \mathscr{A}\} < \infty$, (ii) it is "regular", i.e., for each $A \in \mathscr{A}$, $\epsilon > 0$, there exist B, C in $\mathscr{A}$ with $\operatorname{clos}(B) \subset A \subset \operatorname{int}(C)$ and $\lambda(C) - \lambda(B) < \epsilon$. Then there exists a Radon measure μ on Ω whose completion extends λ (i.e., the conclusion of Theorem 11 holds) iff for each $\epsilon > 0$ there is a compact set $C_\epsilon \subset \Omega$ such that for all $A \subset \Omega - C_\epsilon$, $A \in \mathscr{A}$, we have $\lambda(A) < \epsilon$. [If the conclusion of Theorem 11 holds, then there is a Radon μ, bounded, such that $\hat{\mu}|\mathscr{A} = \lambda$, and for each $\epsilon > 0$, there is $C_\epsilon \subset \Omega$, compact, such that $\mu(\Omega - C_\epsilon) < \epsilon$, by the inner regularity. Conversely, if λ is as given, let $\mathscr{U} = \{\Omega - C: C \subset \Omega \text{ compact}\}$; then if the given condition holds, the regularity of λ implies that for each $A \in \mathscr{A}$ and $\epsilon > 0$ there exist B, C such that $\lambda(B) > \lambda(A) - \epsilon$, $\lambda(C) < \lambda(A) + \epsilon$, and $\operatorname{clos}(B) \subset A \subset \operatorname{int}(C)$. Let $C_0 = C_\epsilon \cap \operatorname{clos}(B)$, $U_0 = (\Omega - C_\epsilon) \cup \operatorname{int}(C)$. So $C_0 \subset A \subset U$, $U \in \mathscr{U}$, and for any $A_1 \in \mathscr{A}$, $C_0 \subset A_1 \subset U$, we get $B - A_1 \subset \Omega - C_\epsilon$ and $A_1 - C \subset U - \operatorname{int}(G) \subset \Omega - C_\epsilon$. Now deduce that $|\lambda(A) - \lambda(A_1)| < \epsilon$. This can be used for another proof of Theorem 6.4.8.]

15. Let Ω_i be locally compact, and μ_i be Radon measures on Ω_i, $i = 1, 2$. If $\mu = \mu_1 \otimes \mu_2$ on $\Omega = \Omega_1 \times \Omega_2$ has the finite subset property, and $f: \Omega \to \mathbb{R}^+$ is a lower semicontinuous function, show that $\omega_1 \mapsto \int_{\Omega_2} f(\omega_1, \omega_2)\mu_1(d\omega_2)$ and $\omega_2 \mapsto \int_{\Omega_2} f(\omega_1, \omega_2)\mu_2(d\omega_1)$ are Borel functions, μ is a Radon measure, and

$$\int_\Omega f\,d\mu = \int_{\Omega_1}\int_{\Omega_2} f\,d\mu_2\,d\mu_1 = \int_{\Omega_2}\int_{\Omega_1} f\,d\mu_1\,d\mu_2.$$

[Verify that $f(\omega) = \sup\{g(\omega): 0 \leqslant g \leqslant f,\ g \in C_c(\Omega)\}$, and apply Exercise 4.3.13, Exercise 6.2.7, and Theorem 6.2.2.]

9.3. LOCAL FUNCTIONALS AND THE RIESZ-MARKOV THEOREM

As applications of the preceding work, we present three important results in this section. First the classical Riesz-Markov theorem is given in two forms (for the positive and the general cases). Then a characterization of local functionals on $C_c(\Omega)$—and the method is applicable for certain other function spaces—is established. From this another proof of the Riesz-Markov theorem is obtained. However, the proof of the stated characterization is exceedingly long, even though we have curtailed some repetitive details. Perhaps most readers should skip it. The reason for its inclusion here is the hope that a simpler argument may be found in future.

Because of the importance of the Riesz-Markov theorem we present an independent demonstration for it, and then an alternative proof will be given. The following form is for positive linear functionals on $C_c(\Omega)$, the space of real continuous functions, on a locally compact Ω, with compact supports.

Theorem 1 (Riesz-Markov; first form). *Let $l: C_c(\Omega) \to \mathbb{R}$ be a positive linear functional, so that for each $f \geqslant 0$ one has $l(f) \geqslant 0$, l additive. Then there exists* *a unique regular Borel measure μ on the Borel σ-ring of Ω such that*

$$l(f) = \int_{\Omega} f(\omega)\mu(d\omega), \qquad f \in C_c(\Omega). \tag{1}$$

Proof. Motivated by the construction of regular Borel measures in Theorem 2.6, we define μ with $l(\cdot)$ and then deduce (1). To simplify our expressions, for each $f \in C_c(\Omega)$, $0 \leqslant f \leqslant 1$, and open set U such that $\operatorname{supp}(f) \subset U$, we write $f \prec U$. Define for each open set U

$$\mu_*(U) = \sup\{l(f) : f \in C_c(\Omega),\, f \prec U\}, \tag{2}$$

$$\mu^*(A) = \inf\{\mu_*(U) : U \supset A,\, U \text{ open}\}, \qquad A \subset \Omega.$$

It is clear that $\mu^*(U) = \mu_*(U)$ for each open set U and μ_*, μ^* are monotone, since by the positivity of l, $f \leqslant g \Rightarrow l(g) = l(f) + l(g - f) \geqslant l(f)$. We wish to show that

(i) μ^* is subadditive on open sets.
(ii) μ^* is additive on disjoint open sets and
(iii) μ^* has the approximation property for compact sets,

and then to apply Theorem 2.11. From these facts (1) will follow immediately.

Regarding (i), let U_1, U_2 be open and choose $f \prec U_1 \cup U_2$. Then by step 1 of the proof of Theorem 2.6, there exist $g_i \prec U_i$ satisfying $g_1 + g_2 = 1$ on the (compact) support of f. Hence $g_i f \prec U_i$, $i = 1, 2$. Since clearly $f = fg_1 + fg_2$, we get from the definition of μ_* in (2)

$$l(f) = l(fg_1) + l(fg_2) \leqslant \mu^*(U_1) + \mu^*(U_2).$$

Taking the supremum over all such f, one has

$$\mu_*(U_1 \cup U_2) = \mu^*(U_1 \cup U_2) \leqslant \mu^*(U_1) + \mu^*(U_2), \tag{3}$$

so that μ^* is finitely subadditive as asserted.

(ii): First observe that $\mu^*(K) < \infty$ for each compact set K. Indeed, writing $K \prec f$ for $f \in C_c(\Omega)$, $0 \leqslant f \leqslant 1$, and $f|_K = 1$, consider any such f. If $U_\alpha = \{\omega : f(\omega) > \alpha\}$, then U_α for $0 < \alpha < 1$ is open, nonempty, and moreover $K \subset U_\alpha$. Let $h \prec U_\alpha$. Then $\alpha h < f$, and so by monotonicity of μ^* we have

$$\mu^*(K) \leqslant \mu^*(U_\alpha) = \sup\{l(h) : h \prec U_\alpha\} \leqslant l(f/\alpha) < \infty,$$

by the linearity of l. Since $0 < \alpha < 1$ is arbitrary, one has $\mu^*(K) \leqslant l(f)$. To prove additivity of μ^* on open sets, we only need to check its superadditivity, because of (3). For this one has to show that

(a) $\mu^*(U) = \sup\{\mu^*(K): K \subset U, \text{compact}\}$, and that
(b) μ^* is additive on compact sets.

By definition of μ^*, for each $0 < \alpha < \mu^*(U)$ there exists an $f \prec U$ such that $\alpha < l(f)$, and if $V \supset \text{supp}(f)$, is any open set, then $\mu_*(V) = \mu^*(V) \geqslant l(f) > \alpha$. If $K = \text{supp}(f)$, then taking the infimum over such V, we get $\mu^*(K) \geqslant l(f) > \alpha$. Since $K \prec f \prec U$ (and such f exist for any $K \subset U$, by Urysohn's lemma), one gets, on taking the sup [since $\mu^*(K) > \alpha$ by letting $\alpha \uparrow \mu^*(V)$], that $\mu^*(V) = \sup\{\mu^*(K): K \subset U,\ K \text{ compact}\}$. Also $\mu^*(K) = \inf\{l(f): f \succ K\}$. This proves (a), with some additional information to be used later.

Let K_1, K_2 be disjoint compact sets. Then by Urysohn's lemma again, there is an $f \in C_c(\Omega)$ such that $f|_{K_1} = 1$, $f|_{K_2} = 0$. By the last statement of the preceding paragraph, for each $\epsilon > 0$, we can find an $f_0 \succ K_1 \cup K_2$ such that $l(f_0) - \epsilon < \mu^*(K_1 \cup K_2)$. But it is evident that $ff_0 \succ K_1$ and $(1-f)f_0 \succ K_2$. Hence

$$\begin{aligned}\mu^*(K_1) + \mu^*(K_2) &\leqslant l(ff_0) + l((1-f)f_0) \\ &= l(f_0) < \mu^*(K_1 \cup K_2) + \epsilon. \end{aligned} \tag{4}$$

But (3) trivially implies with (2) that

$$\mu^*(K_1 \cup K_2) \leqslant \mu^*(U_1 \cup U_2) \leqslant \mu^*(U_1) + \mu^*(U_2)$$

for all $U_i \supset K_i$, so that, taking the infimum over all such open U_i, we get

$$\mu^*(K_1 \cup K_2) \leqslant \mu^*(K_1) + \mu^*(K_2). \tag{5}$$

Since $\epsilon > 0$ is arbitrary, (4) and (5) imply our assertion (b).

If $U_1 \cup U_2 = U$ is a disjoint union of open sets, and if $\mu^*(U) = \infty$, then by (3), the additivity follows on such sets. Suppose that $\mu^*(U_i) < \infty$, $i = 1, 2$. Then by (a) above, for each $\epsilon > 0$, there exist compact sets $K_i \subset U_i$ such that $\mu^*(U_i) < \mu^*(K_i) + \epsilon/2$, $i = 1, 2$. Since U_i are disjoint, so are K_i and we get

$$\begin{aligned}\mu^*(U) &\geqslant \mu^*(K_1 \cup K_2) = \mu^*(K_1) + \mu^*(K_2) \\ &> \mu^*(U_1) + \mu^*(U_2) - \epsilon. \end{aligned} \tag{6}$$

Since $\epsilon > 0$ is arbitrary, (3) and (6) show that μ^* is additive on open sets, proving (ii).

(iii): The approximation property of μ^* is now immediate. In fact, for each $\epsilon > 0$, U open, $\mu^*(U) < \infty$, there is a compact set $K \subset U$ such that $\mu^*(U) - \mu^*(K) < \epsilon$, as noted and used above.

If now we consider $\mathscr{A}$ as the lattice of open sets of Ω, then it clearly satisfies all the conditions of Theorem 2.11, and its approximation for μ^* also holds. Then by that theorem there exists a unique Radon measure (or equivalently, tight measure) on the Borel σ-ring of Ω which agrees with μ^* on $\mathscr{A}$. We denote this by μ, and assert that (1) holds for this measure.

Since $f \in C_c(\Omega)$ can be written as $f = f^+ - f^-$ with $0 \leqslant f^{\pm} \in C_c(\Omega)$, and the integral is linear, it suffices to consider $f \geqslant 0$. But clearly f is μ-integrable. We now use Theorem 4.1.4 and approximate the integral by (Borel) step functions, from which the result will follow. Since f is bounded, for an $\epsilon > 0$ fixed, divide the range of f, $[0, k]$, into $0 = x_0 < x_1 < \cdots < x_n < x_{n+1} = k$, where $k = \sup\{f(\omega): \omega \in \Omega\}$, such that if $U_i = f^{-1}((x_i, x_{i+1}))$ and $C_i = f^{-1}([x_i, x_{i+1}])$, then we have

$$\sum_{i=0}^{n} x_{i+1}\mu(C_i) - \frac{\epsilon}{2} \leqslant \int_\Omega f\,d\mu \leqslant \sum_{i=0}^{n} x_i\mu(U_i) + \frac{\epsilon}{2}. \tag{7}$$

This is precisely the Lebesgue approximation in Theorem 4.1.4. Since $k = 0$ implies $f = 0$, in which case (1) is trivial, we take $k > 0$ and let $\delta = \epsilon/[2(n+1)k]$. Since f has compact support (and is continuous), C_i is compact, and U_i is open, one has $U_i \subset C_i$. But it was shown in step (ii) above that there exist compact sets $K_i \subset U_i$ such that $\mu(U_i) < \mu^*(K_i) + \delta$. Also there is a $g_i \succ C_i$ such that $l(g_i) \geqslant \mu^*(C_i)$ and $l(g_i) < \mu^*(C_i) + \delta$. this is a place where we intend to use the additional information of (ii)(a). By Urysohn's lemma, there is an $h_i \in C_c(\Omega)$, $K_i \prec h_i \prec U_i$, such that $\mu^*(K_i) \leqslant l(h_i)$ and $\mu(U_i) < \mu^*(K_i) + \delta \leqslant l(h_i) + \delta$. But these choices of g_i and h_i imply for the approximations

$$0 \leqslant \sum_{i=0}^{n} x_i h_i \leqslant f \leqslant \sum_{i=0}^{n} x_{i+1} g_i, \tag{8}$$

and they are in $C_c(\Omega)$. Hence, applying l to both sides, we get

$$\sum_{i=0}^{n} x_i l(h_i) \leqslant l(f) \leqslant \sum_{i=0}^{n} x_{i+1} l(g_i)$$

so that

$$\sum_{i=0}^{n} x_i\mu(U_i) - \frac{\epsilon}{2} \leqslant l(f) \leqslant \sum_{i=0}^{n} x_{i+1}\mu(C_i) + \frac{\epsilon}{2}. \tag{9}$$

Thus (7) and (9) imply

$$\int_\Omega f\,d\mu - \epsilon < l(f) < \int_\Omega f\,d\mu + \epsilon. \tag{10}$$

Since $\epsilon > 0$ is arbitrary, this establishes (1).

Finally for uniqueness of such μ, let ν be another regular Borel measure satisfying (1). Since such measures are determined by their values on compact sets of Ω, let K be a compact set. If $\epsilon > 0$ is given, choose an open set $U \supset K$ such that $\mu(U) < \mu(K) + \epsilon$ (as shown in Theorem 2.6). Then there is an $f \in C_c(\Omega)$, $K \prec f \prec U$, such that

$$\nu(K) \leqslant \int_\Omega f\,d\nu = \int_\Omega f\,d\mu, \qquad \text{by assumption}$$

$$\leqslant \mu(U) < \mu(K) + \epsilon.$$

Hence $\nu(K) \leqslant \mu(K)$. Similarly $\mu(K) \leqslant \nu(K)$, so that $\mu = \nu$. The proof of the theorem is complete.

We also can get another proof of this result from Theorem 7.3.4. There we stated the result for a σ-compact space Ω and μ as a Baire measure. Since by Theorem 2.6 each Baire measure on a locally compact space has a unique extension to a regular Borel measure, the result of Theorem 7.3.4 extends to give (1) above, because of the following fact and some easy computation using the uniqueness part (Exercise 2) for such Ω.

Lemma 2. *If μ is a Borel, and ν a Baire, measure on Ω such that μ is the extension of ν (as guaranteed by Theorem 2.6), then*

$$\int_\Omega f\,d\mu = \int_\Omega f\,d\nu, \qquad f \in C_c(\Omega). \tag{11}$$

Proof. We may consider $f \geqslant 0$ again, and evidently f is μ- and ν-integrable. By the structure theorem of measurable functions, there exist Baire (hence Borel) functions $0 \leqslant f_n \uparrow f$, and $\int_\Omega f_n\,d\mu = \int_\Omega f_n\,d\nu$, since μ is an extension of ν. Then (11) follows by the monotone convergence theorem.

The importance of (the proof of) Theorem 1 is that each positive linear functional on $C_c(\Omega)$ generates a unique regular Borel measure on Ω, and this fact can be proved directly without the Daniell-Stone theorem. For this reason, sometimes one defines a Radon measure on a locally compact space as a positive linear functional on $C_c(\Omega)$.

Next we extend Theorem 1 to all *bounded* linear functionals from the positive (not necessarily bounded) case. This is based on the Jordan type decomposition as follows [the norm $\|\cdot\|$ on $C_c(\Omega)$ is the uniform norm, as usual].

Lemma 3. *Let $l\colon C_c(\Omega) \to \mathbb{R}$ be a bounded linear functional. Then there exist positive bounded linear functionals l_i on $C_c(\Omega)$, $i = 1, 2$, such that $l = l_1 - l_2$.*

Proof. If $0 \leqslant f \in C_c(\Omega)$ and $\Omega_0 = \operatorname{supp}(f)$, so that Ω_0 is compact, define

$$L(f) = \sup\{l(g): 0 \leqslant g \leqslant f,\ g \in C_c(\Omega)\}\ (\leqslant \|l\|\,\|f\| < \infty). \tag{12}$$

Then, since $l(0) = 0$, we have that $L(f) \geqslant 0$, $L(0) = 0$, and $L(\cdot)$ is additive on the positive elements of $C_c(\Omega)$. To see the additivity of L, let $0 \leqslant f_1, f_2 \in C_c(\Omega)$ and $0 \leqslant g_i \leqslant f_i$, $g_i \in C_c(\Omega)$, so that $g_1 + g_2 \leqslant f_1 + f_2$ and

$$l(g_1) + l(g_2) = l(g_1 + g_2) \leqslant L(f_1 + f_2).$$

Taking the supremum over all $0 \leqslant g_i \leqslant f_i$, as above, we get $L(f_1) + L(f_2) \leqslant L(f_1 + f_2)$. For the opposite inequality, consider $0 \leqslant g \leqslant f_1 + f_2$, $g \in C_c(\Omega)$. If $g_1 = \min(g, f_1)$, $g_2 = \max((g - f_1), 0)$, then $0 \leqslant g_i \leqslant f_i$, $i = 1, 2$. [For $i = 1$, this is obvious, and for $i = 2$ it is true if $g < f_1$. If $g \geqslant f_1$, then $0 \leqslant g_2 = f - f_1 \leqslant (f_1 + f_2) - f_1 = f_2$.] Since $g_1 + g_2 = g$ and $g_i \in C_c(\Omega)$, we get

$$l(g) = l(g_1) + l(g_2) \leqslant L(f_1) + L(f_2).$$

Taking the supremum over all such g, we have $L(f_1 + f_2) \leqslant L(f_1) + L(f_2)$. This and the earlier inequality imply $L(f_1 + f_2) = L(f_1) + L(f_2)$, and $L(af) = aL(f)$ being evident, for $a \geqslant 0$, the additivity follows.

To see that L has an additive extension to $C_c(\Omega)$ to be a positive linear functional, let $g \in C_c(\Omega)$, and express it as $g = g^+ - g^-$ where $0 \leqslant g^{\pm} \in C_c(\Omega)$. Define $\tilde{L}(g) = L(g^+) - L(g^-)$. This is well defined, since if $g = g_1 - g_2$ for some other $0 \leqslant g_i \in C_c(\Omega)$, then $g^+ - g^- = g_1 - g_2$, so that $0 \leqslant g^+ + g_2 = g_1 + g^- \in C_c(\Omega)$ and $L(g^+) + L(g_2) = L(g^+ + g_2) = L(g_1 + g^-) = L(g_1) + L(g^-) \Rightarrow L(g^+) - L(g^-) = L(g_1) - L(g_2)$, and the value is independent of the decomposition of g. Thus for any $\tilde{f}, \tilde{g} \in C_c(\Omega)$

$$\begin{aligned}
\tilde{L}(\tilde{f} + \tilde{g}) &= \tilde{L}(\tilde{f}^+ - \tilde{f}^- + \tilde{g}^+ - \tilde{g}^-) \\
&= L(\tilde{f}^+ + \tilde{g}^+) - L(\tilde{f}^- + \tilde{g}^-) \\
&= L(\tilde{f}^+) - L(\tilde{f}^-) + L(\tilde{g}^+) - L(\tilde{g}^-) \\
&= \tilde{L}(\tilde{f}) + \tilde{L}(\tilde{g}).
\end{aligned}$$

Similarly $\tilde{L}(a\tilde{f}) = a\tilde{L}(\tilde{f})$, so $\tilde{L}: C_c(\Omega) \to \mathbb{R}$ is linear. Also, $|\tilde{L}(f)| = |L(f^+) - L(f^-)| = |l(f)| \leqslant \|l\|\,\|f\|$, so $\tilde{L}$ is bounded. If $0 \leqslant f \in C_c(\Omega)$, then clearly $\tilde{L}(f) = L(f) \geqslant 0$. Thus $\tilde{L}$ is a positive linear (bounded) functional. Set $\tilde{L} = l^+$.

Define $l^- = l^+ - l$. Then l^- is a bounded linear functional. If $0 \leqslant f \in C_c(\Omega)$, then $l^-(f) = l^+(f) - l(f) \geqslant 0$, by (12). Also,

$$\begin{aligned} l^-(f) &= \sup\{l(g): 0 \leqslant g \leqslant f,\, g \in C_c(\Omega)\} - l(f) \\ &= -\{l(f) - \sup(l(g): 0 \leqslant g \leqslant f)\} \\ &= -\inf\{l(f) - l(g): 0 \leqslant g \leqslant f\} \\ &= -\inf\{l(h): 0 \leqslant h \leqslant f\}. \end{aligned} \tag{13}$$

Clearly $l = l^+ - l^-$, and the lemma follows, with (13) as a bonus.

A linear functional $l: C_c(\Omega) \to \mathbb{R}$ is called *relatively bounded* if for each compact subset $\Omega_0 \subset \Omega$, there is a positive constant K_0 $[= K(\Omega_0)]$ such that $|l(f)| \leqslant K_0 \|f\|$ for all f with supports contained in Ω_0. Thus the above proof and the result are valid for all relatively bounded linear functionals. (See Exercise 3.)

It is now possible to present another form (an extension) of (1) as:

Theorem 4 (Riesz-Markov; general form). *Let $l: C_c(\Omega) \to \mathbb{R}$ be a bounded linear functional. Then there is a unique signed regular Borel measure μ on the Borel σ-ring of Ω such that*

$$l(f) = \int_\Omega f\, d\mu, \qquad f \in C_c(\Omega), \tag{14}$$

and

$$\|l\| \left[= \sup\{|l(f)|: \|f\| \leqslant 1,\, f \in C_c(\Omega)\}\right] = |\mu|(\Omega). \tag{15}$$

Proof. The representation (14) is not difficult, since $l = l_1 - l_2$ for a pair of positive bounded linear functionals $l_i: C_c(\Omega) \to \mathbb{R}$ by Lemma 3, and then by (1) there are positive regular Borel measures μ_i, $i = 1, 2$, such that

$$\begin{aligned} l(f) = l_1(f) - l_2(f) &= \int_\Omega f\, d\mu_1 - \int_\Omega f\, d\mu_2, \qquad f \in C_c(\Omega), \\ &= \int_\Omega f\, d\mu, \qquad \mu = \mu_1 - \mu_2. \end{aligned} \tag{16}$$

To see that μ is a signed Borel measure, we observe that the μ_i vanish outside a σ-compact set Ω_i, $i = 1, 2$. Indeed, consider μ_1. By regularity, for each compact set $C \subset \Omega$, we have $\mu_1(C) < \infty$ and there is an $f \in C_c(\Omega)$ such that

$f: \Omega \to [0,1]$ and $f|_C = 1$. So

$$\mu_1(C) \leqslant \int_\Omega f\, d\mu_1 = l_1(f) \leqslant \sup_{0 \leqslant f \leqslant 1} l_1(f) = \|l_1\| < \infty.$$

Thus if $\alpha = \sup\{\mu_1(C): C \subset \Omega,$ compact$\}$, then $\alpha \leqslant \|l_1\| < \infty$, and hence there exist compact C_n (we can assume $C_n \subset C_{n+1}$) such that $\lim_n \mu_1(C_n) = \mu_1(\cup_{n=1}^\infty C_n) = \alpha$. Now $\Omega_1 = \cup_{n=1}^\infty C_n$ is σ-compact, and $\mu_1(\Omega - \Omega_1) = 0$, for otherwise, by regularity of μ_1, there is a compact set $C_0 \subset \Omega - \Omega_1$ with $\mu_1(C_0) > 0$. But then $\alpha = \mu_1(\Omega_1) < \mu_1(\Omega_1 \cup C_0) = \sup\{\mu_1(C): C$ compact$\}$ $= \alpha$, giving a contradiction. This also shows that $\mu_1(\Omega) = \alpha \leqslant \|l_1\|$. Similarly $\mu_2(\Omega) \leqslant \|l_2\|$, and hence $\mu = \mu_1 - \mu_2$ is a signed regular Borel measure, and $|l(f)| \leqslant \int_\Omega |f|\, d|\mu|$, so that $\|l\| \leqslant |\mu|(\Omega)$ holds.

Since (16) gives (14), and half of (15) [i.e. $\|l\| \leqslant |\mu|(\Omega)$] is proved, we now establish the other half (the opposite inequality). This implies (15), and hence the uniqueness also follows at once.

Evidently $\mu \ll |\mu|$, and the latter is a finite measure. By the Radon-Nikodým theorem, we have $g = d\mu/d|\mu|$, so that g is a Borel function and $|g| = 1$ a.e.($|\mu|$). Hence by Theorem 3.2.9, for each $\epsilon > 0$ there is a compact set K_ϵ such that $g_\epsilon = g|_{K_\epsilon}$ is continuous and $|\mu|(K_\epsilon^c) < \epsilon$. In fact there is a relatively compact open $U \supset K_\epsilon$ and $\tilde{g} \in C_c(\Omega)$ such that $\text{supp}(\tilde{g}) \subset U$ and $\tilde{g} = g\,\text{sgn}(g)$ on K_ϵ. Hence $g' = \tilde{g}\,\text{sgn}(g)$ is in $C_c(\Omega)$, satisfying $(g')^2 = 1$ on K_ϵ. Then

$$\begin{aligned}
|\mu|(\Omega) &= \int_\Omega g \cdot g\, d|\mu| = \int_\Omega g\, d\mu \\
&\leqslant \left| \int_{K_\epsilon} g'\, d\mu \right| + \left| \int_{K_\epsilon^c} (g - g')\, d\mu \right| \\
&\leqslant \left| \int_\Omega g_\epsilon\, d\mu \right| + 2|\mu|(K_\epsilon^c), \qquad g_\epsilon \in C_c(\Omega), \\
&\leqslant |l(g_\epsilon)| + 2\epsilon \leqslant \|l\| + 2\epsilon, \qquad \text{since} \quad |g_\epsilon| \leqslant 1.
\end{aligned}$$

But $\epsilon > 0$ is arbitrary. So we get $|\mu|(\Omega) \leqslant \|l\|$, and (15) follows.

Remark. A further analysis shows that $l = l^+ - l^-$ satisfies the norm relation that $\|l\| = \|l^+\| + \|l^-\|$, and then the argument is simplified without using Luzin's result (Theorem 3.2.9). However, the present method has independent interest, and from it one can deduce, on using the Jordan decomposition of μ, that $l^\pm: f \mapsto \int_\Omega f\, d\mu^\pm$ gives the same norm equation. This alternative method also was recently employed by G. B. Folland (1984).

If Ω is compact, then $1 \in C(\Omega)$ and $l^{\pm}(1) = \mu^{\pm}(\Omega)$, and

$$\|l\| = |\mu|(\Omega) = \mu^{+}(\Omega) + \mu^{-}(\Omega) = l^{+}(1) + l^{-}(1). \tag{17}$$

In this case we can therefore replace (15) by (17). Thus the equations (1), (14), and (15) [or (17)] give different forms of this classical result.

We now consider a representation of local functionals on $C_c(\Omega)$. To describe them precisely, consider a mapping $\Lambda: C_c(\Omega) \to \mathbb{R}$. Then Λ is called a *local functional* if it is continuous on the normed linear space $C_c(\Omega)$ and $\Lambda(f+g) = \Lambda(f) + \Lambda(g)$ for all $f, g \in C_c(\Omega)$ such that $(f \cdot g)(\omega) = 0$, $\omega \in \Omega$. If the last restriction is dropped, then one has the usual situation of a bounded linear functional. Taking, for instance, $\Lambda(f) = |f|^2$, it is clear that a local functional need not be linear. Such mappings arise in many areas, e.g., differential equations and generalized random processes, among others. It is in studying the latter subject that I. Gel'fand and N. Vilenkin (1964) isolated these functionals, under the present name, on the Schwartz space of infinitely differentiable functions on $\mathbb{R}$ with compact supports, and raised the problem of their characterization. We shall present a solution on the space $C_c(\Omega)$, Ω locally compact, for this problem. Similar results on other function spaces are possible and are of interest. An overview of the available work on the subject has been given by the author (1980a).

Using the uniform norm $\|\cdot\|$ on $C_c(\Omega)$ as usual, we have the following characterization of the problem at hand:

Theorem 5. *Let $\Lambda: C_c(\Omega) \to \mathbb{R}$ be a mapping such that:*

(i) (*Sequential continuity*) *If $\{f_n: n \geq 1\} \subset C_c(\Omega)$ is a bounded pointwise convergent sequence, then $\{\Lambda(f_n): n \geq 1\} \subset \mathbb{R}$ is Cauchy.*
(ii) (*Additivity*) *$\Lambda(f_1 + f_2) = \Lambda(f_1) + \Lambda(f_2)$ for all f_1, f_2 satisfying $f_1 f_2 = 0$ for which the equation is defined.*
(iii) (*Bounded uniform continuity*) *For each $\epsilon > 0$, $\gamma > 0$, there is a δ $(= \delta_{\epsilon,\gamma} > 0)$ such that if $\|f_i\| \leq \gamma$, $f_i \in C_c(\Omega)$, $i = 1, 2$, and $\|f_1 - f_2\| < \delta$, then $|\Lambda(f_1) - \Lambda(f_2)| < \epsilon$.*

Then $\Lambda(\cdot)$ is representable as an integral

$$\Lambda(f) = \int_{\Omega} \Phi(f(\omega), \omega)\mu(d\omega), \qquad f \in C_c(\Omega), \tag{18}$$

where μ is a finite regular Borel measure on Ω and $\Phi: \mathbb{R} \times \Omega \to \mathbb{R}$ (sometimes termed a kernel) satisfies the following conditions:

(a) *$\Phi(0, \omega) = 0$, and $\Phi(\cdot, \omega)$ is continuous for a.a. $\omega \in \Omega$,*
(b) *$\Phi(x, \cdot)$ is μ-measurable for all $x \in \mathbb{R}$, and*
(c) *for each $f \in C_c(\Omega)$, $\Phi(f(\omega), \omega)$ is bounded for a.a. $\omega \in \Omega$, and for any sequence $\{f_n: n \geq 1\}$ as in (i), $\{\Phi(f_n, \cdot): n \geq 1\}$ is Cauchy in $L^1(\mu)$.*

Conversely, if (Φ, μ) *satisfies conditions* (a)–(c) *and* $\Lambda(\cdot)$ *is defined by* (18), *then it is a local functional on* $C_c(\Omega)$ *for which statements* (i)–(iii) *hold*.

Proof. The myriad details of the proof will be presented in steps for convenience; no real short cut is available.

1. Let $B_0(\Omega)$ be the space of all functions on Ω which are bounded pointwise limits of sequences from $C_c(\Omega)$. Then $B_0(\Omega)$ is a vector subspace of bounded Baire functions on Ω vanishing at infinity, i.e., for each $\epsilon > 0$ and $f \in B_0(\Omega)$, the set $\{\omega: |f(\omega)| \geqslant \epsilon\}$ is compact. If $K \subset U \subset \Omega$, K compact and U open, then by Urysohn's lemma, for any $h > 0$ there is a $p_{K,U}^h \in C_c(\Omega)$ satisfying $p_{K,U}^h = h$ on K, $= 0$ on $\Omega - U$, with values in $[0, h]$. If A is a compact G_δ in Ω, then there is a sequence $\{p_{A,U_n}^h: n \geqslant 1\} \subset C_c(\Omega)$ such that $p_{A,U_n}^h \downarrow h\chi_A$, by Proposition 1.1. Hence $\chi_A \in B_0(\Omega)$. If A, B are compact G_δ-sets, then $\chi_{A-B} = \chi_A - \chi_{A\cap B}$, $\chi_{A\cup B} = \chi_A + \chi_B - \chi_{A\cap B}$, so that $\{\chi_{A-B}, \chi_{A\cup B}) \subset B_0(\Omega)$, by linearity. Thus the class $\mathscr{C} = \{A \subset \Omega: \chi_A \in B_0(\Omega)\}$ contains the Baire ring $\mathscr{R}$ generated by the compact G_δ-sets of Ω. This structure will be utilized in the ensuing computations.

2. Λ maps bounded sets of $C_c(\Omega)$ into bounded sets of $\mathbb{R}$, and for each $\epsilon > 0$, relatively compact open U, and f, g in $C_c(\Omega)$ with $\operatorname{supp}(f) \subset U$, we have an $f_\epsilon \in C_c(\Omega)$ such that $\|f_\epsilon\| \leqslant \|f\|$ and $|\Lambda(f+g) - \Lambda(f_\epsilon + g)| < \epsilon$. For, let $\tilde{B}$ $[\subset C_c(\Omega)]$ be a ball of radius $d > 0$, and for $\epsilon = 1$, choose $0 < \delta < 1$ such that $f, g \in \tilde{B}$, $\|f - g\| < \delta d \Rightarrow |\Lambda(f) - \Lambda(g)| < 1$, which is possible by (iii) of the hypothesis, since $f, g \in \tilde{B}$ have norms at most d. Let $n = [1/\delta] + 1$, where $[x]$ is the integer part of $x \in \mathbb{R}$. If $\alpha = 1/n$, then for $f \in \tilde{B}$,

$$|\Lambda(f)| \leqslant \sum_{j=1}^{n} |\Lambda(j\alpha f) - \Lambda((j-1)\alpha f)| \leqslant n \leqslant 1 + \frac{1}{\delta}, \tag{19}$$

since $\|j\alpha f - (j-1)\alpha f\| = \alpha\|f\| < d/n \leqslant d\delta$. Thus $\Lambda(\tilde{B}) \subset \mathbb{R}$ is bounded.

So see that Λ has uniformly small changes for small "perturbations", let $\epsilon > 0$ be given, and choose $\delta > 0$ such that for $f, g, h_0 \in C_c(\Omega)$ and $\|(f+g) - (h_0 + g)\| = \|f - h_0\| < \delta$ one has $|\Lambda(f+g) - \Lambda(h_0 + g)| < \epsilon$. We now show that h_0 can be taken to satisfy the additional prescription depending on f and ϵ only, i.e., $h_0 = f_\epsilon$. In fact let $U_1 = \{\omega: |f(\omega)| < \delta\}$, so that $U_1^c \subset U$, since $\operatorname{supp}(f) \subset U$. Thus the closed sets U_1^c and U^c are disjoint. Since U_1^c is also compact, by Corollary 1.2 there is an open F_σ set V such that $U_1^c \subset V \subset U$, and by Urysohn's lemma there is a function $v: \Omega \to [0,1]$, continuous and such that $v|_{U_1^c} = 1$ and $v = 0$ on V^c, so that $\operatorname{supp}(v) \subset V \subset U$. If $f_\epsilon = h_0 = vf$, then $\|h_0\| \leqslant \|f\|$, $\|f - f_\epsilon\| = \|(1 - v)f\| < \delta$, and so $|\Lambda(f+g) - \Lambda(f_\epsilon + g)| < \epsilon$. In particular, if $\|f\| < \delta$ then $|\Lambda(f+g) - \Lambda(g)| < \epsilon$.

3. If $A \subset \Omega$ is closed and $\epsilon > 0$, $h > 0$ are numbers, we can find an open $U \supset A$ such that if $f \in C_c(\Omega)$, $\operatorname{supp}(f) \subset U - A$, $\|f\| \leqslant h$, then $|\Lambda(f)| < \epsilon$.

For, if the conclusion does not hold, then for each open set $U_1 \supset A$ we can find an $\tilde{f}_1 \in C_c(\Omega)$ with $\operatorname{supp}(\tilde{f}_1) \subset U_1 - A$ for which the preceding step fails, i.e.,

$$|\Lambda(\tilde{f}_1)| > \epsilon, \quad \|\tilde{f}_1\| \leqslant h \qquad (\text{with } g = 0 \text{ there}).$$

Then $U_2 = U_1 - \operatorname{supp}(\tilde{f}_1)$ is open $\supset A$, and by the same argument, one can find $\tilde{f}_2 \in C_c(\Omega)$ with $\operatorname{supp}(\tilde{f}_2) \subset U_2 - A$ for which $|\Lambda(\tilde{f}_2)| > \epsilon$. By induction, we can find $\tilde{f}_1, \ldots, \tilde{f}_n$ in $C_c(\Omega)$ with disjoint supports and $\|\tilde{f}_i\| \leqslant h$, but for which $|\Lambda(\tilde{f}_k)| > \epsilon$ for each $k = 1, 2, \ldots, n$. However, $\Sigma_{i=1}^n \tilde{f}_i \in C_c(\Omega)$, $\Lambda(\Sigma_{i=1}^n \tilde{f}_i) = \Sigma_{i=1}^n \Lambda(\tilde{f}_i)$ for each n, and by the same step one has that $\{|\Lambda(\Sigma_{k=1}^n \tilde{f}_k)|: n \geqslant 1\} \subset \mathbb{R}$ is bounded. For a subsequence $\{n_i: i \geqslant 1\}$, $\Sigma_{k=1}^{n_i} \Lambda(f_k)$ converges as $n_i \to \infty$, so that $|\Lambda(f_{n_i})| \to 0$. But this contradicts the fact that $|\Lambda(f_k)| \geqslant \epsilon$ for all k. Hence our supposition is false and the original assertion must hold.

Using the preceding work, we show that there is a set function on the lattice of compact subsets of Ω satisfying the hypothesis of Theorem 2.9, which in turn gives the desired measure μ, and then (18) will be obtained.

4. One extends Λ to $\tilde{\Lambda}$: $B_0(\Omega) \to \mathbb{R}$ as follows. If $A \subset \Omega$ is a compact G_δ, then there exist $f_n \in C_c(\Omega)$ such that $f_n \downarrow \chi_A$ by Proposition 1.1. Hence by (i), $\{\Lambda(f_n): n \geqslant 1\}$ is Cauchy in $\mathbb{R}$, so that $\lim_n \Lambda(f_n) = \tilde{\Lambda}(\chi_A)$ exists, and it is easily seen that the functional $\tilde{\Lambda}$ does not depend on the sequence used, since two such sequences can be combined into one. If $\tilde{f} = \Sigma_{i=1}^n a_i \chi_{A_i}$, $A_i \subset \Omega$, compact G_δ and disjoint, then $\tilde{\Lambda}(\tilde{f}) = \Sigma_{i=1}^n \tilde{\Lambda}(a_i \chi_{A_i})$ and $\tilde{\Lambda}|C_c(\Omega) = \Lambda$. This follows from the fact that $\tilde{\Lambda}(\chi_{A_1} + \chi_{A_2}) = \tilde{\Lambda}(\chi_{A_1}) + \tilde{\Lambda}(\chi_{A_2})$ for disjoint compact G_δ-sets A_1, A_2, since by the complete regularity of Ω, there exist disjoint relatively compact open sets U_{1n}, U_{2n} of A_1, A_2 such that $f_{in} = p^1_{A_i, U_{in}} \in C_c(\Omega)$, $i = 1, 2$, satisfying

$$\begin{aligned}\tilde{\Lambda}(\chi_{A_1} + \chi_{A_2}) &= \lim_n \Lambda(f_{1n} + f_{2n}) \\ &= \lim_n [\Lambda(f_{1n}) + \Lambda(f_{2n})] = \tilde{\Lambda}(\chi_{A_1}) + \tilde{\Lambda}(\chi_{A_2}).\end{aligned}$$

Let $\lambda^h(A) = \tilde{\Lambda}(h\chi_A)$ $[= \lim_n \Lambda(p^h_{A, U_n})]$. Then $\lambda^h(\varnothing) = 0$, and λ^h: $\mathscr{A} \to \mathbb{R}$ is additive, where $\mathscr{A}$ is the lattice of compact G_δ-sets of Ω. We wish to show that $\lambda^h(\cdot)$ extends to the ring $\mathscr{R}$ generated by $\mathscr{A}$ and is additive there. This is a consequence of the next step.

5. The mapping λ^h: $\mathscr{A} \to \mathbb{R}$ is tight, i.e., if $\epsilon > 0$, and if $A \supset B$ are compact G_δ-sets, then there is a $C \in \mathscr{A}$, $C \subset A - B$, such that $|\lambda^h(A) - \lambda^h(B) - \lambda^h(C)| < \epsilon$, where $h > 0$ is a real number. For, we note that $B \subset U \subset \bar{U} \subset A$, where U is open (and $\bar{U}$ its closure) can be assumed. Otherwise we work with $\bar{U}$ (which can be taken G_δ by Corollary 1.2). This is because for $\epsilon > 0$, there is an open $U \supset A$ such that $f \in C_c(\Omega)$, $\operatorname{supp}(f) \subset A - U \Rightarrow |\Lambda(f)| < \epsilon$ by step 3, so that for all $A \subset \tilde{A} \subset U$, $\tilde{A} \in \mathscr{A}$, and for $\tilde{f} \in C_c(\Omega)$ with $\operatorname{supp}(\tilde{f}) \subset$

$U - \tilde{A}$, we have $|\Lambda(\tilde{f})| < \epsilon$. Then by step 4, $|\lambda^h(A) - \Lambda(f_1)| < \epsilon/3$, $\text{supp}(f_1) \subset U$, $f_1|_A = 1$, and similarly $|\lambda^h(\tilde{A}) - \Lambda(\tilde{f}_1)| < \epsilon/3$ for a suitable $\tilde{f}_1$ with $\tilde{f}_1|_{\tilde{A}} = 1$, $\text{supp}(\tilde{f}_1) \subset U$. Also, $f_1 - \tilde{f}_1 \in C_c(\Omega)$, $\|f_1 - \tilde{f}_1\| \leqslant h$, $\text{supp}(\tilde{f}_1 - f_1) \subset U - A$, so that $|\Lambda(f_1) - \Lambda(\tilde{f}_1 - f_1 + f_1)| = |\Lambda(f_1) - \Lambda(\tilde{f}_1)| < \epsilon/3$. Consequently

$$|\lambda^h(A) - \lambda^h(\tilde{A})| \leqslant |\lambda^h(A) - \Lambda(f_1)| + |\Lambda(f_1) - \Lambda(\tilde{f}_1)| + |\Lambda(\tilde{f}_1) - \lambda^h(\tilde{A})| < \epsilon. \quad (20)$$

Hence we may work with $B \subset \tilde{A}$. If $\tilde{C} \subset \tilde{A} - B$, let $C = \tilde{C} \cap A \subset A - B$. Then this C will satisfy our inequality, i.e., $|\lambda^h(C) - \lambda^h(\tilde{C})| < \epsilon$ will obtain.

So consider $B \subset U \subset \bar{U} \subset A$, where $\text{supp}(f) \subset U - B$, $f \in C_c(\Omega) \Rightarrow |\Lambda(f)| < \epsilon$. Let $C = A - U \in \mathscr{A}$ (all compact sets are taken to be in $\mathscr{A}$ which is possible by Corollary 1.2). A considerable amount of repetitive approximation using Urysohn's lemma has to be made to complete this step. Since $C \subset A - B$ also, there are (by normality of C, $B \subset A$) two disjoint open sets V_1, V_2 such that $B \subset V_1$, $C \subset V_2$ and $\bar{V}_2 \subset A$, $\bar{V}_1 \subset A$. Because of the preceding step, one may assume the existence of open V_0, W_0 approximating B and C and such that

$$B \subset V_0 \subset \bar{V}_0 \subset V_1 \quad \text{and} \quad C \subset W_0 \subset \bar{W}_0 \subset V_2.$$

Let v_1, v_2 be a pair of Urysohn functions (i.e. those in $C_c(\Omega)$ with range $[0,1]$) such that $v_1 = 1$ on $\bar{V}_0$, $v_1 = 0$ on V_2, $v_2 = 1$ on $\bar{W}_0$, and $v_2 = 0$ on V_1. If f approximates A in that $f \in C_c(\Omega)$, $f|A = 1$, and $|\lambda^h(A) - \Lambda(f)| < \epsilon$, then let $f_i = v_i f$, $i = 1, 2$, and $g = f - f_1 - f_2$. Now clearly $f_1 f_2 = 0$ and $g = 0$ on $\bar{V}_0 \cup \bar{W}_0$. But by the way U and hence C were chosen, if $\tilde{U} = \text{supp}(f)$, then $\text{supp}(g) \subset \tilde{U} - A$, so that $|\Lambda(g)| < \epsilon$. Since $\Lambda(f - g) = \Lambda(f_1) + \Lambda(f_2)$, we get

$$|\lambda^h(A) - \lambda^h(B) - \lambda^h(C)| \leqslant |\lambda^h(A) - \Lambda(f)| + |\Lambda(f - g) - \Lambda(f)| + |\Lambda(f_1) - \lambda^h(B)| + |\lambda^h(C) - \Lambda(f_2)|. \quad (21)$$

Now by the choice of f and the fact that $B \subset A$, $C \subset A$, $\text{supp}(f_1) \subset \bar{V}_0$, and $\text{supp}(f_2) \subset \bar{W}_0$, the last two terms as well as the first one on the right side of (21) are each less than ϵ, and since $|\Lambda(g)| < \epsilon$, the second term is also $< \epsilon$, by step 3. Hence the asserted inequality of this step holds (with 3ϵ). This C is in reality $\tilde{C}$ of the preceding paragraph. But if C is defined as given there, then $\tilde{C} - C \subset \tilde{A} - A \subset \tilde{U} - A$, where $\tilde{U}$ is an open set approximating A, as noted earlier. Hence if $h_0 \in C_c(\Omega)$, $\text{supp}(h_0) \subset (\tilde{C} - C)$, then $|\Lambda(h_0)| < \epsilon$, so that $|\lambda^h(\tilde{C}) - \lambda^h(C)| < \epsilon$. In other words, the statement of the step holds.

The approximations involved are extensions of the $\epsilon - \delta$ adjustments in the case of a (compact) metric Ω. The latter have been detailed by Friedman and

Katz (1966), and the present case is essentially a rewording for the locally compact spaces.

6. The preceding step allows one to apply Theorem 2.9 with this $\mathscr{A}$ and λ^h, and conclude that λ^h has a unique extension to the ring $\mathscr{R}$ generated by $\mathscr{A}$. Since, by step 1, $\lambda^h(A) = \lim_n \Lambda(p^h_{A,U_n})$, we have $\sup_{A \in \mathscr{A}} |\lambda^h(A)| \leqslant \alpha h$ for some constant $\alpha > 0$. Hence λ^h has the regularity property relative to $\mathscr{A}$, and then by Theorem 2.6 together with Proposition 2.4, we conclude that λ^h has a unique extension to the Borel σ-ring of Ω, to be a regular bounded Borel (signed) measure, which is again denoted by the same symbol.

It is now claimed that λ^h depends continuously on h. Since this is a local property, it suffices to verify it for h in compact intervals. Thus we show that for each $\epsilon > 0$, $k > 0$, there is a δ $(= \delta_{\epsilon,k} > 0)$ such that $|\lambda^h(B) - \lambda^{h'}(B)| < \epsilon$ for each $B \in \mathscr{B}$, the Borel σ-ring of Ω, and $|h - h'| < \delta$, h, h' in $[0, k]$. Actually, by Corollary 4.4.7 due to Nikodým, this implies uniformity in B, but the above statement suffices here. Now by the regularity of λ^h, in conjunction with Theorem 2.6, it is enough to verify the result for each compact G_δ-set. But if the latter assertion is false, then there is a compact Baire set A and a pair h_1, h_2 in $[0, k]$ such that

$$|\lambda^{h_1}(A) - \lambda^{h_2}(A)| > \epsilon \qquad \text{for} \quad |h_1 - h_2| < \delta/2.$$

By step 4, there exist $f_n^j \in C_c(\Omega)$ such that $f_n^j \downarrow h_j \chi_A$ and $\lambda^{h_j}(A) = \lim_n \Lambda(f_n^j)$, $j = 1, 2$. Since $|h_1 - h_2| < \delta/2$, and f_n^j can be taken as $p^{h_j}_{A,U_n}$ in the earlier notation (of step 1), we have $\|f_n^1 - f_n^2\| < \delta$, and then $|\Lambda(f_n^1) - \Lambda(f_n^2)| < \epsilon/3$ for $n \geqslant n_0$ $[n_0 = n_0(\epsilon)]$. Hence $|\Lambda(f_n^j) - \tilde{\Lambda}(h_j \chi_A)| < \epsilon/3$, $j = 1, 2$. Then we get

$$\begin{aligned} \epsilon &< |\lambda^{h_1}(A) - \lambda^{h_2}(A)| \\ &\leqslant |\tilde{\Lambda}(h_1 \chi_A) - \Lambda(f_n^1)| + |\Lambda(f_n^1) - \Lambda(f_n^2)| \quad + |\Lambda(f_n^2) - \tilde{\Lambda}(h_2 \chi_A)| \\ &< \frac{\epsilon}{3} + \frac{\epsilon}{3} + \frac{\epsilon}{3} = \epsilon. \end{aligned}$$

This contradiction shows that $\lambda^h(A)$ depends on h continuously for each such $A \subset \Omega$, and $h \in \mathbb{R}^+$. Clearly $\lambda^0(A) = 0$ for all A. It is noted that all the above work holds if the real $h < 0$.

7. To construct a common measure μ dominating all λ^h, consider an enumeration $h_1, h_2, \ldots$ of the rationals of $\mathbb{R}$. Define

$$\mu = \sum_{n=1}^{\infty} \frac{1}{2^n} \frac{|\lambda^{h_n}|(\cdot)}{1 + |\lambda^{h_n}|(\Omega)}, \tag{22}$$

where $|\lambda^h|(\Omega)$ is the variation norm of the bounded signed Borel measure λ^h. Then μ is a finite Borel measure, and by the preceding step $|\lambda^{h_n}(A)| \leqslant \alpha_n \mu(A)$, $A \in \mathscr{B}$, for some α_n $(= 2^n[1 + |\lambda^{h_n}|(\Omega)]) < \infty$. Hence by the Radon-Nikodým theorem (Theorem 5.3.5, for instance) there is a *bounded* function $\Psi_h(\cdot)$ (a.e. equal to a bounded Borel function for μ) such that $|\Psi_h(\omega)| \leqslant \alpha_n$ for a.a. ω, and

$$\tilde{\Lambda}(h\chi_A) = \lambda^h(A) = \int_A \Psi_h(\omega)\mu(d\omega) = \int_\Omega \Psi(h\chi_A(\omega), \omega)\mu(d\omega), \quad (23)$$

where $\Psi(h, \omega) = \Psi_h(\omega)$, $h \in \mathbb{R}$, $\omega \in A$, and $= 0$ if $\omega \notin A$. Since Ψ_h is only defined uniquely outside a μ-null set that can depend on h, one has to obtain Φ from $\Psi(\cdot, \cdot)$ with some care to establish (18). This will be done in the next step.

8. The function $\Psi(\cdot, \cdot)$, which is uniquely defined on $\mathbb{Q} \times (\Omega - N)$, where $\mu(N) = 0$ and $\mathbb{Q}$ is the set of rationals, is extended to $\mathbb{R} \times \Omega$ in two parts as follows. First, observing that μ may be assumed complete in the above (or by completing it in the opposite case), we have $\Psi(h, \cdot) \in L^\infty(\mu)$, and by Theorem 8.2.3 there is a lifting map ρ: $L^\infty(\mu) \to \mathcal{L}^\infty(\mu)$, so that in (23), $\Psi(h, \cdot)$ may be taken as $\rho(\Psi(h, \cdot))$ everywhere, by modifying it on a μ-negligible set. This would not have been possible if μ given by (22) had been replaced by a (σ-)finite $\tilde{\mu}$ with only $\lambda^h \ll \mu$, since then $\Psi(h, \cdot) \in L^1(\mu)$ (or finite a.e.), and ρ need not exist on such a space. Next, with the function $\Psi(h, \cdot)$ defined everywhere on Ω, and bounded, we have for all h in $\mathbb{Q}$

$$\Phi(a, \omega) = \lim_{h \to a} \Psi(h, \omega), \qquad a \in \mathbb{R}, \quad \omega \in \Omega. \quad (24)$$

Since λ^h is continuous in h ($\in \mathbb{Q}$) and $\mathbb{Q}$ is dense in $\mathbb{R}$, $\Phi(\cdot, \cdot)$ is well defined on $\mathbb{R} \times \Omega \to \mathbb{R}$, and $\Phi(\cdot, \omega)$ is bounded on bounded sets of $\mathbb{R}$ for each $\omega \in \Omega$. Evidently $\Phi(0, \omega) = 0$, $\omega \in \Omega$, and $\Phi(\cdot, \omega)\colon \mathbb{R} \to \mathbb{R}$ is continuous, $\omega \in \Omega$. Also $\Phi(h, \omega) = \Psi(h, \omega)$ for $h \in \mathbb{Q}$, $\Phi(h, \cdot) \in \mathcal{L}^\infty(\mu)$. We assert that $\Phi(\cdot, \cdot)$ satisfies (18).

To see this, if $g = \sum_{i=1}^n h_i \chi_{A_i}$, $A_i \in \mathcal{B}$, disjoint, $h_i \in \mathbb{R}$, then by step 4, $\tilde{\Lambda}$ is local on $B_0(\Omega)$ [and $\tilde{\Lambda}|C_c(\Omega) = \Lambda$], so that

$$\begin{aligned}\tilde{\Lambda}(g) &= \sum_{i=1}^n \tilde{\Lambda}(h_i\chi_{A_i}) = \sum_{i=1}^n \lambda^{h_i}(A_i)\\ &= \sum_{i=1}^n \int_{A_i} \Psi_{h_i}(\omega)\mu(d\omega) = \int_\Omega \Phi\left(\sum_{i=1}^n h_i\chi_{A_i}(\omega), \omega\right)\mu(d\omega)\\ &= \int_\Omega \Phi(g(\omega), \omega)\mu(d\omega).\end{aligned}$$

If $g \in B_0(\Omega)$ is arbitrary, then there exists a sequence of simple Baire functions g_n with coefficients in $\mathbb{Q}$ such that $g_n \to g$ pointwise and boundedly. Hence $\Phi(g_n, \cdot) \to \Phi(g, \cdot)$ pointwise and boundedly, since $\Phi(\cdot, \omega)$ is bounded on bounded sets. But μ is a finite measure. Hence the set $\{\Phi(g_n, \cdot): n \geqslant 1\}$ is uniformly integrable, so that by the Vitali (or even bounded) convergence theorem we get with condition (i)

$$\tilde{\Lambda}(g) = \lim_n \tilde{\Lambda}(g_n) = \lim_n \int_\Omega \Phi(g_n(\omega), \omega)\mu(d\omega)$$

$$= \int_\Omega \Phi(g(\omega), \omega)\mu(d\omega), \qquad \text{by Corollary 4.3.13.} \tag{25}$$

Since $\Lambda = \tilde{\Lambda}|C_c(\Omega)$, (25) reduces to (18) on $C_c(\Omega)$, and $\{\Phi(g_n, \cdot): n \geqslant 1\}$ is evidently Cauchy in $L^1(\mu)$ by the same theorem. This gives the direct part.

9. Conversely, let (Φ, μ) satisfy (a)–(c) of the statement, and Λ be defined by (18). Since $\Phi(0, \omega) = 0$, (ii) clearly holds. Also $\Phi(\cdot, \omega)$ is bounded on bounded sets of $\mathbb{R}$ by (c). $\{\Phi(g_n, \cdot): n \geqslant 1\}$ is a pointwise convergent sequence if $\{g_n: n \geqslant 1\} \subset C_c(\Omega)$ has that property, and it is also uniformly integrable. Hence, by Vitali's theorem again, (i) follows. It remains to verify (iii).

Suppose that (iii) does not hold. Since (18) is true for all $f \in B_0(\Omega)$, we have that $\tilde{\Lambda}: B_0(\Omega) \to \mathbb{R}$ is a local functional. Then there exist positive numbers $\gamma > 0$, $\epsilon > 0$ and a sequence $\{f_n^j: n \geqslant 1\}$, $j = 1, 2$, such that

$$\|f_n^j\| \leqslant \gamma, \qquad \|f_n^1 - f_n^2\| < \frac{1}{n}, \qquad \int_\Omega \left|\Phi(f_n^1, \cdot) - \Phi(f_n^2, \cdot)\right| d\mu > \epsilon. \tag{26}$$

Since $\{\Phi(f_n^j, \cdot): n \geqslant 1\}$ are μ-integrable, there exists a $\delta_1 > 0$, $A \in \mathscr{B}$, $\mu(A) < \delta_1$, and n_1 such that

$$\int_A \left|\Phi(f_{n_1}^1, \cdot) - \Phi(f_{n_1}^2, \cdot)\right| d\mu < \frac{\epsilon}{3}. \tag{27}$$

Here $0 < \delta_1 < \frac{1}{2}\mu(\Omega)$ can be assumed. Since $\Phi(\cdot, \omega)$ is continuous for all $\omega \in \Omega - N$, $\mu(N) = 0$, we get that for any $\eta > 0$ the set Ω_1 defined by

$$\Omega_1 = \left\{\omega: -\gamma \leqslant a_1, a_2 \leqslant \gamma, |a_1 - a_2| \leqslant 1/n \Rightarrow |\Phi(a_1, \omega) - \Phi(a_2, \omega)| < \eta\right\}$$

has full measure, i.e., $\mu(\Omega - \Omega_1) = 0$. Consequently we can find $\Omega_2 \subset \Omega_1$ and $n_2 > n_1$ such that $\mu(\Omega_2^c) < \delta_1$, and $|\Phi(f_{n_2}^1, \cdot)(\omega) - \Phi(f_{n_2}^2, \cdot)(\omega)| < \epsilon/[3\mu(\Omega)]$ for all $\omega \in \Omega_2$. But taking $A = \Omega_2^c$ in (27), we get, on subtraction,

$$\int_{\Omega_2^c} \left|\Phi(f_{n_2}^1, \cdot) - \Phi(f_{n_2}^2, \cdot)\right| d\mu > \epsilon - \frac{\epsilon}{3} = \frac{2\epsilon}{3}. \tag{28}$$

Similarly we can find a $0 < \delta_2 < \delta_1$, $\Omega_3 \subset \Omega_2$, $\mu(\Omega_3^c) < \delta_2$, and an $n_3 > n_2$ such that $|\Phi(f_{n_3}^1, \cdot)(\omega) - \Phi(f_{n_3}^2, \cdot)(\omega)| < \epsilon/[3\mu(\Omega)]$, $\omega \in \Omega_3$, and with $A = \Omega_3^c$

in (27) we get

$$\int_{\Omega_3^c} \left| \Phi\left(f_{n_3}^1, \cdot\right) - \Phi\left(f_{n_3}^2, \cdot\right) \right| d\mu > \frac{2\epsilon}{3}. \tag{29}$$

Proceeding by induction, one finds sequences $\{f_{n_i}^j, \Omega_i: i \geqslant 1\}$ such that $\mu(\Omega_i^c) < \delta_i \ (< \frac{1}{2}\delta_{i-1})$ and

$$\int_{\Omega_i^c} \left| \Phi\left(f_{n_i}^1, \cdot\right) - \Phi\left(f_{n_i}^2, \cdot\right) \right| d\mu > \frac{2\epsilon}{3}, \qquad \int_{\Omega_{i+1}^c} \left| \Phi\left(f_{n_i}^1, \cdot\right) - \Phi\left(f_{n_i}^2, \cdot\right) \right| d\mu < \frac{\epsilon}{3}.$$

If $A_1 = \Omega_1^c$ and $A_k = \Omega_k^c - \bigcup_{i=1}^{k-1}\Omega_i$ for $k > 1$, then A_k, $k \geqslant 1$, are disjoint and $\mu(\bigcup_{i \geqslant n+1}\Omega_i^c) < \frac{1}{2}\delta_n$. So our construction gives $\mu(A_k) < \delta_k$ and then

$$\int_{A_k} \left| \Phi\left(f_{n_k}^1, \cdot\right) - \Phi\left(f_{n_k}^2, \cdot\right) \right| d\mu > \frac{\epsilon}{3}, \qquad k \geqslant 1.$$

If $f^1 = \sum_{k \geqslant 1} f_{n_k}^1 \chi_{A_k}$, $f^2 = \sum_{k \geqslant 1} f_{n_k}^2 \chi_{A_k}$, then $f^i \in B_0(\Omega)$, and hence $\Phi(f^1, \cdot)$, $\Phi(f^2, \cdot)$ are μ-integrable. Consequently

$$\infty > \int_\Omega \left| \Phi(f^1, \cdot) - \Phi(f^2, \cdot) \right| d\mu = \sum_{k \geqslant 1} \int_{A_k} \left| \Phi\left(f_{n_k}^1, \cdot\right) - \Phi\left(f_{n_k}^2, \cdot\right) \right| d\mu$$

$$> \sum_{k \geqslant 1} \frac{\epsilon}{3}.$$

This contradiction shows that $\Lambda(\cdot)$ defined by (18) must satisfy (iii) also.
With this the proof of the theorem is finally completed.

Remark. The preceding argument suggests that if Λ is defined by (18) for a (σ-)finite μ on $B_0(\Omega)$, then it is a local functional satisfying (i)–(iii). This is true, and the σ-finite case needs additional argument, as shown by Mizel (1970); the computation of step 9 above essentially followed the latter. In case $C_c(\Omega)$ is replaced by $L^\infty(\mu)$ to begin with, then our work through the first seven steps of the above proof is not necessary. But step 8 is more detailed, since one has to find a substitute for the lifting map and its properties. This is actually the case considered by Mizel in the above paper. The theorem in its form given here is due to the author (1980b), where applications of this result to probability theory are included.

As a consequence of this theorem let us sketch another proof of the Riesz-Markov representation of continuous linear functionals on the normed linear space $C_c(\Omega)$.

Second Proof of Theorem 4. If $l: C_c(\Omega) \to \mathbb{R}$ is a continuous linear functional, then $l(f + g) = l(f) + l(g)$ for all f, g in $C_c(\Omega)$ and $|l(f)| \leqslant k_0\|f\|$ for some $k_0 < \infty$. It is clear that $l(\cdot)$ is also a local functional which satisfies (i)–(iii) of Theorem 5. Hence (18) holds for this $l(\cdot)$ with some special properties, and then μ is a finite (signed) Borel measure. Using the linearity of l, consider

$$\mathscr{C} = \left\{ A \subset \Omega: \int_A \Phi(f+g,\cdot)\,d\mu = \int_A \Phi(f,\cdot)\,d\mu + \int_A \Phi(g,\cdot)\,d\mu \right\}. \quad (30)$$

Observing that $l_S = l|C_c(S)$ for each compact set $S \subset \Omega$ and that l_S is linear, continuous, and satisfies $l_{S_1} = l_{S_2} = l_{S_1 \cap S_2}$ on $C(S_1 \cap S_2)$ for any compact subsets S_1, S_2, one notes also that the representing μ_S for l_S is $\mu_S = \mu(S \cap \cdot\,)$ in (18). Hence it follows that $\mathscr{C}$ contains each compact subset of Ω. Since l can be extended to $B_0(\Omega)$ also, it follows that $\mathscr{C}$ contains each Baire set, because $l_B(\cdot)$ is again a local functional satisfying (i)–(iii) by the converse part of the theorem. Hence $\mathscr{C} \supset \mathscr{B}$ also. Consequently, the equation in (30) implies

$$\Phi(f(\omega) + g(\omega), \omega) = \Phi(f(\omega), \omega) + \Phi(g(\omega), \omega), \quad (31)$$

for almost all ω in Ω. But, as seen in the construction of Φ, it actually can be taken as $\Phi(\cdot, \omega)$ that is continuous for each $\omega \in \Omega$. But then (f, g being arbitrary) (31) is the classical Cauchy functional equation, for which the solution is $\Phi(f(\omega), \omega) = \beta(\omega)f(\omega)$, $\omega \in \Omega$, for some Borel function $\beta: \Omega \to \mathbb{R}$ which is μ-integrable, as follows from Theorem 5.5.10 [a consequence of which is that $\int_\Omega fg\,d\mu \in \mathbb{R}$ for each $g \in \mathscr{L}^\infty(\mu) \Rightarrow f \in \mathscr{L}^1(\mu)$]. Letting $d\nu = \beta\,d\mu$, it follows that ν is a signed Borel measure on Ω, and (18) becomes

$$l(f) = \int_\Omega f(\omega)\nu(d\omega), \qquad f \in C_c(\Omega), \quad (32)$$

using Proposition 5.3.6 on change of variables. But (32) is the same as (14). Then (15) is proved from this as before. Note that with the Jordan decomposition of ν, we get $l = l^+ - l^-$ from (32), and so Lemma 3 is now a consequence of this proof.

EXERCISES

1. Show that a positive linear functional on $C_c(\Omega)$, Ω locally compact, is automatically relatively bounded. It is actually bounded on the space $C_0(\Omega)$ of real functions which vanish at infinity, i.e., $f \in C_0(\Omega)$ iff $f: \Omega \to \mathbb{R}$ is continuous and $\{\omega: |f(\omega)| \geqslant \alpha\}$ is compact for each $\alpha > 0$. [Note that $(C_0(\Omega), \|\cdot\|)$ is the completion of $C_c(\Omega)$.]

2. Prove the Riesz-Markov theorem on $C_c(\Omega)$ as an extension of the Daniell-Stone theorem (i.e. of Theorem 7.3.4 but with Ω locally compact).

3. If $l: C_c(\Omega) \to \mathbb{R}$ is a relatively bounded linear functional, show that with $l^+(g) = \sup\{l(f): 0 \le f \le g\}$, $g \in C_c(\Omega)$, and $l^- = l^+ - l$, where l^+ and l^- are positive linear functionals, one has a Jordan decomposition of l $(= l^+ - l^-)$.

4. Let l be as in the preceding exercise, and define $|l| = l^+ + l^-$, so that $|l|$ is a positive linear functional. If ν is the regular Borel measure representing $|l|$, show, by Theorem 1, that $\|l\| = \sup\{\nu(A): A \subset \Omega$ is Borel$\}$. (This may be $= +\infty$.) Show further that if l is bounded, then for the representing measure μ given by Theorem 4 we have $|\mu| = \nu$.
5. Using the second proof of Theorem 4 from Theorem 5, show that each continuous linear functional l on $C_0(\Omega)$ admits a Jordan decomposition $l = l^+ - l^-$, and deduce that $\|l\| = \|l^+\| + \|l^-\|$. Verify that l and $|l|$ $(= l^+ + l^-)$ have the same norm $(= \|l\|)$, and this is $l^+(1) + l^-(1)$ if Ω is compact.
6. Let $L: C_c(\Omega) \to \mathbb{C} - \{0\}$ be a mapping, where Ω is locally compact, as usual, and $\mathbb{C}$ is the complex number-space. Suppose that (i) if $f_n \in C_c(\Omega)$, $f_n \to f$ pointwise boundedly, then $\{L(f_n): n \geqslant 1\} \subset \mathbb{C}$ is Cauchy; (ii) $L(f+g) = L(f)L(g)$ if $fg = 0$; and (iii) L maps bounded sets of $C_c(\Omega)$ into bounded sets of $\mathbb{C}$, and also for $\epsilon > 0$, $K > 0$, there is a δ $(= \delta_{\epsilon, K} > 0)$ such that if $\|f_i\| < K$ and $\|f_1 - f_2\| < \delta$, then $|[L(f_1)/L(f_2)] - 1| < \epsilon$. Show that there exist a finite Borel measure on the Borel σ-algebra of Ω, and a kernel $\Phi(\cdot, \cdot): \mathbb{R} \times \Omega \to \mathbb{C}$ satisfying conditions (a)–(c) of Theorem 5, such that

$$L(f) = \exp\left(\int_\Omega \Phi(f(\omega), \omega)\mu(d\omega)\right), \qquad f \in C_c(\Omega).$$

[Verify that $\Lambda = \log L$ is well defined and satisfies the hypothesis of Theorem 5.]
7. If Ω is a σ-compact space and μ is a Borel measure on it, consider $L^\infty(\Omega, \mathbb{B}, \mu)$ in lieu of $C_c(\Omega)$ of Theorem 5. If Λ is a real functional on this space satisfying conditions (i)–(iii) of that theorem, show that there is a kernel $\Phi: \mathbb{R} \times \Omega \to \mathbb{R}$ such that the representation given in that theorem holds. [Here the first seven steps are eliminated in the proof, but the uniform integrability of step 8 needs a detailed argument. Since μ must now be σ-finite (and the Radon-Nikodým densities of λ^h relative to μ need not be bounded), the necessary arguments are as in step 9.]
8. Let μ_i be Radon measures on locally compact spaces Ω_i, $i = 1, 2$, and $\Omega = \Omega_1 \times \Omega_2$. If $\mu = \mu_1 \otimes \mu_2$, show that $C_c(\Omega)$ is dense in $\mathscr{L}^1(\mu)$, and $l: \mapsto \int_\Omega f\, d\mu$, $f \in C_c(\Omega)$, is a positive linear functional. Deduce that $l(f) = \int_\Omega f\, d\tilde{\mu}$ for a unique Radon measure, and $\mathscr{L}^1(\mu) = \mathscr{L}^1(\tilde{\mu})$.

9.4. HAAR MEASURES

It is shown in Theorem 2.2.17(v) that the Lebesgue measure on $\mathbb{R}^n$ is translation invariant. This property plays a fundamental role in Fourier analysis and other classical work. However, such studies as multivariate statistical analysis, mathematical physics, and even some problems of number theory demand that the Lebesgue theory should be generalized to include integration on spaces of nonsingular matrices on $\mathbb{R}^n$ or $\mathbb{C}^n$ etc. The common characteristic is that these are locally compact groups and a translation operation is defined on them. A translation-invariant measure on some compact groups was established by F. Peter and H. Weyl in 1927, and in 1933 A. Haar, in an important advance, established it for *all* separable locally compact cases. Slightly later the existence and essential uniqueness of such an invariant measure on arbitrary local compact groups were proved by A. Weil and J. von Neumann respectively. More historical discussion with relevant references to the above is in Hewitt and Ross (1962). Other proofs of this result have been obtained by various authors. Since the method of existence is different from other proofs of the preceding chapters, we include it here as a brief introduction to the subject of invariant integration.

Recall that a topological (locally compact) group G is algebraically a group and topologically a (locally compact) space such that the group operations $(a, b) \mapsto ab$ and $a \to a^{-1}$ are continuous (the first of $G \times G$ onto G with the product topology for $G \times G$, and the second of G onto G). A typical example to be included in our treatment is that $G = \mathrm{GL}(n)$, all $n \times n$ invertible matrices (with matrix multiplication as group operation) on $\mathbb{R}^n$, in addition to the traditional case that $G = \mathbb{R}^n$ (with addition as group operation). For any subsets A, B of G, we define the sets $gA = \{ga: a \in A\}$, $Ag = \{ag: a \in A\}$, $AB = \{ab: a \in A,\ b \in B\}$, and $A^{-1} = \{a^{-1}: a \in A\}$, where $g \in G$. Thus in general $gA \neq Ag$, $AB \neq BA$, and $A \neq A^{-1}$. In case $A = A^{-1}$, the set A is termed *symmetric*. The identity element of G is generally denoted by e $(= I$ for matrices and $= 0$ for $\mathbb{R}^n)$. Clearly $(AB)^{-1} = B^{-1}A^{-1}$.

To gain some insight, and also for later use, we record a few standard facts. [For a recent survey of the subject, see Comfort (1984).]

Proposition 1. *Let G be an arbitrary topological group, and $U \subset G$ be an open set. Then*:

(a) *gU, Ug are open for each $g \in G$, and more generally, for $A \subset G$ (any set) AU and UA are open*;
(b) *if $A \subset G$ is compact and $B \subset G$ is closed, then AB and BA are closed, and compact if B is also compact*;
(c) *G has a symmetric neighborhood basis at e*;
(d) *a subgroup H of G is open iff it has a nonempty interior, and hence an open subgroup is also closed*; *and*
(e) *if $A(\subset G)$ is compact, U open, and $A \subset U$, then there is a neighborhood V of the identity (of G) such that $(AV) \cup (VA) \subset U$, and if G is locally compact, then V may be chosen so that AV and VA are relatively compact. Again V can be taken to be symmetric here.*

Proof. (a): Let $a \in G$ and $f: a \mapsto ga$. Then f is one to one and onto, and is also continuous, since if W is a neighborhood of $f(a)$ $(= ga)$, and W_1, W_2 are neighborhoods of g, a, then $W_1W_2 \subset W$, so that $f(W_1) \subset W$. Thus f is a homeomorphism. But then $f(U) = gU$ is open. Similarly Ug is open, and $AU = \cup\{aU: a \in A\}$ is the union of open sets, so it is open. Likewise UA is open, and $g\bar{U}, \bar{U}g$ are closed ($\bar{U}$ being the closure of U).

(b): If A, B are compact, then $A \times B \subset G \times G$ is compact in the product topology, and since $h: (a, b) \to ab$ is continuous, $h(A, B) = AB$ is the image of a compact set under h. So it is compact. If B is only closed, then the argument is less simple. It is to be shown that AB is closed, or equivalently, its complement is open. So let $x \in (AB)^c$. Then xB^{-1} is closed [as in (a)], and $A \cap xB^{-1} = \varnothing$. Suppose we can find a neighborhood V of the identity e such that $A \cap (VxB^{-1}) = \varnothing$. This will then imply that $(AB) \cap (Vx) = \varnothing$, so that $Vx \subset (AB)^c$. Since Vx is a neighborhood of x which is open by (a), it follows that each point x in $(AB)^c$ has a neighborhood contained in the complement of AB. So AB is closed. Thus it remains to verify the supposition.

We have the hypothesis that the compact set A and the closed set $F = xB^{-1}$ are disjoint. Then one has to find a neighborhood V (and $\tilde{V}$) of e such that $A \cap FV = \varnothing$ (and $A\tilde{V} \cap F = \varnothing$). Indeed, since F^c is open and $x \in A \subset F^c$ implies the existence of a neighborhood W_x of x such that $W_x^2 \subset F^c$ [as a consequence of the continuity of $(x, y) \mapsto xy$], and since $\{W_x: x \in A\}$ is an open covering of the compact set A, we can find a finite subcover $\{W_{x_i}: i = 1, \ldots, n\}$ such that $W_{x_i}^2 \subset F^c, \bigcup_{i=1}^n W_{x_i} \supset A$. Next consider $x_i^{-1}W_{x_i}$, as neighborhoods of e, and set $V_1 = \bigcap_{i=1}^n x_i^{-1}W_{x_i}$. Then V_1 is a neighborhood of the identity such that $xV_1 \cap A = \varnothing$ for each $x \in A$; and $x \in A \Rightarrow x \in W_{x_i}$ for some $1 \le i \le n$. Hence $xV_1 \subset W_{x_i}^2 \subset F^c$, so that $xV_1 \cap F = \varnothing$. Since $x \in A$ is arbitrary, one has $AV_1 \cap F = \varnothing$. Similarly we can find a $\tilde{V}_1$ such that $A \cap F\tilde{V}_1 = \varnothing$. Taking $\tilde{U} = V_1 \cap \tilde{V}_1$, and U as $U^2 \subset \tilde{U}$. Set $V = U \cap U^{-1}$. Then $AV \cap F = \varnothing = FV \cap A$. Hence (b) holds.

(c): This follows from the continuity of $a \mapsto a^{-1}$. In fact, for an arbitrary U, let $V = U \cap U^{-1}$, so that $V \subset U$ and is symmetric.

(d): If the subgroup H is open, then every point of H is an interior point. Conversely, if $x \in H$ is an interior point, then by definition there is a neighborhood V of e such that $xV \subset H$. If now $z \in H$ is arbitrary, then zV is an open set by (a), and $zV = zx^{-1}(xV) \subset zx^{-1}H = H$, since $zx^{-1} \in H$. Hence z is an interior point and H is open. Finally, if $x \notin H$, then $xH \subset H^c$. Hence $\cup\{xH: x \in H^c\} = H^c$. So if H is open, then xH is open, whence H^c is open, i.e., H is closed.

(e): If $A \subset G$ is compact, then clearly there exist open $U \supset A$ and if $x \in A \subset U$, let W_x be a (symmetric) neighborhood of e such that $xW_x \subset U$. Let V_x be a (symmetric) neighborhood of e with $V_x^2 \subset W_x$, so that $\{xV_x: x \in A\}$ is an open covering of A. By compactness there exist $x_iV_{x_i}$, $1 \le i \le n$, covering A. If $V_0 = \bigcap_{i=1}^n V_{x_i}$, then V_0 is a (symmetric) neighborhood of e, $x \in A \Rightarrow x \in x_iV_i$ for some $i \leqslant n$, and

$$AV_0 \subset \bigcup_{i=1}^n x_iV_iV_0 \subset \bigcup_{i=1}^n x_iV_i^2 \subset \bigcup_{i=1}^n x_iW_{x_i} \subset U. \tag{1}$$

By a similar argument, we can find a $\tilde{V}$ such that $\tilde{V}A \subset U$. Set $V = V_0 \cap \tilde{V}$.

Then this will satisfy the inclusion $(AV) \cup (VA) \subset U$. Finally, if G is locally compact, then by Corollary 1.2, $A \subset \tilde{U} \subset U$ with $\tilde{U}$ relatively compact, so that AV, VA are also, completing the proof.

Hereafter we assume that G is a locally compact Hausdorff group. Two more facts have to be recorded before the main result.

Proposition 2. *Let G be any topological group, and $f \in C_c(G)$. Then we have:*

(a) *f is uniformly continuous, i.e., for each $\epsilon > 0$, there is a neighborhood V of e such that for each $xy^{-1} \in V$ and $y^{-1}x \in V$ one has $|f(x) - f(y)| < \epsilon$;*

(b) *if $0 \leq f \in C_c(G)$, $0 \leq g \in C_c(G)$, $g \not\equiv 0$, and G is also locally compact, then there exists a finite set $\{x_1, \ldots, x_n\} \subset G$ and numbers $a_1, \ldots, a_n > 0$ such that $f(y) \leq \sum_{i=1}^n a_i g(x_i^{-1}y)$ for all $y \in G$.*

Proof. (a): Let $K = \text{supp}(f)$. Then K is compact, and by continuity of f, for each $x \in K$, there is [by Proposition 1(c)] a symmetric neighborhood U_x of e such that $y \in xU_x \Rightarrow |f(x) - f(y)| < \epsilon/2$. By the same proposition we can choose a symmetric neighborhood V_x of e such that $V_x^2 \subset U_x$; and xV_x is then a neighborhood of x. Since $\{xV_x: x \in K\}$ is an open covering of K, by compactness there is a finite subcover $\{x_iV_{x_i}: i = 1, \ldots, n\}$ of K. Let $V = \bigcap_{i=1}^n V_{x_i}$. Then $x \in K \Rightarrow x \in x_iV_{x_i}$ for some i, and so for each $y \in G$ such that $y \in xV$, we have $|f(x) - f(y)| < \epsilon/2$, since $x \in xV \subset x_iV_{x_i}V \subset x_iV_{x_i}^2 \subset x_iU_{x_i}$ and $y \in x_iU_{x_i}$. Thus for $x \in K$ and $y \in G$ such that $x^{-1}y \in V$, we get

$$|f(x) - f(y)| \leq |f(x) - f(x_j)| + |f(x_j) - f(y)| < \frac{\epsilon}{2} + \frac{\epsilon}{2} = \epsilon.$$

If $x \in K^c$ and $x^{-1}y \in V$, then $f(x) = 0$, $f(y) = 0$, and we are done. If $f(y) \neq 0$, let $J = \{y: |f(y)| \geq \epsilon\}$. Then $J \subset \text{int}(K) = W$ (say). However, being a closed subset of K, J is compact and contained in the open set W. By Proposition 1(e), there is a symmetric neighborhood $\tilde{V}$ of e such that $J\tilde{V} \subset W = \text{int}(K) \subset K$. Now $f(y) \neq 0$, but $x^{-1}y = v \in \tilde{V}$. Then $y = xv \notin J$, since otherwise $x = yv^{-1} \in J\tilde{V} \subset K$ because $\tilde{V}^{-1} = \tilde{V}$. Thus $y \in K - J$, so that $|f(y)| < \epsilon$, and $|f(x) - f(y)| = |f(y)| < \epsilon$. This proves the left uniform continuity of f if we take $V_0 = V \cap \tilde{V}$ in the above.

In the same way, f is shown to be right uniformly continuous for a symmetric neighborhood $\tilde{V}_0$, and then, taking $V_0 \cap \tilde{V}_0$ as a new neighborhood, one gets the uniform continuity of f on G, as asserted.

(b): Let now G be locally compact, $K = \text{supp}(f)$. Then K is compact, since $0 \leq f \in C_c(G)$. By hypothesis $0 \leq g \in C_c(G)$ is not identically zero. So there is some $x \in G$ such that $g(x) > 0$. If $s \in G$ is arbitrary, consider a point $u \in G$ such that $s = ux^{-1}$ or $x = s^{-1}u$. Then $g(x) = g(s^{-1}u) > 0$. By the continuity of g, there is a relatively compact open V of u and a number

$0 < a < \infty$ such that $f(u) \le ag(s^{-1}u)$ for all $u \in V_s$. Clearly $\{V_s: s \in G\}$ is an open covering of K. So by its compactness, there is a finite subcover V_{s_i}, $1 \le i \le n$, of K and then numbers $0 < a_i < \infty$ such that $f(x) \le \sum_{i=1}^n a_i g(s_i^{-1}x)$ for each $x \in K$. The inequality trivially holds on $G - K$. This completes the proof of the proposition.

The reason why one assumes that G is Hausdorff is that even if G is taken to be a T_0-space (i.e., for each distinct pair x, y in G there is an open set V that contains one but not both the points), then G is not only Hausdorff but even completely regular. In the locally compact case, the T_0-group is even normal. We shall not prove these statements here; they are discussed extensively in Hewitt and Ross (1963, Section 8). Thus one always takes G to be Hausdorff in the context of invariant integration.

Next we present a motivation for the main result to be proved. Recall that, by Theorem 4.1.4, if $f \ge 0$ is a measurable function on (Ω, Σ, μ) then $\int_\Omega f\,d\mu = \inf\{\sum_i a_i \mu(A_i): \sum_i a_i \chi_{A_i} \ge f\}$, where $h = \sum_i a_i \chi_{A_i}$ is a measurable elementary function. If $\Omega = \mathbb{R}^n$ and μ is the Lebesgue measure with $\{A_i: i \ge 1\}$ as an "equipartition" of Ω, then the right side equals $\{\mu(A) = \sum_{i \ge 1} a_i: h \ge f\}$ [$f \in C_c(\mathbb{R}^n)$], where $\mu(A)$ is the length (or "norm") of each A_j. If $A = A_0$, then $A_i = x_i + A_0$ for suitable x_i and $g = \chi_{A_0} > 0$ on A_0 of positive length. Thus relative to each such g, we set $(f: g) = \inf\{\sum_{i=1}^n a_i: f(\cdot) \le \sum_{i=1}^n a_i g(x_i + \cdot)\}$ as an approximation of $\int_\Omega f\,d\mu$ relative to g. Clearly the procedure improves if $g = \chi_{A_0}$ is made to have smaller support. To achieve this we can take another auxiliary function $f^* \ge 0$, $\int_\Omega f^*\,d\mu > 0$, compute $(f^*: g)$ for it, and consider their relative sizes, i.e., $I_g(f) = (f: g)/(f^*: g)$. Now if supp(g) shrinks, both $(f: g)$ and $(f^*: g)$ approach zero, so that $I_g(f)$ may tend to a limit. This is what one shows to be the case. It will be seen that $I_g(f) = I_g(f(s + \cdot))$, so that it is translation invariant (in the Lebesgue case). If $\tilde{f}$ is another such reference function in place of f^*, then the corresponding functional $J_g(f) = (f: g)/(\tilde{f}: g) = I_g(f)(f^*: g)/(\tilde{f}: g)$, and if $c = \lim_{\text{supp}(g)\downarrow}(f^*: g)/(\tilde{f}: g)$, then I_g and J_g [rather their limits $I_0(\cdot)$ and $J_0(\cdot)$] differ by a constant factor. In the Lebesgue case we are essentially using $f^* = \chi_{[0,1]}$ or a constant multiple of it. The surprising fact is that this idea has been made rigorous by A. Haar, and it will be demonstrated by showing that the $I_g(\cdot)$ has a limit I_0 which is a positive homogeneous subadditive functional on $C_c^+(\Omega)$, and if $I: C_c(\Omega) \to \mathbb{R}$ is defined by $I(f) = I_0(f^+) - I_0(f^-)$, then $I(\cdot)$ is a positive linear functional on $C_c(\Omega)$, which is invariant under translations (to the left). Then Theorem 3.1 will give a unique Radon measure on Ω that is translation invariant, called a *Haar measure* on Ω.

With this general motivating sketch, one has the following:

Theorem 3. *Let G be an arbitrary locally compact* (*Hausdorff*) *group. Then there exists a nonzero positive linear left-invariant functional I: $C_c(G) \to \mathbb{R}$ such that* (i) $I(f) > 0$ *if* $0 \ne f \in C_c^+(G)$, (ii) $I(af + bg) = aI(f) + bI(g)$, *$a, b$ in*

$\mathbb{R}$, (iii) $I(f(s^{-1}\cdot)) = I(f)$ *for all* $s \in G$. *If* $J: C_c(G) \to \mathbb{R}$ *is another such (left-)invariant nonzero positive linear functional, then* $I = aJ$ *for some* $a > 0$. *Thus I is essentially unique.*

Proof. Since there are many details to be proved, we present the argument in steps for convenience. The result will also be given a measure version, and indeed the uniqueness part is given for the latter form. This illuminates the interplay.

1. As noted in the discussion prior to the theorem, we start with $f, g \in C_c^+(G)$, $g \not\equiv 0$. Then by Proposition 2(b) there exist numbers $a_1, \ldots, a_n > 0$ and points $s_1, \ldots, s_n$ of G satisfying

$$f(\cdot) \leq \sum_{i=1}^{n} a_i g(s_i^{-1}\cdot). \tag{$*$}$$

Let $(f: g) = \inf\{\Sigma_{i=1}^n a_i: (*) \text{ holds}\}$. The expression $(f: g)$ has the following evident properties on $C_c^+(G)$:

(a) $(L_s f: g) = (f: L_s g) = (f: g)$, where $(L_s f)(x) = f(s^{-1}x)$, $s \in G$, $x \in g$;
(b) $(af: g) = a(f: g)$ if $a \geq 0$; $(f_1: g) \leq (f_2: g)$ if $0 \leq f_1 \leq f_2$;
(c) $(f_1 + f_2: g) \leq (f_1: g) + (f_2: g)$;
(d) $\|f\| \leq \Sigma_{i=1}^n a_i \|g\| \Rightarrow (f: g) \geq \|f\|/\|g\|$, where $\|f\| = \sup_x |f(x)|$.

Moreover, one also has for $g \neq 0$, $h \neq 0$ in $C_c^+(G)$

(e) $(f: g) \leq (f: h)(h: g)$.

To see this, by Proposition 2(b) again, note that $f \leq \Sigma_{i=1}^n a_i L_{s_i} h$ and similarly $h \leq \Sigma_{j=1}^m b_j L_{t_j} g$, so that $f(x) \leq \Sigma_{i=1}^n a_i L_{s_i}(\Sigma_{j=1}^m b_j L_{t_j} g)(x) = \Sigma_{i=1}^n \Sigma_{j=1}^m a_i b_j (L_{s_i t_j} g)(x)$. Hence

$$(f: g) = \inf\left\{\sum_{i=1}^{n}\sum_{j=1}^{n} a_i b_j: (*) \text{ holds}\right\}$$

$$= \inf\left\{\left(\sum_{i=1}^{n} a_i\right)\left(\sum_{j=1}^{m} b_j\right): (*) \text{ holds}\right\} \leq (f: h)(h: g). \tag{2}$$

Let f^* be a *fixed* nonzero element of $C_c^+(G)$, and define

$$I_g(f) = \frac{(f: g)}{(f^*: g)}, \qquad f \in C_c^+(G). \tag{3}$$

Of course, I_g depends on the reference function f^* even though it is not displayed. Clearly $I_g(0) = 0$, and since $(f: f) = 1$ for $f \neq 0$, we have by (e) above for such f

$$\frac{1}{(f^*: f)} = \frac{(f: g)}{(f^*: f)(f: g)} \leq \frac{(f: g)}{(f^*: g)} = I_g(f)$$

$$\leq \frac{(f: f^*)(f^*: g)}{(f^*: g)} = (f: f^*), \tag{4}$$

so that $I_g(f)$ is bounded independently of g for each $f \neq 0$. Also (a)–(c) above imply that $I_g(\cdot)$ is monotone, positively homogeneous, (left) translation-invariant, and subadditive. We now show that $I_g(\cdot)$ is nearly additive if $\mathrm{supp}(g)$ is small, and then establish that it converges to the desired limit when this support shrinks to a point.

2. For each $\epsilon > 0$ and f_1, f_2 in $C_c^+(G)$, there is a neighborhood V of the identity e such that $\mathrm{supp}(g) \subset V \Rightarrow I_g(f_1) + I_g(f_2) \leq I_g(f_1 + f_2) + \epsilon$. For, let $K = \mathrm{supp}(f_1 + f_2)$ and $1 > \delta > 0$ be given (in addition to $\epsilon > 0$). Select a $v \in C_c^+(G)$ such that $v|_K = 1$, and consider $h = f_1 + f_2 + \delta v$. Since $\mathrm{supp}(f_i) \subset K$, if $h_i = f_i/h$, then $h_i(x) = 0$ for $x \notin \mathrm{supp}(f_i)$, and h_i is continuous, $h_1 + h_2 \leq 1$. By Proposition 2(a), h_i is uniformly continuous. Hence for the $\delta > 0$, there is a (symmetric) neighborhood V of e such that for x, y in G with $x^{-1}y \in V$ we have $|h_i(x) - h_i(y)| < \delta/2$, $i = 1, 2$. If $0 \leq g \in C_c^+(G)$ has its support contained in V (and $g \neq 0$), then $h \leq \sum_{i=1}^n a_i L_{s_i}(g)$ for some $a_i > 0$, $s_i \in G$. This gives

$$f_i(x) = (hh_i)(x) \leq \sum_{j=1}^n a_j h_i(x)(L_{s_j}g)(x)$$

$$\leq \sum_{j=1}^n a_j g(s_j^{-1}x)\left[h_i(s_j) + \frac{\delta}{2}\right], \qquad i = 1, 2, \tag{5}$$

because $g(s_j^{-1}x) = 0$ unless $s_j^{-1}x \in V$, in which case $|h_i(x) - h_i(s_j)| < \delta/2$. Thus (5) implies

$$(f_i: g) \leq \sum_{j=1}^n a_j\left(h_i(s_j) + \frac{\delta}{2}\right), \qquad i = 1, 2.$$

Now summing these over i and using the fact that $h_1 + h_2 \leq 1$, we get

$$(f_1: g) + (f_2: g) \leq \sum_{j=1}^n a_j(1 + \delta). \tag{6}$$

Taking the infimum over all such a_j's, one has from (6)

$$(f_1: g) + (f_2: g) \leq (1 + \delta)(h: g). \tag{7}$$

Dividing (7) by $(f^*: g)$, this yields

$$I_g(f_1) + I_g(f_2) \leq (1 + \delta)I_g(h) \leq (1 + \delta)\left[I_g(f_1 + f_2) + \delta I_g(v)\right],$$

using the subaddivity of I_g. This and (4) give

$$I_g(f_1) + I_g(f_2) \leq I_g(f_1 + f_2) + \delta(f_1 + f_2: f^*) + (1 + \delta)\delta(v: f^*). \tag{8}$$

If we take $\delta = \epsilon/\max([(f_1 + f_2: f^*) + 2(v: f^*)], 1)$ in (8), then $0 < \delta < 1$ and one obtains the stated result.

3. We now assert that there is an $I: C_c(G) \to \mathbb{R}$ which is a positive linear left-invariant functional, as required by the theorem. This does not depend on the auxiliary g's but will depend on the fixed f^*.

Consider, for each $f \in C_c^+(G)$ with $\|f\| > 0$, the compact interval $J_f = [(f^*: f)^{-1}, (f: f^*)]$. By (4), $I_g(f) \in J_f$ for each $\|f\| > 0$ and any $g \in C_c^+(G)$, $g \neq 0$. If $J = \times_f J_f$, then by Tychonov's theorem J is a nonempty compact set in the product topology. So $a_g = \{I_g(f): 0 \neq f \in C_c^+(G)\}$ is an element of J. For each (symmetric) neighborhood V of e, let A_V be the closure in the compact set J of the set of points a_g for all $0 < g \in C_c^+(G)$ such that $\operatorname{supp}(g) \subset V$. Clearly A_V is nonempty and compact, and $\bigcap_{i=1}^n A_{V_i} \supset A_{\bigcap_{i=1}^n V_i}$, so that the family $\{A_V: V \text{ as above}\}$ has the finite intersection property. Hence there is at least one element, denoted I, in $\cap\{A_V: V \text{ a neighborhood of } e\}$. Thus every neighborhood of I intersects $\{I_g: \operatorname{supp}(g) \subset V\}$ for all V, which means, for every $\epsilon > 0$, $f_1, \ldots, f_n$ in $C_c^+(G)$, and $\operatorname{supp}(g) \subset V$, we have $|I(f_i) - I_g(f_i)| < \epsilon$, $i = 1, \ldots, n$. Since $I_g(\cdot)$ is positively homogeneous, left invariant, and subadditive, so is $I(\cdot)$ and by step 2, $I(\cdot)$ is also superadditive and hence I is additive on $C_c^+(G)$. Moreover, for $f \neq 0$, $I(f) \in J_f$, so that $0 < (f^*: f)^{-1} \leq (f: f^*) < \infty$. Set $I(0) = 0$, and note that $I(f^*) = 1$.

Extend I to $C_c(G)$ as follows. If $f \in C_c(G)$, let $f = f^+ - f^-$, so that $0 \leq f^{\pm} \in C_c(G)$. Now let $I(f) = I(f^+) - I(f^-)$. If $f = f_1 - f_2$ is another representation, $0 \leq f_i \in C_c(G)$, then $f^+ - f^- = f = f_1 - f_2 \Rightarrow f^+ + f_2 = f_1 + f^-$, so that $I(f^+ + f_2) = I(f^+) + I(f_2) = I(f_1 + f^-) = I(f_1) + I(f^-)$, implying $I(f^+) - I(f^-) = I(f_1) - I(f_2) = I(f)$. Thus I is uniquely defined, and is a left-invariant positive linear functional. This proves the existence. We translate this to the measure version and then prove its essential uniqueness.

Recall that by Theorem 3.1, there is a unique Radon measure μ on G such that $I(f) = \int_G f\,d\mu$, $f \in C_c(G)$, and μ is a left Haar measure, in the sense that $\mu(sA) = \mu(A)$ for all Borel sets $A \subset G$, $s \in G$. Indeed, the mapping $\tau: x \mapsto s^{-1}x$ being a homeomorphism of G onto G, $sA = \tau^{-1}(A)$ is a Baire (Borel) set if A is such. So the above result follows if it is true for compact Baire sets A. But then there are $0 \leq f_n \in C_c(G)$ such that $f_n \downarrow \chi_A$, by Proposition 1.1, and hence, by the dominated convergence theorem, we get

$$\mu(A) = \lim_n \int_G f_n \, d\mu = \lim_n I(f_n) = \lim_n I(L_s f_n)$$

$$= \lim_n \int_G (L_s f_n) \, d\mu = \mu(sA). \tag{9}$$

Because of the one to one relation between I and μ, we can consider either of these in our work.

4. If I and $\tilde{I}$ (or equivalently μ and $\tilde{\mu}$) are left-invariant Haar functionals (or measures), then $I = a\tilde{I}$ (or $\mu = a\tilde{\mu}$) for some $a > 0$.

We proceed with μ and $\tilde{\mu}$. It suffices to show, for any compact sets C, $\tilde{C}$ of G,

$$\mu(C)\tilde{\mu}(\tilde{C}) = \mu(\tilde{C})\tilde{\mu}(C). \tag{$**$}$$

For then $a = \mu(C)/\tilde{\mu}(C)$ is the same for any compact set, and hence μ, $\tilde{\mu}$ satisfy $\mu = a\tilde{\mu}$ on the lattice of compact sets, so that they agree on all Borel sets. First we assert that for any compact $C \subset G$ and a relatively compact open V,

$$\mu(C)\tilde{\mu}(V) \leq \tilde{\mu}(CV)\mu(V^{-1}). \tag{10}$$

For consider the left side. We have

$$\mu(C)\tilde{\mu}(V) = \int_G \int_G \chi_C(x)\chi_V(y)\mu(dx)\tilde{\mu}(dy)$$

$$= \int_G \int_G \chi_C(x)\chi_V(x^{-1}y)\mu(dx)\tilde{\mu}(dy)$$

(by the left invariance of $\tilde{\mu}$),

$$= \int_G \int_G \chi_{Cy}(xy)\chi_V(x^{-1}y)\tilde{\mu}(dy)\mu(dx)$$

(by Fubini's theorem)

$$\leq \int_G \int_G \chi_{CV}(xy)\chi_V(x^{-1}y)\tilde{\mu}(dy)\mu(dx)$$

(since $CV \supset Cy$ for all $y \in V$)

$$= \int_G \int_G \chi_{CV}(y)\chi_V(x^{-1}y)\tilde{\mu}(dy)\mu(dx)$$

(by the left invariance of $\tilde{\mu}$ for $y \mapsto xy$),

$$= \int_G \int_G \chi_{CV}(y)_{V^{-1}}(y^{-1}x)\mu(dx)\tilde{\mu}(dy)$$

(by Fubini's theorem),

$$= \int_G \int_G \chi_{CV}(y)\chi_{V^{-1}}(x)\mu(dx)\tilde{\mu}(dy)$$

(by the left invariance of μ)

$$= \mu(V^{-1})\tilde{\mu}(CV). \tag{11}$$

Interchanging μ and $\tilde{\mu}$ in the above computation, with $\tilde{C}$ and V^{-1} in place of C and V, we get for (11)

$$\tilde{\mu}(\tilde{C})\mu(V^{-1}) \leqslant \mu(\tilde{C}V^{-1})\tilde{\mu}(V). \tag{12}$$

But by the fact that μ and $\tilde{\mu}$ are Radon measures, they are inner regular, so that V and V^{-1} contain compact sets supporting an $f \in C_c(G)$, $f \neq 0$, implying that $\tilde{\mu}(V)\mu(V^{-1}) > 0$. Multiplying (11) and (12), we have

$$\mu(C)\tilde{\mu}(\tilde{C}) \leqslant \mu(\tilde{C}V^{-1})\tilde{\mu}(CV) \tag{13}$$

(after canceling the positive factor). On the other hand $\mu, \tilde{\mu}$ are also outer regular. Hence for any $\epsilon > 0$, there exist open sets $U, \tilde{U}$ such that $U \supset C$, $\tilde{U} \supset \tilde{C}$, $\mu(\tilde{C}) > \mu(\tilde{U}) - \epsilon$, and $\tilde{\mu}(C) > \tilde{\mu}(U) - \epsilon$. Since V is any relatively compact open set in (13), choose V as such a neighborhood of e satisfying $CV \subset U$ and $\tilde{C}V^{-1} \subset \tilde{U}$. Then (13) becomes

$$\mu(C)\tilde{\mu}(\tilde{C}) \leqslant \mu(U)\tilde{\mu}(\tilde{U}) < [\mu(\tilde{C}) + \epsilon][\tilde{\mu}(C) + \epsilon]. \tag{14}$$

Since $\epsilon > 0$ is arbitrary, (14) implies $\mu(C)\tilde{\mu}(\tilde{C}) \leqslant \mu(\tilde{C})\tilde{\mu}(C)$. By interchanging C and $\tilde{C}$ in this computation we get the opposite inequality, and hence equality holds, i.e., $(**)$ is true. Because of the earlier observation, this shows that $\mu = a\tilde{\mu}$. Hence the theorem is completely demonstrated.

Remark. The existence proof given here is due to A. Weil (1938), and because of the use of Tychonov's theorem, it depends on the axiom of choice. However, with a more involved (and longer) analysis, H. Cartan (1940) showed that one can avoid this axiom in the proof. The details are given in Hewitt and Ross (1963). The uniqueness proof given above is due to S. Kakutani (1948), and again other versions are available. The present demonstration appears to be the shortest one.

A measure ν on G is called a *right-invariant Haar measure* if it is Radon and if $\nu(As) = \nu(A)$ for each Borel set A and $s \in G$. But if μ is a left Haar measure and $\nu(A) = \mu(A^{-1})$, then ν is easily seen to be a right Haar measure, and conversely. Hence the above theorem also demonstrates the existence and essential uniqueness of a right Haar measure. Next note that if μ is a left Haar measure and $s \in G$ is fixed, then $\tilde{\mu}(A) = \mu(As)$, for each Borel set $A \subset G$, again defines a left Haar measure $\tilde{\mu}$. Hence by the above theorem $\tilde{\mu} = a(s)\mu$ for some number $a(s) > 0$. It is immediately noted that $a(st) = a(s)a(t)$, and locally bounded, so that $a: G \to (0, \infty)$ is a continuous multiplicative function determined by G and does not depend on μ. (See Exercise 2.) It is called the *modular function* of G, denoted by $\Delta(\cdot)$, and the group is termed *unimodular* if $\Delta(s) = 1$, $s \in G$.

One has the following observation.

Proposition 4. *Let G be a locally compact group, and μ be a (left) Haar measure on it. Then $\mu(G) < \infty$ iff G is compact, and if either G is compact or abelian, it is unimodular.*

[Also, for compact G, $\mu(G)$ is usually *normalized* so that $\mu(G) = 1$.]

Proof. Since μ is finite on compact sets (being a Radon measure), it is clear that $\mu(G) < \infty$ if G is a compact group. On the other hand, if G is not compact, then there is a relatively compact neighborhood U of the identity e such that a finite number of translates $\{x_iU: 1 \leqslant i \leqslant n\}$ cannot cover G. Let $x_1 = e$, and choose inductively $x_1, \dots, x_n$ from G such that $x_n \in (\bigcup_{i=1}^{n-1} x_iU)^c$. This is possible because $\bigcup_{i=1}^{n-1} x_iU_i$ is relatively compact and G is not. Let V be a symmetric neighborhood of e such that $V^2 \subset U$. Then $\{x_nV: n \geqslant 1\}$ is a disjoint sequence of open sets. For otherwise, $(x_nV) \cap (x_mV) \neq \varnothing$, and if $n > m$, we must have $x_n = x_m v\tilde{v}^{-1}$ for some $v, \tilde{v}$ in V, so that $x_n \in x_mV^2 \subset x_mU$. But this is impossible by the choice of x_n's. Hence $\bigcup_{i=1}^{n} x_iV \subset G$, and by disjointness of these sets and left invariance of μ, we have

$$\mu(G) \geqslant \mu\left(\bigcup_{i=1}^{n} x_iV\right) = n\mu(V), \qquad n \geqslant 1.$$

Since $\mu(V) > 0$ and n is arbitrary, $\mu(G) = \infty$ must hold.

If G is compact, then $Gx = G$ for all $x \in G$. Hence $\mu(G) = \mu(Gx) = \Delta(x)\mu(G)$. Since $\mu(G) > 0$, one has $\Delta(x) = 1$, $x \in G$. Thus G is unimodular. The Abelian case is evident. This finishes the proof.

As a final item of this section we record another property.

Proposition 5. *Let μ be a left Haar measure on G. Then*

$$\int_G f(x^{-1})\mu(dx) = \int_G f(x)\Delta(x^{-1})\mu(dx), \qquad f \in C_c(G). \tag{15}$$

Hence right and left Haar measures are mutually equivalent. If (G_i, μ_i), $i = 1, 2$, *is a pair of (left) Haar measured locally compact groups, then* $\mu_1 \otimes \mu_2$ *is a left Haar measure of* $G_1 \times G_2$ *with modular function* $\Delta_1\Delta_2$, *where the* Δ_i *are the modular functions of* G_i, $i = 1, 2$. *Thus* $G_1 \times G_2$ *is unimodular iff both* G_1 *and* G_2 *are.*

Proof. If $l(f) = \int_G f(x^{-1})\mu(dx)$, then $l(f^s) = l(f)$, so that it is right invariant, where $f^s(x) = f(xs)$. On the other hand, if $\tilde{l}(f)$ is the right side of (15), then it is also right invariant, since

$$\tilde{l}(f^s) = \int_G f(xs)\Delta(x^{-1})\mu(dx)$$

$$= \Delta(s)\int_G f(xs)\Delta\big((xs)^{-1}\big)\mu(dx),$$

(since Δ is a homomorphism)

$$= \int_G f(t)\Delta(t^{-1})\mu(dt) = \tilde{l}(f).$$

Hence $\tilde{l} = a_0 l$ for some $a_0 > 0$. Given $\epsilon > 0$, choose a symmetric relatively compact neighborhood V of e such that $|\Delta(t) - 1| < \epsilon$ for $t \in V$. Then with $f = \chi_V$, since the integrals defining $l, \tilde{l}$ are also valid for Borel functions, we get

$$0 = \int_G f(x)\big[a_0 - \Delta(x^{-1})\big]\mu(dx), \qquad \text{since } f \text{ is symmetric.}$$

Thus $a_0\mu(V) = \int_V \Delta(x^{-1})\mu(dx)$, and hence $\mu(V)(a_0 - 1) = \int_V[\Delta(x^{-1}) - 1]\,\mu(dx)$, so that $\mu(V)|a_0 - 1| \leqslant \epsilon\mu(V) \;\Rightarrow\; a_0 = 1$. This proves (15).

Taking $f = \chi_A$, we get $\mu(A^{-1}) = \int_A \Delta(x^{-1})\mu(dx)$. Since ν: $A \mapsto \mu(A^{-1})$ is a right Haar measure, it follows that $\nu \ll \mu \ll \nu$.

By definition of the product measure, $\mu_1 \otimes \mu_2(A \times B) = \mu_1(A)\mu_2(B)$ for all Borel $A \subset G_1$, $B \subset G_2$. So $\mu_1 \otimes \mu_2$ is a left Haar measure on $G_1 \times G_2$. Also

$$\mu_1 \otimes \mu_2(Aa \times Bb) = \mu_1(Aa)\mu_2(Bb) = \Delta_1(a)\Delta_2(b)\mu_1(A)\mu_2(B)$$

$$= \Delta_1(a)\Delta_2(b)\mu_1 \otimes \mu_2(A \times B)$$

$$\big[= \Delta(a, b)\mu_1 \otimes \mu_2(A \times B)\big].$$

Hence $\Delta(a, b) = \Delta_1(a)\Delta_2(b)$ defines the modular function of $G_1 \times G_2$. The rest is immediate and the proof is finished.

EXERCISES

1. Give an example to show that the product of two closed subsets (even subgroups) H_1 and H_2 of a topological group G need not be closed. [Try $G = \mathbb{R}$, $H_1 = Z$ the integers, and $H_2 = \{\alpha n: n \in Z, \alpha > 0$ an irrational number$\}$.]
2. Show that the modular function Δ of a locally compact group is a continuous homomorphism into $(0, \infty)$.
3. Let $H \subset G$ be an open subgroup of a locally compact group G. If μ and $\tilde{\mu}$ are the left Haar measures on H and G respectively, verify that μ can be taken as the restriction of $\tilde{\mu}$ to the Borel σ-ring of H.
4. Let $L^p(G, \mu)$, $1 \leqslant p < \infty$, be the usual Lebesgue space on a (left) Haar measured locally compact group space (G, μ). Show that the functions $s \mapsto \|f - L_s f\|_p$ and $s \mapsto \|f - R_s f\|_p$ are uniformly continuous, where $(R_s f)(x) = f(xs)$ and $(L_s f)(x) = f(s^{-1}x)$ are the right and left translates of f. [Note that $C_c(G)$ is dense in $L^p(G, \mu)$, and verify the result first for $C_c(G)$, after recalling Proposition 2(a).]
5. Let G be the group of all 2 by 2 nonsingular matrices with the usual matrix multiplication as the group operation. Identifying G as an open subset of $\mathbb{R}^4$, it becomes locally compact. Verify that its Haar measure μ is given by $d\mu(x) = dx_1\, dx_2\, dx_3\, dx_4/[\det(x)]^2$, where

$$x = \begin{pmatrix} x_1 & x_2 \\ x_3 & x_4 \end{pmatrix} \in G$$

and $\det(x)$ is its determinant, and that the group is unimodular. Here $dx_1 \cdots dx_4$ is the Lebesgue measure of $\mathbb{R}^4$.
6. Let $H \subset G$ be the subgroup of the (unimodular) group of the above exercise, where $x \in H$ iff

$$x = \begin{pmatrix} x_1 & x_2 \\ 0 & 1 \end{pmatrix} \quad \text{with} \quad x_1 \neq 0, \quad (x_1, x_2) \in \mathbb{R}^2.$$

Verify that its left and right Haar measures μ and ν are given by $d\mu(x) = dx_1\, dx_2/x_1^2$ and $d\nu(x) = dx_1\, dx_2/|x_1|$. Again $dx_1\, dx_2$ is the Lebesgue measure of $\mathbb{R}^2$. Show that the modular function Δ is given by $\Delta(x) = 1/|x_1|$. (Thus a subgroup of a unimodular group need not be unimodular. However, this cannot happen if the subgroup H is also normal. See Exercise 8 below.)
7. The group H considered in Exercise 6 can be used to show that if A, B are two (noncompact) subsets, of finite (left) Haar measure, then their product AB may have infinite (left) Haar measure. Verify this if

$$A = B = \left\{x \in H: x = \begin{pmatrix} x_1 & x_2 \\ 0 & 1 \end{pmatrix}, 1 \leqslant x_1 < \infty, 0 \leqslant x_2 \leqslant 1\right\},$$

so that $AB = \{x \in H: 1 \leqslant x_1 < \infty \text{ and } 0 \leqslant x_2 \leqslant x_1 + 1\}$.

8. Let H be a closed normal subgroup of a locally compact group G, and consider the quotient group G/H with $\pi: G \to G/H$ as the canonical map. Then a set $A \subset G/H$ is declared open iff $\pi^{-1}(A)$ is an open set in G. This makes G/H also locally compact, and π becomes a continuous open mapping. [So $B \subset G$ open implies $\pi(B)$ open also.] Thus H, G, and G/H carry their own Haar measures, denoted μ_H, μ_G, and $\mu_{G/H}$. Show that given any two of these, the third one can be normalized to satisfy

$$\int_{G/H}\left(\int_H f(xy)\mu_H(dy)\right)\mu_{G/H}(d\dot{x}) = \int_G f(x)\mu_G(dx), \qquad f \in C_c(G),$$

where $\dot{x} = xH \in G/H$. This is called *Weil's formula* and plays a crucial role in harmonic analysis as well as number theory. [Note that $g(\dot{x}) = \int_H f(xy)\mu_H(dy)$ defines $g: G/H \to \mathbb{R}$ and g is constant on the (left) cosets of H. If $\tilde{g} = g \circ \pi$, then $\tilde{g}$ is continuous iff g is, and $\tilde{g}$ is termed an *H-periodic* function. Now $l(f) = \int_{G/H}[\int_H f(xy)\mu_H(dy)]\mu_{G/H}(d\dot{x})$ is a positive linear functional on $C_c(G)$ which is left invariant, and $l(f) \neq 0$ if $f \neq 0$. So by Theorem 3, $l(f) = a_0\int_G f\,d\tilde{\mu}_G$ for some $a_0 > 0$. Take $\mu_G = a_0\tilde{\mu}_G$.] Using this formula, deduce that a closed normal subgroup of a unimodular group is unimodular.

9. A locally compact group G is called *amenable* if there exists a positive linear left-invariant functional l on $C_b(G)$, the space of all bounded continuous real functions on G, such that $l(1) = 1$. All compact and Abelian as well as "solvable" groups are amenable, but there are groups which are not amenable (e.g., the G of Exercise 5). Consider a local functional $\Lambda: C_c(G) \to \mathbb{R}$ satisfying conditions (i)–(iii) of Theorem 3.5 with G amenable. Suppose that $\Lambda(L_x f) = \Lambda(f)$, $x \in G$, so that it is (left-) invariant. Show that in the integral representation there one has

$$\Lambda(f) = \int_G \Phi(f(\omega))\mu(d\omega), \qquad f \in C_c(\Omega),$$

where μ is a left Haar measure, and $\Phi: \mathbb{R} \to \mathbb{R}$ is continuous and satisfies conditions (a)–(c) of that theorem. (In step 5 of the proof of Theorem 3.5, λ^h will be left invariant on a certain algebra, and by a known result, *with* amenability, it can be extended to the Borel σ-ring to be invariant. Then the rest of the argument admits a simplification, and step 7 will be unnecessary.)

10

SOME COMPLEMENTS AND APPLICATIONS

Relations between abstract and topological measures and some applications are considered in this final chapter. Utilizing the lattice properties of measure spaces, a topological measurable space is associated with each measurable space through the Stone isomorphism theorem. With this, it is shown that abstract Lebesgue spaces $L^p(\mu)$ are isometrically isomorphic to the corresponding classes on certain locally compact regular Borel measure spaces. An application to a Stone-Weierstrass theorem on such function spaces is included, and the interactions between the abstract and topological ideas of measures are discussed.

10.1. LATTICE AND HOMOMORPHISM PROPERTIES

A general method of connecting an abstract measure space with a topological one is by means of a measurable mapping of the first to the second with respect to its Borel (or Baire) σ-algebra. Through such a mapping, one induces an image measure, which turns out to be regular or of Radon type. This procedure is common, for instance, in probability theory, where the mapping is a "random variable" and the image measure is termed a distribution function, as noted in Section 3.3. A special case of Theorem 4.2.6 for integrals with image measures will be recalled here to motivate the results to be established.

Lemma 1. *Let* (Ω, Σ, μ) *be a measure space, and* (Ω', Σ') *a measurable space. If* $f: \Omega \to \Omega'$ *is a* (Σ, Σ')*-measurable mapping and* $\mu' = \mu \circ f^{-1}$*, then* $\mu': \Sigma' \to \bar{\mathbb{R}}^+$ *is a measure, and for each* $g: \Omega' \to \bar{\mathbb{R}}$*, measurable for* Σ'*, one has*

$$\int_\Omega g \circ f \, d\mu = \int_{\Omega'} g \, d\mu' \tag{1}$$

in the sense that if either integral exists, so does the other and equality holds.

This is implied by the result already proved for generated outer measures in Theorem 4.2.6. In the present case, it is evident if $g = \chi_A$, $A \in \Sigma'$, and then for simple functions by linearity. The general case follows by approximation with such functions, and the monotone convergence theorem.

The problem here is that f^{-1} is not a point function. But it preserves set operations, $f^{-1}: \Sigma' \to \Sigma$. Such a (set) mapping is a homomorphism. More precisely, a mapping $h: \Sigma' \to \Sigma$ is a (σ-)*homomorphism* if

(i) $h(A \cup B) = h(A) \cup h(B)$,
(ii) $h(A - B) = h(A) - h(B)$,
[(iii) $A_n \in \Sigma'$, $n \geqslant 1 \Rightarrow h(\bigcup_{n=1}^{\infty} A_n) = \bigcup_{n=1}^{\infty} h(A_n)$].

Thus the inverse mapping of a point function is a σ-homomorphism. However, there are other (set) mappings which are not necessarily of the above type, as simple examples show. For instance, let A be an uncountable set with $\mathscr{A}$ as its σ-algebra defined as the class of subsets that either are countable or have countable complements. Let $\mathscr{B} \subset \mathscr{A}$ be the trivial algebra $\{\varnothing, A\}$. If $h: \mathscr{A} \to \mathscr{B}$ is defined as $h(E) = \varnothing$ or $= A$ according as E or E^c is countable, then h is a σ-homomorphism. If there is an $f: A \to A$ such that $f^{-1} = h$, then $f^{-1}(E) = \varnothing$, or $f(\varnothing)$ is a countable set (for E countable). But now f is not a function (it will be a "multifunction"). Thus an immediate question is to find conditions in order for h to be induced by a point mapping. The following result, due to R. Sikorski (1969, Section 15), is in this direction.

Proposition 2. *Let (Ω, Σ, μ) be a complete measure space, and $([0,1] = I, \mathscr{B})$ be the Borelian space of the unit interval. If $h: \mathscr{B} \to \mathscr{A} = \Sigma/\mathscr{N}$ is a given σ-homomorphism such that $h(I) = \Omega$, then there is an essentially unique point mapping $f: \Omega \to I$ such that $h = f^{-1}$. Here $\mathscr{N}$ is the class of μ-null sets, and essential uniqueness means that for any other $g: \Omega \to I$, with $g^{-1} = h$, we have $f(\omega) = g(\omega)$ for almost all ω.*

Proof. Let $h: \mathscr{B} \to \mathscr{A}$ be the given σ-homomorphism, and enumerate the rationals of I as $\{r_1, r_2, \dots\}$. Set $A_k = h([0, r_k])$ and $\Omega = h([0,1])$. Then $A_k \in \mathscr{A} = \Sigma/\mathscr{N}$, so that A_k is the equivalence class of sets $\tilde{A}_k \in \Sigma$ (i.e., $\tilde{A}_k, \bar{A}_k$, in Σ with $\mu(\tilde{A}_k \Delta \bar{A}_k) = 0$, implying $A_k = [\tilde{A}_k]$). Since $h(\cdot)$ preserves lattice operations, for $r_k < r_{k'}$ we get $\tilde{A}_k \subset \tilde{A}_{k'}$ a.e., and so $B_{kk'} = \tilde{A}_k - \tilde{A}_{k'} \in \mathscr{N}$. If $B = \bigcup\{B_{kk'}: r_k \leqslant r_{k'}\}$, then $B \in \mathscr{N}$ and so $\tilde{A}_k \cup B = C_k \in [\tilde{A}_k]$. Also the definition implies $C_k \subset C_{k'}$ for $r_k < r_{k'}$, $C_k \in \Sigma$. Define $f: \Omega \to I$ as follows. For each $\omega \in \Omega$, let $f(\omega)$ be the first index k of C_k, with $\omega \in C_k$: i.e., $f(\omega) = \inf\{k: \omega \in C_k\}$. Then taking $r_\infty = 1$, $r_0 = 0$, one has $\omega \in \Omega = C_\infty$, $\varnothing = C_0$, and $0 \leqslant f(\omega) \leqslant 1$. Evidently f is measurable for $(\Sigma, \mathscr{B})$, since $\{f(\omega) \leqslant r\} = \bigcup\{C_{r_k}: r_k \leqslant r\}$, and $f^{-1}([0,r]) = C_r$ for each $r \in \{r_k: k \geqslant 1\}$. It is clear also that $h([0,r]) = f^{-1}([0,r])$. If $\mathscr{C} = \{B \in \mathscr{B}: f^{-1}(B) = h(B)\}$, then by the fact that f^{-1} and h preserve all the (countable) set operations, $\mathscr{C}$ is a σ-algebra which contains intervals with (rational and then general) end points $0 \leqslant a \leqslant b \leqslant 1$, so that it includes the Borel σ-algebra. Thus $\mathscr{C} = \mathscr{B}$, and $h = f^{-1}$.

For the essential uniqueness, let $g: \Omega \to I$ be another function such that h is also induced by g. Now $A_{rs} = \{\omega: f(\omega) \leqslant r < s < g(\omega)\} = \{\omega: f(\omega) \leqslant r\} \cap \{\omega: g(\omega) > s\} = f^{-1}([0, r]) \cap g^{-1}((s, 1]) = C_r \cap C_s^c \in \mathscr{N}$, so that $\{\omega: f(\omega) < g(\omega)\} = \bigcup\{A_{rs}: r < s, \text{rationals}\} \in \mathscr{N}$. Hence $f = g$ a.e., and the result follows. [The construction of f is similar to that in Lemma 3.1.5.]

The counterexample preceding the statement indicates that the conclusion cannot be strengthened without restricting the hypothesis. However, we shall present an interesting application of this result as an exercise with a sketch. (See Exercise 2.)

Instead of fixing the two spaces (Ω, Σ, μ) and $(I, \mathscr{B})$ on which a (σ-)homomorphism is given, we shall now show that with any given (Ω, Σ, μ), it is possible to associate a topological space $(\tilde{\Omega}, \tilde{\Sigma})$ and a homomorphism $h: \Sigma \to \tilde{\Sigma}$ such that $(\tilde{\Omega}, \tilde{\Sigma}, \tilde{\mu})$ is a regular Borel measure space, where $\tilde{\mu} = \mu \circ h^{-1}$, the image measure. In fact h will be an isomorphism; this fundamental representation is due to M. H. Stone. Some applications using the work of the last chapter will also be given later.

Theorem 3 (Stone). *Let (Ω, Σ) be a pair with Σ as an algebra. Then there exists an extremally disconnected*[†] *compact Hausdorff space $\tilde{\Omega}$ with its algebra $\tilde{\Sigma}$ generated by all of its clopen (= closed-open) subsets such that Σ is isomorphic to $\tilde{\Sigma}$. In case Σ is only a ring, then $\tilde{\Omega}$ will be locally compact, $\tilde{\Sigma}$ will be a ring generated by the compact open sets, and Σ is isomorphic with $\tilde{\Sigma}$.*

Remark. The proof of this important representation explicitly depends on the axiom of choice (or at least the weaker ultrafilter axiom), and no constructive method avoiding this (or an equivalent) axiom is available. Later we include an extension to σ-algebras.

Proof. The demonstration will be given in steps for clarity.

1. The argument depends on Proposition 7.2.8 regarding some properties of ultrafilters. Recall that a subclass $\mathscr{A}_0 \subset \Sigma$ is a filter base if (i) $\varnothing \notin \mathscr{A}_0$, (ii) for $A_1, A_2 \in \mathscr{A}_0$ there is an $A_3 \in \mathscr{A}_0$ such that $A_3 \subset A_1 \cap A_2$. For each such base we let $\mathscr{F} = \{B \subset \Omega: B \supset A \text{ for some } A \in \mathscr{A}_0\}$, so that $\mathscr{F}$ is a filter. But by Proposition 7.2.8 each filter can be refined to an ultrafilter of Ω. Hence each $\varnothing \neq A \in \Sigma$ belongs to an ultrafilter, denoted by $\tilde{\omega}$. Let $\tilde{\Omega}$ be the set of all ultrafilters determined by the elements of Σ. Since by definition of a filter, it is closed under monotone (increasing) nets and finite intersections, we have

[†]As noted in the footnote on page 433, for Theorem 8.3.1, we should perhaps use the word "0-dimensional". Since $\tilde{\Omega}$ is compact, this and extremal disconnectedness are the same, as pointed out there. So we can use the classical concept.

$\{A_i: i \in I\} \subset \tilde{\omega} \Rightarrow \bigcup_i A_i \in \tilde{\omega}$, and since $\varnothing \neq A \in \Sigma \Rightarrow A \in \tilde{\omega}_1$ for some $\tilde{\omega}_1$, we have for any A, B in Σ that $A \cup B \in \tilde{\omega} \Rightarrow A \in \tilde{\omega}$ or $B \in \tilde{\omega}$. But if $A \cap B = \varnothing$, only one of them is in $\tilde{\omega}$ (by Proposition 7.2.8 again), and hence $A \in \Sigma$, $A \notin \tilde{\omega} \Rightarrow A \cap B = \varnothing$ for some $B \in \tilde{\omega}$. For each $A \in \Sigma$, define $h(A) = \{\tilde{\omega} \in \tilde{\Omega}: A \in \tilde{\omega}\}$. We claim that $h: \Sigma \to \mathscr{P}(\tilde{\Omega})$ is an isomorphism of Σ into the power set $\mathscr{P}(\tilde{\Omega})$, and if $\tilde{\Sigma} = h(\Sigma) = \{h(A): A \in \Sigma\}$, then $(\tilde{\Omega}, \tilde{\Sigma})$ will in fact satisfy the conditions of the theorem on introducing a topology in $\tilde{\Omega}$ by the class $\tilde{\Sigma}$. This will then establish the result itself.

2. To see that h is an isomorphism, let A, B be in Σ, $A \neq B$. Then either $A - B$ or $B - A$ is nonvoid. If $A - B \neq \varnothing$, then by definition of a filter base and its ultrafilter, there exists an $\tilde{\omega} \in \tilde{\Omega}$ such that $A - B \in \tilde{\omega}$, but $B \notin \tilde{\omega}$, since $(A - B) \cap B = \varnothing$. Consequently $A - B \subset A \Rightarrow A \in \tilde{\omega}$, so that $\tilde{\omega} \in h(A)$ and $\tilde{\omega} \notin h(B)$. Thus $A - B \neq \varnothing \Rightarrow h(A) \neq h(B)$, and the same holds if $B - A \neq \varnothing$. Also, $A \subset B$ iff $h(A) \subset h(B)$. Indeed, if $A - B \neq \varnothing$, then (as seen above) $h(A) - h(B) \neq \varnothing$, and if $A \subset B$, then clearly $h(A) \subset h(B)$. Since $\tilde{\omega} \in h(A) \Rightarrow A \in \tilde{\omega}$, we get $A \subset B \in \tilde{\omega} \Rightarrow \tilde{\omega} \in h(B)$. Further, $\varnothing \notin \tilde{\omega}$ for any (ultra-)filter, so that $h(\varnothing) = \varnothing$.

Let $A_1, A_2 \in \Sigma$. Since $A_i \subset A_1 \cup A_2$, $i = 1, 2$, it follows from the preceding paragraph that $h(A_1) \cup h(A_2) \subset h(A_1 \cup A_2)$. On the other hand, let $\tilde{\omega} \in h(A_1 \cup A_2)$, so that $A_1 \cup A_2 \in \tilde{\omega}$. Then by Proposition 7.2.8, either A_1 or A_2 is in $\tilde{\omega}$, so that $\tilde{\omega} \in h(A_1) \cup h(A_2)$ and hence $h(A_1 \cup A_2) \subset h(A_1) \cup h(A_2)$. Thus $h(A_1 \cup A_2) = h(A_1) \cup h(A_2)$. Next we assert that $h(A_1 - A_2) = h(A_1) - h(A_2)$ also holds. In fact, $\tilde{\omega} \in h(A_1 - A_2) \Rightarrow A_1 - A_2 \in \tilde{\omega}$, so that $A_1 \supset A_1 - A_2$ and $(A_1 - A_2) \cap A_2 = \varnothing$. Thus $A_1 \in \tilde{\omega}$, $A_2 \notin \tilde{\omega}$. Hence $\tilde{\omega} \in h(A_1) - h(A_2)$, or $h(A_1 - A_2) \subset h(A_1) - h(A_2)$. In the opposite direction, $\tilde{\omega} \in h(A_1) - h(A_2) \Rightarrow A_1 \in \tilde{\omega}$, $A_2 \notin \tilde{\omega}$. By Proposition 7.2.8, since $A_2 \cup A_2^c = \Omega \in \tilde{\omega}$ and $A_2 \cap A_2^c = \varnothing$, we have $A_2^c \in \tilde{\omega}$. So $A_1 \cap A_2^c = A_1 - A_2 \in \tilde{\omega}$ and $\tilde{\omega} \in h(A_1 - A_2)$. Thus $h(A_1) - h(A_2) \subset h(A_1 - A_2)$, or $h(A_1 - A_2) = h(A_1) - h(A_2)$. Finally, $h(A_1 \cap A_2) = h(A_1 - (A_1 - A_2)) = h(A_1) - h(A_1 - A_2) = h(A_1) - [h(A_1) - h(A_2)] = h(A_1) \cap h(A_2)$, so that $h(\cdot)$ is an isomorphism on Σ into $\mathscr{P}(\tilde{\Omega})$.

3. Let us now introduce a topology in $\tilde{\Omega}$. If $\tilde{\Sigma} = h(\Sigma)$, then the preceding properties of h imply that $\tilde{\Sigma}$ is a ring. The fact that $\tilde{\Sigma}$ is closed under finite intersections shows that it qualifies as a base for the topology of $\tilde{\Omega}$. If $\mathscr{T}$ is the topology thus determined by $\tilde{\Sigma}$, then $(\tilde{\Omega}, \mathscr{T})$ is asserted to be a locally compact Hausdorff space, and each member of $\tilde{\Sigma}$ is a compact open set.

The separation property is simple, since $\tilde{\omega}_1 \neq \tilde{\omega}_2$ implies the existence of $\varnothing \neq A_i \in \Sigma$ such that $A_1 \in \tilde{\omega}_1$ and $A_2 \notin \tilde{\omega}_1$, $A_2 \in \tilde{\omega}_2$, $A_1 \cap A_2 = \varnothing$. For otherwise, if $A_1 = (A_1 - A_2) \cup (A_1 \cap A_2) \in \tilde{\omega}_1$ and $A_2 = (A_2 - A_1) \cup (A_1 \cap A_2) \in \tilde{\omega}_2$, then $A_1 - A_2 \in \tilde{\omega}_1$ and $A_2 - A_1 \in \tilde{\omega}_2$. (We cannot have $A_1 \cap A_2 \in \tilde{\omega}_1$, since this implies $A_1, A_2 \in \tilde{\omega}_1 \Rightarrow \tilde{\omega}_1 = \tilde{\omega}_2$, and similarly $A_1 \cap A_2 \notin \tilde{\omega}_2$.) But then $\tilde{A}_1 = A_1 - A_2 \in \Sigma$, $\tilde{A}_2 = A_2 - A_1 \in \Sigma$ and disjoint, $\tilde{\omega}_1 \in h(\tilde{A}_1)$, and $\tilde{\omega}_2 \in h(\tilde{A}_2)$. So $h(\tilde{A}_1) \cap h(\tilde{A}_2) = h(\tilde{A}_1 \cap \tilde{A}_2) = h(\varnothing) = \varnothing$, and $h(\tilde{A}_i) \in \tilde{\Sigma}$, $i = 1, 2$, are open. Thus $(\tilde{\Omega}, \mathscr{T})$ is a Hausdorff space.

Next we assert that $h(A)$ is compact (hence closed and open) for each $A \in \Sigma$. This will imply that each $\tilde{\omega} \in \tilde{\Omega}$ has a compact (open) neighborhood, so that $\tilde{\Omega}$ is locally compact. In proving that $h(A)$ is compact, we use the equivalence (i) $\Leftrightarrow$ (v) of Theorem 1.2.7, whose proof is valid for *any* Hausdorff topological space (no property of $\mathbb{R}^n$ was used). Namely, a subset of a Hausdorff space is compact (i.e., an open covering admits a finite subcover) iff it has the finite intersection property. Thus consider any family of closed sets $\{C_i : i \in I\} \subset h(A)$ with the finite intersection property, i.e., $\bigcap_{j \in J} C_j \neq \varnothing$ for each finite set $J \subset I$. We represent C_i as follows. Since $h(A)$ is open and C_i closed, the set $h(A) - C_i \subset \tilde{\Omega}$ is open in the topology $\mathscr{T}$. Hence it can be written in terms of the basis elements as

$$h(A) - C_i = \bigcup_{\alpha \in \Delta} h(A_{i\alpha}), \quad \{A_{i\alpha} : \alpha \in \Delta\} \subset \Sigma \qquad (\Delta = \text{an index set}).$$

Thus

$$C_i = h(A) - [h(A) - C_i] = h(A) - \bigcup_{\alpha \in \Delta} h(A_{i\alpha})$$

$$= \bigcap_{\alpha \in \Delta} [h(A) - h(A_{i\alpha})] = \bigcap_{\alpha \in \Delta} h(A - A_{i\alpha}).$$

Since the C_i's satisfy $\bigcap_{i \in J} C_i \neq \varnothing$, it follows that $\{h(A - A_{i\alpha}) : i \in I, \alpha \in \Delta\}$ also has the finite intersection property, so that $\{A - A_{i\alpha} : i \in I, \alpha \in \Delta\}$ has the same property because h is an isomorphism. But since this class belongs to a filter base, it is in an ultrafilter $\tilde{\omega} \in \tilde{\Omega}$. Consequently $\tilde{\omega} \in \bigcap\{h(A - A_{i\alpha}) : i \in I, \ \alpha \in \Delta\}$, so that $\bigcap_{i \in I} C_i \neq \varnothing$. This proves that $h(A)$ is compact, $\tilde{\Omega}$ is locally compact, and $\mathscr{T}$ is extremally disconnected.

Note that if $\Omega \in \Sigma$, then $\tilde{\Omega} = h(\Omega) \in \tilde{\Sigma}$ and $\tilde{\Omega}$ is compact by the above proof. In this case the preceding paragraph can be shortened because Ω belongs to every ultrafilter and so $\omega \in \tilde{\Omega} \ \Rightarrow \ \omega \in h(\Omega)$, and $\tilde{\Omega} = h(\Omega)$ follows. Then $h(A^c) = h(\Omega - A) = \tilde{\Omega} - h(A)$ is open, so that $h(A)$ is open and closed. Even if Σ is only a ring, so that Ω is locally compact, each compact open set $\tilde{A} \subset \tilde{\Omega}$ is necessarily in the ring $\tilde{\Sigma}$, since $\tilde{A} = \bigcup_{i \in I} h(A_i)$ for some $\{A_i : i \in I\} \subset \Sigma$, and hence by compactness $\tilde{A} = \bigcup_{j \in J} h(A_j) = h(\bigcup_{j \in J} A_j) \in \tilde{\Sigma}$, where J is a finite index set. This completes the proof.

In the preceding theorem no special properties of the ring Σ of subsets of Ω have been used. Indeed, it could have been an *abstract* Boolean ring with $\varnothing$ as its least element and $\cup, \cap, {}^c$ replaced by $\vee, \wedge, '$, the join, meet, and complement. We state the result for reference.

Proposition 4. *Let $\mathscr{A}$ be an abstract Boolean algebra. Then it is isomorphic to the algebra $\mathscr{B}$ of clopen sets of an extremally disconnected compact Hausdorff space S. Moreover, $\mathscr{B}$ is generated by the clopen G_δ-sets of S, so that it is the Baire algebra of S.*

The last statement follows from the fact that χ_A is a continuous function on S for each clopen set $A \subset S$, i.e., $\chi_A \in C(S)$ [$= C_c(S)$ here]. Thus A is the zero set of the continuous function $1 - \chi_A$, so that it is a Baire set. Since it is compact by the preceding theorem, it is a G_δ (cf. Exercise 9.1.9). A similar argument shows that $\tilde{\Sigma}$ for Theorem 3 is a Baire ring.

The corresponding extension of the representation of σ-algebras is slightly different in form and is due to L. H. Loomis (and independently to R. Sikorski). It is given by the following.

Theorem 5. *Let* (Ω, Σ) *be a measurable space. Let* $\tilde{\Omega}$ *be the extremally disconnected compact space representing* Σ, *as given by Theorem* 3, *and let* Σ' $[= \sigma(\tilde{\Sigma})]$ *be the* σ-*algebra generated by the clopen sets of* $\tilde{\Omega}$. *Then* Σ *is isomorphic to* $\Sigma'/\mathscr{I}$, *where* $\mathscr{I}$ *is the* σ-*ideal of sets in* Σ' *such that* $A \in \mathscr{I}$ *iff* A *is of the first category* (*or meager*) *in* $\tilde{\Omega}$, (*i.e.*, $A = \bigcup_{k=1}^{\infty} A_k$, *with* $\bar{A}_k$ *having an empty interior*). *Explicitly, if* $h: \Sigma \to \tilde{\Sigma}$ *is the isomorphism given by Theorem* 3, *then* $\tilde{h}: \Sigma \to \Sigma'/\mathscr{I}$ *is obtained by setting* $\tilde{h}(B) = [h(B)] \in \Sigma'/\mathscr{I}$, $B \in \Sigma$, ($[h(B)] = h(B) + \mathscr{I}$, *is the coset*).

A proof of this result is outlined in Exercise 4 below.

The Stone isomorphism theorem plays an important role in some applications of considerable interest in analysis. Some of these results will be presented in the next section.

EXERCISES

1. Let (Ω, Σ, P) be a nonatomic complete probability space, and $(I, \mathscr{B})$ be the Borelian unit interval. If there is a σ-homomorphism $h: \mathscr{B} \to \Sigma/\mathscr{N}$, where $\mathscr{N}$ is the σ-ideal of P-null sets of Σ, consider the mapping T defined by $T\chi_A = \chi_{h(A)}$ for $A \in \mathscr{B}$. Show that T extends to a linear transformation on $\mathscr{L}^p(I, \mathscr{B}, \mu) = \mathscr{L}^p(\mu) \to \mathscr{L}^p(\Omega, \Sigma, P) = \mathscr{L}^p(P)$ such that $\|Tf\|_p = \|f\|_p$, $1 \leqslant p < \infty$, for each $f \in \mathscr{L}^p(\mu)$, with μ as the Lebesgue measure. Deduce that $Tf = f \circ \varphi^{-1}$ for a measurable point mapping $\varphi: \Omega \to I$. [Use Proposition 2.]
2. This exercise gives a kind of converse to the above. The full converse is not true, and one has to restrict the result and is therefore more involved. Let $\mathscr{L}^p(P)$, $\mathscr{L}^p(\mu)$ be as in Exercise 1 with $1 \leqslant p < \infty$, $p \neq 2$. Suppose $L^p(P)$, $L^p(\mu)$ are the corresponding Lebesgue spaces and $T: L^p(\mu) \to L^p(P)$ is a linear mapping such that $\|Tf\|_p = \|f\|_p$. Then there exists a measurable $\varphi: \Omega \to I$ and $g \in L^p(P)$ such that $Tf = g(f \circ \varphi)$ for all $f \in L^p(\mu)$. Moreover φ and g are essentially unique, the latter on $\{\omega: g(\omega) \neq 0\}$. [Since the support of $f \in L^p(P)$ is an element of $\Sigma/\mathscr{N}$, define for each $B \in \mathscr{B}$ the set $h(B) = \{\omega: (T\chi_B)(\omega) \neq 0\}$. Next use the inequality given in Exercise 5.5.3, with $\varphi(x) = |x|^p$, $p > 2$, and $1 \leqslant p < 2$, on noting

that there is equality iff $u \cdot v = 0$, i.e., $\|u + v\|_p^p + \|u - v\|_p^p = 2(\|u\|_p^p + \|v\|_p^p)$ iff $u \cdot v = 0$. Verify that $h: \mathscr{B} \to \Sigma/\mathscr{N}$ is additive on disjoint sets, and then that h is a homomorphism. Since $T\chi_A = \lim_n \Sigma_{k=1}^n T\chi_{A_k}$ with $A_k \in \mathscr{B}$, disjoint, and $A = \bigcup_{k=1}^{\infty} A_k$, deduce that h is a σ-homomorphism. Then by Proposition 2, $h = \varphi^{-1}$. Let $g = T1 = T\chi_A + T\chi_{A^c}$, so that $T\chi_A = g\chi_{h(A)} = g \cdot (\chi_A \circ \varphi)$. The rest follows from this computation. This result is due to J. Lamperti. It also has an extension for certain Orlicz spaces.]

3. If Σ is an algebra of Ω, and $(\tilde{\Omega}, \tilde{\Sigma})$ is its Stone representation with $h: \Sigma \to \tilde{\Sigma}$ as the isomorphism, and if $A_n \in \Sigma$, $A = \bigcup_{n=1}^{\infty} A_n \in \Sigma$, then verify that $M = h(A) - \bigcup_{n=1}^{\infty} h(A_n)$ is a meager set. Show the converse: If the set M is meager, $h(A_n) \subset h(A)$, and $\bigcup_n A_n \in \Sigma$, then $A = \bigcup_n A_n$ holds. [Since $h(A_n)$ is open, M is closed in $\tilde{\Omega}$. Also $\operatorname{int}(M) = h(A) - \overline{\bigcup_{n \geqslant 1} h(A_n)}$. But $A = \bigcup_n A_n \Rightarrow h(A_n) \subset h(A) \Rightarrow \overline{\bigcup_{n \geqslant 1} h(A_n)} \subset \overline{h(A)} = h(A)$. Thus unless M is meager [i.e., $\operatorname{int}(M) = \varnothing$], it contains a basis element $h(B)$, $\varnothing \neq B \in \Sigma$. But then $h(B) \subset M$, so $B \subset A$. Since $\varnothing = h(B) \cap h(A_n) = h(B \cap A_n) \Rightarrow B \cap A_n = \varnothing$, this contradicts the relation $A = \bigcup_n A_n$. For the second part, let M be meager. Since $h(A_n) \subset h(A)$ and $B = \bigcup_n A_n \in \Sigma \Rightarrow A \supset A_n$, and so $A \supset B$, it follows that if $A - B = C \neq \varnothing$, then $C \cap A_n = \varnothing$. Hence $h(C) \subset M$. Since $h(C) \neq \varnothing$ and is open, M is not meager.]

4. Prove Theorem 5 using the following sketch. Since Σ is a σ-algebra, $\tilde{\Omega} \in \tilde{\Sigma}$ and is compact. Also the mapping $\tilde{h}: \Sigma \to \tilde{\Sigma}/\mathscr{N}$ is a homomorphism, and if $\tilde{h}(B) \in \mathscr{N}$ for some $B \neq \varnothing$, then it is meager. But $\tilde{h}(B)$ is a compact open set and is nonempty. A classical result of E. Čech says that an open nonempty subset of a compact space is never meager. Hence $\tilde{h}(B) \notin \mathscr{N}$, and $\tilde{h}$ is an isomorphism. To see that $\tilde{h}(\Sigma) = \tilde{\Sigma}/\mathscr{N}$, consider $\Sigma^0 = \{A \in \tilde{\Sigma}: A + \mathscr{N} = \tilde{h}(B) \text{ for some } B \in \Sigma\}$. Then $A \in \Sigma^0 \Rightarrow A^c \in \Sigma'$. Now let $A_n \in \tilde{\Sigma}$, $n \geqslant 1$. So $A_n + \mathscr{N} = \tilde{h}(B_n)$, $B_n \in \Sigma$. Since Σ is a σ-algebra, $\bigcup_n B_n \in \Sigma$, and by Exercise 3 above, $\tilde{h}(\bigcup_n B_n) = h(\bigcup_n B_n) + \mathscr{N}$ $\Rightarrow h(\bigcup_n B_n) - \bigcup_n h(B_n) \in \mathscr{N}$, so that $\bigcup_n A_n + \mathscr{N} \subset \Sigma^0$, or $\bigcup_n A_n \in \Sigma^0/\mathscr{N}$. Hence $\tilde{\Sigma} \subset \Sigma^0 \subset \sigma(\tilde{\Sigma}) = \Sigma'$. Also Σ^0 is a σ-algebra. Thus $\Sigma' = \sigma(\tilde{\Sigma}) \subset \Sigma^0 \subset \Sigma'$ and $\tilde{h}(\Sigma) = \Sigma'/\mathscr{N}$. This argument follows R. Sikorski, and is simpler than L. H. Loomis's.]

5. Show that in Theorem 5, as well as in Exercise 3 above, one may take Σ to be a σ-ring of Ω, and then $\tilde{\Sigma}$ will be the ring of $\tilde{\Omega}$ generated by its compact open sets. (See the method of proof of Theorem 3.)

10.2. SOME APPLICATIONS OF THE STONE ISOMORPHISM THEOREM

Here we present a sample of applications following from Stone's isomorphism theorem that relate to measure and integration theory. If Ω is a set and Σ_0 is an algebra of subsets of Ω, let $\mathrm{ba}(\Omega, \Sigma_0)$ denote the class of all bounded (finitely) additive scalar functions on Σ_0 with the total variation norm. Similarly, if (Ω, Σ) is a measurable space (can be a σ-algebra), let $\text{ca}(\Omega, \Sigma)$ denote the space of bounded σ-additive scalar functions on Σ with variation norm. It is the subset of σ-additive elements of $\text{ba}(\Omega, \Sigma)$. Then both $\text{ba}(\Omega, \Sigma_0)$ and $\text{ca}(\Omega, \Sigma)$ are complete normed linear spaces.

Our first application of Theorem 1.3 is contained in the following result.

Theorem 1. *Let* (Ω, Σ_0) *be a pair with* Σ_0 *as an algebra, and* $(\tilde{\Omega}, \tilde{\Sigma}_0)$ *be its Stone representation space with* $h: \Sigma_0 \to \tilde{\Sigma}_0$ *as its isomorphism. Then the mapping* $T: \mu \mapsto \mu \circ h^{-1}$ *defined by* $(T\mu)(\tilde{A}) = \mu(h^{-1}(\tilde{A}))$, $\tilde{A} \in \tilde{\Sigma}_0$, $\mu \in \text{ba}(\Omega, \Sigma_0)$, *is an isometric isomorphism of* $\text{ba}(\Omega, \Sigma_0)$ *onto* $\text{ba}(\tilde{\Omega}, \tilde{\Sigma}_0)$. *Moreover, each* $\mu_1 = T\mu$ *has a unique extension to* $\mu_2 \in \text{ca}(\tilde{\Omega}, \tilde{\Sigma})$ *where* $\tilde{\Sigma} = \sigma(\tilde{\Sigma}_0)$ *(thus* μ_1 *is* σ*-additive), and each such* μ_2 *is regular, the correspondence* $\tilde{T}: \text{ba}(\tilde{\Omega}, \tilde{\Sigma}_0) \to \text{ca}(\tilde{\Omega}, \tilde{\Sigma})$ *given by* $\tilde{T}\mu_1 = \mu_2$ *being again an isometric isomorphism onto* $\text{ca}(\tilde{\Omega}, \tilde{\Sigma})$. *Moreover*

$$\int_\Omega f\, d\mu = \int_{\tilde{\Omega}} Uf\, d\mu_2, \tag{1}$$

for all bounded $f: \Omega \to \mathbb{R}$ *satisfying* $f^{-1}(B) \in \Sigma_0$, $B \subset \mathbb{R}$, *a Borel set. (Here* $U\chi_A = \chi_{h(A)}$; *it is extended linearly to all bounded measurable* f, *and the integral for finitely additive set functions is defined as in Exercise* 5.5.9.)

Proof. Since $h: \Sigma_0 \to \tilde{\Sigma}_0$ is an isomorphism onto, T is clearly also an isomorphism of $\text{ba}(\Omega, \Sigma_0)$ onto $\text{ba}(\tilde{\Omega}, \tilde{\Sigma}_0)$. This correspondence $T\mu = \mu \circ h^{-1}$ is isometric because

$$\begin{aligned}
|T\mu|(\tilde{\Omega}) &= \sup\left\{ \sum_{i=1}^{n} |(T\mu)(A_i)| : A_i \in \tilde{\Sigma}_0, \text{disjoint}, \bigcup_{i=1}^{n} A_i = \tilde{\Omega} \right\} \\
&= \sup\left\{ \sum_{i=1}^{n} \left|\mu\left(h^{-1}(A_i)\right)\right| : A_i \in \tilde{\Sigma}_0, \text{disjoint}, \bigcup_{i=1}^{n} A_i = \tilde{\Omega} \right\} \\
&= \sup\left\{ \sum_{i=1}^{n} |\mu(B_i)| : B_i \in \Sigma_0, \text{disjoint}, \bigcup_{i=1}^{n} B_i = \Omega \right\} \\
&= |\mu|(\Omega).
\end{aligned} \tag{2}$$

Also, as noted before, $\tilde{\Sigma}_0$ is a Baire algebra of $\tilde{\Omega}$, and since every element $h(B)$ of $\tilde{\Sigma}_0$ is a clopen set (and compact) it follows that each μ_1 in $\text{ba}(\tilde{\Omega}, \tilde{\Sigma}_0)$, being finite, is regular (cf. Definition 9.2.3). Consequently, we obtain that every μ_1 is σ-additive and has a unique (Borel) regular σ-additive extension to $\sigma(\tilde{\Sigma}_0) = \tilde{\Sigma}$. This is an immediate consequence of the Henry extension theorem (cf. Theorem 2.3.10 and exercise 2.3.11), since $\tilde{\Sigma}_0$ is a base of the topology of $\tilde{\Omega}$. For variety, we include an alternative proof of this statement here.

Let $\mu_1 \in \text{ba}(\tilde{\Omega}, \tilde{\Sigma}_0)$. Then the variation measure $|\mu_1|(\cdot)$ is additive. Hence its σ-additivity follows if it is shown to be σ-subadditive, by Lemma 1.3.5. Because of the Jordan decomposition, μ_1 is σ-additive iff $|\mu_1|(\cdot)$ is. But $(\tilde{\Omega}, \tilde{\Sigma})$ is a Baire-measurable space. So if $|\mu_1|(\cdot)$ is σ-additive (and bounded), then Theorem 9.2.5 shows that it is regular, and by Theorem 9.2.6 it has a unique regular Borel extension to all Borel sets of $\tilde{\Omega}$. This implies the same property for μ_1, by Proposition 9.2.4. Hence our statement follows if $|\mu_1|(\cdot)$ is shown to be σ-subadditive. For this let B_n, B be elements of $\tilde{\Sigma}_0$, $B \subset \bigcup_n B_n$. If $\epsilon > 0$ is given, by regularity of $|\mu_1|(\cdot)$, there exist C_n and D in $\tilde{\Sigma}_0$ such that the closure $\bar{D} \subset B$, with $B_n \subset \text{int}(C_n)$, satisfying $|\mu_1|(B - D) < \epsilon/2$ and $|\mu_1|(C_n - B_n) < \epsilon/2^{n+1}$. Hence $\bigcup_n \text{int}(C_n) \supset \bigcup_n B_n \supset B \supset \bar{D}$, compact, so that for a finite m, $\bar{D} \subset \bigcup_{k=1}^m \text{int}(C_k) \subset \bigcup_{k=1}^m C_k$, and we have

$$\sum_{n=1}^{\infty} |\mu_1|(B_n) \geqslant \sum_{n=1}^{\infty} \left(|\mu_1|(C_n) - \frac{\epsilon}{2^{n+1}} \right)$$

$$\geqslant \sum_{n=1}^{m} |\mu_1|(C_n) - \frac{\epsilon}{2} \geqslant |\mu_1|\left(\bigcup_{n=1}^{m} C_n \right) - \frac{\epsilon}{2}$$

$$\geqslant |\mu_1|(D) - \frac{\epsilon}{2} \geqslant |\mu|(B) - \epsilon.$$

Hence $|\mu_1|$ is σ-subadditive, and our assertion is proved.

Thus if μ_2 is an extension of μ_1 from $\tilde{\Sigma}_0$ to $\tilde{\Sigma}$, then it is regular, and if $\mu_2 \in \text{ca}(\tilde{\Omega}, \tilde{\Sigma})$ is given, then $\mu_1 = \mu_2|\tilde{\Sigma}_0$ is regular, so that the mapping $\tilde{T}(\mu_1) = \mu_2$ is an isomorphism of $\text{ba}(\tilde{\Omega}, \tilde{\Sigma}_0)$ onto $\text{ca}(\tilde{\Omega}, \tilde{\Sigma})$. We now verify that $\tilde{T}$ is also an isometry. Since μ_2 is regular on $\tilde{\Sigma}$, this is equivalent to the inner or outer regularity (cf. Exercise 9.2.8). Hence

$$|\mu_2|(\tilde{\Omega}) = \sup\left\{ \sum_{i=1}^{n} |\mu_2(A_i)| : A_i \in \tilde{\Sigma}, \text{disjoint}, \bigcup_{i=1}^{n} A_i = \tilde{\Omega} \right\}$$

$$< \infty.$$

So given $\epsilon > 0$, there is a partition $\{A_i : 1 \leqslant i \leqslant n_\epsilon\} \subset \tilde{\Sigma}$ such that

$$|\mu_2|(\tilde{\Omega}) \leqslant \sum_{i=1}^{n_\epsilon} |\mu_2(A_i)| + \frac{\epsilon}{2}. \tag{3}$$

By the outer regularity of μ_2 and the fact that $\tilde{\Sigma}_0$ is a basis for the topology of $\tilde{\Omega}$, there exist $B_i \in \tilde{\Sigma}_0$, $B_i \supset A_i$ such that

$$\big||\mu_2(A_i)| - |\mu_2(B_i)|\big| \leqslant |\mu_2(A_i) - \mu_2(B_i)|$$

$$< \epsilon/2n_\epsilon, \qquad i = 1, \dots, n_\epsilon.$$

Hence $|\mu_2(A_i)| < |\mu_2(B_i)| + \epsilon/2n_\epsilon$, so that (3) becomes

$$|\mu_2|(\tilde{\Omega}) < \sum_{i=1}^{n_\epsilon}\left(|\mu_2(B_i)| + \frac{\epsilon}{2n_\epsilon}\right) + \frac{\epsilon}{2}$$

$$= \sum_{i=1}^{n_\epsilon} |\mu_1(B_i)| + \epsilon, \qquad \text{since} \quad \mu_1 = \mu_2|\tilde{\Sigma}_0$$

$$= \sum_{i=1}^{n_\epsilon} |T^{-1}(\mu_1)(h^{-1}(B_i))| + \epsilon$$

$$\leqslant |T^{-1}(\mu_1)|(\tilde{\Omega}) + \epsilon = |\mu_1|(\tilde{\Omega}) + \epsilon.$$

The last equality follows from the isomorphism of $\mathrm{ba}(\Omega, \Sigma_0)$ and $\mathrm{ba}(\tilde{\Omega}, \tilde{\Sigma}_0)$ established in (2). Since $\epsilon > 0$ is arbitrary, we get $|\mu_2|(\tilde{\Omega}) \leqslant |\mu_1|(\tilde{\Omega})$. But μ_2 being an extension of μ_1, the opposite inequality is always true, so that $|\mu_1|(\tilde{\Omega}) = |\mu_2|(\tilde{\Omega}) = |(\tilde{T}\mu_1)|(\tilde{\Omega})$.

Finally, let $U: \chi_A \mapsto \chi_{h(A)}$, $A \in \Sigma_0$, and then if $f = \sum_{i=1}^n a_i \chi_{A_i}$, let $f' = \sum_{i=1}^n a_i \chi_{h(A_i)} = Uf$. Then U is linear on the class of all Σ_0-measurable bounded functions, since each such f is a uniform limit of step functions. But for each such function, and for $\mu \in \mathrm{ba}(\Omega, \Sigma_0)$, one has

$$\int_\Omega f\, d\mu = \sum_{i=1}^n a_i \mu(A_i)$$

$$= \sum_{i=1}^n a_i (\mu \circ h^{-1})(B_i), \qquad B_i = h(A_i) \in \tilde{\Sigma}_0$$

$$= \sum_{i=1}^n a_i \mu_1(B_i) = \int_{\tilde{\Omega}} f'\, d\mu_1$$

$$= \int_{\tilde{\Omega}} f'\, d\mu_2, \qquad \text{since} \quad \mu_1 = \mu_2|\tilde{\Sigma}_0. \tag{4}$$

From (4), the definition of the integral on the left (cf. Exercise 5.5.9), and also the bounded convergence theorem on the right, one gets the relation (1) immediately. This establishes the theorem.

Recall that for any elements λ, μ of $\mathrm{ba}(\Omega, \Sigma_0)$, λ is called μ-continuous if $\lim_{|\mu|(A)\to 0}\lambda(A) = 0$, or equivalently with $|\lambda|$ in place of λ here. A useful characterization of this concept and a somewhat stronger conclusion of a part of the preceding result in this context will now be given.

Proposition 2. *Let* (Ω, Σ_0) *and its representation* $(\tilde{\Omega}, \tilde{\Sigma}_0)$ *be as in Theorem* 1. *If* λ, μ *are in* $\mathrm{ba}(\Omega, \Sigma_0)$, λ_1, μ_1 *are their images in* $\mathrm{ba}(\tilde{\Omega}, \tilde{\Sigma}_0)$, *and* λ_2, μ_2 *in* $\mathrm{ca}(\tilde{\Omega}, \tilde{\Sigma})$, *then* $|\lambda_1|(B) = |\lambda_2|(B)$, $B \in \tilde{\Sigma}_0$. *Moreover,* λ *is* μ*-continuous iff* λ_2 *is* μ_2*-continuous or* $\tilde{\lambda}$ *if* $\tilde{\mu}$*-continuous, where* $\tilde{\lambda}$ *and* $\tilde{\mu}$ *are any* σ*-additive extensions of* λ, μ *onto* $\sigma(\Sigma_0)$ *when such exist. (Here* $\lambda_i, \mu_i, \tilde{\Sigma}, \tilde{\Sigma}_0$ *have the same meanings as in Theorem* 1.)

Proof. For each $B \in \tilde{\Sigma}_0$, and $\lambda_1 \in \mathrm{ba}(\tilde{\Omega}, \tilde{\Sigma}_0)$, consider the new set function $\bar{\lambda}_1 = \lambda_1|\tilde{\Sigma}_0(B)$, where $\tilde{\Sigma}_0(B) = \{B \cap C: C \in \tilde{\Sigma}_0\}$, the trace of $\tilde{\Sigma}_0$ on B. Then $\tilde{\Sigma}_0(B) = h(\Sigma_0(h^{-1}(B)))$, so that we can apply Theorem 1 to $\mathrm{ba}(B, \tilde{\Sigma}_0(B))$ and $\mathrm{ca}(B, \tilde{\Sigma}(B))$, where $\tilde{\Sigma}(B) = \sigma(\tilde{\Sigma}_0(B))$. (This is also the restriction of $\tilde{\Sigma}$ to B, as the notation indicates.) But by the preceding theorem, $|\bar{\lambda}_1|(B) = |\tilde{T}\bar{\lambda}_1|(B)$. Since by definition $\bar{\lambda}_1(\cdot) = \lambda_1(B \cap \cdot\,)$, we get $|\lambda_2|(B) = |\tilde{T}\lambda_1|(B) = |\tilde{T}\bar{\lambda}_1|(B) = |\bar{\lambda}_1(B)| = |\lambda_1|(B)$, $B \in \tilde{\Sigma}_0$, as asserted.

For the second part, if λ_2 is μ_2-continuous, then λ_1 is μ_1-continuous, and since $\lambda(h^{-1}(B)) = \lambda_1(B)$, $B \in \tilde{\Sigma}_0$, and h is an isomorphism (and similarly for μ and μ_2), it follows by the preceding paragraph that λ is μ-continuous. If $\tilde{\lambda}$ is $\tilde{\mu}$-continuous, then this implies the same for λ, μ at once. So we only need to prove the converse.

Let us first consider the case of λ_2 and μ_2 if λ is μ-continuous. By the isomorphism (and the first part), λ_1 is μ_1-continuous. But as shown in the alternative argument (see the second paragraph of the preceding proof), λ_1, μ_1 in $\mathrm{ba}(\tilde{\Omega}, \tilde{\Sigma}_0)$ are σ-additive. Consider λ_1^*, μ_1^*, the Carathéodory-generated outer measures by $(\lambda_1, \tilde{\Sigma}_0)$ and $(\mu_1, \tilde{\Sigma}_0)$. If $\mathscr{M}_{\lambda_1^*}, \mathscr{M}_{\mu_1^*}$ are the familiar λ_1^*-, μ_1^*-classes of sets, then they are σ-algebras containing $\tilde{\Sigma}_0$, and hence $\sigma(\tilde{\Sigma}_0) \subset \mathscr{M}_{\lambda_1^*} \cap \mathscr{M}_{\mu_1^*}$ on which λ_1^*, μ_1^* are σ-additive. Also $\lambda_1^*|\tilde{\Sigma}_0 = \lambda_1$, $\mu_1^*|\tilde{\Sigma}_0 = \mu_1$, by Theorem 2.2.10. Note that we can apply this latter result to $|\lambda_1|, |\mu_1|$ or just to the positive and negative parts by their Jordan decompositions. Since these are finite measures, by Theorem 2.3.1, their extensions are unique, and so $\lambda_1^*|\sigma(\tilde{\Sigma}_0) = \lambda_2$, $\mu_1^*|\sigma(\tilde{\Sigma}_0) = \mu_2$. Hence we have for each $A \in \sigma(\tilde{\Sigma}_0)$

$$|\mu_2|(A) = |\mu_1^*|(A)$$

$$= \inf\left\{\sum_{i=1}^{\infty}|\mu_1|(A_i): A \subset \bigcup_{i=1}^{\infty}A_i,\ A_i \in \tilde{\Sigma}_0\right\}. \tag{5}$$

Suppose now that λ is μ-continuous. Then as noted already (by the first part), λ_1 is μ_1-continuous, since $|\lambda|(h^{-1}(E)) = |T\lambda|(E) = |\lambda_1|(E)$, $E \in \tilde{\Sigma}_0$. (Similarly for μ_1.) Thus for each $\epsilon > 0$ there is a $\delta > 0$ such that $|\mu_1|(A) < \delta \Rightarrow |\lambda_1|(A) < \epsilon$, $A \in \tilde{\Sigma}_0$. Since $\tilde{\Sigma}_0$ is an algebra, and $|\mu_1|$ is a measure, we may assume (by disjunctification) that the A_i in (5) are disjoint and cover A. Hence there is such a sequence that $\sum_{i=1}^{\infty}|\mu_1|(A_i) < \infty$, $A \subset \bigcup_{i=1}^{\infty} A_i$, and so for each $n \geqslant 1$

$$\sum_{i=1}^{n} |\mu_1|(A_i) = |\mu_1|\left(\bigcup_{i=1}^{n} A_i\right) < \delta \quad \Rightarrow \quad |\lambda_1|\left(\bigcup_{i=1}^{n} A_i\right) < \epsilon. \tag{6}$$

Since $\mu_2|\tilde{\Sigma}_0 = \mu_1$ and $\lambda_2|\tilde{\Sigma}_0 = \lambda_1$, (6) holds if λ_1, μ_1 are replaced by λ_2, μ_2. Letting $n \to \infty$, this yields, on noting $\bigcup_{i=1}^{\infty} A_i \in \sigma(\tilde{\Sigma}_0)$,

$$|\mu_2|\left(\bigcup_{i=1}^{\infty} A_i\right) \leqslant \delta \quad \Rightarrow \quad |\lambda_2|\left(\bigcup_{i=1}^{\infty} A_i\right) \leqslant \epsilon. \tag{7}$$

It follows from (6) and (7) that $|\mu_2|(A) \leqslant \delta \Rightarrow |\lambda_2|(A) \leqslant \epsilon$, $A \in \sigma(\tilde{\Sigma}_0)$.

If now $\tilde{\lambda}, \tilde{\mu}$ are σ-additive extensions of λ, μ onto $\sigma(\Sigma_0)$, then by definition of an extension, $\lambda = \tilde{\lambda}|\Sigma_0$, $\mu = \tilde{\mu}|\Sigma_0$, so that λ, μ must also be σ-additive. Then the Carathéodory procedure with λ, μ in place of λ_1, μ_1 above shows that $\lambda^*|\sigma(\Sigma_0) = \tilde{\lambda}$ and $\mu^*|\sigma(\Sigma_0) = \tilde{\mu}$ (by Theorem 2.2.10). The same argument leading to (6) and (7) applies to $\tilde{\mu}, \tilde{\lambda}$ in place of μ_2, λ_2 and proves that $\tilde{\lambda}$ is $\tilde{\mu}$-continuous. Thus the result is established in both cases.

As a consequence of the preceding work, one can obtain an analog of the Radon-Nikodým theorem for finitely additive set functions, as originally established by S. Bochner. We include this result, deducing it from the countably additive case via Stone's representation theorem. This has independent interest and is useful in other applications.

Theorem 3 (Bochner). *Let λ, μ be in* $\mathrm{ba}(\Omega, \Sigma_0)$, *where Σ_0 is an algebra of a set Ω, and $\mu \geqslant 0$. If λ is μ-continuous, then for each $\epsilon > 0$ there is a step function $f_\epsilon: \Omega \to \mathbb{R}$, measurable for Σ_0, such that $|\lambda - F_\epsilon|(\Omega) < \epsilon$, where $F_\epsilon(A) = \int_A f_\epsilon \, d\mu$, $A \in \Sigma_0$.*

Proof. Our aim here is to obtain this from Theorem 5.3.3(ii)(b). Thus to get σ-additive functions out of λ, μ, we apply Theorem 10.2.1 and consider $\lambda_2 = V\lambda$ and $\mu_2 = V\mu$, where $V = \tilde{T}T$: $\mathrm{ba}(\Omega, \Sigma_0) \to \mathrm{ca}(\tilde{\Omega}, \tilde{\Sigma})$ is the isometric isomorphism onto the second space. Since λ is μ-continuous, by Proposition 10.2.2 we see that λ_2 is μ_2-continuous and these measures are bounded on the

Baire-measurable space $(\tilde{\Omega}, \tilde{\Sigma})$. Hence by Theorem 5.3.3(ii)(b), there is an essentially unique $[\mu_2]$ measurable function $\tilde{f}: \tilde{\Omega} \to \mathbb{R}$ such that $\lambda_2(A) = \int_A \tilde{f} d\mu_2$, $A \in \tilde{\Sigma}$. Also, $\tilde{f} \in \mathscr{L}^1(\tilde{\Omega}, \tilde{\Sigma}, \mu_2)$, and by Theorem 4.5.7, for each $\epsilon > 0$ there is a step function $\tilde{f}_\epsilon = \sum_{i=1}^n a_i \chi_{\tilde{A}_i}$, $a_i \in \tilde{\Sigma}$, such that $\|\tilde{f} - \tilde{f}_\epsilon\|_1 < \epsilon$, or equivalently $|\lambda_2 - \tilde{F}_\epsilon|(\Omega) < \epsilon$, where $\tilde{F}_\epsilon(A) = \int_A \tilde{f}_\epsilon \, d\mu_2$. But V is an isometric isomorphism. Since $\lambda = V^{-1}\lambda_2$ and $\mu = V^{-1}\mu_2$, if we select F_ϵ in ba(Ω, Σ_0) such that $\tilde{F}_\epsilon = VF_\epsilon$, then from definitions of V and $\tilde{F}_\epsilon$ we get $|\lambda - F_\epsilon|(\Omega) = |\lambda_2 - \tilde{F}_\epsilon|(\tilde{\Omega}) < \epsilon$, by Proposition 2. This completes the proof.

As another application of the isomorphism theorem we show how an arbitrary $\mathscr{L}^p(\mu)$-space can be put into an isometric and isomorphic correspondence with an $\mathscr{L}^p(\tilde{\mu})$ on a Baire measure space $(\tilde{\Omega}, \tilde{\Sigma}, \tilde{\mu})$. Then the adjoint spaces of a general $\mathscr{L}^p(\mu)$ can be understood relatively easily from the corresponding ones with regular topological measures.

Let us start with an auxiliary problem, of special interest.

Proposition 4. *Let $B(\Omega, \Sigma_0)$ be the vector space of functions $f: \Omega \to \mathbb{R}$ which are uniform limits of sequences of Σ_0-measurable real step functions on Ω, where Σ_0 is an algebra of subsets of Ω, as usual. Then there exists an extremally disconnected compact Hausdorff space $\tilde{\Omega}$ (the Stone space of Σ_0) such that $B(\Omega, \Sigma_0)$ is isometrically isomorphic to the space $C(\tilde{\Omega})$ of real continuous functions on $\tilde{\Omega}$, where both spaces are given the uniform norm. Moreover, this isomorphism preserves order in these spaces.*

Proof. If $f = \sum_{i=1}^n a_i \chi_{A_i}$, $A_i \in \Sigma_0$, let $Tf = \sum_{i=1}^n a_i \chi_{h(A_i)}$. Since $h(A_i) \in \tilde{\Sigma}_0$ and is a clopen set, $\chi_{h(A_i)}$ and hence Tf are continuous and one has $Tf \in C(\tilde{\Omega})$. Also $\|Tf\| = \max_i |a_i| = \|f\|$. Since $B(\Omega, \Sigma_0)$ is the uniform closure of step functions, it follows that T: $B(\Omega, \Sigma_0) \to C(\tilde{\Omega})$ is an isometric isomorphism into the latter. The one to one property follows from $\|T(f_1 - f_2)\| = \|f_1 - f_2\| = 0 \;\Rightarrow\; f_1 = f_2$. Moreover, $T(B(\Omega, \Sigma_0))$ is a closed subspace of $C(\tilde{\Omega})$. To see that T is onto needs some work. Note that $T(fg) = (Tf)(Tg)$. This is obvious for step functions, and then the general case follows by (uniform) approximation. If $\tilde{\omega}_1, \tilde{\omega}_2$ are two distinct points of $\tilde{\Omega}$, then by its construction in Theorem 1.3, there exist disjoint sets A, D in Σ_0 such that $\omega_1 \in h(A)$ and $\omega_2 \in h(D)$. Since $h(A \cap D) = \varnothing$, we have $(T\chi_A)(\omega_1) = 1$, $(T\chi_A)(\omega_2) = 0$, so that $T(B(\Omega, \Sigma_0))$ distinguishes points of $\tilde{\Omega}$; it has been shown to be a closed subalgebra of $C(\tilde{\Omega})$. It is clear that the unit function also is in $T(B(\Omega, \Sigma_0))$. Hence by the Stone-Weierstrass Theorem (see the Appendix) $C(\tilde{\Omega}) = T(B(\Omega, \Sigma_0))$, and T is onto. The definition of T gives that $f \geqslant g \Rightarrow Tf \geqslant Tg$ also, and the proof is complete.

We now present an isometric representation of $\mathscr{L}^p(\mu)$ on an abstract measure space (Ω, Σ, μ), on $\mathscr{L}^p(\tilde{\mu})$. For convenience μ is assumed to have the finite subset property (Definition 2.3.2).

Theorem 5. *Let $\mathscr{L}^p(\mu)$ be the Lebesgue space of pth power integrable real measurable functions, $1 \leqslant p < \infty$, on an abstract measure space (Ω, Σ, μ), as above. Then there exists a measure triple $(\tilde{\Omega}, \tilde{\Sigma}, \tilde{\mu})$, where $\tilde{\Omega}$ is a locally compact Hausdorff space, $\tilde{\Sigma}$ is the Baire σ-algebra of $\tilde{\Omega}$, and $\tilde{\mu}$ is a Baire measure on $\tilde{\Sigma}$, such that $\mathscr{L}^p(\tilde{\mu})$ is isometrically (and lattice) isomorphic to $\mathscr{L}^p(\mu)$. Moreover, every element of $\mathscr{L}^p(\tilde{\mu})$ has its support contained in a σ-compact (Baire) set.*

Proof. First we assume that there is a bounded $f_0 > 0$ in $\mathscr{L}^p(\mu)$ and establish the result. Later, an extension of this to the general case will be obtained. Let $\mathscr{A} \subset \mathscr{L}^p(\mu)$ be the algebra of all essentially bounded functions. If Σ_1 is the σ-algebra generated by the functions of $\mathscr{A}$, then $\mathscr{L}^p(\Sigma_1, \mu)$ and $\mathscr{L}^p(\Sigma, \mu)$ are equal in the sense that each element of the one space is a.e. equal to an element of the other with the same norms. To see this, since clearly $\mathscr{L}^p(\Sigma_1, \mu) \subset \mathscr{L}^p(\Sigma, \mu)$ is true ($\Sigma_1 \subset \Sigma$ by construction), consider $f \in \mathscr{L}^p(\Sigma, \mu)$. If $f_n = f\chi_{[|f| \leqslant n]}$, then $f_n \in \mathscr{A}$, so that it is in $\mathscr{L}^p(\Sigma_1, \mu)$ for each n. Letting $n \to \infty$, it follows that f is measurable for Σ_1, and hence $f \in \mathscr{L}^p(\Sigma_1, \mu)$, so that equality holds. Now consider (Ω, Σ_1), and let $(\tilde{\Omega}, \tilde{\Sigma}_1)$ be its Stone representation space. Thus $\tilde{\Omega}$ is a compact extremally disconnected space. Since simple functions are dense in $\mathscr{L}^p(\Sigma, \mu)$, and they belong to $\mathscr{A}$, it follows that for each $A \in \Sigma_1$ with $\mu(A_1) > 0$, there exists $\bar{A}_1 \subset A_1$, $\bar{A}_1 \in \Sigma_1$, and $0 < \mu(\bar{A}_1) < \infty$, $\chi_{\bar{A}_1} \in \mathscr{A}$. So for the mapping $T: \mathscr{A} \to C(\tilde{\Omega})$ of Proposition 4 we have: (i) $T(\mathscr{A})$ separates the points of $\tilde{\Omega}$, (ii) $T(\mathscr{A})$ is an algebra, and (iii) $T(f_0) > 0$. Hence by the Stone-Weierstrass theorem $T(\mathscr{A})$ is uniformly dense in $C(\tilde{\Omega})$.

Let $\tilde{\mu}(\cdot) = \mu \circ h^{-1}(\cdot)$, and consider the space $\mathscr{L}^p(\tilde{\Omega}, \sigma(\tilde{\Sigma}_1), \tilde{\mu})$ $[= \mathscr{L}^p(\tilde{\mu})]$, where $h: \Sigma_1 \to \sigma(\tilde{\Sigma}_1)/\mathscr{N}$ is the σ-isomorphism given by Theorem 1.5, with $\mathscr{N}$ as the σ-ideal of meager sets. Since $\int_{\tilde{\Omega}} |Tf_0|^p \, d\tilde{\mu} = \int_\Omega |f_0|^p \, d\mu < \infty$, and $T(f_0) > 0$ is continuous on the compact space $\tilde{\Omega}$ (hence having a bounded inverse), it follows that $\tilde{\mu}(\tilde{\Omega}) < \infty$. Also $\tilde{\mu}$ is a Baire measure. So $C(\tilde{\Omega}) \subset \mathscr{L}^p(\tilde{\mu})$, and is dense in the latter. It then follows that $T(\mathscr{A})$ is also dense in $\mathscr{L}^p(\tilde{\mu})$, and $\|Tf\|_{p,\tilde{\mu}} = \|f\|_{p,\mu}$, $f \in \mathscr{A}$. Hence for each $\epsilon > 0$, $f \in \mathscr{L}^p(\tilde{\mu})$, there is a $g_\epsilon \in T(\mathscr{A})$ such that $\|f - g_\epsilon\|_{p,\tilde{\mu}} < \epsilon$. But $g_\epsilon = T(g'_\epsilon)$ for a $g'_\epsilon \in \mathscr{A}$, and $\mathscr{A}$ is dense in $\mathscr{L}^p(\mu)$. So if T is extended to $\hat{T}: \mathscr{L}^p(\mu) \to \mathscr{L}^p(\tilde{\mu})$ by setting, for each $f \in \mathscr{L}^p(\mu)$, $f_n = f\chi_{[|f| \leqslant n]}$ and $\hat{T}f = \lim_n Tf_n$, which exists and defines $\hat{T}$ uniquely $[\|Tf_n - Tf_m\|_{p,\tilde{\mu}} = \|f_n - f_m\|_{p,\tilde{\mu}} \to 0$ as $f_n \to f$ in $\mathscr{L}^p(\mu)]$, then we get $\|\hat{T}f\|_{p,\tilde{\mu}} = \lim_n \|Tf_n\|_{p,\tilde{\mu}}$. Thus $\hat{T}(\mathscr{L}^p(\mu)) \subset \mathscr{L}^p(\tilde{\mu})$. But by density of $T(\mathscr{A})(= \hat{T}(\mathscr{A}))$ in $\mathscr{L}^p(\tilde{\mu})$ we have equality in the preceding inclusion. It follows that $\mathscr{L}^p(\tilde{\mu})$ and $\mathscr{L}^p(\mu)$ are isometrically isomorphic, and the result holds in this case.

Next consider the general case that there is no bounded $f_0 > 0$, a.e., in $\mathscr{L}^p(\mu)$. Without changing anything in the latter space, we may assume that μ is also complete. Consider now the classes $\mathscr{C}$ of all sets $\{f_\alpha : \alpha \in J\} \subset \mathscr{L}^p(\mu)$ with the property that for $\alpha \neq \alpha'$, $f_\alpha \wedge f_{\alpha'} = 0$ a.e. (i.e., they have a.e. disjoint supports) and that each $f_\alpha \geqslant 0$ is bounded. Here J is some index set. These collections can be partially ordered by inclusion. Here we use the convention that for any two such classes $\{f_\alpha: \alpha \in J\}$ and $\{f_\beta: \beta \in \tilde{J}\}$, one considers all elements $f_\alpha \wedge f_\beta = 0$ a.e., and discards the excess elements having a.e. equal supports. Thus every linearly ordered subclass of $\mathscr{C}$ will have an upper bound obtained by combining all of them with this procedure. Hence by Zorn's lemma, $\mathscr{C}$ has a maximal collection, denoted $\mathscr{S}_0 = \{f_\alpha: \alpha \in I_0\}$, so that $f_\alpha \geqslant 0$ a.e., essentially bounded, and $f_\alpha \wedge f_{\alpha'} = 0$ a.e. if $\alpha \neq \alpha'$. Note that the collection $\mathscr{S}_0$ can alternatively be described as follows. Consider all the bounded positive elements of $\mathscr{L}^p(\mu)$, and arrange them in a well-ordered sequence $\{g_0, g_1, \ldots, g_\alpha, \ldots, \alpha < \gamma\}$. We use the "numeration theorem" that for any set A there is an ordinal γ equipollent with A, and employing transfinite induction. (See Appendix.) Thus, let $f_0 = g_0$, and if f_β is defined for $\beta < \alpha$ (α being an ordinal number, finite or not), then let $f_\alpha = g_\alpha$ if $g_\alpha \wedge f_\beta = 0$ for $\beta < \alpha$, and $= 0$ otherwise, so that $f_\alpha = 0$ if $g_\beta \wedge f_\alpha > 0$ on a set of positive measure. Thus $\{f_\beta: \beta < \gamma\}$ is defined for all β. Let I denote the index set. Discarding those $f_\beta = 0$, we see that $f_\alpha \wedge f_{\alpha'} = 0$ for $\alpha \neq \alpha'$, and by construction, for any $f \geqslant 0$ one can find an f_α in the class satisfying $f \wedge f_\alpha > 0$ on a set of positive measure. This will be our sequence. Here the transfinite induction takes the role of Zorn's lemma. However, the present procedure is equivalent with that lemma. [See Suppes (1972, Chapter 8); the relevant result is given in the Appendix for completeness.]

If $S_\beta = \operatorname{supp}(f_\beta)$, then $S_\beta \in \Sigma$, $\mu(S_\beta) > 0$. Consider $\Omega_0 = \bigcup_{\beta \in I} S_\beta$. If $(S_\beta, \Sigma(S_\beta), \mu_\beta)$ is the trace of (Ω, Σ, μ) on S_β, $\mu_\beta(\cdot) = \mu(S_\beta \cap \cdot)$, then $\mathscr{L}^p(\mu_\beta)$ $[= \mathscr{L}^p(S_\beta, \Sigma(S_\beta), \mu_\beta)]$ satisfies the hypothesis of the special case considered at the beginning. By that result there is a triple $(\tilde{\Omega}_\alpha, \tilde{\Sigma}_\alpha, \tilde{\mu}_\alpha)$, such that $\mathscr{L}^p(\mu_\alpha)$ is isometrically isomorphic to $\mathscr{L}^p(\tilde{\mu}_\alpha)$. Also it is easy to see that $\mathscr{L}^p(\Omega, \Sigma, \mu)$ and $\mathscr{L}^p(\Omega_0, \Sigma_0, \mu_0) = \mathscr{L}^p(\mu_0)$ contain the same elements, and the corresponding members have the same norm, where Ω_0 is as above, Σ_0 is the σ-ring generated by $\bigcup_{\beta \in I} \Sigma(S_\beta)$, and μ_0 is given by

$$\mu_0(A) = \sum_{\beta \in I} \mu_\beta(A) = \sup\left\{ \sum_{\beta \in I_1} \mu(S_\beta \cap A) : I_1 \subset I \right\}. \tag{8}$$

Here I_1 ranges over all finite subsets of I. If $\mu(A) < \infty$, the sum is at most countable. (Cf. Exercise 2.2.12.)

We now define $(\tilde{\Omega}, \tilde{\Sigma}, \tilde{\mu}_0)$. Let $\tilde{\Omega} = \bigcup_{\alpha \in I} \{\tilde{\Omega}_\alpha \times \{\alpha\}\}$, and if $i_\alpha: \tilde{\Omega}_\alpha \to \tilde{\Omega}$ is the canonical injection mapping, endow $\tilde{\Omega}$ with the finest topology $\mathscr{T}$ for which each $\{i_\alpha: \alpha \in I\}$ is continuous. (This $\tilde{\Omega}$ is often called the *topological set sum*.) Identifying each $\tilde{\Omega}_\alpha$ with $i_\alpha(\tilde{\Omega}_\alpha)$, we get that a set $A \subset \tilde{\Omega}$ is open (or closed) iff $A \cap \tilde{\Omega}_\alpha$ is open (or closed) for each $\alpha \in I$. So $\tilde{\Omega}_\alpha$ is both open and closed in $\tilde{\Omega}$. Since each point of $\tilde{\Omega}$ thus has a relatively compact neighborhood, it is a locally compact space, and by this construction, $\mathcal{T}$ is clearly a Hausdorff topology. Then $\tilde{\Sigma}$ is the σ-ring generated by $\bigcup_\alpha h(\Sigma(\Omega_\alpha))$, and $\tilde{\mu}_0(A) = \mu_0(h^{-1}(A))$, $A \in \tilde{\Sigma}$. It follows that $\tilde{\mu}_0$ is a Baire measure, and $\mathcal{L}^p(\mu_0)$ and $\mathcal{L}^p(\tilde{\mu}_0)$ are easily seen to be isometrically isomorphic. Moreover, the lattice operations are also preserved in both cases. Finally, if $\hat{f} \in \mathcal{L}^p(\mu_0)$, then $\hat{f}$ vanishes outside of a set of σ-finite measure. Since every set of finite $\tilde{\mu}$-measure is contained in the union of a countable collection of $\tilde{\Omega}_\alpha$'s which are compact, it is clear that the support of $\hat{f}$ is contained in a σ-compact Baire set of $\tilde{\Omega}$. This concludes the proof of the theorem.

The first part of the preceding result (especially its proof) yields the following, which is termed a *Stone-Weierstrass theorem for function spaces.*

Proposition 6. *Let (Ω, Σ, μ) be a measure space and $\mathcal{L}^p(\mu)$ be the corresponding scalar Lebesgue space on it, and let $1 \leqslant p < \infty$. Suppose $\mathcal{A} \subset \mathcal{L}^p(\mu)$ is an algebra of essentially bounded functions satisfying the following conditions*:

(i) *there is an $f_0 \in \mathcal{A}$ such that $f_0 > 0$ a.e.*;
(ii) *for every pair of disjoint measurable sets A_1, A_2 such that $\mu(A_i) > 0$ for $i = 1, 2$, there is an $f \in \mathcal{A}$ such that $f > 0$ a.e. on A_1 and $f \leqslant 0$ a.e. on A_2*;
[(iii) *if $\mathcal{L}^p(\mu)$ is a complex function space, then with every $f \in \mathcal{A}$, its conjugate $\bar{f} \in \mathcal{A}$*].

Then $\mathcal{A}$ is norm dense in $\mathcal{L}^p(\mu)$.

Proof. Let $\mathcal{A}$ be as in the statement, and $\mathcal{A}_1$ be the algebra of essentially bounded elements of $\mathcal{L}^p(\mu)$. Then $\mathcal{A} \subset \mathcal{A}_1$, and if Σ_1 is the σ-algebra generated by $\mathcal{A}_1$, we have $\mathcal{L}^p(\Sigma_1, \mu) = \mathcal{L}^p(\Sigma, \mu)$, as seen at the beginning of the preceding proof.

First suppose that the space $\mathcal{L}^p(\mu)$ is real. Then by the first part of the last theorem, there is a Stone space representation $\mathcal{L}^p(\tilde{\mu})$ which is isometrically isomorphic to $\mathcal{L}^p(\mu)$, where $(\tilde{\Omega}, \tilde{\Sigma}, \tilde{\mu})$ is a Baire measure space with $\tilde{\Omega}$ as a compact extremally disconnected Hausdorff space, $\tilde{\Sigma}$ as the σ-algebra generated by the clopen sets of $\tilde{\Omega}$, and $\tilde{\mu}$ as a (finite) Baire measure on $\tilde{\Sigma}$. Thus $T(\mathcal{A}_1)$ is norm dense in $\mathcal{L}^p(\tilde{\mu})$ by the last proof, T being the isomorphism map. But then conditions (i) and (ii) imply that $T(\mathcal{A})$ [$\subset T(\mathcal{A}_1) \subset C(\tilde{\Omega})$] distinguishes the points of $\tilde{\Omega}$ and is an algebra with $T(f_0)$ in it. Hence $T(\mathcal{A})$ is uniformly dense [in $T(\mathcal{A}_1)$ and so] in $C(\tilde{\Omega})$. It follows that $T(\mathcal{A})$ is norm dense in $\mathcal{L}^p(\tilde{\mu})$, $1 \leqslant p < \infty$ [cf. Exercise 5(a)]. Since T is an isometry onto, the same is true of $\mathcal{A}$ in $\mathcal{L}^p(\mu)$. Thus the result holds if the space is real.

In the complex case, hypotheses (i)–(iii) imply that the real elements of $\mathcal{A}$ satisfy (i) and (ii). Since $\mathcal{L}^p(\mu) = \{f = f_1 + if_2 : f_1, f_2 \in \mathcal{L}^p_{\mathbb{R}}(\mu)\}$, where

$\mathscr{L}^p_{\mathbb{R}}(\mu)$ is the real space on (Ω, Σ, μ), then by the preceding paragraph the real and imaginary elements of $\mathscr{A}$ are dense in $\mathscr{L}^p_{\mathbb{R}}(\mu)$, and then it follows immediately that $\mathscr{A}$ is dense in $\mathscr{L}^p(\mu)$ itself. In other words, we have used the complex form of the usual Stone-Weierstrass theorem in disguise. This completes the proof.

As an immediate consequence of this result, we have the following, which is essentially due to R. H. Farrell (1962). He proved it using a different argument.

Corollary 7. *Let $\mathscr{L}^p(\mu)$ be the Lebesgue space on (Ω, Σ, μ), where $\Omega = \mathbb{R}$; $\Sigma = \mathscr{B}$, the Borel σ-algebra; and μ is a finite Borel measure. If $\mathscr{A}$ is the algebra generated by 1 and a bounded strictly increasing $f: \mathbb{R} \to \mathbb{R}$, then $\mathscr{A}$ is norm dense in $\mathscr{L}^p(\mu)$, $1 \leqslant p < \infty$.*

The preceding isomorphism results give the following description of the adjoint spaces of $\mathscr{L}^p(\mu)$, $1 \leqslant p < \infty$, on arbitrary (but with the finite subset property, to use the preceding) measure spaces (Ω, Σ, μ). If $1 < p < \infty$, we already know that the adjoint space $[\mathscr{L}^p(\mu)]^*$ is representable as $\mathscr{L}^q(\mu)$, $q = p/(p-1)$, and no such result holds if $p = 1$. In fact $[\mathscr{L}^1(\mu)]^* = \mathscr{L}^\infty(\mu)$ iff μ is a localizable measure. (See Theorems 5.5.5 and 5.5.9.) The difficulty there was that the underlying measure space is *held fixed* in these representations. However, by Theorem 5 above, $\mathscr{L}^p(\Omega, \Sigma, \mu)$ is isometrically isomorphic to $\mathscr{L}^p(\tilde{\Omega}, \tilde{\Sigma}, \tilde{\mu})$ on a locally compact $\tilde{\Omega}$ with $\tilde{\mu}$ as the Baire measure, which thus is (strictly) localizable. Hence the adjoint space of the latter is representable on $\mathscr{L}^q(\tilde{\Omega}, \tilde{\Sigma}, \tilde{\mu})$. The new insight here is that $[\mathscr{L}^1(\Omega, \Sigma, \mu)]^*$ is isometrically isomorphic with $\mathscr{L}^\infty(\tilde{\Omega}, \tilde{\Sigma}, \tilde{\mu})$ on a *different measure space* when μ is not necessarily localizable. Using Proposition 4 above, we can go a step further in this direction. Since the last space is an algebra also, it is isometrically (lattice) isomorphic to the Banach space of continuous real functions $C(S_1)$ on a (possibly still different) compact Stone space S_1. But by Theorem 9.3.4 the adjoint space of $C(S_1)$ is representable as $\text{rca}(S_1)$, the space of regular (signed) Borel measures on S_1. Hence on combining these isomorphisms (each of which is an isometry and preserves order), we can conclude that the second adjoint of an arbitrary $\mathscr{L}^1(\mu)$ [on (Ω, Σ, μ) without restrictions (except for the convenient finite subset property) on measures] is isometrically isomorphic to $\text{rca}(S_1)$ based on another compact Hausdorff space S_1. But an $\text{rca}(S_1)$ admits the following further analysis.

Let μ, ν, ξ be any elements of $M = \text{rca}(S_1)$. We have a partial order in M. Namely, let $\mu \leqslant \nu$ iff $\mu(A) \leqslant \nu(A)$ for all Borel sets $A \subset S_1$. With this $(M, \leqslant)$ becomes a vector lattice, and its norm $\|\mu\| = |\mu|(S_1)$ (the total variation) makes it a Banach lattice. Thus the following eight conditions are satisfied for M:

(i) $\mu \leqslant \nu,\ \nu \leqslant \mu \ \Rightarrow\ \mu = \nu$,
(ii) $\mu \leqslant \nu,\ \nu \leqslant \xi \ \Rightarrow\ \mu \leqslant \xi$,
(iii) $\mu \leqslant \nu,\ a \in \mathbb{R}^+ \ \Rightarrow\ a\mu \leqslant a\nu$,
(iv) $\mu \leqslant \nu \ \Rightarrow\ \mu + \xi \leqslant \nu + \xi,\ \xi \in M$,
(v) $\mu_n \leqslant \nu_n,\ \|\mu_n - \mu\| \to 0,\ \|\nu_n - \nu\| \to 0 \ \Rightarrow\ \mu \leqslant \nu$,
(vi) $\mu, \nu \in M \ \Rightarrow\ \mu \vee \nu = \max(\mu, \nu) \in M,\ \mu \wedge \nu = \min(\mu, \nu) \in M$,
(vii) $\mu \geqslant 0,\ \nu \geqslant 0 \ \Rightarrow\ \|\mu + \nu\| = \|\mu\| + \|\nu\|$, and
(viii) $\mu, \nu \in M,\ \mu \wedge \nu = 0 \ \Rightarrow\ \|\mu + \nu\| = \|\mu - \nu\|$.

The last three conditions are perhaps not obvious. For a proof, let $\lambda = |\mu| + |\nu|$. Then $\lambda \in M$ and dominates both μ and ν. Since it is a finite measure, by the Radon-Nikodým theorem (Theorem 5.3.3)

$$\mu(A) = \int_A f\,d\lambda, \quad \nu(A) = \int_A g\,d\lambda, \qquad A \subset S_1 \text{ (Borel)}. \tag{9}$$

Then $\mu \wedge \nu$ and $\mu \vee \nu$ are Borel measures given by

$$(\mu \wedge \nu)(A) = \int_A (f \wedge g)\,d\lambda, \quad (\mu \vee \nu)(A) = \int_A (f \vee g)\,d\lambda, \qquad (A \text{ Borel}) \tag{10}$$

(cf. Exercise 5.3.13). Similarly

$$\begin{aligned} \|\mu + \nu\| = |\mu + \nu|(S_1) &= \int_{S_1} (f + g)\,d\lambda \\ &= \int_{S_1} f\,d\lambda + \int_{S_1} g\,d\lambda = \|\mu\| + \|\nu\|, \end{aligned} \tag{11}$$

and $(\mu \wedge \nu) = 0 \ \Leftrightarrow\ f \cdot g = 0$, so that

$$\begin{aligned} \|\mu + \nu\| = \int_{S_1} |f + g|\,d\lambda &= \int_{\operatorname{supp}(f)} |f|\,d\lambda + \int_{\operatorname{supp}(g)} |g|\,d\lambda \\ &= \int_{S_1} |f - g|\,d\lambda = \|\mu - \nu\|. \end{aligned} \tag{12}$$

Any given Banach lattice $\mathscr{X}$ satisfying conditions (i)–(viii) is called an *abstract L*, or *an* (AL), *space*). Thus $M = \text{rca}(S_1)$ is an (AL) space. Every such (AL) space is isometrically (lattice) isomorphic to a concrete space $\mathscr{L}^1(S_2, \mathscr{B}, \xi)$ on some locally compact Hausdorff space S_2 with its Baire σ-ring $\mathscr{B}$ and ξ as a Baire measure on $\mathscr{B}$. This representation was proved by S. Kakutani (1941). Here S_2 may have nothing in common with S_1. Thus the second adjoint of an $\mathscr{L}^1(\mu)$ on an abstract measure space is isometrically isomorphic with $\mathscr{L}^1(S_2, \mathscr{B}, \xi)$, and by change of spaces one can study some properties of (even) higher adjoint spaces. This is the useful part of such isomorphism studies. On the other hand, one does not have any control over these new measure spaces. Thus the representation of $[\mathscr{L}^p(\mu)]^*$, obtained in Chapter 5, allows *finer* analysis of these spaces, since the basic (Ω, Σ, μ) is not altered. Here we omit further discussion.

EXERCISES

1. Let (Ω, Σ) be a pair with Σ an algebra of subsets of Ω, and $\text{ba}(\Omega, \Sigma)$ the space of bounded additive functions $\mu: \Sigma \to \mathbb{R}$. Let $\text{ca}(\Omega, \Sigma)$ be the subspace of σ-additive elements of $\text{ba}(\Omega, \Sigma)$. Using the total variation norm, show that both these spaces are weakly sequentially complete. [By Proposition 2, $\text{ba}(\Omega, \Sigma)$ is isometrically isomorphic to $\text{rca}(S_1)$ on some compact Stone space, and the latter is an (AL) space, and hence is isometric and isomorphic to an $\mathscr{L}^1(S_2, \mathscr{B}, \xi)$, by Kakutani's representation. See Theorem 5.5.13 on weak sequential completeness.]
2. Let $\Phi: \mathbb{R}^+ \to \mathbb{R}^+$ be a convex function with $\Phi(x) = 0$ iff $x = 0$. If (Ω, Σ, μ) is a measure space, let $\mathscr{L}^\Phi(\mu)$ be the set of $(\Sigma\text{-})$measurable $f: \Omega \to \mathbb{R}$ such that $N_\Phi(f) < \infty$, where $N_\Phi(f) = \inf\{\alpha > 0: \int_\Omega \Phi(|f|/\alpha)\, d\mu \leqslant 1\}$. As noted before, $(\mathscr{L}^\Phi(\mu), N_\Phi(\cdot))$ is a Banach space. If $\mathscr{A} \subset \mathscr{L}^\Phi(\mu)$ is an algebra of essentially bounded functions, and Σ_0 is the σ-algebra generated by the elements of $\mathscr{A}$, and conditions (i) and (ii) of Proposition 6 hold for it, show that for each $f \in \mathscr{L}^\Phi(\mu)$ there are Σ_0-measurable bounded functions f_n in $\mathscr{L}^\Phi(\mu)$ such that $f_n \to f$ a.e. and $N_\Phi(f_n) \to N_\Phi(f)$ as $n \to \infty$.
3. Let $\mathscr{P}$ be the set of all extended real valued measurable functions on a measure space (Ω, Σ, μ), and $\rho: \mathscr{P} \to \overline{\mathbb{R}}^+$ be a mapping which satisfies the following conditions: (i) $\rho(f) = 0$ iff $f = 0$ a.e., $\rho(f_1 + af_2) \leqslant \rho(f_1) + a\rho(f_2)$ for all $a \in \mathbb{R}^+$, (ii) $0 \leqslant f_1 \leqslant f_2 \leqslant \cdots \to f$ a.e. $\Rightarrow$ $\rho(f_n) \uparrow \rho(f)$, and (iii) $0 < \rho(f) < \infty$ for at least one $f \in \mathscr{P}$. Let M^ρ be the vector lattice of all real measurable functions on (Ω, Σ, μ) such that if $\Sigma_0 = \{A \in \Sigma: \rho(\chi_A) < \infty\}$, then for each $f \in M^\rho$ and $\epsilon > 0$ there is a Σ_0 step function g_ϵ with $\rho(|g_\epsilon|) < \infty$ and $\rho(|f - g_\epsilon|) < \epsilon$. ($M^\rho$ is thus a Banach space.) If $\mathscr{A} \subset M^\rho$ is an algebra of bounded functions satisfying conditions (i) and (ii) of Proposition 6, show that $\mathscr{A}$ is dense in M^ρ. [If $\rho(\cdot) = N_\Phi(\cdot)$ and $\Phi(2x) \leqslant c\Phi(x)$, $x > 0$, $0 < c < \infty$, in Exercise 2 above, then it is an example of M^ρ.]

4. Let (Ω, Σ, μ) be a Radon measure space with Ω a Hausdorff space. If $\mathscr{F}$ is a family of additive functions $\nu: \Sigma \to \mathbb{R}$, of p-bounded variation relative to μ in the sense of Definition 5.1.1, with $1 < p < \infty$, suppose that $\mathscr{F}$ satisfies the following conditions: (i) $\nu \in \mathscr{F} \Rightarrow |\nu(A)| \leqslant k_\nu \mu(A)$, $A \in \Sigma$, where $0 < k_\nu < \infty$, (ii) there exists a $\nu_0 \in \mathscr{F}$ such that $\nu_0 \geqslant 0$ and μ is ν_0-continuous, (iii) if A_1, A_2 are disjoint sets of positive μ-measure from Σ, then there is a $\nu \in \mathscr{F}$ such that $\nu(E) \leqslant 0$ for all $E \subset A_1$, $E \in \Sigma$, and $\nu(B) > 0$ for all $B \subset A_2$, $B \in \Sigma$. Suppose that $f: \Omega \to \mathbb{R}$ is measurable for Σ, and $\int_\Omega |f|\, d|\nu| < \infty$ but $\int_\Omega f\, d\nu = 0$ for all $\nu \in \mathscr{F}$. Then show that $f = 0$ a.e.$[\mu]$. [This problem plays an important role in statistical estimation theory. Note that $\mathscr{F}$ is equivalent to a set of bounded functions in $\mathscr{L}^p(\mu)$ via Theorem 9.1.3 and Exercise 5.1.2, and if $\mathscr{A}$ is the algebra generated by this equivalent set (of $\mathscr{F}$), contained in $\mathscr{L}^p(\mu)$, then it satisfies the conditions of Proposition 2. So the functional $l_f: g \mapsto \int_\Omega fg\, d\nu_0$, $g \in \tilde{\mathscr{A}}$, is linear and continuous on $\mathscr{L}^\infty(\nu_0)$, where $\tilde{\mathscr{A}}$ is the algebra generated by 1 and $\mathscr{A}$. Verify that $l_f(\tilde{\mathscr{A}}) = 0$, and then, by Proposition 6 (as well as Theorem 5.5.10), $l_f = 0$, so that $f = 0$ a.e.$[\mu]$. This and the preceding two exercises, together with the results in the latter half of this section, are taken with simplifications from the author's paper (1969).]

5. Let $(\Omega, \mathscr{B}, \mu)$ be a Baire measure space with Ω locally compact Hausdorff. If $\mathscr{A}$ is the space of real continuous functions on Ω with compact supports, then $\mathscr{A} \subset \mathscr{L}^p(\mu)$, $1 \leqslant p < \infty$, and is norm dense in the latter space. Prove this statement in the following two different ways:

(a) As a consequence of Theorem 9.1.1. [This is used already in the earlier work. Since Baire simple functions are dense in $\mathscr{L}^p(\mu)$, it suffices to show that each (indicator) function χ_A in $\mathscr{L}^p(\mu)$ can be approximated by an element of $\mathscr{A}$. By the inner regularity of μ, A is approximable by a compact G_δ-set $C \subset A$, and then Theorem 9.1.1 gives an $f \in \mathscr{A}$ which approximates χ_A. We have used this in Proposition 6.]

(b) Assume that Proposition 6 can be proved directly when (Ω, Σ, μ) there is a Baire measure space on a locally compact space as here. [Such a proof in this (Baire measure) case was given by Farrell (1962).] With this, prove the density statement from Proposition 6, so that there is a deeper "equivalence" between these two statements. [Indeed, let Ω be σ-compact, at first. Then there exists an increasing sequence of compact sets C_n, $n \geqslant 1$, $C_n \subset \operatorname{int}(C_{n+1})$, such that $\Omega = \bigcup_{n=1}^\infty C_n$ and $\hat{\mu}(C_n) < \infty$, where $\hat{\mu}$ is the Borel extension of μ. Then we can obtain a bounded positive f_0 as $f_0 = \sum_{n \geqslant 1} f_n / n^2 \hat{\mu}(C_{n+1})$, $f_0 \in \mathscr{L}^p(\mu)$, where f_n is continuous, $0 \leqslant f_n \leqslant 1$, $f_n|_{C_n} = 1$, and $f_n|_{C_{n+1}^c} = 0$ by Urysohn's lemma. If $\mathscr{A}_1$ is the algebra generated by f_0 and $\mathscr{A}$, then it satisfies the hypothesis of Proposition 6, so that its conclusion gives the norm density of $\mathscr{A}_1$ in $\mathscr{L}^p(\mu)$. Since $f_n \in \mathscr{A}$ and f_0 can be approximated in norm by a finite linear combination of these f_n, the result holds in this case. If Ω is merely locally compact, then since $f \in \mathscr{L}^p(\mu)$ has σ-compact support, the result can be reduced to the preceding case by replacing $(\Omega, \mathscr{B}, \mu)$

with $(S_f, \mathscr{B}(S_f), \mu(S_f \cap \cdot\,))$, where $S_f = \operatorname{supp}(f)$, $\mathscr{B}(S_f)$ is the trace of $\mathscr{B}$ on S_f, and $\mu(S_f \cap \cdot\,)$ is the restriction of μ to $\mathscr{B}(S_f)$.]

(c) Deduce that, in either form, the class $C_0(\Omega)$ is also norm dense in $\mathscr{L}^p(\mu)$, where $C_0(\Omega)$ is the space of continuous functions on $\Omega \to \mathbb{R}$ vanishing at infinity. [If $(\Omega, \mathscr{B}, \mu)$ is a Radon measure space with Ω any Hausdorff space, we have shown in Theorem 4.5.9 that $C_c(\Omega)$ is dense in the metric of $\mathscr{L}^p(\mu)$, $0 < p < \infty$. The present case is similar, but a more specialized version.]

10.3. REMARKS ON TOPOLOGY THROUGH MEASURE

The main idea of the preceding section is to associate a topological measure space with an abstract one through the Stone isomorphism or some of its variants. However, for a number of problems it is desirable to introduce a topology in the given space itself and study the properties of the same measure in this topology. We have seen how this might be done for strictly localizable spaces through a lifting map. It was shown in Theorem 8.3.1 that the resulting topology is again extremally disconnected, but generally it is not Hausdorff—so that, for instance, a compact set may not be closed. Can this situation be improved if one considers a suitable subclass of (abstract) measure spaces? The problem is discussed briefly in this concluding section.

A natural place to look for specialized measure spaces in which the preceding problem is likely to have an affirmative answer is to consider (Ω, Σ, μ), with Ω as a group, Σ as a σ-ring containing all the translates sA $(= \{sa\colon a \in A\})$ of $A \in \Sigma$, and μ as (left) translation invariant, i.e., $\mu(sA) = \mu(A)$, but μ is not identically zero. Since for such measure spaces one should also be able to define the convolution operation and hence the Fubini-Stone theorem (Theorem 6.2.1) as well as Tonelli's theorem should be applicable, one needs to assume that in (Ω, Σ, μ), μ is σ-finite on the σ-ring Σ. With this and the further condition that the mapping $T\colon \Omega \times \Omega \to \Omega \times \Omega$ defined by $T(x, y) = (y^{-1}x, y)$ is measurable relative to $\Sigma \otimes \Sigma$, the product σ-ring (as in Definition 6.1.4), A. Weil (1938, Appendix I), showed that it is possible to introduce a topology, hereafter called the *Weil topology*, in Ω, relative to which it becomes a dense subgroup of a locally compact group $\tilde{\Omega}$ with the given μ as the restriction of its Haar measure. We now describe this result in more precise terms.

Let (Ω, Σ, μ) be a measurable space with Ω a group and Σ a σ-ring having all left translates of each of its members. Suppose μ is a σ-finite translation-invariant measure which is not identically zero and such that the mapping $T\colon \Omega \times \Omega \to \Omega \times \Omega$, defined above, is measurable for $\Sigma \otimes \Sigma$. Such a triple is termed, for short, a *measurable group*. Let $\rho\colon \Sigma \times \Sigma \to \overline{\mathbb{R}}^+$ be the (Fréchet) "distance function," i.e., $\rho(A, B) = \mu(A \Delta B)$, and consider the class $N(A; \epsilon) = \{\omega\colon \rho(\omega A, A) < \epsilon\}$, $A \in \Sigma$, $0 < \mu(A) < \infty$, and $0 < \epsilon < 2\mu(A)$. The idea here is to show that the class $\mathscr{N} = \{N(A; \epsilon)\colon A \in \Sigma\}$ forms a neighborhood base at e for a topology of Ω. One terms a measurable group *separated* if for each $\omega \neq e$, there exists a set $A \in \Sigma$, $0 < \mu(A) < \infty$, such that $\rho(\omega A, A) > 0$. [Here again e denotes the identity of the group Ω.]

With these concepts we can state the desired result as:

Theorem 1. *Let* (Ω, Σ, μ) *be a measurable group which is separated. Then the class* $\mathcal{N}$, *introduced above, can be taken as a base of the Weil topology at e with respect to which* Ω *becomes a topological group. Moreover, there is a locally compact group* $\tilde{\Omega}$ *with a Haar measure* $\tilde{\mu}$ *on its* σ*-ring* $\mathcal{B}$ *of all Baire sets, having* Ω *as a dense subgroup such that* $\Sigma \supset \mathcal{B}(\Omega)$ *and* $\mu(A) = \tilde{\mu}(\tilde{A})$ *for all* $A = \tilde{A} \cap \Omega$. *Here* $\mathcal{B}(\Omega)$, *as usual, is the trace of* $\mathcal{B}$ *on* Ω. *(In the terminology introduced for Corollary* 2.3.8, Ω *becomes a thick subgroup of* $\tilde{\Omega}$.)

To show that $\mathcal{N}$ is a base of e, one has to verify: (i) if $\omega \neq e$, there is a member $N(A; \epsilon)$ of $\mathcal{N}$ excluding ω, (ii) given $U_i \in \mathcal{N}$, $i = 1, 2$, there is a $V \in \mathcal{N}$ satisfying $V \subset U_1 \cap U_2$, (iii) for each $U \in \mathcal{N}$ there is a $V \in \mathcal{N}$ with $VV^{-1} \subset U$, (iv) for $U \in \mathcal{N}$, $\omega \in U$, there is a $V \in \mathcal{N}$ such that $V\omega \subset U$, and (v) $\omega \in \Omega$, $U \in \mathcal{N} \Rightarrow V \subset \omega U \omega^{-1}$ for some $V \in \mathcal{N}$. In this topology, which thus will be Hausdorff, one shows that each point ω $(\in \Omega)$ has a relatively compact neighborhood and that $\tilde{\Omega}$ will be the completion of Ω. But there are many details to be filled in, and we refer the reader to Halmos (1950, Section 62) for a complete account. Weil's original proof is carried out in the context of $\mathcal{L}^2(\mu)$, starting with the result: Let $f \in \mathcal{L}^2(\mu)$. Then for each $E \in \Sigma$, $0 < \mu(E) < \infty$, and $\epsilon > 0$, there is an $A \in \Sigma$, $0 < \mu(A) < \infty$, $A \subset E$, such that for any $(\omega, \omega') \in A \times A$ we have $\|L_\omega f - L_{\omega'} f\|_2 < \epsilon$, where L_ω is the left translation operator. Then consider $W(A, \epsilon) = \{\omega: \|L_\omega \chi_A - \chi_A\|_2 < \epsilon\}$ for each $A \in \Sigma$, $0 < \mu(A) < \infty$. Since $|L_\omega \chi_A - \chi_A| = \chi_{(\omega A) \Delta A}$, we see that $W(A; \epsilon)$ and $N(A; \epsilon)$ above both determine the same topology. The argument using the properties of the $\mathcal{L}^2(\mu)$-space uses more integration theory than the one indicated above with the Fréchet space (Σ, ρ). Even though one can be translated into the other, it is instructive to study them both, and we recommend this effort to the interested reader.

Another related question is to specialize further to see whether we can actually assert $\Omega = \tilde{\Omega}$ in the above theorem. Such a result was established by G. W. Mackey (1957, Theorem 7.1) and is stated here for completeness of our discussion. It is shown to be the case for a subclass of groups which are called *analytic*. Recall that by Proposition 7.1.3, given a measurable space (Ω, Σ), a set $A \subset \Omega$ is analytic relative to Σ iff it is obtained by the Souslin operation given in Definition 7.1.1. Another equivalent statement is the following. The abstract space Ω is analytic iff there is an isomorphism $\varphi: \Omega \to J$, an analytic subset of $\mathbb{R}$, such that φ and φ^{-1} are measurable relative to Σ and the Borel σ-algebra $\mathcal{B}(J)$ of J. (Here the statement that $J \subset \mathbb{R}$ is analytic can be understood simply as J being the range of a continuous mapping of the complete separable metric space $\mathbb{N}^{\mathbb{N}}$, with the product topology. These equivalences can be proved and are not obvious.) Then we have Mackey's result as:

Theorem 2. *Let (Ω, Σ, μ) be a measurable group with $\Omega \in \Sigma$, and be also analytic. Then Ω can be endowed uniquely with a locally compact topology, so that it becomes a topological group with μ as its (left) Haar measure.*

The proof of this result is to show that the Weil topology is again the appropriate one, but this time the extension $\tilde{\Omega}$ coincides with Ω, using various properties of analytic spaces. Again the reader is referred to the original source, since the details are technical and also depend on the preceding result.

EXERCISES

1. Let (Ω, Σ, μ) be a (σ-finite) measurable group. Show that the functions $\omega \mapsto \mu(\omega A)$ and $\omega \mapsto \mu(A\omega)$ are measurable (for Σ) for each $A \in \Sigma$, $\mu(A) < \infty$. [Use Fubini's theorem.]
2. Suppose (Ω, Σ, μ) is a (σ-finite) measurable group in which Σ is also a complete σ-algebra. Let M_μ be the set of all σ-finite measures on Σ which are equivalent to μ. If $T: \Omega \to \Omega$ is an isomorphism and is measurable (i.e., $T^{-1}(\Sigma) \subset \Sigma$), verify that the measures $\nu: A \mapsto \mu[T^{-1}(A)]$, $A \in \Sigma$, form a set having the same properties as M_μ. Hence it is termed a *measure class*, and denoted $T(M_\mu)$. If $T = L_\omega$ $[R_\omega]$, the left [right] translation for $\omega \in \Omega$, and $L_\omega(M_\mu) = M_\mu$ [or $R_\omega(M_\mu) = M_\mu$], $\omega \in \Omega$, then call M_μ left [right] invariant. Show that if Ω has a left-invariant measure class, then it has a measure class which is both right and left invariant.
3. Let (Ω, Σ, μ) be a σ-finite measurable group. As in Proposition 2.6, suppose that for every pair of distinct points ω_1, ω_2 of Ω there is a set $A \in \Sigma$ with $\omega_1 \in A$, $\omega_2 \notin A$. (Thus Σ separates points of Ω.) If there exists a countable sequence of sets $\{A_n: n \geqslant 1\} \subset \Sigma$ which separate points of Ω, (in particular the measure space is separable), verify that for each $\omega \in \Omega$ distinct from e, there exists an $A \in \Sigma$ such that $\mu(A \cap \omega A) < \mu(A) < \infty$. [Assume the opposite and derive a contradiction.] (These are some of the properties needed to prove Theorem 2, and they show what type of argument is needed.)

APPENDIX

We recall here some equivalent formulations of Zorn's lemma and the axiom of choice which have been used freely in the text. Also, some topological results which are given in standard texts are recorded for ready reference.

Let R be a partial order relation in a set S. Thus R is reflexive, antisymmetric, and transitive. It is a linear (or simple) ordering if moreover, for each pair x, y in S, either xRy or yRx holds, i.e., they are related by the order R.

Zorn's lemma (Z_1). *If S is nonempty and has a partial order R, and if the sum (or upper bound) of each nonempty chain (i.e., linearly ordered subset) of S is in S, then S has a maximal element.*

Axiom of choice (AC_1). *For any set A, there is a choice function $f: 2^A \to A$ such that for any nonempty $B \subset A$, $f(B) \in B$.*

The form which we employ in showing the existence of a Lebesgue-nonmeasurable set in $\mathbb{R}$ is the following.

Axiom of choice (AC_2). *If $\mathscr{A}$ is a collection of nonempty pairwise disjoint sets of a set S, then there exists a set B whose intersection with each member of $\mathscr{A}$ has exactly one element.*

Another result which we use occasionally is the next one.

Numeration theorem. *For every set S there is an ordinal number α such that S and α are equipollent.*

The following topological result is related to the above.

Tychonov's product theorem. *If $\{\Omega_i: i \in I\}$ is an arbitrary family of compact topological spaces, then $\Omega = \times_{i \in I} \Omega_i$ is compact in its product topology.*

We have the following equivalences:

$$\begin{aligned}\text{axiom of choice } (AC_1) &\Leftrightarrow \text{axiom of choice } (AC_2)\\ &\Leftrightarrow \text{Zorn's lemma } (Z_1)\\ &\Leftrightarrow \text{numeration theorem.}\end{aligned}$$

Also, the axiom of choice (AC_1) is employed in proving the Tychonov product theorem. But in 1950 J. L. Kelley showed that conversely the latter implies the axiom of choice, when the Ω_i are not Hausdorff. (Cf. Davis (1977, p. 83.)

A good place for proofs of the above assertions (except the last one) is P. Suppes (1972, Chapter 8); for the last result and for other topological discussion, Kelley (1955) is standard.

We have used Urysohn's lemma and the Stone-Weierstrass theorem also several times. They are as follows.

Urysohn's lemma

(1) *If Ω is a locally compact Hausdorff space, let $C \subset \Omega$ be a compact set. Then there exists a relatively compact open set $V \supset C$ and a continuous function $f: \Omega \to \mathbb{R}^+$ such that $0 \leqslant f \leqslant 1$, $f|_C = 1$, and $f = 0$ outside $\bar{V}$, the (compact) closure of V.*

(2) *If Ω is a normal space and A_1, A_2 are disjoint closed subsets of Ω, then there is a continuous $f: \Omega \to \mathbb{R}^+$ such that $0 \leqslant f \leqslant 1$, $f|_{A_1} = 1$, and $f|_{A_2} = 0$.*

The more general form (2) implies (1), since $\bar{V}$ is normal. But we use the first case most of the time in our work.

Stone-Weierstrass theorem

(1) *Let Ω be a compact Hausdorff space, and $C(\Omega)$ be the space of all real continuous functions on Ω. If $\mathscr{A} \subset C(\Omega)$ is a subalgebra which contains constants and separates points of Ω [i.e., if $\omega_1 \neq \omega_2$ are points of Ω, then $f(\omega_1) \neq f(\omega_2)$ for some $f \in \mathscr{A}$], then $\mathscr{A}$ is uniformly dense in $C(\Omega)$.*

(2) *If, in the above, $C(\Omega)$ is the space of complex continuous functions, and $\mathscr{A} \subset C(\Omega)$ is as in (1) and also contains the complex conjugates of each of its elements, then again $\mathscr{A}$ is uniformly dense in $C(\Omega)$.*

(3) *If Ω is locally compact (but noncompact) in the above, and $C(\Omega)$ is replaced by $C_0(\Omega)$, the space of continuous functions vanishing at infinity, then let $\mathscr{A} \subset C_0(\Omega)$ be a subalgebra which separates points of Ω. Let it contain complex conjugates of each of its elements, and assume that for each $\omega \in \Omega$ there is an $f \in \mathscr{A}$ such that $f(\omega) \neq 0$. Then $\mathscr{A}$ is again uniformly dense in $C_0(\Omega)$.*

A proof of this result may be found in Loomis (1953) or in Dunford and Schwartz (1958).

Recall that a topological space Ω is completely regular if points are closed and for each closed set $A \subset \Omega$ and point $\omega \in \Omega - A$, there is a continuous $0 \leqslant f \leqslant 1$ such that $f(\omega) = 0$, $f(A) = \{1\}$. (Ω is also termed a *Tychonov space*.) The following properties are of interest in our work.

Theorem

(a) *The Cartesian product of completely regular spaces is completely regular.*

(b) (*Stone-Čech compactification*) *If Ω is completely regular, then it can be densely embedded in a compact Hausdorff space $\tilde{\Omega}$, such that each continuous $f: \Omega \to Y$, a compact space, has a unique continuous extension to $\tilde{\Omega} \to Y$, i.e., $f \circ i: \tilde{\Omega} \to Y$ has such an extension, where $i: \Omega \to \tilde{\Omega}$ is the embedding.*

Proofs of these results can be found in Kelley's book noted above.

REFERENCES

Alexandrov, A. D. (1940–1943). Additive set functions, I–III, *Mat. Sb. (N.S.)*, I, **8** (50) (1940), 307–348; II, **9** (51) (1941), 563–628; III, **13** (55) (1943), 169–238.

Bartle, R. G. (1976). *The Elements of Real Analysis* (2nd ed.), Wiley, New York.

Bingham, M. S., and K. R. Parthasarathy (1968). A probabilistic proof of Bochner's theorem on positive definite functions, *J. London Math. Soc.* **43**, 626–632.

Bourbaki, N. (1965). *Intégration* (2nd ed.), Hermann, Paris, Chapters I–V.

Calderón, A. P. (1966). Spaces between L^1 and L^∞ and the theorem of Marcinkiewicz, *Studia Math.* **26**, 273–299.

Carathéodory, C. (1927). *Vorlesungen über reele Functionen* (2nd ed.), Teuber, Leipzig.

Cartan, H. (1940). Sur la mesure de Haar, *C. R. Acad. Sci.* **211**, 759–762.

Choquet, G. (1955). Theory of capacities, *Ann. Inst. Fourier Grenoble* **5**, 131–295.

Comfort, W. W. (1984). Topological groups, in *Handbook of Set-Theoretic Topology*, North-Holland, Amsterdam, 1143–1263.

Daniell, P. J. (1918, 1922). A general form of integral, *Ann. Math.* **19**, 279–294; *ibid.* **22**, 203–220.

Davis, M. (1977). *Applied Nonstandard Analysis*, Wiley-Interscience, New York.

Dellacherie, C., and P. A. Meyer (1978). *Probabilité et Potentiel, Part* A, Hermann, Paris.

Dinculeanu, N. (1974). *Integration on Locally Compact Spaces*, Noordhoff, Leyden.

Dinculeanu, N. (1967). *Vector Measures*, Pergamon Press, London.

Doob, J. L. (1953). *Stochastic Processes*, Wiley, New York.

Dunford, N., and J. T. Schwartz (1958). *Linear Operators, Part I: General Theory*, Wiley-Interscience, New York.

Fan, S. C. (1941) Integration with respect to an outer measure function, *Am. J. Math.* **63**, 319–338.

Farrell, R. H. (1962). Dense algebras of functions in L^p, *Proc. Am. Math. Soc.* **13**, 324–328.

Federer, H. (1969). *Geometric Measure Theory*, Springer, Berlin.

Filmore, P. A. (1966). On topology induced by measure, *Proc. Am. Math. Soc.* **17**, 854–857.

Folland, G. B. (1984). *Real Analysis: Modern Techniques and Their Applications*, Wiley-Interscience, New York.

Friedman, N., and M. Katz (1966). A representation theorem for additive functionals, *Arch. Rat. Mech. Anal.* **21**, 49–57.

Gardner, R. J., and W. F. Pfeffer (1984). Borel measures, in *Handbook of Set-Theoretic Topology*, North-Holland, Amsterdam.

Gel'fand, I., and N. Vilenkin (1964). *Generalized Functions, Vol. 4, Harmonic Analysis on Hilbert Spaces* (translation), Academic Press, New York.

Gillman, L., and M. Jerison (1960). *Rings of Continuous Functions*, Van Nostrand, New York.

Halmos, P. R. (1950). *Measure Theory*, Van Nostrand, New York.

Hardy, G. H., J. E. Littlewood, and G. Pólya (1952). *Inequalities* (2nd ed.), Cambridge Univ. Press, London.

Hausdorff, F. (1962). *Set Theory* (3rd ed.) (translation), Chelsea, New York.

Hayes, C. A., and C. Y. Pauc (1970). *Derivation and Martingales*, Springer, Berlin.

Henry, J.-P. (1969). Prolongement des mesures de Radon, *Ann. Inst. Fourier Grenoble* **19**, 237–247.

Hewitt, E. (1960). Integration by parts for Stieltjes integrals, *Am. Math. Monthly*, **67**, 419–423.

Hewitt, E., and K. A. Ross (1963). *Abstract Harmonic Analysis*—I, Springer, Berlin.

Hewitt, E., and K. Stromberg (1965). *Real and Abstract Analysis*, Springer, Berlin.

Huff, R. E. (1973). The Yoshida-Hewitt decomposition as an ergodic theorem, in *Vector and Operator Valued Measures and Applications*, Academic Press, New York, pp. 133–139.

Hunt, R. A. (1966). On L^{pq} spaces, *L'Enseign. Math.* **22**, 249–276.

Ionescu Tulcea, A. and C., (1969). *Topics in the Theory of Lifting*, Springer, Berlin.

Jacobs, K. (1978). *Measure and Integral*, Academic Press, New York.

Kakutani, S. (1941). Concrete representation of abstract (L)-spaces and the mean ergodic theorem, *Ann. Math.* **42**, 523–537.

Kakutani, S. (1948). A proof of the uniqueness of Haar's measure, *Ann. Math.* **49**, 225–226.

Kelley, J. L. (1950), Tychonov's infinite product theorem implies the axiom of choice, *Fund. Math.* **37**, 75–78.

Kelley, J. L. (1955). *General Topology*, Van Nostrand, New York.

Kelley, J. L., and T. P. Srinivasan (1971). Premeasures on lattices of sets, *Math. Ann.* **190**, 233–241.

Kisyński, J. (1968). On the generation of tight measures, *Studia Math.* **30**, 141–151.

Lebesgue, H. (1902). Intégrale, longuer, aire, *Ann. Mat. Pura Appl.* (*3*) **7**, 231–359.

Loomis, L. H. (1947). On the representation of σ-complete Boolean algebras, *Bull. Am. Math. Soc.* **53**, 257–260.

Loomis, L. H. (1953). *An Introduction to Abstract Harmonic Analysis*, Van Nostrand, New York.

Luther, N. Y. (1967). Unique extension and product measures, *Canad. J. Math.* **19**, 757–763.

Mackey, G. W. (1957). Borel structures in groups and their duals, *Trans. Am. Math. Soc.* **85**, 134–165.

Maharam, D. (1947). An algebraic characterization of measure algebras, *Ann. Math.* **48**, 154–167.

Maharam, D. (1958). On a theorem of von Neumann, *Proc. Am. Math. Soc.* **9**, 987–994.

McShane, E. J. (1950). Linear functionals on certain Banach spaces, *Proc. Am. Math. Soc.* **1**, 402–408.

McShane, E. J. (1962). Families of measures and representations of algebras of operators, *Trans. Am. Math. Soc.* **102**, 328–345.

Meyer, P. A. (1966). *Probability and Potentials*, Blaisdell, Waltham, MA.

Mizel, V. J. (1970). Characterization of nonlinear transformations possessing kernels, *Canad. J. Math.* **22**, 449–471.

Nachbin, L. (1965). *The Haar Integral*, Van Nostrand, New York.

Pettis, B. J. (1951). On the extensions of measures, *Ann. Math.* **54**, 186–197.

Rao, B. V. (1969). On discrete Borel spaces and projective sets, *Bull. Am. Math. Soc.* **75**, 614–617.

Rao, M. M. (1969). Stone-Weierstrass theorems for function spaces, *J. Math. Anal. Appl.* **25**, 362–371.

Rao, M. M. (1975). Conditional measures and operators, *J. Multivar. Anal.* **5**, 330–413.

Rao, M. M. (1976). Two characterizations of conditional probability, *Proc. Am. Math. Soc.* **59**, 75–80.

Rao, M. M. (1979). *Stochastic Processes and Integration*, Sijthoff and Noordhoff, Alphen aan den Rijn, The Netherlands.

Rao, M. M. (1980a). Local functionals, Oberwolfach Proceedings on Measure Theory, *Lecture Notes Math.* **794**, 484–496.

Rao, M. M. (1980b). Local functionals on $C_\infty(G)$ and probability, *J. Functional Anal.* **39**, 23–41.

Rao, M. M. (1981). *Foundations of Stochastic Analysis*, Academic Press, New York.

Rao, M. M. (1984). *Probability Theory with Applications*, Academic Press, New York.

Reiter, H. (1968). *Classical Harmonic Analysis and Locally Compact Groups*, Oxford Math. Monographs, London.

Rogers, C. A. (1970). *Hausdorff Measures*, Cambridge Univ. Press, London.

Royden, H. L. (1968). *Real Analysis* (2nd ed.), Macmillan, New York.

Rudin, W. (1974). *Real and Complex Analysis* (2nd ed.), McGraw-Hill, New York.

Rudin, W. (1976). *Principles of Mathematical Analysis* (3rd ed.), McGraw-Hill, New York.

Saks, S. (1964). *Theory of the Integral* (2nd ed.), Dover, New York.

Sastry, A. S., and K. P. S. Sastry (1977). Measure extensions of set functions over lattices of sets, *J. Indian Math. Soc.* **41**, 317–330.

Schwartz, L. (1973). *Radon Measures on Arbitrary Topological Spaces and Cylindrical Measures*, Oxford Univ. Press, London.

Segal, I. E. (1954). Equivalence of measure spaces, *Am. J. Math.* **73**, 275–313.

Sikorski, R. (1969). *Boolean Algebras* (3rd ed.), Springer, Berlin.

Sion, M. (1963). On capacity and measurability, *Ann. Inst. Fourier Grenoble* **13**, 88–99.

Sion, M. (1968). *Introduction to the Methods of Real Analysis*, Holt, Rinehart and Winston, New York.

Sion, M. (1973). *A Theory of Group Valued Measures*, Lect. Notes in Math., 355, Springer, Berlin.

Skorokhod, A. V. (1970). On admissible translates of measures in Hilbert space, *Theor. Prob. Appl.* **15**, 557–580.

Solovay, R. M. (1970). A model of set theory in which every set of reals is Lebesgue measurable, *Ann. Math.* **92**, 1–56.

Stone, M. H. (1936). The theory of representations for Boolean algebras, *Trans. Am. Math. Soc.* **40**, 37–111.

Stone, M. H. (1948, 1949). Notes on integration, I–IV, *Proc. Nat. Acad. Sci.* **34**, 336–342, 447–455, 483–490; **35**, 50–58.

Suppes, P. (1972). *Axiomatic Set Theory* (2nd ed.) Dover, New York.

Traynor, T. (1974). An elementary proof of the lifting theorem, *Pacific J. Math.* **53**, 267–272.

Uhl, J. J., Jr. (1967). Orlicz spaces of finitely additive set functions, *Studia Math.* **29**, 19–58.

Varadarajan, V. S. (1961). Measures on topological spaces, *Mat. Sb. (N.S.)* **55** (97), 35–100. (AMS translation, Ser. 2, *Math. USSR-Sb.* **48**, 161–228.)

Varberg, D. E. (1965). On absolutely continuous functions, *Am. Math. Monthly* **72**, 831–841.

Vitali, G. (1905). *Sul problema della misura dei gruppi di punta di una retta*, Bologna.

Vitali, G. (1907). Sul integrazione per serie, *Rend. Circ. Math. Palermo* **23**, 137–155.

Weil, A. (1938). *L'Intégration dans les groupes topologique et ses applications* (2nd ed., 1953), Hermann, Paris.

Wheeler, R. F. (1983). A survey of Baire measures on topological spaces, *Expositions Math.* **1**, 97–190.

Zaanen, A. C. (1961). The Radon-Nikodým theorem, I–II, *Indag. Math.* **23**, 157–187.

Zaanen, A. C. (1967). *Integration* (2nd ed.), North-Holland, Amsterdam.

Zakon, E. (1966). Integration of nonmeasurable functions, *Canad. Math. Bull.* **9**, 307–330.

INDEX OF SYMBOLS AND NOTATION

$A \times \bigtimes_{i \in I-\beta} \Omega_i$: a cylinder with base A in Ω_β, 344

$\{(\Omega_\alpha, \Sigma_\alpha, \mu_\alpha, g_{\alpha\beta})$: $\alpha < \beta$ in $\mathscr{F}\}$: a projective system of measures, 354

$\underleftarrow{\lim}\ (\Omega_\alpha, \Sigma_\alpha, \mu_\alpha, g_{\alpha\beta})$: projective limit of the system, 354

$\mathbb{Z}^k$: the integer lattice of $\mathbb{R}^k$, 373

$\mathbb{T}^k$: the k-torus, 373

CHAPTER 7

$N(\mathscr{D})$: nucleus of a system $\mathscr{D}$, 384

$\mathscr{A}(\mathscr{C})$: the class of $\mathscr{C}$-analytic sets, 384

$I^*(\cdot)$: outer capacity, 392

$f \vee g$: maximum of f and g, 402

$f \wedge g$: minimum of f and g, 403

$C_c(\Omega)$: continuous real functions with compact supports, 411

CHAPTER 8

$F = G/H$: factor space, 413

$\rho(\cdot)$: lifting map, 414

$\lambda(\cdot)$: set lifting, 418

$\tilde{\lambda}(\cdot)$: lower density, 418

$C_b(S)$: bounded real continuous functions on S, 419

$f \overset{w}{\equiv} g$: weakly equivalent f and g, 420

$\mathscr{M}(S)$: the space of continuous functions for the lifting topology, 437

CHAPTER 9

$\mathscr{B}$ ($\tilde{\mathscr{B}}$): Baire (Borel) σ-algebra, 441

$\mathscr{B}_0$ ($\tilde{\mathscr{B}}_0$): restrictedly Baire (Borel) σ-algebra, 441

$\mathscr{I}(K, O)$: intervals for classes of sets, 448

$\lambda(\cdot)$: a content, 466

$f \prec U$: support of f is in U, 472

$\Lambda(\cdot)$: local functional, 479

gU (Ug): the left (right) translate of U, 489

$\Delta(\cdot)$: modular function, 498

$\dot{x} = xH$: left coset of H, 501

CHAPTER 10

$\tilde{\Omega}$: set of ultrafilters, 504

$\tilde{\Sigma}$: ring of clopen sets of $\tilde{\Omega}$, 504

rca(S): regular countably additive scalar set functions, 518

(AL): abstract L-space, 520

Φ: convex (Young) function, 520

L_ω (R_ω): left (right) translation operator, 523

AUTHOR INDEX

SUBJECT INDEX